Prealgebra

SIXTH EDITION

Charles P. McKeague
CUESTA COLLEGE

Australia • Brazil • Japan • Korea • Mexico • Singapore • Spain • United Kingdom • United States

***Prealgebra*, Sixth Edition**
Charles P. McKeague

Mathematics Editor: Marc Bove
Publisher: Charlie Van Wagner
Consulting Editor: Richard T. Jones
Assistant Editor: Shaun Williams
Editorial Assistant: Mary De La Cruz
Media Editor: Maureen Ross
Marketing Manager: Joe Rogove
Marketing Assistant: Angela Kim
Marketing Communications Manager: Katy Malatesta
Project Manager, Editorial Production: Hal Humphrey
Art Director: Vernon Boes
Print Buyer: Judy Inouye
Permissions Editor: Bob Kauser
Production Service: XYZ Textbooks
Text Designer: Diane Beasley
Photo Researcher: Kathleen Olson
Cover Designer: Irene Morris
Cover Image: Pete McArthur
Compositor: Devin Christ/XYZ Textbooks
Illustrator: Kristina Chung/XYZ Textbooks

ISBN-13: 978-0-495-55991-7

ISBN-10: 0-495-55991-1

Brooks/Cole
10 Davis Drive
Belmont, CA 94002-3098
USA

Cengage Learning is a leading provider of customized learning solutions with office locations around the globe, including Singapore, the United Kingdom, Australia, Mexico, Brazil, and Japan. Locate your local office at:
www.cengage.com/international

Cengage Learning products are represented in Canada by Nelson Education, Ltd.

For your course and learning solutions, visit **www.academic.cengage.com**

To learn more about Brooks/Cole, visit **www.cengage.com/brookscole**
Purchase any of our products at your local college store or at our preferred online store **www.ichapters.com**

Printed in the United States of America
1 2 3 4 5 6 7 11 10 09 08

Brief Contents

Contents

1 Whole Numbers 1

2 Introduction to Algebra 101

3 Fractions and Mixed Numbers 161

4 Solving Equations 255

5 Decimals 351

6 Ratio and Proportion 447

Preface to the Instructor

I have a passion for teaching mathematics. That passion carries through to my textbooks. My goal is a textbook that is user-friendly for both students and instructors. For students, this book forms a bridge to beginning algebra with clear, concise writing, continuous review, and interesting applications. For the instructor, I build features into the text that reinforce the habits and study skills we know will bring success to our students.

The sixth edition of Prealgebra builds upon these strengths.

Applying the Concepts Students are always curious about how the mathematics they are learning can be applied, so we have included applied problems in most of the problem sets in the book and have labeled them to show students the array of uses of mathematics. These applied problems are written in an inviting way, many times accompanied by new interesting illustrations to help students overcome some of the apprehension associated with application problems.

Getting Ready for the Next Section Many students think of mathematics as a collection of discrete, unrelated topics. Their instructors know that this is not the case. The new Getting Ready for the Next Section problems reinforce the cumulative, connected nature of this course by showing how the concepts and techniques flow one from another throughout the course. These problems review all of the material that students will need in order to be successful, forming a bridge to the next section, gently preparing students to move forward.

Maintaining Your Skills One of the major themes of our book is continuous review. We strive to continuously hone techniques learned earlier by keeping the important concepts in the forefront of the course. The Maintaining Your Skills problems review material from the previous chapter, or they review problems that form the foundation of the course—the problems that you expect students to be able to solve when they get to the next course.

The Prealgebra Course as a Bridge to Further Success

Prealgebra is a bridge course. The course and its syllabus bring the student to the level of ability required of college students, while getting them ready to make a successful start in beginning algebra. The algebraic concepts are bridges linked cumulatively to other and more advanced algebra concepts.

Our Proven Commitment to Student Success

After five successful editions, we have developed several interlocking, proven features that will improve students' chances of success in the course. We place practical, easily understood study skills in the first five chapters scattered throughout the sections. Here are some of the other, important success features of the book.

Chapter Pretest These are meant as a diagnostic test taken before the starting work in the chapter. Much of the material here is learned in the chapter so proficiency on the pretests is not necessary.

Getting Ready for Chapter X This is a set of problems from previous chapters that students need in order to be successful in the current chapter. These are review problems intended to reinforce the idea that all topics in the course are built on previous topics.

Getting Ready for Class Just before each problem set is a list of four questions under the heading Getting Ready for Class. These problems require written responses from students and are to be done before students come to class. The answers can be found by reading the preceding section. These questions reinforce the importance of reading the section before coming to class.

Blueprint for Problem Solving Found in the main text, this feature is a detailed outline of steps required to successfully attempt application problems. Intended as a guide to problem solving in general, the blueprint takes the student through the solution process to various kinds of applications.

End-of-Chapter Summary, Review, and Assessment

We have learned that students are more comfortable with a chapter that sums up what they have learned thoroughly and accessibly, and reinforces concepts and techniques well. To help students grasp concepts and get more practice, each chapter ends with the following features that together give a comprehensive reexamination of the chapter.

Chapter Summary The chapter summary recaps all main points from the chapter in a visually appealing grid. In the margin, next to each topic, is an example that illustrates the type of problem associated with the topic being reviewed. Our way of summarizing shows students that concepts in mathematics do relate—and that mastering one concept is a bridge to the next. When students prepare for a test, they can use the chapter summary as a guide to the main concepts of the chapter.

Chapter Review Following the chapter summary in each chapter is the chapter review. It contains an extensive set of problems that review all the main topics in the chapter. This feature can be used flexibly, as assigned review, as a recommended self-test for students as they prepare for examinations, or as an in-class quiz or test.

Cumulative Review Starting in Chapter 2, following the chapter review in each chapter is a set of problems that reviews material from all preceding chapters. This keeps students current with past topics and helps them retain the information they study.

Chapter Test A set of problems representative of all the main points of the chapter. These don't contain as many problems as the chapter review, and should be completed in 50 minutes.

Chapter Projects Each chapter closes with a pair of projects. One is a group project, suitable for students to work on in class. Group projects list details about number of participants, equipment, and time, so that instructors can determine how well the project fits into their classroom. The second project is a research project for students to do outside of class and tends to be open ended.

Additional Features of the Book

Facts from Geometry Many of the important facts from geometry are listed under this heading. In most cases, an example or two accompanies each of the facts to give students a chance to see how topics from geometry are related to the algebra they are learning.

Chapter Openings Each chapter opens with an introduction in which a real-world application is used to stimulate interest in the chapter. We expand on these opening applications later in the chapter.

Descriptive Statistics Beginning in Chapter 1 and then continuing through the rest of the book, students are introduced to descriptive statistics. In Chapter 1 we cover tables and bar charts, as well as mean, median, and mode. These topics are carried through the rest of the book. Along the way we add to the list of descriptive statistics by including scatter diagrams and line graphs. In Chapter 4, we move on to graph ordered pairs and linear equations on a rectangular coordinate system.

Supplements for the Instructor

Please contact your sales representative.

Annotated Instructor's Edition ISBN-10: 0495828254 | ISBN-13: 9780495828259
This special instructor's version of the text contains answers next to all exercises and instructor notes at the appropriate location.

Complete Solutions Manual ISBN-10: 0495828815 | ISBN-13: 9780495828815
This manual contains complete solutions for all problems in the text.

Enhanced WebAssign ISBN-10: 0495828955 | ISBN-13: 9780495828952

Enhanced WebAssign 1-Semester Printed Access Card for Lower Level Math
ISBN-10: 0495390801 | ISBN-13: 9780495390800
Enhanced WebAssign, used by over one million students at more than 1,100 institutions, allows you to assign, collect, grade, and record homework assignments via the web. This proven and reliable homework system includes thousands of algorithmically generated homework problems, links to relevant textbook sections, video examples, problem-specific tutorials, and more.

ExamView® Algorithmic Equation
ISBN-10: 0495828890 | ISBN-13: 9780495828891
Create, deliver, and customize tests and study guides (both print and online) in minutes with this easy-to-use assessment and tutorial system. ExamView offers both a Quick Test Wizard and an Online Test Wizard that guide you step-by-step through the process of creating tests—you can even see the test you are creating on the screen exactly as it will print or display online.

PowerLecture with JoinIn™ Student Response System, ExamView®, Lecture Video ISBN-10: 0495828904 | ISBN-13: 9780495828907

Test Bank ISBN-10: 0495828823 | ISBN-13: 9780495828822
This test bank contains several tests per chapter as well as final exams. The test bank is made up of a combination of multiple choice, free response, true/false, and fill in the blank questions.

Text-Specific Videos ISBN-10: 0495828874 | ISBN-13: 9780495828877
This set of text-specific videos features segments taught by the author, worked-out solutions to many examples in the book. Available to instructors only.

Supplements for the Student

Enhanced WebAssign 1-Semester Printed Access Card for Lower Level Math

ISBN-10: 0495390801 | ISBN-13: 9780495390800
Enhanced WebAssign, used by over one million students at more than 1,100 institutions, allows you to assign, collect, grade, and record homework assignments via the web. This proven and reliable homework system includes thousands of algorithmically generated homework problems, links to relevant textbook sections, video examples, problem-specific tutorials, and more.

Student Solutions Manual ISBN-10: 0495828807
This manual contains complete annotated solutions to all odd problems in the problem sets and all chapter review and chapter test exercises.

Acknowledgments

I would like to thank my editor at Cengage Learning, Marc Bove, for his help and encouragement with this project. Many thanks also to Rich Jones, my developmental editor, for his suggestions on content, and his availability for consulting. Ellena Reda contributed both new ideas and exercises to this revision. Devin Christ, the head of production at our office, was a tremendous help in organizing and planning the details of putting this book together. Mary Gentilucci, Michael Landrum and Tammy Fisher-Vasta assisted with error checking and proofreading. Special thanks to my other friends at Cengage Learning: Sam Subity and Shaun Williams for handling the media and ancillary packages on this project, and Hal Humphrey, my project manager, who did a great job of coordinating everyone and everything in order to publish this book.

Finally, I am grateful to the following instructors for their suggestions and comments: Patricia Clark, Sinclair CC; Matthew Hudock, St. Phillip's College; Bridget Young, Suffolk County CC; Bettie Truitt, Black Hawk College; Armando Perez, Laredo CC; Diane Allen, College of Technology Idaho State; Jignasa Rami, CCBC Catonsville; Yon Kim, Passaic Community College; Elizabeth Chu, Suffolk County CC, Ammerman; Marilyn Larsen, College of the Mainland; Sherri Ucravich, University of Wisconsin; Scott Beckett, Jacksonville State University; Nimisha Raval, Macon Technical Institute; Gary Franchy, Davenport University, Warren; Debbi Loeffler, CC of Baltimore County; Scott Boman, Wayne County CC; Dayna Coker, Southwestern Oklahoma State; Annette Wiesner, University of Wisconsin; Anne Kmet, Grossmont College; Mary Wagner-Krankel, St. Mary's University; Joseph Deguzman, Riverside CC, Norco; Deborah McKee, Weber State University; Gail Burkett, Palm Beach CC; Lee Ann Spahr, Durham Technical CC; Randall Mills, KCTCS Big Sandy CC/Tech; Jana Bryant, Manatee CC; Fred Brown, University of Maine, Augusta; Jeff Waller, Grossmont College; Robert Fusco, Broward CC, FL; Larry Perez, Saddleback College, CA; Victoria Anemelu, San Bernardino Valley, CA; John Close, Salt Lake CC, UT; Randy Gallaher, Lewis and Clark CC; Julia Simms, Southern Illinois U; Julianne Labbiento, Lehigh Carbon CC; Joanne Kendall, Cy-Fair College; Ann Davis, Northeastern Tech.

Pat. McKeague
November 2008

Preface to the Student

I often find my students asking themselves the question "Why can't I understand this stuff the first time?" The answer is "You're not expected to." Learning a topic in mathematics isn't always accomplished the first time around. There are many instances when you will find yourself reading over new material a number of times before you can begin to work problems. That's just the way things are in mathematics. If you don't understand a topic the first time you see it, that doesn't mean there is something wrong with you. Understanding mathematics takes time. The process of understanding requires reading the book, studying the examples, working problems, and getting your questions answered.

How to Be Successful in Mathematics

1. If you are in a lecture class, be sure to attend all class sessions on time. You cannot know exactly what goes on in class unless you are there. Missing class and then expecting to find out what went on from someone else is not the same as being there yourself.

2. Read the book. It is best to read the section that will be covered in class beforehand. Reading in advance, even if you do not understand everything you read, is still better than going to class with no idea of what will be discussed.

3. Work problems every day and check your answers. The key to success in mathematics is working problems. The more problems you work, the better you will become at working them. The answers to the odd-numbered problems are given in the back of the book. When you have finished an assignment, be sure to compare your answers with those in the book. If you have made a mistake, find out what it is, and correct it.

4. Do it on your own. Don't be misled into thinking someone else's work is your own. Having someone else show you how to work a problem is not the same as working the same problem yourself. It is okay to get help when you are stuck. As a matter of fact, it is a good idea. Just be sure you do the work yourself.

5. Review every day. After you have finished the problems your instructor has assigned, take another 15 minutes and review a section you have already completed. The more you review, the longer you will retain the material you have learned.

6. Don't expect to understand every new topic the first time you see it. Sometimes you will understand everything you are doing, and sometimes you won't. That's just the way things are in mathematics. Expecting to understand each new topic the first time you see it can lead to disappointment and frustration. The process of understanding takes time. It requires that you read the book, work problems, and get your questions answered.

7. Spend as much time as it takes for you to master the material. No set formula exists for the exact amount of time you need to spend on mathematics to master it. You will find out as you go along what is or isn't enough time for you. If you end up spending 2 or more hours on each section in order to master the material there, then that's how much time it takes; trying to get by with less will not work.

8. Relax. It's probably not as difficult as you think.

Whole Numbers

1

Chapter Outline

Introduction

The Hoover Dam, as shown in an image from Google Earth, sits on the border of Nevada and Arizona and was the largest producer of hydroelectric power in the United States when it was completed in 1935. Hydroelectric power is the most widely used form of renewable energy today, accounting for about 19% of the world's electricity. Hydroelectricity is a very clean source of power as it does not produce carbon dioxide or any waste products.

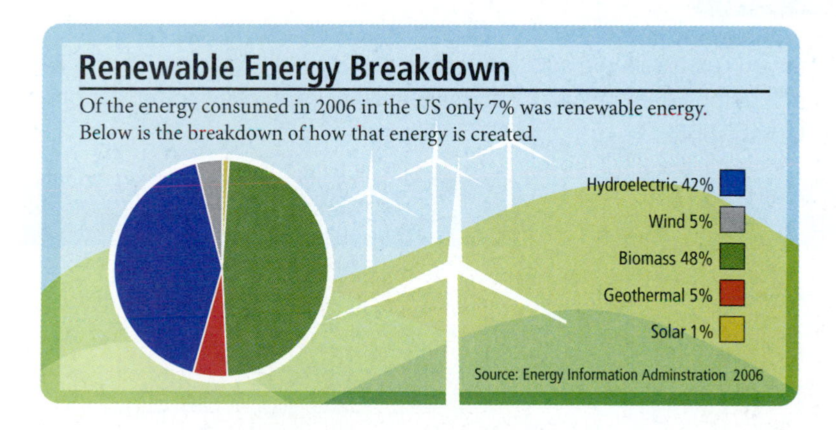

As the chart indicates, the demand for energy is soaring, and developing new sources of energy production is more important than ever. In this chapter we will begin reading and understanding this type of chart.

Chapter Pretest

The pretest below contains problems that are representative of the problems you will find in the chapter. Those of you studying on your own, or working in a self-paced course, can use the pretest to determine which parts of the chapter will require the most work on your part.

1. Write 7,062 in expanded form.

2. Write 3,409,021 in words.

3. Write eighteen thousand, five hundred seven with digits instead of words.

4. Add.
$$\begin{array}{r} 341 \\ +256 \\ \hline \end{array}$$

5. Add.
$$\begin{array}{r} 1{,}029 \\ +4{,}381 \\ \hline \end{array}$$

6. Subtract.
$$\begin{array}{r} 512 \\ -301 \\ \hline \end{array}$$

7. Subtract.
$$\begin{array}{r} 1{,}700 \\ -1{,}436 \\ \hline \end{array}$$

8. Multiply.
$$\begin{array}{r} 27 \\ \times\ 8 \\ \hline \end{array}$$

9. Multiply.
$$\begin{array}{r} 536 \\ \times\ 40 \\ \hline \end{array}$$

10. Divide.
$$18\overline{)576}$$

11. Divide.
$$23\overline{)4{,}018}$$

Round.

12. 513 to the nearest ten

13. 6,798 to the nearest hundred

Simplify.

14. $7 + 3 \cdot 2^3$

15. $4 + 5[2 + 6(9 - 7)]$

16. Find the mean, the median, and range for 4, 5, 7, 9, 15.

17. Write the expression using symbols, then simplify.
6 times the difference of 12 and 8.

Getting Ready for Chapter 1

To get started in this book, we assume that you can do simple addition and multiplication problems. To check to see that you are ready for the chapter, fill in each of the tables below. If you have difficulty, you can find further practice in Appendix A and Appendix B at the back of the book.

TABLE 1
Addition Facts

+	0	1	2	3	4	5	6	7	8	9
0										
1										
2										
3										
4										
5										
6										
7										
8										
9										

TABLE 2
Multiplication Facts

×	0	1	2	3	4	5	6	7	8	9
0										
1										
2										
3										
4										
5										
6										
7										
8										
9										

Place Value and Names for Numbers

1.1

Objectives

A State the place value for a digit in a number written in standard notation.

B Write a whole number in expanded form.

C Write a number in words.

D Write a number from words.

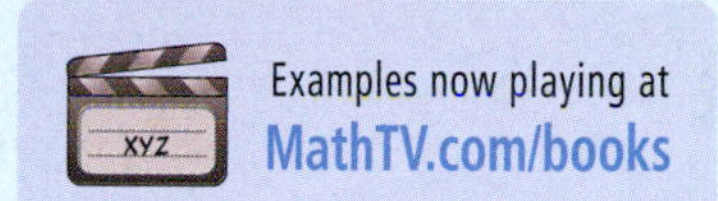

Introduction . . .

The two diagrams below are known as Pascal's triangle, after the French mathematician and philosopher Blaise Pascal (1623–1662). Both diagrams contain the same information. The one on the left contains numbers in our number system; the one on the right uses numbers from Japan in 1781.

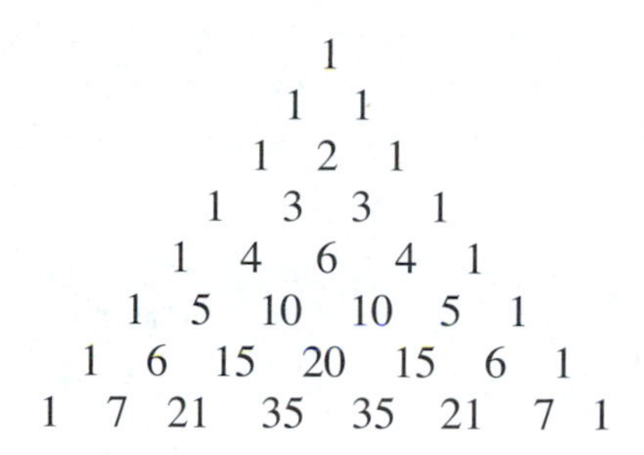

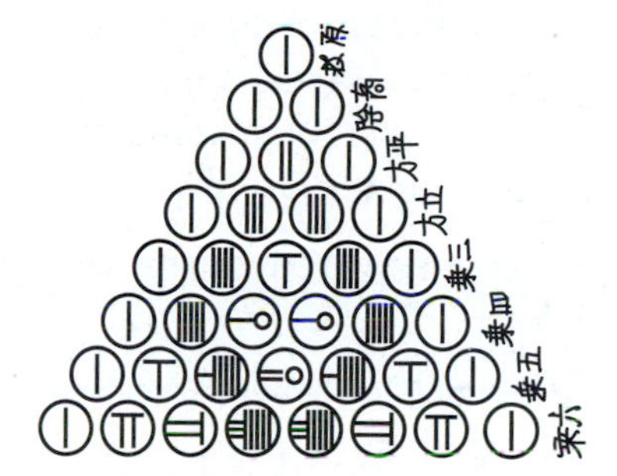

PASCAL'S TRIANGLE IN JAPAN
From Mural Chūzen's *Sampō Dōshi-mon* (1781)

A Place Value

Our number system is based on the number 10 and is therefore called a "base 10" number system. We write all numbers in our number system using the *digits* 0, 1, 2, 3, 4, 5, 6, 7, 8, and 9. The positions of the digits in a number determine the values of the digits. For example, the 5 in the number 251 has a different value from the 5 in the number 542.

The *place values* in our number system are as follows: The first digit on the right is in the *ones column*. The next digit to the left of the ones column is in the *tens column*. The next digit to the left is in the *hundreds column*. For a number like 542, the digit 5 is in the hundreds column, the 4 is in the tens column, and the 2 is in the ones column.

If we keep moving to the left, the columns increase in value. The table shows the name and value of each of the first seven columns in our number system:

Millions Column	Hundred Thousands Column	Ten Thousands Column	Thousands Column	Hundreds Column	Tens Column	Ones Column
1,000,000	100,000	10,000	1,000	100	10	1

Note Next to each Example in the text is a Practice Problem with the same number. After you read through an Example, try the Practice Problem next to it. The answers to the Practice Problems are at the bottom of the page. Be sure to check your answers as you work these problems. The worked-out solutions to all Practice Problems with more than one step are given in the back of the book. So if you find a Practice Problem that you cannot work correctly, you can look up the correct solution to that problem in the back of the book.

EXAMPLE 1 Give the place value of each digit in the number 305,964.

SOLUTION Starting with the digit at the right, we have:

4 in the ones column, 6 in the tens column, 9 in the hundreds column, 5 in the thousands column, 0 in the ten thousands column, and 3 in the hundred thousands column.

PRACTICE PROBLEMS

1. Give the place value of each digit in the number 46,095.

Answer

1. 5 ones, 9 tens, 0 hundreds, 6 thousands, 4 ten thousands

NASA

Large Numbers

The photograph shown here was taken by the Hubble telescope in April 2002. The object in the photograph is called the *Cone Nebula.* In astronomy, distances to objects like the Cone Nebula are given in light-years, the distance light travels in a year. If we assume light travels 186,000 miles in one second, then a light-year is 5,865,696,000,000 miles; that is

5 trillion, 865 billion, 696 million miles

To find the place value of digits in large numbers, we can use Table 1. Note how the Hundreds, Thousands, Millions, Billions, and Trillions categories are each broken into Ones, Tens, and Hundreds. Note also that we have written the digits for our light-year in the last row of the table.

TABLE 1

Trillions			Billions			Millions			Thousands			Hundreds		
Hundreds	Tens	Ones	Hundreds	Tens	Ones	Hundreds	Tens	Ones	Hundreds	Tens	Ones	Hundreds	Tens	Ones
		5	8	6	5	6	9	6	0	0	0	0	0	0

2. Give the place value of each digit in the number 21,705,328,456.

EXAMPLE 2 Give the place value of each digit in the number 73,890,672,540.

SOLUTION The following diagram shows the place value of each digit.

Ten Billions	*Billions*	*Hundred Millions*	*Ten Millions*	*Millions*	*Hundred Thousands*	*Ten Thousands*	*Thousands*	*Hundreds*	*Tens*	*Ones*
7	3,	8	9	0,	6	7	2,	5	4	0

B Expanded Form

We can use the idea of place value to write numbers in *expanded form*. For example, the number 542 can be written in expanded form as

$$542 = 500 + 40 + 2$$

because the 5 is in the hundreds column, the 4 is in the tens column, and the 2 is in the ones column.

Here are more examples of numbers written in expanded form.

3. Write 3,972 in expanded form.

EXAMPLE 3 Write 5,478 in expanded form.

SOLUTION $5,478 = 5,000 + 400 + 70 + 8$

We can use money to make the results from Example 3 more intuitive. Suppose you have $5,478 in cash as follows:

Answers

2. 6 ones, 5 tens, 4 hundreds, 8 thousands, 2 ten thousands, 3 hundred thousands, 5 millions, 0 ten millions, 7 hundred millions, 1 billion, 2 ten billions

3. 3,000 + 900 + 70+ 2

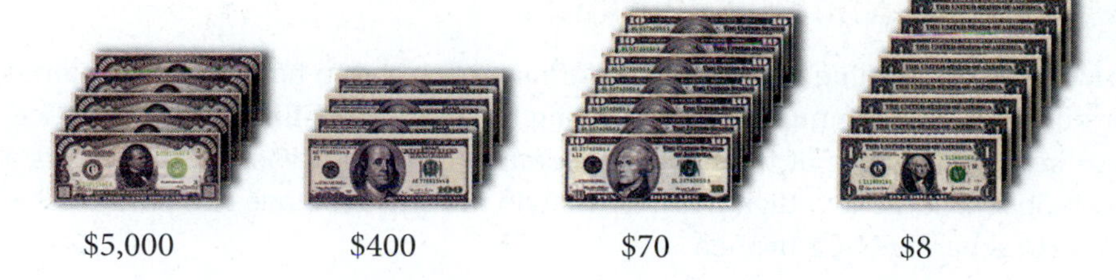

Using this diagram as a guide, we can write

$$\$5,478 = \$5,000 + \$400 + \$70 + \$8$$

which shows us that our work writing numbers in expanded form is consistent with our intuitive understanding of the different denominations of money.

EXAMPLE 4 Write 354,798 in expanded form.

SOLUTION $354,798 = 300,000 + 50,000 + 4,000 + 700 + 90 + 8$

EXAMPLE 5 Write 56,094 in expanded form.

SOLUTION Notice that there is a 0 in the hundreds column. This means we have 0 hundreds. In expanded form we have

$$56,094 = 50,000 + 6,000 + 90 + 4$$

↑ Note that we don't have to include the 0 hundreds

EXAMPLE 6 Write 5,070,603 in expanded form.

SOLUTION The columns with 0 in them will not appear in the expanded form.

$$5,070,603 = 5,000,000 + 70,000 + 600 + 3$$

4. Write 271,346 in expanded form.

5. Write 71,306 in expanded form.

6. Write 4,003,560 in expanded form.

STUDY SKILLS

Some of the students enrolled in my mathematics classes develop difficulties early in the course. Their difficulties are not associated with their ability to learn mathematics; they all have the potential to pass the course. Research has identified three variables that affect academic achievement. These are (1) how much math you know before entering a course, (2) the quality of instruction (classroom atmosphere, teaching style, textbook content and format), and (3) your academic self concept, attitude, anxiety, and study habits. As a student, you have the most control over the last variable. Your academic self concept is a significant predictor of mathematics achievement. Students who get off to a poor start do so because they have not developed the study skills necessary to be successful in mathematics. Throughout this textbook you will find tips and things you can do to begin to develop effective study skills and improve your academic self concept.

Put Yourself on a Schedule

The general rule is that you spend two hours on homework for every hour you are in class. Make a schedule for yourself in which you set aside two hours each day to work on this course. Once you make the schedule, stick to it. Don't just complete your assignments and then stop. Use all the time you have set aside. If you complete the assignment and have time left over, read the next section in the book, and then work more problems. As the course progresses you may find that two hours a day is not enough time to master the material in this course. If it takes you longer than two hours a day to reach your goals for this course, then that's how much time it takes. Trying to get by with less will not work.

Answers

4. 200,000 + 70,000 + 1,000 + 300 + 40 + 6

5. 70,000 + 1,000 + 300 + 6

6. 4,000,000 + 3,000 + 500 + 60

C Writing Numbers in Words

The idea of place value and expanded form can be used to help write the names for numbers. Naming numbers and writing them in words takes some practice. Let's begin by looking at the names of some two-digit numbers. Table 2 lists a few. Notice that the two-digit numbers that do not end in 0 have two parts. These parts are separated by a hyphen.

TABLE 2

Number	In English	Number	In English
25	*Twenty-five*	30	*Thirty*
47	*Forty-seven*	62	*Sixty-two*
93	*Ninety-three*	77	*Seventy-seven*
88	*Eighty-eight*	50	*Fifty*

The following examples give the names for some larger numbers. In each case the names are written according to the place values given in Table 1.

EXAMPLE 7 Write each number in words.

a. 452 **b.** 397 **c.** 608

SOLUTION **a.** Four hundred fifty-two

b. Three hundred ninety-seven

c. Six hundred eight

EXAMPLE 8 Write each number in words.

a. 3,561 **b.** 53,662 **c.** 547,801

SOLUTION **a.** Three thousand, five hundred sixty-one

↑

Notice how the comma separates the thousands from the hundreds

b. Fifty-three thousand, six hundred sixty-two

c. Five hundred forty-seven thousand, eight hundred one

EXAMPLE 9 Write each number in words.

a. 507,034,005

b. 739,600,075

c. 5,003,007,006

SOLUTION **a.** Five hundred seven million, thirty-four thousand, five

b. Seven hundred thirty-nine million, six hundred thousand, seventy-five

c. Five billion, three million, seven thousand, six

7. Write each number in words.
a. 724
b. 595
c. 307

8. Write each number in words.
a. 4,758
b. 62,779
c. 305,440

9. Write each number in words.
a. 707,044,002
b. 452,900,008
c. 4,008,002,001

Answers

7. **a.** Seven hundred twenty-four
b. Five hundred ninety-five
c. Three hundred seven

8. **a.** Four thousand, seven hundred fifty-eight
b. Sixty-two thousand, seven hundred seventy-nine
c. Three hundred five thousand, four hundred forty

9. **a.** Seven hundred seven million, forty-four thousand, two
b. Four hundred fifty-two million, nine hundred thousand, eight
c. Four billion, eight million, two thousand, one

STUDY SKILLS
Find Your Mistakes and Correct Them

There is more to studying mathematics than just working problems. You must always check your answers with the answers in the back of the book. When you have made a mistake, find out what it is, and then correct it. Making mistakes is part of the process of learning mathematics. I have never had a successful student who didn't make mistakes—lots of them. Your mistakes are your guides to understanding; look forward to them.

Here is a practical reason for being able to write numbers in word form.

Michael Smith
1221 Main Street
Anytown, NY 11001

1001

DATE 7/8/08

PAY TO THE ORDER OF Campus Book Store $ 423.00

Four hundred twenty-three and no cents DOLLARS

Michael Smith

⑆010011⑆ 0111332200233⑈ 1142232⑈

D Writing Numbers from Words

The next examples show how we write a number given in words as a number written with digits.

EXAMPLE 10 Write five thousand, six hundred forty-two, using digits instead of words.

SOLUTION *Five thousand, six hundred forty-two*

5, 6 42

EXAMPLE 11 Write each number with digits instead of words.

a. Three million, fifty-one thousand, seven hundred
b. Two billion, five
c. Seven million, seven hundred seven

SOLUTION

a. 3,051,700

b. 2,000,000,005

c. 7,000,707

10. Write six thousand, two hundred twenty-one using digits instead of words.

11. Write each number with digits instead of words.
a. Eight million, four thousand, two hundred
b. Twenty-five million, forty
c. Nine million, four hundred thirty-one

Answers
10. 6,221
11. **a.** 8,004,200
b. 25,000,040
c. 9,000,431

Sets and the Number Line

In mathematics a collection of numbers is called a *set*. In this chapter we will be working with the set of *counting numbers* and the set of *whole numbers*, which are defined as follows:

> **Note** Counting numbers are also called natural numbers.

$$\text{Counting numbers} = \{1, 2, 3, \ldots\}$$
$$\text{Whole numbers} = \{0, 1, 2, 3, \ldots\}$$

The dots mean "and so on," and the braces { } are used to group the numbers in the set together.

Another way to visualize the whole numbers is with a *number line*. To draw a number line, we simply draw a straight line and mark off equally spaced points along the line, as shown in Figure 1. We label the point at the left with 0 and the rest of the points, in order, with the numbers 1, 2, 3, 4, 5, and so on.

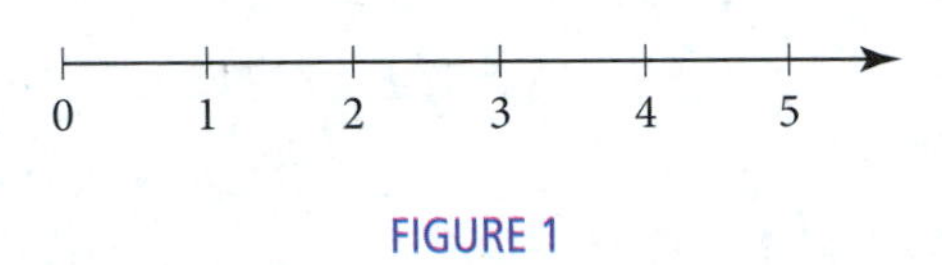

FIGURE 1

The arrow on the right indicates that the number line can continue in that direction forever. When we refer to numbers in this chapter, we will always be referring to the whole numbers.

STUDY SKILLS
Gather Information on Available Resources

You need to anticipate that you will need extra help sometime during the course. There is a form to fill out in Appendix A to help you gather information on resources available to you. One resource is your instructor; you need to know your instructor's office hours and where the office is located. Another resource is the math lab or study center, if they are available at your school. It also helps to have the phone numbers of other students in the class, in case you miss class. You want to anticipate that you will need these resources, so now is the time to gather them together.

Getting Ready for Class

After reading through the preceding section, respond in your own words and in complete sentences.

1. Give the place value of the 9 in the number 305,964.
2. Write the number 742 in expanded form.
3. Place a comma and a hyphen in the appropriate place so that the number 2,345 is written correctly in words below:

   two thousand three hundred forty five

4. Is there a largest whole number?

Problem Set 1.1

A Give the place value of each digit in the following numbers. [Examples 1, 2]

1. 78
2. 93
3. 45
4. 79
5. 348
6. 789
7. 608
8. 450
9. 2,378
10. 6,481
11. 273,569
12. 768,253

Give the place value of the 5 in each of the following numbers.

13. 458,992
14. 75,003,782
15. 507,994,787
16. 320,906,050
17. 267,894,335
18. 234,345,678,789
19. 4,569,000
20. 50,000

B Write each of the following numbers in expanded form. [Examples 3–6]

21. 658
22. 479
23. 68
24. 71
25. 4,587
26. 3,762
27. 32,674
28. 54,883
29. 3,462,577
30. 5,673,524
31. 407
32. 508
33. 30,068
34. 50,905
35. 3,004,008
36. 20,088,060

C Write each of the following numbers in words. [Examples 7–9]

37. 29

38. 75

39. 40

40. 90

41. 573

42. 895

43. 707

44. 405

45. 770

46. 450

47. 23,540

48. 56,708

49. 3,004

50. 5,008

51. 3,040

52. 5,080

53. 104,065,780

54. 637,008,500

55. 5,003,040,008

56. 7,050,800,001

57. 2,546,731

58. 6,998,454

D Write each of the following numbers with digits instead of words. [Examples 10, 11]

59. Three hundred twenty-five

60. Forty-eight

61. Five thousand, four hundred thirty-two

62. One hundred twenty-three thousand, sixty-one

63. Eighty-six thousand, seven hundred sixty-two

64. One hundred million, two hundred thousand, three hundred

65. Two million, two hundred

66. Two million, two

67. Two million, two thousand, two hundred

68. Two billion, two hundred thousand, two hundred two

Applying the Concepts

69. The illustration shows the average income of workers 18 and older by education.

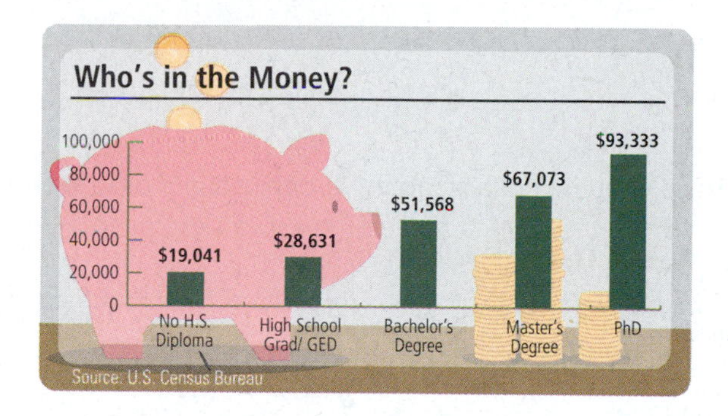

Write the following numbers in words:

a. the average income of someone with only a high school education

b. the average income of someone with a Ph.D.

70. Write the following numbers in words from the information in the given illustration:

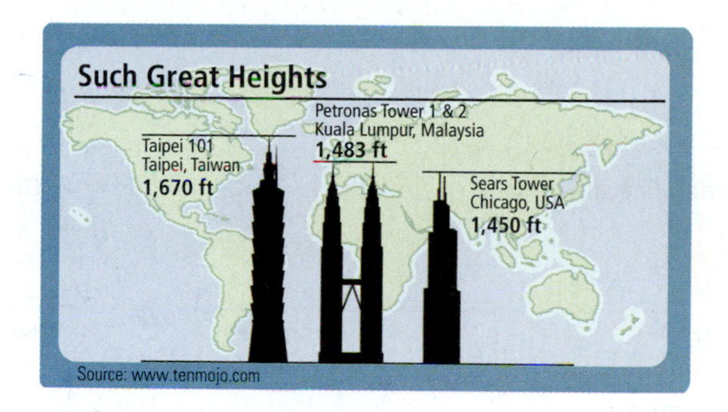

a. the height in feet of the Taipei 101 building in Taipei, Taiwan

b. the height in feet of the Sears Tower in Chicago, Illinois

71. MP3s A new MP3 player has the ability to hold over 125,000 songs. Write the place value of the 1 in the number of songs.

72. Music Downloads The top three downloaded songs for one month on Amazon.com had a combined 450,320 downloads. Write the place value of the 3 in the number of downloads.

73. Baseball Salaries According to mlb.com, major league baseball's 2008 average player salary was $3,173,403, representing an increase of 7% from the previous season's average. Write 3,173,403 in words.

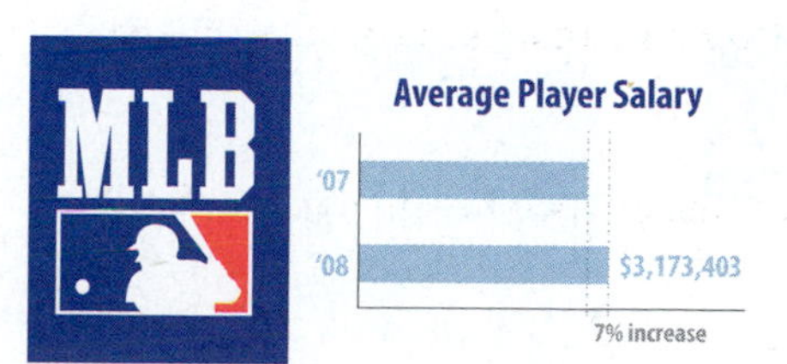

74. Astronomy The distance from the sun to the earth is 92,897,416 miles. Write this number in expanded form.

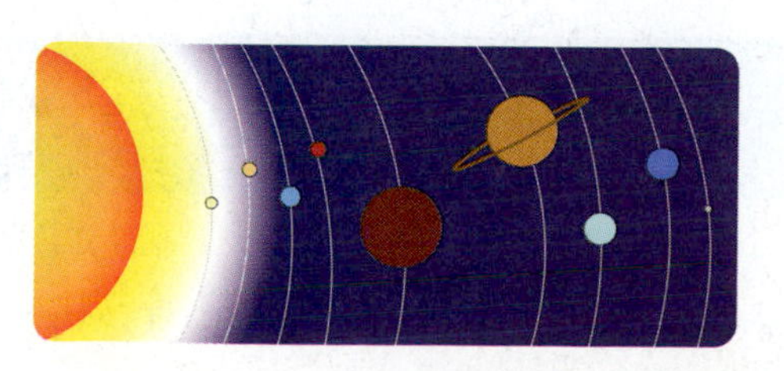

75. Web Searches The phrase "math help" was searched approximately 21,480 times in one month in 2008 from Google. Write this number in words.

76. Web Searches The phrase "math help" was searched approximately 6,180 times in one month in 2008 from Yahoo. Write this number in words.

Writing Checks In each of the checks below, fill in the appropriate space with the dollar amount in either digits or in words.

77.

Michael Smith
1221 Main Street
Anytown, NY 11001 1002

DATE 7/8/08

PAY TO THE ORDER OF Sunshine Apartment Complex $ 750.00

DOLLARS

Michael Smith

⑆01001⑆ 0111332200233⑈ 1142232⑈

78.

Michael Smith
1221 Main Street
Anytown, NY 11001 1003

DATE 7/8/08

PAY TO THE ORDER OF Electric and Gas Company $

Two hundred sixteen dollars and no cents DOLLARS

Michael Smith

⑆01001⑆ 0111332200233⑈ 1142232⑈

Populations of Countries The table below gives estimates of the populations of some countries for mid-year 2008. The first column under *Population* gives the population in digits. The second column gives the population in words. Fill in the blanks.

Country	Population	
	Digits	Words
79. United States	________	Three hundred four million
80. People's Republic of China	________	One billion, three hundred thirty million
81. Japan	127,000,000	________________
82. United Kingdom	61,000,000	________________

(From U.S. Census Bureau, International Data Base)

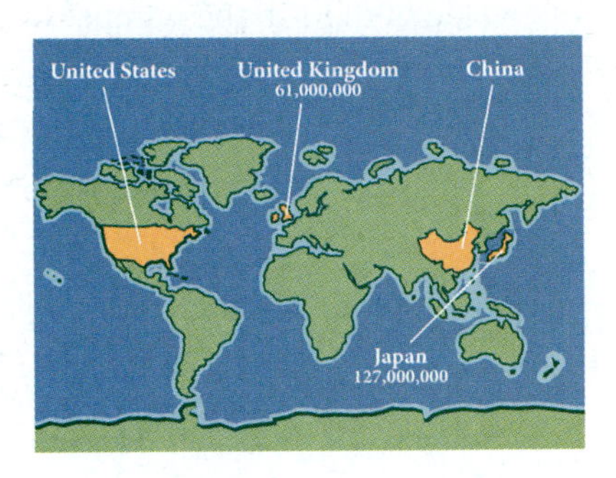

Populations of Cities The table below gives estimates of the populations of some cities for mid-year 2008. The first column under *Population* gives the population in digits. The second column gives the population in words. Fill in the blanks.

City	Population	
	Digits	Words
83. Tokyo	________	Thirty-six million
84. Los Angeles	________	Eighteen million
85. Paris	10,900,000	________________
86. London	7,500,000	________________

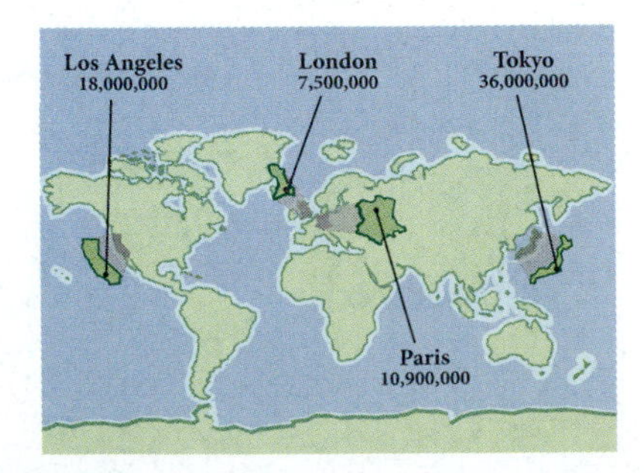

Addition with Whole Numbers, and Perimeter

1.2

Objectives

A Add whole numbers.

B Understand the notation and vocabulary of addition.

C Use the properties of addition.

D Find a solution to an equation by inspection.

E Find the perimeter of a figure.

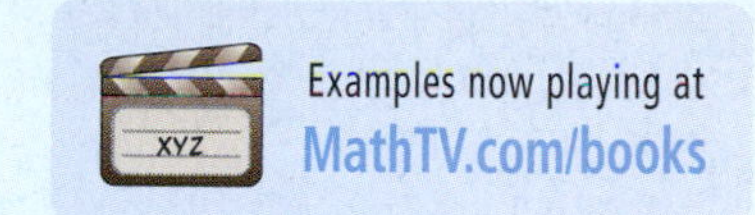

Introduction . . .

The chart shows the number of babies born in 2006, grouped together according to the age of mothers.

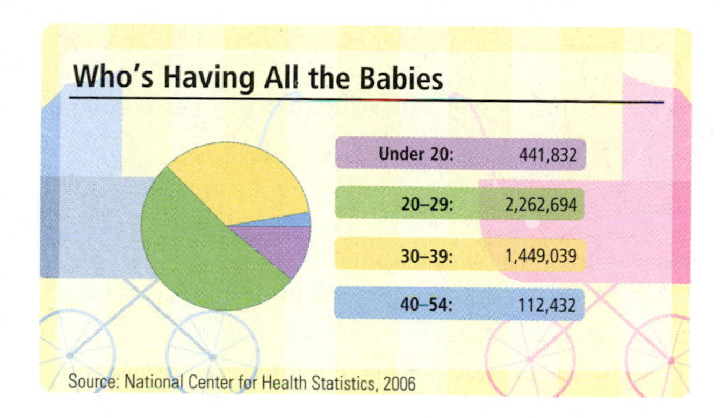

There is much more information available from the table than just the numbers shown. For instance, the chart tells us how many babies were born to mothers less than 30 years of age. But to find that number, we need to be able to do addition with whole numbers. Let's begin by visualizing addition on the number line.

Facts of Addition

Using lengths to visualize addition can be very helpful. In mathematics we generally do so by using the number line. For example, we add 3 and 5 on the number line like this: Start at 0 and move to 3, as shown in Figure 1. From 3, move 5 more units to the right. This brings us to 8. Therefore, $3 + 5 = 8$.

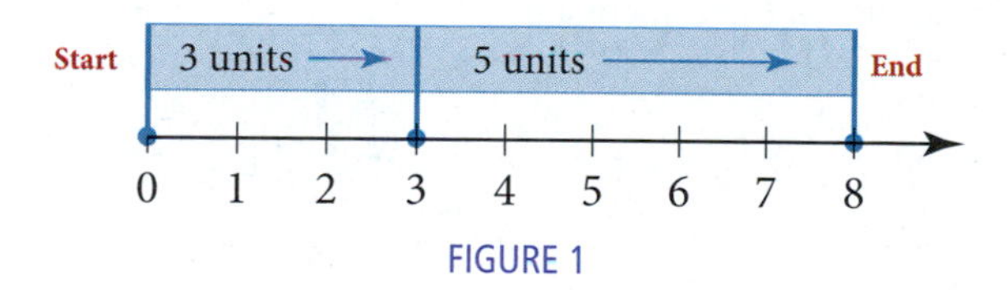

FIGURE 1

If we do this kind of addition on the number line with all combinations of the numbers 0 through 9, we get the results summarized in Table 1 on the next page.

We call the information in Table 1 our basic addition facts. Your success with the examples and problems in this section depends on knowing the basic addition facts.

Note Table 1 is a summary of the addition facts that you *must* know in order to make a successful start in your study of prealgebra. You *must* know how to add any pair of numbers that come from the list. You *must* be fast and accurate. You don't want to have to think about the answer to 7 + 9. You should know it's 16. Memorize these facts now. Don't put it off until later.

Appendix B at the back of the book has 100 problems on the basic addition facts for you to practice. You may want to go there now and work those problems.

TABLE 1

ADDITION TABLE

+	0	1	2	3	4	5	6	7	8	9
0	0	1	2	3	4	5	6	7	8	9
1	1	2	3	4	5	6	7	8	9	10
2	2	3	4	5	6	7	8	9	10	11
3	3	4	5	6	7	8	9	10	11	12
4	4	5	6	7	8	9	10	11	12	13
5	5	6	7	8	9	10	11	12	13	14
6	6	7	8	9	10	11	12	13	14	15
7	7	8	9	10	11	12	13	14	15	16
8	8	9	10	11	12	13	14	15	16	17
9	9	10	11	12	13	14	15	16	17	18

We read Table 1 in the following manner: Suppose we want to use the table to find the answer to 3 + 5. We locate the 3 in the column on the left and the 5 in the row at the top. We read *across* from the 3 and *down* from the 5. The entry in the table that is across from 3 and below 5 is 8.

A Adding Whole Numbers

To add whole numbers, we add digits within the same place value. First we add the digits in the ones place, then the tens place, then the hundreds place, and so on.

PRACTICE PROBLEMS

1. Add: 63 + 25

EXAMPLE 1 Add: 43 + 52

SOLUTION This type of addition is best done vertically. First we add the digits in the ones place.

$$\begin{array}{r} 43 \\ +\ 52 \\ \hline 5 \end{array}$$

Then we add the digits in the tens place.

$$\begin{array}{r} 43 \\ +\ 52 \\ \hline 95 \end{array}$$

Note To show *why* we add digits with the same place value, we can write each number showing the place value of the digits:

$$\begin{array}{rcl} 43 &=& 4 \text{ tens} + 3 \text{ ones} \\ +\ 52 &=& 5 \text{ tens} + 2 \text{ ones} \\ \hline && 9 \text{ tens} + 5 \text{ ones} \end{array}$$

2. Add: 342 + 605

EXAMPLE 2 Add: 165 + 801

SOLUTION Writing the sum vertically, we have

$$\begin{array}{r} 165 \\ +\ 801 \\ \hline 966 \end{array}$$

966 ← Add ones place
↑ Add tens place
↑ Add hundreds place

Answers
1. 88
2. 947

A Addition with Carrying

In Examples 1 and 2, the sums of the digits with the same place value were always 9 or less. There are many times when the sum of the digits with the same place value will be a number larger than 9. In these cases we have to do what is called *carrying* in addition. The following examples illustrate this process.

EXAMPLE 3 Add: 197 + 213 + 324

SOLUTION We write the sum vertically and add digits with the same place value.

$$\begin{array}{r} \overset{1}{1}97 \\ 213 \\ +\ 324 \\ \hline 4 \end{array}$$ When we add the ones, we get 7 + 3 + 4 = 14. We write the 4 and carry the 1 to the tens column

$$\begin{array}{r} \overset{1}{1}\overset{1}{9}7 \\ 213 \\ +\ 324 \\ \hline 34 \end{array}$$ We add the tens, including the 1 that was carried over from the last step. We get 13, so we write the 3 and carry the 1 to the hundreds column

$$\begin{array}{r} \overset{1}{1}\overset{1}{9}7 \\ 213 \\ +\ 324 \\ \hline 734 \end{array}$$ We add the hundreds, including the 1 that was carried over from the last step

3. Add.
 a. 375 + 121 + 473
 b. 495 + 699 + 978

Notice that Practice Problem 3 has two parts. Part a is similar to the problem shown in Example 3. Part b is similar also, but a little more challenging in nature. We will do this from time to time throughout the text. If a practice problem contains more parts than the example to which it corresponds, then the additional parts cover the same concept, but are more challenging than Part a.

EXAMPLE 4 Add: 46,789 + 2,490 + 864

SOLUTION We write the sum vertically—with the digits with the same place value aligned—and then use the shorthand form of addition.

$$\begin{array}{rrrrr} \overset{1}{4} & \overset{2}{6}, & \overset{2}{7} & \overset{1}{8} & 9 \\ & 2, & 4 & 9 & 0 \\ & & 8 & 6 & 4 \\ \hline 5 & 0, & 1 & 4 & 3 \end{array}$$

These are the numbers that have been carried

Write the 3; carry the 1 — Ones
Write the 4; carry the 2 — Tens
Write the 1; carry the 2 — Hundreds
Write the 0; carry the 1 — Thousands
No carrying necessary — Ten thousands

4. Add.
 a. 57,904 + 7,193 + 655
 b. 68,495 + 7,236 + 878 + 29 + 5

Adding numbers as we are doing here takes some practice. Most people don't make mistakes in carrying. Most mistakes in addition are made in adding the numbers in the columns. That is why it is so important that you are accurate with the basic addition facts given in this chapter.

B Vocabulary

The word we use to indicate addition is the word *sum*. If we say "the sum of 3 and 5 is 8," what we mean is 3 + 5 = 8. The word *sum* always indicates addition. We can state this fact in symbols by using the letters a and b to represent numbers.

Answers
3. a. 969 b. 2,172
4. a. 65,752 b. 76,643

Definition

If a and b are any two numbers, then the **sum** of a and b is $a + b$. To find the sum of two numbers, we add them.

Table 2 gives some phrases and sentences in English and their mathematical equivalents written in symbols.

Note When mathematics is used to solve everyday problems, the problems are almost always stated in words. The translation of English to symbols is a very important part of mathematics.

TABLE 2

In English	In Symbols
The sum of 4 and 1	$4 + 1$
4 added to 1	$1 + 4$
8 more than m	$m + 8$
x increased by 5	$x + 5$
The sum of x and y	$x + y$
The sum of 2 and 4 is 6	$2 + 4 = 6$

C Properties of Addition

Once we become familiar with addition, we may notice some facts about addition that are true regardless of the numbers involved. The first of these facts involves the number 0 (zero).

Whenever we add 0 to a number, the result is the original number. For example,

$$7 + 0 = 7 \quad \text{and} \quad 0 + 3 = 3$$

Because this fact is true no matter what number we add to 0, we call it a property of 0.

Addition Property of 0

If we let a represent any number, then it is always true that

$$a + 0 = a \quad \text{and} \quad 0 + a = a$$

In words: Adding 0 to any number leaves that number unchanged.

Note When we use letters to represent numbers, as we do when we say "If a and b are any two numbers," then a and b are called variables, because the values they take on vary. We use the variables a and b in the definitions and properties on this page because we want you to know that the definitions and properties are true for all numbers that you will encounter in this book.

A second property we notice by becoming familiar with addition is that the order of two numbers in a sum can be changed without changing the result.

$$3 + 5 = 8 \quad \text{and} \quad 5 + 3 = 8$$
$$4 + 9 = 13 \quad \text{and} \quad 9 + 4 = 13$$

This fact about addition is true for *all* numbers. The order in which you add two numbers doesn't affect the result. We call this fact the *commutative property of addition,* and we write it in symbols as follows.

Commutative Property of Addition

If a and b are any two numbers, then it is always true that

$$a + b = b + a$$

In words: Changing the order of two numbers in a sum doesn't change the result.

STUDY SKILLS
Accept Definitions

It is important that you don't overcomplicate definitions. When I tell my students that my name is Mr. McKeague, they don't ask "why?" You should approach definitions in the same way. Just accept them as they are, and memorize them if you have to. If someone asks you what the commutative property is, you should be able to respond, "With addition, the commutative property says that if a and b are two numbers then $a + b = b + a$. In other words, you can change the order of two numbers you are adding without changing the result."

EXAMPLE 5 Use the commutative property of addition to rewrite each sum.

a. $4 + 6$ **b.** $5 + 9$ **c.** $3 + 0$ **d.** $7 + n$

SOLUTION The commutative property of addition indicates that we can change the order of the numbers in a sum without changing the result. Applying this property we have:

a. $4 + 6 = 6 + 4$
b. $5 + 9 = 9 + 5$
c. $3 + 0 = 0 + 3$
d. $7 + n = n + 7$

Notice that we did not actually add any of the numbers. The instructions were to use the commutative property, and the commutative property involves only the order of the numbers in a sum.

5. Use the commutative property of addition to rewrite each sum.
a. $7 + 9$
b. $6 + 3$
c. $4 + 0$
d. $5 + n$

The last property of addition we will consider here has to do with sums of more than two numbers. Suppose we want to find the sum of 2, 3, and 4. We could add 2 and 3 first, and then add 4 to what we get:

$$(2 + 3) + 4 = 5 + 4 = 9$$

Or, we could add the 3 and 4 together first and then add the 2:

$$2 + (3 + 4) = 2 + 7 = 9$$

The result in both cases is the same. If we try this with any other numbers, the same thing happens. We call this fact about addition the *associative property of addition,* and we write it in symbols as follows.

Note This discussion is here to show why we write the next property the way we do. Sometimes it is helpful to look ahead to the property itself (in this case, the associative property of addition) to see what it is that is being justified.

Associative Property of Addition

If a, b, and c represent any three numbers, then

$$(a + b) + c = a + (b + c)$$

In words: Changing the grouping of three numbers in a sum doesn't change the result.

Answer
5. a. $9 + 7$ **b.** $3 + 6$ **c.** $0 + 4$ **d.** $n + 5$

EXAMPLE 6 Use the associative property of addition to rewrite each sum.

a. $(5 + 6) + 7$ **b.** $(3 + 9) + 1$ **c.** $6 + (8 + 2)$ **d.** $4 + (9 + n)$

SOLUTION The associative property of addition indicates that we are free to regroup the numbers in a sum without changing the result.

a. $(5 + 6) + 7 = 5 + (6 + 7)$
b. $(3 + 9) + 1 = 3 + (9 + 1)$
c. $6 + (8 + 2) = (6 + 8) + 2$
d. $4 + (9 + n) = (4 + 9) + n$

The commutative and associative properties of addition tell us that when adding whole numbers, we can use any order and grouping. When adding several numbers, it is sometimes easier to look for pairs of numbers whose sums are 10, 20, and so on.

EXAMPLE 7 Add: $9 + 3 + 2 + 7 + 1$

SOLUTION We find pairs of numbers that we can add quickly:

$$9 + 3 + 2 + 7 + 1$$
$$= 10 + 10 + 2$$
$$= 22$$

D Solving Equations

We can use the addition table to help solve some simple equations. If n is used to represent a number, then the equation

$$n + 3 = 5$$

will be true if n is 2. The number 2 is therefore called a *solution* to the equation, because, when we replace n with 2, the equation becomes a true statement:

$$2 + 3 = 5$$

Equations like this are really just puzzles, or questions. When we say, "Solve the equation $n + 3 = 5$," we are asking the question, "What number do we add to 3 to get 5?"

When we solve equations by reading the equation to ourselves and then stating the solution, as we did with the equation above, we are solving the equation by inspection.

EXAMPLE 8 Find the solution to each equation by inspection.

a. $n + 5 = 9$
b. $n + 6 = 12$
c. $4 + n = 5$
d. $13 = n + 8$

SOLUTION We find the solution to each equation by using the addition facts given in Table 1.

a. The solution to $n + 5 = 9$ is 4, because $4 + 5 = 9$.
b. The solution to $n + 6 = 12$ is 6, because $6 + 6 = 12$.
c. The solution to $4 + n = 5$ is 1, because $4 + 1 = 5$.
d. The solution to $13 = n + 8$ is 5, because $13 = 5 + 8$.

---

6. Use the associative property of addition to rewrite each sum.
a. $(3 + 2) + 9$
b. $(4 + 10) + 1$
c. $5 + (9 + 1)$
d. $3 + (8 + n)$

7. Add.
a. $6 + 2 + 4 + 8 + 3$
b. $24 + 17 + 36 + 13$

Note The letter n as we are using it here is a variable, because it represents a number. In this case it is the number that is a solution to an equation.

8. Use inspection to find the solution to each equation.
a. $n + 9 = 17$
b. $n + 2 = 10$
c. $8 + n = 9$
d. $16 = n + 10$

Answers
6. a. $3 + (2 + 9)$ **b.** $4 + (10 + 1)$ **c.** $(5 + 9) + 1$ **d.** $(3 + 8) + n$
7. a. 23 **b.** 90
8. a. 8 **b.** 8 **c.** 1 **d.** 6

E Perimeter

FACTS FROM GEOMETRY Perimeter

We end this section with an introduction to perimeter. Here we will find the perimeter of several different shapes called *polygons*. A *polygon* is a closed geometric figure, with at least three sides, in which each side is a straight line segment.

The most common polygons are squares, rectangles, and triangles. Examples of these are shown in Figure 2.

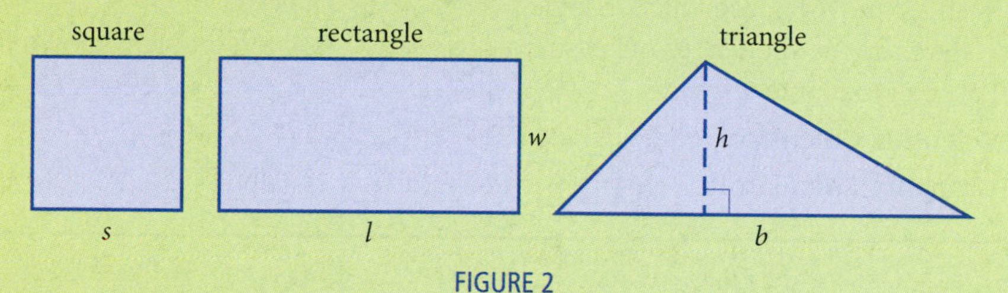

FIGURE 2

In the square, s is the length of the side, and each side has the same length. In the rectangle, l stands for the length, and w stands for the width. The width is usually the lesser of the two. The b and h in the triangle are the base and height, respectively. The height is always perpendicular to the base. That is, the height and base form a 90°, or right, angle where they meet.

Note In the triangle, the small square where the broken line meets the base is the notation we use to show that the two line segments meet at right angles. That is, the height h and the base b are perpendicular to each other; the angle between them is 90°.

Definition

The **perimeter** of any polygon is the sum of the lengths of the sides, and it is denoted with the letter P.

EXAMPLE 9 Find the perimeter of each geometric figure.

a. 15 inches

b. 24 feet, 37 feet

c. 36 yards, 23 yards, 24 yards, 24 yards, 12 yards

SOLUTION In each case we find the perimeter by adding the lengths of all the sides.

a. The figure is a square. Because the length of each side in the square is the same, the perimeter is

$$P = 15 + 15 + 15 + 15 = 60 \text{ inches}$$

b. In the rectangle, two of the sides are 24 feet long, and the other two are 37 feet long. The perimeter is the sum of the lengths of the sides.

$$P = 24 + 24 + 37 + 37 = 122 \text{ feet}$$

c. For this polygon, we add the lengths of the sides together. The result is the perimeter.

$$P = 36 + 23 + 24 + 12 + 24 = 119 \text{ yards}$$

9. Find the perimeter of each geometric figure.

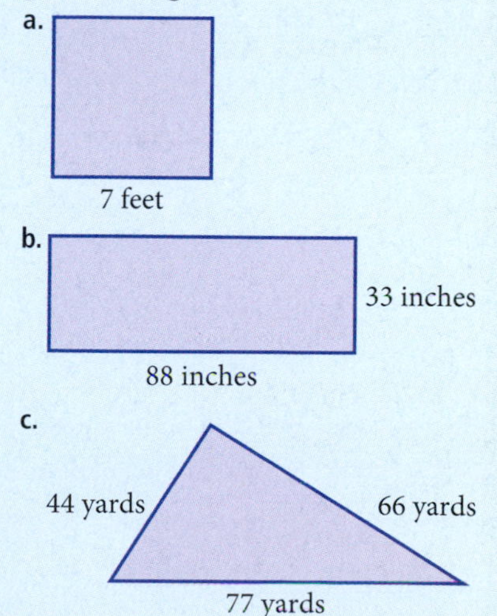

Answer

9. a. 28 feet b. 242 inches c. 187 yards

USING TECHNOLOGY

Calculators

From time to time we will include some notes like this one, which show how a calculator can be used to assist us with some of the calculations in the book. Most calculators on the market today fall into one of two categories: those with algebraic logic and those with function logic. Calculators with algebraic logic have a key with an equals sign on it. Calculators with function logic do not have an equals key. Instead they have a key labeled ENTER or EXE (for execute). Scientific calculators use algebraic logic, and graphing calculators, such as the TI-83, use function logic.

Here are the sequences of keystrokes to use to work the problem shown in Part c of Example 9.

Scientific Calculator: 36 [+] 23 [+] 24 [+] 12 [+] 24 [=]

Graphing Calculator: 36 [+] 23 [+] 24 [+] 12 [+] 24 [ENT]

Getting Ready for Class

After reading through the preceding section, respond in your own words and in complete sentences.

1. What number is the sum of 6 and 8?
2. Make up an addition problem using the number 456 that does not involve carrying.
3. Make up an addition problem using the number 456 that involves carrying from the ones column to the tens column only.
4. What is the perimeter of a geometric figure?

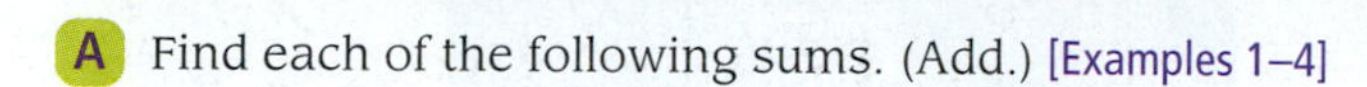

Problem Set 1.2

A Find each of the following sums. (Add.) [Examples 1–4]

1. 3 + 5 + 7

2. 2 + 8 + 6

3. 1 + 4 + 9

4. 2 + 8 + 3

5. 5 + 9 + 4 + 6

6. 8 + 1 + 6 + 2

7. 1 + 2 + 3 + 4 + 5

8. 5 + 6 + 7 + 8 + 9

9. 9 + 1 + 8 + 2

10. 7 + 3 + 6 + 4

A Add each of the following. (There is no carrying involved in these problems.) [Examples 1, 2]

11.	**12.**	**13.**	**14.**	**15.**	**16.**
43	56	81	37	4,281	2,749
25	23	17	22	3,016	1,250

17.	**18.**	**19.**	**20.**	**21.**	**22.**
3,482	2,496	32	521	6,245	27
3,005	7,503	21	340	203	4,510
		43	135	1,001	342

A Add each of the following. (All problems involve carrying in at least one column.) [Examples 3, 4]

23.	**24.**	**25.**	**26.**	**27.**	**28.**
49	85	74	36	682	439
16	29	28	46	193	270

29.	**30.**	**31.**	**32.**	**33.**	**34.**
638	444	4,963	8,291	6,205	8,888
191	595	5,428	7,489	9,999	9,999

35.	**36.**	**37.**	**38.**	**39.**	**40.**
56,789	45,678	52,468	13,579	4,296	5,637
98,765	87,654	58,642	97,531	8,720	481
				4,375	7,899

41.	**42.**	**43.**	**44.**	**45.**	**46.**
4,994	6,824	12	21	999	646
449	371	34	43	444	464
9,449	4,857	56	65	555	525
		78	87	222	252

47.	**48.**
9,245	45
672	9,876
8,341	54
27	6,789

B Complete the following tables.

49.

First Number a	Second Number b	Their Sum $a + b$
61	38	
63	36	
65	34	
67	32	

50.

First Number a	Second Number b	Their Sum $a + b$
10	45	
20	35	
30	25	
40	15	

51.

First Number a	Second Number b	Their Sum $a + b$
9	16	
36	64	
81	144	
144	256	

52.

First Number a	Second Number b	Their Sum $a + b$
25	75	
24	76	
23	77	
22	78	

C Rewrite each of the following using the commutative property of addition. [Example 5]

53. $5 + 9$ **54.** $2 + 1$ **55.** $3 + 8$ **56.** $9 + 2$ **57.** $6 + 4$ **58.** $1 + 7$

C Rewrite each of the following using the associative property of addition. [Example 6]

59. $(1 + 2) + 3$ **60.** $(4 + 5) + 9$ **61.** $(2 + 1) + 6$ **62.** $(2 + 3) + 8$

63. $1 + (9 + 1)$ **64.** $2 + (8 + 2)$ **65.** $(4 + n) + 1$ **66.** $(n + 8) + 1$

D Find a solution for each equation. [Example 8]

67. $n + 6 = 10$ **68.** $n + 4 = 7$ **69.** $n + 8 = 13$ **70.** $n + 6 = 15$

71. $4 + n = 12$ **72.** $5 + n = 7$ **73.** $17 = n + 9$ **74.** $13 = n + 5$

B Write each of the following expressions in words. Use the word *sum* in each case. [Table 2]

75. $4 + 9$ **76.** $9 + 4$ **77.** $8 + 1$

78. $9 + 9$ **79.** $2 + 3 = 5$ **80.** $8 + 2 = 10$

B Write each of the following in symbols. [Table 2]

81. **a.** The sum of 5 and 2
b. 3 added to 8

82. **a.** The sum of a and 4
b. 6 more than x

83. **a.** m increased by 1
b. The sum of m and n

84. **a.** The sum of 4 and 8 is 12.
b. The sum of a and b is 6.

E Find the perimeter of each figure. The first four figures are squares. [Example 9]

85.

86.

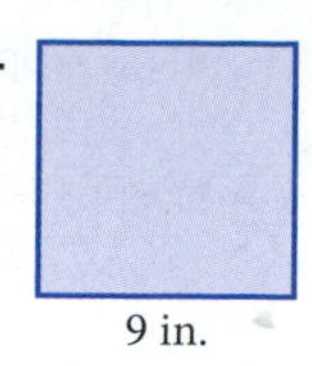

87.

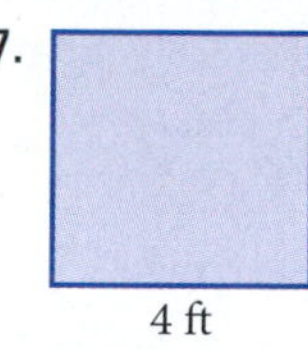

88.

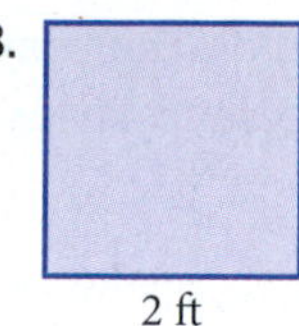

89.

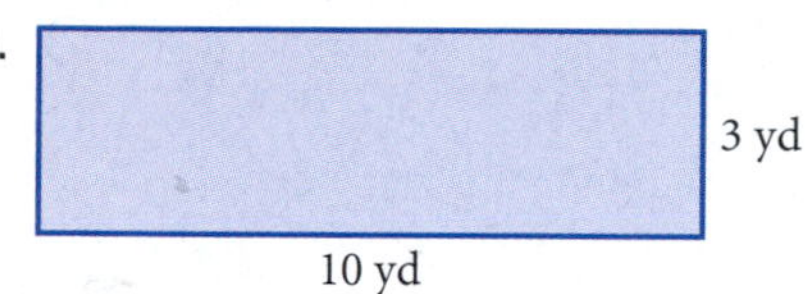

90.

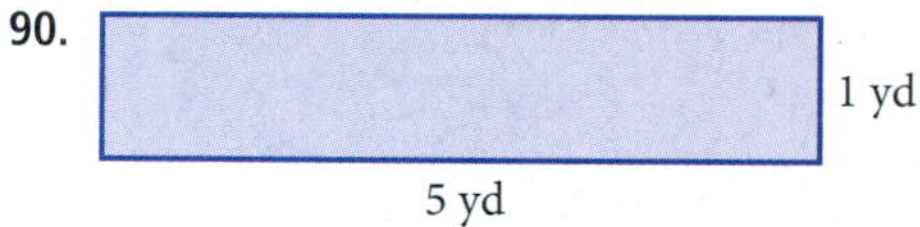

91.

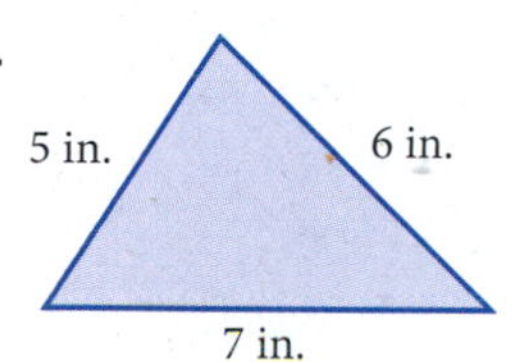

92.

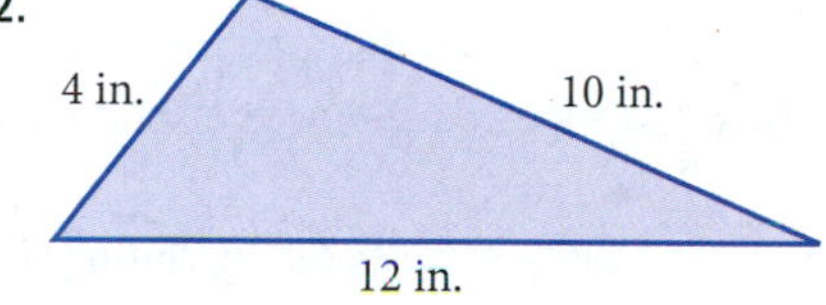

E Applying the Concepts

93. Classroom appliances use a lot of energy. You can save energy by unplugging or turning of unused appliances. Use the information in the given illustration to find the following:

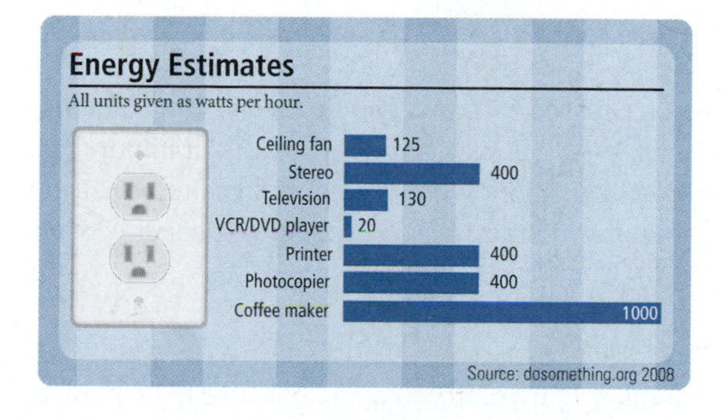

a. the number of watts/hour saved by unplugging a DVD player and a television

b. the number of watts/hour saved by unplugging a ceiling fan and a coffee maker

94. The information in the illustration represents the number of picture messages sent for the first nine months of the year, in millions. Use the information to find the following:

a. the number of picture messages sent in all nine months

b. the number of picture messages sent in March and April

95. Checkbook Balance On Monday Bob had a balance of $241 in his checkbook. On Tuesday he made a deposit of $108, and on Thursday he wrote a check for $24. What was the balance in his checkbook on Wednesday?

RECORD ALL CHARGES OR CREDITS THAT AFFECT YOUR ACCOUNT

NUMBER	DATE	DESCRIPTION OF TRANSACTION	PAYMENT/DEBIT (-)	DEPOSIT/CREDIT (+)	BALANCE
					$241 00
	11/06	Deposit		$108 00	?
1401	11/18	Postage Stamps	$24 00		?

96. Number of Passengers A plane flying from Los Angeles to New York left Los Angeles with 67 passengers on board. The plane stopped in Bakersfield and picked up 28 passengers, and then it stopped again in Dallas where 57 more passengers came on board. How many passengers were on the plane when it landed in New York?

97. College Costs According to data from *The Chronicle of Higher Education*, the most expensive college in the country is George Washington University in Washington, D.C. According to the university's website, a student entering as a freshman during the 2008 – 09 academic year can expect to pay the expenses shown in the chart below:

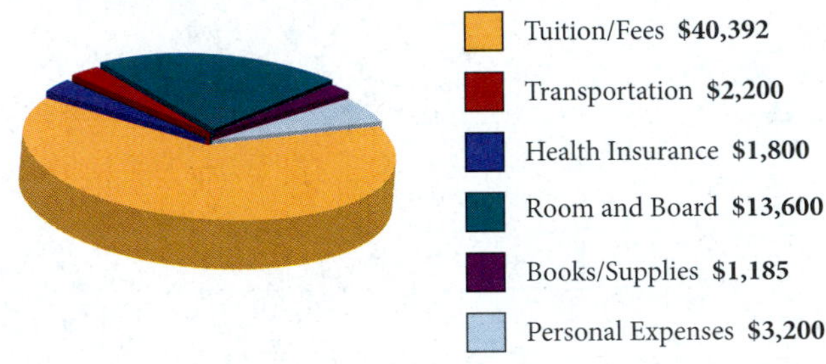

a. What are the total of the expenses for one year at George Washington University?

b. How much of these total expenses are college related?

c. What is the total amount for expenses that are not directly related to attending this college?

98. Improving Your Quantitative Literacy Quantitative literacy is a subject discussed by many people involved in teaching mathematics. The person they are concerned with when they discuss it is you. We are going to work at improving your quantitative literacy, but before we do that we should answer the question, what is quantitative literacy? Lynn Arthur Steen, a noted mathematics educator, has stated that quantitative literacy is "the capacity to deal effectively with the quantitative aspects of life."

a. Give a definition for the word *quantitative.*

b. Give a definition for the word *literacy.*

c. Are there situations that occur in your life that you find distasteful or that you try to avoid because they involve numbers and mathematics? If so, list some of them here. (For example, some people find the process of buying a car particularly difficult because they feel that the numbers and details of the financing are beyond them.)

Rounding Numbers, Estimating Answers, and Displaying Information

1.3

Objectives

A Round whole numbers.

B Estimate the answer to a problem.

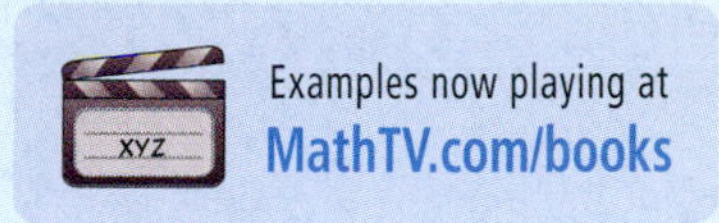

Introduction . . .

Many times when we talk about numbers, it is helpful to use numbers that have been *rounded off*, rather than exact numbers. For example, the city where I live has a population of 43,704. But when I tell people how large the city is, I usually say, "The population is about 44,000." The number 44,000 is the original number rounded to the nearest thousand. The number 43,704 is closer to 44,000 than it is to 43,000, so it is rounded to 44,000. We can visualize this situation on the number line.

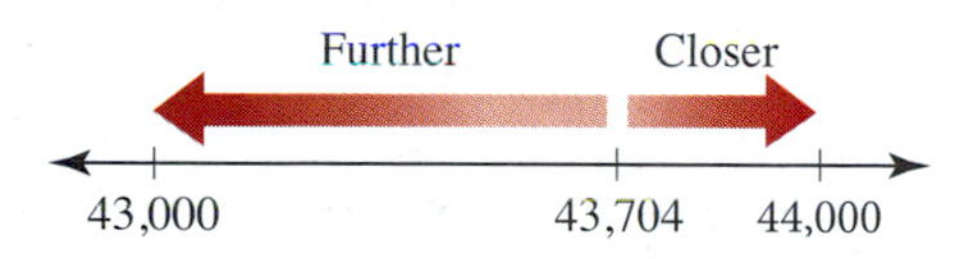

A Rounding

The steps used in rounding numbers are given below.

> **Strategy** Rounding Whole Numbers
>
> To summarize, we list the following steps:
>
> **Step 1:** Locate the digit just to the right of the place you are to round to.
>
> **Step 2:** If that digit is less than 5, replace it and all digits to its right with zeros.
>
> **Step 3:** If that digit is 5 or more, replace it and all digits to its right with zeros, and add 1 to the digit to its left.

Note After you have used the steps listed here to work a few problems, you will find that the procedure becomes almost automatic.

You can see from these rules that in order to round a number you must be told what column (or place value) to round to.

EXAMPLE 1 Round 5,382 to the nearest hundred.

SOLUTION There is a 3 in the hundreds column. We look at the digit just to its right, which is 8. Because 8 is greater than 5, we add 1 to the 3, and we replace the 8 and 2 with zeros:

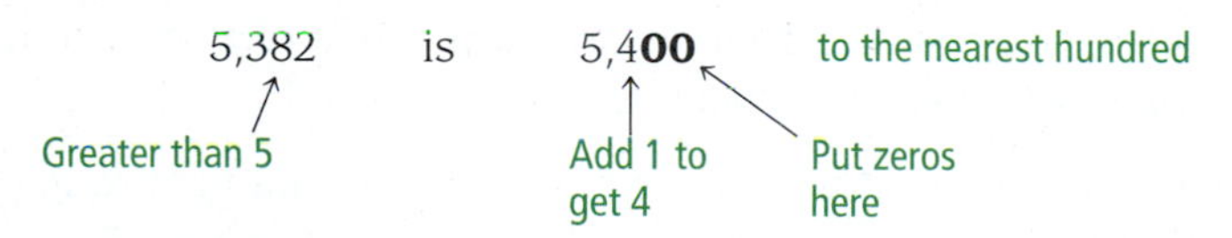

EXAMPLE 2 Round 94 to the nearest ten.

SOLUTION There is a 9 in the tens column. To its right is 4. Because 4 is less than 5, we simply replace it with 0:

PRACTICE PROBLEMS

1. Round 5,742 to the nearest
   a. hundred
   b. thousand

2. Round 87 to the nearest
   a. ten
   b. hundred

Answers

1. a. 5,700 b. 6,000
2. a. 90 b. 100

3. Round 980 to the nearest
   a. hundred
   b. thousand

EXAMPLE 3 Round 973 to the nearest hundred.

SOLUTION We have a 9 in the hundreds column. To its right is 7, which is greater than 5. We add 1 to 9 to get 10, and then replace the 7 and 3 with zeros:

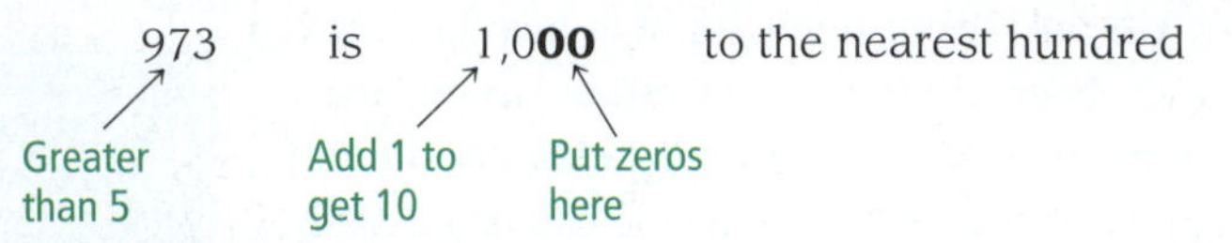

4. Round 376,804,909 to the nearest
   a. million
   b. ten thousand

EXAMPLE 4 Round 47,256,344 to the nearest million.

SOLUTION We have 7 in the millions column. To its right is 2, which is less than 5. We simply replace all the digits to the right of 7 with zeros to get:

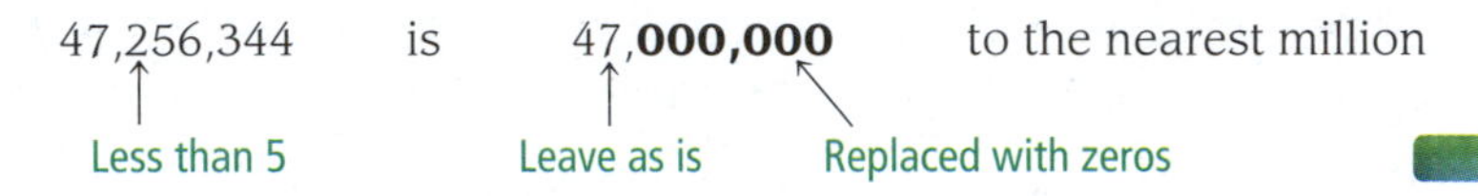

Table 1 gives more examples of rounding.

TABLE 1

Original Number	Rounded to the Nearest Ten	Rounded to the Nearest Hundred	Rounded to the Nearest Thousand
6,914	6,910	6,900	7,000
8,485	8,490	8,500	8,000
5,555	5,560	5,600	6,000
1,234	1,230	1,200	1,000

Rule
If we are doing calculations and are asked to round our answer, we do all our arithmetic first and then round the result. That is, the last step is to round the answer; we don't round the numbers first and then do the arithmetic.

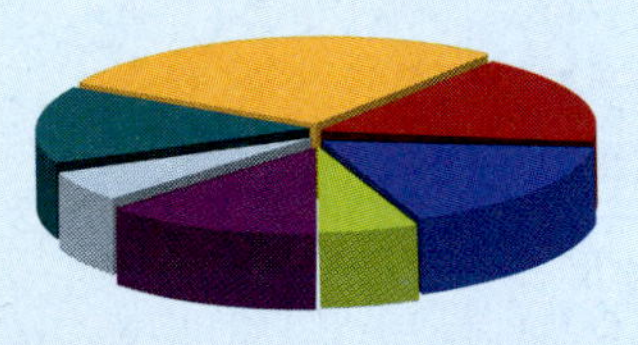

House Payments **$10,200**
Taxes **$6,137**
Miscellaneous **$6,142**
Entertainment **$2,142**
Car Expenses **$4,847**
Savings **$2,149**
Food **$5,296**

5. Use the pie chart above to answer these questions.
   a. To the nearest ten dollars, what is the total amount spent on food and car expenses?
   b. To the nearest hundred dollars, how much is spent on savings and taxes?
   c. To the nearest thousand dollars, how much is spent on items other than food and entertainment?

EXAMPLE 5 The pie chart in the margin shows how a family earning $36,913 a year spends their money.

a. To the nearest hundred dollars, what is the total amount spent on food and entertainment?
b. To the nearest thousand dollars, how much of their income is spent on items other than taxes and savings?

SOLUTION In each case we add the numbers in question and then round the sum to the indicated place.

a. We add the amounts spent on food and entertainment and then round that result to the nearest hundred dollars.

Food	$5,296
Entertainment	2,142
Total	$7,438 = $7,400 to the nearest hundred dollars

Answers
3. a. 1,000 b. 1,000
4. a. 377,000,000 b. 376,800,000

b. We add the numbers for all items except taxes and savings.

House payments	$10,200
Food	5,296
Car expenses	4,847
Entertainment	2,142
Miscellaneous	6,142
Total	$28,627 = $29,000 to the nearest thousand dollars

B Estimating

When we *estimate* the answer to a problem, we simplify the problem so that an approximate answer can be found quickly. There are a number of ways of doing this. One common method is to use rounded numbers to simplify the arithmetic necessary to arrive at an approximate answer, as our next example shows.

EXAMPLE 6 Estimate the answer to the following problem by rounding each number to the nearest thousand.

$$\begin{array}{r} 4{,}872 \\ 1{,}691 \\ 777 \\ +\ 6{,}124 \\ \hline \end{array}$$

SOLUTION We round each of the four numbers in the sum to the nearest thousand. Then we add the rounded numbers.

4,872	rounds to	5,000
1,691	rounds to	2,000
777	rounds to	1,000
+ 6,124	rounds to	+ 6,000
		14,000

We estimate the answer to this problem to be approximately 14,000. The actual answer, found by adding the original unrounded numbers, is 13,464.

Here is a practical application for which the ability to estimate can be a useful tool.

EXAMPLE 7 On the way home from classes you stop at the local grocery store to pick up a few things. You know that you have a $20.00 bill in your wallet. You pick up the following items: a loaf of wheat bread for $2.29, a gallon of milk for $3.96, a dozen eggs for $2.18, a pound of apples for $1.19, and a box of your favorite cereal for $4.59. Use estimation to determine if you will have enough to pay for your groceries when you get to the cashier.

SOLUTION We round the items in our grocery cart off to the nearest dollar:

wheat bread for $2.29	rounds to	$2.00
milk for $3.96	rounds to	$4.00
eggs for $2.18	rounds to	$2.00
apples for $1.19	rounds to	$1.00
+ cereal for $4.59	rounds to	+ $5.00
		$14.00

We estimate our total to be $14.00. Thus, $20.00 will be enough to pay for the groceries. (The actual cost of the groceries is $14.21.)

6. Estimate the answer by first rounding each number to the nearest thousand.

a. 5,287 / 2,561 / 888 / +4,898

b. 702 / 3,944 / 1,001 / +3,500

Note In Example 6 we are asked to *estimate* an answer, so it is okay to round the numbers in the problem before adding them. In Example 5 we are asked for a rounded answer, meaning that we are to find the exact answer to the problem and then round to the indicated place. In that case we must not round the numbers in the problem before adding.

Answer

5. a. $10,140 **b.** $8,300 **c.** $29,000

6. a. 14,000 **b.** 10,000

DESCRIPTIVE STATISTICS

Bar Charts

The table and chart below give two representations for the amount of caffeine in five different drinks, one numeric and the other visual.

TABLE 2

Beverage (6-ounce cup)	Caffeine (in milligrams)
Brewed coffee	100
Instant coffee	70
Tea	50
Cocoa	5
Decaffeinated coffee	4

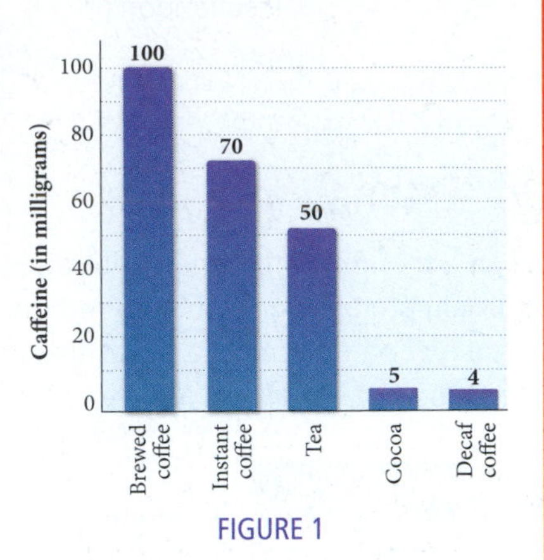

FIGURE 1

The diagram in Figure 1 is called a *bar chart*. The horizontal line below which the drinks are listed is called the *horizontal axis*, while the vertical line that is labeled from 0 to 100 is called the *vertical axis*.

Getting Ready for Class

After reading through the preceding section, respond in your own words and in complete sentences.

1. Describe the process you would use to round the number 5,382 to the nearest thousand.
2. Describe the process you would use to round the number 47,256,344 to the nearest ten thousand.
3. Find a number not containing the digit 7 that will round to 700 when rounded to the nearest hundred.
4. When I ask a class of students to round the number 7,499 to the nearest thousand, a few students will give the answer as 8,000. In what way are these students using the rule for rounding numbers incorrectly?

Problem Set 1.3

A Round each of the numbers to the nearest ten. [Examples 1–5]

1. 42 **2.** 44 **3.** 46 **4.** 48 **5.** 45 **6.** 73

7. 77 **8.** 75 **9.** 458 **10.** 455 **11.** 471 **12.** 680

13. 56,782 **14.** 32,807 **15.** 4,504 **16.** 3,897

Round each of the numbers to the nearest hundred. [Examples 1–5]

17. 549 **18.** 954 **19.** 833 **20.** 604 **21.** 899 **22.** 988

23. 1090 **24.** 6,778 **25.** 5,044 **26.** 56,990 **27.** 39,603 **28.** 31,999

Round each of the numbers to the nearest thousand. [Examples 1–5]

29. 4,670 **30.** 9,054 **31.** 9,760 **32.** 4,444 **33.** 978 **34.** 567

35. 657,892 **36.** 688,909 **37.** 509,905 **38.** 608,433 **39.** 3,789,345 **40.** 5,744,500

A Complete the following table by rounding the numbers on the left as indicated by the headings in the table. [Examples 1–5]

Original Number	Rounded to the Nearest		
	Ten	Hundred	Thousand
41. 7,821			
42. 5,945			
43. 5,999			
44. 4,353			
45. 10,985			
46. 11,108			
47. 99,999			
48. 95,505			

B **Estimating** Estimate the answer to each of the following problems by rounding each number to the indicated place value and then adding. [Example 6]

49. hundred
750
275
+ 120

50. thousand
1,891
765
+ 3,223

51. hundred
472
422
536
+ 511

52. hundred
399
601
744
+ 298

53. thousand
25,399
7,601
18,744
+ 6,298

54. thousand
9,999
8,888
7,777
+ 6,666

55. hundred
9,999
8,888
7,777
+ 6,666

56. ten thousand
127,675
72,560
+ 219,065

57. ten thousand
65,000
31,000
15,555
+ 72,000

58. ten
10,061
10,044
10,035
+ 10,025

59. hundred
20,150
18,250
12,350
+ 30,450

60. hundred
1,950
2,849
3,750
+ 4,649

Applying the Concepts

61. Age of Mothers About 4 million babies were born in 2006. The chart shows the breakdown by mothers' age and number of babies. Use the chart to answer the following questions.

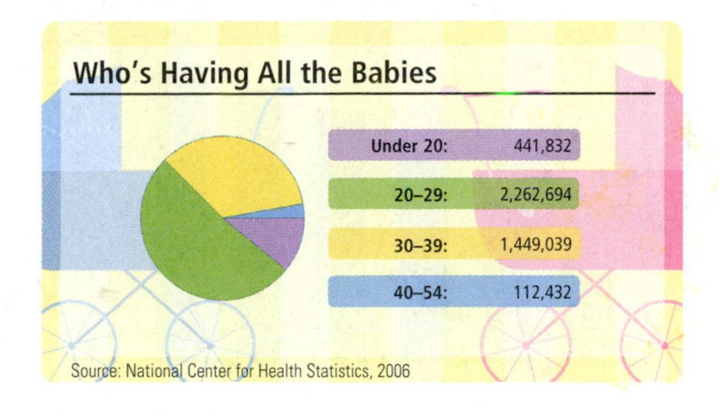

a. What is the exact number of babies born in 2006?

b. Using your answer from Part a, is the statement "About 4 million babies were born in 2006" correct?

c. To the nearest hundred thousand, how many babies were born to mothers aged 20 to 29 in 2006?

d. To the nearest thousand, how many babies were born to mothers 40 years old or older?

62. Business Expenses The pie chart shows one year's worth of expenses for a small business. Use the chart to answer the following questions.

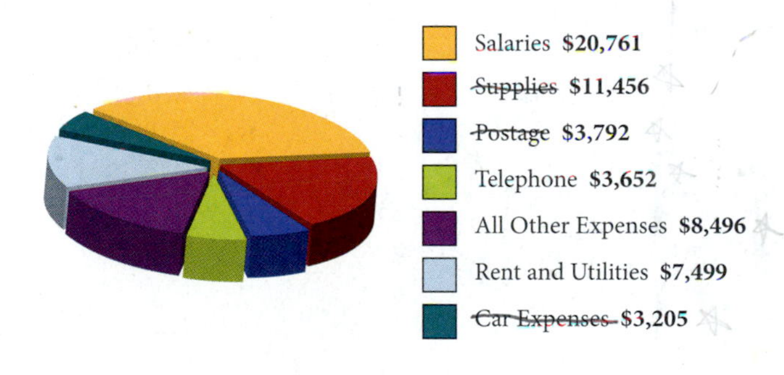

a. To the nearest hundred dollars, how much was spent on postage and supplies?

b. Find the total amount spent, to the nearest hundred dollars, on rent and utilities and car expenses.

c. To the nearest thousand dollars, how much was spent on items other than salaries and rent and utilities?

d. To the nearest thousand dollars, how much was spent on items other than postage, supplies, and car expenses?

The bar chart below is similar to the one we studied in this section. It was given to me by a friend who owns and operates an alcohol dragster. The dragster contains a computer that gives information about each of his races. This particular race was run during the 1993 Winternationals. The bar chart gives the speed of a race car in a quarter-mile drag race every second during the race. The horizontal lines have been added to assist you with Problems 63–66.

63. Is the speed of the race car after 3 seconds closer to 160 miles per hour or 190 miles per hour?

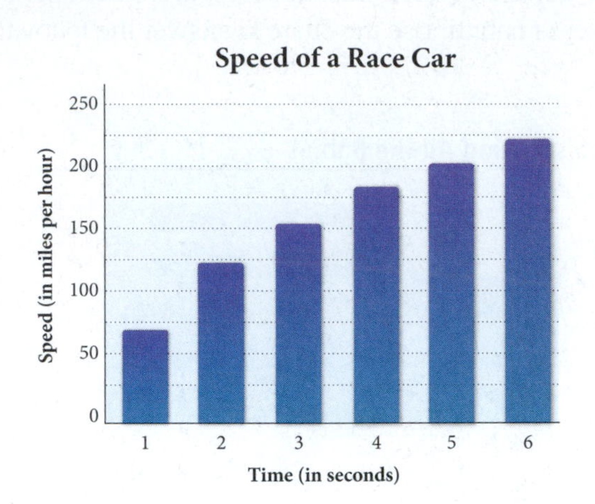

64. After 4 seconds, is the speed of the race car closer to 150 miles per hour or 190 miles per hour?

65. Estimate the speed of the car after 1 second.

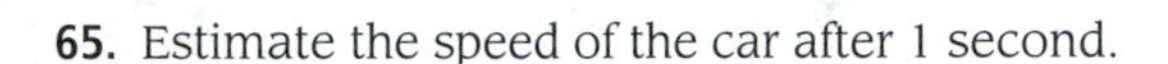

66. Estimate the speed of the car after 6 seconds.

67. Fast Food The following table lists the number of calories consumed by eating some popular fast foods. Use the axes in the figure below to construct a bar chart from the information in the table.

CALORIES IN FAST FOOD

Food	Calories
McDonald's Hamburger	270
Burger King Hamburger	260
Jack in the Box Hamburger	280
McDonald's Big Mac	510
Burger King Whopper	630
Jack in the Box Colossus Burger	940

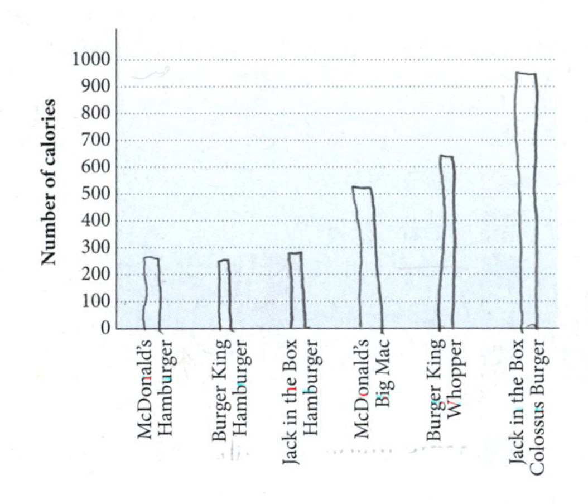

68. Exercise The following table lists the number of calories burned in 1 hour of exercise by a person who weighs 150 pounds. Use the axes in the figure below to construct a bar chart from the information in the table.

CALORIES BURNED BY A 150-POUND PERSON IN ONE HOUR

Activity	Calories
Bicycling	374
Bowling	265
Handball	680
Jazzercise	340
Jogging	680
Skiing	544

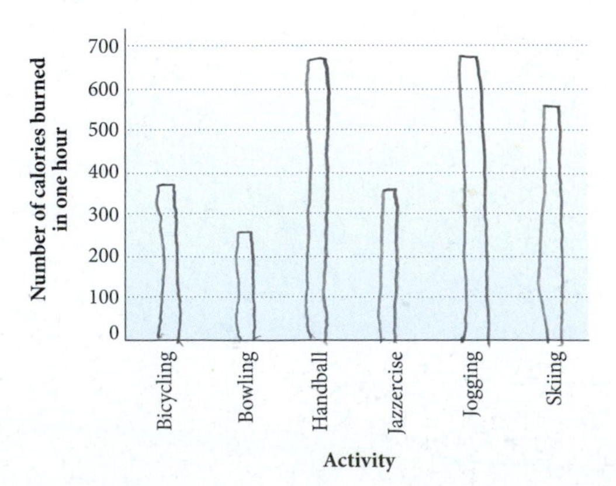

Subtraction with Whole Numbers

1.4

Objectives

A Understand the notation and vocabulary of subtraction.

B Subtract whole numbers.

C Subtraction with borrowing.

D Solving problems with subtraction.

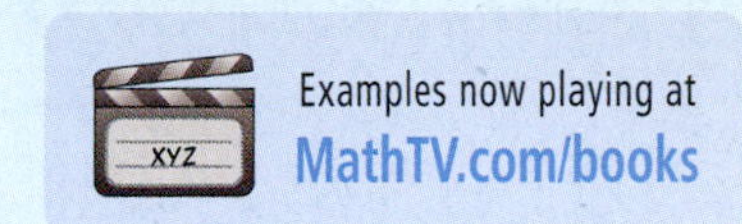

Introduction . . .

In business, subtraction is used to calculate profit. Profit is found by subtracting costs from revenue. The following double bar chart shows the costs and revenue of the Baby Steps Shoe Company during one 4-week period.

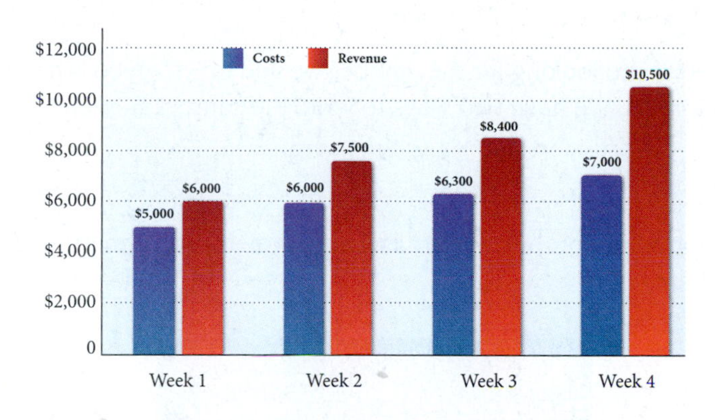

To find the profit for Week 1, we subtract the costs from the revenue, as follows:

Profit = \$6,000 − \$5,000
Profit = \$1,000

Subtraction is the opposite operation of addition. If you understand addition and can work simple addition problems quickly and accurately, then subtraction shouldn't be difficult for you.

A Vocabulary

The word *difference* always indicates subtraction. We can state this in symbols by letting the letters a and b represent numbers.

Definition

The **difference** of two numbers a and b is $a - b$

Table 1 gives some word statements involving subtraction and their mathematical equivalents written in symbols.

TABLE 1

In English	In Symbols
The difference of 9 and 1	$9 - 1$
The difference of 1 and 9	$1 - 9$
The difference of m and 4	$m - 4$
The difference of x and y	$x - y$
3 subtracted from 8	$8 - 3$
2 subtracted from t	$t - 2$
The difference of 7 and 4 is 3	$7 - 4 = 3$
The difference of 9 and 3 is 6	$9 - 3 = 6$

B The Meaning of Subtraction

When we want to subtract 3 from 8, we write

$8 - 3$, 8 subtract 3, or 8 minus 3

The number we are looking for here is the difference between 8 and 3, or the number we add to 3 to get 8. That is:

$8 - 3 = ?$ is the same as $? + 3 = 8$

In both cases we are looking for the number we add to 3 to get 8. The number we are looking for is 5. We have two ways to write the same statement.

Subtraction		Addition
$8 - 3 = 5$	or	$5 + 3 = 8$

For every subtraction problem, there is an equivalent addition problem. Table 2 lists some examples.

TABLE 2

Subtraction		Addition
$7 - 3 = 4$	because	$4 + 3 = 7$
$9 - 7 = 2$	because	$2 + 7 = 9$
$10 - 4 = 6$	because	$6 + 4 = 10$
$15 - 8 = 7$	because	$7 + 8 = 15$

To subtract numbers with two or more digits, we align the numbers vertically and subtract in columns.

PRACTICE PROBLEMS

1. Subtract.
   a. $684 - 431$
   b. $7{,}406 - 3{,}405$

2. a. Subtract 405 from 6,857.
   b. Subtract 234 from 345.

EXAMPLE 1 Subtract: $376 - 241$

SOLUTION We write the problem vertically, aligning digits with the same place value. Then we subtract in columns.

$$\begin{array}{r} 376 \\ -\ 241 \\ \hline 135 \end{array}$$

← Subtract the bottom number in each column from the number above it

EXAMPLE 2 Subtract 503 from 7,835.

SOLUTION In symbols this statement is equivalent to

$7{,}835 - 503$

To subtract we write 503 below 7,835 and then subtract in columns.

$$\begin{array}{rrrr} 7, & 8 & 3 & 5 \\ - & 5 & 0 & 3 \\ \hline 7, & 3 & 3 & 2 \end{array}$$

$5 - 3 = 2$ Ones

$3 - 0 = 3$ Tens

$8 - 5 = 3$ Hundreds

$7 - 0 = 7$ Thousands

Answers

1. a. 253 b. 4,001
2. a. 6,452 b. 111

As you can see, subtraction problems like the ones in Examples 1 and 2 are fairly simple. We write the problem vertically, lining up the digits with the same place value, and subtract in columns. We always subtract the bottom number from the top number.

C Subtraction with Borrowing

Subtraction must involve *borrowing* when the bottom digit in any column is larger than the digit above it. In one sense, borrowing is the reverse of the carrying we did in addition.

EXAMPLE 3 Subtract: $92 - 45$

SOLUTION We write the problem vertically with the place values of the digits showing:

$$\begin{array}{rl} 92 = & 9 \text{ tens} + 2 \text{ ones} \\ -\ 45 = & 4 \text{ tens} \quad\ 5 \text{ ones} \\ \hline \end{array}$$

Look at the ones column. We cannot subtract immediately, because 5 is larger than 2. Instead, we borrow 1 ten from the 9 tens in the tens column. We can rewrite the number 92 as

$$\begin{aligned} & 9 \text{ tens} + 2 \text{ ones} \\ &= 8 \text{ tens} + 1 \text{ ten} + 2 \text{ ones} \\ &= 8 \text{ tens} + 12 \text{ ones} \end{aligned}$$

Now we are in a position to subtract.

$$\begin{array}{rlll} 92 = & 9 \text{ tens} + 2 \text{ ones} & = & 8 \text{ tens} + 12 \text{ ones} \\ -\ 45 = & 4 \text{ tens} \quad\ 5 \text{ ones} & = & 4 \text{ tens} \quad\ \ 5 \text{ ones} \\ \hline & & & 4 \text{ tens} + \ 7 \text{ ones} \end{array}$$

The result is 4 tens + 7 ones, which can be written in standard form as 47.

Writing the problem out in this way is more trouble than is actually necessary. The shorthand form of the same problem looks like this:

$$\begin{array}{rr} \overset{8}{\cancel{9}} & \overset{12}{\cancel{2}} \\ -\ 4 & 5 \\ \hline 4 & 7 \end{array}$$

← This shows we have borrowed 1 ten to go with the 2 ones

$12 - 5 = 7$ Ones

$8 - 4 = 4$ Tens

This shortcut form shows all the necessary work involved in subtraction with borrowing. We will use it from now on.

3. Subtract.
 a. $63 - 47$
 b. $532 - 403$

Note The discussion here shows why borrowing is necessary and how we go about it. To understand borrowing you should pay close attention to this discussion.

Answer
3. a. 16 b. 129

The borrowing that changed 9 tens + 2 ones into 8 tens + 12 ones can be visualized with money.

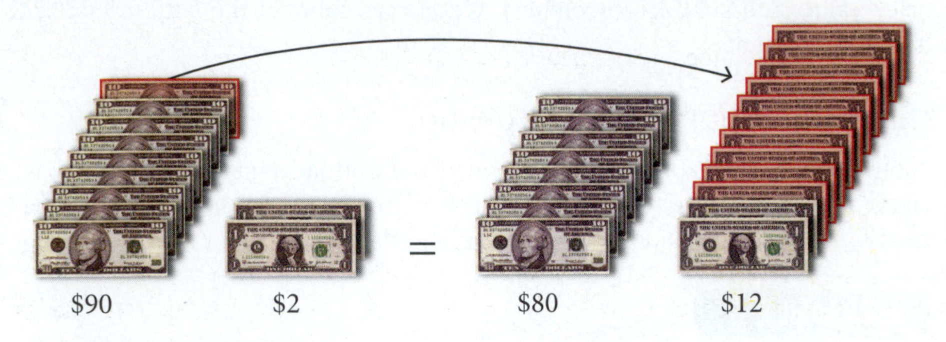

4. a. Find the difference of 656 and 283.
   b. Find the difference of 3,729 and 1,749.

EXAMPLE 4 Find the difference of 549 and 187.

SOLUTION In symbols the difference of 549 and 187 is written

$$549 - 187$$

Writing the problem vertically so that the digits with the same place value are aligned, we have

$$\begin{array}{r} 549 \\ -\,187 \\ \hline \end{array}$$

The top number in the tens column is smaller than the number below it. This means that we will have to borrow from the next larger column.

$$\begin{array}{rrr} \overset{4}{\cancel{5}} & \overset{14}{\cancel{4}} & 9 \\ -\,1 & 8 & 7 \\ \hline 3 & 6 & 2 \end{array}$$

Borrow 1 hundred to go with the 4 tens

$9 - 7 = 2$ Ones

$14 - 8 = 6$ Tens

$4 - 1 = 3$ Hundreds

The actual work we did in borrowing looks like this:

5 hundreds + 4 tens + 9 ones

= 4 hundreds + 1 hundred + 4 tens + 9 ones

= 4 hundreds + 14 tens + 9 ones

Getting Ready for Class

After reading through the preceding section, respond in your own words and in complete sentences.

1. Which sentence below describes the problem shown in Example 1?
   a. The difference of 241 and 376 is 135.
   b. The difference of 376 and 241 is 135.
2. Write a subtraction problem using the number 234 that involves borrowing from the tens column to the ones column.
3. Write a subtraction problem using the number 234 in which the answer is 111.
4. Describe how you would subtract the number 56 from the number 93.

Answers
4. a. 373 b. 1,980

Problem Set 1.4

A Perform the indicated operation. [Examples 1, 2, 4]

1. Subtract 24 from 56.

2. Subtract 71 from 89.

3. Subtract 23 from 45.

4. Subtract 97 from 98.

5. Find the difference of 29 and 19.

6. Find the difference of 37 and 27.

7. Find the difference of 126 and 15.

8. Find the difference of 348 and 32.

B Work each of the following subtraction problems. [Examples 1, 2]

9. $\begin{array}{r} 975 \\ -\ 663 \\ \hline \end{array}$

10. $\begin{array}{r} 480 \\ -\ 260 \\ \hline \end{array}$

11. $\begin{array}{r} 904 \\ -\ 501 \\ \hline \end{array}$

12. $\begin{array}{r} 657 \\ -\ 507 \\ \hline \end{array}$

13. $\begin{array}{r} 9{,}876 \\ -\ 8{,}765 \\ \hline \end{array}$

14. $\begin{array}{r} 5{,}008 \\ -\ 3{,}002 \\ \hline \end{array}$

15. $\begin{array}{r} 7{,}976 \\ -\ 3{,}432 \\ \hline \end{array}$

16. $\begin{array}{r} 6{,}980 \\ -\ \ \ 470 \\ \hline \end{array}$

C Find the difference in each case. (These problems all involve borrowing.) [Example 3]

17. 52 − 37

18. 65 − 48

19. 70 − 37

20. 90 − 21

21. 74 − 69

22. 31 − 28

23. 51 − 18

24. 64 − 58

25. 329 − 234

26. 518 − 492

27. 348 − 196

28. 759 − 661

29. $\begin{array}{r} 932 \\ -\ 658 \\ \hline \end{array}$

30. $\begin{array}{r} 895 \\ -\ 597 \\ \hline \end{array}$

31. $\begin{array}{r} 647 \\ -\ 159 \\ \hline \end{array}$

32. $\begin{array}{r} 842 \\ -\ 199 \\ \hline \end{array}$

33. $\begin{array}{r} 905 \\ -\ 367 \\ \hline \end{array}$

34. $\begin{array}{r} 804 \\ -\ 238 \\ \hline \end{array}$

35. $\begin{array}{r} 600 \\ -\ 437 \\ \hline \end{array}$

36. $\begin{array}{r} 800 \\ -\ 342 \\ \hline \end{array}$

37. $\begin{array}{r} 4{,}583 \\ -\ 2{,}973 \\ \hline \end{array}$

38. $\begin{array}{r} 7{,}849 \\ -\ 2{,}957 \\ \hline \end{array}$

39. $\begin{array}{r} 79{,}040 \\ -\ 32{,}957 \\ \hline \end{array}$

40. $\begin{array}{r} 86{,}492 \\ -\ 78{,}506 \\ \hline \end{array}$

A Complete the following tables.

41.

First Number a	Second Number b	The Difference of a and b $a - b$
25	15	
24	16	
23	17	
22	18	

42.

First Number a	Second Number b	The Difference of a and b $a - b$
90	79	
80	69	
70	59	
60	49	

43.

First Number a	Second Number b	The Difference of a and b $a - b$
400	256	
400	144	
225	144	
225	81	

44.

First Number a	Second Number b	The Difference of a and b $a - b$
100	36	
100	64	
25	16	
25	9	

A Write each of the following expressions in words. Use the word *difference* in each case.

45. $10 - 2$

46. $9 - 5$

47. $a - 6$

48. $7 - x$

49. $8 - 2 = 6$

50. $m - 1 = 4$

51. What number do you subtract from 8 to get 5?

52. What number do you subtract from 6 to get 0?

53. What number do you subtract from 15 to get 7?

54. What number do you subtract from 21 to get 14?

55. What number do you subtract from 35 to get 12?

56. What number do you subtract from 41 to get 11?

A Write each of the following sentences as mathematical expressions.

57. The difference of 8 and 3

58. The difference of x and 2

59. 9 subtracted from y

60. a subtracted from b

61. The difference of 3 and 2 is 1.

62. The difference of 10 and y is 5.

63. The difference of 37 and $9x$ is 10.

64. The difference of $3x$ and $2y$ is 15.

65. The difference of $2y$ and $15x$ is 24.

66. The difference of $25x$ and $9y$ is 16.

67. The difference of $(x + 2)$ and $(x + 1)$ is 1.

68. The difference of $(x - 2)$ and $(x - 4)$ is 2.

D Applying the Concepts

Not all of the following application problems involve only subtraction. Some involve addition as well. Be sure to read each problem carefully.

69. **Checkbook Balance** Diane has $504 in her checking account. If she writes five checks for a total of $249, how much does she have left in her account?

70. **Checkbook Balance** Larry has $763 in his checking account. If he writes a check for each of the three bills listed below, how much will he have left in his account?

Item	Amount
Rent	$418
Phone	25
Car repair	117

71. **Home Prices** In 1985, Mr. Hicks paid $137,500 for his home. He sold it in 2008 for $310,000. What is the difference between what he sold it for and what he bought it for?

72. **Oil Spills** In March 1977, an oil tanker hit a reef off Taiwan and spilled 3,134,500 gallons of oil. In March 1989, an oil tanker hit a reef off Alaska and spilled 10,080,000 gallons of oil. How much more oil was spilled in the 1989 disaster?

73. **Wind Speeds** On April 12, 1934, the wind speed on top of Mount Washington was recorded at 231 miles per hour. When Hurricane Katrina struck on August 28, 2005, the highest recorded wind speed was 140 miles per hour. How much faster was the wind on top of Mount Washington, than the winds from Hurricane Katrina?

74. **Concert Attendance** Eleven thousand, seven hundred fifty-two people attended a recent concert at the Pepsi Arena in Albany, New York. If the arena holds 17,500 people, how many empty seats were there at the concert?

75. Computer Hard Drive You purchase a new computer with 320 gigabytes of hard drive capacity. (A gigabyte is roughly a billion bytes). After loading a variety of programs you discover that you have used 147 gigabytes of your hard drive's capacity. How much hard drive capacity do you still have available?

76. State Size Alaska is the largest state in the United States with an area of 663,267 square miles. Rhode Island is the smallest state with an area of 1,545 square miles. How many more square miles does Alaska have when compared to Rhode Island?

77. Wind Energy The bar chart below shows the states producing the most wind energy in 2006.

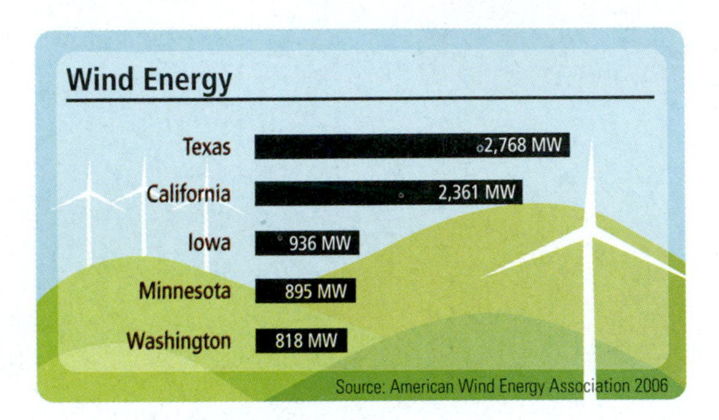

a. Use the information in the bar chart to fill in the missing entries in the table.

State	Energy (megawatts)
Texas	
California	
Iowa	
	818

b. How much more wind energy is produced in Texas than in California?

78. Auto Insurance Costs The bar chart below shows the cities with the highest annual insurance rates in 2006.

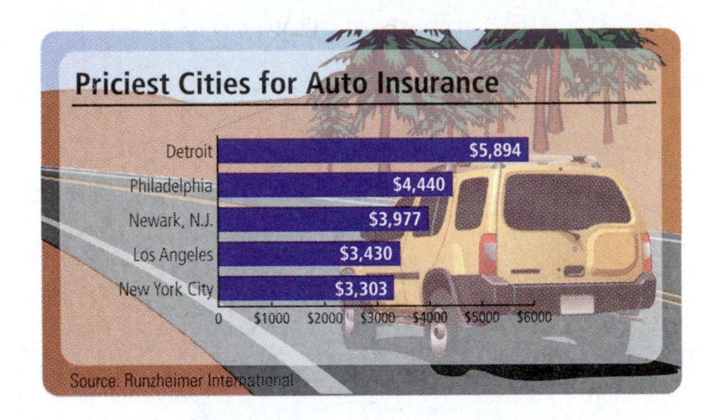

a. Use the information in the bar chart to fill in the missing entries in the table.

City	Cost (dollars)
Detroit	
Philadelphia	
Los Angeles	
	3,303

b. How much more does auto insurance cost in Detroit than in Los Angeles?

Multiplication with Whole Numbers

1.5

Objectives

A Multiply whole numbers.

B Understand the notation and vocabulary of multiplication.

C Identify properties of multiplication.

D Solve equations with multiplication.

E Solve applications with multiplication.

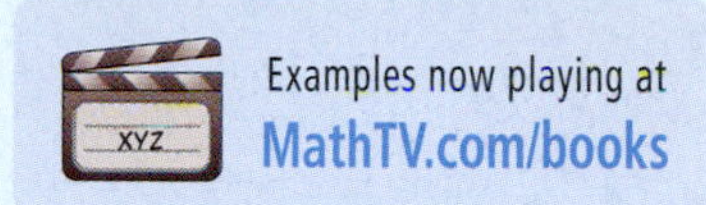

Introduction . . .

A supermarket orders 35 cases of a certain soft drink. If each case contains 12 cans of the drink, how many cans were ordered?

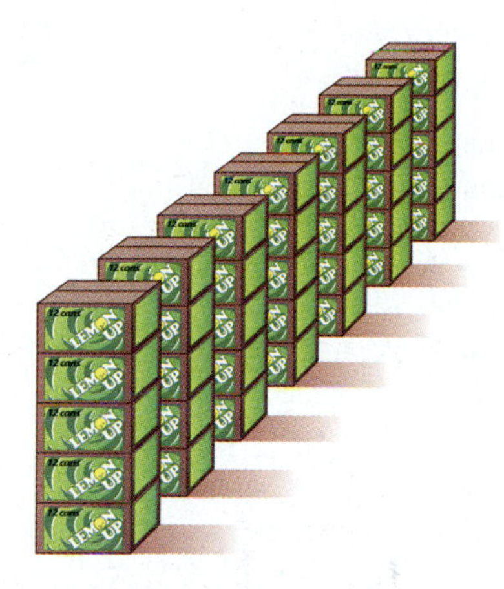

To solve this problem and others like it, we must use multiplication. Multiplication is what we will cover in this section.

A Multiplying Whole Numbers

To begin, we can think of multiplication as shorthand for repeated addition. That is, multiplying 3 times 4 can be thought of this way:

$$3 \text{ times } 4 = 4 + 4 + 4 = 12$$

Multiplying 3 times 4 means to add three 4's. We can write 3 times 4 as 3×4, or $3 \cdot 4$.

EXAMPLE 1 Multiply: $3 \cdot 4{,}000$

SOLUTION Using the definition of multiplication as repeated addition, we have

$$\begin{aligned} 3 \cdot 4{,}000 &= 4{,}000 + 4{,}000 + 4{,}000 \\ &= 12{,}000 \end{aligned}$$

Here is one way to visualize this process.

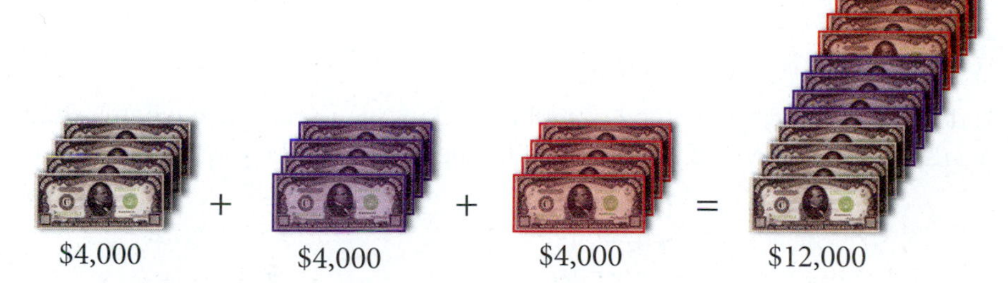

Notice that if we had multiplied 3 and 4 to get 12 and then attached three zeros on the right, the result would have been the same.

PRACTICE PROBLEMS

1. Multiply.
   a. $4 \cdot 70$
   b. $4 \cdot 700$
   c. $4 \cdot 7{,}000$

Answer

1. a. 280
   b. 2,800
   c. 28,000

Note The kind of notation we will use to indicate multiplication will depend on the situation. For example, when we are solving equations that involve letters, it is not a good idea to indicate multiplication with the symbol ×, since it could be confused with the letter *x*. The symbol we will use to indicate multiplication most often in this book is the multiplication dot.

Note We are assuming that you know the basic multiplication facts given in the table below. If you need some practice with these facts, go to Appendix 2 at the back of the book.

BASIC MULTIPLICATION FACTS

×	1	2	3	4	5	6	7	8	9
1	1	2	3	4	5	6	7	8	9
2	2	4	6	8	10	12	14	16	18
3	3	6	9	12	15	18	21	24	27
4	4	8	12	16	20	24	28	32	36
5	5	10	15	20	25	30	35	40	45
6	6	12	18	24	30	36	42	48	54
7	7	14	21	28	35	42	49	56	63
8	8	16	24	32	40	48	56	64	72
9	9	18	27	36	45	54	63	72	81

2. Identify the products and factors in the statement $6 \cdot 7 = 42$

3. Identify the products and factors in the statement $70 = 2 \cdot 5 \cdot 7$

B Notation

There are many ways to indicate multiplication. All the following statements are equivalent. They all indicate multiplication with the numbers 3 and 4.

$$3 \cdot 4, \quad 3 \times 4, \quad 3(4), \quad (3)4, \quad (3)(4), \quad \begin{array}{r} 4 \\ \times\, 3 \\ \hline \end{array}$$

If one or both of the numbers we are multiplying are represented by letters, we may also use the following notation:

$5n$	means	5 times n
ab	means	a times b

B Vocabulary

We use the word *product* to indicate multiplication. If we say "The product of 3 and 4 is 12," then we mean

$$3 \cdot 4 = 12$$

Both $3 \cdot 4$ and 12 are called the product of 3 and 4. The 3 and 4 are called *factors*.

TABLE 1

In English	In Symbols
The product of 2 and 5	$2 \cdot 5$
The product of 5 and 2	$5 \cdot 2$
The product of 4 and n	$4n$
The product of x and y	xy
The product of 9 and 6 is 54	$9 \cdot 6 = 54$
The product of 2 and 8 is 16	$2 \cdot 8 = 16$

EXAMPLE 2 Identify the products and factors in the statement

$$9 \cdot 8 = 72$$

SOLUTION The factors are 9 and 8, and the products are $9 \cdot 8$ and 72.

EXAMPLE 3 Identify the products and factors in the statement

$$30 = 2 \cdot 3 \cdot 5$$

SOLUTION The factors are 2, 3, and 5. The products are $2 \cdot 3 \cdot 5$ and 30.

C Distributive Property

To develop an efficient method of multiplication, we need to use what is called the *distributive property*. To begin, consider the following two problems:

Problem 1	Problem 2
$3(4 + 5)$	$3(4) + 3(5)$
$= 3(9)$	$= 12 + 15$
$= 27$	$= 27$

The result in both cases is the same number, 27. This indicates that the original two expressions must have been equal also. That is,

$$3(4 + 5) = 3(4) + 3(5)$$

Answers

2. Factors: 6, 7; products: $6 \cdot 7$ and 42
3. Factors: 2, 5, 7; products: $2 \cdot 5 \cdot 7$ and 70

This is an example of the distributive property. We say that multiplication *distributes* over addition.

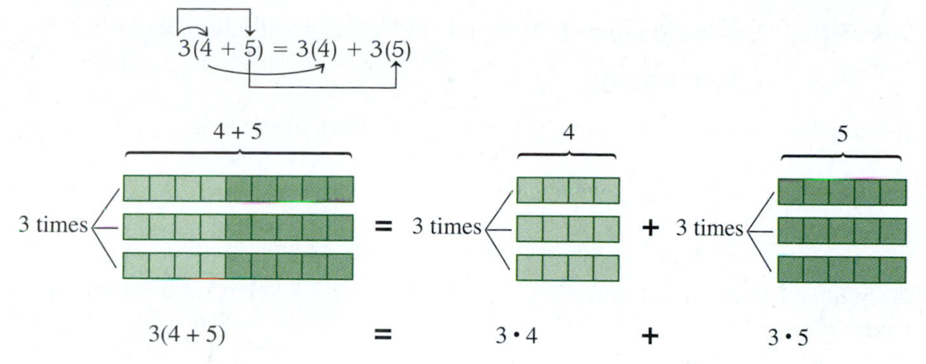

We can write this property in symbols using the letters a, b, and c to represent any three whole numbers.

Distributive Property

If a, b, and c represent any three whole numbers, then

$$a(b + c) = a(b) + a(c)$$

A Multiplication with Whole Numbers

Suppose we want to find the product 7(65). By writing 65 as 60 + 5 and applying the distributive property, we have:

$$\begin{aligned} 7(65) &= 7(60 + 5) && 65 = 60 + 5 \\ &= 7(60) + 7(5) && \text{Distributive property} \\ &= 420 + 35 && \text{Multiplication} \\ &= 455 && \text{Addition} \end{aligned}$$

We can write the same problem vertically like this:

$$\begin{array}{rl} 60 + 5 & \\ \times \quad 7 & \\ \hline 35 & \leftarrow \quad 7(5) = 35 \\ + \quad 420 & \leftarrow \quad 7(60) = 420 \\ \hline 455 & \end{array}$$

This saves some space in writing. But notice that we can cut down on the amount of writing even more if we write the problem this way:

$$\begin{array}{r} \overset{3}{6}5 \\ \times\ 7 \\ \hline 455 \end{array}$$

STEP 2: 7(6) = 42; add the 3 we carried to 42 to get 45

STEP 1: 7(5) = 35; write the 5 in the ones column, and then carry the 3 to the tens column

This shortcut notation takes some practice.

EXAMPLE 4 Multiply: 9(43)

$$\begin{array}{r} \overset{2}{4}3 \\ \times\ 9 \\ \hline 387 \end{array}$$

STEP 2: 9(4) = 36; add the 2 we carried to 36 to get 38

STEP 1: 9(3) = 27; write the 7 in the ones column, and then carry the 2 to the tens column

4. Multiply.
a. 8(57)
b. 8(570)

Answer
4. a. 456 **b.** 4,560

5. Multiply.
   a. 45(62)
   b. 45(620)

Note This discussion is to show why we multiply the way we do. You should go over it in detail, so you will understand the reasons behind the process of multiplication. Besides being able to do multiplication, you should understand it.

6. Multiply.
   a. 356(641)
   b. 3,560(641)

Answers
5. a. 2,790 b. 27,900
6. a. 228,196 b. 2,281,960

EXAMPLE 5

Multiply: 52(37)

SOLUTION This is the same as 52(30 + 7) or by the distributive property

$$52(30) + 52(7)$$

We can find each of these products by using the shortcut method:

$$\begin{array}{r} 52 \\ \times\ 30 \\ \hline 1{,}560 \end{array} \qquad \begin{array}{r} \overset{1}{52} \\ \times\ 7 \\ \hline 364 \end{array}$$

The sum of these two numbers is 1,560 + 364 = 1,924. Here is a summary of what we have so far:

$$\begin{aligned} 52(37) &= 52(30 + 7) && 37 = 30 + 7 \\ &= 52(30) + 52(7) && \text{Distributive property} \\ &= 1{,}560 + 364 && \text{Multiplication} \\ &= 1{,}924 && \text{Addition} \end{aligned}$$

The shortcut form for this problem is

$$\begin{array}{r} 52 \\ \times\ 37 \\ \hline 364 \\ +\ 1{,}560 \\ \hline 1{,}924 \end{array} \quad \begin{array}{l} \\ \\ \leftarrow 7(52) = 364 \\ \leftarrow 30(52) = 1{,}560 \\ \\ \end{array}$$

In this case we have not shown any of the numbers we carried, simply because it becomes very messy.

EXAMPLE 6

Multiply: 279(428)

SOLUTION

$$\begin{array}{r} 279 \\ \times\ 428 \\ \hline 2{,}232 \\ 5{,}580 \\ +\ 111{,}600 \\ \hline 119{,}412 \end{array} \quad \begin{array}{l} \\ \\ \leftarrow 8(279) = 2{,}232 \\ \leftarrow 20(279) = 5{,}580 \\ \leftarrow 400(279) = 111{,}600 \\ \\ \end{array}$$

USING TECHNOLOGY

Calculators

Here is how we would work the problem shown in Example 6 on a calculator:

Scientific Calculator: 279 [×] 428 [=]

Graphing Calculator: 279 [×] 428 [ENT]

Estimating

One way to estimate the answer to the problem shown in Example 6 is to round each number to the nearest hundred and then multiply the rounded numbers. Doing so would give us this:

$$300(400) = 120{,}000$$

Our estimate of the answer is 120,000, which is close to the actual answer, 119,412. Making estimates is important when we are using calculators; having an estimate of the answer will keep us from making major errors in multiplication.

E Applications

EXAMPLE 7 A supermarket orders 35 cases of a certain soft drink. If each case contains 12 cans of the drink, how many cans were ordered?

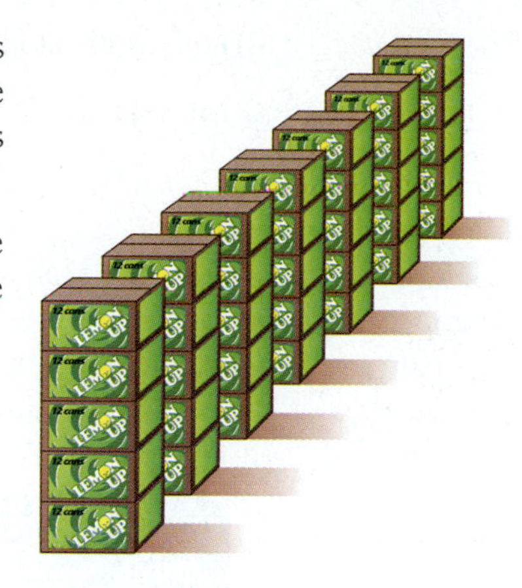

SOLUTION We have 35 cases, and each case has 12 cans. The total number of cans is the product of 35 and 12, which is 35(12):

$$\begin{array}{rll} 12 & & \\ \times\ 35 & & \\ \hline 60 & \longleftarrow & 5(12) = 60 \\ +\ 360 & \longleftarrow & 30(12) = 360 \\ \hline 420 & & \end{array}$$

There is a total of 420 cans of the soft drink.

7. If each tablet of vitamin C contains 550 milligrams of vitamin C, what is the total number of milligrams of vitamin C in a bottle that contains 365 tablets?

EXAMPLE 8 Shirley earns \$12 an hour for the first 40 hours she works each week. If she has \$109 deducted from her weekly check for taxes and retirement, how much money will she take home if she works 38 hours this week?

SOLUTION To find the amount of money she earned for the week, we multiply 12 and 38. From that total we subtract 109. The result is her take-home pay. Without showing all the work involved in the calculations, here is the solution:

$$\begin{array}{rl} 38(\$12) = \$456 & \text{Her total weekly earnings} \\ \$456 - \$109 = \$347 & \text{Her take-home pay} \end{array}$$

8. If Shirley works 36 hours the next week and has the same amount deducted from her check for taxes and retirement, how much will she take home?

EXAMPLE 9 In 1993, the government standardized the way in which nutrition information is presented on the labels of most packaged food products. Figure 1 shows one of these standardized food labels. It is from a package of Fritos Corn Chips that I ate the day I was writing this example. Approximately how many chips are in the bag, and what is the total number of calories consumed if all the chips in the bag are eaten?

Nutrition Facts
Serving Size 1 oz. (28g/About 32 chips)
Servings Per Container: 3

Amount Per Serving	
Calories 160	Calories from fat 90
	% Daily Value*
Total Fat 10 g	**16%**
Saturated Fat 1.5g	**8%**
Cholesterol 0mg	**0%**
Sodium 160mg	**7%**
Total Carbohydrate 15g	**5%**
Dietary Fiber 1g	**4%**
Sugars 0g	
Protein 2g	
Vitamin A 0% •	Vitamin C 0%
Calcium 2% •	Iron 0%

*Percent Daily Values are based on a 2,000 calorie diet

FIGURE 1

SOLUTION Reading toward the top of the label, we see that there are about 32 chips in one serving, and approximately 3 servings in the bag. Therefore, the total number of chips in the bag is

$$3(32) = 96 \text{ chips}$$

9. The amounts given in the middle of the nutrition label in Figure 1 are for one serving of chips. If all the chips in the bag are eaten, how much fat has been consumed? How much sodium?

Note The letter g that is shown after some of the numbers in the nutrition label in Figure 1 stands for grams, a unit used to measure weight. The unit mg stands for milligrams, another, smaller unit of weight. We will have more to say about these units later in the book.

Answers
7. 200,750 milligrams
8. \$323

10. If a 150-pound person bowls for 3 hours, will he or she burn all the calories consumed by eating two bags of the chips mentioned in Example 9?

This is an approximate number, because each serving is approximately 32 chips. Reading further we find that each serving contains 160 calories. Therefore, the total number of calories consumed by eating all the chips in the bag is

$$3(160) = 480 \text{ calories}$$

As we progress through the book, we will study more of the information in nutrition labels.

EXAMPLE 10 The table below lists the number of calories burned in 1 hour of exercise by a person who weighs 150 pounds. Suppose a 150-pound person goes bowling for 2 hours after having eaten the bag of chips mentioned in Example 9. Will he or she burn all the calories consumed from the chips?

Activity	Calories Burned in 1 Hour by a 150-Pound Person
Bicycling	374
Bowling	265
Handball	680
Jazzercize	340
Jogging	680
Skiing	544

SOLUTION Each hour of bowling burns 265 calories. If the person bowls for 2 hours, a total of

$$2(265) = 530 \text{ calories}$$

will have been burned. Because the bag of chips contained only 480 calories, all of them have been burned with 2 hours of bowling.

C More Properties of Multiplication

Multiplication Property of 0
If a represents any number, then

$$a \cdot 0 = 0 \quad \text{and} \quad 0 \cdot a = 0$$

In words: Multiplication by 0 always results in 0.

Multiplication Property of 1
If a represents any number, then

$$a \cdot 1 = a \quad \text{and} \quad 1 \cdot a = a$$

In words: Multiplying any number by 1 leaves that number unchanged.

Commutative Property of Multiplication
If a and b are any two numbers, then

$$ab = ba$$

In words: The order of the numbers in a product doesn't affect the result.

Answers
9. 30 g of fat, 480 mg of sodium
10. No

Associative Property of Multiplication

If a, b, and c represent any three numbers, then

$$(ab)c = a(bc)$$

In words: We can change the grouping of the numbers in a product without changing the result.

To visualize the commutative property, we can think of an instructor with 12 students.

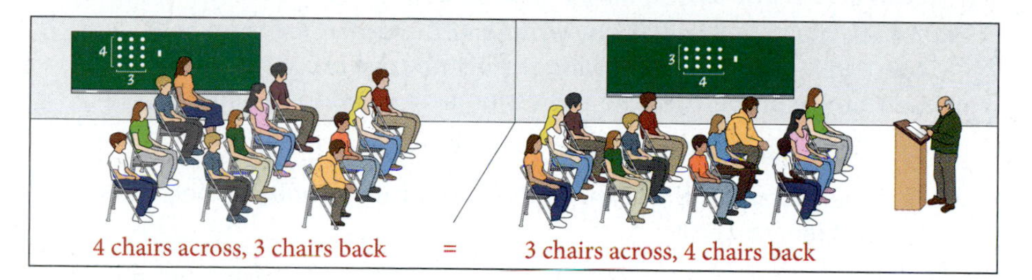

EXAMPLE 11 Use the commutative property of multiplication to rewrite each of the following products:

a. $7 \cdot 9$ b. $4(6)$

SOLUTION Applying the commutative property to each expression, we have:

a. $7 \cdot 9 = 9 \cdot 7$ b. $4(6) = 6(4)$

11. Use the commutative property of multiplication to rewrite each of the following products.
a. $5 \cdot 8$
b. $7(2)$

EXAMPLE 12 Use the associative property of multiplication to rewrite each of the following products:

a. $(2 \cdot 7) \cdot 9$ b. $3 \cdot (8 \cdot 2)$

SOLUTION Applying the associative property of multiplication, we regroup as follows:

a. $(2 \cdot 7) \cdot 9 = 2 \cdot (7 \cdot 9)$ b. $3 \cdot (8 \cdot 2) = (3 \cdot 8) \cdot 2$

12. Use the associative property of multiplication to rewrite each of the following products.
a. $(5 \cdot 7) \cdot 4$
b. $4 \cdot (6 \cdot 4)$

D Solving Equations

If n is used to represent a number, then the equation

$$4 \cdot n = 12$$

is read "4 times n is 12," or "The product of 4 and n is 12." This means that we are looking for the number we multiply by 4 to get 12. The number is 3. Because the equation becomes a true statement if n is 3, we say that 3 is the solution to the equation.

EXAMPLE 13 Find the solution to each of the following equations:

a. $6 \cdot n = 24$ b. $4 \cdot n = 36$ c. $15 = 3 \cdot n$ d. $21 = 3 \cdot n$

SOLUTION
a. The solution to $6 \cdot n = 24$ is 4, because $6 \cdot 4 = 24$.
b. The solution to $4 \cdot n = 36$ is 9, because $4 \cdot 9 = 36$.
c. The solution to $15 = 3 \cdot n$ is 5, because $15 = 3 \cdot 5$.
d. The solution to $21 = 3 \cdot n$ is 7, because $21 = 3 \cdot 7$.

13. Use multiplication facts to find the solution to each of the following equations.
a. $5 \cdot n = 35$
b. $8 \cdot n = 72$
c. $49 = 7 \cdot n$
d. $27 = 9 \cdot n$

Answers
11. a. $8 \cdot 5$ b. $2(7)$
12. a. $5 \cdot (7 \cdot 4)$ b. $(4 \cdot 6) \cdot 4$
13. a. 7 b. 9 c. 7 d. 3

Getting Ready for Class

After reading through the preceding section, respond in your own words and in complete sentences.

1. Use the numbers 7, 8, and 9 to give an example of the distributive property.

2. When we write the distributive property in words, we say "multiplication distributes over addition." It is also true that multiplication distributes over subtraction. Use the variables *a*, *b*, and *c* to write the distributive property using multiplication and subtraction.

3. We can multiply 8 and 487 by writing 487 in expanded form as $400 + 80 + 7$ and then applying the distributive property. Apply the distributive property to the expression below and then simplify.

$$8(400 + 80 + 7) =$$

4. Find the mistake in the following multiplication problem. Then work the problem correctly.

$$\begin{array}{r} 43 \\ \times\ 68 \\ \hline 344 \\ +\ 258 \\ \hline 602 \end{array}$$

Problem Set 1.5

A Multiply each of the following. [Example 1]

1. $3 \cdot 100$ **2.** $7 \cdot 100$ **3.** $3 \cdot 200$ **4.** $4 \cdot 200$ **5.** $6 \cdot 500$ **6.** $8 \cdot 400$

7. $5 \cdot 1{,}000$ **8.** $8 \cdot 1{,}000$ **9.** $3 \cdot 7{,}000$ **10.** $6 \cdot 7{,}000$ **11.** $9 \cdot 9{,}000$ **12.** $7 \cdot 7{,}000$

A Find each of the following products. (Multiply.) In each case use the shortcut method. [Examples 4–6]

13. $\begin{array}{r} 25 \\ \times 4 \\ \hline \end{array}$ **14.** $\begin{array}{r} 43 \\ \times 9 \\ \hline \end{array}$ **15.** $\begin{array}{r} 38 \\ \times 6 \\ \hline \end{array}$ **16.** $\begin{array}{r} 45 \\ \times 7 \\ \hline \end{array}$ **17.** $\begin{array}{r} 18 \\ \times 2 \\ \hline \end{array}$ **18.** $\begin{array}{r} 29 \\ \times 3 \\ \hline \end{array}$

19. $\begin{array}{r} 72 \\ \times 20 \\ \hline \end{array}$ **20.** $\begin{array}{r} 68 \\ \times 30 \\ \hline \end{array}$ **21.** $\begin{array}{r} 19 \\ \times 50 \\ \hline \end{array}$ **22.** $\begin{array}{r} 24 \\ \times 40 \\ \hline \end{array}$ **23.** $\begin{array}{r} 69 \\ \times 25 \\ \hline \end{array}$ **24.** $\begin{array}{r} 27 \\ \times 36 \\ \hline \end{array}$

25. $\begin{array}{r} 11 \\ \times 11 \\ \hline \end{array}$ **26.** $\begin{array}{r} 12 \\ \times 21 \\ \hline \end{array}$ **27.** $\begin{array}{r} 97 \\ \times 16 \\ \hline \end{array}$ **28.** $\begin{array}{r} 24 \\ \times 39 \\ \hline \end{array}$ **29.** $\begin{array}{r} 168 \\ \times 25 \\ \hline \end{array}$ **30.** $\begin{array}{r} 452 \\ \times 34 \\ \hline \end{array}$

31. 728×91

32. 680×76

33. 698×400

34. 879×600

35. 111×111

36. 123×321

37. 532×200

38. 277×900

39. 856×232

40. 455×248

41. 976×628

42. 432×555

43. $2,468 \times 135$

44. $2,725 \times 324$

45. $24,563 \times 735$

46. $56,728 \times 852$

47. $44,777 \times 5,888$

48. $33,999 \times 2,555$

B Complete the following tables.

49.

First Number a	Second Number b	Their Product ab
11	11	
11	22	
22	22	
22	44	

50.

First Number a	Second Number b	Their Product ab
25	15	
25	30	
50	15	
50	30	

51.

First Number a	Second Number b	Their Product ab
25	10	
25	100	
25	1,000	
25	10,000	

52.

First Number a	Second Number b	Their Product ab
11	111	
11	222	
22	111	
22	222	

53.

First Number a	Second Number b	Their Product ab
12	20	
36	20	
12	40	
36	40	

54.

First Number a	Second Number b	Their Product ab
10	12	
100	12	
1,000	12	
10,000	12	

B Write each of the following expressions in words, using the word *product*.

55. $6 \cdot 7$

56. $9(4)$

57. $2 \cdot n$

58. $5 \cdot x$

59. $9 \cdot 7 = 63$

60. $(5)(6) = 30$

B Write each of the following in symbols.

61. The product of 7 and n

62. The product of 9 and x

63. The product of 6 and 7 is 42.

64. The product of 8 and 9 is 72.

65. The product of 0 and 6 is 0.

66. The product of 1 and 6 is 6.

B Identify the products in each statement.

67. $9 \cdot 7 = 63$

68. $2(6) = 12$

69. $4(4) = 16$

70. $5 \cdot 5 = 25$

B Identify the factors in each statement.

71. $2 \cdot 3 \cdot 4 = 24$

72. $6 \cdot 1 \cdot 5 = 30$

73. $12 = 2 \cdot 2 \cdot 3$

74. $42 = 2 \cdot 3 \cdot 7$

C Rewrite each of the following using the commutative property of multiplication. [Example 11]

75. $5(9)$

76. $4(3)$

77. $6 \cdot 7$

78. $8 \cdot 3$

C Rewrite each of the following using the associative property of multiplication. [Example 12]

79. $2 \cdot (7 \cdot 6)$

80. $4 \cdot (8 \cdot 5)$

81. $3 \times (9 \times 1)$

82. $5 \times (8 \times 2)$

C Use the distributive property to rewrite each expression, then simplify.

83. $7(2 + 3)$ **84.** $4(5 + 8)$ **85.** $9(4 + 7)$ **86.** $6(9 + 5)$

87. $3(x + 1)$ **88.** $5(x + 8)$ **89.** $2(x + 5)$ **90.** $4(x + 3)$

D Find a solution for each equation. [Example 13]

91. $4 \cdot n = 12$ **92.** $3 \cdot n = 12$ **93.** $9 \cdot n = 81$ **94.** $6 \cdot n = 36$

95. $0 = n \cdot 5$ **96.** $6 = 1 \cdot n$

E Applying the Concepts

Most, but not all, of the application problems that follow require multiplication. Read the problems carefully before trying to solve them.

97. Planning a Trip A family decides to drive their compact car on their vacation. They figure it will require a total of about 130 gallons of gas for the vacation. If each gallon of gas will take them 22 miles, how long is the trip they are planning?

98. Rent A student pays $675 rent each month. How much money does she spend on rent in 2 years?

99. Downloading Songs You receive a gift card for the Apple™ iTunes™ store for $25.00 and download 18 songs at $0.99 per song. How much is left on your gift card?

100. Cost of Building a Home When you consider building a new home it is helpful to be able to estimate the cost of building that house. A simple way to do this is to multiply the number of square feet under the roof of the house by the average building cost per square foot. Suppose you contact a builder who estimates that, on average, he charges $142.00 per square foot. Determine the cost to build a 2,067 square foot house.

101. World's Busiest Airport Atlanta, Georgia is home to the world's busiest airport, Hartsfield-Jackson Atlanta International Airport. According to the Federal Aviation Administration about 50 jets can land and take off every 15 minutes which is about 200 jets an hour. About how many jets land and take off in the month of July?

102. Flowers It is probably no surprise that Valentine's Day is the busiest day of the year for florists. It is estimated that 214 million roses were produced for Valentine's Day in 2007 (Source: Society of American Florists). If a single rose costs a consumer $2.50, what was the total revenue for the roses produced?

Exercise and Calories The table below is an extension of the table we used in Example 10 of this section. It gives the amount of energy expended during 1 hour of various activities for people of different weights. The accompanying figure is a nutrition label from a bag of Doritos tortilla chips. Use the information from the table and the nutrition label to answer Problems 103–108.

CALORIES BURNED THROUGH EXERCISE

Activity	Calories Per Hour 120 Pounds	150 Pounds	180 Pounds
Bicycling	299	374	449
Bowling	212	265	318
Handball	544	680	816
Jazzercise	272	340	408
Jogging	544	680	816
Skiing	435	544	653

Nutrition Facts

Serving Size 1 oz. (28g/About 12 chips)
Servings Per Container About 2

Amount Per Serving	
Calories 140	Calories from fat 60
	% Daily Value*
Total Fat 7g	**11%**
Saturated Fat 1g	**6%**
Cholesterol 0mg	**0%**
Sodium 170mg	**7%**
Total Carbohydrate 18g	**6%**
Dietary Fiber 1g	**4%**
Sugars less than 1g	
Protein 2g	
Vitamin A 0% •	Vitamin C 0%
Calcium 4% •	Iron 2%

*Percent Daily Values are based on a 2,000 calorie diet

103. Suppose you weigh 180 pounds. How many calories would you burn if you play handball for 2 hours and then ride your bicycle for 1 hour?

104. How many calories are burned by a 120-lb person who jogs for 1 hour and then goes bike riding for 2 hours?

105. How many calories would you consume if you ate the entire bag of chips?

106. Approximately how many chips are in the bag?

107. If you weigh 180 pounds, will you burn off the calories consumed by eating 3 servings of tortilla chips if you ride your bike 1 hour?

108. If you weigh 120 pounds, will you burn off the calories consumed by eating 3 servings of tortilla chips if you ride your bike for 1 hour?

Estimating

Mentally estimate the answer to each of the following problems by rounding each number to the indicated place and then multiplying.

109. 750 hundred
$\times$ 12 ten

110. 591 hundred
$\times$ 323 hundred

111. 3,472 thousand
$\times$ 511 hundred

112. 399 hundred
$\times$ 298 hundred

113. 2,399 thousand
$\times$ 698 hundred

114. 9,999 thousand
$\times$ 666 hundred

Extending the Concepts: Number Sequences

A *geometric sequence* is a sequence of numbers in which each number is obtained from the previous number by multiplying by the same number each time. For example, the sequence 3, 6, 12, 24, . . . is a geometric sequence, starting with 3, in which each number comes from multiplying the previous number by 2. Find the next number in each of the following geometric sequences.

115. 5, 10, 20, . . .

116. 10, 50, 250, . . .

117. 2, 6, 18, . . .

118. 12, 24, 48, . . .

Division with Whole Numbers

1.6

Objectives

A Understand the notation and vocabulary of division.

B Divide whole numbers.

C Solve applications using division.

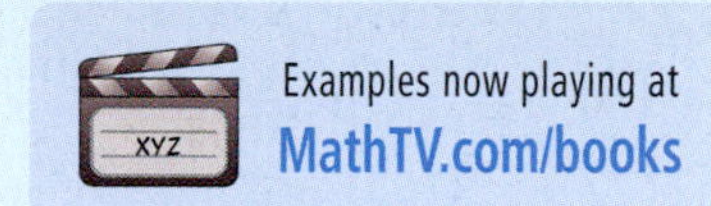

Introduction . . .

Darlene is planning a party and would like to serve 8-ounce glasses of soda. The glasses will be filled from 32-ounce bottles of soda. In order to know how many bottles of soda to buy, she needs to find out how many of the 8-ounce glasses can be filled by one of the 32-ounce bottles. One way to solve this problem is with division: dividing 32 by 8. A diagram of the problem is shown in Figure 1.

FIGURE 1

As a division problem:	As a multiplication problem:
$32 \div 8 = 4$	$4 \cdot 8 = 32$

A Notation

As was the case with multiplication, there are many ways to indicate division. All the following statements are equivalent. They all mean 10 divided by 5.

$$10 \div 5, \quad \frac{10}{5}, \quad 10/5, \quad 5\overline{)10}$$

The kind of notation we use to write division problems will depend on the situation. We will use the notation $5\overline{)10}$ mostly with the long-division problems found in this chapter. The notation $\frac{10}{5}$ will be used in the chapter on fractions and in later chapters. The horizontal line used with the notation $\frac{10}{5}$ is called the *fraction bar.*

A Vocabulary

The word *quotient* is used to indicate division. If we say "The quotient of 10 and 5 is 2," then we mean

$$10 \div 5 = 2 \quad \text{or} \quad \frac{10}{5} = 2$$

The 10 is called the *dividend,* and the 5 is called the *divisor.* All the expressions, $10 \div 5$, $\frac{10}{5}$, and 2, are called the *quotient* of 10 and 5.

TABLE 1

In English	In Symbols
The quotient of 15 and 3	$15 \div 3$, or $\frac{15}{3}$, or 15/3
The quotient of 3 and 15	$3 \div 15$, or $\frac{3}{15}$, or 3/15
The quotient of 8 and n	$8 \div n$, or $\frac{8}{n}$, or $8/n$
x divided by 2	$x \div 2$, or $\frac{x}{2}$, or $x/2$
The quotient of 21 and 3 is 7	$21 \div 3 = 7$, or $\frac{21}{3} = 7$

The Meaning of Division

One way to arrive at an answer to a division problem is by thinking in terms of multiplication. For example, if we want to find the quotient of 32 and 8, we may ask, "What do we multiply by 8 to get 32?"

$$32 \div 8 = ? \qquad \text{means} \qquad 8 \cdot ? = 32$$

Because we know from our work with multiplication that $8 \cdot 4 = 32$, it must be true that

$$32 \div 8 = 4$$

Table 2 lists some additional examples.

TABLE 2

Division		Multiplication
$18 \div 6 = 3$	because	$6 \cdot 3 = 18$
$32 \div 8 = 4$	because	$8 \cdot 4 = 32$
$10 \div 2 = 5$	because	$2 \cdot 5 = 10$
$72 \div 9 = 8$	because	$9 \cdot 8 = 72$

B Division by One-Digit Numbers

Consider the following division problem:

$$465 \div 5$$

We can think of this problem as asking the question, "How many fives can we subtract from 465?" To answer the question we begin subtracting multiples of 5. One way to organize this process is shown below:

```
   90    ⟵ We first guess that there are at least 90 fives in 465
 5)465
 − 450   ⟵ 90(5) = 450
 -----
    15   ⟵ 15 is left after we subtract 90 fives from 465
```

What we have done so far is subtract 90 fives from 465 and found that 15 is still left. Because there are 3 fives in 15, we continue the process.

```
     3   ⟵ There are 3 fives in 15
    90
 5)465
  −450
    15
  − 15   ⟵ 3 · 5 = 15
     0   ⟵ The difference is 0
```

The total number of fives we have subtracted from 465 is

$90 + 3 = 93$

We now summarize the results of our work.

$465 \div 5 = 93$ which we check with multiplication ⟶ $\begin{array}{r} \overset{1}{9}3 \\ \times\ 5 \\ \hline 465 \end{array}$

A Notation

The division problem just shown can be shortened by eliminating the subtraction signs, eliminating the zeros in each estimate, and eliminating some of the numbers that are repeated in the problem.

```
The shorthand       3     looks like      93     The arrow
form for this      90     this.        5)465     indicates that
problem         5)465                    45↓     we bring down
                  450                     15     the 5 after
                   15                     15     we subtract.
                   15                      0
                    0
```

The problem shown above on the right is the shortcut form of what is called *long division.* Here is an example showing this shortcut form of long division from start to finish.

EXAMPLE 1 Divide: $595 \div 7$

SOLUTION Because $7(8) = 56$, our first estimate of the number of sevens that can be subtracted from 595 is 80:

```
     8   ⟵ The 8 is placed above the tens column
 7)595      so we know our first estimate is 80
   56↓   ⟵ 8(7) = 56
    35   ⟵ 59 − 56 = 3; then bring down the 5
```

Since $7(5) = 35$, we have

```
    85   ⟵ There are 5 sevens in 35
 7)595
   56↓
    35
    35   ⟵ 5(7) = 35
     0   ⟵ 35 − 35 = 0
```

Our result is $595 \div 7 = 85$, which we can check with multiplication:

$$\begin{array}{r} \overset{3}{8}5 \\ \times\ 7 \\ \hline 595 \end{array}$$

PRACTICE PROBLEMS

1. Divide.

a. $296 \div 4$

b. $2{,}960 \div 4$

Answer

1. a. 74 **b.** 740

2. Divide.
   a. 6,792 ÷ 24
   b. 67,920 ÷ 24

B Division by Two-Digit Numbers

EXAMPLE 2 Divide: 9,380 ÷ 35

SOLUTION In this case our divisor, 35, is a two-digit number. The process of division is the same. We still want to find the number of thirty-fives we can subtract from 9,380.

```
      2     ← The 2 is placed above the hundreds column
35)9,380
   7 0↓     ← 2(35) = 70
   2 38     ← 93 − 70 = 23; then bring down the 8
```

We can make a few preliminary calculations to help estimate how many thirty-fives are in 238:

$$5 \times 35 = 175 \qquad 6 \times 35 = 210 \qquad 7 \times 35 = 245$$

Because 210 is the closest to 238 without being larger than 238, we use 6 as our next estimate:

```
     26     ← 6 in the tens column means this estimate is 60
35)9,380
   7 0
   2 38
   2 10↓    ← 6(35) = 210
     280    ← 238 − 210 = 28; bring down the 0
```

Because 35(8) = 280, we have

```
     268
35)9,380
   7 0
   2 38
   2 10
     280
     280    ← 8(35) = 280
       0    ← 280 − 280 = 0
```

We can check our result with multiplication:

```
    268
  ×  35
  1,340
  8,040
  9,380
```

3. Divide.
   1,872 ÷ 9

EXAMPLE 3 Divide: 1,872 by 18.

SOLUTION Here is the first step.

```
      1     ← 1 is placed above hundred column
18)1,872
   1 8      ← Multiply 1(18) to get 18
     0      ← Subtract to get 0
```

Answer
2. a. 283 b. 2,830

The next step is to bring down the 7 and divide again.

```
     10   ⟵ 0 is placed above tens column. 0 is the largest number
18)1,872     we can multiply by 18 and not go over 7
   1 8↓
   ---
    07
     0    ⟵ Multiply 0(18) to get 0
   ---
     7    ⟵ Subtract to get 7
```

Here is the complete problem.

```
     104
18)1,872
   1 8↓|
   ---
    07 |
     0↓
   ---
     72
     72
     --
      0
```

To show our answer is correct, we multiply.

$$18(104) = 1{,}872$$

B Division with Remainders

Suppose Darlene was planning to use 6-ounce glasses instead of 8-ounce glasses for her party. To see how many glasses she could fill from the 32-ounce bottle, she would divide 32 by 6. If she did so, she would find that she could fill 5 glasses, but after doing so she would have 2 ounces of soda left in the bottle. A diagram of this problem is shown in Figure 2.

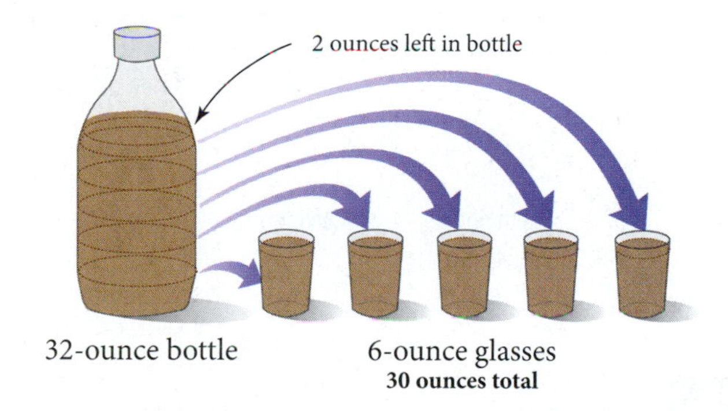

FIGURE 2

Writing the results in the diagram as a division problem looks like this:

```
                  5  ⟵ Quotient
Divisor ⟶      6)32  ⟵ Dividend
                 30
                 --
                  2  ⟵ Remainder
```

3. 208

4. Divide.
   a. 1,883 ÷ 27
   b. 1,883 ÷ 18

EXAMPLE 4 Divide: 1,690 ÷ 67

SOLUTION Dividing as we have previously, we get

```
     25
67)1,690
   1 34↓
     350
     335
      15 ⟵ 15 is left over
```

We have 15 left, and because 15 is less than 67, no more sixty-sevens can be subtracted. In a situation like this we call 15 the *remainder* and write

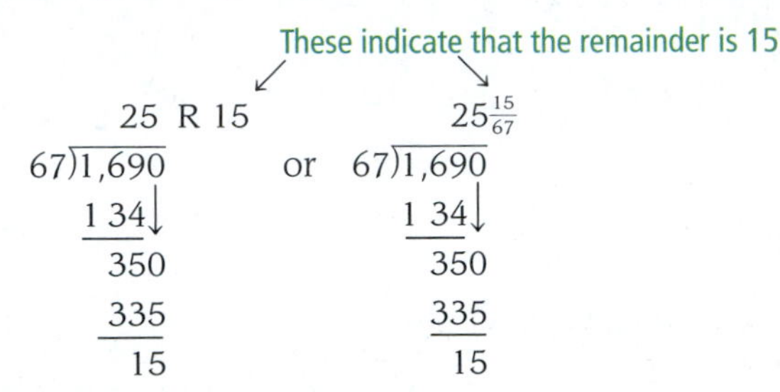

CALCULATOR NOTE

Here is how we would work the problem shown in Example 4 on a calculator:

SCIENTIFIC CALCULATOR:

1690 [÷] 67 [=]

GRAPHING CALCULATOR:

1690 [÷] 67 [ENT]

In both cases the calculator will display 25.223881 (give or take a few digits at the end), which gives the remainder in decimal form. We will discuss decimals later in the book.

Both forms of notation shown above indicate that 15 is the remainder. The notation R 15 is the notation we will use in this chapter. The notation $\frac{15}{67}$ will be useful in the chapter on fractions.

To check a problem like this, we multiply the divisor and the quotient as usual, and then add the remainder to this result:

```
   67
 × 25
  335
1,340
1,675 ⟵ Product of divisor and quotient
```

1,675 + 15 = 1,690

(↑ Remainder, ↖ Dividend)

5. A family spends $1,872 on a 12-day vacation. How much did they spend each day on average?

C Applications

EXAMPLE 5 A family has an annual income of $35,880. How much is their average monthly income?

SOLUTION Because there are 12 months in a year and the yearly (annual) income is $35,880, we want to know what $35,880 divided into 12 equal parts is. Therefore we have

```
    2 990
12)35,880
   24↓
   11 8
   10 8↓
    1 08
    1 08
       00
```

Because 35,880 ÷ 12 = 2,990, the monthly income for this family is $2,990.

Note To estimate the answer to Example 5 quickly, we can replace 35,880 with 36,000 and mentally calculate

36,000 ÷ 12

which gives an estimate of 3,000. Our actual answer, 2,990, is close enough to our estimate to convince us that we have not made a major error in our calculation.

Answers

4. a. 69 R 20, or $69\frac{20}{27}$
   b. 104 R 11, or $104\frac{11}{18}$
5. $156

Division by Zero

We cannot divide by 0. That is, we cannot use 0 as a divisor in any division problem. Here's why.

Suppose there was an answer to the problem

$$\frac{8}{0} = ?$$

That would mean that

$$0 \cdot ? = 8$$

But we already know that multiplication by 0 always produces 0. There is no number we can use for the ? to make a true statement out of

$$0 \cdot ? = 8$$

Because this was equivalent to the original division problem

$$\frac{8}{0} = ?$$

we have no number to associate with the expression $\frac{8}{0}$. It is undefined.

Rule

Division by 0 is undefined. Any expression with a divisor of 0 is undefined. We cannot divide by 0.

Getting Ready for Class

After reading through the preceding section, respond in your own words and in complete sentences.

1. Which sentence below describes the problem shown in Example 1?

a. The quotient of 7 and 595 is 85.

b. Seven divided by 595 is 85.

c. The quotient of 595 and 7 is 85.

2. In Example 2, we divide 9,380 by 35 to obtain 268. Suppose we add 35 to 9,380, making it 9,415. What will our answer be if we divide 9,415 by 35?

3. Example 4 shows that 1,690 ÷ 67 gives a quotient of 25 with a remainder of 15. If we were to divide 1,692 by 67, what would the remainder be?

4. Explain why division by 0 is undefined in mathematics.

Problem Set 1.6

A Write each of the following in symbols.

1. The quotient of 6 and 3
2. The quotient of 3 and 6
3. The quotient of 45 and 9
4. The quotient of 12 and 4
5. The quotient of r and s
6. The quotient of s and r
7. The quotient of 20 and 4 is 5.
8. The quotient of 20 and 5 is 4.

Write a multiplication statement that is equivalent to each of the following division statements.

9. $6 \div 2 = 3$
10. $6 \div 3 = 2$
11. $\frac{36}{9} = 4$
12. $\frac{36}{4} = 9$
13. $\frac{48}{6} = 8$
14. $\frac{35}{7} = 5$
15. $28 \div 7 = 4$
16. $81 \div 9 = 9$

B Find each of the following quotients. (Divide.) [Examples 1–3]

17. $25 \div 5$
18. $72 \div 8$
19. $40 \div 5$
20. $12 \div 2$
21. $9 \div 0$
22. $7 \div 1$
23. $360 \div 8$
24. $285 \div 5$
25. $\frac{138}{6}$
26. $\frac{267}{3}$
27. $5\overline{)7{,}650}$
28. $5\overline{)5{,}670}$
29. $5\overline{)6{,}750}$
30. $5\overline{)6{,}570}$
31. $3\overline{)54{,}000}$
32. $3\overline{)50{,}400}$
33. $3\overline{)50{,}040}$
34. $3\overline{)50{,}004}$

Estimating

Work Problems 35 through 38 mentally, without using a calculator.

35. The quotient $845 \div 93$ is closest to which of the following numbers?
a. 10 **b.** 100 **c.** 1,000 **d.** 10,000

36. The quotient $762 \div 43$ is closest to which of the following numbers?
a. 2 **b.** 20 **c.** 200 **d.** 2,000

37. The quotient $15{,}208 \div 771$ is closest to which of the following numbers?
a. 2 **b.** 20 **c.** 200 **d.** 2,000

38. The quotient $24{,}471 \div 523$ is closest to which of the following numbers?
a. 5 **b.** 50 **c.** 500 **d.** 5,000

Mentally give a one-digit estimate for each of the following quotients. That is, for each quotient, mentally estimate the answer using one of the digits 1, 2, 3, 4, 5, 6, 7, 8, or 9.

39. $316 \div 289$ **40.** $662 \div 289$ **41.** $728 \div 355$ **42.** $728 \div 177$

43. $921 \div 243$ **44.** $921 \div 442$ **45.** $673 \div 109$ **46.** $673 \div 218$

B Divide. You shouldn't have any wrong answers because you can always check your results with multiplication. [Examples 1–3]

47. $1{,}440 \div 32$ **48.** $1{,}206 \div 67$ **49.** $\dfrac{2{,}401}{49}$ **50.** $\dfrac{4{,}606}{49}$

51. $28\overline{)12{,}096}$ **52.** $28\overline{)96{,}012}$ **53.** $63\overline{)90{,}594}$ **54.** $45\overline{)17{,}595}$

55. $87\overline{)61{,}335}$ **56.** $79\overline{)48{,}032}$ **57.** $45\overline{)135{,}900}$ **58.** $56\overline{)227{,}920}$

B Complete the following tables.

59.

First Number a	Second Number b	The Quotient of a and b $\frac{a}{b}$
100	25	
100	26	
100	27	
100	28	

60.

First Number a	Second Number b	The Quotient of a and b $\frac{a}{b}$
100	25	
101	25	
102	25	
103	25	

B The following division problems all have remainders. [Example 4]

61. $6\overline{)370}$

62. $8\overline{)390}$

63. $3\overline{)271}$

64. $3\overline{)172}$

65. $26\overline{)345}$

66. $26\overline{)543}$

67. $71\overline{)16,620}$

68. $71\overline{)33,240}$

69. $23\overline{)9,250}$

70. $23\overline{)20,800}$

71. $169\overline{)5,950}$

72. $391\overline{)34,450}$

C Applying the Concepts [Example 5]

The application problems that follow may involve more than merely division. Some may require addition, subtraction, or multiplication, whereas others may use a combination of two or more operations.

73. Monthly Income A family has an annual income of $42,300. How much is their monthly income?

74. Hourly Wages If a man works an 8-hour shift and is paid $96, how much does he make for 1 hour?

75. Price per Pound If 6 pounds of a certain kind of fruit cost $4.74, how much does 1 pound cost?

76. Cost of a Dress A dress shop orders 45 dresses for a total of $2,205. If they paid the same amount for each dress, how much was each dress?

77. Filling Glasses How many 32-ounce bottles of Coke will be needed to fill sixteen 6-ounce glasses?

78. Filling Glasses How many 8-ounce glasses can be filled from three 32-ounce bottles of soda?

79. Filling Glasses How many 5-ounce glasses can be filled from a 32-ounce bottle of milk? How many ounces of milk will be left in the bottle when all the glasses are full?

80. Filling Glasses How many 3-ounce glasses can be filled from a 28-ounce bottle of milk? How many ounces of milk will be left in the bottle when all the glasses are filled?

81. Boston Red Sox The annual payroll for the Boston Red Sox for the 2007 season was about $156 million dollars. If there are 40 players on the roster what is the average salary per player for the Boston Red Sox?

82. Miles per Gallon A traveling salesman kept track of his mileage for 1 month. He found that he traveled 1,104 miles and used 48 gallons of gas. How many miles did he travel on each gallon of gas?

83. Milligrams of Calcium Suppose one egg contains 25 milligrams of calcium, a piece of toast contains 40 milligrams of calcium, and a glass of milk contains 215 milligrams of calcium. How many milligrams of calcium are contained in a breakfast that consists of three eggs, two glasses of milk, and four pieces of toast?

84. Milligrams of Iron Suppose a glass of juice contains 3 milligrams of iron and a piece of toast contains 2 milligrams of iron. If Diane drinks two glasses of juice and has three pieces of toast for breakfast, how much iron is contained in the meal?

85. Fitness Walking The guidelines for fitness now indicate that a person who walks 10,000 steps daily is physically fit. According to *The Walking Site* on the Internet, it takes just over 2,000 steps to walk one mile. If that is the case, how many miles do you need to walk in order to take 10,000 steps?

86. Fundraiser As part of a fundraiser for the Earth Day activities on your campus, three volunteers work to stuff 3,210 envelopes with information about global warming. How many envelopes did each volunteer stuff?

Exponents, Order of Operations, and Averages

1.7

Objectives

A Identify the base and exponent of an expression.

B Simplify expressions with exponents.

C Use the rule for order of operations.

D Find the mean, median, mode, and range of a set of numbers.

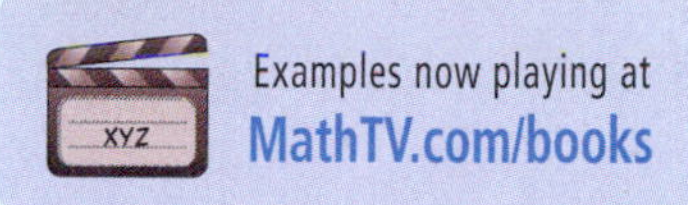

Exponents are a shorthand way of writing repeated multiplication. In the expression 2^3, 2 is called the *base* and 3 is called the *exponent*. The expression 2^3 is read "2 to the third power" or "2 cubed." The exponent 3 tells us to use the base 2 as a multiplication factor three times.

$$2^3 = 2 \cdot 2 \cdot 2 \quad \text{2 is used as a factor three times}$$

We can simplify the expression by multiplication:

$$\begin{aligned} 2^3 &= 2 \cdot 2 \cdot 2 \\ &= 4 \cdot 2 \\ &= 8 \end{aligned}$$

The expression 2^3 is equal to the number 8. We can summarize this discussion with the following definition.

Definition

An **exponent** is a whole number that indicates how many times the base is to be used as a factor. Exponents indicate repeated multiplication.

A Exponents

In the expression 5^2, 5 is the base and 2 is the exponent. The meaning of the expression is

$$\begin{aligned} 5^2 &= 5 \cdot 5 \quad \text{5 is used as a factor two times} \\ &= 25 \end{aligned}$$

The expression 5^2 is read "5 to the second power" or "5 squared."

Here are some more examples.

EXAMPLE 1 3^2 The base is 3, and the exponent is 2. The expression is read "3 to the second power" or "3 squared."

EXAMPLE 2 3^3 The base is 3, and the exponent is 3. The expression is read "3 to the third power" or "3 cubed."

EXAMPLE 3 2^4 The base is 2, and the exponent is 4. The expression is read "2 to the fourth power."

As you can see from these examples, a base raised to the second power is also said to be *squared,* and a base raised to the third power is also said to be *cubed.* These are the only two exponents (^{2} and ^{3}) that have special names. All other exponents are referred to only as "fourth powers," "fifth powers," "sixth powers," and so on.

PRACTICE PROBLEMS

For each expression, name the base and the exponent, and write the expression in words.

1. 5^2

2. 2^3

3. 1^4

Answers

1–3. See solutions section.

B Expressions with Exponents

The next examples show how we can simplify expressions involving exponents by using repeated multiplication.

Simplify each of the following by using repeated multiplication.

4. 5^2

5. 9^2

6. 2^3

7. 1^4

8. 2^5

EXAMPLE 4 $3^2 = 3 \cdot 3 = 9$

EXAMPLE 5 $4^2 = 4 \cdot 4 = 16$

EXAMPLE 6 $3^3 = 3 \cdot 3 \cdot 3 = 9 \cdot 3 = 27$

EXAMPLE 7 $3^4 = 3 \cdot 3 \cdot 3 \cdot 3 = 9 \cdot 9 = 81$

EXAMPLE 8 $2^4 = 2 \cdot 2 \cdot 2 \cdot 2 = 4 \cdot 4 = 16$

USING TECHNOLOGY

Calculators

Here is how we use a calculator to evaluate exponents, as we did in Example 8:

Scientific Calculator: 2 [x^y] 4 [=]

Graphing Calculator: 2 [^] 4 [ENT] or 2 [x^y] 4 [ENT]
(depending on the calculator)

Finally, we should consider what happens when the numbers 0 and 1 are used as exponents. First of all, any number raised to the first power is itself. That is, if we let the letter a represent any number, then

$$a^1 = a$$

To take care of the cases when 0 is used as an exponent, we must use the following definition:

Definition

Any number other than 0 raised to the 0 power is 1. That is, if a represents any nonzero number, then it is always true that

$$a^0 = 1$$

Simplify each of the following expressions.

9. 7^1

10. 4^1

11. 9^0

12. 1^0

EXAMPLE 9 $5^1 = 5$

EXAMPLE 10 $9^1 = 9$

EXAMPLE 11 $4^0 = 1$

EXAMPLE 12 $8^0 = 1$

Answers
4. 25 **5.** 81 **6.** 8 **7.** 1 **8.** 32
9. 7 **10.** 4 **11.** 1 **12.** 1

C Order of Operations

The symbols we use to specify operations, +, −, ·, ÷, along with the symbols we use for grouping, ( ) and [ ], serve the same purpose in mathematics as punctuation marks in English. They may be called the punctuation marks of mathematics.

Consider the following sentence:

Bob said John is tall.

It can have two different meanings, depending on how we punctuate it:

1. "Bob," said John, "is tall."
2. Bob said, "John is tall."

Without the punctuation marks we don't know which meaning the sentence has.

Now, consider the following mathematical expression:

$$4 + 5 \cdot 2$$

What should we do? Should we add 4 and 5 first, or should we multiply 5 and 2 first? There seem to be two different answers. In mathematics we want to avoid situations in which two different results are possible. Therefore we follow the rule for order of operations.

Definition

Order of Operations When evaluating mathematical expressions, we will perform the operations in the following order:

1. If the expression contains grouping symbols, such as parentheses ( ), brackets [ ], or a fraction bar, then we perform the operations inside the grouping symbols, or above and below the fraction bar, first.
2. Then we evaluate, or simplify, any numbers with exponents.
3. Then we do all multiplications and divisions in order, starting at the left and moving right.
4. Finally, we do all additions and subtractions, from left to right.

Note To help you to remember the order of operations you can use the popular sentence

Please **E**xcuse **M**y **D**ear **A**unt **S**ally, or the acronym **PEMDAS**

Parentheses (or grouping)
Exponents
Multiplication and
Division, from left to right
Addition and
Subtraction, from left to right

According to our rule, the expression $4 + 5 \cdot 2$ would have to be evaluated by multiplying 5 and 2 first, and then adding 4. The correct answer—and the only answer—to this problem is 14.

$$\begin{aligned} 4 + 5 \cdot 2 &= 4 + 10 && \text{Multiply first} \\ &= 14 && \text{Then add} \end{aligned}$$

Here are some more examples that illustrate how we apply the rule for order of operations to simplify (or evaluate) expressions.

EXAMPLE 13 Simplify: $4 \cdot 8 - 2 \cdot 6$

SOLUTION We multiply first and then subtract:

$$\begin{aligned} 4 \cdot 8 - 2 \cdot 6 &= 32 - 12 && \text{Multiply first} \\ &= 20 && \text{Then subtract} \end{aligned}$$

13. Simplify.
 a. $5 \cdot 7 - 3 \cdot 6$
 b. $5 \cdot 70 - 3 \cdot 60$

Answer
13. a. 17 **b.** 170

14. Simplify: $7 + 3(6 + 4)$

EXAMPLE 14 Simplify: $5 + 2(7 - 1)$

SOLUTION According to the rule for the order of operations, we must do what is inside the parentheses first:

$$\begin{aligned} 5 + 2(7 - 1) &= 5 + 2(6) && \text{Inside parentheses first} \\ &= 5 + 12 && \text{Then multiply} \\ &= 17 && \text{Then add} \end{aligned}$$

15. Simplify.
a. $28 \div 7 - 3$
b. $6 \cdot 3^2 + 64 \div 2^4 - 2$

EXAMPLE 15 Simplify: $9 \cdot 2^3 + 36 \div 3^2 - 8$

SOLUTION

$$\begin{aligned} 9 \cdot 2^3 + 36 \div 3^2 - 8 &= 9 \cdot 8 + 36 \div 9 - 8 && \text{Exponents first} \\ &= 72 + 4 - 8 && \text{Then multiply and divide, left to right} \\ &= 76 - 8 && \text{Add and subtract,} \\ &= 68 && \text{left to right} \end{aligned}$$

USING TECHNOLOGY

Calculators

Here is how we use a calculator to work the problem shown in Example 14:

Scientific Calculator: 5 [+] 2 [×] [(] 7 [−] 1 [)] [=]

Graphing Calculator: 5 [+] 2 [(] 7 [−] 1 [)] [ENT]

Example 15 on a calculator looks like this:

Scientific Calculator: 9 [×] 2 [x^y] 3 [+] 36 [÷] 3 [x^y] 2 [−] 8 [=]

Graphing Calculator: 9 [×] 2 [^] 3 [+] 36 [÷] 3 [^] 2 [−] 8 [ENT]

16. Simplify.
a. $5 + 3[24 - 5(6 - 2)]$
b. $50 + 30[240 - 50(6 - 2)]$

EXAMPLE 16 Simplify: $3 + 2[10 - 3(5 - 2)]$

SOLUTION The brackets, [ ], are used in the same way as parentheses. In a case like this we move to the innermost grouping symbols first and begin simplifying:

$$\begin{aligned} 3 + 2[10 - 3(5 - 2)] &= 3 + 2[10 - 3(3)] \\ &= 3 + 2[10 - 9] \\ &= 3 + 2[1] \\ &= 3 + 2 \\ &= 5 \end{aligned}$$

Table 1 lists some English expressions and their corresponding mathematical expressions written in symbols.

TABLE 1

In English	Mathematical Equivalent
5 times the sum of 3 and 8	$5(3 + 8)$
Twice the difference of 4 and 3	$2(4 - 3)$
6 added to 7 times the sum of 5 and 6	$6 + 7(5 + 6)$
The sum of 4 times 5 and 8 times 9	$4 \cdot 5 + 8 \cdot 9$
3 subtracted from the quotient of 10 and 2	$10 \div 2 - 3$

Answers
14. 37 15. a. 1 b. 56
16. a. 17 b. 1,250

DESCRIPTIVE STATISTICS

D Average

Next we turn our attention to averages. If we go online to the Merriam-Webster dictionary at www.m-w.com, we find the following definition for the word *average* when it is used as a noun:

av · er · age *noun* 1a: a single value (as a mean, mode, or median) that summarizes or represents the general significance of a set of unequal values . . .

In everyday language, the word *average* can refer to the mean, the median, or the mode. The mean is probably the most common average.

Mean

Definition

To find the **mean** for a set of numbers, we add all the numbers and then divide the sum by the number of numbers in the set. The mean is sometimes called the **arithmetic mean**.

EXAMPLE 17 An instructor at a community college earned the following salaries for the first five years of teaching. Find the mean of these salaries.

$35,344 $38,290 $39,199 $40,346 $42,866

SOLUTION We add the five numbers and then divide by 5, the number of numbers in the set.

$$\text{Mean} = \frac{35{,}344 + 38{,}290 + 39{,}199 + 40{,}346 + 42{,}866}{5} = \frac{196{,}045}{5} = 39{,}209$$

The instructor's mean salary for the first five years of work is $39,209 per year.

Median

The table below shows the median weekly wages for a number of professions for the first quarter of 2008.

WEEKLY WAGES	
All Americans	$719
Butchers	$495
Dietitians	$734
Social workers	$757
Electricians	$805
Clergy	$797
Special ed teachers	$881
Lawyers	$1591

Source: U.S. Bureau of Labor Statistics (all wages are median figures for 2008)

17. A woman traveled the following distances on a 5-day business trip: 187 miles, 273 miles, 150 miles, 173 miles, and 227 miles. What was the mean distance the woman traveled each day?

Answer

17. 202 miles

If you look at the type at the bottom of the table, you can see that the numbers are the *median* figures for 2008. The median for a set of numbers is the number such that half of the numbers in the set are above it and half are below it. Here is the exact definition.

Definition

To find the **median** for a set of numbers, we write the numbers in order from smallest to largest. If there is an odd number of numbers, the median is the middle number. If there is an even number of numbers, then the median is the mean of the two numbers in the middle.

18. Find the median for the distances in Practice Problem 17.

EXAMPLE 18 Find the median of the numbers given in Example 17.

SOLUTION The numbers in Example 17, written from smallest to largest, are shown below. Because there are an odd number of numbers in the set, the median is the middle number.

35,344 38,290 39,199 40,346 42,866
↑
median

The instructor's median salary for the first five years of teaching is $39,199.

19. A teacher earns the following amounts for the first 4 years he teaches. Find the median.
$40,770 $42,635 $44,475 $46,320

EXAMPLE 19 A teacher at a community college in California will make the following salaries for the first four years she teaches.

$51,890 $53,745 $55,601 $57,412

Find the mean and the median for the four salaries.

SOLUTION To find the mean, we add the four numbers and then divide by 4:

$$\frac{51{,}890 + 53{,}745 + 55{,}601 + 57{,}412}{4} = \frac{218{,}648}{4} = 54{,}662$$

To find the median, we write the numbers in order from smallest to largest. Then, because there is an even number of numbers, we average the middle two numbers to obtain the median.

51,890 53,745 55,601 57,412
median
↓

$$\frac{53{,}745 + 55{,}601}{2} = 54{,}673$$

The mean is $54,662, and the median is $54,673.

Mode

The mode is best used when we are looking for the most common eye color in a group of people, the most popular breed of dog in the United States, and the movie that was seen the most often. When we have a set of numbers in which one number occurs more often than the rest, that number is the *mode*.

Definition

The **mode** for a set of numbers is the number that occurs most frequently. If all the numbers in the set occur the same number of times, there is no mode.

Answers
18. 187 miles **19.** $43,555

For example, consider this set of iPods:

Given the set of iPods the most popular color is red. We call this the mode.

EXAMPLE 20 A math class with 18 students had the grades shown below on their first test. Find the mean, the median, and the mode.

77	87	100	65	79	87
79	85	87	95	56	87
56	75	79	93	97	92

SOLUTION To find the mean, we add all the numbers and divide by 18:

$$\text{mean} = \frac{77+87+100+65+79+87+79+85+87+95+56+87+56+75+79+93+97+92}{18}$$

$$= \frac{1,476}{18} = 82$$

To find the median, we must put the test scores in order from smallest to largest; then, because there are an even number of test scores, we must find the mean of the middle two scores.

56 56 65 75 77 79 79 79 85 87 87 87 87 92 93 95 97 100

$$\text{Median} = \frac{85 + 87}{2} = 86$$

The mode is the most frequently occurring score. Because 87 occurs 4 times, and no other scores occur that many times, 87 is the mode.

The mean is 82, the median is 86, and the mode is 87.

20. The students in a small math class have the following scores on their final exam. Find the mode.

56 89 74 68 97 74 68
74 88 45

More Vocabulary

When we used the word *average* for the first time in this section, we used it as a noun. It can also be used as an adjective and a verb. Below is the definition of the word *average* when it is used as a verb.

av · er · age *verb* . . . 2 : to find the arithmetic mean of (a series of unequal quantities) . . .

In everyday language, if you are asked for, or given, the *average* of a set of numbers, the word *average* can represent the mean, the median, or the mode. When used in this way, the word *average* is a noun. However, if you are asked to *average* a set of numbers, then the word *average* is a verb, and you are being asked to find the mean of the numbers.

Answer
20. 74

Range

The range of a set of data is the difference between the greatest and least values. While the range of scores on the latest math test may be high, the difference between the highest and lowest gas prices around town will be much smaller.

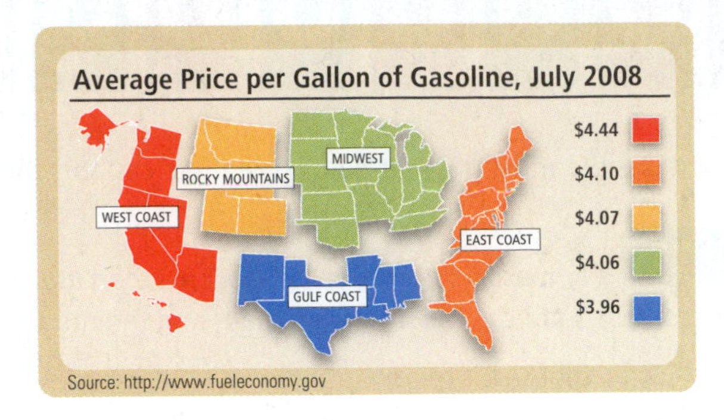

From the information on average gas prices around the country, we see that the lowest average price was found in Gulf Coast states at \$3.96 per gallon with the highest prices being paid on the West Coast at \$4.44 per gallon. The range of this set of data is the difference between these two numbers:

$$\$4.44 - \$3.96 = \$0.48$$

We say the gas prices in July of 2008 had a range of \$0.48.

Definition

The **range** for a set of numbers is the difference between the largest number and the smallest number in the sample.

STUDY SKILLS
Read the Book Before Coming to Class

As we mentioned in the Preface, it is best to have read the section to be covered in class before getting to class. Even if you don't understand everything that you have read, you are still better off reading ahead than not. The Getting Ready for Class questions at the end of each section are intended to give you things to look for in the reading that will be important in understanding what is in the section.

Getting Ready for Class

After reading through the preceding section, respond in your own words and in complete sentences.

1. In the expression 5^3, which number is the base?
2. Give a written description of the process you would use to simplify the expression $3 + 4(5 + 6)$.
3. What is the first step in simplifying the expression $8 + 6 \div 3 - 1$?
4. What number must we use for x, if the mean of 6, 8, and x is to be 8?

Problem Set 1.7

A For each of the following expressions, name the base and the exponent. [Examples 1–3]

1. 4^5 **2.** 5^4 **3.** 3^6 **4.** 6^3 **5.** 8^2 **6.** 2^8

7. 9^1 **8.** 1^9 **9.** 4^0 **10.** 0^4

B Use the definition of exponents as indicating repeated multiplication to simplify each of the following expressions. [Examples 4–12]

11. 6^2 **12.** 7^2 **13.** 2^3 **14.** 2^4 **15.** 1^4 **16.** 5^1

17. 9^0 **18.** 27^0 **19.** 9^2 **20.** 8^2 **21.** 10^1 **22.** 8^1

23. 12^1 **24.** 16^0 **25.** 45^0 **26.** 3^4

C Use the rule for the order of operations to simplify each expression. [Examples 13–16]

27. $16 - 8 + 4$ **28.** $16 - 4 + 8$ **29.** $20 \div 2 \cdot 10$ **30.** $40 \div 4 \cdot 5$

31. $20 - 4 \cdot 4$ **32.** $30 - 10 \cdot 2$ **33.** $3 + 5 \cdot 8$ **34.** $7 + 4 \cdot 9$

35. $3 \cdot 6 - 2$ **36.** $5 \cdot 1 + 6$ **37.** $6 \cdot 2 + 9 \cdot 8$ **38.** $4 \cdot 5 + 9 \cdot 7$

39. $4 \cdot 5 - 3 \cdot 2$ **40.** $5 \cdot 6 - 4 \cdot 3$ **41.** $5^2 + 7^2$ **42.** $4^2 + 9^2$

43. $480 + 12(32)^2$ **44.** $360 + 14(27)^2$ **45.** $3 \cdot 2^3 + 5 \cdot 4^2$ **46.** $4 \cdot 3^2 + 5 \cdot 2^3$

47. $8 \cdot 10^2 - 6 \cdot 4^3$ **48.** $5 \cdot 11^2 - 3 \cdot 2^3$ **49.** $2(3 + 6 \cdot 5)$ **50.** $8(1 + 4 \cdot 2)$

51. $19 + 50 \div 5^2$

52. $9 + 8 \div 2^2$

53. $9 - 2(4 - 3)$

54. $15 - 6(9 - 7)$

55. $4 \cdot 3 + 2(5 - 3)$

56. $6 \cdot 8 + 3(4 - 1)$

57. $4[2(3) + 3(5)]$

58. $3[2(5) + 3(4)]$

59. $(7 - 3)(8 + 2)$

60. $(9 - 5)(9 + 5)$

61. $3(9 - 2) + 4(7 - 2)$

62. $7(4 - 2) - 2(5 - 3)$

63. $18 + 12 \div 4 - 3$

64. $20 + 16 \div 2 - 5$

65. $4(10^2) + 20 \div 4$

66. $3(4^2) + 10 \div 5$

67. $8 \cdot 2^4 + 25 \div 5 - 3^2$

68. $5 \cdot 3^4 + 16 \div 8 - 2^2$

69. $5 + 2[9 - 2(4 - 1)]$

70. $6 + 3[8 - 3(1 + 1)]$

71. $3 + 4[6 + 8(2 - 0)]$

72. $2 + 5[9 + 3(4 - 1)]$

73. $\dfrac{15 + 5(4)}{17 - 12}$

74. $\dfrac{20 + 6(2)}{11 - 7}$

Translate each English expression into an equivalent mathematical expression written in symbols. Then simplify.

75. 8 times the sum of 4 and 2

76. 3 times the difference of 6 and 1

77. Twice the sum of 10 and 3

78. 5 times the difference of 12 and 6

79. 4 added to 3 times the sum of 3 and 4

80. 25 added to 4 times the difference of 7 and 5

81. 9 subtracted from the quotient of 20 and 2

82. 7 added to the quotient of 6 and 2

83. The sum of 8 times 5 and 5 times 4

84. The difference of 10 times 5 and 6 times 2

D Find the mean and the range for each set of numbers. [Examples 17–20]

85. 1, 2, 3, 4, 5

86. 2, 4, 6, 8, 10

87. 1, 3, 9, 11

88. 5, 7, 9, 12, 12

D Find the median and the range for each set of numbers. [Examples 18–20]

89. 5, 9, 11, 13, 15

90. 42, 48, 50, 64

91. 10, 20, 50, 90, 100

92. 700, 900, 1100

D Find the mode and the range for each set of numbers. [Example 20]

93. 14, 18, 27, 36, 18, 73

94. 11, 27, 18, 11, 72, 11

Applying the Concepts

Nutrition Labels Use the three nutrition labels below to work Problems 95–100.

SPAGHETTI

Nutrition Facts
Serving Size 2 oz. (56g/l/8 of pkg) dry
Servings Per Container: 8

Amount Per Serving	
Calories 210	Calories from fat 10
	% Daily Value*
Total Fat 1g	**2%**
Saturated Fat 0g	**0%**
Poly unsaturated Fat 0.5g	
Monounsaturated Fat 0g	
Cholesterol 0mg	**0%**
Sodium 0mg	**0%**
Total Carbohydrate 42g	**14%**
Dietary Fiber 2g	**7%**
Sugars 3g	
Protein 7g	
Vitamin A 0% •	Vitamin C 0%
Calcium 0% •	Iron 10%

*Percent Daily Values are based on a 2,000 calorie diet

CANNED ITALIAN TOMATOES

Nutrition Facts
Serving Size 1/2 cup (121g)
Servings Per Container: about 3 1/2

Amount Per Serving	
Calories 25	Calories from fat 0
	% Daily Value*
Total Fat 0g	**0%**
Saturated Fat 0g	**0%**
Cholesterol 0mg	**0%**
Sodium 300mg	**12%**
Potassium 145mg	**4%**
Total Carbohydrate 4g	**2%**
Dietary Fiber 1g	**4%**
Sugars 4g	
Protein 1g	
Vitamin A 20% •	Vitamin C 15%
Calcium 4% •	Iron 15%

*Percent Daily Values are based on a 2,000 calorie diet. Your daily values may be higher or lower depending on your calorie needs.

SHREDDED ROMANO CHEESE

Nutrition Facts
Serving Size 2 tsp (5g)
Servings Per Container: 34

Amount Per Serving	
Calories 20	Calories from fat 10
	% Daily Value*
Total Fat 1.5g	**2%**
Saturated Fat 1g	**5%**
Cholesterol 5mg	**2%**
Sodium 70mg	**3%**
Total Carbohydrate 0g	**0%**
Fiber 0g	**0%**
Sugars 0g	
Protein 2g	
Vitamin A 0% •	Vitamin C 0%
Calcium 4% •	Iron 0%

*Percent Daily Values (DV) are based on a 2,000 calorie diet

Find the total number of calories in each of the following meals.

95. Spaghetti 1 serving
Tomatoes 1 serving
Cheese 1 serving

96. Spaghetti 1 serving
Tomatoes 2 servings
Cheese 1 serving

97. Spaghetti 2 servings
Tomatoes 1 serving
Cheese 1 serving

98. Spaghetti 2 servings
Tomatoes 1 serving
Cheese 2 servings

Find the number of calories from fat in each of the following meals.

99. Spaghetti 2 servings
Tomatoes 1 serving
Cheese 1 serving

100. Spaghetti 2 servings
Tomatoes 1 serving
Cheese 2 servings

The following table lists the number of calories consumed by eating some popular fast foods. Use the table to work Problems 101 and 102.

CALORIES IN FOOD

Food	Calories
McDonald's hamburger	270
Burger King hamburger	260
Jack in the Box hamburger	280
McDonald's Big Mac	510
Burger King Whopper	630
Jack in the Box Colossus burger	940

101. Compare the total number of calories in the meal in Problem 95 with the number of calories in a McDonald's Big Mac.

102. Compare the total number of calories in the meal in Problem 98 with the number of calories in a Burger King hamburger.

103. Average If a basketball team has scores of 61, 76, 98, 55, 76, and 102 in their first six games, find

a. the mean score **b.** the median score **c.** the mode of the scores **d.** the range of scores

104. Home Sales Below are listed the prices paid for 10 homes that sold during the month of February in a city in Texas.

$210,000	$139,000	$122,000	$145,000	$120,000
$540,000	$167,000	$125,000	$125,000	$950,000

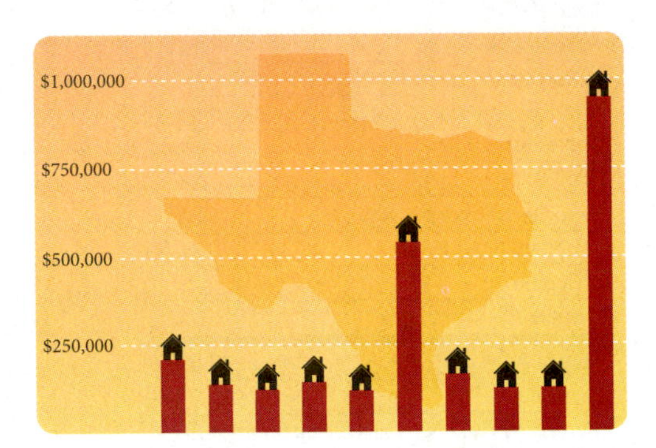

a. Find the mean housing price for the month.

b. Find the median housing price for the month.

c. Find the mode of the housing prices for the month.

d. Which measure of "average" best describes the average housing price for the month? Explain your answer.

105. Average Enrollment The number of students enrolled in a community college during a 5-year period was as follows:

Year	Enrollment
1999	6,789
2000	6,970
2001	7,242
2002	6,981
2003	6,423

Find the mean enrollment and the range of enrollments for this 5-year period.

106. Car Prices The following prices were listed for Volkswagen Jettas on the ebay.com car auction site. Use the table to find each of the following:

a. the mean car price

b. the median car price

c. the mode for the car prices

d. the range of car prices

CAR PRICES

Year	Price
1998	$10,000
1999	$14,500
1999	$10,500
1999	$11,700
1999	$15,500
2000	$10,500
2000	$18,200
2001	$19,900

107. Blood Pressure Screening When you have your blood pressure measured, it is written down as two numbers, one over the other. The top number, which is called the systolic pressure, shows the pressure in your arteries when your heart is forcing blood through them. The bottom number, called the diastolic pressure, shows the pressure in your arteries when your heart relaxes. Blood pressure screening is a part of the annual health fair held on your campus. The systolic reading (measured in mmHg) of 10 students were recorded:

140 112 118 120 138 119 130 130 125 128

Use this information to find

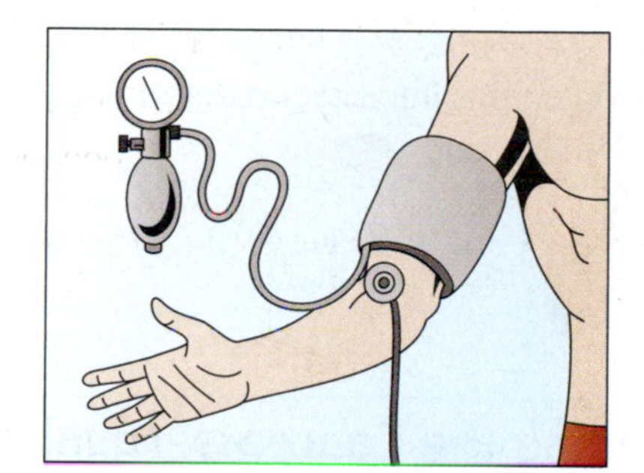

a. the mean systolic pressure.

b. the median systolic pressure.

c. the mode for the systolic pressure.

d. the range in the values for systolic pressure.

108. California Counties The table below shows those California counties which had a population of more than 1,000,000 in 2006. Use this information to find the following:

a. the mean population for these counties

b. the median population

c. the mode population

d. the range in the values for the population in these counties

COUNTY POPULATION

County	Population
Los Angeles	10,294,280
San Diego	3,120,088
Orange	3,098,183
Riverside	2,070,315
San Bernardino	2,039,467
Santa Clara	1,820,176
Alameda	1,530,620
Sacramento	1,415,117
Contra Costa	1,044,201

e. Which measure seems to best describe the average population? Explain your choice.

109. Gasoline Prices The Energy Information Administration (EIA) was created by Congress in 1977 and is a statistical agency of the U.S. Department of Energy. According to the EIA, the average retail prices for regular gasoline in California can be seen in the table below. Use this information to find

a. the median price for a gallon of regular gas.

b. the mode price.

c. the range in the price of regular gas between March 17th and May 5th.

AVERAGE PRICE FOR REGULAR GAS IN CALIFORNIA

Date	Price per Gallon
3/17/2008	$3.60
3/24/2008	$3.60
3/31/2008	$3.61
4/07/2008	$3.69
4/14/2008	$3.77
4/21/2008	$3.85
4/28/2008	$3.89
5/05/2008	$3.90

110. Cell Phones The following table shows the total voice minutes and number of calls sent and received for different age groups in 2005 in the US (*Telephia Customer Value Metrics, Q3 2005*). Use this information to find

a. the mean number of minutes used and the mean number of calls sent and received.

b. the range of minutes used and calls sent and received.

c. Based on this information determine the average length of a cell phone call.

CELL PHONE USAGE

Age	Total Voice Minutes Used	Number of Calls Sent/Received
18-24	1,304	340
25-36	970	246
37-55	726	197
56+	441	119

Extending the Concepts: Number Sequences

There is a relationship between the two sequences below. The first sequence is the *sequence of odd numbers*. The second sequence is called the *sequence of squares*.

1, 3, 5, 7, . . . The sequence of odd numbers
1, 4, 9, 16, . . . The sequence of squares

111. Add the first two numbers in the sequence of odd numbers.

112. Add the first three numbers in the sequence of odd numbers.

113. Add the first four numbers in the sequence of odd numbers.

114. Add the first five numbers in the sequence of odd numbers.

1.8 Area and Volume

Objectives

A Find the area of a polygon.
B Find the volume of an object.
C Find the surface area of an object.

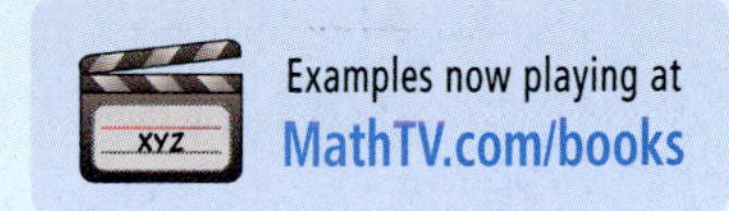

A Area

The area of a flat object is a measure of the amount of surface the object has. The area of the rectangle below is 6 square inches, because it takes 6 square inches to cover it.

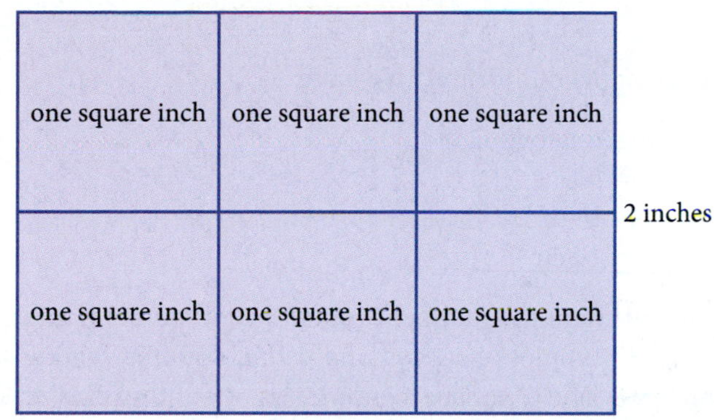

A rectangle with an area of 6 square inches

The area of this rectangle can also be found by multiplying the length and the width.

$$\begin{aligned}\text{Area} &= (\text{length}) \cdot (\text{width}) \\ &= (3 \text{ inches}) \cdot (2 \text{ inches}) \\ &= (3 \cdot 2) \cdot (\text{inches} \cdot \text{inches}) \\ &= 6 \text{ square inches}\end{aligned}$$

From this example, and others, we conclude that the area of any rectangle is the product of the length and width.

Here are three common geometric figures along with the formula for the area of each one.

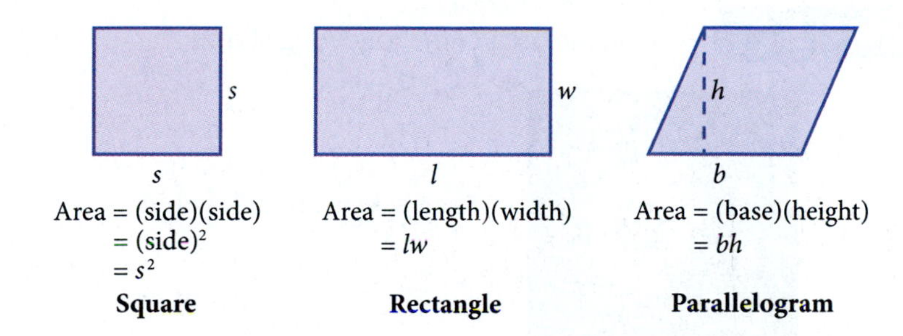

PRACTICE PROBLEMS

1. Find the area.

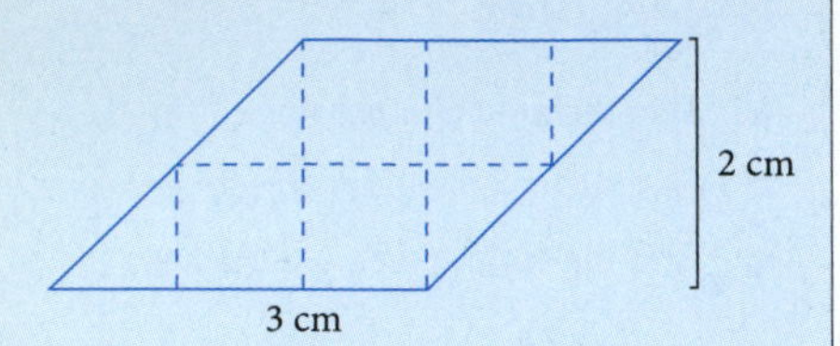

2. Find the area of a rectangular stamp if it is 35 mm wide and 70 mm long.

Answers

1. 6 cm² **2.** 2,450 mm²

EXAMPLE 1 The parallelogram below has a base of 5 centimeters and a height of 2 centimeters. Find the area.

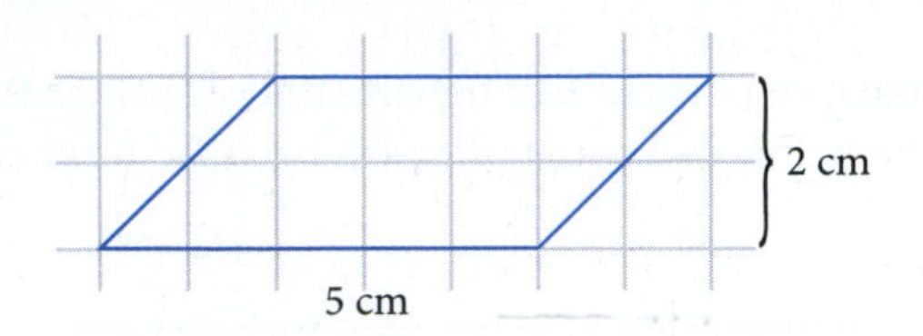

SOLUTION If we apply our formula we have

$$\begin{aligned} \text{Area} &= (\text{base})(\text{height}) \\ A &= bh \\ &= 5 \cdot 2 \\ &= 10 \text{ cm}^2 \end{aligned}$$

Or, we could simply count the number of square centimeters it takes to cover the object. There are 8 complete squares and 4 half-squares, giving a total of 10 squares for an area of 10 square centimeters. Counting the squares in this manner helps us see why the formula for the area of a parallelogram is the product of the base and the height.

To justify our formula in general, we simply rearrange the parts to form a rectangle.

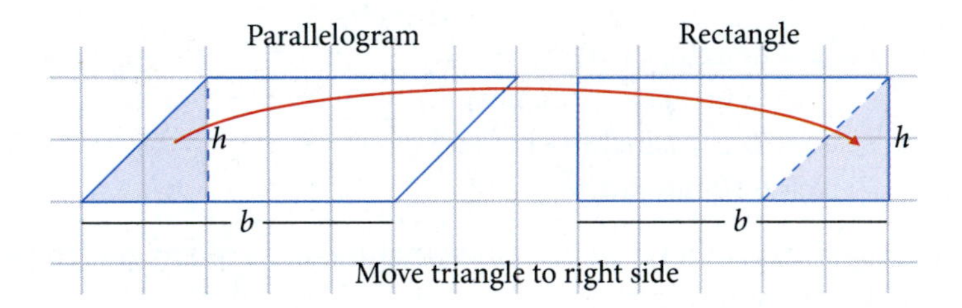

EXAMPLE 2 Find the area of the following stamp.

Each side is 35 millimeters

SOLUTION Applying our formula for area we have

$$A = s^2 = (35 \text{ mm})^2 = 1{,}225 \text{ mm}^2$$

EXAMPLE 3 Find the total area of the house and deck shown below.

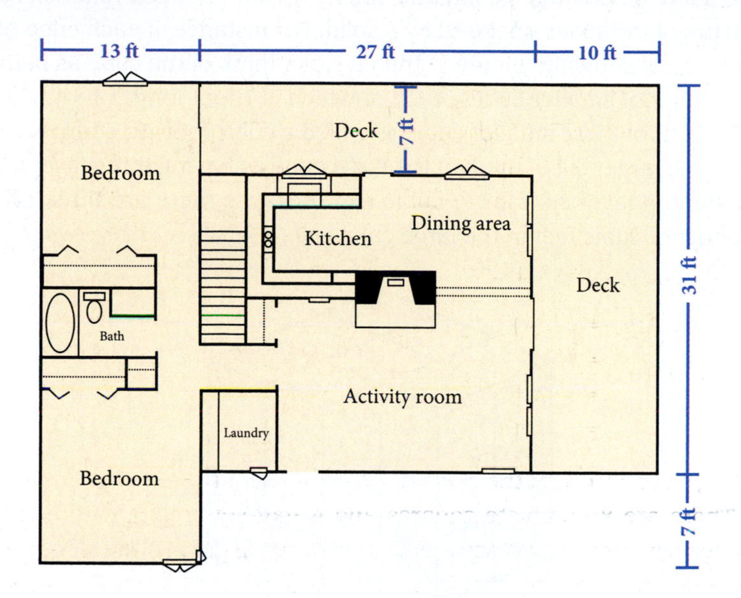

SOLUTION We begin by drawing an additional line, so that the original figure is now composed of two rectangles. Next, we fill in the missing dimensions on the two rectangles.

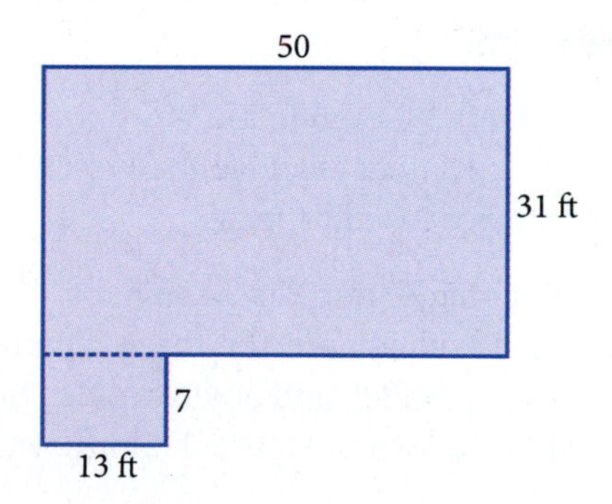

Finally, we calculate the area of the original figure by adding the areas of the individual figures:

$$\begin{aligned} \text{Area} &= \text{Area of the small rectangle} + \text{Area of the large rectangle} \\ &= 13 \cdot 7 + 50 \cdot 31 \\ &= 91 + 1{,}550 \\ &= 1{,}641 \text{ square feet} \end{aligned}$$

3. Find the area of the house without the deck.

Answer
3. 1,142 sq ft

B Volume

Next, we move up one dimension and consider what is called *volume*. Volume is the measure of the space enclosed by a solid. For instance, if each edge of a cube is 3 feet long, as shown in Figure 1, then we can think of the cube as being made up of a number of smaller cubes, each of which is 1 foot long, 1 foot wide, and 1 foot high. Each of these smaller cubes is called a cubic foot. To count the number of them in the larger cube, think of the large cube as having three layers. You can see that the top layer contains 9 cubic feet. Because there are three layers, the total number of cubic feet in the large cube is $9 \cdot 3 = 27$.

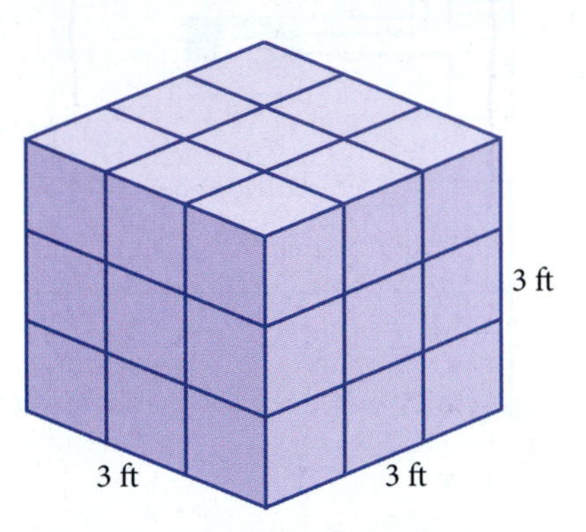

FIGURE 1 A cube in which each edge is 3 feet long

On the other hand, if we multiply the length, the width, and the height of the cube, we have the same result:

$$\begin{aligned} \text{Volume} &= (3\text{ feet})(3\text{ feet})(3\text{ feet}) \\ &= (3 \cdot 3 \cdot 3)(\text{feet} \cdot \text{feet} \cdot \text{feet}) \\ &= 27\text{ ft}^3 \text{ or } 27 \text{ cubic feet} \end{aligned}$$

For the present we will confine our discussion of volume to volumes of *rectangular solids.* Rectangular solids are the three-dimensional equivalents of rectangles: Opposite sides are parallel, and any two sides that meet, meet at right angles. A rectangular solid is shown in Figure 2, along with the formula used to calculate its volume.

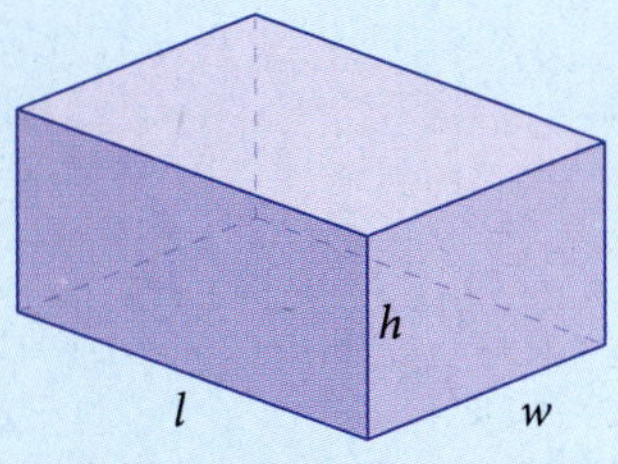

Volume = (length)(width)(height)
$V = lwh$

FIGURE 2 A Rectangular Solid

EXAMPLE 4 Find the volume of a rectangular solid with length 15 inches, width 3 inches, and height 5 inches.

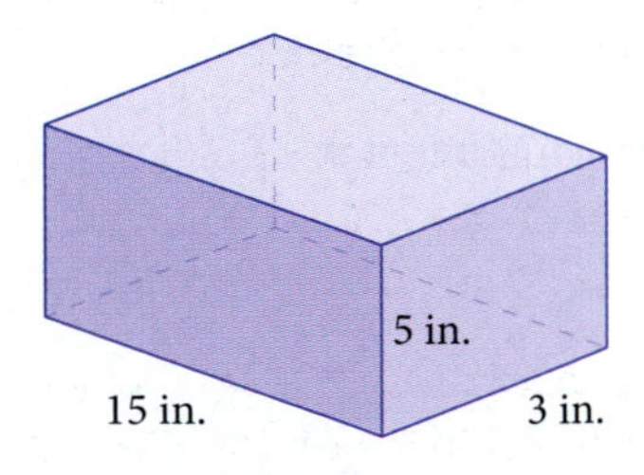

SOLUTION To find the volume we apply the formula shown in Figure 2:

$$\begin{aligned} V &= l \cdot w \cdot h \\ &= (15\text{ in.})(3\text{ in.})(5\text{ in.}) \\ &= 225\text{ in}^3 \end{aligned}$$

4. A home has a dining room that is 12 feet wide and 15 feet long. If the ceiling is 8 feet high, find the volume of the dining room.

Answer
4. 1,440 cubic feet

C Surface Area

Figure 3 shows a closed box with length l, width w, and height h. The surfaces of the box are labeled as sides, top, bottom, front, and back.

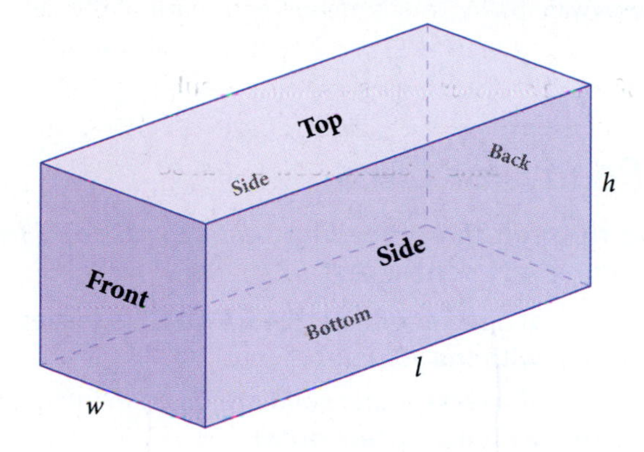

FIGURE 3 A box with dimensions l, w, and h

To find the *surface area* of the box, we add the areas of each of the six surfaces that are labeled in Figure 3.

$$\text{Surface area} = \text{side} + \text{side} + \text{front} + \text{back} + \text{top} + \text{bottom}$$
$$S = l \cdot h + l \cdot h + h \cdot w + h \cdot w + l \cdot w + l \cdot w$$
$$= 2lh + 2hw + 2lw$$

EXAMPLE 5 Find the surface area of the box shown in Figure 4.

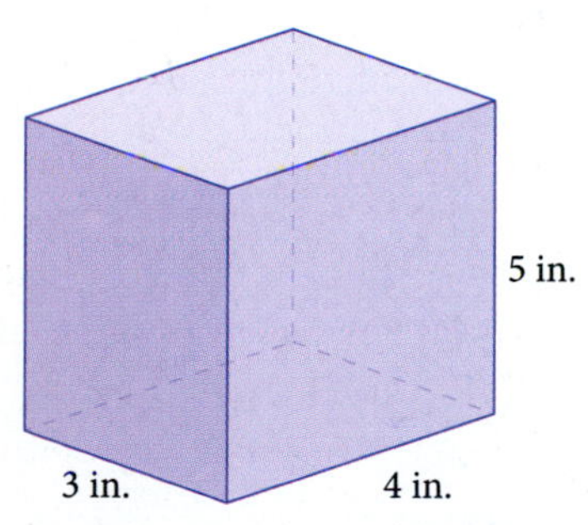

FIGURE 4 A box 4 inches long, 3 inches wide, and 5 inches high

SOLUTION To find the surface area we find the area of each surface individually, and then we add them together:

$$\text{Surface area} = 2(3 \text{ in.})(4 \text{ in.}) + 2(3 \text{ in.})(5 \text{ in.}) + 2(4 \text{ in.})(5 \text{ in.})$$
$$= 24 \text{ in}^2 + 30 \text{ in}^2 + 40 \text{ in}^2$$
$$= 94 \text{ in}^2$$

The total surface area is 94 square inches. If we calculate the volume enclosed by the box, it is $V = (3 \text{ in.})(4 \text{ in.})(5 \text{ in.}) = 60 \text{ in}^3$. The surface area measures how much material it takes to make the box, whereas the volume measures how much space the box will hold.

5. A family is painting a dining room that is 12 feet wide and 15 feet long.
   a. If the ceiling is 8 feet high, find the surface area of the walls and the ceiling, but not the floor.
   b. If a gallon of paint will cover 400 square feet, how many gallons should they buy to paint the walls and the ceiling?

Answer

5. **a.** 612 square feet
   b. 2 gallons will cover everything, with some paint left over.

STUDY SKILLS

List Difficult Problems

Begin to make lists of problems that give you the most difficulty—those that you are repeatedly making mistakes with.

Getting Ready for Class

After reading through the preceding section, respond in your own words and in complete sentences.

1. If the dimensions of a rectangular solid are given in inches, what units will be associated with the volume?
2. If the dimensions of a rectangular solid are given in inches, what units will be associated with the surface area?
3. How do you find the area of a square?
4. How do you find the area of a parallelogram?

Problem Set 1.8

A Find the area enclosed by each figure. [Examples 1–3]

1. 5 cm, 5 cm

2. 10 ft, 10 ft

3. 14 m, 24 m

4. 9 in., 4 in.

5. 6 ft, 10 ft

6. 6 ft, 8 ft

7. 3 m, 3 m, 6 m, 7 m, 4 m, 9 m

8. 10 in., 5 in., 10 in., 5 in., 5 in., 5 in.

9. 1 cm, 4 cm, 2 cm, 7 cm

10. 10 m, 9 m, 4 m, 3 m

11.

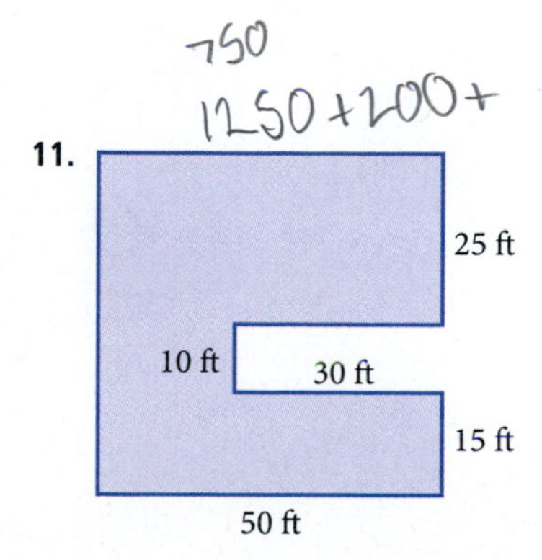

12.

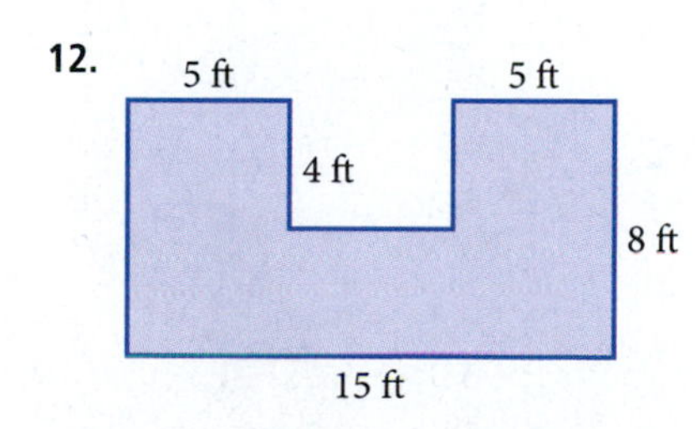

13.

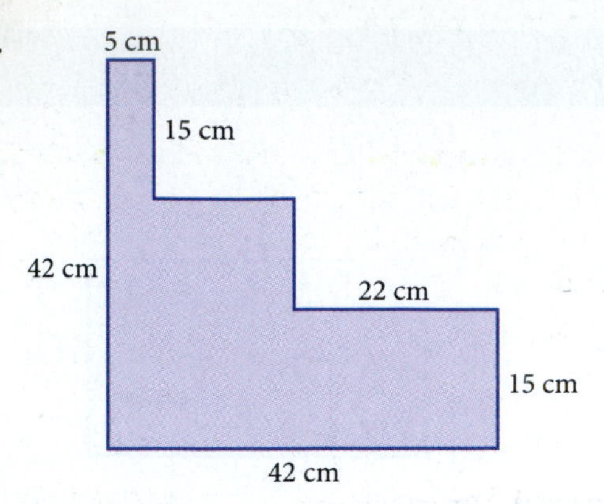

14.

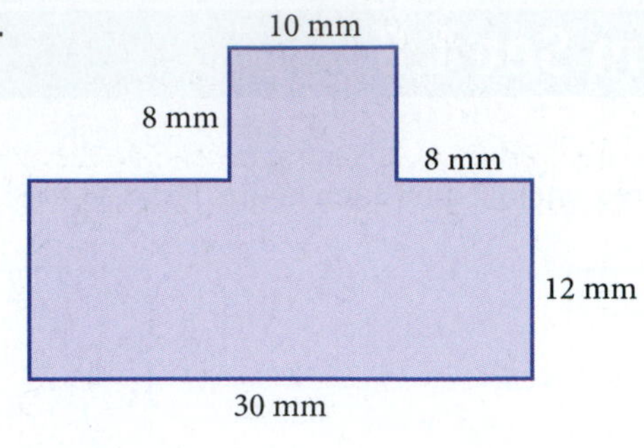

15. Find the area of a square with side 10 inches.

16. Find the area of a square with side 6 centimeters.

B **C** Find the volume and surface area of each figure. [Example 4, 5]

17.

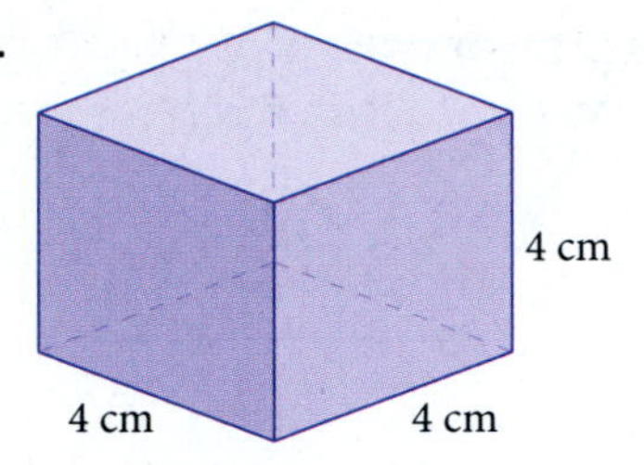

18.

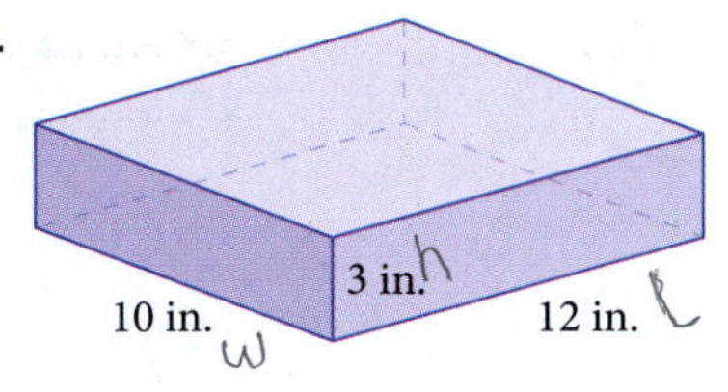

19.

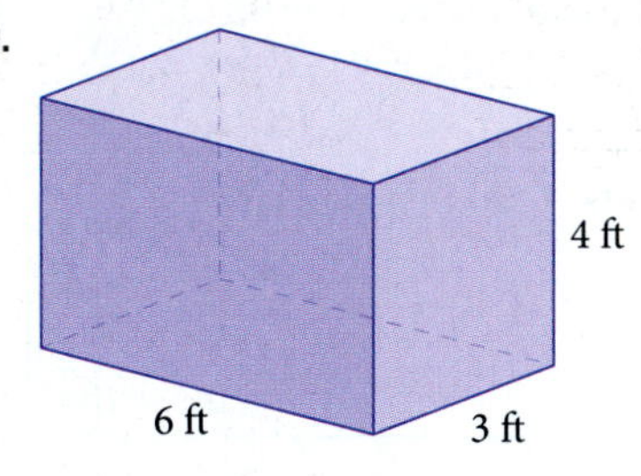

20.

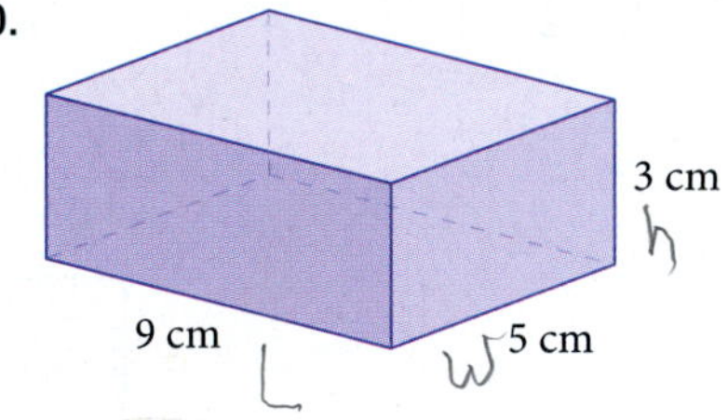

B Find the volume of each figure. [Example 4]

21.

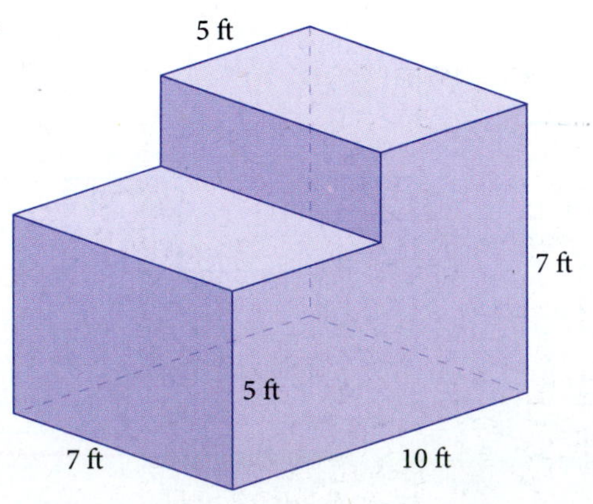

22.

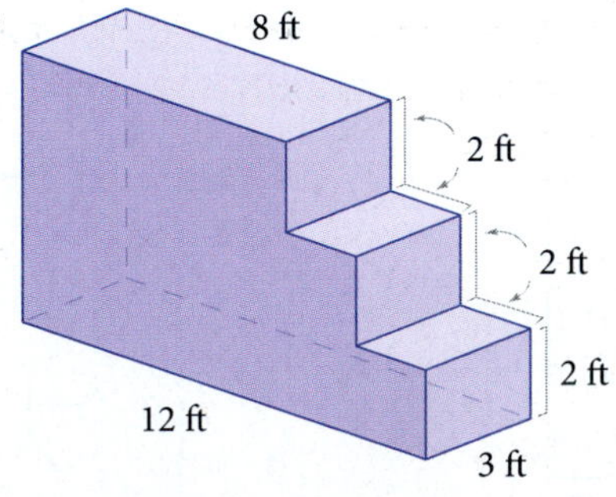

Applying the Concepts

23. Area A swimming pool is 20 feet wide and 40 feet long. If it is surrounded by square tiles, each of which is 1 foot by 1 foot, how many tiles are there surrounding the pool?

24. Area A garden is rectangular with a width of 8 feet and a length of 12 feet. If it is surrounded by a walkway 2 feet wide, how many square feet of area does the walkway cover?

25. Comparing Areas The side of a square is 5 feet long. If all four sides are increased by 2 feet, by how much is the area increased?

26. Comparing Areas The length of a side in a square is 20 inches. If all four sides are decreased by 4 inches, by how much is the area decreased?

27. Area of a Euro A 10 euro banknote has a width of 67 millimeters and a length of 127 millimeters. Find the area.

28. Area of a Dollar A $10 bill has a width of 65 millimeters and a length of 156 millimeters. Find the area.

29. Area of a Stamp The stamp here shows the Mexican artist Frida Kahlo. The image area of the stamp has a width of 20 millimeters and a length of 36 millimeter. Find the area of the image.

© 2004 Banco de México

30. Area of a Stamp The stamp shown here was issued in 2001 to honor the Italian scientist Enrico Fermi. The image area of the stamp has a width of 21 millimeters and a length of 35 millimeters. Find the area of the image.

31. Hot Air Balloon The woodcut shows the giant hot air balloon known as "Le Geant de Nadar" when it was displayed in the Crystal Palace in England in 1868. The wicker car of the balloon was two stories, consisting of a 6-compartment cottage with a viewing deck on top. If the car was 8 feet high with a square base 13 feet on each side, find the volume.

Science Museum/Science and Society Picture Library

32. Reading House Plans Find the area of the floor of the house shown here if the garage is not included with the house and if the garage is included with the house.

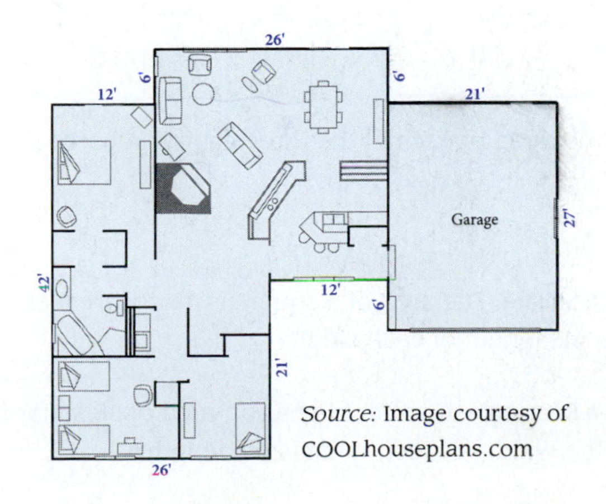

Source: Image courtesy of COOLhouseplans.com

Extending the Concepts

33. a. Each side of the red square in the corner is 1 centimeter, and all squares are the same size. On the grid below, draw three more squares. Each side of the first one will be 2 centimeters, each side of the second square will be 3 centimeters, and each side of the third square will be 4 centimeters.

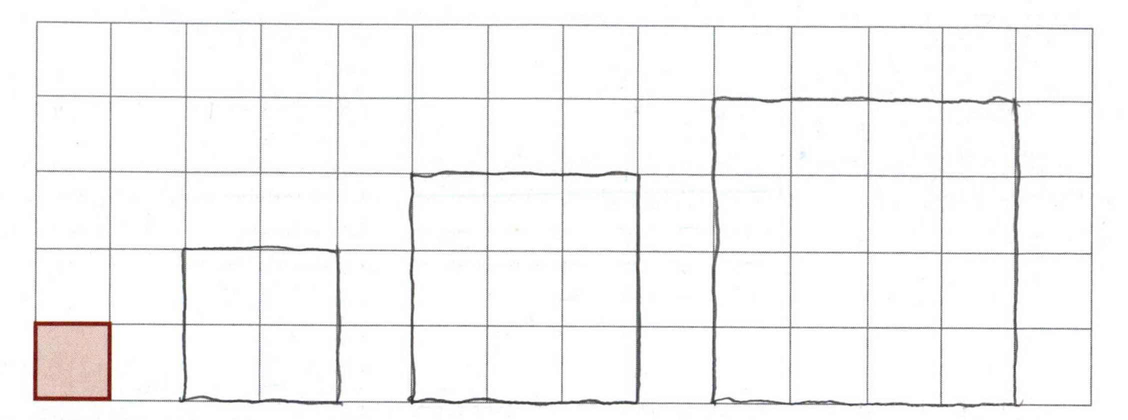

b. Use the squares you have drawn above to complete each of the following tables.

PERIMETERS OF SQUARES

Length of each Side (in Centimeters)	Perimeter (in Centimeters)
1	4
2	8
3	12
4	16

AREAS OF SQUARES

Length of each Side (in Centimeters)	Area (in Square Centimeters)
1	1
2	4
3	9
4	16

34. a. The lengths of the sides of the squares in the grid below are all 1 centimeter. The red square has a perimeter of 12 centimeters. On the grid below, draw two different rectangles, each with a perimeter of 12 centimeters.

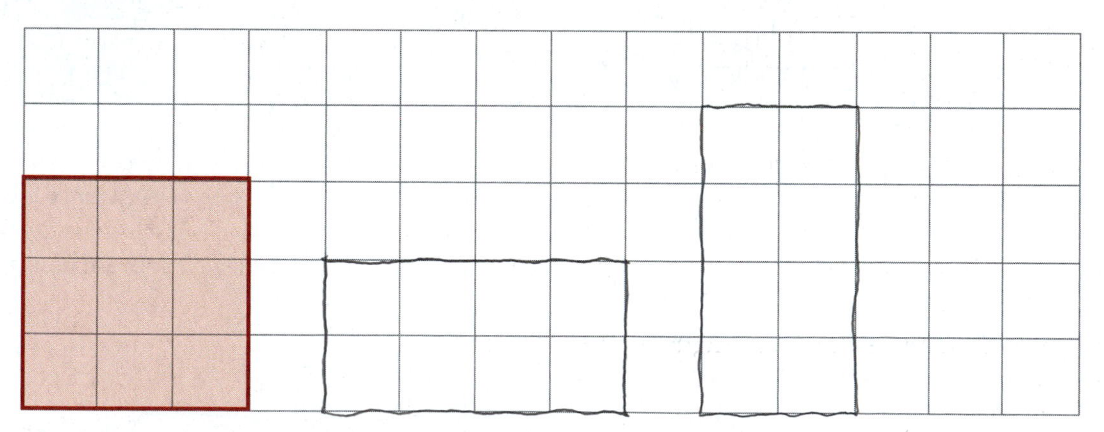

b. Find the area of each of the three figures in part a.

35. Area of a Square The area of a square is 49 square feet. What is the length of each side?

36. Area of a Square The area of a square is 144 square feet. How long is each side?

37. Area of a Rectangle A rectangle has an area of 36 square feet. If the width is 4 feet, what is the length?

39. Area of a Rectangle A rectangle has an area of 39 square feet. If the length is 13 feet, what is the width?

Chapter 1 Summary

The numbers in brackets indicate the sections in which the topics were discussed.

EXAMPLES

The margins of the chapter summaries will be used for examples of the topics being reviewed, whenever convenient.

Place Values for Decimal Numbers [1.1]

The place values for the digits of any base 10 number are as follows:

Trillions			Billions			Millions			Thousands			Hundreds		
Hundreds	Tens	Ones	Hundreds	Tens	Ones	Hundreds	Tens	Ones	Hundreds	Tens	Ones	Hundreds	Tens	Ones

1. The number 42,103,045 written in words is "forty-two million, one hundred three thousand, forty-five."

The number 5,745 written in expanded form is
5,000 + 700 + 40 + 5

Vocabulary Associated with Addition, Subtraction, Multiplication, and Division [1.2, 1.4, 1.5, 1.6]

The word *sum* indicates addition.

The word *difference* indicates subtraction.

The word *product* indicates multiplication.

The word *quotient* indicates division.

2. The sum of 5 and 2 is $5 + 2$.
The difference of 5 and 2 is $5 - 2$.
The product of 5 and 2 is $5 \cdot 2$.
The quotient of 10 and 2 is $10 \div 2$.

Properties of Addition and Multiplication [1.2, 1.5]

If a, b, and c represent any three numbers, then the properties of addition and multiplication used most often are:

Commutative property of addition: $a + b = b + a$
Commutative property of multiplication: $a \cdot b = b \cdot a$
Associative property of addition: $(a + b) + c = a + (b + c)$
Associative property of multiplication: $(a \cdot b) \cdot c = a \cdot (b \cdot c)$
Distributive property: $a(b + c) = a(b) + a(c)$

3. $3 + 2 = 2 + 3$
$3 \cdot 2 = 2 \cdot 3$
$(x + 3) + 5 = x + (3 + 5)$
$(4 \cdot 5) \cdot 6 = 4 \cdot (5 \cdot 6)$
$3(4 + 7) = 3(4) + 3(7)$

Perimeter of a Polygon [1.2]

The *perimeter* of any polygon is the sum of the lengths of the sides, and it is denoted with the letter P.

4. The perimeter of the rectangle below is
$P = 37 + 37 + 24 + 24$
$= 122$ feet

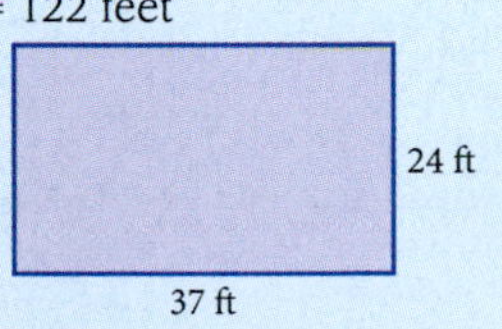

Steps for Rounding Whole Numbers [1.3]

1. Locate the digit just to the right of the place you are to round to.
2. If that digit is less than 5, replace it and all digits to its right with zeros.
3. If that digit is 5 or more, replace it and all digits to its right with zeros, and add 1 to the digit to its left.

5. 5,482 to the nearest ten is 5,480

5,482 to the nearest hundred is 5,500

5,482 to the nearest thousand is 5,000

Division by 0 (Zero) [1.6]

Division by 0 is undefined. We cannot use 0 as a divisor in any division problem.

6. Each expression below is undefined.
$5 \div 0 \quad \frac{7}{0} \quad 4/0$

Order of Operations [1.7]

To simplify a mathematical expression:

1. We simplify the expression inside the grouping symbols first. Grouping symbols are parentheses ( ), brackets [ ], or a fraction bar.

2. Then we evaluate any numbers with exponents.

3. We then perform all multiplications and divisions in order, starting at the left and moving right.

4. Finally, we do all the additions and subtractions, from left to right.

7. $4 + 6(8 - 2)$
$= 4 + 6(6)$ Inside parentheses first
$= 4 + 36$ Then multiply
$= 40$ Then add

Average [1.7]

The *average* for a set of numbers can be the mean, the median, or the mode.

8. The mean of 4, 7, 9 and 12 is
$(4 + 7 + 9 + 12) \div 4 = 32 \div 4$
$= 8$

Exponents [1.7]

In the expression 2^3, 2 is the *base* and 3 is the *exponent.* An exponent is a shorthand notation for repeated multiplication. The exponent 0 is a special exponent. Any nonzero number to the 0 power is 1.

9. $2^3 = 2 \cdot 2 \cdot 2 = 8$
$5^0 = 1$
$3^1 = 3$

Formulas for Area [1.8]

Below are two common geometric figures, along with the formulas for their areas.

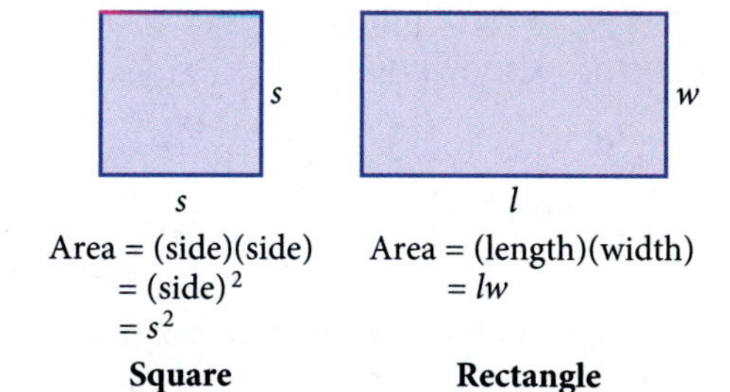

Formulas for Volume and Surface Area [1.8]

The object below is a rectangular solid.

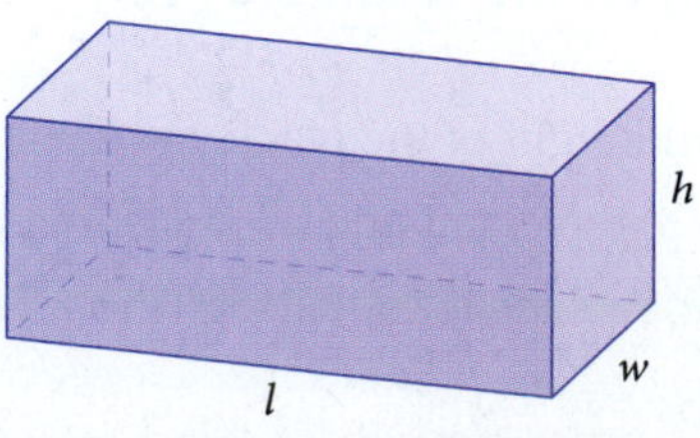

Volume $= V = lwh$ Surface Area $= S = 2lh + 2hw + 2lw$

Chapter 1 Review

The numbers in brackets indicate the sections in which problems of a similar type can be found.

1. One of the largest Pacific blue marlins was caught near Hawaii in 1982. It weighed 1,376 pounds. Write 1,376 in words. [1.1]

2. In 2003 the New York Yankees had the highest home attendance in major league baseball. The attendance that year was 3,465,600. Write 3,465,600 in words. [1.1]

For Problems 3 and 4, write each number with digits instead of words. [1.1]

3. Five million, two hundred forty-five thousand, six hundred fifty-two

4. Twelve million, twelve thousand, twelve

5. In 2003 the Montreal Expos had the lowest attendance in major league baseball. The attendance that year was 1,025,639. Write 1,025,639 in expanded form. [1.1]

6. According to the American Medical Association, in 2002, there were 215,005 female physicians practicing medicine in the United States. Write 215,005 in expanded form. [1.1]

Identify each of the statements in Problems 7–14 as an example of one of the following properties. [1.2, 1.5]

a. Addition property of 0
b. Multiplication property of 0
c. Multiplication property of 1
d. Commutative property of addition
e. Commutative property of multiplication
f. Associative property of addition
g. Associative property of multiplication

7. $5 + 7 = 7 + 5$

8. $(4 + 3) + 2 = 4 + (3 + 2)$

9. $6 \cdot 1 = 6$

10. $8 + 0 = 8$

11. $5 \cdot 0 = 0$

12. $4 \cdot 6 = 6 \cdot 4$

13. $5 \cdot (3 \cdot 2) = (5 \cdot 3) \cdot 2$

14. $(6 + 2) + 3 = (2 + 6) + 3$

Find each of the following sums. (Add.) [1.2]

15. $\begin{array}{r} 498 \\ +\ 251 \\ \hline \end{array}$

16. $\begin{array}{r} 784 \\ +\ 598 \\ \hline \end{array}$

17. $\begin{array}{r} 7{,}384 \\ 251 \\ +\ 637 \\ \hline \end{array}$

18. $\begin{array}{r} 4{,}901 \\ 648 \\ +3{,}592 \\ \hline \end{array}$

Find each of the following differences. (Subtract.) [1.4]

19. $\begin{array}{r} 789 \\ -475 \\ \hline \end{array}$

20. $\begin{array}{r} 792 \\ -178 \\ \hline \end{array}$

21. $\begin{array}{r} 5,908 \\ -2,759 \\ \hline \end{array}$

22. $\begin{array}{r} 3,527 \\ -1,789 \\ \hline \end{array}$

Find each of the following products. (Multiply.) [1.5]

23. 8(73)

24. 7(984)

25. 63(59)

26. 49(876)

Find each of the following quotients. (Divide.) [1.6]

27. $692 \div 4$

28. $1,020 \div 15$

29. $36\overline{)15,408}$

30. $286\overline{)21,736}$

Round the number 3,781,092 to the nearest: [1.3]

31. Ten

32. Hundred

33. Hundred thousand

34. Million

Use the rule for the order of operations to simplify each expression as much as possible. [1.7]

35. $4 + 3 \cdot 5^2$

36. $7(9)^2 - 6(4)^3$

37. $3(2 + 8 \cdot 9)$

38. $7 - 2(6 - 4)$

39. $24 \div 6 \cdot 2$

40. $20 \cdot 3 \div 12 \cdot 2$

41. $4(3 - 1)^3$

42. $36 \div 9 \cdot 3^2$

43. A first-year math student had grades of 80, 67, 78, and 91 on the first four tests. What is the student's mean test grade and median test grade? [1.7]

44. If a person has scores of 205, 222, 197, 236, 185, and 215 for six games of bowling, what is the mean score for the six games and the range of scores for the six games? [1.7]

Write an expression using symbols that is equivalent to each of the following expressions; then simplify. [1.7]

45. 3 times the sum of 4 and 6

46. 9 times the difference of 5 and 3

47. Twice the difference of 17 and 5

48. The product of 5 and the sum of 8 and 2

Applying the Concepts

49. Income and Expenses A person has a monthly income of \$1,783 and monthly expenses of \$1,295. What is the difference between the monthly income and the expenses? [1.4]

50. Number of Sheep A rancher bought 395 sheep and then sold 197 of them. How many were left? [1.4]

Area and Perimeter The rules for soccer state that the playing field must be from 100 to 120 yards long and 55 to 75 yards wide. The 1999 Women's World Cup was played at the Rose Bowl on a playing field 116 yards long and 72 yards wide. The diagram below shows the smallest possible soccer field, the largest possible soccer field, and the soccer field at the Rose Bowl. [1.2, 1.8]

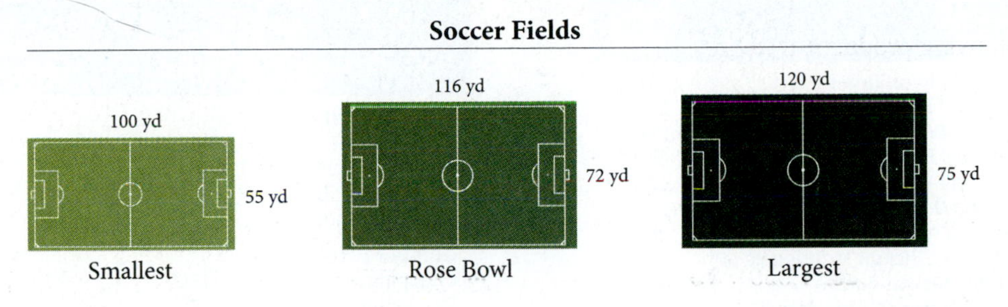

51. Find the perimeter of each soccer field.

52. Find the area of each soccer field.

53. Monthly Budget Each month a family budgets $1,150 for rent, $625 for food, and $257 for entertainment. What is the sum of these numbers? [1.2]

54. Checking Account If a person wrote 23 checks in January, 37 checks in February, 40 checks in March, and 27 checks in April, what is the total number of checks written in the 4-month period? [1.2]

55. Yearly Income A person has a yearly income of $23,256. What is the person's monthly income? [1.6]

56. Jogging It takes a jogger 126 minutes to run 14 miles. At that rate, how long does it take the jogger to run 1 mile? [1.6]

57. Take-Home Pay Jeff makes $16 an hour for the first 40 hours he works in a week and $24 an hour for every hour after that. Each week he has $228 deducted from his check for income taxes and retirement. If he works 45 hours in one week, how much is his take-home pay? [1.5]

58. Take-Home Pay Barbara earns $8 an hour for the first 40 hours she works in a week and $12 an hour for every hour after that. Each week she has $123 deducted from her check for income taxes and retirement. What is her take-home pay for a week in which she works 50 hours? [1.5]

Exercise and Calories The tables below are similar to two of the tables we have worked with in this chapter. Use the information in the tables to work the problems below. [1.2, 1.4, 1.5]

NUMBER OF CALORIES IN FAST FOOD	
Food	**Calories**
McDonald's hamburger	270
Burger King hamburger	260
Jack in the Box hamburger	280
McDonald's Big Mac	510
Burger King Whopper	630
Jack in the Box Colossus burger	940
Roy Rogers roast beef sandwich	260
McDonald's Chicken McNuggets (6)	300
Taco Bell chicken burrito	345
McDonald's french fries (large)	450
Burger King BK Broiler	540
Burger King chicken sandwich	700

NUMBER OF CALORIES BURNED IN 30 MINUTES		
	130-Pound Person	**170-Pound Person**
Indoor Activities		
Vacuuming	100	130
Mopping floors	105	135
Shopping for food	110	135
Ironing clothes	115	145
Outdoor Activities		
Chopping wood	150	200
Ice skating	170	235
Cross-country skiing	250	330
Shoveling snow	260	350

59. How many calories do you consume if you eat one large order of McDonald's french fries and 2 Big Macs?

60. How many calories do you consume if you eat one order of Chicken McNuggets and a McDonald's hamburger?

61. How many more calories are in one Colossus burger than in two Taco Bell chicken burritos?

62. What is the difference in calories between a Whopper and a BK Broiler?

63. If you weigh 170 pounds and ice skate for 1 hour, will you burn all the calories consumed by eating one Whopper?

64. If you weigh 130 pounds and go cross-country skiing for 1 hour, will you burn all the calories consumed by eating one large order of McDonald's french fries?

65. Suppose you eat a Big Mac and a large order of fries for lunch. If you weigh 130 pounds, what combination of 30-minute activities could you do to burn all the calories you consumed at lunch?

66. Suppose you weigh 170 pounds and you eat two Taco Bell chicken burritos for lunch. What combination of 30-minute activities could you do to burn all the calories in the burritos?

67. Find the volume and surface area of the rectangular solid given. [1.8]

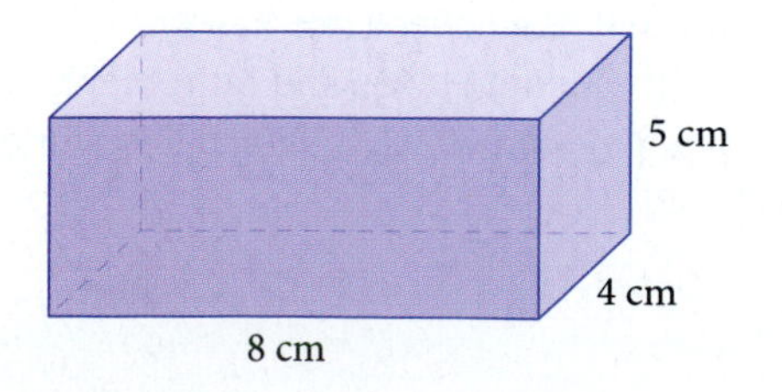

68. Find the volume and surface area of the rectangular solid given. [1.8]

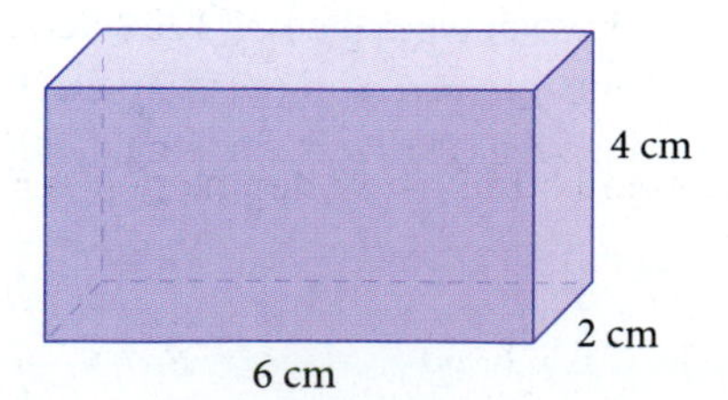

Chapter 1 Test

1. Write the number 20,347 in words.

2. Write the number two million, forty-five thousand, six with digits instead of words.

3. Write the number 123,407 in expanded form.

Identify each of the statements in Problems 4–7 as an example of one of the following properties.

a. Addition property of 0
b. Multiplication property of 0
c. Multiplication property of 1
d. Commutative property of addition
e. Commutative property of multiplication
f. Associative property of addition
g. Associative property of multiplication

4. $(5 + 6) + 3 = 5 + (6 + 3)$ **5.** $7 \cdot 1 = 7$ **6.** $9 + 0 = 9$ **7.** $5 \cdot 6 = 6 \cdot 5$

Find each of the following sums. (Add.)

8. $\begin{array}{r} 135 \\ +\ 741 \\ \hline \end{array}$

9. $\begin{array}{r} 5{,}401 \\ 329 \\ +\ 10{,}653 \\ \hline \end{array}$

Find each of the following differences. (Subtract.)

10. $\begin{array}{r} 937 \\ -\ 413 \\ \hline \end{array}$

11. $\begin{array}{r} 7{,}052 \\ -\ 3{,}967 \\ \hline \end{array}$

Find each of the following products. (Multiply.)

12. 9(186)

13. 62(359)

Find each of the following quotients. (Divide.)

14. $1{,}105 \div 13$

15. $583\overline{)12{,}243}$

16. Round the number 516,249 to the nearest ten thousand.

Use the rule for the order of operations to simplify each expression as much as possible.

17. $8(5)^2 - 7(3)^3$

18. $8 - 2(5 - 3)$

19. $7 + 2(53 - 3)$

20. $3(x - 2)$

21. Home Sales Below are listed the prices paid for 10 homes that sold during the month of February in the city of White Bear Lake. Find the mean, median, and mode from these prices.

\$210,000	\$139,000	\$122,000	\$145,000	\$120,000
\$540,000	\$167,000	\$125,000	\$125,000	\$950,000

Translate into symbols, then simplify.

22. Twice the sum of 11 and 7

23. The quotient of 20 and 5 increased by 9

24. Hours of Commuting In 2001 the Texas Transportation Institute conducted a study of the number of hours a year commuters spent in gridlock in the country's 68 largest urban areas. The top five areas from that study are listed in the bar chart below. Use the information in the bar chart to complete the table.

Time Spent in Gridlock

Hours: 0, 10, 20, 30, 40, 50, 60, 70, 80, 90

Los Angeles 52, Washington 35, Seattle-Everett 32, Atlanta 34, Boston 29

Urban Area	Average Hours in Gridlock Per Year
Los Angeles	52
Washington	
Seattle-Everett	
	34
Boston	

25. Geometry Find the perimeter and the area of the rectangle below.

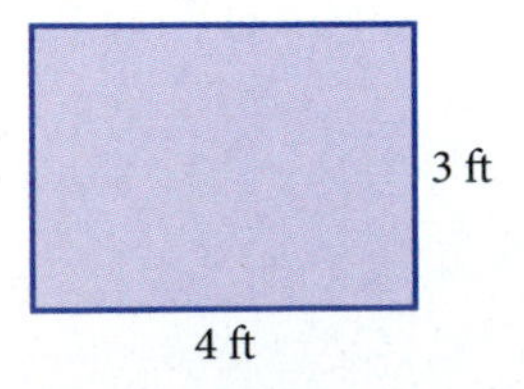

26. Geometry Find the volume and surface area of the rectangular solid given.

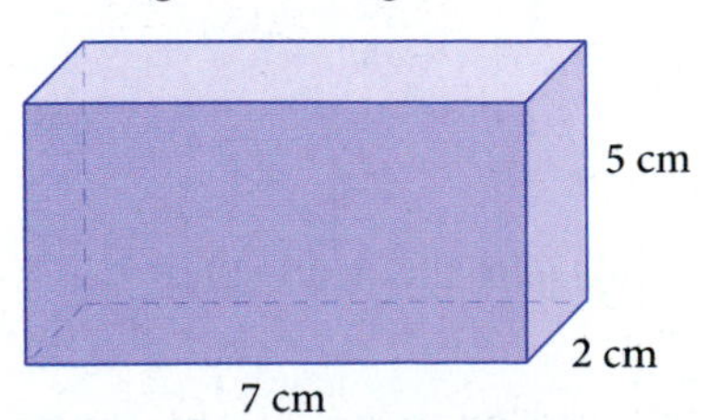

Chapter 1 Projects

WHOLE NUMBERS

GROUP PROJECT

Egyptian Numbers

Number of People 3

Time Needed 10 minutes

Equipment Pencil and paper

Background The Egyptians had a fully developed number system as early as 3500 B.C. They recorded very large numbers in the macehead of Narmer, which boasts of the spoils taken during wars, and the *Book of the Dead,* a collection of religious texts. The Egyptians used a base-ten system. A special pictograph was used to represent each power of ten. Here are some pictographs used.

1	10	100	1,000	10,000	100,000	1,000,000
staff	horseshoe	rope	lotus flower	bent finger	tadpole or frog	astonished person

Example Usually the direction of writing was from right to left, with the larger units first. Symbols were placed in rows to save lateral space. Writing the number 132,146 in Egyptian hieroglyphics looks like this:

132,146 =

Procedure Write each of the following Egyptian numbers in our system.

1.

2.

Express each of the given numbers in Egyptian hieroglyphics.

3. 1,842

4. 4,310,175

Students and Instructors: The end of each chapter in this book will have two projects. The group projects are intended to be done in class. The research projects are to be completed outside of class. They can be done in groups or individually.

RESEARCH PROJECT

Leopold Kronecker

Leopold Kronecker (1823–1891) was a German mathematician and logician who thought that arithmetic should be based on whole numbers. He is known for the quote, "God made the natural numbers; all else is the work of man." He was openly critical of the efforts of his contemporaries. Kronecker's primary work was in the field of algebraic number theory. Research the life of Leopold Kronecker, or discuss the work of a mathematician who was criticized by Kronecker.

Courtesy of Wolfram Research/ National Science Foundation

Introduction to Algebra

Chapter Outline

Introduction

The Grand Canyon, located in the state of Arizona, is a large gorge created by the Colorado River over millions of years. Much of the Grand Canyon is located in the Grand Canyon National Park, which receives over four million visitors per year. Visitors come to hike trails and view the magnificent rock formations.

Many of the hiking trails have significant changes in altitude. We sometimes represent changes in altitude with negative numbers. In this chapter we will work problems involving both negative numbers and some of the trails found in the Grand Canyon.

Chapter Pretest

The pretest below contains problems that are representative of the problems you will find in the chapter.

Give the opposite of each number.

1. 10 **2.** -3

Place either < or > between the numbers so that the resulting statement is true.

3. $-3 \quad -5$ **4.** $2 \quad |-4|$

Simplify each expression.

5. $-|-8|$ **6.** $-(-1)$

Perform the indicated operations.

7. $9 - 19$ **8.** $3 + (-4)$ **9.** $-6 - 8$ **10.** $-5 + (-15)$ **11.** $-40 - (-7)$ **12.** $5(-4)$

13. $(-2)(-8)$ **14.** $21 \div (-7)$ **15.** $\frac{-40}{-5}$ **16.** $\frac{0}{-1}$

Simplify the following expressions as much as possible.

17. $(-2)^3$ **18.** $-10 + 2(1 - 3)$ **19.** $\frac{7 + 4(-1)}{2 - 3}$

20. $(3x + 4) + 9$ **21.** $5(3y - 2)$ **22.** $-4b + 3b$

23. On a certain day, the temperature reaches a high of 30° above 0 and a low of 5° below 0. What is the difference between the high and low temperatures for the day?

Getting Ready for Chapter 2

The problems below review material covered previously that you need to know in order to be successful in Chapter 2. If you have any difficulty with the problems here, you need to go back and review before going on to Chapter 2.

1. Use the associative property to rewrite the expression: $5(3 \cdot 2)$

2. Use the distributive property to rewrite the expression: $4 \cdot 7 - 4 \cdot 2$

Perform the indicated operation.

3. $5 - 3$ **4.** $9 - 0$ **5.** $6 - 1$ **6.** $7 - 7$ **7.** 5^3 **8.** $12 + 6$

9. $14 - 8$ **10.** $6 \div 3$ **11.** $60 \div 20$ **12.** $8 \div 8$ **13.** $9 \div 3$ **14.** $7 \div 1$

15. $12 \div 4$ **16.** $12 \div 3$

17. Find the perimeter of the square.

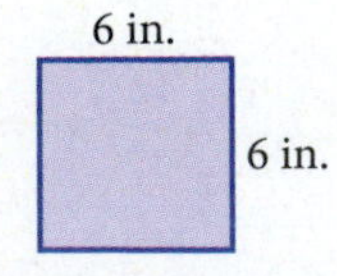

18. Find the area of the rectangle.

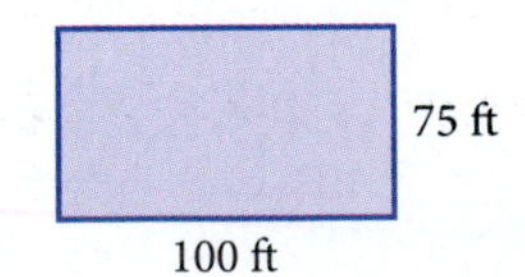

Positive and Negative Numbers

2.1

Objectives

A Use the number line and inequality symbols to compare numbers.

B Find the absolute value of a number.

C Find the opposite of a number.

D Solve applications involving negative numbers.

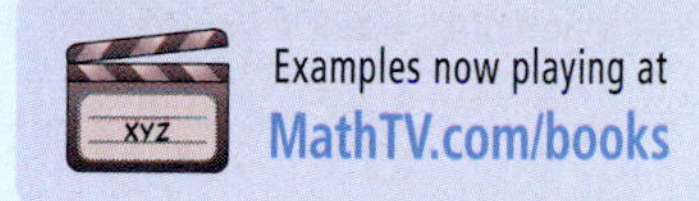

Introduction . . .

Before the late nineteenth century, time zones did not exist. Each town would set their clocks according to the motions of the Sun. It was not until the late 1800s that a system of worldwide time zones was developed. This system divides the earth into 24 time zones with Greenwich, England designated as the center of the time zones (GMT). This location is assigned a value of zero. Each of the World Time Zones is assigned a number ranging from -12 to $+12$ depending on its position east or west of Greenwich, England.

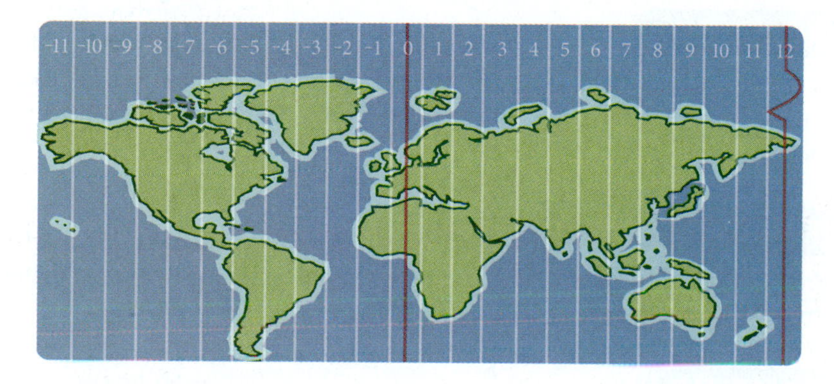

If New York is 5 time zones to the left of GMT, this would be noted as $-5{:}00$ GMT.

A Comparing Numbers

To see the relationship between negative and positive numbers, we can extend the number line as shown in Figure 1. We first draw a straight line and label a convenient point with 0. This is called the *origin*, and it is usually in the middle of the line. We then label positive numbers to the right (as we have done previously), and negative numbers to the left.

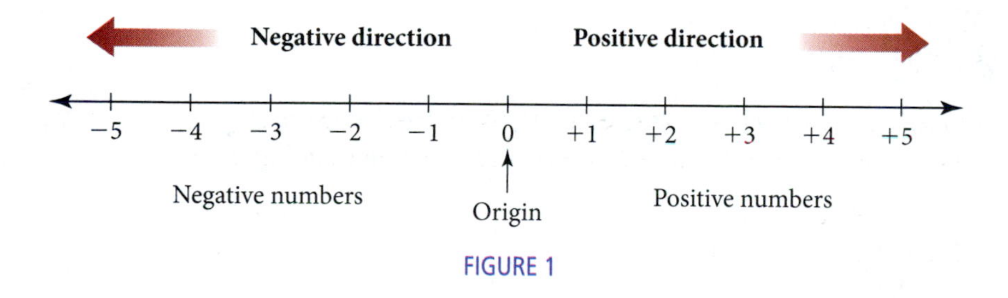

FIGURE 1

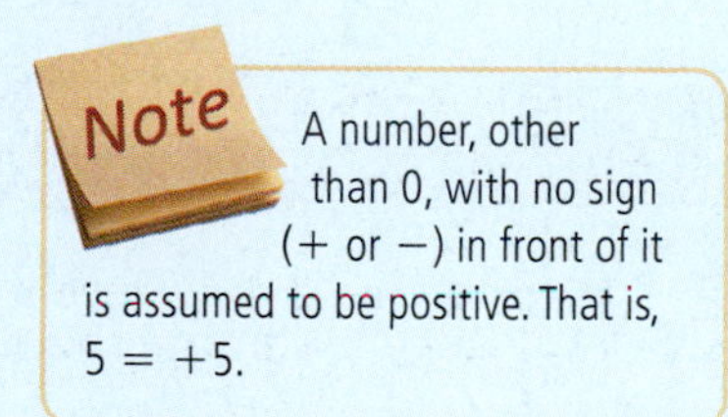

Note A number, other than 0, with no sign (+ or −) in front of it is assumed to be positive. That is, $5 = +5$.

The numbers increase going from left to right. If we move to the right, we are moving in the positive direction. If we move to the left, we are moving in the negative direction. *Any number to the left of another number is considered to be smaller than the number to its right.*

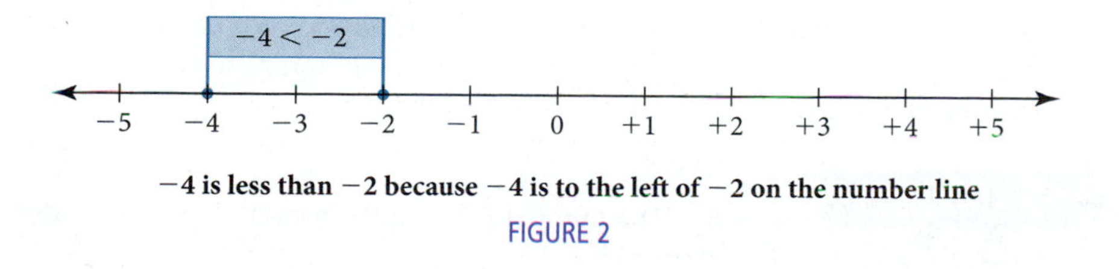

−4 is less than −2 because −4 is to the left of −2 on the number line

FIGURE 2

We see from the line that every negative number is less than every positive number.

In algebra we can use inequality symbols when comparing numbers.

Notation

If a and b are any two numbers on the number line, then

$a < b$ is read "a is less than b"

$a > b$ is read "a is greater than b"

As you can see, the inequality symbols always point to the smaller of the two numbers being compared. Here are some examples that illustrate how we use the inequality symbols.

PRACTICE PROBLEMS

Write each statement in words.

1. $2 < 8$
2. $5 > 10$ (Is this a true statement?)
3. $-4 < 4$
4. $-7 < -2$

EXAMPLE 1 $3 < 5$ is read "3 is less than 5." Note that it would also be correct to write $5 > 3$. Both statements, "3 is less than 5" and "5 is greater than 3," have the same meaning. The inequality symbols always point to the smaller number.

EXAMPLE 2 $0 > 100$ is a false statement, because 0 is less than 100, not greater than 100. To write a true inequality statement using the numbers 0 and 100, we would have to write either $0 < 100$ or $100 > 0$.

EXAMPLE 3 $-3 < 5$ is a true statement, because -3 is to the left of 5 on the number line, and, therefore, it must be less than 5. Another statement that means the same thing is $5 > -3$.

EXAMPLE 4 $-5 < -2$ is a true statement, because -5 is to the left of -2 on the number line, meaning that -5 is less than -2. Both statements $-5 < -2$ and $-2 > -5$ have the same meaning; they both say that -5 is a smaller number than -2.

B Absolute Value

It is sometimes convenient to talk about only the numerical part of a number and disregard the sign (+ or −) in front of it. The following definition gives us a way of doing this.

Definition

The **absolute value** of a number is its distance from 0 on the number line. We denote the absolute value of a number with vertical lines. For example, the absolute value of -3 is written $|-3|$.

The absolute value of a number is never negative because it is a distance, and a distance is always measured in positive units (unless it happens to be 0).

Here are some examples of absolute value problems.

Give the absolute value of each of the following.

5. $|6|$
6. $|-5|$

EXAMPLE 5 $|5| = 5$ The number 5 is 5 units from 0.

EXAMPLE 6 $|-3| = 3$ The number -3 is 3 units from 0.

Answers

1. 2 is less than 8.
2. 5 is greater than 10. (No.)
3. -4 is less than 4.
4. -7 is less than -2.
5. 6 6. 5

EXAMPLE 7 $|-7| = 7$ The number -7 is 7 units from 0.

Give the absolute value.

7. $|-8|$

C Opposites

Definition

Two numbers that are the same distance from 0 but in opposite directions from 0 are called **opposites.*** The notation for the opposite of a is $-a$.

EXAMPLE 8 Give the opposite of each of the following numbers:

5, 7, 1, −5, −8

SOLUTION The opposite of 5 is −5.
The opposite of 7 is −7.
The opposite of 1 is −1.
The opposite of −5 is −(−5), or 5.
The opposite of −8 is −(−8), or 8.

8. Give the opposite of each of the following numbers: 8, 10, 0, −4.

We see from this example that the opposite of every positive number is a negative number, and likewise, the opposite of every negative number is a positive number. The last two parts of Example 8 illustrate the following property:

Property

If a represents any positive number, then it is always true that

$$-(-a) = a$$

In other words, this property states that the opposite of a negative number is a positive number.

It should be evident now that the symbols + and − can be used to indicate several different ideas in mathematics. In the past we have used them to indicate addition and subtraction. They can also be used to indicate the direction a number is from 0 on the number line. For instance, the number +3 (read "positive 3") is the number that is 3 units from zero in the positive direction. On the other hand, the number −3 (read "negative 3") is the number that is 3 units from 0 in the negative direction. The symbol − can also be used to indicate the opposite of a number, as in $-(-2) = 2$. The interpretation of the symbols + and − depends on the situation in which they are used. For example:

3 + 5	The + sign indicates addition.
7 − 2	The − sign indicates subtraction.
−7	The − sign is read "negative" 7.
−(−5)	The first − sign is read "the opposite of." The second − sign is read "negative" 5.

This may seem confusing at first, but as you work through the problems in this chapter you will get used to the different interpretations of the symbols + and −.

We should mention here that the set of whole numbers along with their opposites forms the set of *integers*. That is:

$$\text{Integers} = \{\ldots, -3, -2, -1, 0, 1, 2, 3, \ldots\}$$

*In some books opposites are called *additive inverses*.

Answers

7. 8 8. −8, −10, 0, 4

DESCRIPTIVE STATISTICS

D Displaying Negative Numbers

In the table below, the temperatures below zero are represented by negative numbers.

EXAMPLE 9 Use the information in Table 1 and the template below to draw a scatter diagram and a line graph representing the information in Table 1.

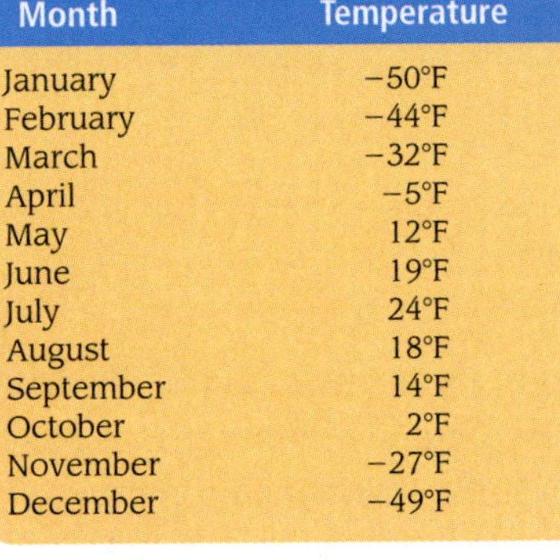

TABLE 1

RECORD LOW TEMPERATURES FOR JACKSON HOLE, WYOMING

Month	Temperature
January	−50°F
February	−44°F
March	−32°F
April	−5°F
May	12°F
June	19°F
July	24°F
August	18°F
September	14°F
October	2°F
November	−27°F
December	−49°F

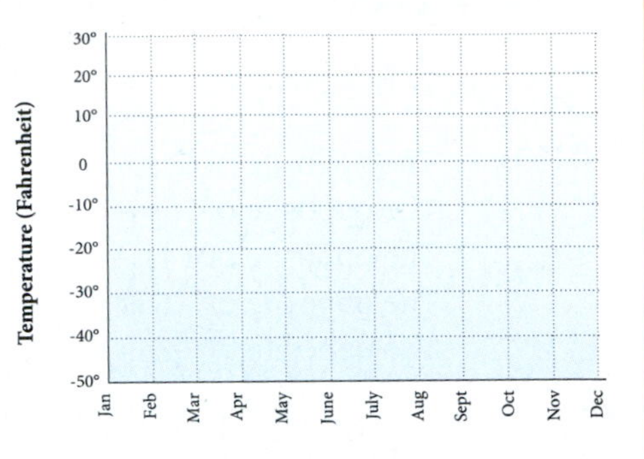

SOLUTION Notice that the vertical axis in the template looks like the number line we have been using. To produce the scatter diagram, we place a dot above each month, across from the temperature for that month. For example, the dot above July will be across from 24°. Doing the same for each of the months, we have the scatter diagram shown in Figure 3. To produce the line graph in Figure 4, we simply connect the dots in Figure 3 with line segments.

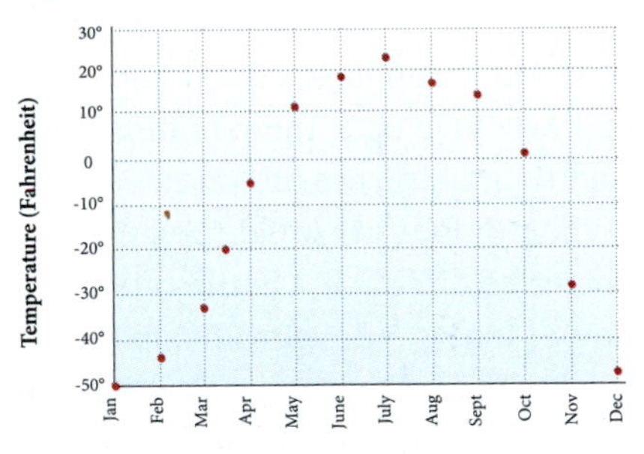

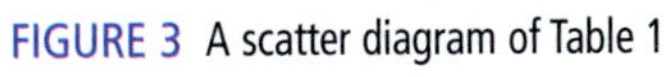

FIGURE 3 A scatter diagram of Table 1

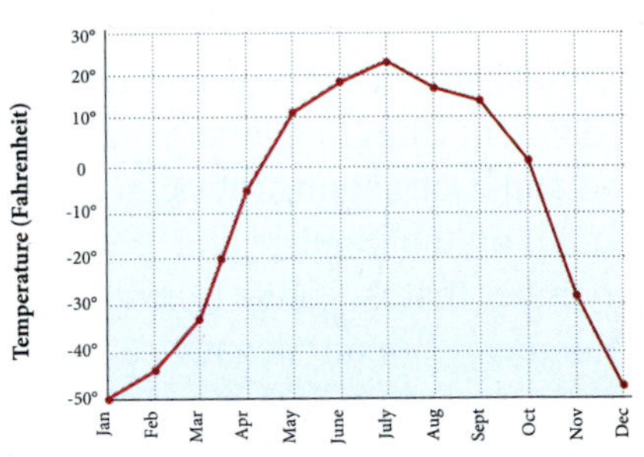

FIGURE 4 A line graph of Table 1

Getting Ready for Class

After reading through the preceding section, respond in your own words and in complete sentences.

1. Write the statement "3 is less than 5" in symbols.
2. What is the absolute value of a number?
3. Describe what we mean by numbers that are "opposites" of each other.
4. If you locate two different numbers on the number line, which one will be the smaller number?

9. Use the information in the table below to make both a scatter diagram and a line graph.

AVERAGE MONTHLY PRECIPITATION SAN LUIS OBISPO, CALIFORNIA

Month	Precipitation (mm)
Jan	134.1
Mar	113.8
May	11.9
July	0.8
Sept	11.2
Nov	55.1

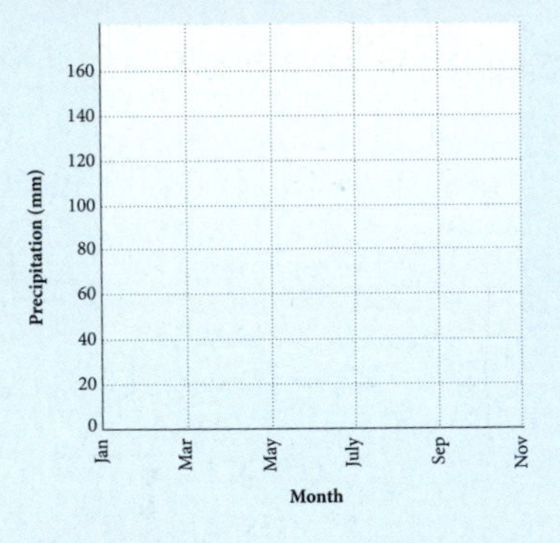

Note In the United States, temperature is measured on the Fahrenheit temperature scale. On this scale, water boils at 212 degrees and freezes at 32 degrees. To denote a temperature of 32 degrees on the Fahrenheit scale, we write 32°F, which is read "32 degrees Fahrenheit."

Answer

9. See solutions section.

Problem Set 2.1

A Write each of the following in words. [Example 1]

1. $4 < 7$
2. $0 < 10$
3. $5 > -2$
4. $8 > -8$
5. $-10 < -3$.
6. $-20 < -5$
7. $0 > -4$
8. $0 > -100$

Write each of the following in symbols.

9. 30 is greater than -30.
10. -30 is less than 30.
11. -10 is less than 0.
12. 0 is greater than -10.
13. -3 is greater than -15.
14. -15 is less than -3.

A Place either $<$ or $>$ between each of the following pairs of numbers so that the resulting statement is true. [Examples 2–4]

15. $3 \quad 7$
16. $17 \quad 0$
17. $7 \quad -5$
18. $2 \quad -13$
19. $-6 \quad 0$
20. $-14 \quad 0$
21. $-12 \quad -2$
22. $-20 \quad -1$
23. $-\frac{1}{2} \quad -\frac{3}{4}$
24. $-\frac{6}{7} \quad \frac{5}{6}$
25. $-0.75 \quad 0.25$
26. $-1 \quad -3.5$
27. $-0.1 \quad -0.01$
28. $-0.04 \quad -0.4$
29. $-3 \quad |6|$
30. $|8| \quad -2$
31. $15 \quad |-4|$
32. $20 \quad |-6|$
33. $|-2| \quad |-7|$
34. $|-3| \quad |-1|$

B Find each of the following absolute values. [Examples 5–7]

35. $|2|$
36. $|7|$
37. $|100|$
38. $|10{,}000|$
39. $|-8|$
40. $|-9|$
41. $|-231|$
42. $|-457|$
43. $\left|-\frac{3}{4}\right|$
44. $\left|-\frac{1}{10}\right|$
45. $|-200|$
46. $|-350|$
47. $|8|$
48. $|9|$
49. $|231|$
50. $|457|$

C Give the opposite of each of the following numbers. [Example 8]

51. 3 **52.** -5 **53.** -2 **54.** 15 **55.** 75 **56.** -32

57. 0 **58.** 1 **59.** -121 **60.** -200 **61.** 555 **62.** 125

Simplify each of the following.

63. $-(-2)$ **64.** $-(-5)$ **65.** $-(-8)$ **66.** $-(-3)$

67. $-|-2|$ **68.** $-|-5|$ **69.** $-|-8|$ **70.** $-|-3|$

71. What number is its own opposite?

72. Is $|a| = a$ always a true statement?

73. If n is a negative number, is $-n$ positive or negative?

74. If n is a positive number, is $-n$ positive or negative?

Estimating

Work Problems 75–80 mentally, without pencil and paper or a calculator.

75. Is -60 closer to 0 or -100?

76. Is -20 closer to 0 or -30?

77. Is -10 closer to -20 or 20?

78. Is -20 closer to -40 or 10?

79. Is -362 closer to -360 or -370?

80. Is -368 closer to -360 or -370?

D Applying the Concepts [Example 9]

81. The London Eye has a height of 450 feet. Describe the location of someone standing on the ground in relation to someone at the top of the London Eye.

82. The Eiffel Tower has several levels visitors can walk around on. The first is 57 meters above the ground, the second is 115 meters high, and the third level is 276 meters high. What is the location of someone standing on the first level in relation to someone standing on the third level?

83. The Bright Angel trail at Grand Canyon National Park ends at Indian Garden, 3,060 feet below the trailhead. Write this as a negative number with respect to the trailhead.

84. The South Kaibab Trail at Grand Canyon National Park ends at Cedar Ridge, 1,140 feet below the trailhead. Write this as a negative number with respect to the trailhead.

85. Car Depreciation Depreciation refers to the decline in a car's market value during the time you own the car. According to sources such as Kelley Blue Book and Edmunds.com, not all cars depreciate at the same rate. Suppose you pay $25,000 for a new car which has a high rate of depreciation. Your car loses about $5,000 in value per year. Represent this loss in value as a negative number. A car with a low rate of depreciation loses about $2,750 in value each year. Represent this loss as a negative number.

86. Census Figures In June, 2007 the U.S. Census Bureau released population estimates for the twenty-five cities with the largest population loss between July 1, 2005 and July 1, 2006. New Orleans had the largest population loss. The city's population fell by 228,782 people. Detroit, Michigan experienced a population loss of 12,344 people during the same time period. Represent the loss of population for New Orleans and for Detroit as a negative number.

87. Temperature and Altitude Yamina is flying from Phoenix to San Francisco on a Boeing 737 jet. When the plane reaches an altitude of 33,000 feet, the temperature outside the plane is 61 degrees below zero Fahrenheit. Represent this temperature with a negative number. If the temperature outside the plane gets warmer by 10 degrees, what will the new temperature be?

88. Temperature Change At 11:00 in the morning in Superior, Wisconsin, Jim notices the temperature is 15 degrees below zero Fahrenheit. Write this temperature as a negative number. At noon it has warmed up by 8 degrees. What is the temperature at noon?

89. Temperature Change At 10:00 in the morning in White Bear Lake, Minnesota, Zach notices the temperature is 5 degrees below zero Fahrenheit. Write this temperature as a negative number. By noon the temperature has dropped another 10 degrees. What is the temperature at noon?

90. Snorkeling Steve is snorkeling in the ocean near his home in Maui. At one point he is 6 feet below the surface. Represent this situation with a negative number. If he descends another 6 feet, what negative number will represent his new position?

91. Time Zones New Orleans, Louisiana, is 1 time zone west of New York City. Represent this time zone as a negative number, as discussed in the introduction to this chapter.

92. Time Zones Seattle, Washington, is 2 time zones west of New Orleans, Louisiana. Represent this time zone as a negative number, as discussed in the introduction to this chapter.

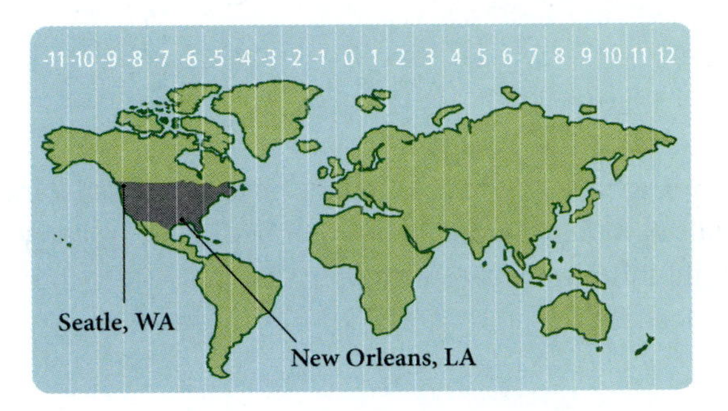

Table 2 lists various wind chill temperatures. The top row gives air temperature, while the first column gives wind speed in miles per hour. The numbers within the table indicate how cold the weather will feel. For example, if the thermometer reads 30°F and the wind is blowing at 15 miles per hour, the wind chill temperature is 9°F.

TABLE 2

WIND CHILL TEMPERATURES

	Air temperatures (°F)							
Wind Speed	30°	25°	20°	15°	10°	5°	0°	−5°
10 mph	16°	10°	3°	−3°	−9°	−15°	−22°	−27°
15 mph	9°	2°	−5°	−11°	−18°	−25°	−31°	−38°
20 mph	4°	−3°	−10°	−17°	−24°	−31°	−39°	−46°
25 mph	1°	−7°	−15°	−22°	−29°	−36°	−44°	−51°
30 mph	−2°	−10°	−18°	−25°	−33°	−41°	−49°	−56°

93. Wind Chill Find the wind chill temperature if the thermometer reads 25°F and the wind is blowing at 25 miles per hour.

94. Wind Chill Find the wind chill temperature if the thermometer reads 10°F and the wind is blowing at 25 miles per hour.

95. Wind Chill Which will feel colder: a day with an air temperature of 10°F and a 25-mph wind, or a day with an air temperature of −5°F and a 10-mph wind?

96. Wind Chill Which will feel colder: a day with an air temperature of 15°F and a 20-mph wind, or a day with an air temperature of 5°F and a 10-mph wind?

Table 3 lists the record low temperatures for each month of the year for Lake Placid, New York. Table 4 lists the record high temperatures for the same city.

TABLE 3

RECORD LOW TEMPERATURES FOR LAKE PLACID, NEW YORK

Month	Temperature
January	−36°F
February	−30°F
March	−14°F
April	−2°F
May	19°F
June	22°F
July	35°F
August	30°F
September	19°F
October	15°F
November	−11°F
December	−26°F

TABLE 4

RECORD HIGH TEMPERATURES FOR LAKE PLACID, NEW YORK

Month	Temperature
January	54°F
February	59°F
March	69°F
April	82°F
May	90°F
June	93°F
July	97°F
August	93°F
September	90°F
October	87°F
November	67°F
December	60°F

97. Temperature Figure 5 is a bar chart of the information in Table 3. Use the template in Figure 6 to construct a scatter diagram of the same information. Then connect the dots in the scatter diagram to obtain a line graph of that same information. (Notice that we have used the numbers 1 through 12 to represent the months January through December.)

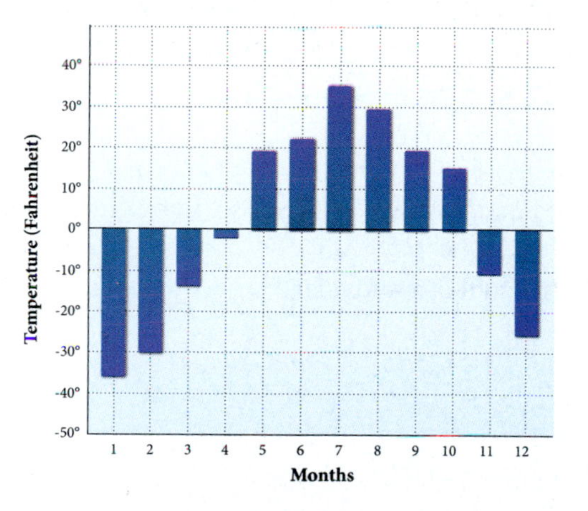

FIGURE 5 A bar chart of Table 3

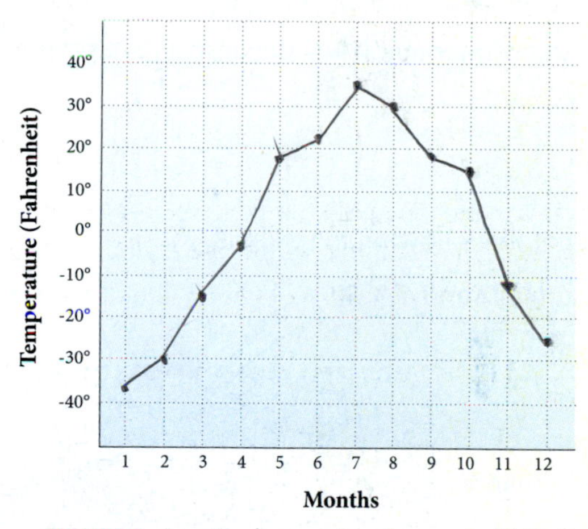

FIGURE 6 A scatter diagram, then line graph of Table 3

98. Temperature Figure 7 is a bar chart of the information in Table 4. Use the template in Figure 8 to construct a scatter diagram of the same information. Then connect the dots in the scatter diagram to obtain a line graph of that same information. (Again, we have used the numbers 1 through 12 to represent the months January through December.)

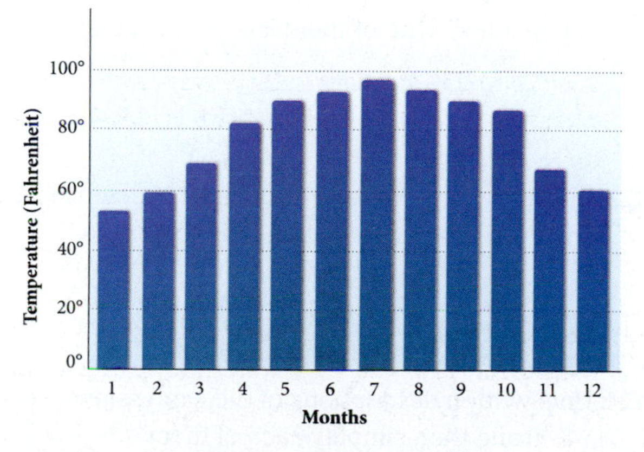

FIGURE 7 A bar chart of Table 4

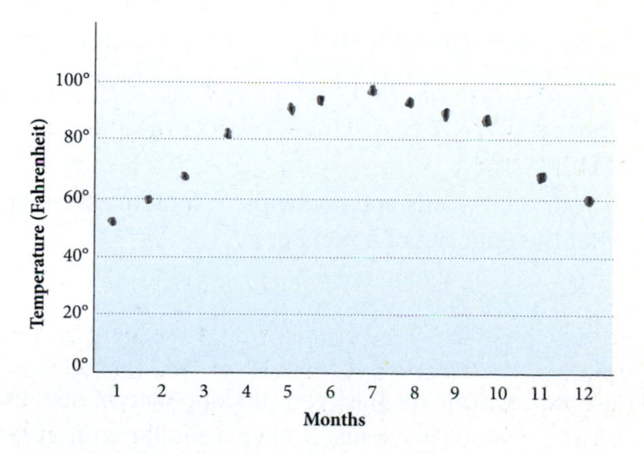

FIGURE 8 A scatter diagram, then line graph of Table 4

Getting Ready for the Next Section

Add or subtract.

99. $10 + 15$

100. $12 + 15$

101. $15 - 10$

102. $15 - 12$

103. $10 - 5 - 3 + 4$

104. $12 - 3 - 7 + 5$

105. $[3 + 10] + [8 - 2]$

106. $[2 + 12] + [7 - 5]$

107. $276 + 32 + 4{,}005$

108. $17 + 3 + 152 + 1{,}200$

109. $635 - 579$

110. $2{,}987 - 1{,}130$

Maintaining Your Skills

Complete each statement using the commutative property of addition.

111. $3 + 5 =$

112. $9 + x =$

Complete each statement using the associative property of addition.

113. $7 + (2 + 6) =$

114. $(x + 3) + 5 =$

Write each of the following in symbols.

115. The sum of x and 4

116. The sum of x and 4 is 9.

117. 5 more than y

118. x increased by 8

Extending the Concepts

119. There are two numbers that are 5 units from 2 on the number line. One of them is 7. What is the other one?

120. There are two numbers that are 5 units from -2 on the number line. One of them is 3. What is the other one?

121. In your own words and in complete sentences, explain what the opposite of a number is.

122. In your own words and in complete sentences, explain what the absolute value of a number is.

123. The expression $-(-3)$ is read "the opposite of negative 3," and it simplifies to just 3. Give a similar written description of the expression $-|-3|$, and then simplify it.

124. Give written descriptions of the expressions $-(-4)$ and $-|-4|$ and then simplify each of them.

2.2 Addition with Negative Numbers

Objectives

A Use the number line to add positive and negative numbers.

B Add positive and negative numbers using a rule.

C Solve applications involving addition with positive and negative numbers.

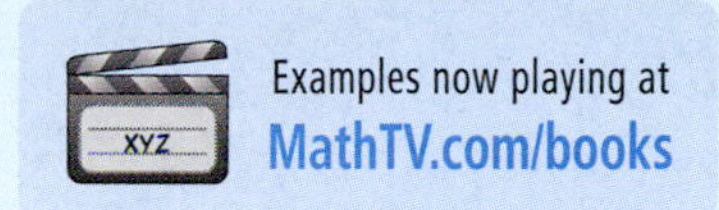

Introduction . . .

Suppose you are in Las Vegas playing blackjack and you lose \$3 on the first hand and then you lose \$5 on the next hand. If you represent winning with positive numbers and losing with negative numbers, how will you represent the results from your first two hands? Since you lost \$3 and \$5 for a total of \$8, one way to represent the situation is with addition of negative numbers:

$$(-\$3) + (-\$5) = -\$8$$

From this example we see that the sum of two negative numbers is a negative number. To generalize addition of positive and negative numbers, we can use the number line.

A Adding with a Number Line

We can think of each number on the number line as having two characteristics: (1) a *distance* from 0 (absolute value) and (2) a *direction* from 0 (positive or negative). The distance from 0 is represented by the numerical part of the number (like the 5 in the number -5), and its direction is represented by the + or − sign in front of the number.

We can visualize addition of numbers on the number line by thinking in terms of distance and direction from 0. Let's begin with a simple problem we know the answer to. We interpret the sum $3 + 5$ on the number line as follows:

1. The first number is 3, which tells us "start at the origin, and move 3 units in the positive direction."
2. The + sign is read "and then move."
3. The 5 means "5 units in the positive direction."

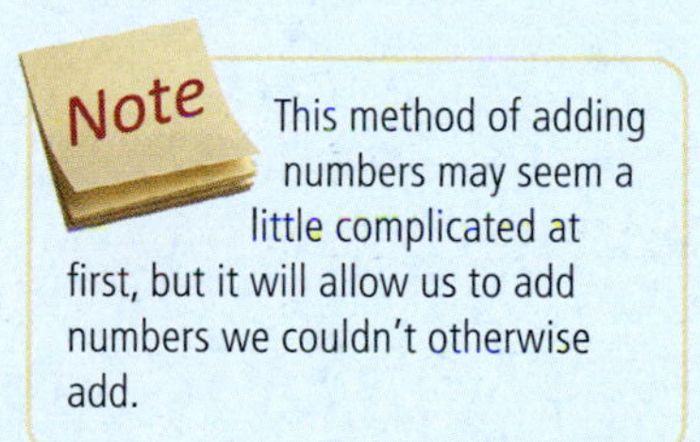

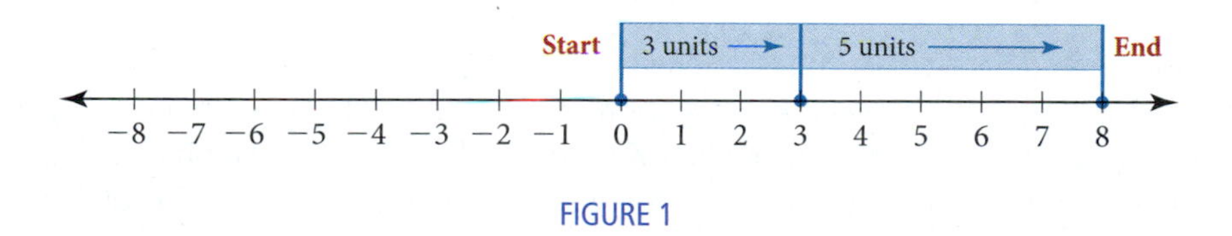

FIGURE 1

Figure 1 shows these steps. To summarize, $3 + 5$ means to start at the origin (0), move 3 units in the *positive* direction, and then move 5 units in the *positive* direction. We end up at 8, which is the sum we are looking for: $3 + 5 = 8$.

EXAMPLE 1 Add $3 + (-5)$ using the number line.

SOLUTION We start at the origin, move 3 units in the positive direction, and then move 5 units in the negative direction, as shown in Figure 2. The last arrow ends at -2, which must be the sum of 3 and -5. That is:

$$3 + (-5) = -2$$

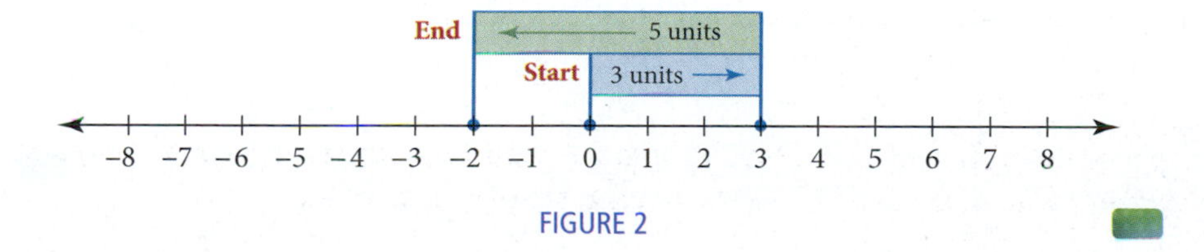

FIGURE 2

PRACTICE PROBLEMS

1. Add: $2 + (-5)$

Answer

1. -3

2. Add: $-2 + 5$

EXAMPLE 2 Add $-3 + 5$ using the number line.

SOLUTION We start at the origin, move 3 units in the negative direction, and then move 5 units in the positive direction, as shown in Figure 3. We end up at 2, which is the sum of -3 and 5. That is:

$$-3 + 5 = 2$$

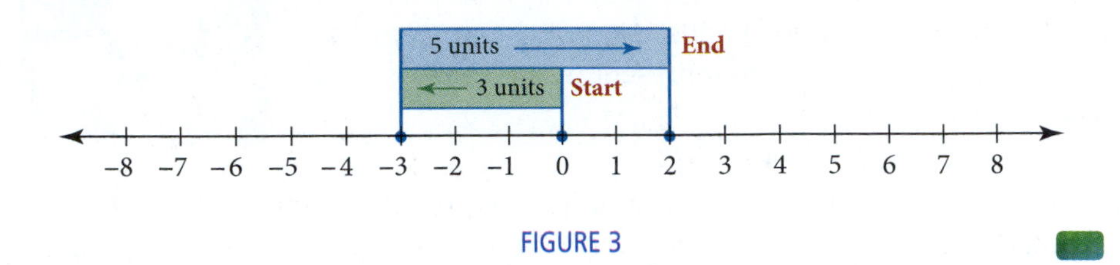

FIGURE 3

3. Add: $-2 + (-5)$

EXAMPLE 3 Add $-3 + (-5)$ using the number line.

SOLUTION We start at the origin, move 3 units in the negative direction, and then move 5 more units in the negative direction. This is shown on the number line in Figure 4. As you can see, the last arrow ends at -8. We must conclude that the sum of -3 and -5 is -8. That is:

$$-3 + (-5) = -8$$

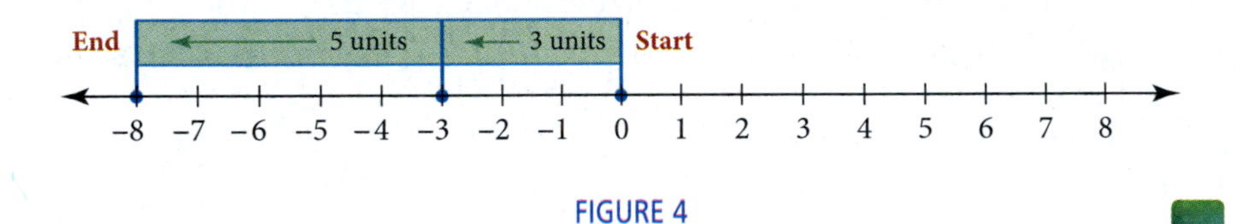

FIGURE 4

Adding numbers on the number line as we have done in these first three examples gives us a way of visualizing addition of positive and negative numbers. We eventually want to be able to write a rule for addition of positive and negative numbers that doesn't involve the number line. The number line is a way of justifying the rule we will eventually write. Here is a summary of the results we have so far:

$$3 + 5 = 8 \qquad -3 + 5 = 2$$
$$3 + (-5) = -2 \qquad -3 + (-5) = -8$$

Examine these results to see if you notice any pattern in the answers.

4. Add: $2 + 6$

EXAMPLE 4 $4 + 7 = 11$

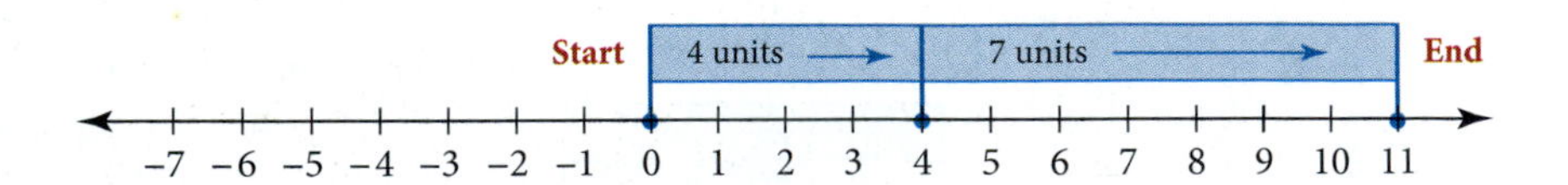

Answers
2. 3 **3.** −7 **4.** 8

EXAMPLE 5 $4 + (-7) = -3$

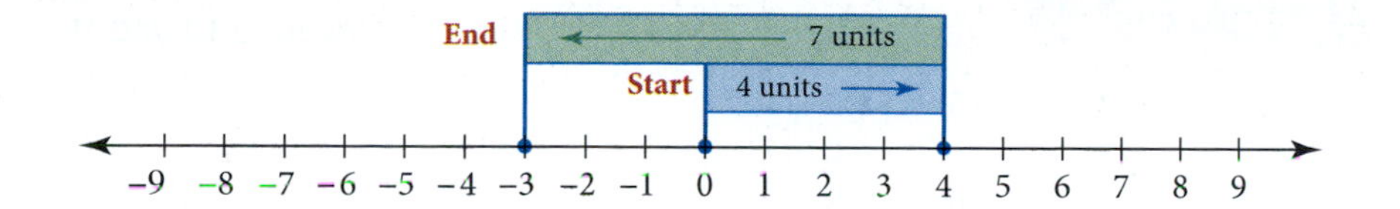

5. Add: $2 + (-6)$

EXAMPLE 6 $-4 + 7 = 3$

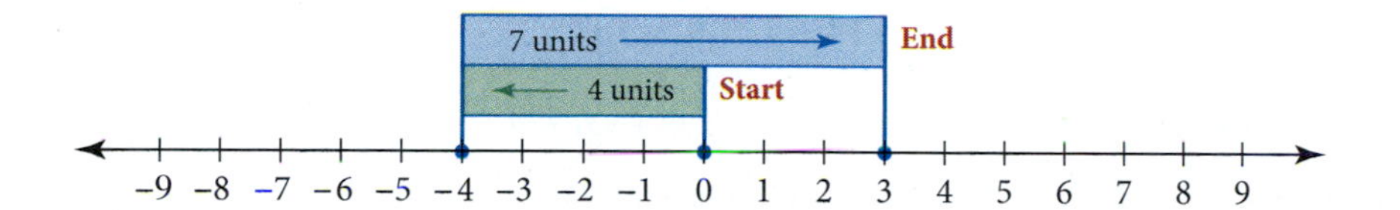

6. Add: $-2 + 6$

EXAMPLE 7 $-4 + (-7) = -11$

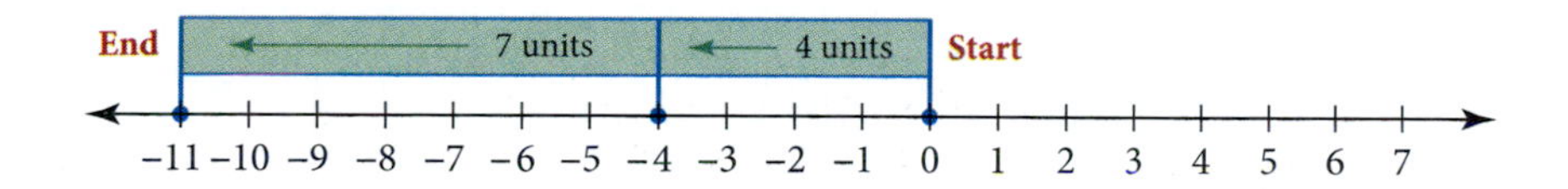

7. Add: $-2 + (-6)$

B Addition

A summary of the results of these last four examples looks like this:

$$
\begin{aligned}
4 + \quad 7 &= 11 \\
4 + (-7) &= -3 \\
-4 + \quad 7 &= 3 \\
-4 + (-7) &= -11
\end{aligned}
$$

Looking over all the examples in this section, and noticing how the results in the problems are related, we can write the following rule for adding any two numbers:

Rule

1. To add two numbers with the *same* sign: Simply add their absolute values, and use the common sign. If both numbers are positive, the answer is positive. If both numbers are negative, the answer is negative.
2. To add two numbers with *different* signs: Subtract the smaller absolute value from the larger absolute value. The answer will have the sign of the number with the larger absolute value.

This rule covers all possible addition problems involving positive and negative numbers. You *must* memorize it. After you have worked some problems, the rule will seem almost automatic.

Answers

5. −4 **6.** 4 **7.** −8

The following examples show how the rule is used. You will find that the rule for addition is consistent with all the results obtained using the number line.

8. Add all combinations of positive and negative 12 and 15.

EXAMPLE 8 Add all combinations of positive and negative 10 and 15.

SOLUTION

$$\begin{aligned} 10 + 15 &= 25 \\ 10 + (-15) &= -5 \\ -10 + 15 &= 5 \\ -10 + (-15) &= -25 \end{aligned}$$

Notice that when we add two numbers with the same sign, the answer also has that sign. When the signs are not the same, the answer has the sign of the number with the larger absolute value.

Once you have become familiar with the rule for adding positive and negative numbers, you can apply it to more complicated sums.

9. Simplify: $12 + (-3) + (-7) + 5$

EXAMPLE 9 Simplify: $10 + (-5) + (-3) + 4$

SOLUTION Adding left to right, we have:

$$\begin{aligned} 10 + (-5) + (-3) + 4 &= \mathbf{5} + (-3) + 4 && 10 + (-5) = 5 \\ &= \mathbf{2} + 4 && 5 + (-3) = 2 \\ &= 6 \end{aligned}$$

10. Simplify: $[-2 + (-12)] + [7 + (-5)]$

EXAMPLE 10 Simplify: $[-3 + (-10)] + [8 + (-2)]$

SOLUTION We begin by adding the numbers inside the brackets.

$$\begin{aligned} [-3 + (-10)] + [8 + (-2)] &= [-13] + [6] \\ &= -7 \end{aligned}$$

USING TECHNOLOGY

Calculator Note

There are a number of different ways in which calculators display negative numbers. Some calculators use a key labeled [+/−], whereas others use a key labeled [(−)]. You will need to consult with the manual that came with your calculator to see how your calculator does the job.

Here are a couple of ways to find the sum $-10 + (-15)$ on a calculator:

Scientific Calculator: 10 [+/−] [+] 15 [+/−] [=]

Graphing Calculator: [(−)] 10 [+] [(−)] 15 [ENT]

Getting Ready for Class

After reading through the preceding section, respond in your own words and in complete sentences.

1. Explain how you would use the number line to add 3 and 5.
2. If two numbers are negative, such as −3 and −5, what sign will their sum have?
3. If you add two numbers with different signs, how do you determine the sign of the answer?
4. With respect to addition with positive and negative numbers, does the phrase "two negatives make a positive" make any sense?

Answers
8. See solutions section.
9. 7 **10.** −12

Problem Set 2.2

A Draw a number line from -10 to $+10$ and use it to add the following numbers. [Examples 1–7]

1. $2 + 3$ **2.** $2 + (-3)$ **3.** $-2 + 3$ **4.** $-2 + (-3)$ **5.** $5 + (-7)$ **6.** $-5 + 7$

7. $-4 + (-2)$ **8.** $-8 + (-2)$ **9.** $10 + (-6)$ **10.** $-9 + 3$ **11.** $7 + (-3)$ **12.** $-7 + 3$

13. $-4 + (-5)$ **14.** $-2 + (-7)$

B Combine the following by using the rule for addition of positive and negative numbers. (Your goal is to be fast and accurate at addition, with the latter being more important.) [Example 8]

15. $7 + 8$ **16.** $9 + 12$ **17.** $5 + (-8)$ **18.** $4 + (-11)$

19. $-6 + (-5)$ **20.** $-7 + (-2)$ **21.** $-10 + 3$ **22.** $-14 + 7$

23. $-1 + (-2)$ **24.** $-5 + (-4)$ **25.** $-11 + (-5)$ **26.** $-16 + (-10)$

27. $4 + (-12)$ **28.** $9 + (-1)$ **29.** $-85 + (-42)$ **30.** $-96 + (-31)$

31. $-121 + 170$ **32.** $-130 + 158$ **33.** $-375 + 409$ **34.** $-765 + 213$

Complete the following tables.

35.

First Number a	Second Number b	Their Sum $a+b$
5	−3	
5	−4	
5	−5	
5	−6	
5	−7	

36.

First Number a	Second Number b	Their Sum $a+b$
−5	3	
−5	4	
−5	5	
−5	6	
−5	7	

37.

First Number x	Second Number y	Their Sum $x+y$
−5	−3	
−5	−4	
−5	−5	
−5	−6	
−5	−7	

38.

First Number x	Second Number y	Their Sum $x+y$
30	−20	
−30	20	
−30	−20	
30	20	
−30	0	

B Add the following numbers left to right. [Example 9]

39. $24 + (-6) + (-8)$

40. $35 + (-5) + (-30)$

41. $-201 + (-143) + (-101)$

42. $-27 + (-56) + (-89)$

43. $-321 + 752 + (-324)$

44. $-571 + 437 + (-502)$

45. $-2 + (-5) + (-6) + (-7)$

46. $-8 + (-3) + (-4) + (-7)$

47. $15 + (-30) + 18 + (-20)$

48. $20 + (-15) + 30 + (-18)$

49. $-78 + (-42) + 57 + 13$

50. $-89 + (-51) + 65 + 17$

B Use the rule for order of operations to simplify each of the following. [Example 10]

51. $(-8 + 5) + (-6 + 2)$

52. $(-3 + 1) + (-9 + 4)$

53. $(-10 + 4) + (-3 + 12)$

54. $(-11 + 5) + (-3 + 2)$

55. $20 + (-30 + 50) + 10$

56. $30 + (-40 + 20) + 50$

57. $108 + (-456 + 275)$

58. $106 + (-512 + 318)$

59. $[5 + (-8)] + [3 + (-11)]$

60. $[8 + (-2)] + [5 + (-7)]$

61. $[57 + (-35)] + [19 + (-24)]$

62. $[63 + (-27)] + [18 + (-24)]$

63. Find the sum of -8, -10, and -3.

64. Find the sum of -4, 17, and -6.

65. What number do you add to 8 to get 3?

66. What number do you add to 10 to get 4?

67. What number do you add to -3 to get -7?

68. What number do you add to -5 to get -8?

69. What number do you add to -4 to get 3?

70. What number do you add to -7 to get 2?

71. If the sum of -3 and 5 is increased by 8, what number results?

72. If the sum of -9 and -2 is increased by 10, what number results?

Estimating

Work Problems 73–80 mentally, without pencil and paper or a calculator.

73. The answer to the problem $251 + 249$ is closest to which of the following numbers?
a. 500 **b.** 0 **c.** -500

74. The answer to the problem $251 + (-249)$ is closest to which of the following numbers?
a. 500 **b.** 0 **c.** -500

75. The answer to the problem $-251 + 249$ is closest to which of the following numbers?
a. 500 **b.** 0 **c.** -500

76. The answer to the problem $-251 + (-249)$ is closest to which of the following numbers?
a. 500 **b.** 0 **c.** -500

77. The sum of 77 and 22 is closest to which of the following numbers?
a. -100 **b.** -60 **c.** 60 **d.** 100

78. The sum of -77 and 22 is closest to which of the following numbers?
a. -100 **b.** -60 **c.** 60 **d.** 100

79. The sum of 77 and -22 is closest to which of the following numbers?
a. -100 **b.** -60 **c.** 60 **d.** 100

80. The sum of -77 and -22 is closest to which of the following numbers?
a. -100 **b.** -60 **c.** 60 **d.** 100

C Applying the Concepts

81. One of the trails at the Grand Canyon starts at Bright Angel Trailhead and then drops 4,060 feet to the Colorado River and then climbs 4,440 feet to Yaki Point. What is the trail's ending position in relation to the Bright Angel Trailhead? If the trail ends below the starting position write the answer as a negative number.

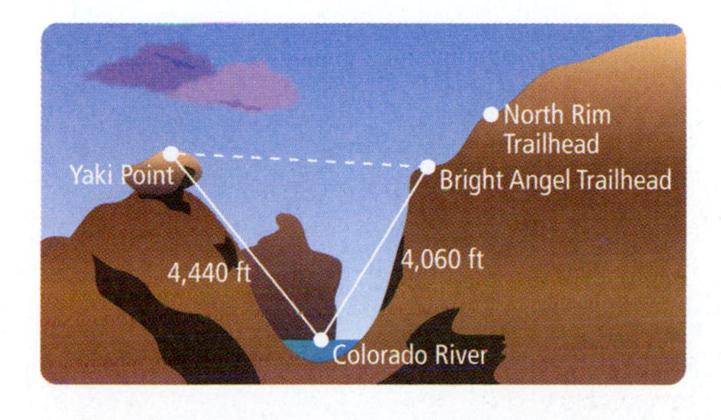

82. One of the trails in the Grand Canyon starts at the North Rim trailhead and drops 5,490 feet to the Colorado River. The trail then climbs 4,060 feet to the Bright Angel Trailhead. What is the Bright Angel Trailhead's position in relation to the North Rim Trailhead? If the trail ends below the starting position write the answer as a negative number.

83. Checkbook Balance Ethan has a balance of −$40 in his checkbook. If he deposits $100 and then writes a check for $50, what is the new balance in his checkbook?

RECORD ALL CHARGES OR CREDITS THAT AFFECT YOUR ACCOUNT

NUMBER	DATE	DESCRIPTION OF TRANSACTION	PAYMENT/DEBIT (-)	DEPOSIT/CREDIT (+)	BALANCE
					-$40 00
	9/20	Deposit		$100 00	
1502	9/21	Vons Market	$50 00		

84. Checkbook Balance Kendra has a balance of −$20 in her checkbook. If she deposits $45 and then writes a check for $15, what is the new balance in her checkbook?

RECORD ALL CHARGES OR CREDITS THAT AFFECT YOUR ACCOUNT

NUMBER	DATE	DESCRIPTION OF TRANSACTION	PAYMENT/DEBIT (−)	DEPOSIT/CREDIT (+)	BALANCE
					-$20 00
	9/25	Deposit		$45 00	
1504	9/28	SLO Soccer	$15 00		

Getting Ready for the Next Section

Give the opposite of each number.

85. 2

86. 3

87. −4

88. −5

89. $\frac{2}{5}$

90. $\frac{3}{8}$

91. −30

92. −15

93. 60.3

94. 70.4

95. Subtract 3 from 5.

96. Subtract 2 from 8.

97. Find the difference of 7 and 4.

98. Find the difference of 8 and 6.

Maintaining Your Skills

The problems below review subtraction with whole numbers.

Subtract.

99. 763 − 159

100. 1,007 − 136

101. 465 − 462 − 3

102. 481 − 479 − 2

Write each of the following statements in symbols.

103. The difference of 10 and x.

104. The difference of x and 10.

105. 17 subtracted from y.

106. y subtracted from 17.

2.3 Subtraction with Negative Numbers

Objectives

A Subtract numbers by thinking of subtraction as addition of the opposite.

B Solve applications involving subtraction with positive and negative numbers.

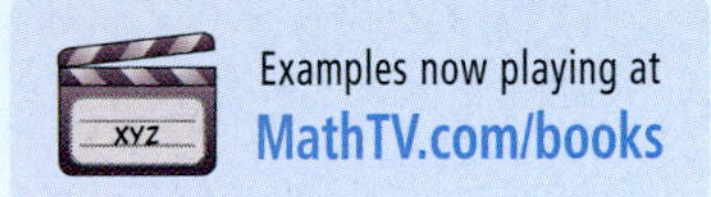

Introduction . . .

How would we represent the final balance in a checkbook if the original balance was \$20 and we wrote a check for \$30? The final balance would be $-\$10$. We can summarize the whole situation with subtraction:

$$\$20 - \$30 = -\$10$$

RECORD ALL CHARGES OR CREDITS THAT AFFECT YOUR ACCOUNT

NUMBER	DATE	DESCRIPTION OF TRANSACTION	PAYMENT/DEBIT (-)	DEPOSIT/CREDIT (+)	BALANCE
					\$20 00
1501	9/15	Campus Bookstore	\$30 00		-\$10 00

From this we see that subtracting 30 from 20 gives us -10. Another example that gives the same answer but involves addition is this:

$$20 + (-30) = -10$$

A Subtraction

From the two examples above, we find that subtracting 30 gives the same result as adding -30. We use this kind of reasoning to give a definition for subtraction that will allow us to use the rules we developed for addition to do our subtraction problems. Here is that definition:

> **Definition**
>
> **Subtraction** If a and b represent any two numbers, then it is always true that
>
> $$a - b = a + (-b)$$
>
> To subtract b Add its opposite, $-b$
>
> *In words:* Subtracting a number is equivalent to adding its opposite.

Note This definition of subtraction may seem a little strange at first. In Example 1 you will notice that using the definition gives us the same results we are used to getting with subtraction. As we progress further into the section, we will use the definition to subtract numbers we haven't been able to subtract before.

Let's see if this definition conflicts with what we already know to be true about subtraction.

EXAMPLE 1 Subtract: $5 - 2$

SOLUTION From previous experience we know that

$$5 - 2 = 3$$

We can get the same answer by using the definition we just gave for subtraction. Instead of subtracting 2, we can add its opposite, -2. Here is how it looks:

$$5 - 2 = 5 + (-2) \quad \text{Change subtraction to addition of the opposite}$$

$$= 3 \quad \text{Apply the rule for addition of positive and negative numbers}$$

The result is the same whether we use our previous knowledge of subtraction or the new definition. The new definition is essential when the problems begin to get more complicated.

PRACTICE PROBLEMS

1. Subtract: $7 - 3$

Answer

1. 4

2. Subtract: $-7 - 3$

Note A real-life analogy to Example 2 would be: "If the temperature were 7° below 0 and then it dropped another 2°, what would the temperature be then?"

3. Subtract: $-8 - 6$

4. Subtract: $10 - (-6)$

5. Subtract: $-10 - (-15)$

Note Examples 4 and 5 may give results you are not used to getting. But you must realize that the results are correct. That is, $12 - (-6)$ is 18, and $-20 - (-30)$ is 10. If you think these results should be different, then you are not thinking of subtraction correctly.

6. Subtract each of the following.
   a. $8 - 5$
   b. $-8 - 5$
   c. $8 - (-5)$
   d. $-8 - (-5)$
   e. $12 - 10$
   f. $-12 - 10$
   g. $12 - (-10)$
   h. $-12 - (-10)$

Answers
2. -10 3. -14 4. 16 5. 5
6. a. 3 b. -13 c. 13 d. -3 e. 2 f. -22 g. 22 h. -2

EXAMPLE 2

Subtract: $-7 - 2$

SOLUTION We have never subtracted a positive number from a negative number before. We must apply our definition of subtraction:

$$-7 - 2 = -7 + (-2)$$ Instead of subtracting 2, we add its opposite, -2

$$= -9$$ Apply the rule for addition

EXAMPLE 3

Subtract: $-10 - 5$

SOLUTION We apply the definition of subtraction (if you don't know the definition of subtraction yet, go back and read it) and add as usual.

$$-10 - 5 = -10 + (-5)$$ Definition of subtraction

$$= -15$$ Addition

EXAMPLE 4

Subtract: $12 - (-6)$

SOLUTION The first $-$ sign is read "subtract," and the second one is read "negative." The problem in words is "12 subtract negative 6." We can use the definition of subtraction to change this to the addition of positive 6:

$$12 - (-6) = 12 + 6$$ Subtracting -6 is equivalent to adding $+6$

$$= 18$$ Addition

EXAMPLE 5

Subtract: $-20 - (-30)$

SOLUTION Instead of subtracting -30, we can use the definition of subtraction to write the problem again as the addition of 30:

$$-20 - (-30) = -20 + 30$$ Definition of subtraction

$$= 10$$ Addition

Examples 1–5 illustrate all the possible combinations of subtraction with positive and negative numbers. There are no new rules for subtraction. We apply the definition to change each subtraction problem into an equivalent addition problem. The rule for addition can then be used to obtain the correct answer.

EXAMPLE 6

The following table shows the relationship between subtraction and addition:

Subtraction	Addition of the Opposite	Answer
$7 - 9$	$7 + (-9)$	-2
$-7 - 9$	$-7 + (-9)$	-16
$7 - (-9)$	$7 + 9$	16
$-7 - (-9)$	$-7 + 9$	2
$15 - 10$	$15 + (-10)$	5
$-15 - 10$	$-15 + (-10)$	-25
$15 - (-10)$	$15 + 10$	25
$-15 - (-10)$	$-15 + 10$	-5

EXAMPLE 7 Combine: $-3 + 6 - 2$

SOLUTION The first step is to change subtraction to addition of the opposite. After that has been done, we add left to right.

$$-3 + 6 - 2 = -3 + 6 + (-2) \quad \text{Subtracting 2 is equivalent to adding } -2$$
$$= 3 + (-2) \quad \text{Add left to right}$$
$$= 1$$

EXAMPLE 8 Combine: $10 - (-4) - 8$

SOLUTION Changing subtraction to addition of the opposite, we have

$$10 - (-4) - 8 = 10 + 4 + (-8)$$
$$= 14 + (-8)$$
$$= 6$$

EXAMPLE 9 Subtract 3 from -5.

SOLUTION Subtracting 3 is equivalent to adding -3.

$$-5 - 3 = -5 + (-3) = -8$$

Subtracting 3 from -5 gives us -8.

EXAMPLE 10 Subtract -4 from 9.

SOLUTION Subtracting -4 is the same as adding $+4$:

$$9 - (-4) = 9 + 4 = 13$$

Subtracting -4 from 9 gives us 13.

EXAMPLE 11 Find the difference of -7 and -4.

SOLUTION Subtracting -4 from -7 looks like this:

$$-7 - (-4) = -7 + 4 = -3$$

The difference of -7 and -4 is -3.

USING TECHNOLOGY

Calculator Note

Here is how we work the subtraction problem shown in Example 11 on a calculator.

Scientific Calculator: 7 [+/−] [−] 4 [+/−] [=]

Graphing Calculator: [(−)] 7 [−] [(−)] 4 [ENT]

7. Combine: $-4 + 6 - 7$

8. Combine: $15 - (-5) - 8$

9. Subtract 2 from -8.

10. Subtract -5 from 7.

11. Find the difference of -8 and -6.

Answers
7. -5 **8.** 12 **9.** -10 **10.** 12 **11.** -2

STUDY SKILLS

Don't Let Your Intuition Fool You

As you become more experienced and more successful in mathematics, you will be able to trust your mathematical intuition. For now, though, it can get in the way of success. For example, if you ask a beginning algebra student to "subtract 3 from -5" many will answer -2 or 2. Both answers are incorrect, even though they may seem intuitively true.

B Application

EXAMPLE 12 Many of the planes used by the United States during World War II were not pressurized or sealed from outside air. As a result, the temperature inside these planes was the same as the surrounding air temperature outside. Suppose the temperature inside a B-17 Flying Fortress is 50°F at takeoff and then drops to -30°F when the plane reaches its cruising altitude of 28,000 feet. Find the difference in temperature inside this plane at takeoff and at 28,000 feet.

Courtesy of the U.S. Air Force Museum

SOLUTION The temperature at takeoff is 50°F, whereas the temperature at 28,000 feet is -30°F. To find the difference we subtract, with the numbers in the same order as they are given in the problem:

$$50 - (-30) = 50 + 30 = 80$$

The difference in temperature is 80°F.

12. Suppose the temperature is 42°F at takeoff and then drops to -42°F when the plane reaches its cruising altitude. Find the difference in temperature at takeoff and at cruising altitude.

Subtraction and Taking Away

Some people may believe that the answer to $-5 - 9$ should be -4 or 4, not -14. If this is happening to you, you are probably thinking of subtraction in terms of taking one number away from another. Thinking of subtraction in this way works well with positive numbers if you always subtract the smaller number from the larger. In algebra, however, we encounter many situations other than this. The definition of subtraction, that $a - b = a + (-b)$ clearly indicates the correct way to use subtraction. That is, when working subtraction problems, you should think "addition of the opposite," not "taking one number away from another."

Getting Ready for Class

After reading through the preceding section, respond in your own words and in complete sentences.

1. Write the subtraction problem $5 - 3$ as an equivalent addition problem.
2. Explain the process you would use to subtract 2 from -7.
3. Write an addition problem that is equivalent to the subtraction problem $-20 - (-30)$.
4. To find the difference of -7 and -4 we subtract what number from -7?

Answer

12. 84°F

Problem Set 2.3

A Subtract. [Examples 1–5]

1. $7 - 5$
2. $5 - 7$
3. $8 - 6$
4. $6 - 8$
5. $-3 - 5$
6. $-5 - 3$
7. $-4 - 1$
8. $-1 - 4$
9. $5 - (-2)$
10. $2 - (-5)$
11. $3 - (-9)$
12. $9 - (-3)$
13. $-4 - (-7)$
14. $-7 - (-4)$
15. $-10 - (-3)$
16. $-3 - (-10)$
17. $15 - 18$
18. $20 - 32$
19. $100 - 113$
20. $121 - 21$
21. $-30 - 20$
22. $-50 - 60$
23. $-79 - 21$
24. $-86 - 31$
25. $156 - (-243)$
26. $292 - (-841)$
27. $-35 - (-14)$
28. $-29 - (-4)$

Complete the following tables.

29.

First number x	second number y	the difference of x and y $x - y$
8	6	
8	7	
8	8	
8	9	
8	10	

30.

First number x	second number y	the difference of x and y $x - y$
10	12	
10	11	
10	10	
10	9	
10	8	

31.

First number x	second number y	the difference of x and y $x - y$
8	−6	
8	−7	
8	−8	
8	−9	
8	−10	

32.

First number x	second number y	the difference of x and y $x - y$
−10	−12	
−10	−11	
−10	−10	
−10	−9	
−10	−8	

A Simplify as much as possible by first changing all subtractions to addition of the opposite and then adding left to right. [Examples 7, 8]

33. $4 - 5 - 6$

34. $7 - 3 - 2$

35. $-8 + 3 - 4$

36. $-10 - 1 + 16$

37. $-8 - 4 - 2$

38. $-7 - 3 - 6$

39. $12 - 30 - 47$

40. $-29 - 53 - 37$

41. $33 - (-22) - 66$

42. $44 - (-11) + 55$

43. $101 - (-95) + 6$

44. $-211 - (-207) + 3$

45. $-900 + 400 - (-100)$

46. $-300 + 600 - (-200)$

A Translate each of the following and simplify the result. [Examples 9–11]

47. Subtract -6 from 5.

48. Subtract 8 from -2.

49. Find the difference of -5 and -1.

50. Find the difference of -7 and -3.

51. Subtract -4 from the sum of -8 and 12.

52. Subtract -7 from the sum of 7 and -12.

53. What number do you subtract from -3 to get -9?

54. What number do you subtract from 5 to get 8?

Estimating

Work Problems 55–60 mentally, without pencil and paper or a calculator.

55. The answer to the problem $52 - 49$ is closest to which of the following numbers?
a. 100 **b.** 0 **c.** -100

56. The answer to the problem $-52 - 49$ is closest to which of the following numbers?
a. 100 **b.** 0 **c.** -100

57. The answer to the problem $52 - (-49)$ is closest to which of the following numbers?
a. 100 **b.** 0 **c.** -100

58. The answer to the problem $-52 - (-49)$ is closest to which of the following numbers?
a. 100 **b.** 0 **c.** -100

59. Is $-161 - (-62)$ closer to -200 or -100?

60. Is $-553 - 50$ closer to -600 or -500?

B Applying the Concepts [Example 12]

61. The graph shows the record low temperatures for the Grand Canyon. What is the temperature difference between January and July?

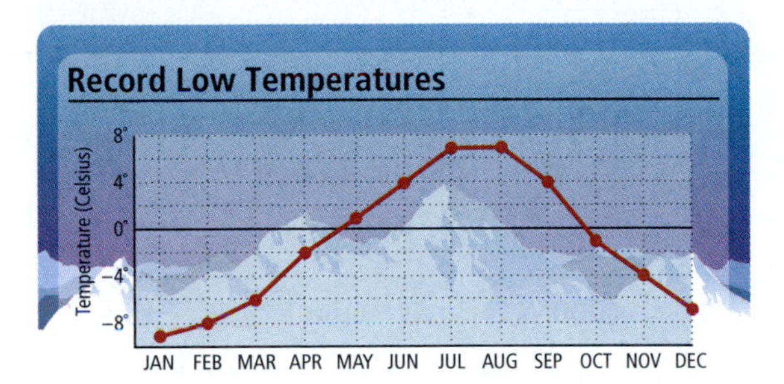

Source: National Park Service

62. The graph shows the lowest and highest points in the Grand Canyon and Death Valley. What is the difference between the lowest point in the Grand Canyon and the lowest point in Death Valley?

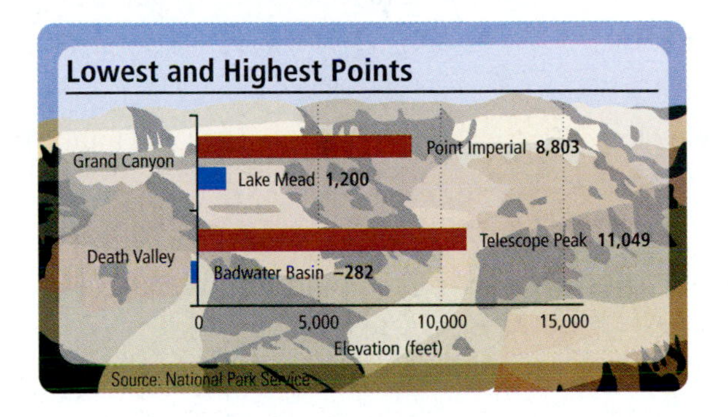

63. The highest point in Grand Canyon National Park is at Point Imperial with an elevation of 8,803 feet. The lowest point in the park is at Lake Mead at 1,200 feet. What is the difference between the highest and the lowest points?

64. Temperature On Monday the temperature reached a high of 28° above 0. That night it dropped to 16° below 0. What is the difference between the high and the low temperatures for Monday?

65. Tracking Inventory By definition, inventory is the total amount of goods contained in a store or warehouse at any given time. It is helpful for store owners to know the number of items they have available for sale in order to accommodate customer demand. This table shows the beginning inventory on May 1st and tracks the number of items bought and sold for one month. Determine the number of items in inventory at the end of the month.

Date	Transaction	Number of Units Available	Number of Units Sold
May 1	Beginning Inventory	400	
May 3	Purchase	100	
May 8	Sale		700
May 15	Purchase	600	
May 19	Purchase	200	
May 25	Sale		400
May 27	Sale		300
May 31	Ending Inventory		

66. Profit and Loss You own a small business which provides computer support to homeowners who wish to create their own in-house computer network. In addition to setting up the network you also maintain and troubleshoot home PCs. Business gets off to a slow start. You record a profit of $2,298 during the first quarter of the year, a loss of $2,854 during the second quarter, a profit of $3,057 during the third quarter, and a profit of $1,250 for the last quarter of the year. Do you end the year with a net profit or a net loss? Represent that profit or loss as a positive or negative value.

Tuition Cost The chart shows the cost of college tuition and fees at public four-year universities. Because of tax breaks, along with federal and state grants, the actual cost per student is much less than the total cost of tuition and fees. Use the information in this chart to answers Questions 67 through 70.

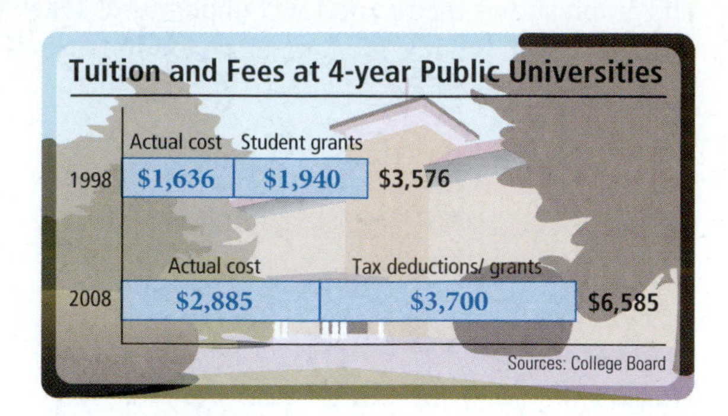

67. Find the difference in student grants in 1998 and student grants and tax deductions in 2008.

68. Find the difference in actual costs in 1998 and actual costs in 2008.

69. Find the difference in total costs in 1998 and total costs in 2008.

70. What has increased more from 1998 to 2008, student grants and tax deductions or actual student costs?

Repeated below is the table of wind chill temperatures that we used previously. Use it for Problems 71–74.

	Air Temperature (°F)							
Wind speed	30°	25°	20°	15°	10°	5°	0°	−5°
10 mph	16°	10°	3°	−3°	−9°	−15°	−22°	−27°
15 mph	9°	2°	−5°	−11°	−18°	−25°	−31°	−38°
20 mph	4°	−3°	−10°	−17°	−24°	−31°	−39°	−46°
25 mph	1°	−7°	−15°	−22°	−29°	−36°	−44°	−51°
30 mph	−2°	−10°	−18°	−25°	−33°	−41°	−49°	−56°

71. Wind Chill If the temperature outside is 15°F, what is the difference in wind chill temperature between a 15-mile-per-hour wind and a 25-mile-per-hour wind?

72. Wind Chill If the temperature outside is 0°F, what is the difference in wind chill temperature between a 15-mile-per-hour wind and a 25-mile-per-hour wind?

73. Wind Chill Find the difference in temperature between a day in which the air temperature is 20°F and the wind is blowing at 10 miles per hour and a day in which the air temperature is 10°F and the wind is blowing at 20 miles per hour.

74. Wind Chill Find the difference in temperature between a day in which the air temperature is 0°F and the wind is blowing at 10 miles per hour and a day in which the air temperature is −5°F and the wind is blowing at 20 miles per hour.

Use the tables below to work Problems 75–78.

RECORD LOW TEMPERATURES FOR LAKE PLACID, NEW YORK	
Month	Temperature
January	−36°F
February	−30°F
March	−14°F
April	−2°F
May	19°F
June	22°F
July	35°F
August	30°F
September	19°F
October	15°F
November	−11°F
December	−26°F

RECORD HIGH TEMPERATURES FOR LAKE PLACID, NEW YORK	
Month	Temperature
January	54°F
February	59°F
March	69°F
April	82°F
May	90°F
June	93°F
July	97°F
August	93°F
September	90°F
October	87°F
November	67°F
December	60°F

75. Temperature Difference Find the difference between the record high temperature and the record low temperature for the month of December.

76. Temperature Difference Find the difference between the record high temperature and the record low temperature for the month of March.

77. Temperature Difference Find the difference between the record low temperatures of March and December.

78. Temperature Difference Find the difference between the record high temperatures of March and December.

Getting Ready for the Next Section

Perform the indicated operations.

79. $3(2)(5)$

80. $5(2)(4)$

81. 6^2

82. 8^2

83. 4^3

84. 3^3

85. $6(3 + 5)$

86. $2(5 + 8)$

87. $3(9 - 2) + 4(7 - 2)$

88. $2(5 - 3) - 7(4 - 2)$

89. $(3 + 7)(6 - 2)$

90. $(6 + 1)(9 - 4)$

Simplify each of the following.

91. $2 + 3(4 + 1)$

92. $6 + 5(2 + 3)$

93. $(6 + 2)(6 - 2)$

94. $(7 + 1)(7 - 1)$

95. 5^2

96. 2^3

97. $2^3 \cdot 3^2$

98. $2^3 + 3^2$

Maintaining Your Skills

Write each of the following in symbols.

99. The product of 3 and 5.

100. The product of 5 and 3.

101. The product of 7 and x.

102. The product of 2 and y.

Rewrite the following using the commutative property of multiplication.

103. $3(5) =$

104. $7(x) =$

Rewrite the following using the associative property of multiplication.

105. $5(7 \cdot 8) =$

106. $4(6 \cdot y)$

Apply the distributive property to each expression and then simplify the result.

107. $2(3 + 4)$

108. $5(6 + 7)$

Extending the Concepts

109. Give an example that shows that subtraction is not a commutative operation.

110. Why is the expression "two negatives make a positive" not correct?

111. Give an example of an everyday situation that is modeled by the subtraction problem $\$10 - \$12 = -\$2$.

112. Give an example of an everyday situation that is modeled by the subtraction problem $-\$10 - \$12 = -\$22$.

In Chapter 1 we defined an arithmetic sequence as a sequence of numbers in which each number, after the first number, is obtained from the previous number by adding the same amount each time.

Find the next two numbers in each arithmetic sequence below.

113. $10, 5, 0, \ldots$

114. $8, 3, -2, \ldots$

115. $-10, -6, -2, \ldots$

116. $-4, -1, 2, \ldots$

Multiplication with Negative Numbers

2.4

Objectives

A Multiply positive and negative numbers.

B Apply the rule for order of operations to expressions containing positive and negative numbers.

C Solve applications involving multiplication with positive and negative numbers.

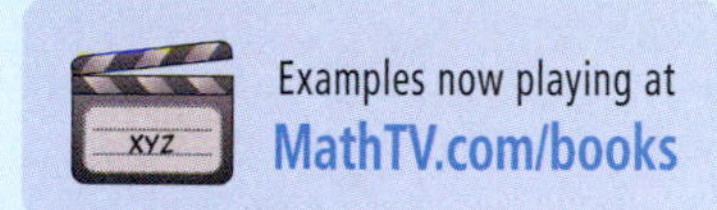

Introduction . . .

Suppose you buy three shares of a certain stock on Monday, and by Friday the price per share has dropped \$5. How much money have you lost? The answer is \$15. Because it is a loss, we can express it as −\$15. The multiplication problem below can be used to describe the relationship among the numbers.

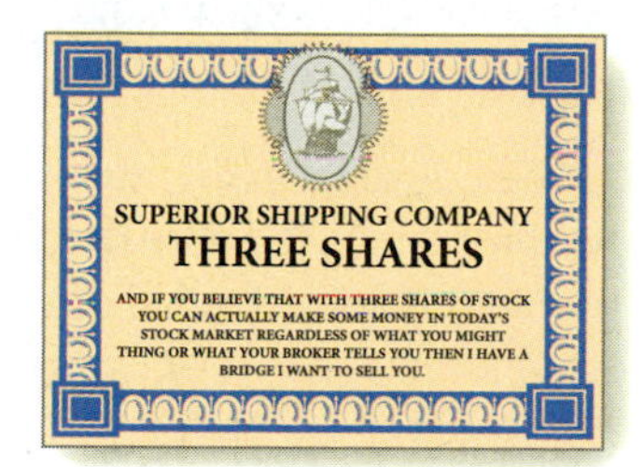

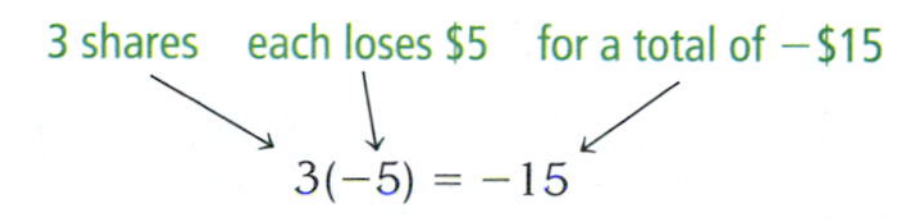

$$3(-5) = -15$$

From this we conclude that it is reasonable to say that the product of a positive number and a negative number is a negative number.

A Multiplication

In order to generalize multiplication with negative numbers, recall that we first defined multiplication by whole numbers to be repeated addition. That is:

$$\underset{\text{Multiplication}}{3 \cdot 5} = \underset{\text{Repeated addition}}{5 + 5 + 5}$$

This concept is very helpful when it comes to developing the rule for multiplication problems that involve negative numbers. For the first example we look at what happens when we multiply a negative number by a positive number.

EXAMPLE 1 Multiply: $3(-5)$

SOLUTION Writing this product as repeated addition, we have

$$\begin{aligned} 3(-5) &= (-5) + (-5) + (-5) \\ &= -10 + (-5) \\ &= -15 \end{aligned}$$

The result, −15, is obtained by adding the three negative 5's.

EXAMPLE 2 Multiply: $-3(5)$

SOLUTION In order to write this multiplication problem in terms of repeated addition, we will have to reverse the order of the two numbers. This is easily done, because multiplication is a commutative operation.

$$\begin{aligned} -3(5) &= 5(-3) && \text{Commutative property} \\ &= (-3) + (-3) + (-3) + (-3) + (-3) && \text{Repeated addition} \\ &= -15 && \text{Addition} \end{aligned}$$

The product of −3 and 5 is −15.

EXAMPLE 3 Multiply: $-3(-5)$

SOLUTION It is impossible to write this product in terms of repeated addition. We will find the answer to $-3(-5)$ by solving a different problem. Look at the following problem:

$$-3[5 + (-5)] = -3[0] = 0$$

PRACTICE PROBLEMS

1. Multiply: $2(-6)$
2. Multiply: $-2(6)$
3. Multiply: $-2(-6)$

Answers
1. −12 2. −12

Note The discussion here explains why $-3(-5) = 15$. We want to be able to justify everything we do in mathematics. The discussion tells *why* $-3(-15) = 15$.

The result is 0, because multiplying by 0 always produces 0. Now we can work the same problem another way, and in the process find the answer to $-3(-5)$. Applying the distributive property to the same expression, we have

$$\begin{aligned} -3[5 + (-5)] &= -3(5) + (-3)(-5) && \text{Distributive property} \\ &= -15 + (?) && -3(5) = -15 \end{aligned}$$

The question mark must be $+15$, because we already know that the answer to the problem is 0, and $+15$ is the only number we can add to -15 to get 0. So, our problem is solved:

$$-3(-5) = +15$$

Table 1 gives a summary of what we have done so far in this section.

TABLE 1

Original Numbers Have	For Example	The Answer Is
Same signs	$3(5) = 15$	Positive
Different signs	$-3(5) = -15$	Negative
Different signs	$3(-5) = -15$	Negative
Same signs	$-3(-5) = 15$	Positive

From the examples we have done so far in this section and their summaries in Table 1, we write the following rule for multiplication of positive and negative numbers:

Rule

To multiply any two numbers, we multiply their absolute values.

1. The answer is *positive* if both the original numbers have the same sign. That is, the product of two numbers with the same sign is positive.
2. The answer is *negative* if the original two numbers have different signs. The product of two numbers with different signs is negative.

This rule should be memorized. By the time you have finished reading this section and working the problems at the end of the section, you should be fast and accurate at multiplication with positive and negative numbers.

Multiply.

4. $3(2)$

5. $-3(-2)$

6. $3(-2)$

7. $-3(2)$

8. $8(-9)$

9. $-6(-4)$

10. $-5(2)(-4)$

EXAMPLE 4 $2(4) = 8$ Like signs; positive answer

EXAMPLE 5 $-2(-4) = 8$ Like signs; positive answer

EXAMPLE 6 $2(-4) = -8$ Unlike signs; negative answer

EXAMPLE 7 $-2(4) = -8$ Unlike signs; negative answer

EXAMPLE 8 $7(-6) = -42$ Unlike signs; negative answer

EXAMPLE 9 $-5(-8) = 40$ Like signs; positive answer

EXAMPLE 10

$$\begin{aligned} -3(2)(-5) &= -6(-5) && \text{Multiply } -3 \text{ and 2 to get } -6 \\ &= 30 \end{aligned}$$

Answers

3. 12 **4.** 6 **5.** 6 **6.** −6 **7.** −6 **8.** −72 **9.** 24 **10.** 40

EXAMPLE 11 Use the definition of exponents to expand each expression. Then simplify by multiplying.

a. $(-6)^2 = (-6)(-6)$ — Definition of exponents
$= 36$ — Multiply

b. $-6^2 = -6 \cdot 6$ — Definition of exponents
$= -36$ — Multiply

c. $(-4)^3 = (-4)(-4)(-4)$ — Definition of exponents
$= -64$ — Multiply

d. $-4^3 = -4 \cdot 4 \cdot 4$ — Definition of exponents
$= -64$ — Multiply

In Example 11, the base is a negative number in Parts a and c, but not in Parts b and d. We know this is true because of the use of parentheses.

B Order of Operations

EXAMPLE 12 Simplify: $-6[3 + (-5)]$

SOLUTION We begin inside the brackets and work our way out:

$$-6[3 + (-5)] = -6[-2]$$
$$= 12$$

EXAMPLE 13 Simplify: $-4 + 5(-6 + 2)$

SOLUTION Simplifying inside the parentheses first, we have

$-4 + 5(-6 + 2) = -4 + 5(-4)$ — Simplify inside parentheses
$= -4 + (-20)$ — Multiply
$= -24$ — Add

EXAMPLE 14 Simplify: $-2(7) + 3(-6)$

SOLUTION Multiplying left to right before we add gives us

$$-2(7) + 3(-6) = -14 + (-18)$$
$$= -32$$

EXAMPLE 15 Simplify: $-3(2 - 9) + 4(-7 - 2)$

SOLUTION We begin by subtracting inside the parentheses:

$$-3(2 - 9) + 4(-7 - 2) = -3(-7) + 4(-9)$$
$$= 21 + (-36)$$
$$= -15$$

EXAMPLE 16 Simplify: $(-3 - 7)(2 - 6)$

SOLUTION Again, we begin by simplifying inside the parentheses:

$$(-3 - 7)(2 - 6) = (-10)(-4)$$
$$= 40$$

11. Use the definition of exponents to expand each expression. Then simplify by multiplying.
a. $(-8)^2$
b. -8^2
c. $(-3)^3$
d. -3^3

12. Simplify: $-2[5 + (-8)]$

13. Simplify: $-3 + 4(-7 + 3)$

14. Simplify: $-3(5) + 4(-4)$

15. Simplify: $-2(3 - 5) - 7(-2 - 4)$

16. Simplify: $(-6 - 1)(4 - 9)$

Answers
11. a. 64 **b.** −64 **c.** −27 **d.** −27
12. 6 **13.** −19 **14.** −31 **15.** 46
16. 35

STUDY SKILLS
Continue to Set and Keep a Schedule

Sometimes I find students do well at first and then become overconfident. They begin to put in less time with their homework. Don't do it. Keep to the same schedule.

USING TECHNOLOGY

Calculator Note

Here is how we work the problem shown in Example 16 on a calculator. (The [×] key on the first line may, or may not, be necessary. Try your calculator without it and see.)

Scientific Calculator: [(] 3 [+/−] [−] 7 [)] [×] [(] 2 [−] 6 [)] [=]

Graphing Calculator: [(] [(−)] 3 [−] 7 [)] [(] 2 [−] 6 [)] [ENT]

Getting Ready for Class

After reading through the preceding section, respond in your own words and in complete sentences.

1. Write the multiplication problem $3(-5)$ as an addition problem.
2. Write the multiplication problem $2(4)$ as an addition problem.
3. If two numbers have the same sign, then their product will have what sign?
4. If two numbers have different signs, then their product will have what sign?

Problem Set 2.4

A Find each of the following products. (Multiply.) [Examples 1–10]

1. $7(-8)$
2. $-3(5)$
3. $-6(10)$
4. $4(-8)$
5. $-7(-8)$
6. $-4(-7)$
7. $-9(-9)$
8. $-6(-3)$
9. $4(-6)$
10. $5(-2)$
11. $-6(-5)$
12. $-8(-3)$
13. $-5(0)(10)$
14. $8(-4)(0)$
15. $3(-2)(4)$
16. $5(-1)(3)$
17. $-4(3)(-2)$
18. $-4(5)(-6)$
19. $-1(-2)(-3)$
20. $-2(-3)(-4)$

Use the definition of exponents to expand each of the following expressions. Then multiply according to the rule for multiplication. [Example 11]

21. a. $(-4)^2$
    b. -4^2
22. a. $(-5)^2$
    b. -5^2
23. a. $(-5)^3$
    b. -5^3
24. a. $(-4)^3$
    b. -4^3
25. a. $(-2)^4$
    b. -2^4
26. a. $(-1)^4$
    b. -1^4

Complete the following tables. Remember, if $x = -5$, then $x^2 = (-5)^2 = 25$. [Example 11]

27.

Number x	Square x^2
−3	
−2	
−1	
0	
1	
2	
3	

28.

Number x	Cube x^3
−3	
−2	
−1	
0	
1	
2	
3	

29.

First Number x	Second Number y	Their Product xy
6	2	
6	1	
6	0	
6	−1	
6	−2	

30.

First Number a	Second Number b	Their Product ab
−5	3	
−5	2	
−5	1	
−5	0	
−5	−1	
−5	−2	
−5	−3	

B Use the rule for order of operations along with the rules for addition, subtraction, and multiplication to simplify each of the following expressions. [Examples 12–16]

31. $4(-3 + 2)$

32. $7(-6 + 3)$

33. $-10(-2 - 3)$

34. $-5(-6 - 2)$

35. $-3 + 2(5 - 3)$

36. $-7 + 3(6 - 2)$

37. $-7 + 2[-5 - 9]$

38. $-8 + 3[-4 - 1]$

39. $2(-5) + 3(-4)$

40. $6(-1) + 2(-7)$

41. $3(-2)4 + 3(-2)$

42. $2(-1)(-3) + 4(-6)$

43. $(8 - 3)(2 - 7)$

44. $(9 - 3)(2 - 6)$

45. $(2 - 5)(3 - 6)$

46. $(3 - 7)(2 - 8)$

47. $3(5 - 8) + 4(6 - 7)$

48. $2(3 - 7) + 3(5 - 6)$

49. $-2(8 - 10) + 3(4 - 9)$

50. $-3(6 - 9) + 2(3 - 8)$

51. $-3(4 - 7) - 2(-3 - 2)$

52. $-5(-2 - 8) - 4(6 - 10)$

53. $3(-2)(6 - 7)$

54. $4(-3)(2 - 5)$

55. Find the product of -3, -2, and -1.

56. Find the product of -7, -1, and 0.

57. What number do you multiply by -3 to get 12?

58. What number do you multiply by -7 to get -21?

59. Subtract -3 from the product of -5 and 4.

60. Subtract 5 from the product of -8 and 1.

Work Problems 61–68 mentally, without pencil and paper or a calculator.

61. The product $-32(-522)$ is closest to which of the following numbers?
a. 15,000 **b.** -500 **c.** $-1{,}500$ **d.** $-15{,}000$

62. The product $32(-522)$ is closest to which of the following numbers?
a. 15,000 **b.** -500 **c.** $-1{,}500$ **d.** $-15{,}000$

63. The product $-47(470)$ is closest to which of the following numbers?
a. 25,000 **b.** 420 **c.** $-2{,}500$ **d.** $-25{,}000$

64. The product $-47(-470)$ is closest to which of the following numbers?
a. 25,000 **b.** 420 **c.** $-2{,}500$ **d.** $-25{,}000$

65. The product $-222(-987)$ is closest to which of the following numbers?
a. 200,000 **b.** 800 **c.** -800 **d.** $-1{,}200$

66. The sum $-222 + (-987)$ is closest to which of the following numbers?
a. 200,000 **b.** 800 **c.** -800 **d.** $-1{,}200$

67. The difference $-222 - (-987)$ is closest to which of the following numbers?
a. 200,000 **b.** 800 **c.** -800 **d.** $-1{,}200$

68. The difference $-222 - 987$ is closest to which of the following numbers?
a. 200,000 **b.** 800 **c.** -800 **d.** $-1{,}200$

C Applying the Concepts

69. The chart shows the record low temperatures for Grand Canyon National Park, by month. Write the record low temperature for March.

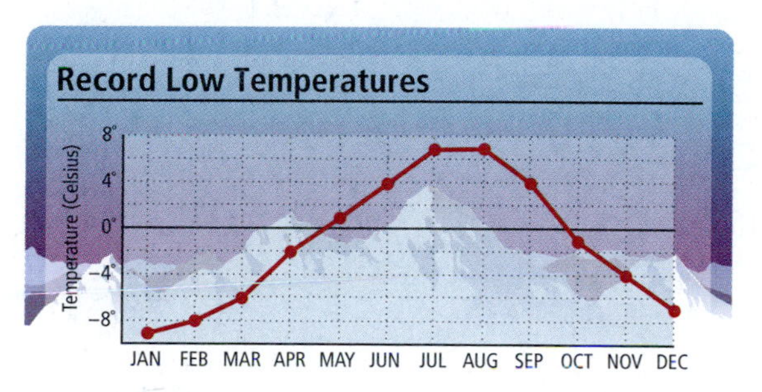

Source: National Park Service

70. The chart shows the cities with the highest annual insurance rates.

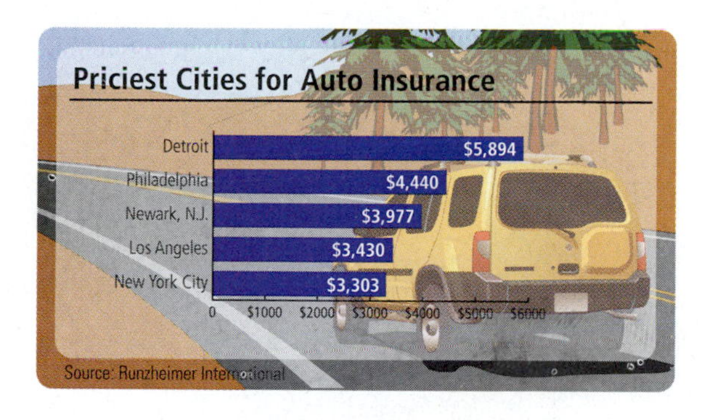

a. What is the monthly payment for a driver in Philadelphia?

b. Use negative numbers to write an expression for the cost of three months of auto insurance for a driver living in Philadelphia.

71. Temperature Change A hot-air balloon is rising to its cruising altitude. Suppose the air temperature around the balloon drops 4 degrees each time the balloon rises 1,000 feet. What is the net change in air temperature around the balloon as it rises from 2,000 feet to 6,000 feet?

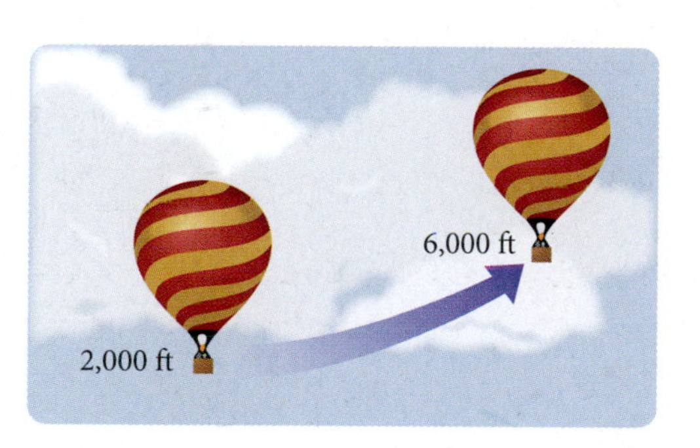

72. Temperature Change A small airplane is rising to its cruising altitude. Suppose the air temperature around the plane drops 4 degrees each time the plane increases its altitude by 1,000 feet. What is the net change in air temperature around the plane as it rises from 5,000 feet to 12,000 feet?

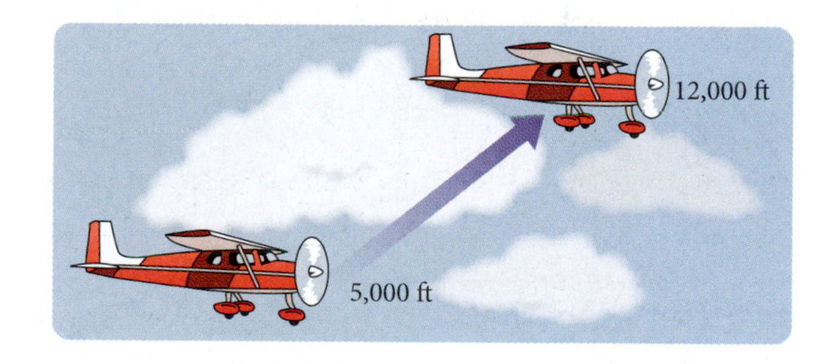

73. Expense Account A business woman has a travel expense account of $1,000. If she spends $75 a week for 8 weeks what will the balance of her expense account be at the end of this time.

74. Gas Prices Two local gas stations offer different prices for a gallon of regular gasoline. The Exxon Mobil station is currently selling their gas at $3.99 per gallon. The Getty station is currently selling their gas for $3.85 per gallon. Represent the net savings to you on a purchase of 15 gallons of regular gas if you buy gas from the Getty gas station.

Getting Ready for the Next Section

Perform the indicated operations.

75. $35 \div 5$

76. $32 \div 4$

77. $\frac{20}{4}$

78. $\frac{30}{5}$

79. $12 - 17$

80. $7 - 11$

81. $(6 \cdot 3) \div 2$

82. $(8 \cdot 5) \div 4$

83. $80 \div 10 \div 2$

84. $80 \div 2 \div 10$

85. $15 + 5(4) \div 10$

86. $[20 + 6(2)] \div (11 - 7)$

87. $4(10^2) + 20 \div 4$

88. $3(4^2) + 10 \div 5$

Maintaining Your Skills

Write each of the following statements in symbols.

89. The quotient of 12 and 6

90. The quotient of x and 5

Rewrite each of the following multiplication problems as an equivalent division problem.

91. $2(3) = 6$

92. $5 \cdot 4 = 20$

Rewrite each of the following division problems as an equivalent multiplication problem.

93. $10 \div 5 = 2$

94. $\frac{63}{9} = 7$

Divide.

95. $4{,}984 \div 56$

96. $4{,}994 \div 56$

Extending the Concepts

In Chapter 1 we defined a geometric sequence to be a sequence of numbers in which each number, after the first number, is obtained from the previous number by multiplying by the same amount each time.

Find the next two terms in each of the following geometric sequences.

97. $2, -6, 18, \ldots$

98. $1, -4, 16, \ldots$

99. $-2, 6, -18, \ldots$

100. $-1, 4, -16, \ldots$

Simplify each of the following according to the rule for order of operations.

101. $5(-2)^2 - 3(-2)^3$

102. $8(-1)^3 - 6(-3)^2$

103. $7 - 3(4 - 8)$

104. $6 - 2(9 - 11)$

105. $5 - 2[3 - 4(6 - 8)]$

106. $7 - 4[6 - 3(2 - 9)]$

2.5 Division with Negative Numbers

Objectives

A Divide positive and negative numbers.

B Apply the rule for order of operations to expressions that contain positive and negative numbers.

C Solve applications involving division with positive and negative numbers.

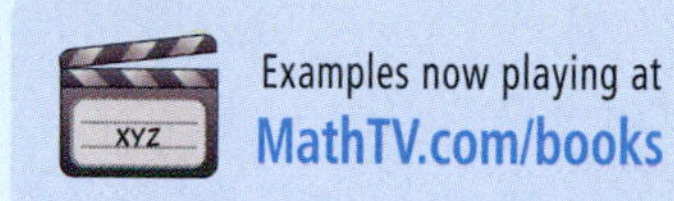

Introduction . . .

Suppose four friends invest equal amounts of money in a moving truck to start a small business. After 2 years the truck has dropped \$10,000 in value. If we represent this change with the number $-\$10{,}000$, then the loss to each of the four partners can be found with division:

$10,000 drop in 2 years

$$(-\$10{,}000) \div 4 = -\$2{,}500$$

From this example it seems reasonable to assume that a negative number divided by a positive number will give a negative answer.

To cover all the possible situations we can encounter with division of negative numbers, we use the relationship between multiplication and division. If we let n be the answer to the problem $12 \div (-2)$, then we know that

$$12 \div (-2) = n \quad \text{and} \quad -2(n) = 12$$

From our work with multiplication, we know that n must be -6 in the multiplication problem above, because -6 is the only number we can multiply -2 by to get 12. Because of the relationship between the two problems above, it must be true that 12 divided by -2 is -6.

The following pairs of problems show more quotients of positive and negative numbers. In each case the multiplication problem on the right justifies the answer to the division problem on the left.

$$\begin{aligned} 6 \div 3 &= 2 && \text{because} && 3(2) = 6 \\ 6 \div (-3) &= -2 && \text{because} && -3(-2) = 6 \\ -6 \div 3 &= -2 && \text{because} && 3(-2) = -6 \\ -6 \div (-3) &= 2 && \text{because} && -3(2) = -6 \end{aligned}$$

The results given above can be used to write the rule for division with negative numbers.

A Division

Rule

To divide two numbers, we divide their absolute values.

1. The answer is *positive* if both the original numbers have the same sign. That is, the quotient of two numbers with the same signs is positive.
2. The answer is *negative* if the original two numbers have different signs. That is, the quotient of two numbers with different signs is negative.

EXAMPLE 1 $-12 \div 4 = -3$ Unlike signs, negative answer

EXAMPLE 2 $12 \div (-4) = -3$ Unlike signs; negative answer

EXAMPLE 3 $-12 \div (-4) = 3$ Like signs; positive answer

EXAMPLE 4 $\dfrac{12}{-4} = -3$ Unlike signs; negative answer

EXAMPLE 5 $\dfrac{-20}{-4} = 5$ Like signs; positive answer

PRACTICE PROBLEMS

Divide.

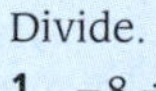

1. $-8 \div 2$
2. $8 \div (-2)$
3. $-8 \div (-2)$

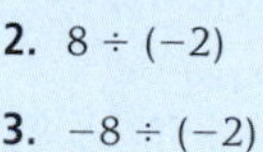

4. $\dfrac{20}{-5}$

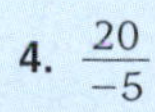

5. $\dfrac{-30}{-5}$

Answers

1. -4 2. -4 3. 4 4. -4 5. 6

From the examples we have done so far, we can make the following generalization about quotients that contain negative signs:

If a and b are numbers and b is not equal to 0, then

$$\frac{-a}{b} = \frac{a}{-b} = -\frac{a}{b} \quad \text{and} \quad \frac{-a}{-b} = \frac{a}{b}$$

B Order of Operations

The last examples in this section involve more than one operation. We use the rules developed previously in this chapter and the rule for order of operations to simplify each.

EXAMPLE 6 Simplify: $\frac{6(-3)}{-2}$

SOLUTION We begin by multiplying 6 and −3:

$$\frac{6(-3)}{-2} = \frac{-18}{-2} \quad \text{Multiplication; } 6(-3) = -18$$

$$= 9 \quad \text{Like signs; positive answer}$$

EXAMPLE 7 Simplify: $\frac{-15 + 5(-4)}{12 - 17}$

SOLUTION Simplifying above and below the fraction bar, we have

$$\frac{-15 + 5(-4)}{12 - 17} = \frac{-15 + (-20)}{-5} = \frac{-35}{-5} = 7$$

EXAMPLE 8 Simplify: $-4(10^2) + 20 \div (-4)$

SOLUTION Applying the rule for order of operations, we have

$$-4(10^2) + 20 \div (-4) = -4(100) + 20 \div (-4) \quad \text{Exponents first}$$

$$= -400 + (-5) \quad \text{Multiply and divide}$$

$$= -405 \quad \text{Add}$$

EXAMPLE 9 Simplify: $-80 \div 10 \div 2$

SOLUTION In a situation like this, the rule for order of operations states that we are to divide left to right.

$$-80 \div 10 \div 2 = -8 \div 2 \quad \text{Divide } -80 \text{ by } 10$$

$$= -4$$

6. Simplify: $\frac{8(-5)}{-4}$

7. Simplify: $\frac{-20 + 6(-2)}{7 - 11}$

8. Simplify: $-3(4^2) + 10 \div (-5)$

9. Simplify: $-80 \div 2 \div 10$

Answers
6. 10 **7.** 8 **8.** −50 **9.** −4

Getting Ready for Class

After reading through the preceding section, respond in your own words and in complete sentences.

1. Write a multiplication problem that is equivalent to the division problem $-12 \div 4 = -3$.
2. Write a multiplication problem that is equivalent to the division problem $-12 \div (-4) = 3$.
3. If two numbers have the same sign, then their quotient will have what sign?
4. Dividing a negative number by 0 always results in what kind of expression?

Problem Set 2.5

A Find each of the following quotients. (Divide.) [Examples 1–5]

1. $-15 \div 5$

2. $15 \div (-3)$

3. $20 \div (-4)$

4. $-20 \div 4$

5. $-30 \div (-10)$

6. $-50 \div (-25)$

7. $\frac{-14}{-7}$

8. $\frac{-18}{-6}$

9. $\frac{12}{-3}$

10. $\frac{12}{-4}$

11. $-22 \div 11$

12. $-35 \div 7$

13. $\frac{0}{-3}$

14. $\frac{0}{-5}$

15. $125 \div (-25)$

16. $-144 \div (-9)$

Complete the following tables.

17.

First Number a	Second Number b	The Quotient of a and b $\frac{a}{b}$
100	−5	
100	−10	
100	−25	
100	−50	

18.

First Number a	Second Number b	The Quotient of a and b $\frac{a}{b}$
24	−4	
24	−3	
24	−2	
24	−1	

19.

First Number a	Second Number b	The Quotient of a and b $\frac{a}{b}$
−100	−5	
−100	5	
100	−5	
100	5	

20.

First Number a	Second Number b	The Quotient of a and b $\frac{a}{b}$
−24	−2	
−24	−4	
−24	−6	
−24	−8	

21. Find the quotient of -25 and 5.

22. Find the quotient of -38 and -19.

23. What number do you divide by -5 to get -7?

24. What number do you divide by 6 to get -7?

25. Subtract -3 from the quotient of 27 and 9.

26. Subtract -7 from the quotient of -72 and -9.

B Use any of the rules developed in this chapter and the rule for order of operations to simplify each of the following expressions as much as possible. [Examples 6–9]

27. $\frac{4(-7)}{-28}$

28. $\frac{6(-3)}{-18}$

29. $\frac{-3(-10)}{-5}$

30. $\frac{-4(-12)}{-6}$

31. $\frac{2(-3)}{6-3}$

32. $\frac{2(-3)}{3-6}$

33. $\frac{4-8}{8-4}$

34. $\frac{9-5}{5-9}$

35. $\frac{2(-3)+10}{-4}$

36. $\frac{7(-2)-6}{-10}$

37. $\frac{2+3(-6)}{4-12}$

38. $\frac{3+9(-1)}{5-7}$

39. $\frac{6(-7)+3(-2)}{20-4}$

40. $\frac{9(-8)+5(-1)}{12-1}$

41. $\frac{3(-7)(-4)}{6(-2)}$

42. $\frac{-2(4)(-8)}{(-2)(-2)}$

43. $(-5)^2 + 20 \div 4$

44. $6^2 + 36 \div 9$

45. $100 \div (-5)^2$

46. $400 \div (-4)^2$

47. $-100 \div 10 \div 2$

48. $-500 \div 50 \div 10$

49. $-100 \div (10 \div 2)$

50. $-500 \div (50 \div 10)$

51. $(-100 \div 10) \div 2$

52. $(-500 \div 50) \div 10$

Estimating

Work Problems 53–60 mentally, without pencil and paper or a calculator.

53. Is $397 \div (-401)$ closer to 1 or -1?

54. Is $-751 \div (-749)$ closer to 1 or -1?

55. The quotient $-121 \div 27$ is closest to which of the following numbers?

a. -150 **b.** -100 **c.** -4 **d.** 6

56. The quotient $1{,}000 \div (-337)$ is closest to which of the following numbers?

a. 663 **b.** -3 **c.** -30 **d.** -663

57. Which number is closest to the sum $-151 + (-49)$?

a. -200 **b.** -100 **c.** 3 **d.** 7,500

58. Which number is closest to $-151 - (-49)$?

a. -200 **b.** -100 **c.** 3 **d.** 7,500

59. Which number is closest to the product $-151(-49)$?

a. -200 **b.** -100 **c.** 3 **d.** 7,500

60. Which number is closest to the quotient $-151 \div (-49)$?

a. -200 **b.** -100 **c.** 3 **d.** 7,500

C Applying the Concepts

61. The chart shows the most expensive cities to live in. Expenses can also be written as negative numbers. Find the monthly cost to live in Los Angeles. Use negative numbers.

Priciest Cities to Inhabit in the U.S.

Manhattan $146,060
San Francisco $133,887
Los Angeles $117,726
San Jose $108,506
Washington, D.C. $102,589
Annual Cost (dollars)
Source: Runzheimer

62. The chart shows the cities with the most expensive auto insurance. Because insurance is an expense, it can be written as a negative number. What is the monthly cost of insurance in New York City? Use negative numbers and round to the nearest cent.

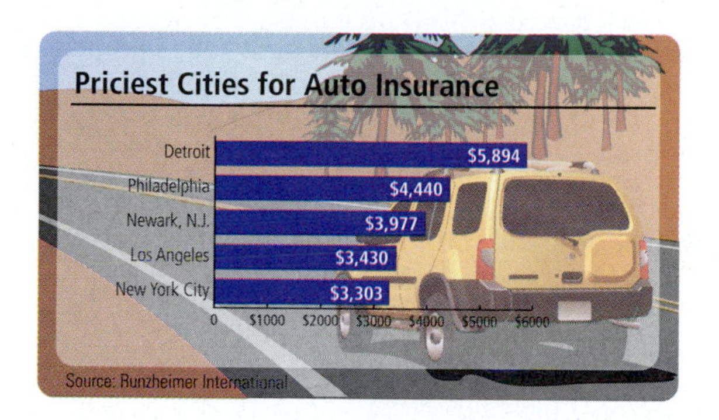

63. Temperature Line Graph The table below gives the low temperature for each day of one week in White Bear Lake, Minnesota. Use the diagram in the figure to draw a line graph of the information in the table.

LOW TEMPERATURES IN WHITE BEAR LAKE, MINNESOTA	
Day	**Temperature**
Monday	10 °F
Tuesday	8 °F
Wednesday	−5 °F
Thursday	−3 °F
Friday	−8 °F
Saturday	5 °F
Sunday	7 °F

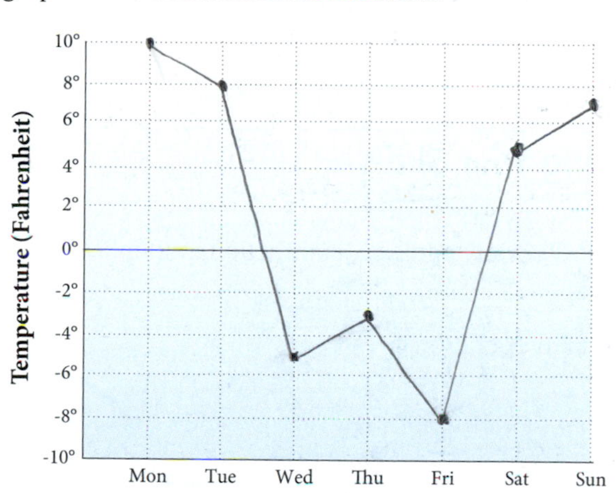

64. Temperature Line Graph The table below gives the low temperature for each day of one week in Fairbanks, Alaska. Use the diagram in the figure to draw a line graph of the information in the table.

LOW TEMPERATURES IN FAIRBANKS, ALASKA	
Day	**Temperature**
Monday	−26 °F
Tuesday	−5 °F
Wednesday	9 °F
Thursday	12 °F
Friday	3 °F
Saturday	−15 °F
Sunday	−20 °F

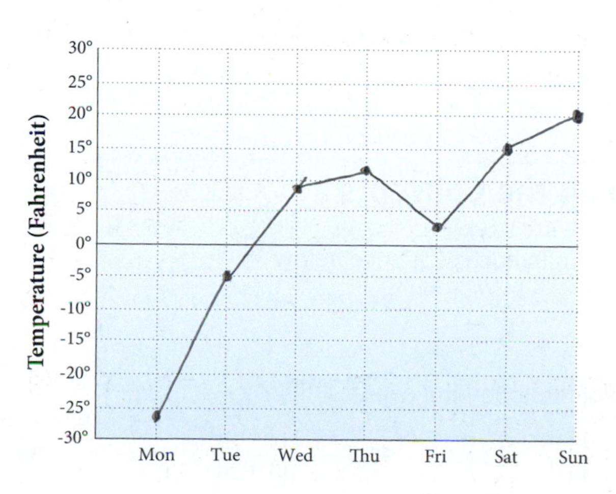

Getting Ready for the Next Section

The problems below review some of the properties of addition and multiplication we covered in Chapter 1.

Rewrite each expression using the commutative property of addition or multiplication.

65. $3 + x$

66. $4y$

Rewrite each expression using the associative property of addition or multiplication.

67. $5 + (7 + a)$

68. $(x + 4) + 6$

69. $3(4y)$

70. $(3y)8$

Apply the distributive property to each expression.

71. $5(3 + 7)$

72. $8(4 + 2)$

Simplify.

73. 6^2

74. 12^2

75. 4^3

76. 5^2

77. $2(100) + 2(75)$

78. $2(100) + 2(53)$

79. $100(75)$

80. $100(53)$

Maintaining Your Skills

The problems below review addition, subtraction, multiplication, and division of positive and negative numbers, as covered in this chapter.

Perform the indicated operations.

81. $8 + (-4)$

82. $-8 + 4$

83. $-8 + (-4)$

84. $-8 - 4$

85. $8 - (-4)$

86. $-8 - (-4)$

87. $8(-4)$

88. $-8(4)$

89. $-8(-4)$

90. $8 \div (-4)$

91. $-8 \div 4$

92. $-8 \div (-4)$

Extending the Concepts

Find the next term in each sequence below.

93. $32, -16, 8, \ldots$

94. $243, -81, 27, \ldots$

95. $-32, 16, -8, \ldots$

96. $-243, 81, -27, \ldots$

Simplify each of the following expressions.

97. $\dfrac{6 - 3(2 - 11)}{6 - 3(2 + 11)}$

98. $\dfrac{8 + 4(3 - 5)}{8 - 4(3 + 5)}$

99. $\dfrac{6 - (3 - 4) - 3}{1 - 2 - 3}$

100. $\dfrac{7 - (3 - 6) - 4}{-1 - 2 - 3}$

Simplifying Algebraic Expressions

2.6

Objectives

A Simplify expressions by using the associative property.

B Apply the distributive property to expressions containing numbers and variables.

C Use the distributive property to combine similar terms.

D Use the formulas for area and perimeter of squares and rectangles.

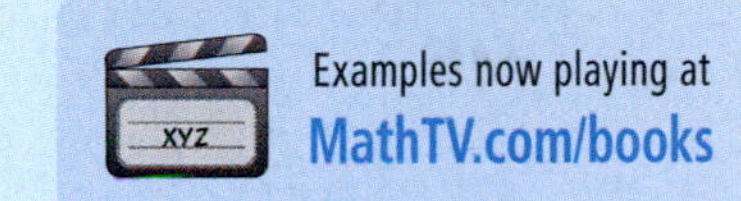

Introduction . . .

The woodcut shown here depicts Queen Dido of Carthage around 900 B.C., having an ox hide cut into small strips that will be tied together to make a long rope. The rope will be used to enclose her territory. The question, which has become known as the Queen Dido problem, is: what shape will enclose the largest territory?

To translate the problem into something we are more familiar with, suppose we have 24 yards of fencing that we are to use to build a rectangular dog run. If we want the dog run to have the largest area possible then we want the rectangle, with perimeter 24 yards, that encloses the largest area. The diagram below shows six dog runs, each of which has a perimeter of 24 yards. Notice how the length decreases as the width increases.

Dog Runs with Perimeter = 24 yards

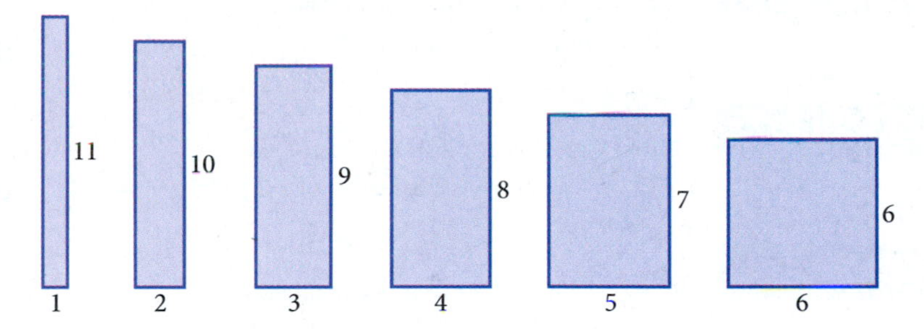

Since area is length times width, we can build a table and a line graph that show how the area changes as we change the width of the dog run.

AREA ENCLOSED BY RECTANGLE OF PERIMETER 24 YARDS

Width (Yards)	Area (Square Yards)
1	11
2	20
3	27
4	32
5	35
6	36

Area Enclosed by Fixed Perimeter

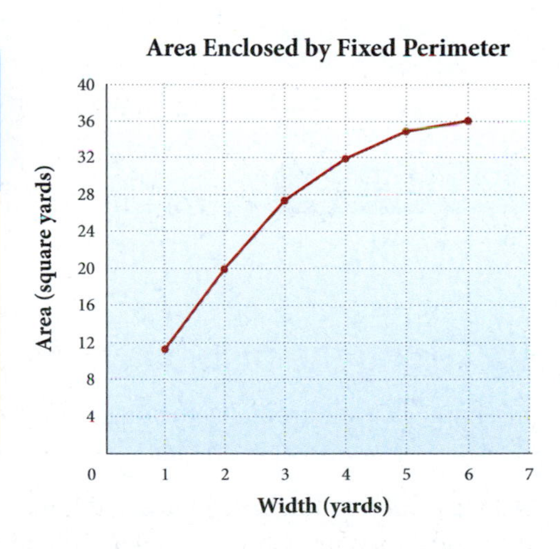

STUDY SKILLS
Increase Effectiveness

You want to become more and more effective with the time you spend on your homework. You want to increase the amount of learning you obtain in the time you have set aside. Increase those activities that you feel are the most beneficial and decrease those that have not given you the results you want.

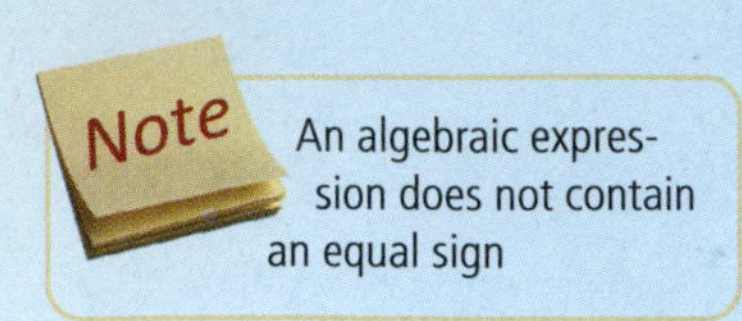

An algebraic expression does not contain an equal sign

In this section we want to simplify expressions containing variables—that is, algebraic expressions. An algebraic expression is a combination of constants and variables joined by arithmetic operations such as addition, subtraction, multiplication and division.

A Using the Associative Property

To begin let's review how we use the associative properties for addition and multiplication to simplify expressions.

Consider the expression $4(5x)$. We can apply the associative property of multiplication to this expression to change the grouping so that the 4 and the 5 are grouped together, instead of the 5 and the x. Here's how it looks:

$$4(5x) = (4 \cdot 5)x \quad \text{Associative property}$$
$$= 20x \quad \text{Multiplication: } 4 \cdot 5 = 20$$

We have simplified the expression to $20x$, which in most cases in algebra will be easier to work with than the original expression.

Here are some more examples.

EXAMPLE 1

$$7(3a) = (7 \cdot 3)a \quad \text{Associative property}$$
$$= 21a \quad \text{7 times 3 is 21}$$

EXAMPLE 2

$$-2(5x) = (-2 \cdot 5)x \quad \text{Associative property}$$
$$= -10x \quad \text{The product of } -2 \text{ and 5 is } -10$$

EXAMPLE 3

$$3(-4y) = [3(-4)]y \quad \text{Associative property}$$
$$= -12y \quad \text{3 times } -4 \text{ is } -12$$

We can use the associative property of addition to simplify expressions also.

EXAMPLE 4

$$3 + (8 + x) = (3 + 8) + x \quad \text{Associative property}$$
$$= 11 + x \quad \text{The sum of 3 and 8 is 11}$$

EXAMPLE 5

$$(2x + 5) + 10 = 2x + (5 + 10) \quad \text{Associative property}$$
$$= 2x + 15 \quad \text{Addition}$$

B Using the Distributive Property

In Chapter 1 we introduced the distributive property. In symbols it looks like this:

$$a(b + c) = ab + ac$$

Because subtraction is defined as addition of the opposite, the distributive property holds for subtraction as well as addition. That is,

$$a(b - c) = ab - ac$$

We say that multiplication distributes over addition and subtraction. Here are some examples that review how the distributive property is applied to expressions that contain variables.

EXAMPLE 6

$$4(x + 5) = 4(x) + 4(5) \quad \text{Distributive property}$$
$$= 4x + 20 \quad \text{Multiplication}$$

PRACTICE PROBLEMS

Multiply.

1. $5(7a)$
2. $-3(9x)$
3. $5(-8y)$

Simplify.

4. $6 + (9 + x)$
5. $(3x + 7) + 4$

Apply the distributive property.

6. $6(x + 4)$

Answers

1. $35a$ **2.** $-27x$ **3.** $-40y$
4. $15 + x$ **5.** $3x + 11$ **6.** $6x + 24$

EXAMPLE 7

$$2(a - 3) = 2(a) - 2(3) \quad \text{Distributive property}$$
$$= 2a - 6 \quad \text{Multiplication}$$

7. $7(a - 5)$

In Examples 1–3 we simplified expressions such as $4(5x)$ by using the associative property. Here are some examples that use a combination of the associative property and the distributive property.

EXAMPLE 8

$$4(5x + 3) = 4(5x) + 4(3) \quad \text{Distributive property}$$
$$= (4 \cdot 5)x + 4(3) \quad \text{Associative property}$$
$$= 20x + 12 \quad \text{Multiplication}$$

8. $6(4x + 5)$

EXAMPLE 9

$$7(3a - 6) = 7(3a) - 7(6) \quad \text{Distributive property}$$
$$= 21a - 42 \quad \text{Associative property and multiplication}$$

9. $3(8a - 4)$

EXAMPLE 10

$$5(2x + 3y) = 5(2x) + 5(3y) \quad \text{Distributive property}$$
$$= 10x + 15y \quad \text{Associative property and multiplication}$$

10. $8(3x + 4y)$

We can also use the distributive property to simplify expressions like $4x + 3x$. Because multiplication is a commutative operation, we can also rewrite the distributive property like this:

$$b \cdot a + c \cdot a = (b + c)a$$

Applying the distributive property in this form to the expression $4x + 3x$, we have

$$4x + 3x = (4 + 3)x \quad \text{Distributive property}$$
$$= 7x \quad \text{Addition}$$

C Similar Terms

Expressions like $4x$ and $3x$ are called *similar terms* because the variable parts are the same. Some other examples of similar terms are $5y$ and $-6y$ and the terms $7a$, $-13a$, and $\frac{3}{4}a$. To simplify an algebraic expression (an expression that involves both numbers and variables), we combine similar terms by applying the distributive property. Table 1 shows several pairs of similar terms and how they can be combined using the distributive property.

TABLE 1

Original Expression		Apply Distributive Property		Simplified Expression
$4x + 3x$	=	$(4 + 3)x$	=	$7x$
$7a + a$	=	$(7 + 1)a$	=	$8a$
$-5x + 7x$	=	$(-5 + 7)x$	=	$2x$
$8y - y$	=	$(8 - 1)y$	=	$7y$
$-4a - 2a$	=	$(-4 - 2)a$	=	$-6a$
$3x - 7x$	=	$(3 - 7)x$	=	$-4x$

As you can see from the table, the distributive property can be applied to any combination of positive and negative terms so long as they are similar terms.

Answers
7. $7a - 35$ 8. $24x + 30$
9. $24a - 12$ 10. $24x + 32y$

D Algebraic Expressions Representing Area and Perimeter

Below are a square with a side of length s and a rectangle with a length of l and a width of w. The table that follows the figures gives the formulas for the area and perimeter of each.

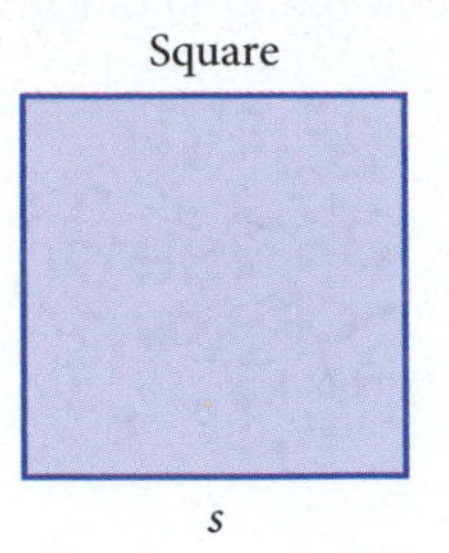

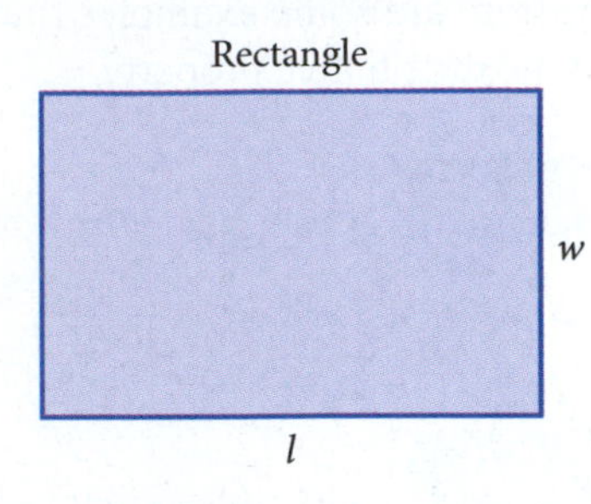

	Square	Rectangle
Area A	s^2	lw
Perimeter P	$4s$	$2l + 2w$

11. Find the area and perimeter of a square if its side is 12 feet long.

EXAMPLE 11 Find the area and perimeter of a square with a side 6 inches long.

SOLUTION Substituting 6 for s in the formulas for area and perimeter of a square, we have

$$\text{Area} = A = s^2 = 6^2 = 36 \text{ square inches}$$

$$\text{Perimeter} = P = 4s = 4(6) = 24 \text{ inches}$$

12. A football field is 100 yards long and approximately 53 yards wide. Find the area and perimeter.

EXAMPLE 12 A soccer field is 100 yards long and 75 yards wide. Find the area and perimeter.

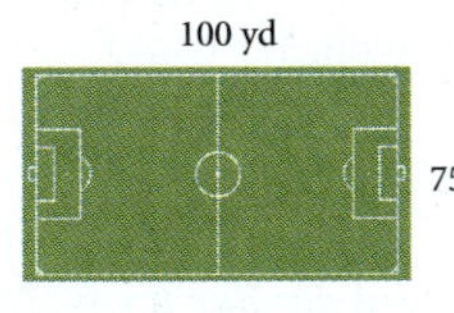

SOLUTION Substituting 100 for l and 75 for w in the formulas for area and perimeter of a rectangle, we have

$$\text{Area} = A = lw = 100(75) = 7{,}500 \text{ square yards}$$

$$\text{Perimeter} = P = 2l + 2w = 2(100) + 2(75) = 200 + 150 = 350 \text{ yards}$$

Getting Ready for Class

After reading through the preceding section, respond in your own words and in complete sentences.

1. Without actually multiplying, how do you apply the associative property to the expression $4(5x)$?
2. What are similar terms?
3. Explain why $2a - a$ is a, rather than 1.
4. Can two rectangles with the same perimeter have different areas? Explain your answer.

Answers

11. $A = 144$ sq ft, $P = 48$ ft
12. $A = 5{,}300$ sq yd, $P = 306$ yd

Problem Set 2.6

A Apply the associative property to each expression, and then simplify the result. [Examples 1–5]

1. $5(4a)$
2. $8(9a)$
3. $6(8a)$
4. $3(2a)$
5. $-6(3x)$
6. $-2(7x)$
7. $-3(9x)$
8. $-4(6x)$
9. $5(-2y)$
10. $3(-8y)$
11. $6(-10y)$
12. $5(-5y)$
13. $2 + (3 + x)$
14. $9 + (6 + x)$
15. $5 + (8 + x)$
16. $3 + (9 + x)$
17. $4 + (6 + y)$
18. $2 + (8 + y)$
19. $7 + (1 + y)$
20. $4 + (1 + y)$
21. $(5x + 2) + 4$
22. $(8x + 3) + 10$
23. $(6y + 4) + 3$
24. $(3y + 7) + 8$
25. $(12a + 2) + 19$
26. $(6a + 3) + 14$
27. $(7x + 8) + 20$
28. $(14x + 3) + 15$

B Apply the distributive property to each expression, and then simplify. [Examples 6–10]

29. $7(x + 5)$
30. $8(x + 3)$
31. $6(a - 7)$
32. $4(a - 9)$
33. $2(x - y)$
34. $5(x - a)$
35. $4(5 + x)$
36. $8(3 + x)$
37. $3(2x + 5)$
38. $8(5x + 4)$
39. $6(3a + 1)$
40. $4(8a + 3)$
41. $2(6x - 3y)$
42. $7(5x - y)$
43. $5(7 - 4y)$
44. $8(6 - 3y)$

C Use the distributive property to combine similar terms. (See Table 1.)

45. $3x + 5x$
46. $7x + 8x$
47. $3a + a$
48. $8a + a$
49. $-2x + 6x$
50. $-3x + 9x$
51. $6y - y$
52. $3y - y$
53. $-8a - 2a$
54. $-7a - 5a$
55. $4x - 9x$
56. $5x - 11x$

Applying the Concepts

57. A farmer is replacing several turbines on his windmills. He plans to replace x turbines, and he is going to get \$300 off each turbine he buys. Also, he'll get a \$250 rebate on his entire purchase. Write an expression that describes this situation and then simplify.

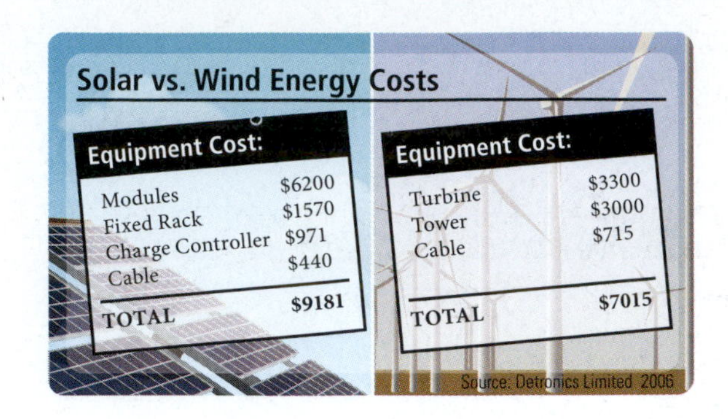

58. A homeowner is replacing 4 solar modules. She is going to receive a discount of some amount x off each module and a \$350 mail-in rebate. Write an expression that describes this situation and then simplify.

D **Area and Perimeter** Find the area and perimeter of each square if the length of each side is as given below. [Example 11]

59. $s = 6$ feet **60.** $s = 14$ yards **61.** $s = 9$ inches **62.** $s = 15$ meters

D **Area and Perimeter** Find the area and perimeter for a rectangle if the length and width are as given below. [Example 12]

63. $l = 20$ inches, $w = 10$ inches **64.** $l = 40$ yards, $w = 20$ yards

65. $l = 25$ feet, $w = 12$ feet **66.** $l = 210$ meters, $w = 120$ meters

Temperature Scales In the metric system, the scale we use to measure temperature is the Celsius scale. On this scale water boils at 100 degrees and freezes at 0 degrees. When we write 100 degrees measured on the Celsius scale, we use the notation 100°C, which is read "100 degrees Celsius." If we know the temperature in degrees Fahrenheit, we can convert to degrees Celsius by using the formula

$$C = \frac{5(F - 32)}{9}$$

where F is the temperature in degrees Fahrenheit. Use this formula to find the temperature in degrees Celsius for each of the following Fahrenheit temperatures.

67. 68°F **68.** 59°F **69.** 41°F **70.** 23°F **71.** 14°F **72.** 32°F

Chapter 2 Summary

Absolute Value [2.1]

The absolute value of a number is its distance from 0 on the number line. It is the numerical part of a number. The absolute value of a number is never negative.

EXAMPLES

1. $|3| = 3$ and $|-3| = 3$

Opposites [2.1]

Two numbers are called opposites if they are the same distance from 0 on the number line but in opposite directions from 0. The opposite of a positive number is a negative number, and the opposite of a negative number is a positive number.

2. $-(5) = -5$ and $-(-5) = 5$

Addition of Positive and Negative Numbers [2.2]

1. To add two numbers with *the same sign:* Simply add absolute values and use the common sign. If both numbers are positive, the answer is positive. If both numbers are negative, the answer is negative.
2. To add two numbers with *different signs:* Subtract the smaller absolute value from the larger absolute value. The answer has the same sign as the number with the larger absolute value.

3.

$$3 + 5 = 8$$
$$-3 + (-5) = -8$$

$$5 + (-3) = 2$$
$$-5 + 3 = -2$$

Subtraction [2.3]

Subtracting a number is equivalent to adding its opposite. If a and b represent numbers, then subtraction is defined in terms of addition as follows:

$$a - b = a + (-b)$$

↑ Subtraction ↑ Addition of the opposite

4.

$$3 - 5 = 3 + (-5) = -2$$
$$-3 - 5 = -3 + (-5) = -8$$
$$3 - (-5) = 3 + 5 = 8$$
$$-3 - (-5) = -3 + 5 = 2$$

Multiplication with Positive and Negative Numbers [2.4]

To multiply two numbers, multiply their absolute values.

1. The answer is *positive* if both numbers have the same sign.
2. The answer is *negative* if the numbers have different signs.

5.

$$3(5) = 15$$
$$3(-5) = -15$$
$$-3(5) = -15$$
$$-3(-5) = 15$$

Division [2.5]

The rule for assigning the correct sign to the answer in a division problem is the same as the rule for multiplication. That is, like signs give a positive answer, and unlike signs give a negative answer.

6.

$$\frac{12}{4} = 3$$
$$\frac{-12}{4} = -3$$
$$\frac{12}{-4} = -3$$
$$\frac{-12}{-4} = 3$$

Simplifying Expressions [2.6]

We simplify algebraic expressions by applying the commutative, associative, and distributive properties.

7. Simplify.

a. $-2(5x) = (-2 \cdot 5)x = -10x$

b. $4(2a - 8) = 4(2a) - 4(8)$
$= 8a - 32$

Combining Similar Terms [2.6]

We combine similar terms by applying the distributive property.

8. Combine similar terms.

a. $5x + 7x = (5 + 7)x = 12x$

b. $2y - 8y = (2 - 8)y = -6y$

Chapter 2 Review

Give the opposite of each number. [2.1]

1. 17 **2.** -32 **3.** -4.6 **4.** $\frac{3}{5}$

For each pair of numbers, name the smaller number. [2.1]

5. 6; -6 **6.** -8; -3 **7.** $|-3|$; 2 **8.** $|-4|$; $|6|$

Simplify each expression. [2.1]

9. $-(-4)$ **10.** $-|-4|$ **11.** $|-6|$ **12.** $|19|$

Perform the indicated operations. [2.2, 2.3, 2.4, 2.5]

13. $5 + (-7)$ **14.** $-3 + 8$ **15.** $-345 + (-626)$ **16.** $-23 + 58$

17. $7 - 9 - 4 - 6$ **18.** $-7 - 5 - 2 - 3$ **19.** $4 - (-3)$ **20.** $30 - 42$

21. $5(-4)$ **22.** $-4(-3)$ **23.** $(56)(-31)$ **24.** $(20)(-4)$

25. $\frac{48}{-16}$ **26.** $\frac{-20}{5}$ **27.** $\frac{-14}{-7}$ **28.** $\frac{-25}{5}$

Simplify the following expressions as much as possible. [2.2, 2.3, 2.4, 2.5]

29. $(-6)^2$ **30.** $\left(-\frac{3}{4}\right)^2$ **31.** $(-2)^3$ **32.** $(-0.2)^4$

33. $7 + 4(6 - 9)$ **34.** $(-3)(-4) + 2(-5)$ **35.** $(7 - 3)(7 - 9)$ **36.** $3(-6) + 8(2 - 5)$

37. $\frac{8 - 4}{-8 + 4}$ **38.** $\frac{-4 + 2(-5)}{6 - 4}$ **39.** $\frac{8(-2) + 5(-4)}{12 - 3}$ **40.** $\frac{-2(5) + 4(-3)}{10 - 8}$

41. Give the sum of -19 and -23. [2.2]

42. Give the sum of -78 and -51. [2.2]

43. Find the difference of -6 and 5. [2.3]

44. Subtract -8 from -10. [2.3]

45. What is the product of -9 and 3? [2.4]

46. What is -3 times the sum of -9 and -4? [2.2, 2.4]

47. Divide the product of 8 and -4 by -16. [2.4, 2.5]

48. Give the quotient of -38 and 2. [2.5]

Indicate whether each statement is *True* or *False*. [2.2, 2.3, 2.4, 2.5]

49. $\frac{-10}{-5} = -2$

50. $10 - (-5) = 15$

51. $2(-3) = -3 + (-3)$

52. $-6 - (-2) = -8$

53. $3 - 5 = 5 - 3$

54. Reaction Distance The table below shows how many feet your car will travel from the time you decide you want to stop to the time it takes you to hit the brake pedal. Use the template to construct a line graph of the information in the table. [2.1]

REACTION DISTANCES

Speed (mi/hr)	Distance (ft)
0	0
10	11
20	22
30	33
40	44
50	55
60	66
70	77
80	88

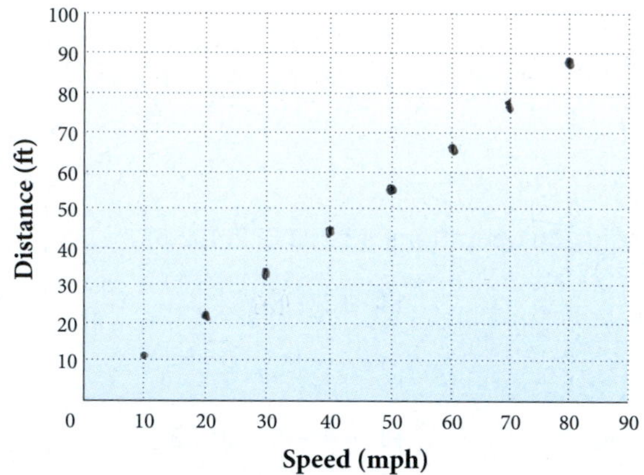

55. Gambling A gambler wins \$58 Saturday night and then loses \$86 on Sunday. Use positive and negative numbers to describe this situation. Then give the gambler's net loss or gain as a positive or negative number. [2.2]

56. Name two numbers that are 7 units from -8 on the number line. [2.1]

57. Temperature On Wednesday, the temperature reaches a high of 17° above 0 and a low of 7° below 0. What is the difference between the high and low temperatures for Wednesday? [2.3]

58. If the difference between two numbers is -3, and one of the numbers is 5, what is the other number? [2.3]

Use the associative properties to simplify each expression. [2.6]

59. $(3x + 4) + 8$

60. $8(3x)$

61. $-3(7a)$

62. $6(-5y)$

Apply the distributive property and then simplify if possible. [2.6]

63. $4(x + 3)$

64. $2(x - 5)$

65. $7(3y - 8)$

66. $3(2a + 5b)$

Combine similar terms. [2.6]

67. $7x - 4x$

68. $-8a + 10a$

69. $5y - y$

70. $12x + 4x$

Chapter 2 Cumulative Review

Add.

1. $\begin{array}{r} 6{,}801 \\ 539 \\ \underline{+\ 374} \end{array}$

2. $-675 + 892$

Subtract.

3. $\begin{array}{r} 5{,}038 \\ \underline{-2{,}769} \end{array}$

4. $675 - 892$

Multiply.

5. $52(867)$

6. $(-3)(-14)$

7. $5(2x - 8)$

8. $-3(8x + 9)$

Divide.

9. $1023 \div 15$

10. $473\overline{)15{,}609}$

11. $\dfrac{-24}{-6}$

12. $\dfrac{-75}{15}$

Simplify.

13. $9 - 3(8 - 5)$

14. $7(4)^2 - 5(2)^3$

15. $-|-11|$

16. $-(-9)$

17. $(-3)^3$

18. $(-7)^2$

19. $(6 - 3)(8 - 13)$

20. $\dfrac{-7 + 3(-5)}{9 - 20}$

21. $\dfrac{-8(6)}{-4}$

Place an inequality symbol (< or >) between the following pairs of numbers so that the resulting statement is true.

22. $-4 \quad -6$

23. $|3| \quad |-5|$

24. Identify the property or properties used in the following: $3y + (5 + 2y) = (3y + 2y) + 5$

25. Translate into symbols; then simplify: twice the difference of 19 and 7

26. Baseball The table below shows the top four records for the highest attended opening-day games in major league baseball. Complete the table by rounding each number to the nearest hundred.

OPENING DAY ATTENDANCE RECORDS

Game	Date	Attendance	To the Nearest Hundred
Montreal at Colorado	4/9/93	80,227	
San Francisco at Los Angeles	4/7/58	78,672	
Detroit at Cleveland	4/7/73	74,420	
St. Louis at Cleveland	4/20/48	73,163	

27. Stopping Distances The bar chart below shows how many feet it takes to stop a car traveling at different rates of speed, once the brakes are applied. Use this information in the bar chart to fill in the table.

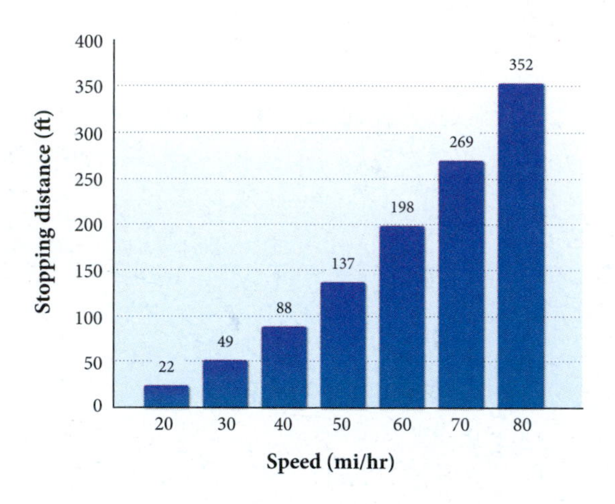

Speed (mi/hr)	Distance (ft)
20	22
30	49
40	88
50	137
60	198
70	269
80	352

28. Stock Market A stock gains 3 points on Tuesday, then loses 4 points on Wednesday, and gains 2 points on Thursday. What is the net gain or loss of the stock for this 3-day period?

29. Hourly Pay Jean tutors in the math lab and earns $56 in one week. If she works 8 hours that week, what is her hourly pay?

30. Average If a basketball team has scores of 64, 76, 98, 55, and 102 in their first five games, what is the average (mean) number of points scored for the five games?

31. Temperature Change On Monday, Jack notices that the temperature reaches a high of 18° above 0 and a low of 5° below 0. What is the difference between the high and low temperatures for Monday?

32. Geometry Find the perimeter and area of a square with side 8 inches.

33. Movie Tickets A movie theater has a total of 250 seats. If they have a sell-out crowd for a matinee, and each ticket costs $7, how much money will ticket sales bring in that afternoon?

34. Number Problem The product of 6 and 8 is how much larger than the sum of 6 and 8?

Chapter 2 Test

Give the opposite of each number.

1. 14
2. -5

Place an inequality symbol (< or >) between each pair of numbers so that the resulting statement is true.

3. $-1 \quad -4$
4. $|-4| \quad |2|$

Simplify each expression.

5. $-(-7)$
6. $-|-2|$

Perform the indicated operations.

7. $8 + (-17)$
8. $-4 - 2$
9. $-9 + (-12)$
10. $-65 - (-29)$
11. $(-6)(-7)$
12. $-3(-18)$
13. $\frac{-80}{16}$
14. $\frac{-35}{-7}$

Simplify the following expressions as much as possible.

15. $(-3)^2$
16. $(-2)^3$
17. $(-7)(3) + (-2)(-5)$
18. $(8 - 5)(6 - 11)$
19. $\frac{-5 + 3(-3)}{5 - 7}$
20. $\frac{-3(2) + 5(-2)}{7 - 3}$
21. Give the sum of -15 and -46.
22. Subtract -5 from -12.
23. What is the product of -8 and -3?
24. Give the quotient of 45 and -9.

25. Garbage Production The table and bar chart below give the annual production of garbage in the United States for some specific years.

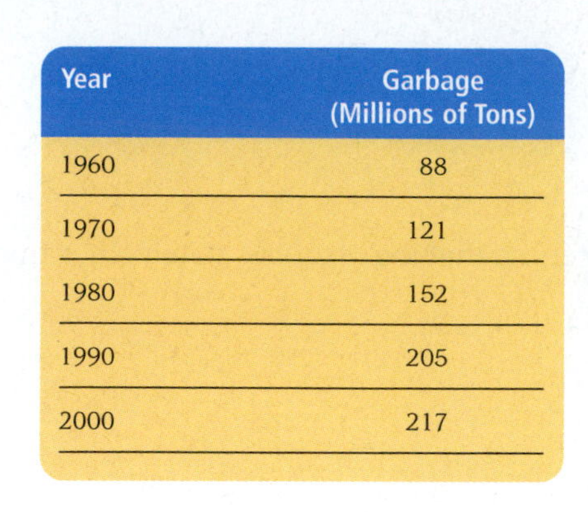

Year	Garbage (Millions of Tons)
1960	88
1970	121
1980	152
1990	205
2000	217

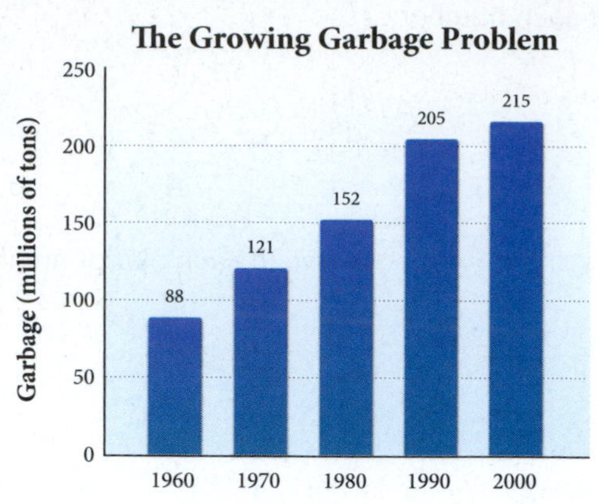

Use the information from the table and bar chart to construct a line graph using the template below.

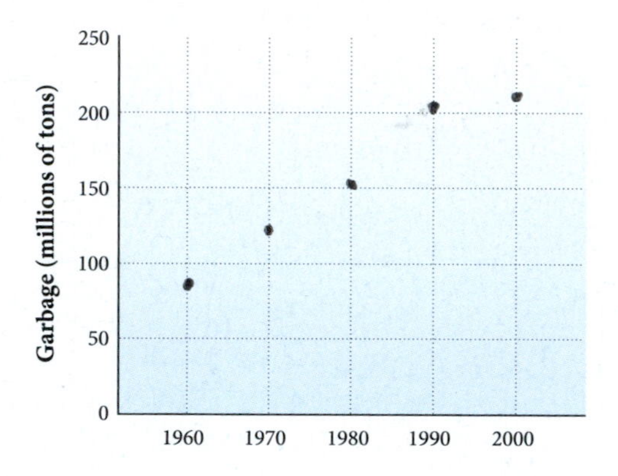

26. Gambling A gambler loses \$100 Saturday night and wins \$65 on Sunday. Give the gambler's net loss or gain as a positive or negative number.

27. Temperature On Friday, the temperature reaches a high of 21° above 0 and a low of 4° below 0. What is the difference between the high and low temperatures for Friday?

Apply the distributive property and simplify.

28. $7(x - 5)$

29. $4(5x - 1)$

30. $2(9x - 8y)$

Combine similar terms.

31. $12x + 20x$

32. $9a - a$

Chapter 2 Projects

INTRODUCTION TO ALGEBRA

GROUP PROJECT

Random Motion

Number of People 3

Time Needed 15 minutes

Equipment Coins, dice, pencil, and paper

Background Microscopic atoms and molecules move randomly. We use random movement models to help us understand their motion. Random motion also helps us understand things like the stock market and computer science.

In a random walk, an ant starts at a lamppost and takes steps of equal length along the street. We can think of the lamppost as the origin. The ant either takes a step in the negative or positive direction. Mathematicians have studied questions such as where the ant is likely to end up after taking a certain number of steps.

Stage	Coin	Die	Position of Ant
0	—	—	0
1			
2			
3			
4			
5			
6			
7			
8			
9			
10			

Procedure You will use a coin and die to simulate random motion. The ant will start at 0 on the number line. Roll the die and flip the coin. The ant will move the number of steps shown on the die. If the coin comes up heads, the ant moves in the positive direction. If the coin comes up tails, the ant moves in the negative direction. Repeat this process 10 times. Start each stage from the ending position of the previous stage. For example, if the ant ends up at -3 after Stage 1, then in Stage 2 the ant starts at -3. Record your results in the table.

RESEARCH PROJECT

David Harold Blackwell

At age 22, David Blackwell earned his doctorate, becoming the seventh African American to earn a Ph.D. in mathematics. In high school, Blackwell did not care for algebra and trigonometry. When he took a course in analysis, he really became interested in math. Although Blackwell faced a good deal of racism during his career, he became a successful teacher, author, and mathematician. Research the life and work of Dr. Blackwell, and then present your results in an essay.

Courtesy of David Harold Blackwell

Fractions and Mixed Numbers

Chapter Outline

Introduction

Crater Lake, located in the Cascade Mountain range in Southern Oregon, is 594 meters deep, making it the deepest lake in the United States. Here is a chart showing the depth of Crater Lake and the location of some lakes that are deeper than Crater Lake.

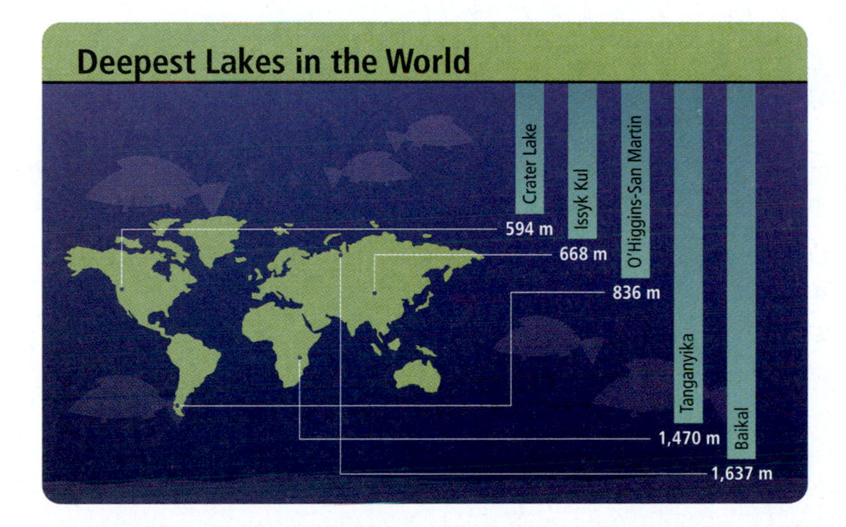

As you can see from the chart, although Crater lake is the deepest lake in the United States, it is far from being the deepest lake in the world. We can use fractions to compare the depths of these lakes. For example, Crater Lake is approximately $\frac{2}{5}$ as deep as Lake Tanganyika. In this chapter, we begin our work with fractions.

Chapter Pretest

The pretest below contains problems that are representative of the problems you will find in the chapter.

1. Reduce to lowest terms: $\frac{16xy^2}{20xy}$

2. Factor 112 into a product of prime factors.

Perform the indicated operations. Reduce all answers to lowest terms.

3. $\frac{3}{8} \cdot 16$

4. $10 \div \left(-\frac{5}{6}\right)$

5. $\frac{32}{45} \div \frac{40}{63}$

6. $\frac{x^2}{y} \div \frac{x}{y}$

7. $\frac{5}{16} + \frac{7}{20}$

8. $\frac{5}{8} - \left(-\frac{1}{6}\right)$

9. Write $4\frac{3}{5}$ as an improper fraction.

10. Write $\frac{21}{8}$ as a mixed number.

Perform the indicated operations. Reduce all answers to lowest terms.

11. $1\frac{3}{8} \cdot 2\frac{4}{5}$

12. $12 \div 3\frac{1}{6}$

13. $4\frac{1}{5} + 1\frac{1}{10}$

14. $6\frac{1}{6} - 3\frac{1}{3}$

Simplify each of the following as much as possible.

15. $\left(-\frac{1}{4}\right)^2$

16. $\left(\frac{1}{3}\right)^2 \cdot 27 + \left(\frac{3}{2}\right)^2 \cdot 4$

17. $12 + \frac{9}{10} \div \frac{10}{9}$

18. $\left(1 - \frac{1}{4}\right)\left(1\frac{2}{3} - \frac{1}{3}\right)$

19. $\dfrac{\frac{3}{5}}{\frac{9}{10}}$

20. $\dfrac{2 + \frac{1}{4}}{2 - \frac{1}{4}}$

Getting Ready for Chapter 3

The problems below review material covered previously that you need to know in order to be successful in Chapter 3. If you have any difficulty with the problems here, you need to go back and review before going on to Chapter 3.

1. Place either $<$ or $>$ between the two numbers so that the resulting statement is true.
a. 9 5 **b.** 10 0 **c.** 0 1 **d.** 2001 201

Simplify.

2. $2 \cdot 5 \cdot x$

3. $a + 5 - 3$

4. $13 + 12 - 9$

5. $5 \cdot 4 + 3$

6. $64 \div 8 \cdot 2$

7. $17 - (3 \cdot 5 + 2)$

8. $(3 + 5)(2 + 1)$

9. $3 + 2(3 + 4)^2$

10. $32 \div 4^2 + 75 \div 5^2$

The following division problems all have remainders. Divide.

11. $11 \div 4$

12. $208 \div 24$

13. $8{,}648 \div 43$

14. $14{,}713 \div 29$

15. Use the distributive property to rewrite $2 \cdot 7 + 3 \cdot 7$.

16. Rewrite using exponents: $2 \cdot 2 \cdot 3 \cdot 3 \cdot 3$.

3.1 The Meaning and Properties of Fractions

Objectives

A Identify the numerator and denominator of a fraction.

B Identify proper and improper fractions.

C Write equivalent fractions.

D Simplify fractions with division.

E Compare the size of fractions.

Examples now playing at MathTV.com/books

Introduction . . .

The information in the table below was taken from the website for Cal Poly. The pie chart was created from the table. Both the table and pie chart use fractions to specify how the students at Cal Poly are distributed among the different schools within the university.

CAL POLY ENROLLMENT FOR FALL

School	Fraction Of Students
Agriculture	$\frac{11}{50}$
Architecture and Environmental Design	$\frac{1}{10}$
Business	$\frac{3}{20}$
Engineering	$\frac{1}{4}$
Liberal Arts	$\frac{4}{25}$
Science and Mathematics	$\frac{3}{25}$

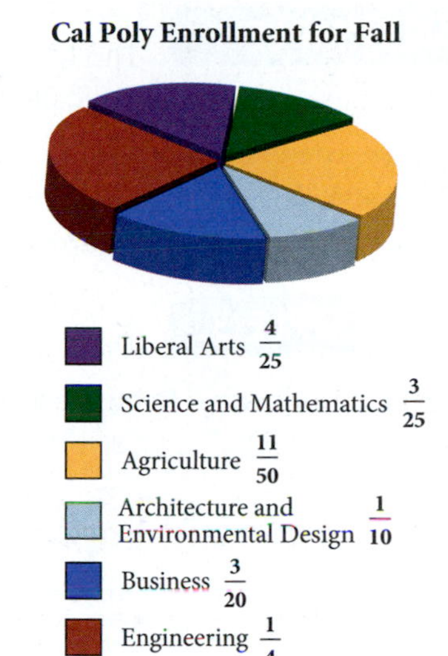

From the table, we see that $\frac{1}{4}$ (one-fourth) of the students are enrolled in the School of Engineering. This means that one out of every four students at Cal Poly is studying engineering. The fraction $\frac{1}{4}$ tells us we have 1 part of 4 equal parts.

Figure 1 at the right shows a rectangle that has been divided into equal parts in four different ways. The shaded area for each rectangle is $\frac{1}{2}$ the total area.

Now that we have an intuitive idea of the meaning of fractions, here are the more formal definitions and vocabulary associated with fractions.

Note As we mentioned in Chapter 1, when we use a letter to represent a number, or a group of numbers, that letter is called a variable. In the definition below, we are restricting the numbers that the variable b can represent to numbers other than 0. As you will see later in the chapter, we do this to avoid writing an expression that would imply division by the number 0.

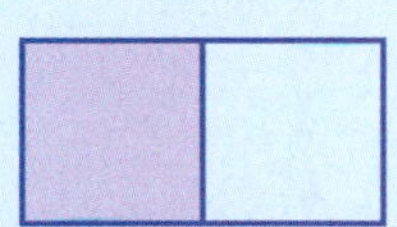

a. $\frac{1}{2}$ is shaded

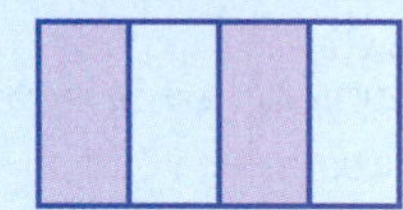

b. $\frac{2}{4}$ are shaded

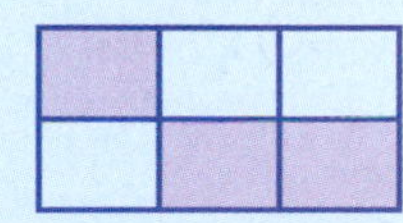

c. $\frac{3}{6}$ are shaded

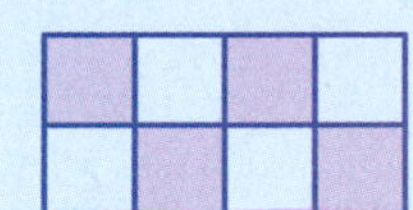

d. $\frac{4}{8}$ are shaded

FIGURE 1 Four Ways to Visualize $\frac{1}{2}$

A The Numerator and Denominator

Definition

A **fraction** is any number that can be put in the form $\frac{a}{b}$ (also sometimes written a/b), where a and b are numbers and b is not 0.

Some examples of fractions are:

$\frac{1}{2}$	$\frac{3}{4}$	$\frac{7}{8}$	$\frac{9}{5}$
One-half	*Three-fourths*	*Seven-eighths*	*Nine-fifths*

STUDY SKILLS
Intend to Succeed

I always have a few students who simply go through the motions of studying without intending to master the material. It is more important to them to look like they are studying than to actually study. You need to study with the intention of being successful in the course no matter what it takes.

Definition

For the fraction $\frac{a}{b}$, a and b are called the **terms** of the fraction. More specifically, a is called the **numerator,** and b is called the **denominator.**

$$\text{fraction } \frac{a}{b} \begin{array}{l} \leftarrow \textbf{numerator} \\ \leftarrow \textbf{denominator} \end{array}$$

PRACTICE PROBLEMS

1. Name the terms of the fraction $\frac{5}{6}$. Which is the numerator and which is the denominator?
2. Name the the numerator and the denominator of the fraction $\frac{x}{3}$.
3. Why is the number 9 considered to be a fraction?

EXAMPLE 1 The terms of the fraction $\frac{3}{4}$ are 3 and 4. The 3 is called the numerator, and the 4 is called the denominator.

EXAMPLE 2 The numerator of the fraction $\frac{a}{5}$ is a. The denominator is 5. Both a and 5 are called terms.

EXAMPLE 3 The number 7 may also be put in fraction form, because it can be written as $\frac{7}{1}$. In this case, 7 is the numerator and 1 is the denominator.

B Proper and Improper Fractions

Definition

A **proper fraction** is a fraction in which the numerator is less than the denominator. If the numerator is greater than or equal to the denominator, the fraction is called an **improper fraction.**

4. Which of the following are proper fractions?

   $\frac{1}{6}$ $\frac{2}{3}$ $\frac{8}{5}$

5. Which of the following are improper fractions?

   $\frac{5}{9}$ $\frac{6}{5}$ $\frac{4}{3}$ 7

EXAMPLE 4 The fractions $\frac{3}{4}$, $\frac{1}{8}$, and $\frac{9}{10}$ are all proper fractions, because in each case the numerator is less than the denominator.

EXAMPLE 5 The numbers $\frac{9}{5}$, $\frac{10}{10}$, and 6 are all improper fractions, because in each case the numerator is greater than or equal to the denominator. (Remember that 6 can be written as $\frac{6}{1}$, in which case 6 is the numerator and 1 is the denominator.)

Fractions on the Number Line

We can give meaning to the fraction $\frac{2}{3}$ by using a number line. If we take that part of the number line from 0 to 1 and divide it into *three equal parts,* we say that we have divided it into *thirds* (see Figure 2). Each of the three segments is $\frac{1}{3}$ (one third) of the whole segment from 0 to 1.

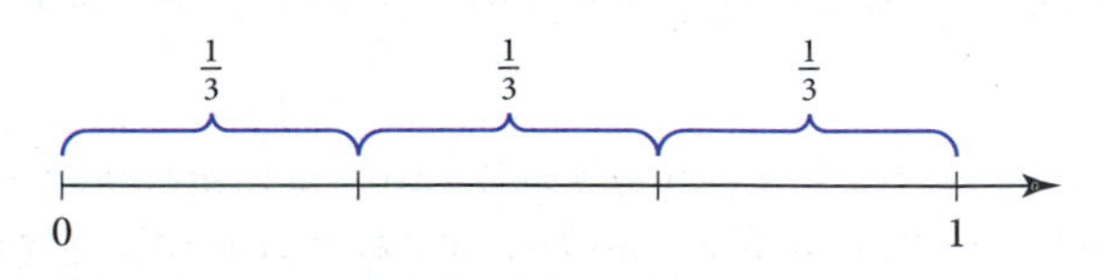

FIGURE 2

Note There are many ways to give meaning to fractions like $\frac{2}{3}$ other than by using the number line. One popular way is to think of cutting a pie into three equal pieces, as shown below. If you take two of the pieces, you have taken $\frac{2}{3}$ of the pie.

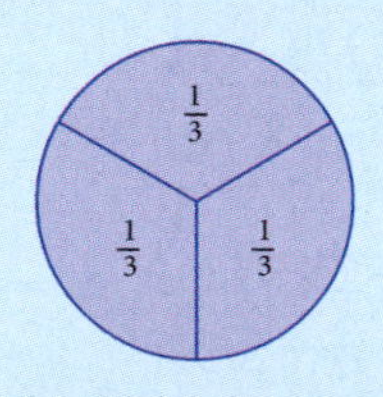

Answers

1. Terms: 5 and 6; numerator: 5; denominator: 6
2. Numerator: x; denominator: 3
3. Because it can be written $\frac{9}{1}$
4. $\frac{1}{6}, \frac{2}{3}$ 5. $\frac{6}{5}, \frac{4}{3}, 7$

Two of these smaller segments together are $\frac{2}{3}$ (two thirds) of the whole segment. And three of them would be $\frac{3}{3}$ (three thirds), or the whole segment, as indicated in Figure 3.

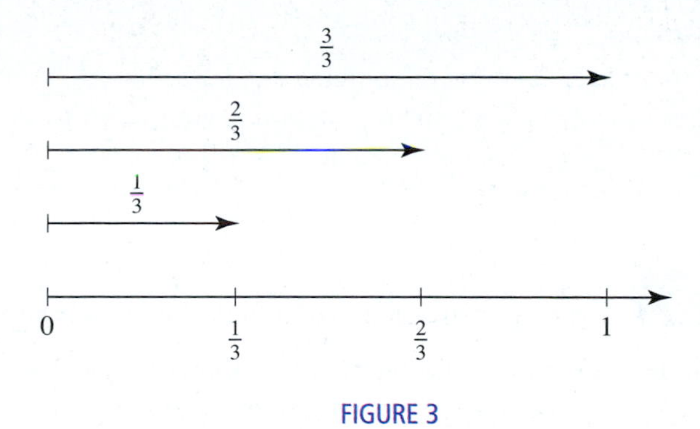

FIGURE 3

Let's do the same thing again with six and twelve equal divisions of the segment from 0 to 1 (see Figure 4).

The same point that we labeled with $\frac{1}{3}$ in Figure 3 is now labeled with $\frac{2}{6}$ and with $\frac{4}{12}$. It must be true then that

$$\frac{4}{12} = \frac{2}{6} = \frac{1}{3}$$

Although these three fractions look different, each names the same point on the number line, as shown in Figure 4. All three fractions have the same *value*, because they all represent the same number.

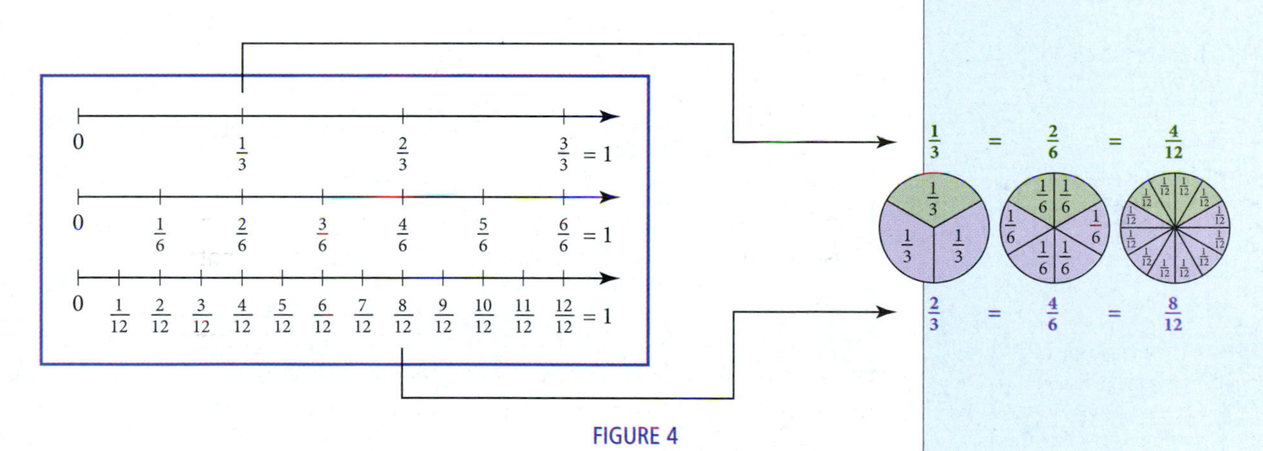

FIGURE 4

C Equivalent Fractions

Definition

Fractions that represent the same number are said to be **equivalent.** Equivalent fractions may look different, but they must have the same value.

It is apparent that every fraction has many different representations, each of which is equivalent to the original fraction. The next two properties give us a way of changing the terms of a fraction without changing its value.

Property 1 for Fractions

If a, b, and c are numbers and b and c are not 0, then it is always true that

$$\frac{a}{b} = \frac{a \cdot c}{b \cdot c}$$

In words: If the numerator and the denominator of a fraction are multiplied by the same nonzero number, the resulting fraction is equivalent to the original fraction.

6. Write $\frac{2}{3}$ as an equivalent fraction with denominator 12.

EXAMPLE 6 Write $\frac{3}{4}$ as an equivalent fraction with denominator 20.

SOLUTION The denominator of the original fraction is 4. The fraction we are trying to find must have a denominator of 20. We know that if we multiply 4 by 5, we get 20. Property 1 indicates that we are free to multiply the denominator by 5 so long as we do the same to the numerator.

$$\frac{3}{4} = \frac{3 \cdot \mathbf{5}}{4 \cdot \mathbf{5}} = \frac{15}{20}$$

The fraction $\frac{15}{20}$ is equivalent to the fraction $\frac{3}{4}$.

7. Write $\frac{2}{3}$ as an equivalent fraction with denominator $12x$.

EXAMPLE 7 Write $\frac{3}{4}$ as an equivalent fraction with denominator $12x$.

SOLUTION If we multiply 4 by $3x$, we will have $12x$:

$$\frac{3}{4} = \frac{3 \cdot \mathbf{3x}}{4 \cdot \mathbf{3x}} = \frac{9x}{12x}$$

Property 2 for Fractions

If a, b, and c are integers and b and c are not 0, then it is always true that

$$\frac{a}{b} = \frac{a \div c}{b \div c}$$

In words: If the numerator and the denominator of a fraction are divided by the same nonzero number, the resulting fraction is equivalent to the original fraction.

8. Write $\frac{15}{20}$ as an equivalent fraction with denominator 4.

EXAMPLE 8 Write $\frac{10}{12}$ as an equivalent fraction with denominator 6.

SOLUTION If we divide the original denominator 12 by 2, we obtain 6. Property 2 indicates that if we divide both the numerator and the denominator by 2, the resulting fraction will be equal to the original fraction:

$$\frac{10}{12} = \frac{10 \div \mathbf{2}}{12 \div \mathbf{2}} = \frac{5}{6}$$

Answers

6. $\frac{8}{12}$ **7.** $\frac{8x}{12x}$ **8.** $\frac{3}{4}$

D The Number 1 and Fractions

There are two situations involving fractions and the number 1 that occur frequently in mathematics. The first is when the denominator of a fraction is 1. In this case, if we let a represent any number, then

$$\frac{a}{1} = a \qquad \text{for any number } a$$

The second situation occurs when the numerator and the denominator of a fraction are the same nonzero number:

$$\frac{a}{a} = 1 \qquad \text{for any nonzero number } a$$

EXAMPLE 9 Simplify each expression.

a. $\frac{24}{1}$ **b.** $\frac{24}{24}$ **c.** $\frac{48x}{24x}$ **d.** $\frac{72y}{24y}$

SOLUTION In each case we divide the numerator by the denominator:

a. $\frac{24}{1} = 24$ **b.** $\frac{24}{24} = 1$ **c.** $\frac{48x}{24x} = 2$ **d.** $\frac{72y}{24y} = 3$

E Comparing Fractions

We can compare fractions to see which is larger or smaller when they have the same denominator.

EXAMPLE 10 Write each fraction as an equivalent fraction with denominator 24. Then write them in order from smallest to largest.

$$\frac{5}{8} \quad \frac{5}{6} \quad \frac{3}{4} \quad \frac{2}{3}$$

SOLUTION We begin by writing each fraction as an equivalent fraction with denominator 24.

$$\frac{5}{8} = \frac{15}{24} \quad \frac{5}{6} = \frac{20}{24} \quad \frac{3}{4} = \frac{18}{24} \quad \frac{2}{3} = \frac{16}{24}$$

Now that they all have the same denominator, the smallest fraction is the one with the smallest numerator and the largest fraction is the one with the largest numerator. Writing them in order from smallest to largest we have:

$$\frac{15}{24} < \frac{16}{24} < \frac{18}{24} < \frac{20}{24}$$

or

$$\frac{5}{8} < \frac{2}{3} < \frac{3}{4} < \frac{5}{6}$$

STUDY SKILLS
Be Focused, Not Distracted

I have students who begin their assignments by asking themselves, "Why am I taking this class?" or, "When am I ever going to use this stuff?" If you are asking yourself similar questions, you may be distracting yourself from doing the things that will produce the results you want in this course. Don't dwell on questions and evaluations of the class that can be used as excuses for not doing well. If you want to succeed in this course, focus your energy and efforts toward success.

9. Simplify.

a. $\frac{18}{1}$ **b.** $\frac{18}{18}$

c. $\frac{36}{18}$ **d.** $\frac{72}{18}$

10. Write each fraction as an equivalent fraction with denominator 12. Then write in order from smallest to largest.
$\frac{1}{3}, \frac{1}{6}, \frac{1}{4}, \frac{5}{12}$

Answers

9. a. 18 **b.** 1 **c.** 2 **d.** 4

10. $\frac{2}{12}, \frac{3}{12}, \frac{4}{12}, \frac{5}{12}$

DESCRIPTIVE STATISTICS

Scatter Diagrams and Line Graphs

The table and bar chart give the daily gain in the price of a certain stock for one week, when stock prices were given in terms of fractions instead of decimals.

Change in Stock Price	
Day	**Gain**
Monday	$\frac{3}{4}$
Tuesday	$\frac{9}{16}$
Wednesday	$\frac{3}{32}$
Thursday	$\frac{7}{32}$
Friday	$\frac{1}{16}$

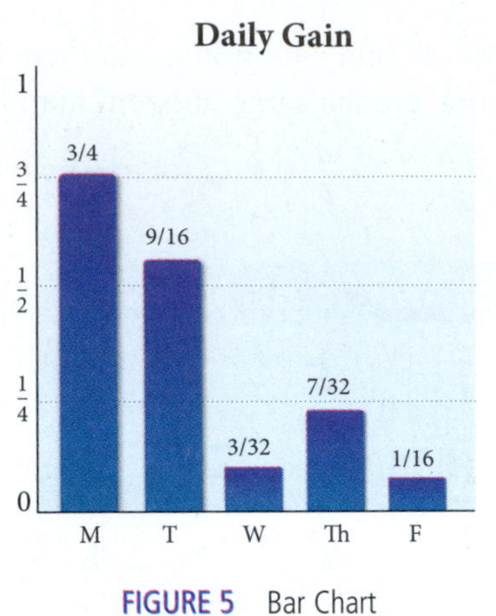

FIGURE 5 Bar Chart

Figure 6 below shows another way to visualize the information in the table. It is called a scatter diagram. In the *scatter diagram,* dots are used instead of the bars shown in Figure 5 to represent the gain in stock price for each day of the week. If we connect the dots in Figure 6 with straight lines, we produce the diagram in Figure 7, which is known as a *line graph.*

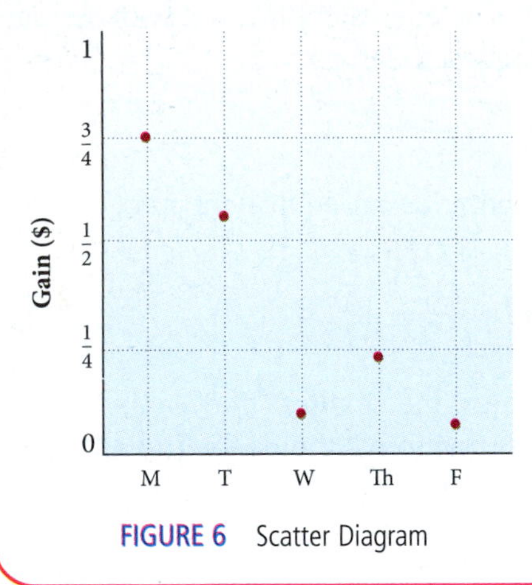

FIGURE 6 Scatter Diagram

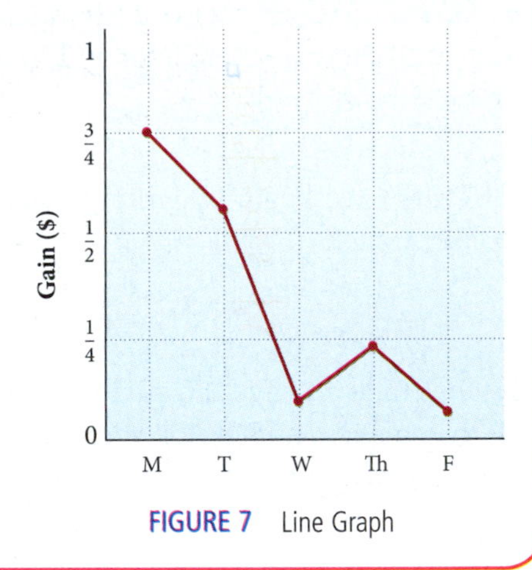

FIGURE 7 Line Graph

Getting Ready for Class

After reading through the preceding section, respond in your own words and in complete sentences. An answer of yes or no should always be accompanied by a sentence explaining why the answer is yes or no.

1. Explain what a fraction is.
2. Which term in the fraction $\frac{7}{8}$ is the numerator?
3. Is the fraction $\frac{3}{9}$ a proper fraction?
4. What word do we use to describe fractions such as $\frac{1}{5}$ and $\frac{4}{20}$, which look different, but have the same value?

Problem Set 3.1

A Name the numerator of each fraction. [Examples 1–3]

1. $\frac{1}{3}$
2. $\frac{1}{4}$
3. $\frac{2}{3}$
4. $\frac{2}{4}$
5. $\frac{x}{8}$
6. $\frac{y}{10}$
7. $\frac{a}{b}$
8. $\frac{x}{y}$

A Name the denominator of each fraction. [Examples 1–3]

9. $\frac{2}{5}$
10. $\frac{3}{5}$
11. 6
12. 2
13. $\frac{a}{12}$
14. $\frac{b}{14}$

A Complete the following tables.

15.

Numerator a	Denominator b	Fraction $\frac{a}{b}$
3	5	$\frac{3}{5}$
1	7	$\frac{1}{7}$
x	y	$\frac{x}{y}$
$x + 1$	x	$\frac{x+1}{x}$

16.

Numerator a	Denominator b	Fraction $\frac{a}{b}$
2	9	$\frac{2}{9}$
4	3	$\frac{4}{3}$
1	x	$\frac{1}{x}$
x	$x + 1$	$\frac{x}{x+1}$

B

17. For the set of numbers $\left\{\frac{3}{4}, \frac{6}{5}, \frac{12}{3}, \frac{1}{2}, \frac{9}{10}, \frac{20}{10}\right\}$, list all the proper fractions.

18. For the set of numbers $\left\{\frac{1}{8}, \frac{7}{9}, \frac{6}{3}, \frac{18}{6}, \frac{3}{5}, \frac{9}{8}\right\}$, list all the improper fractions.

Indicate whether each of the following is *True* or *False*.

19. Every whole number greater than 1 can also be expressed as an improper fraction.

20. Some improper fractions are also proper fractions.

C

21. Adding the same number to the numerator and the denominator of a fraction will not change its value.

22. The fractions $\frac{3}{4}$ and $\frac{9}{16}$ are equivalent.

C Divide the numerator and the denominator of each of the following fractions by 2. [Examples 6–8]

23. $\frac{6}{8}$ **24.** $\frac{10}{12}$ **25.** $\frac{86}{94}$ **26.** $\frac{106}{142}$

C Divide the numerator and the denominator of each of the following fractions by 3. [Examples 6–8]

27. $\frac{12}{9}$ **28.** $\frac{33}{27}$ **29.** $\frac{39}{51}$ **30.** $\frac{57}{69}$

C Write each of the following fractions as an equivalent fraction with denominator 6. [Examples 6–8]

31. $\frac{2}{3}$ **32.** $\frac{1}{2}$ **33.** $\frac{55}{66}$ **34.** $\frac{65}{78}$

C Write each of the following fractions as an equivalent fraction with denominator 12. [Examples 6–8]

35. $\frac{2}{3}$ **36.** $\frac{5}{6}$ **37.** $\frac{56}{84}$ **38.** $\frac{143}{156}$

C Write each fraction as an equivalent fraction with denominator $12x$. [Example 7]

39. $\frac{1}{6}$ **40.** $\frac{3}{4}$

C Write each number as an equivalent fraction with denominator $24a$. [Example 7]

41. 2 **42.** 1 **43.** 5 **44.** 8

45. One-fourth of the first circle below is shaded. Use the other three circles to show three other ways to shade one-fourth of the circle.

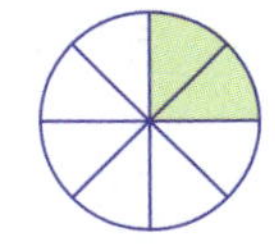 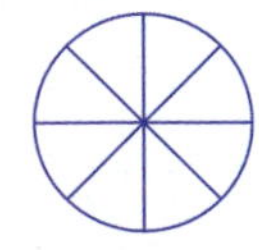 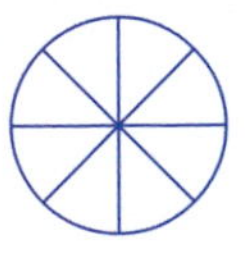 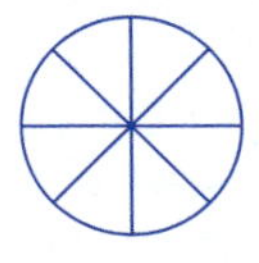

46. The objects below are hexagons, six-sided figures. One-third of the first hexagon is shaded. Shade the other three hexagons to show three other ways to represent one-third.

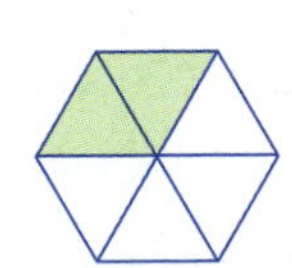  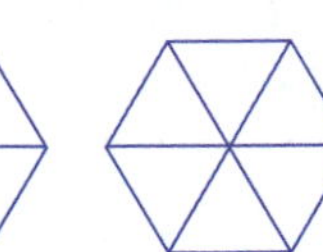 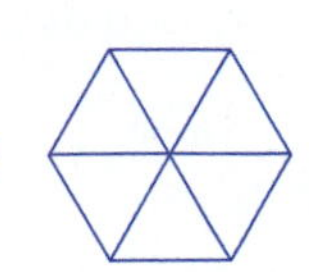

D Simplify by dividing the numerator by the denominator.

47. $\frac{3}{1}$ **48.** $\frac{3}{3}$ **49.** $\frac{6}{3}$ **50.** $\frac{12}{3}$ **51.** $\frac{37}{1}$ **52.** $\frac{37}{37}$

53. For each square below, what fraction of the area is given by the shaded region?

a. 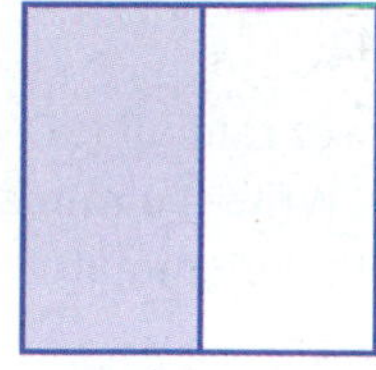

b. 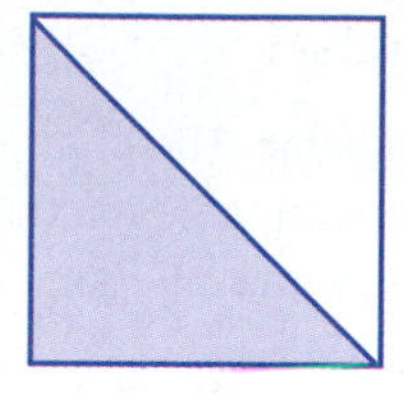

c. 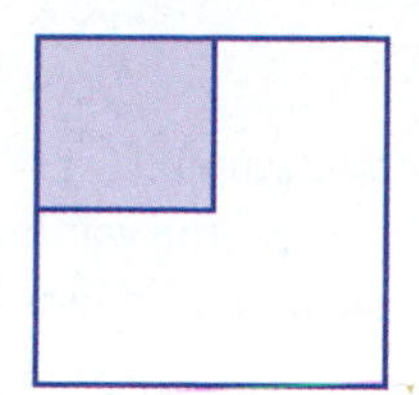

d. 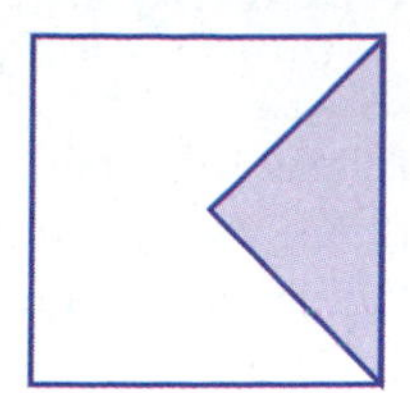

54. For each square below, what fraction of the area is given by the shaded region?

a. 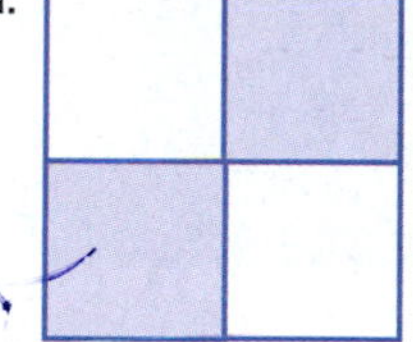

b. 

c. 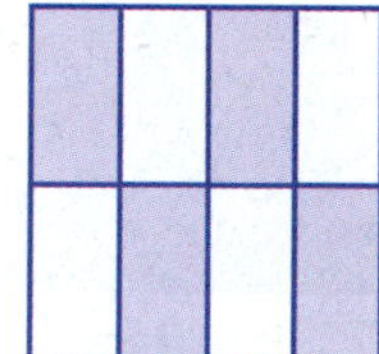

d. 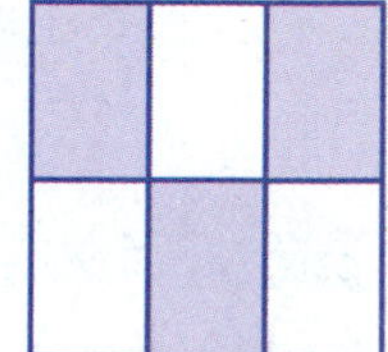

The number line below extends from 0 to 2, with the segment from 0 to 1 and the segment from 1 to 2 each divided into 8 equal parts. Locate each of the following numbers on this number line.

55. $\frac{1}{4}$ **56.** $\frac{1}{8}$ **57.** $\frac{1}{16}$ **58.** $\frac{5}{8}$ **59.** $\frac{3}{4}$

60. $\frac{15}{16}$ **61.** $\frac{3}{2}$ **62.** $\frac{5}{4}$ **63.** $\frac{31}{16}$ **64.** $\frac{15}{8}$

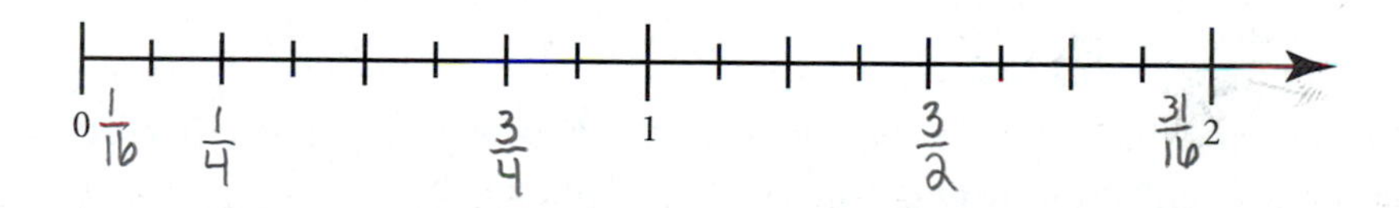

E [Example 10]

65. Write each fraction as an equivalent fraction with denominator 100. Then write them in order from smallest to largest.

$$\frac{3}{10} \quad \frac{1}{20} \quad \frac{4}{25} \quad \frac{2}{5}$$

66. Write each fraction as an equivalent fraction with denominator 30. Then write them in order from smallest to largest.

$$\frac{1}{15} \quad \frac{5}{6} \quad \frac{7}{10} \quad \frac{1}{2}$$

Applying the Concepts

67. Rainfall The chart shows the average rainfall for Death Valley in the given months. Write the rainfall for January as an equivalent fraction with denominator 12.

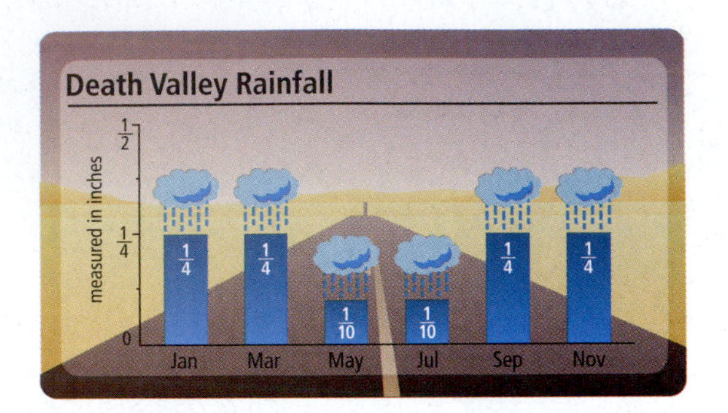

68. Rainfall The chart shows the average rainfall for Death Valley in the given months. Write the rainfall for April as an equivalent fraction with denominator 75.

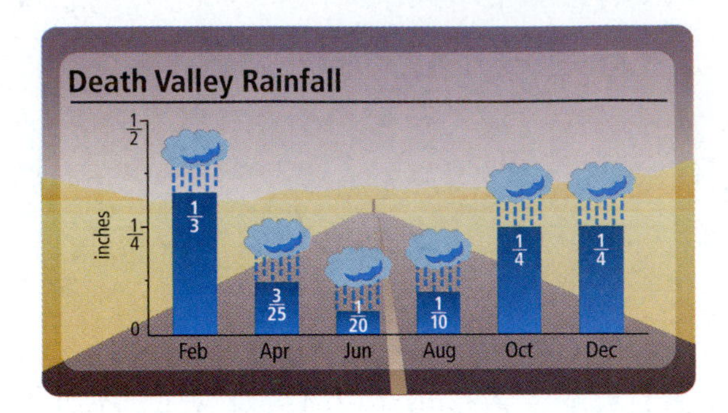

69. Sending E-mail The pie chart below shows the fraction of workers who responded to a survey about sending non-work-related e-mail from the office. Use the pie chart to fill in the table.

Workers sending personal e-mail from the office

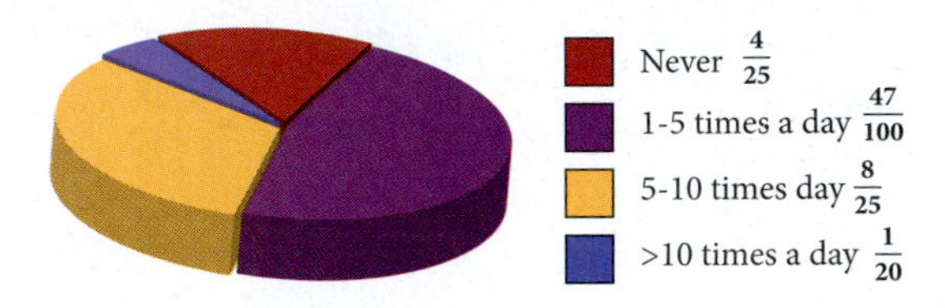

How Often Workers Send Non-Work-Related E-Mail From the Office	Fraction of Respondents Saying Yes
never	$\frac{4}{25}$
1 to 5 times a day	$\frac{47}{100}$
5 to 10 times a day	$\frac{8}{25}$
more than 10 times a day	$\frac{1}{20}$

70. Surfing the Internet The pie chart below shows the fraction of workers who responded to a survey about viewing non-work-related sites during working hours. Use the pie chart to fill in the table.

Workers surfing the net from the office

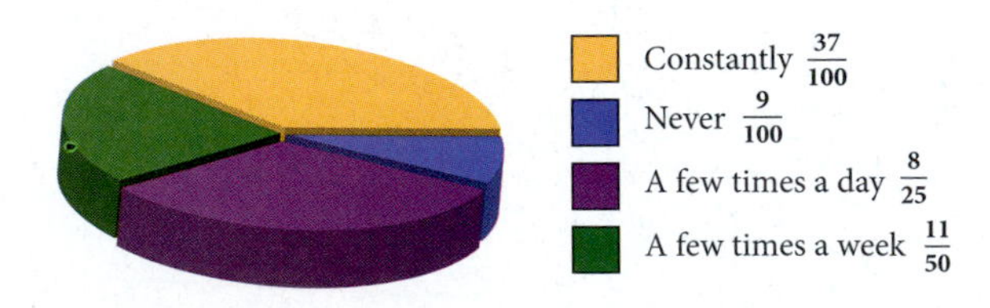

How Often Workers View Non-Work-Related Sites From the Office	Fraction of Respondents Saying Yes
never	
a few times a week	
a few times a day	
constantly	

71. Number of Children If there are 3 girls in a family with 5 children, then we say that $\frac{3}{5}$ of the children are girls. If there are 4 girls in a family with 5 children, what fraction of the children are girls?

72. Medical School If 3 out of every 7 people who apply to medical school actually get accepted, what fraction of the people who apply get accepted?

73. Downloaded Songs The new iPod™ Shuffle will hold up to 500 songs. You load 311 of your favorite tunes onto your iPod. Represent the number of songs on your iPod as a fraction of the total number of songs it can hold.

74. Cell Phones In a survey of 1,000 cell phone subscribers it was determined that 160 subscribers owned more than one cell phone and used different carriers for each phone. Represent the number of cell phone subscribers with more than one carrier as a fraction.

75. College Basketball Recently the men's basketball team at the University of Maryland won 19 of the 33 games they played. What fraction represents the number of games won?

76. Score on a Test Your math teacher grades on a point system. You take a test worth 75 points and score a 67 on the test. Represent your score as a fraction.

77. Circles A circle measures 360 degrees, which is commonly written as 360°. The shaded region of each of the circles below is given in degrees. Write a fraction that represents the area of the shaded region for each of these circles.

a.

b.

c.

d.

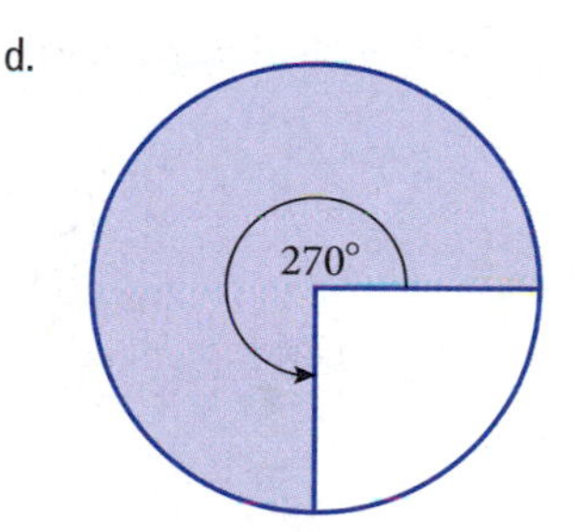

78. Carbon Dating All living things contain a small amount of carbon-14, which is radioactive and decays. The half-life of carbon-14 is 5,600 years. During the lifetime of an organism, the carbon-14 is replenished, but after its death the carbon-14 begins to disappear. By measuring the amount left, the age of the organism can be determined with surprising accuracy. The line graph below shows the fraction of carbon-14 remaining after the death of an organism. Use the line graph to complete the table.

Concentration of Carbon-14

Years Since Death of Organism	Fraction of Carbon-14 Remaining
0	1
	$\frac{1}{2}$
11,200	
16,800	
	$\frac{1}{16}$

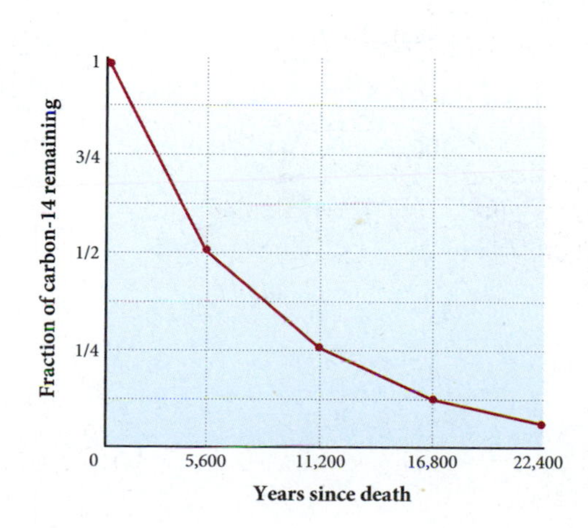

Estimating

79. Which of the following fractions is closest to the number 0?

a. $\frac{1}{2}$ **b.** $\frac{1}{3}$ **c.** $\frac{1}{4}$ **d.** $\frac{1}{5}$

80. Which of the following fractions is closest to the number 1?

a. $\frac{1}{2}$ **b.** $\frac{1}{3}$ **c.** $\frac{1}{4}$ **d.** $\frac{1}{5}$

81. Which of the following fractions is closest to the number 0?

a. $\frac{1}{8}$ **b.** $\frac{3}{8}$ **c.** $\frac{5}{8}$ **d.** $\frac{7}{8}$

82. Which of the following fractions is closest to the number 1?

a. $\frac{1}{8}$ **b.** $\frac{3}{8}$ **c.** $\frac{5}{8}$ **d.** $\frac{7}{8}$

Getting Ready for the Next Section

Multiply.

83. $2 \cdot 2 \cdot 3 \cdot 3 \cdot 3$

84. $2^2 \cdot 3^3$

85. $2^2 \cdot 3 \cdot 5$

86. $2 \cdot 3^2 \cdot 5$

Divide.

87. $12 \div 3$

88. $15 \div 3$

89. $20 \div 4$

90. $24 \div 4$

91. $42 \div 6$

92. $72 \div 8$

93. $102 \div 2$

94. $105 \div 7$

Maintaining Your Skills

The problems below review material covered previously.

Simplify.

95. $3 + 4 \cdot 5$

96. $20 - 8 \cdot 2$

97. $5 \cdot 2^4 - 3 \cdot 4^2$

98. $7 \cdot 8^2 + 2 \cdot 5^2$

99. $4 \cdot 3 + 2(5 - 3)$

100. $6 \cdot 8 + 3(4 - 1)$

101. $18 + 12 \div 4 - 3$

102. $20 + 16 \div 2 - 5$

Prime Numbers, Factors, and Reducing to Lowest Terms

3.2

Objectives

A Identify numbers as prime or composite.

B Factor a number into a product of prime factors.

C Write a fraction in lowest terms.

D Solve applications involving reducing fractions to lowest terms.

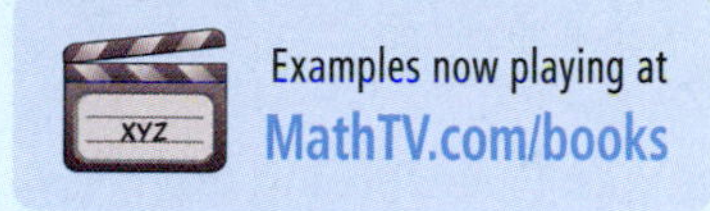

Introduction . . .

Suppose you and a friend decide to split a medium-sized pizza for lunch. When the pizza is delivered you find that it has been cut into eight equal pieces. If you eat four pieces, you have eaten $\frac{4}{8}$ of the pizza, but you also know that you have eaten $\frac{1}{2}$ of the pizza. The fraction $\frac{4}{8}$ is equivalent to the fraction $\frac{1}{2}$; that is, they both have the same value. The mathematical process we use to rewrite $\frac{4}{8}$ as $\frac{1}{2}$ is called *reducing to lowest terms.* Before we look at that process, we need to define some new terms. Here is our first one:

A Prime Numbers

Definition

A **prime number** is any whole number greater than 1 that has exactly two divisors—itself and 1. (A number is a divisor of another number if it divides it without a remainder.)

$$\text{Prime numbers} = \{2, 3, 5, 7, 11, 13, 17, 19, 23, 29, 31, 37, \ldots\}$$

The list goes on indefinitely. Each number in the list has exactly two distinct divisors—itself and 1.

Definition

Any whole number greater than 1 that is not a prime number is called a **composite number.** A composite number always has at least one divisor other than itself and 1.

Identify each of the numbers below as either a prime number or a composite number. For those that are composite, give two divisors other than the number itself or 1.

a. 43 **b.** 12

SOLUTION

a. 43 is a prime number, because the only numbers that divide it without a remainder are 43 and 1.

b. 12 is a composite number, because it can be written as $12 = 4 \cdot 3$, which means that 4 and 3 are divisors of 12. (These are not the only divisors of 12; other divisors are 1, 2, 6, and 12.)

PRACTICE PROBLEMS

1. Which of the numbers below are prime numbers, and which are composite? For those that are composite, give two divisors other than the number itself and 1.
37, 39, 51, 59

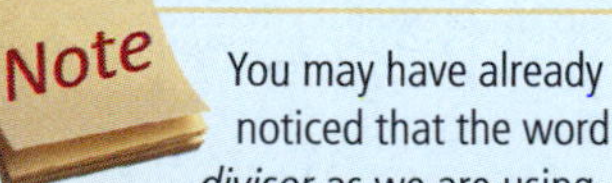

You may have already noticed that the word *divisor* as we are using it here means the same as the word *factor.* A divisor and a factor of a number are the same thing. A number can't be a divisor of another number without also being a factor of it.

B Factoring

Every composite number can be written as the product of prime factors. Let's look at the composite number 108. We know we can write 108 as $2 \cdot 54$. The number 2 is a prime number, but 54 is not prime. Because 54 can be written as $2 \cdot 27$, we have

$$\begin{aligned} 108 &= 2 \cdot 54 \\ &= 2 \cdot 2 \cdot 27 \end{aligned}$$

Answer

1. See solutions section.

Note This process works by writing the original composite number as the product of any two of its factors and then writing any factor that is not prime as the product of any two of its factors. The process is continued until all factors are prime numbers. You do not have to start with the smallest prime factor, as shown in Example 1. No matter which factors you start with you will always end up with the same prime factorization of a number.

2. Factor into a product of prime factors.
   a. 90
   b. 900

Note There are some "shortcuts" to finding the divisors of a number. For instance, if a number ends in 0 or 5, then it is divisible by 5. If a number ends in an even number (0, 2, 4, 6, or 8), then it is divisible by 2. A number is divisible by 3 if the sum of its digits is divisible by 3. For example, 921 is divisible by 3 because the sum of its digits is $9 + 2 + 1 = 12$, which is divisible by 3.

3. Which of the following fractions are in lowest terms?
   $\frac{1}{6}, \frac{2}{8}, \frac{15}{25}, \frac{9}{13}$

4. Reduce $\frac{12}{18}$ to lowest terms by dividing the numerator and the denominator by 6.

Answers
2. a. $2 \cdot 3^2 \cdot 5$ b. $2^2 \cdot 3^2 \cdot 5^2$
3. $\frac{1}{6}, \frac{9}{13}$ 4. $\frac{2}{3}$

Now the number 27 can be written as $3 \cdot 9$ or $3 \cdot 3 \cdot 3$ (because $9 = 3 \cdot 3$), so

$$108 = 2 \cdot 54$$
$$108 = 2 \cdot 2 \cdot 27$$
$$108 = 2 \cdot 2 \cdot 3 \cdot 9$$
$$108 = 2 \cdot 2 \cdot 3 \cdot 3 \cdot 3$$

This last line is the number 108 written as the product of prime factors. We can use exponents to rewrite the last line:

$$108 = 2^2 \cdot 3^3$$

EXAMPLE 2 Factor 60 into a product of prime factors.

SOLUTION We begin by writing 60 as $6 \cdot 10$ and continue factoring until all factors are prime numbers:

$$\begin{aligned} 60 &= 6 \cdot 10 \\ &= 2 \cdot 3 \cdot 2 \cdot 5 \\ &= 2^2 \cdot 3 \cdot 5 \end{aligned}$$

Notice that if we had started by writing 60 as $3 \cdot 20$, we would have achieved the same result:

$$\begin{aligned} 60 &= 3 \cdot 20 \\ &= 3 \cdot 2 \cdot 10 \\ &= 3 \cdot 2 \cdot 2 \cdot 5 \\ &= 2^2 \cdot 3 \cdot 5 \end{aligned}$$

C Reducing Fractions

We can use the method of factoring numbers into prime factors to help reduce fractions to lowest terms. Here is the definition for lowest terms.

Definition

A fraction is said to be in **lowest terms** if the numerator and the denominator have no factors in common other than the number 1.

EXAMPLE 3 The fractions $\frac{1}{2}, \frac{1}{3}, \frac{2}{3}, \frac{1}{4}, \frac{3}{4}, \frac{1}{5}, \frac{2}{5}, \frac{3}{5}$, and $\frac{4}{5}$ are all in lowest terms, because in each case the numerator and the denominator have no factors other than 1 in common. That is, in each fraction, no number other than 1 divides both the numerator and the denominator exactly (without a remainder).

EXAMPLE 4 The fraction $\frac{6}{8}$ is not written in lowest terms, because the numerator and the denominator are both divisible by 2. To write $\frac{6}{8}$ in lowest terms, we apply Property 2 from Section 3.1 and divide both the numerator and the denominator by 2:

$$\frac{6}{8} = \frac{6 \div \mathbf{2}}{8 \div \mathbf{2}} = \frac{3}{4}$$

The fraction $\frac{3}{4}$ is in lowest terms, because 3 and 4 have no factors in common except the number 1.

Reducing a fraction to lowest terms is simply a matter of dividing the numerator and the denominator by all the factors they have in common. We know from Property 2 of Section 3.1 that this will produce an equivalent fraction.

EXAMPLE 5 Reduce the fraction $\frac{12}{15}$ to lowest terms by first factoring the numerator and the denominator into prime factors and then dividing both the numerator and the denominator by the factor they have in common.

SOLUTION The numerator and the denominator factor as follows:

$$12 = 2 \cdot 2 \cdot 3 \quad \text{and} \quad 15 = 3 \cdot 5$$

The factor they have in common is 3. Property 2 tells us that we can divide both terms of a fraction by 3 to produce an equivalent fraction. So

$$\begin{aligned} \frac{12}{15} &= \frac{2 \cdot 2 \cdot 3}{3 \cdot 5} && \text{Factor the numerator and the denominator completely} \\ &= \frac{2 \cdot 2 \cdot 3 \div \mathbf{3}}{3 \cdot 5 \div \mathbf{3}} && \text{Divide by 3} \\ &= \frac{2 \cdot 2}{5} = \frac{4}{5} \end{aligned}$$

The fraction $\frac{4}{5}$ is equivalent to $\frac{12}{15}$ and is in lowest terms, because the numerator and the denominator have no factors other than 1 in common.

We can shorten the work involved in reducing fractions to lowest terms by using a slash to indicate division. For example, we can write the above problem as:

$$\frac{12}{15} = \frac{2 \cdot 2 \cdot \not{3}}{\not{3} \cdot 5} = \frac{4}{5}$$

So long as we understand that the slashes through the 3's indicate that we have divided both the numerator and the denominator by 3, we can use this notation.

D Applications

EXAMPLE 6 Laura is having a party. She puts 4 six-packs of soda in a cooler for her guests. At the end of the party she finds that only 4 sodas have been consumed. What fraction of the sodas are left? Write your answer in lowest terms.

SOLUTION She had 4 six-packs of soda, which is 4(6) = 24 sodas. Only 4 were consumed at the party, so 20 are left. The fraction of sodas left is

$$\frac{20}{24}$$

Factoring 20 and 24 completely and then dividing out both the factors they have in common gives us

$$\frac{20}{24} = \frac{\not{2} \cdot \not{2} \cdot 5}{\not{2} \cdot \not{2} \cdot 2 \cdot 3} = \frac{5}{6}$$

EXAMPLE 7 Reduce $\frac{6}{42}$ to lowest terms.

SOLUTION We begin by factoring both terms. We then divide through by any factors common to both terms:

$$\frac{6}{42} = \frac{\not{2} \cdot \not{3}}{\not{2} \cdot \not{3} \cdot 7} = \frac{1}{7}$$

5. Reduce the fraction $\frac{15}{20}$ to lowest terms by first factoring the numerator and the denominator into prime factors and then dividing out the factors they have in common.

6. Reduce to lowest terms.
   a. $\frac{30}{35}$ b. $\frac{300}{350}$

The slashes in Example 6 indicate that we have divided both the numerator and the denominator by $2 \cdot 2$, which is equal to 4. With some fractions it is apparent at the start what number divides the numerator and the denominator. For instance, you may have recognized that both 20 and 24 in Example 6 are divisible by 4. We can divide both terms by 4 without factoring first, just as we did in Section 3.1. Property 2 guarantees that dividing both terms of a fraction by 4 will produce an equivalent fraction:

$$\frac{20}{24} = \frac{20 \div \mathbf{4}}{24 \div \mathbf{4}} = \frac{5}{6}$$

7. Reduce to lowest terms.
   a. $\frac{8}{72}$ b. $\frac{16}{144}$

Answers

5. $\frac{3}{4}$ 6. Both are $\frac{6}{7}$.

We must be careful in a problem like this to remember that the slashes indicate division. They are used to indicate that we have divided both the numerator and the denominator by $2 \cdot 3 = 6$. The result of dividing the numerator 6 by $2 \cdot 3$ is 1. It is a very common mistake to call the numerator 0 instead of 1 or to leave the numerator out of the answer.

Reduce each fraction to lowest terms.

8. $\frac{5}{50}$

9. $\frac{120}{25}$

10. $\frac{54x}{90xy}$

11. $\frac{306a^2}{228a}$

EXAMPLE 8 Reduce to lowest terms:

$$\frac{4}{40} = \frac{\cancel{2} \cdot \cancel{2} \cdot 1}{\cancel{2} \cdot \cancel{2} \cdot 2 \cdot 5} = \frac{1}{10}$$

EXAMPLE 9 Reduce to lowest terms:

$$\frac{105}{30} = \frac{\cancel{3} \cdot \cancel{5} \cdot 7}{2 \cdot \cancel{3} \cdot \cancel{5}} = \frac{7}{2}$$

If a fraction contains variables (letters) in its numerator and/or denominator, we treat the variables in the same way we treat the other factors in the numerator or denominator. If a variable factor is common to the numerator and the denominator, we can reduce to lowest terms by dividing the numerator and the denominator by that variable. (Remember also that any variable that appears in a denominator is assumed to be nonzero.)

EXAMPLE 10 Reduce $\frac{36xy}{120x}$ to lowest terms.

SOLUTION First we factor 36 and 120 completely. Then we divide out all the factors that are common to the numerator and the denominator, including any variable factors.

$$\frac{36xy}{120x} = \frac{\cancel{2} \cdot \cancel{2} \cdot \cancel{3} \cdot 3 \cdot \cancel{x} \cdot y}{\cancel{2} \cdot \cancel{2} \cdot 2 \cdot \cancel{3} \cdot 5 \cdot \cancel{x}} = \frac{3y}{10}$$

EXAMPLE 11 Reduce $\frac{204a^2}{342a}$ to lowest terms.

SOLUTION We factor both the numerator and the denominator completely and then divide out any factors common to both.

$$\frac{204a^2}{342a} = \frac{\cancel{2} \cdot 2 \cdot \cancel{3} \cdot 17 \cdot \cancel{a} \cdot a}{\cancel{2} \cdot \cancel{3} \cdot 3 \cdot 19 \cdot \cancel{a}} = \frac{34a}{57}$$

Getting Ready for Class

After reading through the preceding section, respond in your own words and in complete sentences.

A. What is a prime number?

B. Why is the number 22 a composite number?

C. Factor 120 into a product of prime factors.

D. What is meant by the phrase "a fraction in lowest possible terms"?

Answers

7. Both are $\frac{1}{9}$. 8. $\frac{1}{10}$ 9. $\frac{24}{5}$

10. $\frac{3}{5y}$ 11. $\frac{51a}{38}$

Problem Set 3.2

A Identify each of the numbers below as either a prime number or a composite number. For those that are composite, give at least one divisor (factor) other than the number itself or the number 1. [Example 1]

1. 11 **2.** 23 **3.** 105 **4.** 41

5. 81 **6.** 50 **7.** 13 **8.** 219

B Factor each of the following into a product of prime factors. [Example 2]

9. 12 **10.** 8 **11.** 81 **12.** 210

13. 215 **14.** 75 **15.** 15 **16.** 42

C Reduce each fraction to lowest terms. [Examples 4, 5, 7–11]

17. $\frac{5}{10}$ **18.** $\frac{3}{6}$ **19.** $\frac{4}{6}$ **20.** $\frac{4}{10}$

21. $\frac{8x}{10x}$ **22.** $\frac{6x}{10x}$ **23.** $\frac{36}{20}$ **24.** $\frac{32}{12}$

25. $\frac{42}{66}$ **26.** $\frac{36}{60}$ **27.** $\frac{24xy}{40y}$ **28.** $\frac{50xy}{75x}$

29. $\frac{14}{98}$ **30.** $\frac{12}{84}$ **31.** $\frac{70}{90}$ **32.** $\frac{80}{90}$

33. $\frac{42x^2}{30x}$ **34.** $\frac{60x^2}{36x}$ **35.** $\frac{18a^2b}{90ab}$ **36.** $\frac{150ab^2}{210ab}$

37. $\frac{110}{70}$ **38.** $\frac{45}{75}$ **39.** $\frac{180xyz}{108xy}$ **40.** $\frac{105xyz}{30yz}$

41. $\frac{96x^2y}{108xy^2}$ **42.** $\frac{66x^2y}{84xy^2}$ **43.** $\frac{126}{165}$ **44.** $\frac{210}{462}$

45. $\frac{102a^2bc^3}{114ab^2c^2}$ **46.** $\frac{255a^3b^2c}{285ab^2c^3}$ **47.** $\frac{294}{693}$ **48.** $\frac{273}{385}$

49. Reduce each fraction to lowest terms.

a. $\frac{6}{51}$ b. $\frac{6}{52}$ c. $\frac{6}{54}$ d. $\frac{6}{56}$ e. $\frac{6}{57}$

50. Reduce each fraction to lowest terms.

a. $\frac{6}{42}$ b. $\frac{6}{44}$ c. $\frac{6}{45}$ d. $\frac{6}{46}$ e. $\frac{6}{48}$

51. Reduce each fraction to lowest terms.

a. $\frac{2}{90}$ b. $\frac{3}{90}$ c. $\frac{5}{90}$ d. $\frac{6}{90}$ e. $\frac{9}{90}$

52. Reduce each fraction to lowest terms.

a. $\frac{3}{105}$ b. $\frac{5}{105}$ c. $\frac{7}{105}$ d. $\frac{15}{105}$ e. $\frac{21}{105}$

53. The answer to each problem below is wrong. Give the correct answer.

a. $\frac{5}{15} = \frac{\cancel{5}}{3 \cdot \cancel{5}} = \frac{0}{3}$ b. $\frac{5}{6} = \frac{3 + \cancel{2}}{4 + \cancel{2}} = \frac{3}{4}$ c. $\frac{6}{30} = \frac{\cancel{2} \cdot \cancel{3}}{\cancel{2} \cdot \cancel{3} \cdot 5} = 5$

54. The answer to each problem below is wrong. Give the correct answer.

a. $\frac{10}{20} = \frac{7 + \cancel{3}}{17 + \cancel{3}} = \frac{7}{17}$ b. $\frac{9}{36} = \frac{\cancel{3} \cdot \cancel{3}}{2 \cdot 2 \cdot \cancel{3} \cdot \cancel{3}} = \frac{0}{4}$ c. $\frac{4}{12} = \frac{\cancel{2} \cdot \cancel{2}}{\cancel{2} \cdot \cancel{2} \cdot 3} = 3$

55. Which of the fractions $\frac{6}{8}$, $\frac{15}{20}$, $\frac{9}{16}$, and $\frac{21}{28}$ does not reduce to $\frac{3}{4}$?

56. Which of the fractions $\frac{4}{9}$, $\frac{10}{15}$, $\frac{8}{12}$, and $\frac{6}{12}$ do not reduce to $\frac{2}{3}$?

The number line below extends from 0 to 2, with the segment from 0 to 1 and the segment from 1 to 2 each divided into 8 equal parts. Locate each of the following numbers on this number line.

57. $\frac{1}{2}$, $\frac{2}{4}$, $\frac{4}{8}$, and $\frac{8}{16}$

58. $\frac{3}{2}$, $\frac{6}{4}$, $\frac{12}{8}$, and $\frac{24}{16}$

59. $\frac{5}{4}$, $\frac{10}{8}$, and $\frac{20}{16}$

60. $\frac{1}{4}$, $\frac{2}{8}$, and $\frac{4}{16}$

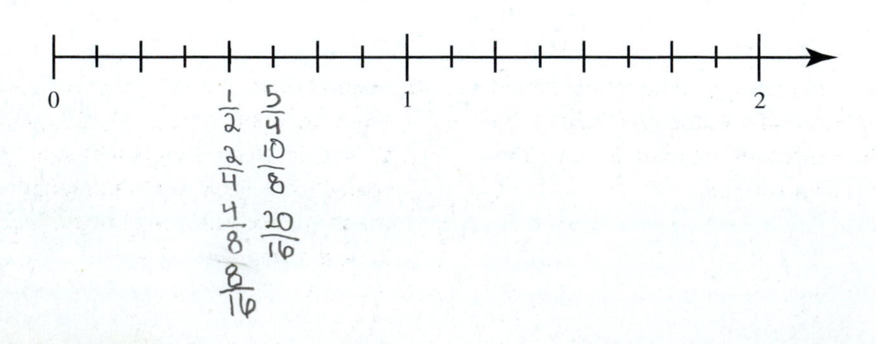

D Applying the Concepts [Example 6]

61. **Tower Heights** The Eiffel Tower is 1,060 feet tall and the Stratosphere Tower in Las Vegas is 1,150 feet tall. Write the height of the Eiffel tower over the height of the Stratosphere Tower and then reduce to lowest terms.

62. **Car Insurance** The chart below shows the annual cost of auto insurance for some major U.S. cities. Write the price of auto insurance in Los Angeles over the price of insurance in Detroit and then reduce to lowest terms.

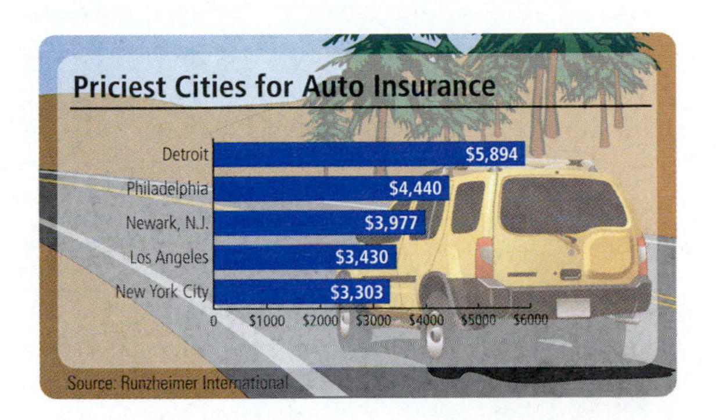

63. **Hours and Minutes** There are 60 minutes in 1 hour. What fraction of an hour is 20 minutes? Write your answer in lowest terms.

64. **Final Exam** Suppose 33 people took the final exam in a math class. If 11 people got an A on the final exam, what fraction of the students did not get an A on the exam? Write your answer in lowest terms.

65. **Driving Distractions** Many of us focus our attention on things other than driving when we are behind the wheel of our car. In a survey of 150 drivers, it was noted that 48 drivers spend time reading or writing while they are driving. Represent the number of drivers who spend time reading or writing while driving as a fraction in lowest terms.

66. **Watching Television** According to the U.S. Census Bureau, it is estimated that the average person watches 4 hours of TV each day. Represent the number of hours of TV watched each day as a fraction in lowest terms.

67. **Hurricanes** Over a recent five-year period, 9 hurricanes struck the mainland of the United States. Three of these hurricanes were classified as a category 3, 4 or 5. Represent the number of major hurricanes that struck the mainland U.S. over this time period as a fraction in lowest terms.

68. **Gasoline Tax** Suppose a gallon of regular gas costs \$3.99, and 54 cents of this goes to pay state gas taxes. What fractional part of the cost of a gallon of gas goes to state taxes? Write your answer in lowest terms.

69. **On-Time Record** A random check of Delta airline flights for the past month showed that of the 350 flights scheduled 185 left on time. Represent the number of on time flights as a fraction in lowest terms.

70. **Internet Users** Based on the most recent data available, there are approximately 1,320,000,000 Internet users in the world. North America makes up about 240,000,000 of this total. Represent the number of Internet users in North America as a fraction of the total expressed in lowest terms.

Nutrition The nutrition labels below are from two different granola bars.

71. What fraction of the calories in Bar 1 comes from fat?

72. What fraction of the calories in Bar 2 comes from fat?

73. For Bar 1, what fraction of the total fat is from saturated fat?

74. What fraction of the total carbohydrates in Bar 1 is from sugar?

GRANOLA BAR 1

Nutrition Facts

Serving Size 2 bars (47g)
Servings Per Container: 6

Amount Per Serving	
Calories 210	Calories from fat 70
	% Daily Value*
Total Fat 8g	**12%**
Saturated Fat 1g	**5%**
Cholesterol 0mg	**0%**
Sodium 150mg	**6%**
Total Carbohydrate 32g	**11%**
Fiber 2g	**10%**
Sugars 12g	
Protein 4g	

*Percent Daily Values are based on a 2,000 calorie diet. Your daily values may be higher or lower depending on your calorie needs.

GRANOLA BAR 2

Nutrition Facts

Serving Size 1 bar (21g)
Servings Per Container: 8

Amount Per Serving	
Calories 80	Calories from fat 15
	% Daily Value*
Total Fat 1.5g	**2%**
Saturated Fat 0g	**0%**
Cholesterol 0mg	**0%**
Sodium 60mg	**3%**
Total Carbohydrate 16g	**5%**
Fiber 1g	**4%**
Sugars 5g	
Protein 2g	

*Percent Daily Values are based on a 2,000 calorie diet. Your daily values may be higher or lower depending on your calorie needs.

Getting Ready for the Next Section

Multiply.

75. $1 \cdot 3 \cdot 1$

76. $2 \cdot 4 \cdot 5$

77. $3 \cdot 5 \cdot 3$

78. $1 \cdot 4 \cdot 1$

79. $5 \cdot 5 \cdot 1$

80. $6 \cdot 6 \cdot 2$

Factor into prime factors.

81. 60

82. 72

83. $15 \cdot 4$

84. $8 \cdot 9$

Expand and multiply.

85. 3^2

86. 4^2

87. 5^2

88. 6^2

Maintaining Your Skills

Simplify.

89. $16 - 8 + 4$

90. $16 - 4 + 8$

91. $24 - 14 + 8$

92. $24 - 16 + 6$

93. $36 - 6 + 12$

94. $36 - 9 + 20$

95. $48 - 12 + 17$

96. $48 - 13 + 15$

Multiplication with Fractions, and the Area of a Triangle

3.3

Objectives

A Multiply fractions.

B Find the area of a triangle.

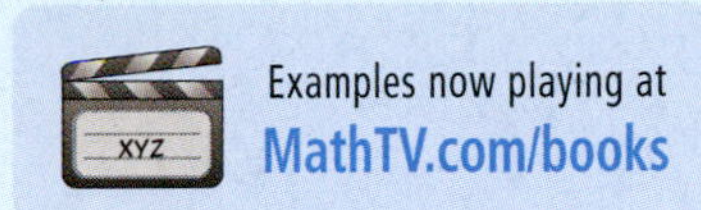

Introduction . . .

A recipe calls for $\frac{3}{4}$ cup of flour. If you are making only $\frac{1}{2}$ the recipe, how much flour do you use? This question can be answered by multiplying $\frac{1}{2}$ and $\frac{3}{4}$. Here is the problem written in symbols:

$$\frac{1}{2} \cdot \frac{3}{4} = \frac{3}{8}$$

As you can see from this example, to multiply two fractions, we multiply the numerators and then multiply the denominators. We begin this section with the rule for multiplication of fractions.

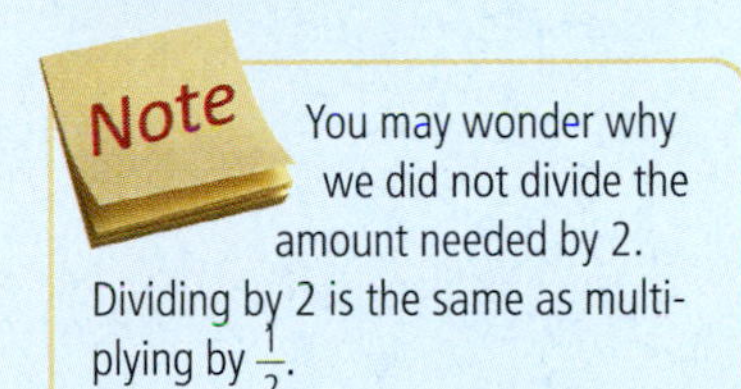

Note You may wonder why we did not divide the amount needed by 2. Dividing by 2 is the same as multiplying by $\frac{1}{2}$.

A Multiplying Fractions

Rule

The product of two fractions is the fraction whose numerator is the product of the two numerators and whose denominator is the product of the two denominators. We can write this rule in symbols as follows:

If a, b, c, and d represent any numbers and b and d are not zero, then

$$\frac{a}{b} \cdot \frac{c}{d} = \frac{a \cdot c}{b \cdot d}$$

EXAMPLE 1 Multiply: $\frac{3}{5} \cdot \frac{2}{7}$

SOLUTION Using our rule for multiplication, we multiply the numerators and multiply the denominators:

$$\frac{3}{5} \cdot \frac{2}{7} = \frac{3 \cdot 2}{5 \cdot 7} = \frac{6}{35}$$

The product of $\frac{3}{5}$ and $\frac{2}{7}$ is the fraction $\frac{6}{35}$. The numerator 6 is the product of 3 and 2, and the denominator 35 is the product of 5 and 7.

EXAMPLE 2 Multiply: $\frac{3}{8} \cdot 5$

SOLUTION The number 5 can be written as $\frac{5}{1}$. That is, 5 can be considered a fraction with numerator 5 and denominator 1. Writing 5 this way enables us to apply the rule for multiplying fractions.

$$\begin{aligned} \frac{3}{8} \cdot 5 &= \frac{3}{8} \cdot \frac{5}{1} \\ &= \frac{3 \cdot 5}{8 \cdot 1} \\ &= \frac{15}{8} \end{aligned}$$

PRACTICE PROBLEMS

1. Multiply: $\frac{2}{3} \cdot \frac{5}{9}$

2. Multiply: $\frac{2}{5} \cdot 7$

Answers

1. $\frac{10}{27}$ 2. $\frac{14}{5}$

3. Multiply: $\frac{1}{3}\left(\frac{4}{5} \cdot \frac{1}{3}\right)$

EXAMPLE 3 Multiply: $\frac{1}{2}\left(\frac{3}{4} \cdot \frac{1}{5}\right)$

SOLUTION We find the product inside the parentheses first and then multiply the result by $\frac{1}{2}$:

$$\begin{aligned}\frac{1}{2}\left(\frac{3}{4} \cdot \frac{1}{5}\right) &= \frac{1}{2}\left(\frac{3 \cdot 1}{4 \cdot 5}\right) \\ &= \frac{1}{2}\left(\frac{3}{20}\right) \\ &= \frac{1 \cdot 3}{2 \cdot 20} = \frac{3}{40}\end{aligned}$$

The properties of multiplication that we developed in Chapter 1 for whole numbers apply to fractions as well. That is, if a, b, and c are fractions, then

$a \cdot b = b \cdot a$ — Multiplication with fractions is commutative

$a \cdot (b \cdot c) = (a \cdot b) \cdot c$ — Multiplication with fractions is associative

To demonstrate the associative property for fractions, let's do Example 3 again, but this time we will apply the associative property first:

$$\begin{aligned}\frac{1}{2}\left(\frac{3}{4} \cdot \frac{1}{5}\right) &= \left(\frac{1}{2} \cdot \frac{3}{4}\right) \cdot \frac{1}{5} && \text{Associative property} \\ &= \left(\frac{1 \cdot 3}{2 \cdot 4}\right) \cdot \frac{1}{5} \\ &= \left(\frac{3}{8}\right) \cdot \frac{1}{5} \\ &= \frac{3 \cdot 1}{8 \cdot 5} = \frac{3}{40}\end{aligned}$$

The result is identical to that of Example 3.

Here is another example that involves the associative property. Problems like this will be useful when we solve equations.

4. Multiply: $\frac{1}{4}(4y)$

EXAMPLE 4 Multiply: $\frac{1}{3}(3x)$

SOLUTION Remember that $3x$ means 3 times x. The 3 and the x are not combined. If we apply the associative property so that we can group the $\frac{1}{3}$ and 3 together, we will be able to multiply them:

$$\begin{aligned}\frac{1}{3}(3x) &= \left(\frac{1}{3} \cdot 3\right)x && \text{Associative property} \\ &= 1x && \text{The product of } \tfrac{1}{3} \text{ and 3 is 1} \\ &= x && \text{The product of 1 and } x \text{ is } x\end{aligned}$$

5. Multiply: $\frac{1}{2}(2x - 4)$

EXAMPLE 5 Multiply: $\frac{1}{3}(3x - 12)$

SOLUTION

$$\begin{aligned}\frac{1}{3}(3x - 12) &= \frac{1}{3}(3x) - \frac{1}{3}(12) && \text{Distributive property} \\ &= 1x - \frac{12}{3} && \text{Simplify} \\ &= x - 4 && \text{Divide}\end{aligned}$$

The answers to all the examples so far in this section have been in lowest terms. Let's see what happens when we multiply two fractions to get a product that is not in lowest terms.

Answers

3. $\frac{4}{45}$ 4. y 5. $x - 2$

EXAMPLE 6 Multiply: $\frac{15}{8} \cdot \frac{4}{9}$

SOLUTION Multiplying the numerators and multiplying the denominators, we have

$$\frac{15}{8} \cdot \frac{4}{9} = \frac{15 \cdot 4}{8 \cdot 9} = \frac{60}{72}$$

The product is $\frac{60}{72}$, which can be reduced to lowest terms by factoring 60 and 72 and then dividing out any factors they have in common:

$$\frac{60}{72} = \frac{\cancel{2} \cdot \cancel{2} \cdot \cancel{3} \cdot 5}{\cancel{2} \cdot \cancel{2} \cdot 2 \cdot \cancel{3} \cdot 3} = \frac{5}{6}$$

We can actually save ourselves some time by factoring before we multiply. Here's how it is done:

$$\frac{15}{8} \cdot \frac{4}{9} = \frac{15 \cdot 4}{8 \cdot 9} = \frac{(3 \cdot 5) \cdot (2 \cdot 2)}{(2 \cdot 2 \cdot 2) \cdot (3 \cdot 3)} = \frac{\cancel{3} \cdot 5 \cdot \cancel{2} \cdot \cancel{2}}{\cancel{2} \cdot \cancel{2} \cdot 2 \cdot \cancel{3} \cdot 3} = \frac{5}{6}$$

The result is the same in both cases. Reducing to lowest terms before we actually multiply takes less time.

Here are some additional examples.

EXAMPLE 7 Multiply: $-\frac{9}{2}\left(-\frac{8}{18}\right)$

SOLUTION

$$-\frac{9}{2}\left(-\frac{8}{18}\right) = \frac{9 \cdot 8}{2 \cdot 18} = \frac{(3 \cdot 3) \cdot (2 \cdot 2 \cdot 2)}{2 \cdot (2 \cdot 3 \cdot 3)} = \frac{\cancel{3} \cdot \cancel{3} \cdot \cancel{2} \cdot \cancel{2} \cdot 2}{\cancel{2} \cdot \cancel{2} \cdot \cancel{3} \cdot \cancel{3}} = \frac{2}{1} = 2$$

Like signs give a positive product

EXAMPLE 8 Multiply: $\frac{x^2y}{z} \cdot \frac{z^3}{xy}$

SOLUTION

$$\frac{x^2y}{z} \cdot \frac{z^3}{xy} = \frac{x^2y \cdot z^3}{z \cdot xy} = \frac{(x \cdot x \cdot y)(z \cdot z \cdot z)}{z \cdot x \cdot y} = \frac{\cancel{x} \cdot x \cdot \cancel{y} \cdot \cancel{z} \cdot z \cdot z}{\cancel{z} \cdot \cancel{x} \cdot \cancel{y}} = \frac{x \cdot z \cdot z}{1} = xz^2$$

6. Multiply.
   a. $\frac{12}{25} \cdot \frac{5}{6}$
   b. $\frac{12}{25} \cdot \frac{50}{60}$

7. Multiply.
   a. $\frac{8}{3} \cdot \frac{9}{24}$
   b. $\frac{8}{30} \cdot \frac{90}{24}$

Although $\frac{2}{1}$ is in lowest terms, it is still simpler to write the answer as just 2. We will always do this when the denominator is the number 1.

8. Find the product: $\frac{yz^2}{x} \cdot \frac{x^3}{yz}$

Remember, we are assuming that any variables that appear in a denominator are not 0.

Answers

6. Both are $\frac{2}{5}$ 7. Both are 1 8. x^2z

9. Multiply: $\frac{3}{4} \cdot \frac{8}{3} \cdot \frac{1}{6}$

EXAMPLE 9 Multiply: $\frac{2}{3} \cdot \frac{6}{5} \cdot \frac{5}{8}$

SOLUTION

$$\frac{2}{3} \cdot \frac{6}{5} \cdot \frac{5}{8} = \frac{2 \cdot 6 \cdot 5}{3 \cdot 5 \cdot 8}$$

$$= \frac{2 \cdot (2 \cdot 3) \cdot 5}{3 \cdot 5 \cdot (2 \cdot 2 \cdot 2)}$$

$$= \frac{\cancel{2} \cdot \cancel{2} \cdot \cancel{3} \cdot \cancel{5}}{\cancel{3} \cdot \cancel{5} \cdot \cancel{2} \cdot \cancel{2} \cdot 2}$$

$$= \frac{1}{2}$$

Apply the definition of exponents, and then multiply.

10. $\left(\frac{2}{3}\right)^2$

In Chapter 1 we did some work with exponents. We can extend our work with exponents to include fractions, as the following examples indicate.

EXAMPLE 10 Expand and multiply: $\left(-\frac{3}{4}\right)^2$

SOLUTION

$$\left(-\frac{3}{4}\right)^2 = \left(-\frac{3}{4}\right)\left(-\frac{3}{4}\right)$$

$$= \frac{3 \cdot 3}{4 \cdot 4} \qquad \text{Like signs give a positive product}$$

$$= \frac{9}{16}$$

11. a. $\left(\frac{3}{4}\right)^2 \cdot \frac{1}{2}$

b. $\left(\frac{2}{3}\right)^3 \cdot \frac{9}{8}$

EXAMPLE 11 Expand and multiply: $\left(\frac{5}{6}\right)^2$

SOLUTION

$$\left(\frac{5}{6}\right)^2 \cdot \frac{1}{2} = \frac{5}{6} \cdot \frac{5}{6} \cdot \frac{1}{2}$$

$$= \frac{5 \cdot 5 \cdot 1}{6 \cdot 6 \cdot 2}$$

$$= \frac{25}{72}$$

The word *of* used in connection with fractions indicates multiplication. If we want to find $\frac{1}{2}$ of $\frac{2}{3}$, then what we do is multiply $\frac{1}{2}$ and $\frac{2}{3}$.

12. a. Find $\frac{2}{3}$ of $\frac{1}{2}$.

b. Find $\frac{3}{5}$ of 15.

EXAMPLE 12 Find $\frac{1}{2}$ of $\frac{2}{3}$.

SOLUTION Knowing the word *of,* as used here, indicates multiplication, we have

$$\frac{1}{2} \text{ of } \frac{2}{3} = \frac{1}{2} \cdot \frac{2}{3}$$

$$= \frac{1 \cdot \cancel{2}}{\cancel{2} \cdot 3} = \frac{1}{3}$$

This seems to make sense. Logically, $\frac{1}{2}$ of $\frac{2}{3}$ should be $\frac{1}{3}$, as Figure 1 shows.

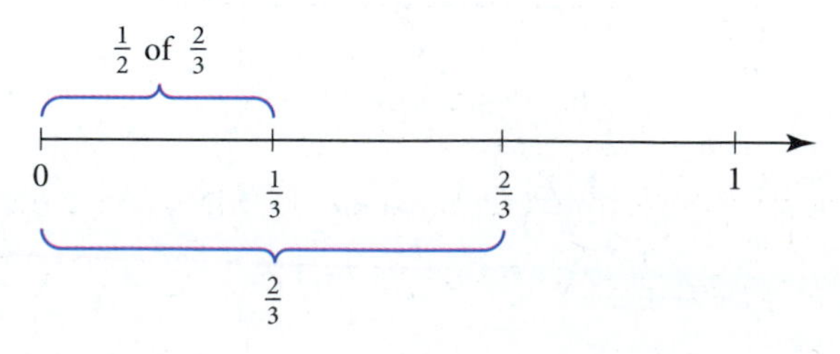

FIGURE 1

Answers

9. $\frac{1}{3}$ 10. $\frac{4}{9}$ 11. a. $\frac{9}{32}$ b. $\frac{1}{3}$
12. a. $\frac{1}{3}$ b. 9

EXAMPLE 13 What is $\frac{3}{4}$ of 12?

SOLUTION Again, *of* means multiply.

$$\begin{aligned}\frac{3}{4} \text{ of } 12 &= \frac{3}{4}(12) \\ &= \frac{3}{4}\left(\frac{12}{1}\right) \\ &= \frac{3 \cdot 12}{4 \cdot 1} \\ &= \frac{3 \cdot \cancel{2} \cdot \cancel{2} \cdot 3}{\cancel{2} \cdot \cancel{2} \cdot 1} \\ &= \frac{9}{1} = 9\end{aligned}$$

B The Area of a Triangle

FACTS FROM GEOMETRY The Area of a Triangle

The formula for the area of a triangle is one application of multiplication with fractions. Figure 2 shows a triangle with base b and height h. Below the triangle is the formula for its area. As you can see, it is a product containing the fraction $\frac{1}{2}$.

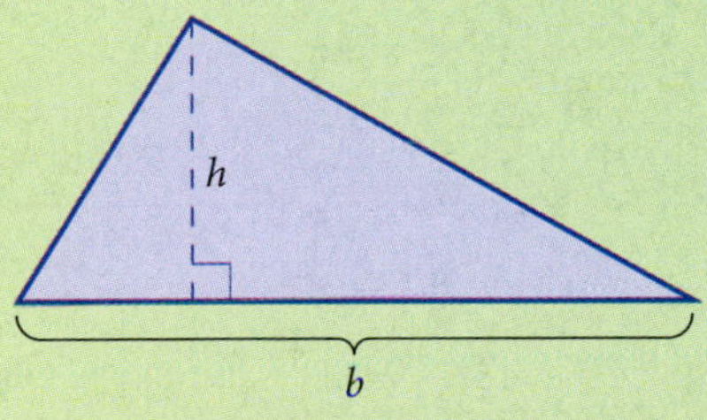

$$\text{Area} = \frac{1}{2}\,(\text{base})(\text{height})$$

$$A = \frac{1}{2}\,bh$$

FIGURE 2 The area of a triangle

EXAMPLE 14 Find the area of the triangle in Figure 3.

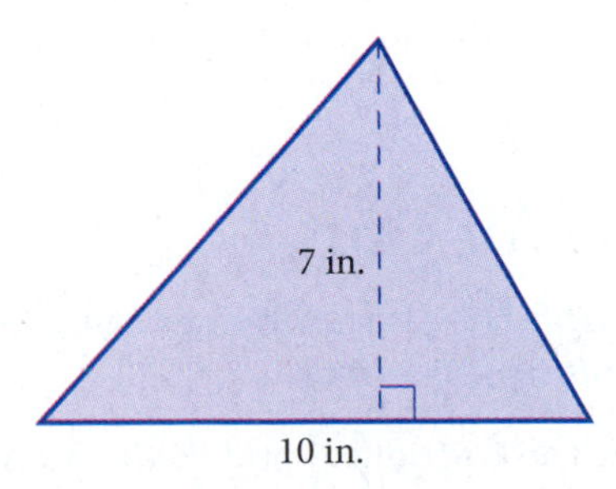

FIGURE 3 A triangle with base 10 inches and height 7 inches

SOLUTION Applying the formula for the area of a triangle, we have

$$A = \frac{1}{2}bh = \frac{1}{2} \cdot 10 \cdot 7 = 5 \cdot 7 = 35 \text{ in}^2$$

13. a. What is $\frac{2}{3}$ of 12?

b. What is $\frac{2}{3}$ of 120?

As you become familiar with multiplying fractions, you may notice shortcuts that reduce the number of steps in the problems. It's okay to use these shortcuts if you understand why they work and are consistently getting correct answers. If you are using shortcuts and not consistently getting correct answers, then go back to showing all the work until you completely understand the process.

14. Find the area of the triangle below.

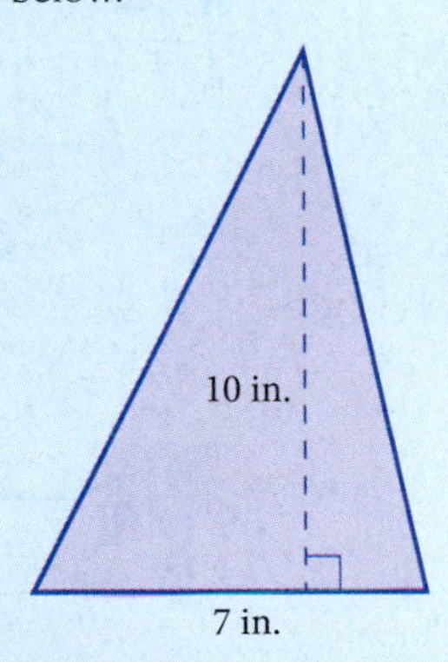

How did we get in^2 as the final units in Example 14? In this problem

$$\begin{aligned}A &= \frac{1}{2}bh \\ &= \frac{1}{2} \cdot 10 \textbf{ inches} \cdot 7 \textbf{ inches} \\ &= 5 \textbf{ in.} \cdot 7 \textbf{ in.} = 35 \textbf{ in}^2\end{aligned}$$

Answers

13. a. 8 **b.** 80 **14.** 35 in^2

15. Find the total area enclosed by the figure.

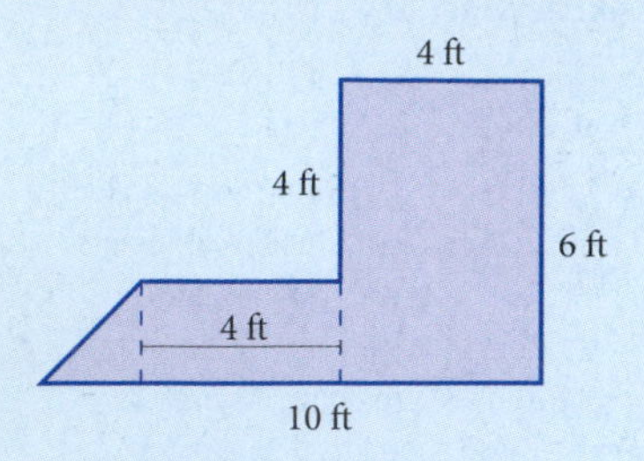

EXAMPLE 15 Find the area of the figure in Figure 4.

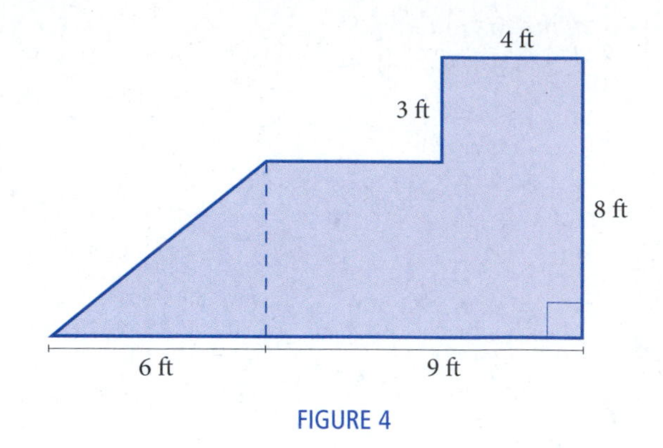

FIGURE 4

SOLUTION We divide the figure into three parts and then find the area of each part (see Figure 5). The area of the whole figure is the sum of the areas of its parts.

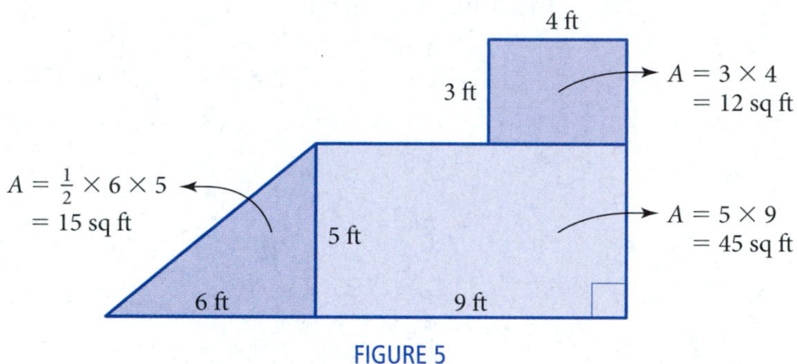

FIGURE 5

$$\begin{aligned} \text{Total area} &= 12 + 45 + 15 \\ &= 72 \text{ sq ft} \end{aligned}$$

Note This is just a reminder about unit notation. In Example 14 we wrote our final units as **in²** but could have just as easily written them as **sq in**. In Example 15 we wrote our final units as **sq ft** but could have just as easily written them as **ft²**.

STUDY SKILLS
Be Resilient

Don't let setbacks keep you from your goals. You want to put yourself on the road to becoming someone who can succeed in this class or any class in college. Failing a test or quiz or having a difficult time on some topics is normal. No one goes through college without some setbacks. A low grade on a test or quiz is simply a signal that some reevaluation of your study habits needs to take place.

Getting Ready for Class

After reading through the preceding section, respond in your own words and in complete sentences.

1. When we multiply the fractions $\frac{3}{5}$ and $\frac{2}{7}$, the numerator in the answer will be what number?
2. When we ask for $\frac{1}{2}$ of $\frac{2}{3}$, are we asking for an addition problem or a multiplication problem?
3. True or false? Reducing to lowest terms before you multiply two fractions will give the same answer as if you were to reduce after you multiply.
4. Write the formula for the area of a triangle with base x and height y.

Answer

15. 34 ft²

Problem Set 3.3

A Find each of the following products. (Multiply.) [Examples 1–4, 6–9]

1. $\frac{2}{3} \cdot \frac{4}{5}$ **2.** $\frac{5}{6} \cdot \frac{7}{4}$ **3.** $\frac{1}{2} \cdot \frac{7}{4}$ **4.** $\frac{3}{5} \cdot \frac{4}{7}$ **5.** $-\frac{5}{3} \cdot \frac{3}{5}$ **6.** $-\frac{4}{7} \cdot \frac{7}{4}$

7. $\frac{3}{4} \cdot 9$ **8.** $\frac{2}{3} \cdot 5$ **9.** $\frac{1}{x} \cdot 3$ **10.** $\frac{1}{y} \cdot 8$ **11.** $-\frac{6}{7}\left(-\frac{7}{6}\right)$ **12.** $\frac{2}{9}\left(\frac{9}{2}\right)$

13. $\frac{1}{2} \cdot \frac{1}{3} \cdot \frac{1}{4}$ **14.** $\frac{2}{3} \cdot \frac{4}{5} \cdot \frac{1}{3}$ **15.** $\frac{2}{5} \cdot \frac{3}{5} \cdot \frac{4}{5}$ **16.** $\frac{1}{4} \cdot \frac{3}{4} \cdot \frac{3}{4}$ **17.** $\frac{x}{y} \cdot \frac{y}{z} \cdot \frac{z}{x}$ **18.** $\frac{y}{x} \cdot \frac{x}{z} \cdot \frac{z}{y}$

A Complete the following tables.

19.

First Number x	Second Number y	Their Product xy
$\frac{1}{2}$	$\frac{2}{3}$	
$\frac{2}{3}$	$\frac{3}{4}$	
$\frac{3}{4}$	$\frac{4}{5}$	
$\frac{5}{a}$	$\frac{a}{6}$	

20.

First Number x	Second Number y	Their Product xy
12	$\frac{1}{2}$	
12	$\frac{1}{3}$	
12	$\frac{1}{4}$	
12	$\frac{1}{6}$	

21.

First Number x	Second Number y	Their Product xy
$\frac{1}{2}$	30	
$\frac{1}{5}$	30	
$\frac{1}{6}$	30	
$\frac{1}{15}$	30	

22.

First Number x	Second Number y	Their Product xy
$\frac{1}{3}$	$\frac{3}{5}$	
$\frac{3}{5}$	$\frac{5}{7}$	
$\frac{5}{7}$	$\frac{7}{9}$	
$\frac{7}{b}$	$\frac{b}{11}$	

A Multiply each of the following. Be sure all answers are written in lowest terms. [Examples 1–4, 6–9]

23. $\frac{9}{20} \cdot \frac{4}{3}$

24. $\frac{135}{16} \cdot \frac{2}{45}$

25. $\frac{3}{4} \cdot 12$

26. $\frac{3}{4} \cdot 20$

27. $-\frac{1}{3}(-3)$

28. $-\frac{1}{5}(-5)$

29. $\frac{2}{5} \cdot 20$

30. $\frac{3}{5} \cdot 15$

31. $\frac{72}{35} \cdot \frac{55}{108} \cdot \frac{7}{110}$

32. $\frac{32}{27} \cdot \frac{72}{49} \cdot \frac{1}{40}$

33. $\frac{a^2b}{c} \cdot \frac{c^3}{ab^2}$

34. $\frac{ab^2}{c} \cdot \frac{c^2}{a^2b}$

Use the associative property to rewrite each of the following expressions, and then simplify as much as possible.

35. $\frac{1}{2}(2x)$

36. $\frac{1}{4}(4x)$

37. $\frac{1}{5}(5y)$

38. $\frac{1}{6}(6y)$

39. $\frac{1}{8}(8a)$

40. $\frac{1}{3}(3a)$

A Expand and simplify each of the following. [Examples 10, 11]

41. $\left(\frac{2}{3}\right)^2$

42. $\left(\frac{3}{5}\right)^2$

43. $\left(-\frac{3}{4}\right)^2$

44. $\left(-\frac{2}{7}\right)^2$

45. $\left(-\frac{1}{2}\right)^2$

46. $\left(-\frac{1}{3}\right)^2$

47. $\left(-\frac{2}{3}\right)^3$

48. $\left(-\frac{3}{5}\right)^3$

49. $\left(\frac{3}{4}\right)^2 \cdot \frac{8}{9}$

50. $\left(\frac{5}{6}\right)^2 \cdot \frac{12}{15}$

51. $\left(\frac{1}{2}\right)^2\left(\frac{3}{5}\right)^2$

52. $\left(\frac{3}{8}\right)^2\left(\frac{4}{3}\right)^2$

53. $\left(\frac{1}{2}\right)^2 \cdot 8 + \left(\frac{1}{3}\right)^2 \cdot 9$

54. $\left(\frac{2}{3}\right)^2 \cdot 9 + \left(\frac{1}{2}\right)^2 \cdot 4$

A Apply the distributive property, then simplify. [Example 5]

55. $\frac{1}{2}(12x + 6)$

56. $\frac{3}{5}(15x - 10)$

57. $\frac{2}{3}(-3x - 6)$

58. $\frac{3}{4}(-4x - 12)$

59. $3\left(\frac{5}{6}a + \frac{4}{9}\right)$

60. $2\left(\frac{3}{4}a - \frac{5}{6}\right)$

61. $-3\left(\frac{2}{3}y + \frac{5}{6}\right)$

62. $-4\left(\frac{5}{12}y + \frac{3}{8}\right)$

63. $\frac{4}{5}\left(\frac{5}{6}x - 10\right)$

64. $\frac{3}{7}\left(\frac{7}{9}x - 21\right)$

A [Examples 12, 13]

65. Find $\frac{3}{8}$ of 64.

66. Find $\frac{2}{3}$ of 18.

67. What is $\frac{1}{3}$ of the sum of 8 and 4?

68. What is $\frac{3}{5}$ of the sum of 8 and 7?

69. Find $\frac{1}{2}$ of $\frac{3}{4}$ of 24.

70. Find $\frac{3}{5}$ of $\frac{1}{3}$ of 15.

Find the mistakes in Problems 71 and 72. Correct the right-hand side of each one.

71. $\frac{1}{2} \cdot \frac{3}{5} = \frac{4}{10}$

72. $\frac{2}{7} \cdot \frac{3}{5} = \frac{5}{35}$

73. a. Complete the following table.

Number x	Square x^2
1	1
2	4
3	9
4	16
5	25
6	36
7	49
8	64

b. Using the results of part a, fill in the blank in the following statement:

For numbers larger than 1, the square of the number is _____ than the number.

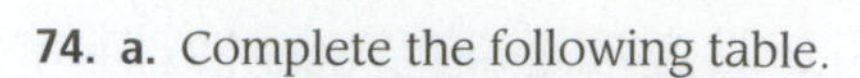

74. a. Complete the following table.

Number x	Square x^2
$\frac{1}{2}$	
$\frac{1}{3}$	
$\frac{1}{4}$	
$\frac{1}{5}$	
$\frac{1}{6}$	
$\frac{1}{7}$	
$\frac{1}{8}$	

b. Using the results of part a, fill in the blank in the following statement:

For numbers between 0 and 1, the square of the number is _____ than the number.

B [Examples 14, 15]

75. Find the area of the triangle with base 19 inches and height 14 inches.

76. Find the area of the triangle with base 13 inches and height 8 inches.

77. The base of a triangle is $\frac{4}{3}$ feet and the height is $\frac{2}{3}$ feet. Find the area.

78. The base of a triangle is $\frac{8}{7}$ feet and the height is $\frac{14}{5}$ feet. Find the area.

Find the area of each figure.

79.

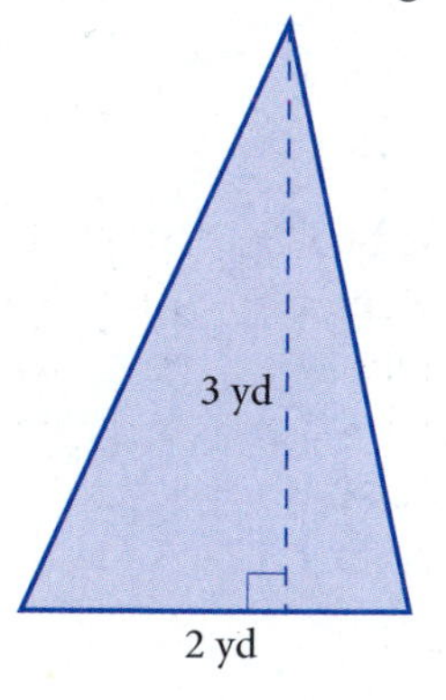

80.

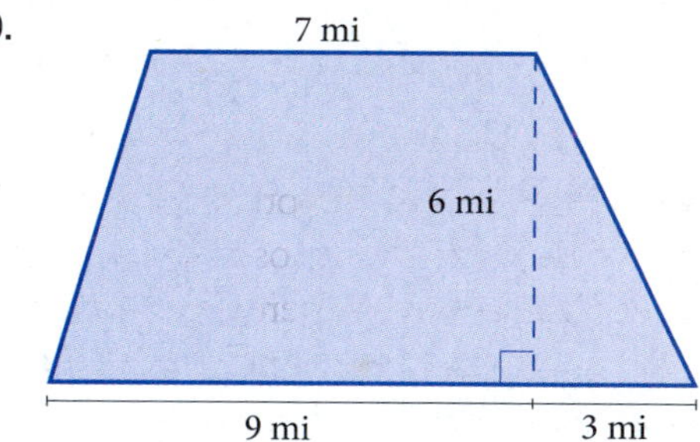

81.

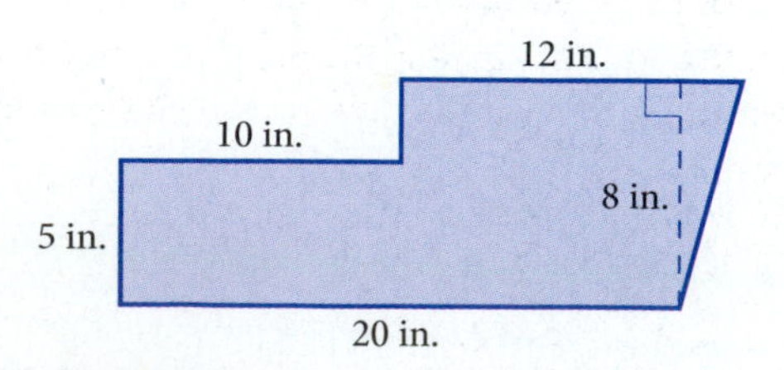

82.

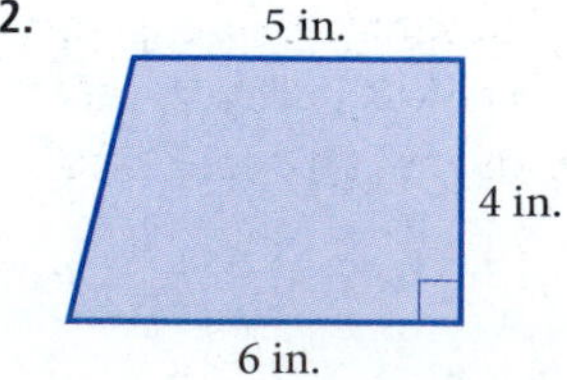

Applying the Concepts

83. Rainfall The chart shows the average rainfall for Death Valley in the given months. Use this chart to answer the questions below.

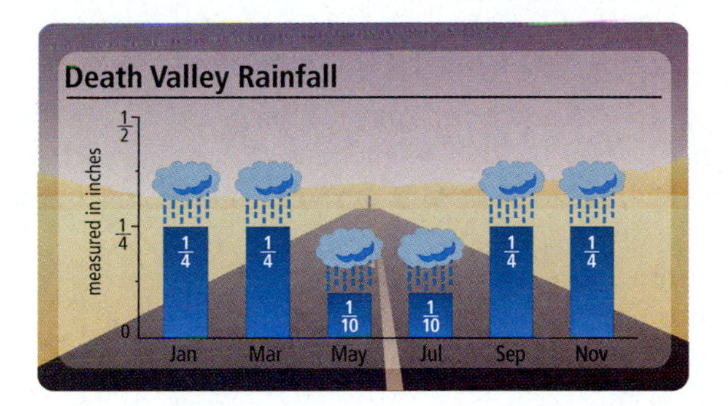

a. How many inches of rain is 5 times the average for January?

b. How many inches of rain is 7 times the average for May?

c. How many inches of rain is 12 times the average for September?

84. Rainfall The chart shows the average rainfall for Death Valley in the given months. Use this chart to answer the questions below.

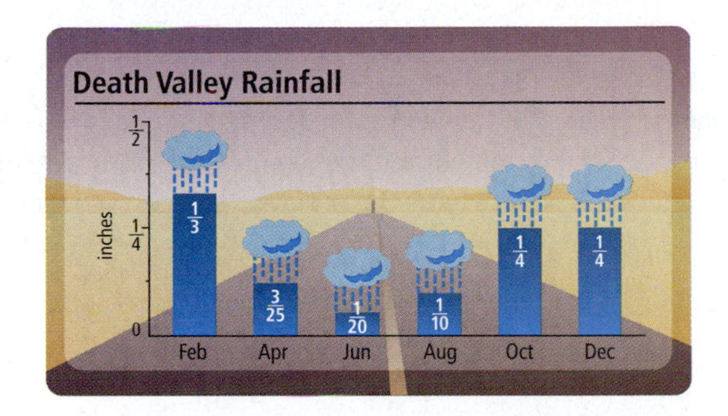

a. How many inches of rain is 6 times the average for February?

b. How many inches of rain is 5 times the average for April?

c. How many inches of rain is 8 times the average for October?

85. Hot Air Balloon Aerostar International makes a hot air balloon called the Rally 105 that has a volume of 105,400 cubic feet. Another balloon, the Rally 126, was designed with a volume that is approximately $\frac{6}{5}$ the volume of the Rally 105. Find the volume of the Rally 126 to the nearest hundred cubic feet.

86. Bicycle Safety The National Safe Kids Campaign and Bell Sports sponsored a study that surveyed 8,159 children ages 5 to 14 who were riding bicycles. Approximately $\frac{2}{5}$ of the children were wearing helmets, and of those, only $\frac{13}{20}$ were wearing the helmets correctly. About how many of the children were wearing helmets correctly?

87. Health Care According to a study reported on MSNBC, almost one-third of the people diagnosed with diabetes don't seek proper medical care. If there are 12 million Americans with diabetes, about how many of them are seeking proper medical care?

88. Working Students Studies indicate that approximately $\frac{3}{4}$ of all undergraduate college students work while attending school. A local community college has a student enrollment of 8,500 students. How many of these students work while attending college?

89. Cigarette Tax In a recent survey of 1,410 adults, it was determined that $\frac{3}{5}$ of those surveyed favored raising the tax on cigarettes as a way to discourage young people from smoking. What number of adults believes that this would reduce the number of young people who smoke?

90. Shared Rent You and three friends decide to rent an apartment for the academic year rather than to live in the dorms. The monthly rent is \$1250. If you and your friends split the rent equally, what is your share of the monthly rent?

91. Importing Oil According to the U.S. Department of Energy, we imported approximately 8,340,000 barrels of oil in November 2007, which represents a typical month. We import a little over $\frac{1}{5}$ of our oil from Canada, approximately $\frac{3}{20}$ of our oil from Venezuela, and less than $\frac{1}{10}$ of our oil from Iraq. Determine the amount of oil we imported from each of these countries.

92. Improving Your Quantitative Literacy MSNBC reported that at least three-fourths of the 55 companies that advertise nationally on television will cut spending on commercials because of electronics that let viewers record programs and edit out commercials. Does this mean at least 41, or at least 42, of the companies will cut spending on commercials?

Geometric Sequences Recall that a geometric sequence is a sequence in which each term comes from the previous term by multiplying by the same number each time. For example, the sequence $1, \frac{1}{2}, \frac{1}{4}, \frac{1}{8}, \ldots$ is a geometric sequence in which each term is found by multiplying the previous term by $\frac{1}{2}$. By observing this fact, we know that the next term in the sequence will be $\frac{1}{8} \cdot \frac{1}{2} = \frac{1}{16}$.

Find the next number in each of the geometric sequences below.

93. $1, \frac{1}{3}, \frac{1}{9}, \ldots$

94. $1, \frac{1}{4}, \frac{1}{16}, \ldots$

95. $\frac{3}{2}, 1, \frac{2}{3}, \frac{4}{9}, \ldots$

96. $\frac{2}{3}, 1, \frac{3}{2}, \frac{9}{4}, \ldots$

Estimating For each problem below, mentally estimate which of the numbers 0, 1, 2, or 3 is closest to the answer. Make your estimate without using pencil and paper or a calculator.

97. $\frac{11}{5} \cdot \frac{19}{20}$

98. $\frac{3}{5} \cdot \frac{1}{20}$

99. $\frac{16}{5} \cdot \frac{23}{24}$

100. $\frac{9}{8} \cdot \frac{31}{32}$

Getting Ready for the Next Section

In the next section we will do division with fractions. As you already know, division and multiplication are closely related. These review problems are intended to let you see more of the relationship between multiplication and division.

Perform the indicated operations.

101. $8 \div 4$

102. $8 \cdot \frac{1}{4}$

103. $15 \div 3$

104. $15 \cdot \frac{1}{3}$

105. $18 \div 6$

106. $18 \cdot \frac{1}{6}$

For each number below, find a number to multiply it by to obtain 1.

107. $\frac{3}{4}$

108. $\frac{9}{5}$

109. $\frac{1}{3}$

110. $\frac{1}{4}$

111. 7

112. 2

Maintaining Your Skills

Simplify.

113. $20 \div 2 \cdot 10$

114. $40 \div 4 \cdot 5$

115. $24 \div 8 \cdot 3$

116. $24 \div 4 \cdot 6$

117. $36 \div 6 \cdot 3$

118. $36 \div 9 \cdot 2$

119. $48 \div 12 \cdot 2$

120. $48 \div 8 \cdot 3$

Division with Fractions

3.4

Objectives

A Divide fractions.

B Simplify order of operation problems involving division of fractions.

C Solve application problems involving division of fractions.

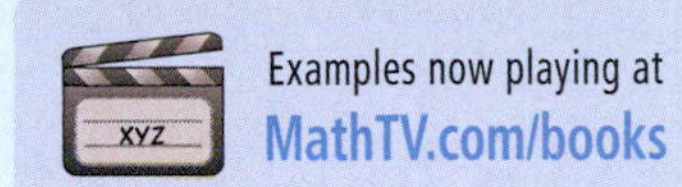

Introduction . . .

A few years ago our 4-H club was making blankets to keep their lambs clean at the county fair. Each blanket required $\frac{3}{4}$ yard of material. We had 9 yards of material left over from the year before. To see how many blankets we could make, we divided 9 by $\frac{3}{4}$. The result was 12, meaning that we could make 12 lamb blankets for the fair.

Before we define division with fractions, we must first introduce the idea of *reciprocals.* Look at the following multiplication problems:

$$\frac{3}{4} \cdot \frac{4}{3} = \frac{12}{12} = 1 \qquad \frac{7}{8} \cdot \frac{8}{7} = \frac{56}{56} = 1$$

In each case the product is 1. Whenever the product of two numbers is 1, we say the two numbers are *reciprocals.*

Definition

Two numbers whose product is 1 are said to be **reciprocals.** In symbols, the reciprocal of $\frac{a}{b}$ is $\frac{b}{a}$, because

$$\frac{a}{b} \cdot \frac{b}{a} = \frac{a \cdot b}{b \cdot a} = \frac{a \cdot b}{a \cdot b} = 1 \qquad (a \neq 0, b \neq 0)$$

Every number has a reciprocal except 0. The reason that 0 does not have a reciprocal is because the product of *any* number with 0 is 0. It can never be 1. Reciprocals of whole numbers are fractions with 1 as the numerator. For example, the reciprocal of 5 is $\frac{1}{5}$, because

$$5 \cdot \frac{1}{5} = \frac{5}{1} \cdot \frac{1}{5} = \frac{5}{5} = 1$$

Table 1 lists some numbers and their reciprocals.

TABLE 1

Number	Reciprocal	Reason
$\frac{3}{4}$	$\frac{4}{3}$	Because $\frac{3}{4} \cdot \frac{4}{3} = \frac{12}{12} = 1$
$\frac{9}{5}$	$\frac{5}{9}$	Because $\frac{9}{5} \cdot \frac{5}{9} = \frac{45}{45} = 1$
$\frac{1}{3}$	3	Because $\frac{1}{3} \cdot 3 = \frac{1}{3} \cdot \frac{3}{1} = \frac{3}{3} = 1$
7	$\frac{1}{7}$	Because $7 \cdot \frac{1}{7} = \frac{7}{1} \cdot \left(\frac{1}{7}\right) = \frac{7}{7} = 1$

A Dividing Fractions

Division with fractions is accomplished by using reciprocals. More specifically, we can define division by a fraction to be the same as multiplication by its reciprocal. Here is the precise definition:

Definition

If a, b, c, and d are numbers and b, c, and d are all not equal to 0, then

$$\frac{a}{b} \div \frac{c}{d} = \frac{a}{b} \cdot \frac{d}{c}$$

Note Defining division to be the same as multiplication by the reciprocal does make sense. If we divide 6 by 2, we get 3. On the other hand, if we multiply 6 by $\frac{1}{2}$ (the reciprocal of 2), we also get 3. Whether we divide by 2 or multiply by $\frac{1}{2}$, we get the same result.

This definition states that dividing by the fraction $\frac{c}{d}$ is exactly the same as multiplying by its reciprocal $\frac{d}{c}$. Because we developed the rule for multiplying fractions in Section 3.3, we do not need a new rule for division. We simply *replace the divisor by its reciprocal* and multiply. Here are some examples to illustrate the procedure.

PRACTICE PROBLEMS

1. Divide.
   a. $\frac{1}{3} \div \frac{1}{6}$
   b. $\frac{1}{30} \div \frac{1}{60}$

2. Divide: $\frac{5}{9} \div \frac{10}{3}$

3. Divide.
   a. $\frac{3}{4} \div 3$
   b. $\frac{3}{5} \div 3$
   c. $\frac{3}{7} \div 3$

4. Divide: $4 \div \frac{1}{5}$

Answers

1. Both are 2 2. $\frac{1}{6}$

3. a. $\frac{1}{4}$ b. $\frac{1}{5}$ c. $\frac{1}{7}$ 4. 20

EXAMPLE 1 Divide: $\frac{1}{2} \div \frac{1}{4}$

SOLUTION The divisor is $\frac{1}{4}$, and its reciprocal is $\frac{4}{1}$. Applying the definition of division for fractions, we have

$$\begin{aligned} \frac{1}{2} \div \frac{1}{4} &= \frac{1}{2} \cdot \frac{4}{1} \\ &= \frac{1 \cdot 4}{2 \cdot 1} \\ &= \frac{1 \cdot \cancel{2} \cdot 2}{\cancel{2} \cdot 1} \\ &= \frac{2}{1} \\ &= 2 \end{aligned}$$

The quotient of $\frac{1}{2}$ and $\frac{1}{4}$ is 2. Or, $\frac{1}{4}$ "goes into" $\frac{1}{2}$ two times. Logically, our definition for division of fractions seems to be giving us answers that are consistent with what we know about fractions from previous experience. Because 2 times $\frac{1}{4}$ is $\frac{2}{4}$ or $\frac{1}{2}$, it seems logical that $\frac{1}{2}$ divided by $\frac{1}{4}$ should be 2.

EXAMPLE 2 Divide: $\frac{3}{8} \div \frac{9}{4}$

SOLUTION Dividing by $\frac{9}{4}$ is the same as multiplying by its reciprocal, which is $\frac{4}{9}$:

$$\begin{aligned} \frac{3}{8} \div \frac{9}{4} &= \frac{3}{8} \cdot \frac{4}{9} \\ &= \frac{\cancel{3} \cdot \cancel{2} \cdot \cancel{2}}{\cancel{2} \cdot \cancel{2} \cdot 2 \cdot \cancel{3} \cdot 3} \\ &= \frac{1}{6} \end{aligned}$$

The quotient of $\frac{3}{8}$ and $\frac{9}{4}$ is $\frac{1}{6}$.

EXAMPLE 3 Divide: $-\frac{2}{3} \div 2$

SOLUTION The reciprocal of 2 is $\frac{1}{2}$. Applying the definition for division of fractions, we have

$$\begin{aligned} -\frac{2}{3} \div 2 &= -\frac{2}{3} \cdot \frac{1}{2} \\ &= -\frac{\cancel{2} \cdot 1}{3 \cdot \cancel{2}} \\ &= -\frac{1}{3} \end{aligned}$$

EXAMPLE 4 Divide: $2 \div \left(-\frac{1}{3}\right)$

SOLUTION We replace $-\frac{1}{3}$ by its reciprocal, which is -3, and multiply:

$$\begin{aligned} 2 \div \left(-\frac{1}{3}\right) &= 2(-3) \\ &= -6 \end{aligned}$$

Here are some further examples of division with fractions. Notice in each case that the first step is the only new part of the process.

EXAMPLE 5 Divide: $\frac{4}{27} \div \frac{16}{9}$

SOLUTION
$$\frac{4}{27} \div \frac{16}{9} = \frac{4}{27} \cdot \frac{9}{16}$$
$$= \frac{\not{4} \cdot \not{9}}{3 \cdot \not{9} \cdot \not{4} \cdot 4}$$
$$= \frac{1}{12}$$

In Example 5 we did not factor the numerator and the denominator completely in order to reduce to lowest terms because, as you have probably already noticed, it is not necessary to do so. We need to factor only enough to show what numbers are common to the numerator and the denominator. If we factored completely in the second step, it would look like this:

$$= \frac{\not{2} \cdot \not{2} \cdot \not{3} \cdot \not{3}}{\not{3} \cdot \not{3} \cdot 3 \cdot \not{2} \cdot \not{2} \cdot 2 \cdot 2}$$
$$= \frac{1}{12}$$

The result is the same in both cases. From now on we will factor numerators and denominators only enough to show the factors we are dividing out.

EXAMPLE 6 Divide: $\frac{16}{35} \div 8$

SOLUTION
$$\frac{16}{35} \div 8 = \frac{16}{35} \cdot \frac{1}{8}$$
$$= \frac{2 \cdot \not{8} \cdot 1}{35 \cdot \not{8}}$$
$$= \frac{2}{35}$$

EXAMPLE 7 Divide: $-27 \div \left(-\frac{3}{2}\right)$

SOLUTION
$$-27 \div \left(-\frac{3}{2}\right) = -27 \cdot \left(-\frac{2}{3}\right)$$
$$= \frac{\not{3} \cdot 9 \cdot 2}{\not{3}}$$
$$= 18$$

EXAMPLE 8 Divide: $\frac{x^2}{y} \div \frac{x}{y^3}$

SOLUTION
$$\frac{x^2}{y} \div \frac{x}{y^3} = \frac{x^2}{y} \cdot \frac{y^3}{x}$$
$$= \frac{\not{x} \cdot x \cdot \not{y} \cdot y \cdot y}{\not{y} \cdot \not{x}}$$
$$= \frac{x \cdot y \cdot y}{1}$$
$$= xy^2$$

Find each quotient.

5. a. $\frac{5}{32} \div \frac{10}{42}$
   b. $\frac{15}{32} \div \frac{30}{42}$

6. a. $\frac{12}{25} \div 6$
   b. $\frac{24}{25} \div 6$

7. a. $12 \div \frac{4}{3}$
   b. $12 \div \frac{4}{5}$
   c. $12 \div \frac{4}{7}$

8. $\frac{x^3}{y} \div \frac{x^2}{y^2}$

Answers
5. Both are $\frac{21}{32}$ 6. a. $\frac{2}{25}$ b. $\frac{4}{25}$
7. a. 9 b. 15 c. 21 8. xy

B Fractions and the Order of Operations

The next two examples combine what we have learned about division of fractions with the rule for order of operations.

EXAMPLE 9 The quotient of $\frac{8}{3}$ and $\frac{1}{6}$ is increased by 5. What number results?

SOLUTION Translating to symbols, we have

$$\frac{8}{3} \div \frac{1}{6} + 5 = \frac{8}{3} \cdot \frac{6}{1} + 5$$
$$= 16 + 5$$
$$= 21$$

EXAMPLE 10 Simplify: $32 \div \left(\frac{4}{3}\right)^2 + 75 \div \left(\frac{5}{2}\right)^2$

SOLUTION According to the rule for order of operations, we must first evaluate the numbers with exponents, then divide, and finally, add.

$$32 \div \left(\frac{4}{3}\right)^2 + 75 \div \left(\frac{5}{2}\right)^2 = 32 \div \frac{16}{9} + 75 \div \frac{25}{4}$$
$$= 32 \cdot \frac{9}{16} + 75 \cdot \frac{4}{25}$$
$$= 18 + 12$$
$$= 30$$

C Applications

EXAMPLE 11 A 4-H club is making blankets to keep their lambs clean at the county fair. If each blanket requires $\frac{3}{4}$ yard of material, how many blankets can they make from 9 yards of material?

SOLUTION To answer this question we must divide 9 by $\frac{3}{4}$.

$$9 \div \frac{3}{4} = 9 \cdot \frac{4}{3}$$
$$= 3 \cdot 4$$
$$= 12$$

They can make 12 blankets from the 9 yards of material.

Getting Ready for Class

After reading through the preceding section, respond in your own words and in complete sentences.

1. What do we call two numbers whose product is 1?
2. True or false? The quotient of $\frac{3}{5}$ and $\frac{3}{8}$ is the same as the product of $\frac{3}{5}$ and $\frac{8}{3}$.
3. How are multiplication and division of fractions related?
4. Dividing by $\frac{19}{9}$ is the same as multiplying by what number?

---

9. The quotient of $\frac{5}{4}$ and $\frac{1}{8}$ is increased by 8. What number results?

10. Simplify: $18 \div \left(\frac{3}{5}\right)^2 + 48 \div \left(\frac{2}{5}\right)^2$

11. How many blankets can the 4-H club make with 12 yards of material?

Answers
9. 18 **10.** 350 **11.** 16 blankets

Problem Set 3.4

A Find the quotient in each case by replacing the divisor by its reciprocal and multiplying. [Examples 1–8]

1. $\frac{3}{4} \div \frac{1}{5}$
2. $\frac{1}{3} \div \frac{1}{2}$
3. $-\frac{2}{3} \div \frac{1}{2}$
4. $-\frac{5}{8} \div \frac{1}{4}$
5. $6 \div \left(-\frac{2}{3}\right)$
6. $8 \div \left(-\frac{3}{4}\right)$
7. $20 \div \frac{1}{10}$
8. $16 \div \frac{1}{8}$
9. $\frac{3}{4} \div (-2)$
10. $\frac{3}{5} \div (-2)$
11. $\frac{7}{8} \div \frac{7}{8}$
12. $\frac{4}{3} \div \frac{4}{3}$
13. $\frac{7}{8} \div \frac{8}{7}$
14. $\frac{4}{3} \div \frac{3}{4}$
15. $\frac{9}{16} \div \left(-\frac{3}{4}\right)$
16. $-\frac{25}{36} \div \left(-\frac{5}{6}\right)$
17. $\frac{25}{46} \div \frac{40}{69}$
18. $\frac{25}{24} \div \frac{15}{36}$
19. $\frac{13}{28} \div \frac{39}{14}$
20. $\frac{28}{125} \div \frac{5}{2}$
21. $\frac{27}{196} \div \frac{9}{392}$
22. $\frac{16}{135} \div \frac{2}{45}$
23. $\frac{x^2}{y^3} \div \frac{x}{y^2}$
24. $\frac{x}{y^2} \div \frac{x^3}{y}$
25. $\frac{ab^2}{c} \div \frac{b}{c}$
26. $\frac{a^2b}{c^2} \div \frac{a}{c^3}$
27. $\frac{10a^2}{3b} \div \frac{5a}{6b}$
28. $\frac{12a^2}{5b} \div \frac{6a}{5b^2}$
29. $\frac{3}{4} \div \frac{1}{2} \cdot 6$
30. $12 \div \frac{6}{7} \cdot 7$
31. $\frac{2}{3} \cdot \frac{3}{4} \div \frac{5}{8}$
32. $4 \cdot \frac{7}{6} \div 7$
33. $\frac{35}{110} \cdot \frac{80}{63} \div \frac{16}{27}$
34. $\frac{20}{72} \cdot \frac{42}{18} \div \frac{20}{16}$

B Simplify each expression as much as possible. [Examples 9, 10]

35. $10 \div \left(\frac{1}{2}\right)^2$

36. $12 \div \left(\frac{1}{4}\right)^2$

37. $\frac{18}{35} \div \left(\frac{6}{7}\right)^2$

38. $\frac{48}{55} \div \left(\frac{8}{11}\right)^2$

39. $\frac{4}{5} \div \frac{1}{10} + 5$

40. $\frac{3}{8} \div \frac{1}{16} + 4$

41. $10 + \frac{11}{12} \div \frac{11}{24}$

42. $15 + \frac{13}{14} \div \frac{13}{42}$

43. $24 \div \left(\frac{2}{5}\right)^2 + 25 \div \left(\frac{5}{6}\right)^2$

44. $18 \div \left(\frac{3}{4}\right)^2 + 49 \div \left(\frac{7}{9}\right)^2$

45. $100 \div \left(\frac{5}{7}\right)^2 + 200 \div \left(\frac{2}{3}\right)^2$

46. $64 \div \left(\frac{8}{11}\right)^2 + 81 \div \left(\frac{9}{11}\right)^2$

47. What is the quotient of $\frac{3}{8}$ and $\frac{5}{8}$?

48. Find the quotient of $\frac{4}{5}$ and $\frac{16}{25}$.

49. If the quotient of 18 and $\frac{3}{5}$ is increased by 10, what number results?

50. If the quotient of 50 and $\frac{5}{3}$ is increased by 8, what number results?

51. Show that multiplying 3 by 5 is the same as dividing 3 by $\frac{1}{5}$.

52. Show that multiplying 8 by $\frac{1}{2}$ is the same as dividing 8 by 2.

C Applying the Concepts [Example 11]

53. Pyramids The Luxor Hotel in Las Vegas is $\frac{5}{7}$ the original height of the Great Pyramid of Giza. If the hotel is 350 feet tall, what was the original height of the Great Pyramid of Giza?

54. Skyscrapers The Bloomberg tower in New York City is $\frac{3}{5}$ the height of the Sears Tower. How tall is the Bloomberg tower?

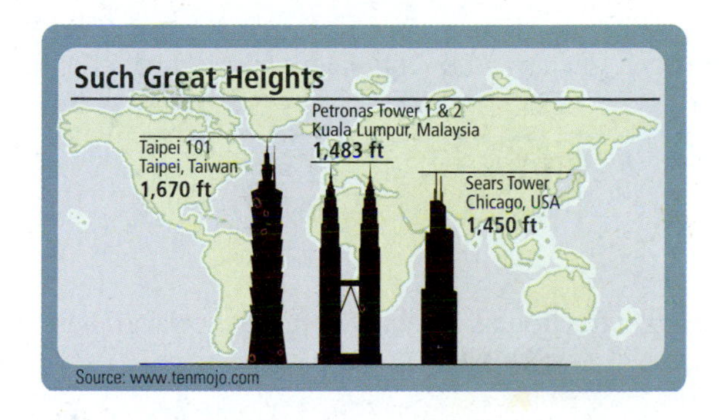

55. Sewing If $\frac{6}{7}$ yard of material is needed to make a blanket, how many blankets can be made from 12 yards of material?

56. Manufacturing A clothing manufacturer is making scarves that require $\frac{3}{8}$ yard of material each. How many can be made from 27 yards of material?

57. Cooking A man is making cookies from a recipe that calls for $\frac{3}{4}$ teaspoon of oil. If the only measuring spoon he can find is a $\frac{1}{8}$ teaspoon, how many of these will he have to fill with oil in order to have a total of $\frac{3}{4}$ teaspoon of oil?

58. Cooking A cake recipe calls for $\frac{1}{2}$ cup of sugar. If the only measuring cup available is a $\frac{1}{8}$ cup, how many of these will have to be filled with sugar to make a total of $\frac{1}{2}$ cup of sugar?

59. Cartons of Milk If a small carton of milk holds exactly $\frac{1}{2}$ pint, how many of the $\frac{1}{2}$-pint cartons can be filled from a 14-pint container?

60. Pieces of Pipe How many pieces of pipe that are $\frac{2}{3}$ foot long must be laid together to make a pipe 16 feet long?

61. Lot Size A land developer wants to subdivide 5 acres of property into lots suitable for building a home. If each lot is to be $\frac{1}{4}$ of an acre in size how many lots can be made?

62. House Plans If $\frac{1}{8}$ inch represents 1 ft on a drawing of a new home, determine the dimensions of a bedroom that measures 2 inches by 2 inches on the drawing.

Getting Ready for the Next Section

Write each fraction as an equivalent fraction with denominator 6.

63. $\frac{1}{2}$ **64.** $\frac{1}{3}$ **65.** $\frac{3}{2}$ **66.** $\frac{2}{3}$

Write each fraction as an equivalent fraction with denominator 12.

67. $\frac{1}{3}$ **68.** $\frac{1}{2}$ **69.** $\frac{2}{3}$ **70.** $\frac{3}{4}$

Write each fraction as an equivalent fraction with denominator 30.

71. $\frac{7}{15}$ **72.** $\frac{3}{10}$ **73.** $\frac{3}{5}$ **74.** $\frac{1}{6}$

Write each fraction as an equivalent fraction with denominator 24.

75. $\frac{1}{2}$ **76.** $\frac{1}{4}$ **77.** $\frac{1}{6}$ **78.** $\frac{1}{8}$

Write each fraction as an equivalent fraction with denominator 36.

79. $\frac{5}{12}$ **80.** $\frac{7}{18}$ **81.** $\frac{1}{4}$ **82.** $\frac{1}{6}$

Maintaining Your Skills

83. Fill in the table by rounding the numbers.

Number	Rounded to the Nearest		
	Ten	Hundred	Thousand
74	70	100	
747	750	700	1000
474	470	500	

84. Fill in the table by rounding the numbers.

Number	Rounded to the Nearest		
	Ten	Hundred	Thousand
63			
636			
363			

85. Estimating The quotient $-253 \div 24$ is closer to which of the following?

a. 5 **b.** -10 **c.** 15 **d.** -20

86. Estimating The quotient $1{,}000 \div -47$ is closer to which of the following?

a. 5 **b.** -10 **c.** 15 **d.** -20

Addition and Subtraction with Fractions

3.5

Objectives

A Add and subtract fractions with the same denominator.

B Add and subtract fractions with different denominators.

C Solve applications involving addition and subtraction of fractions.

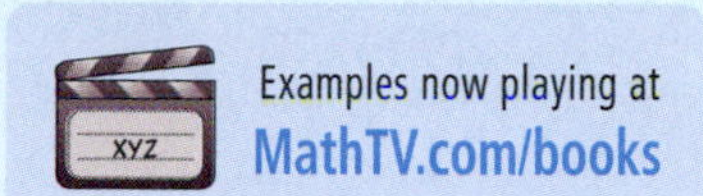

Most people who have done any work with adding fractions know that you add fractions that have the same denominator by adding their numerators, but not their denominators. However, most people don't know why this works. The reason why we add numerators but not denominators is because of the distributive property. And that is what the discussion at the left is all about. If you really want to understand addition of fractions, pay close attention to this discussion.

Introduction . . .

Adding and subtracting fractions is actually just another application of the distributive property. The distributive property looks like this:

$$a(b + c) = a(b) + a(c)$$

where a, b, and c may be whole numbers or fractions. We will want to apply this property to expressions like

$$\frac{2}{7} + \frac{3}{7}$$

But before we do, we must make one additional observation about fractions.

The fraction $\frac{2}{7}$ can be written as $2 \cdot \frac{1}{7}$, because

$$2 \cdot \frac{1}{7} = \frac{2}{1} \cdot \frac{1}{7} = \frac{2}{7}$$

Likewise, the fraction $\frac{3}{7}$ can be written as $3 \cdot \frac{1}{7}$, because

$$3 \cdot \frac{1}{7} = \frac{3}{1} \cdot \frac{1}{7} = \frac{3}{7}$$

In general, we can say that the fraction $\frac{a}{b}$ can always be written as $a \cdot \frac{1}{b}$, because

$$a \cdot \frac{1}{b} = \frac{a}{1} \cdot \frac{1}{b} = \frac{a}{b}$$

To add the fractions $\frac{2}{7}$ and $\frac{3}{7}$, we simply rewrite each of them as we have done above and apply the distributive property. Here is how it works:

$$\frac{2}{7} + \frac{3}{7} = 2 \cdot \frac{1}{7} + 3 \cdot \frac{1}{7} \qquad \text{Rewrite each fraction}$$

$$= (2 + 3) \cdot \frac{1}{7} \qquad \text{Apply the distributive property}$$

$$= 5 \cdot \frac{1}{7} \qquad \text{Add 2 and 3 to get 5}$$

$$= \frac{5}{7} \qquad \text{Rewrite } 5 \cdot \frac{1}{7} \text{ as } \frac{5}{7}$$

We can visualize the process shown above by using circles that are divided into 7 equal parts:

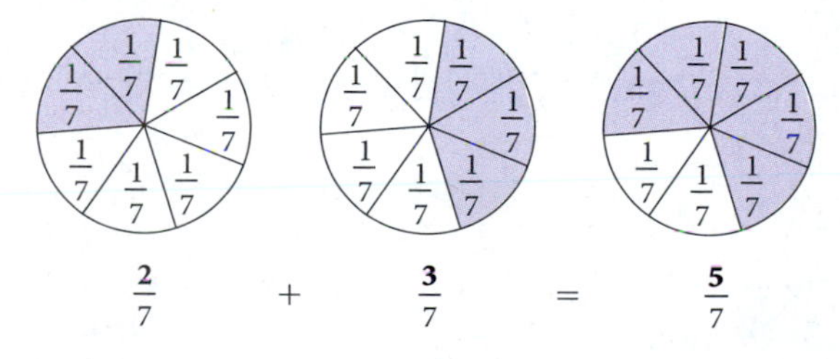

The fraction $\frac{5}{7}$ is the sum of $\frac{2}{7}$ and $\frac{3}{7}$. The steps and diagrams above show why we add numerators *but do not add denominators.* Using this example as justification, we can write a rule for adding two fractions that have the same denominator.

A Combining Fractions with the Same Denominator

Rule

To add two fractions that have the same denominator, we add their numerators to get the numerator of the answer. The denominator in the answer is the same denominator as in the original fractions.

What we have here is the sum of the numerators placed over the *common denominator.* In symbols we have the following:

Addition and Subtraction of Fractions

If a, b, and c are numbers, and c is not equal to 0, then

$$\frac{a}{c} + \frac{b}{c} = \frac{a+b}{c}$$

This rule holds for subtraction as well. That is,

$$\frac{a}{c} - \frac{b}{c} = \frac{a-b}{c}$$

EXAMPLE 1 Add: $\frac{3}{8} + \frac{1}{8}$

SOLUTION

$\frac{3}{8} + \frac{1}{8} = \frac{3+1}{8}$ Add numerators; keep the same denominator

$= \frac{4}{8}$ The sum of 3 and 1 is 4

$= \frac{1}{2}$ Reduce to lowest terms

EXAMPLE 2 Subtract: $\frac{a+5}{8} - \frac{3}{8}$

SOLUTION

$\frac{a+5}{8} - \frac{3}{8} = \frac{a+5-3}{8}$ Combine numerators; keep the same denominator

$= \frac{a+2}{8}$ The difference of 5 and 3 is 2

EXAMPLE 3 Subtract: $\frac{9}{x} - \frac{3}{x}$

SOLUTION

$\frac{9}{x} - \frac{3}{x} = \frac{9-3}{x}$ Subtract numerators; keep the same denominator

$= \frac{6}{x}$ The difference of 9 and 3 is 6

EXAMPLE 4 Add: $\frac{3}{7} + \frac{2}{7} + \frac{9}{7}$

SOLUTION

$\frac{3}{7} + \frac{2}{7} + \frac{9}{7} = \frac{3+2+9}{7}$

$= \frac{14}{7}$

$= 2$

As Examples 1–4 indicate, addition and subtraction are simple, straightforward processes when all the fractions have the same denominator. We will now turn

PRACTICE PROBLEMS

Find the sum or difference. Reduce all answers to lowest terms.

1. $\frac{3}{10} + \frac{1}{10}$
2. $\frac{a+5}{12} + \frac{3}{12}$
3. $\frac{8}{7} - \frac{5}{7}$
4. $\frac{5}{9} + \frac{8}{9} + \frac{5}{9}$

Answers

1. $\frac{2}{5}$ 2. $\frac{a+8}{12}$ 3. $\frac{3}{7}$ 4. 2

our attention to the process of adding fractions that have different denominators. In order to get started, we need the following definition.

B The Least Common Denominator or LCD

Definition

The **least common denominator** (LCD) for a set of denominators is the smallest number that is exactly divisible by each denominator. (Note that, in some books, the least common denominator is also called the **least common multiple.**)

In other words, all the denominators of the fractions involved in a problem must divide into the least common denominator exactly. That is, they divide it without leaving a remainder.

EXAMPLE 5 Find the LCD for the fractions $\frac{5}{12}$ and $\frac{7}{18}$.

SOLUTION The least common denominator for the denominators 12 and 18 must be the smallest number divisible by both 12 and 18. We can factor 12 and 18 completely and then build the LCD from these factors. Factoring 12 and 18 completely gives us

$$12 = 2 \cdot 2 \cdot 3 \qquad 18 = 2 \cdot 3 \cdot 3$$

Now, if 12 is going to divide the LCD exactly, then the LCD must have factors of $2 \cdot 2 \cdot 3$. If 18 is to divide it exactly, it must have factors of $2 \cdot 3 \cdot 3$. We don't need to repeat the factors that 12 and 18 have in common:

$$\left.\begin{aligned} 12 &= 2 \cdot 2 \cdot 3 \\ 18 &= 2 \cdot 3 \cdot 3 \end{aligned}\right\} \qquad \text{LCD} = 2 \cdot 2 \cdot 3 \cdot 3 = 36$$

12 divides the LCD

18 divides the LCD

The LCD for 12 and 18 is 36. It is the smallest number that is divisible by both 12 and 18; 12 divides it exactly three times, and 18 divides it exactly two times.

We can visualize the results in Example 5 with the diagram below. It shows that 36 is the smallest number that both 12 and 18 divide evenly. As you can see, 12 divides 36 exactly 3 times, and 18 divides 36 exactly 2 times.

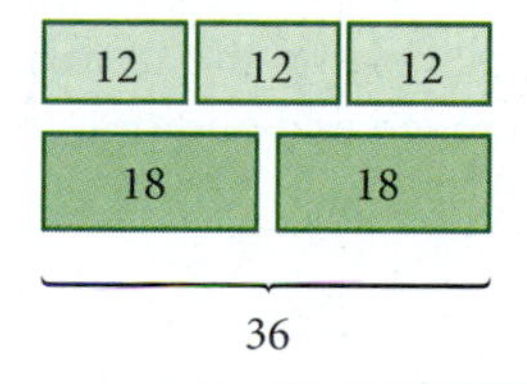

EXAMPLE 6 Add: $\frac{5}{12} + \frac{7}{18}$

SOLUTION We can add fractions only when they have the same denominators. In Example 5, we found the LCD for $\frac{5}{12}$ and $\frac{7}{18}$ to be 36. We change $\frac{5}{12}$ and $\frac{7}{18}$ to equivalent fractions that have 36 for a denominator by applying Property 1 for fractions:

$$\frac{5}{12} = \frac{5 \cdot \mathbf{3}}{12 \cdot \mathbf{3}} = \frac{15}{36}$$

$$\frac{7}{18} = \frac{7 \cdot \mathbf{2}}{18 \cdot \mathbf{2}} = \frac{14}{36}$$

5. a. Find the LCD for the fractions: $\frac{5}{18}$ and $\frac{3}{14}$

b. Find the LCD for the fractions: $\frac{5}{36}$ and $\frac{3}{28}$

The ability to find least common denominators is very important in mathematics. The discussion here is a detailed explanation of how to find an LCD.

6. Add.

a. $\frac{5}{18} + \frac{3}{14}$

b. $\frac{5}{36} + \frac{3}{28}$

Answer

5. a. 126 b. 252

The fraction $\frac{15}{36}$ is equivalent to $\frac{5}{12}$, because it was obtained by multiplying both the numerator and the denominator by 3. Likewise, $\frac{14}{36}$ is equivalent to $\frac{7}{18}$, because it was obtained by multiplying the numerator and the denominator by 2. All we have left to do is to add numerators.

$$\frac{15}{36} + \frac{14}{36} = \frac{29}{36}$$

The sum of $\frac{5}{12}$ and $\frac{7}{18}$ is the fraction $\frac{29}{36}$. Let's write the complete problem again step by step.

$$\frac{5}{12} + \frac{7}{18} = \frac{5 \cdot \mathbf{3}}{12 \cdot \mathbf{3}} + \frac{7 \cdot \mathbf{2}}{18 \cdot \mathbf{2}}$$ Rewrite each fraction as an equivalent fraction with denominator 36

$$= \frac{15}{36} + \frac{14}{36}$$

$$= \frac{29}{36}$$ Add numerators; keep the common denominator

7. a. Find the LCD for $\frac{2}{9}$ and $\frac{4}{15}$.
   b. Find the LCD for $\frac{2}{27}$ and $\frac{4}{45}$.

EXAMPLE 7

Find the LCD for $\frac{3}{4}$ and $\frac{1}{6}$.

SOLUTION We factor 4 and 6 into products of prime factors and build the LCD from these factors.

$$\left.\begin{array}{l} 4 = 2 \cdot 2 \\ 6 = 2 \cdot 3 \end{array}\right\} \text{LCD} = 2 \cdot 2 \cdot 3 = 12$$

The LCD is 12. Both denominators divide it exactly; 4 divides 12 exactly 3 times, and 6 divides 12 exactly 2 times.

8. Add.
   a. $\frac{2}{9} + \frac{4}{15}$
   b. $\frac{2}{27} + \frac{4}{45}$

EXAMPLE 8

Add: $\frac{3}{4} + \frac{1}{6}$

SOLUTION In Example 7, we found that the LCD for these two fractions is 12. We begin by changing $\frac{3}{4}$ and $\frac{1}{6}$ to equivalent fractions with denominator 12:

$$\frac{3}{4} = \frac{3 \cdot \mathbf{3}}{4 \cdot \mathbf{3}} = \frac{9}{12}$$

$$\frac{1}{6} = \frac{1 \cdot \mathbf{2}}{6 \cdot \mathbf{2}} = \frac{2}{12}$$

The fraction $\frac{9}{12}$ is equal to the fraction $\frac{3}{4}$, because it was obtained by multiplying the numerator and the denominator of $\frac{3}{4}$ by 3. Likewise, $\frac{2}{12}$ is equivalent to $\frac{1}{6}$, because it was obtained by multiplying the numerator and the denominator of $\frac{1}{6}$ by 2. To complete the problem we add numerators:

$$\frac{9}{12} + \frac{2}{12} = \frac{11}{12}$$

The sum of $\frac{3}{4}$ and $\frac{1}{6}$ is $\frac{11}{12}$. Here is how the complete problem looks:

$$\frac{3}{4} + \frac{1}{6} = \frac{3 \cdot \mathbf{3}}{4 \cdot \mathbf{3}} + \frac{1 \cdot \mathbf{2}}{6 \cdot \mathbf{2}}$$ Rewrite each fraction as an equivalent fraction with denominator 12

$$= \frac{9}{12} + \frac{2}{12}$$

$$= \frac{11}{12}$$ Add numerators; keep the same denominator

Note We can visualize the work in Example 8 using circles and shading:

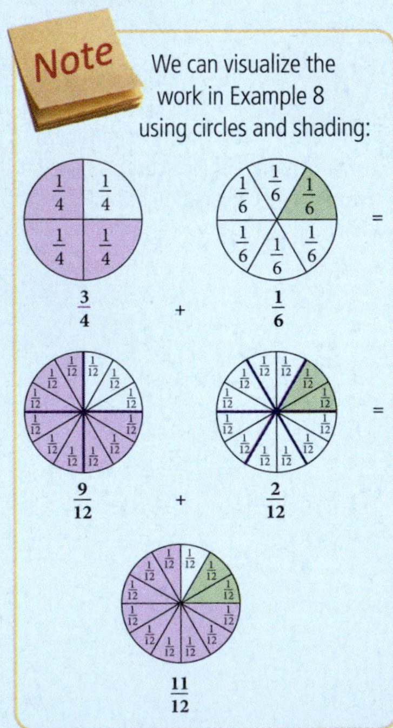

Answers

6. a. $\frac{31}{63}$ b. $\frac{31}{126}$ 7. a. 45 b. 135
8. a. $\frac{22}{45}$ b. $\frac{22}{135}$

EXAMPLE 9 Subtract: $\frac{7}{15} - \frac{3}{10}$

SOLUTION Let's factor 15 and 10 completely and use these factors to build the LCD:

$$\left.\begin{aligned} 15 &= 3 \cdot 5 \\ 10 &= 2 \cdot 5 \end{aligned}\right\} \text{LCD} = 2 \cdot 3 \cdot 5 = 30$$

15 divides the LCD

10 divides the LCD

Changing to equivalent fractions and subtracting, we have

$$\frac{7}{15} - \frac{3}{10} = \frac{7 \cdot \mathbf{2}}{15 \cdot \mathbf{2}} - \frac{3 \cdot \mathbf{3}}{10 \cdot \mathbf{3}}$$ Rewrite as equivalent fractions with the LCD for the denominator

$$= \frac{14}{30} - \frac{9}{30}$$

$$= \frac{5}{30}$$ Subtract numerators; keep the LCD

$$= \frac{1}{6}$$ Reduce to lowest terms

As a summary of what we have done so far, and as a guide to working other problems, we now list the steps involved in adding and subtracting fractions with different denominators.

Strategy Adding or Subtracting Any Two Fractions

Step 1 Factor each denominator completely, and use the factors to build the LCD. (Remember, the LCD is the smallest number divisible by each of the denominators in the problem.)

Step 2 Rewrite each fraction as an equivalent fraction that has the LCD for its denominator. This is done by multiplying both the numerator and the denominator of the fraction in question by the appropriate whole number.

Step 3 Add or subtract the numerators of the fractions produced in Step 2. This is the numerator of the sum or difference. The denominator of the sum or difference is the LCD.

Step 4 Reduce the fraction produced in Step 3 to lowest terms if it is not already in lowest terms.

The idea behind adding or subtracting fractions is really very simple. We can only add or subtract fractions that have the same denominators. If the fractions we are trying to add or subtract do not have the same denominators, we rewrite each of them as an equivalent fraction with the LCD for a denominator.

Here are some additional examples of sums and differences of fractions.

EXAMPLE 10 Subtract: $\frac{x}{5} - \frac{1}{6}$

SOLUTION The LCD for 5 and 6 is their product, 30. We begin by rewriting each fraction with this common denominator:

$$\frac{x}{5} - \frac{1}{6} = \frac{x \cdot \mathbf{6}}{5 \cdot \mathbf{6}} - \frac{1 \cdot \mathbf{5}}{6 \cdot \mathbf{5}}$$

$$= \frac{6x}{30} - \frac{5}{30}$$

$$= \frac{6x - 5}{30}$$

9. Subtract: $\frac{8}{25} - \frac{3}{20}$

10. Subtract: $\frac{3}{4} - \frac{x}{5}$

Answers

9. $\frac{17}{100}$ **10.** $\frac{15 - x}{20}$

11. Add.

a. $\frac{1}{9} + \frac{1}{4} + \frac{1}{6}$

b. $\frac{1}{90} + \frac{1}{40} + \frac{1}{60}$

EXAMPLE 11 Add: $\frac{1}{6} + \frac{1}{8} + \frac{1}{4}$

SOLUTION We begin by factoring the denominators completely and building the LCD from the factors that result:

$$\left.\begin{aligned} 6 &= 2 \cdot 3 \\ 8 &= 2 \cdot 2 \cdot 2 \\ 4 &= 2 \cdot 2 \end{aligned}\right\} \quad \text{LCD} = 2 \cdot 2 \cdot 2 \cdot 3 = 24$$

8 divides the LCD

4 divides the LCD

6 divides the LCD

We then change to equivalent fractions and add as usual:

$$\frac{1}{6} + \frac{1}{8} + \frac{1}{4} = \frac{1 \cdot \mathbf{4}}{6 \cdot \mathbf{4}} + \frac{1 \cdot \mathbf{3}}{8 \cdot \mathbf{3}} + \frac{1 \cdot \mathbf{6}}{4 \cdot \mathbf{6}} = \frac{4}{24} + \frac{3}{24} + \frac{6}{24} = \frac{13}{24}$$

12. Subtract: $2 - \frac{3}{4}$

EXAMPLE 12 Subtract: $3 - \frac{5}{6}$

SOLUTION The denominators are 1 (because $3 = \frac{3}{1}$) and 6. The smallest number divisible by both 1 and 6 is 6.

$$3 - \frac{5}{6} = \frac{3}{1} - \frac{5}{6} = \frac{3 \cdot \mathbf{6}}{1 \cdot \mathbf{6}} - \frac{5}{6} = \frac{18}{6} - \frac{5}{6} = \frac{13}{6}$$

13. Add: $\frac{5}{x} + \frac{2}{3}$

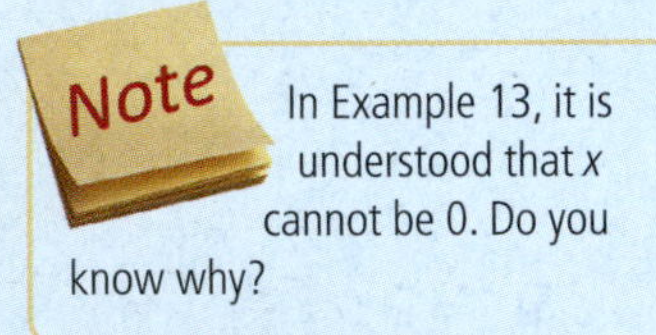

In Example 13, it is understood that x cannot be 0. Do you know why?

EXAMPLE 13 Add: $\frac{4}{x} + \frac{2}{3}$

SOLUTION The LCD for x and 3 is $3x$. We multiply the numerator and the denominator of the first fraction by 3 and the numerator and the denominator of the second fraction by x to get two fractions with the same denominator. We then add the numerators:

$$\frac{4}{x} + \frac{2}{3} = \frac{4 \cdot \mathbf{3}}{x \cdot \mathbf{3}} + \frac{2 \cdot \mathbf{x}}{3 \cdot \mathbf{x}} \quad \text{Change to equivalent fractions}$$

$$= \frac{12}{3x} + \frac{2x}{3x}$$

$$= \frac{12 + 2x}{3x} \quad \text{Add the numerators}$$

14. Simplify: $\frac{x}{4} + \frac{x}{2}$

EXAMPLE 14 Simplify: $\frac{x}{2} + \frac{3x}{4}$

SOLUTION

$$\frac{x}{2} + \frac{3x}{4} = \left(\frac{1}{2} + \frac{3}{4}\right)x \quad \text{Distributive property}$$

$$= \left(\frac{2}{4} + \frac{3}{4}\right)x = \frac{5}{4}x$$

Getting Ready for Class

After reading through the preceding section, respond in your own words and in complete sentences.

1. When adding two fractions with the same denominators, we always add their __________, but we never add their __________.
2. What does the abbreviation *LCD* stand for?
3. What is the first step when finding the LCD for the fractions $\frac{5}{12}$ and $\frac{7}{18}$?
4. When adding fractions, what is the last step?

Answers

11. a. $\frac{19}{36}$ **b.** $\frac{19}{360}$ **12.** $\frac{5}{4}$

13. $\frac{15 + 2x}{3x}$ **14.** $\frac{3x}{4}$

Problem Set 3.5

A Find the following sums and differences, and reduce to lowest terms. (Add or subtract as indicated.) [Examples 1–4]

1. $\frac{3}{6} + \frac{1}{6}$

2. $\frac{2}{5} + \frac{3}{5}$

3. $\frac{3}{8} - \frac{5}{8}$

4. $\frac{1}{7} - \frac{6}{7}$

5. $-\frac{1}{4} + \frac{3}{4}$

6. $-\frac{4}{9} + \frac{7}{9}$

7. $\frac{x}{3} - \frac{1}{3}$

8. $\frac{x}{8} - \frac{1}{8}$

9. $\frac{1}{4} + \frac{2}{4} + \frac{3}{4}$

10. $\frac{2}{5} + \frac{3}{5} + \frac{4}{5}$

11. $\frac{x+7}{2} - \frac{1}{2}$

12. $\frac{x+5}{4} - \frac{3}{4}$

13. $\frac{1}{10} - \frac{3}{10} - \frac{4}{10}$

14. $\frac{3}{20} - \frac{1}{20} - \frac{4}{20}$

15. $\frac{1}{a} + \frac{4}{a} + \frac{5}{a}$

16. $\frac{5}{a} + \frac{4}{a} + \frac{3}{a}$

B Complete the following tables.

17.

First Number a	Second Number b	The Sum of a and b $a + b$
$\frac{1}{2}$	$\frac{1}{3}$	
$\frac{1}{3}$	$\frac{1}{4}$	
$\frac{1}{4}$	$\frac{1}{5}$	
$\frac{1}{5}$	$\frac{1}{6}$	

18.

First Number a	Second Number b	The Sum of a and b $a + b$
1	$\frac{1}{2}$	
1	$\frac{1}{3}$	
1	$\frac{1}{4}$	
1	$\frac{1}{5}$	

19.

First Number a	Second Number b	The Sum of a and b $a + b$
$\frac{1}{12}$	$\frac{1}{2}$	
$\frac{1}{12}$	$\frac{1}{3}$	
$\frac{1}{12}$	$\frac{1}{4}$	
$\frac{1}{12}$	$\frac{1}{6}$	

20.

First Number a	Second Number b	The Sum of a and b $a + b$
$\frac{1}{8}$	$\frac{1}{2}$	
$\frac{1}{8}$	$\frac{1}{4}$	
$\frac{1}{8}$	$\frac{1}{16}$	
$\frac{1}{8}$	$\frac{1}{24}$	

B Find the LCD for each of the following; then use the methods developed in this section to add or subtract as indicated. [Examples 5–14]

21. $\frac{4}{9} + \frac{1}{3}$

22. $\frac{1}{2} + \frac{1}{4}$

23. $2 + \frac{1}{3}$

24. $3 + \frac{1}{2}$

25. $\frac{3}{4} + 1$

26. $\frac{3}{4} + 2$

27. $\frac{1}{2} + \frac{2}{3}$

28. $\frac{1}{8} + \frac{3}{4}$

29. $\frac{x}{4} + \frac{1}{5}$

30. $\frac{x}{3} + \frac{1}{5}$

31. $\frac{2}{x} + \frac{3}{5}$

32. $\frac{3}{x} - \frac{2}{5}$

33. $\frac{5}{12} - \left(-\frac{3}{8}\right)$

34. $\frac{9}{16} - \left(-\frac{7}{12}\right)$

35. $-\frac{1}{20} + \frac{8}{30}$

36. $-\frac{1}{30} + \frac{9}{40}$

37. $\frac{a}{10} + \frac{1}{100}$

38. $\frac{a}{100} + \frac{7}{10}$

39. $\frac{3}{7} + \frac{4}{x}$

40. $\frac{2}{9} + \frac{5}{x}$

41. $\frac{17}{30} + \frac{11}{42}$

42. $\frac{19}{42} + \frac{13}{70}$

43. $\frac{25}{84} + \frac{41}{90}$

44. $\frac{23}{70} + \frac{29}{84}$

45. $\frac{13}{126} - \frac{13}{180}$

46. $\frac{17}{84} - \frac{17}{90}$

47. $\frac{3}{4} + \frac{1}{8} + \frac{5}{6}$

48. $\frac{3}{8} + \frac{2}{5} + \frac{1}{4}$

49. $\frac{4}{y} + \frac{2}{3} + \frac{1}{2}$

50. $\frac{3}{y} + \frac{3}{4} + \frac{1}{5}$

51. $\frac{1}{2} + \frac{1}{3} + \frac{1}{4} + \frac{1}{6}$

52. $\frac{1}{8} + \frac{1}{4} + \frac{1}{5} + \frac{1}{10}$

53. $10 - \frac{2}{9}$

54. $9 - \frac{3}{5}$

55. $\frac{1}{10} + \frac{4}{5} - \frac{3}{20}$

56. $\frac{1}{2} + \frac{3}{4} - \frac{5}{8}$

57. $\frac{1}{4} - \frac{1}{8} + \frac{1}{2} - \frac{3}{8}$

58. $\frac{7}{8} - \frac{3}{4} + \frac{5}{8} - \frac{1}{2}$

There are two ways to work the problems below. You can combine the fractions inside the parentheses first and then multiply, or you can apply the distributive property first, then add.

59. $15\left(\frac{2}{3} + \frac{3}{5}\right)$

60. $15\left(\frac{4}{5} - \frac{1}{3}\right)$

61. $4\left(\frac{1}{2} + \frac{1}{4}\right)$

62. $6\left(\frac{1}{3} + \frac{1}{2}\right)$

63. Find the sum of $\frac{3}{7}$, 2, and $\frac{1}{9}$.

64. Find the sum of 6, $\frac{6}{11}$, and 11.

65. Give the difference of $\frac{7}{8}$ and $\frac{1}{4}$.

66. Give the difference of $\frac{9}{10}$ and $\frac{1}{100}$.

B Apply the distributive property, then find the LCD and simplify.

67. $\frac{1}{2}x + \frac{1}{6}x$

68. $\frac{2}{3}x + \frac{5}{6}x$

69. $\frac{1}{2}x - \frac{3}{4}x$

70. $\frac{2}{3}x - \frac{5}{6}x$

71. $\frac{1}{3}x + \frac{3}{5}x$

72. $\frac{2}{3}x - \frac{3}{5}x$

73. $\frac{3x}{4} + \frac{x}{6}$

74. $\frac{3x}{4} - \frac{2x}{3}$

75. $\frac{2x}{5} + \frac{5x}{8}$

76. $\frac{3x}{5} - \frac{3x}{8}$

C Applying the Concepts

Some of the application problems below involve multiplication or division, while others involve addition or subtraction.

77. Rainfall How much total rainfall did Death Valley get during the months of July and September?

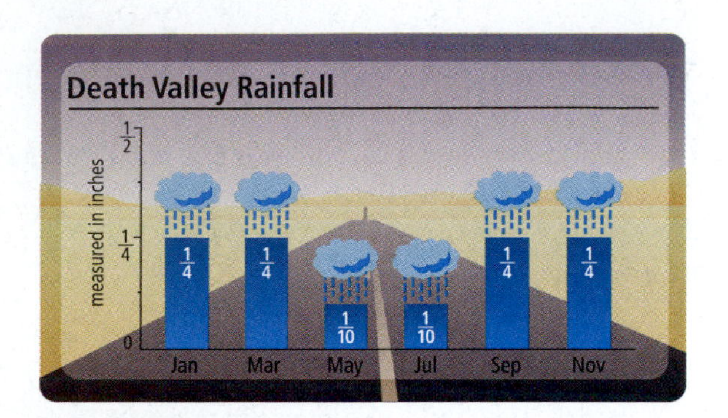

78. Rainfall How much more rainfall did Death Valley get in February than in December?

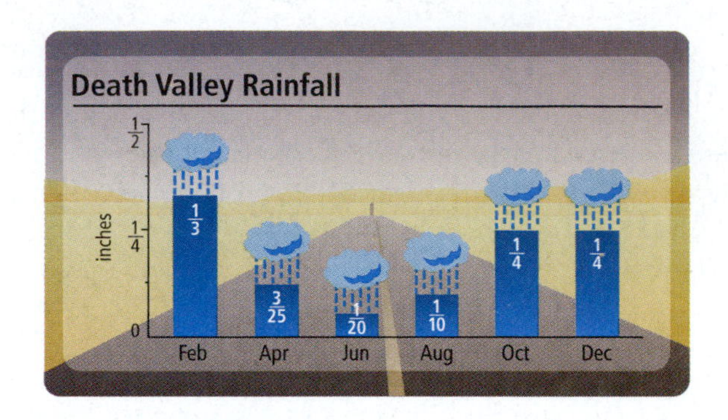

79. Capacity One carton of milk contains $\frac{1}{2}$ pint while another contains 4 pints. How much milk is contained in both cartons?

80. Baking A recipe calls for $\frac{2}{3}$ cup of flour and $\frac{3}{4}$ cup of sugar. What is the total amount of flour and sugar called for in the recipe?

81. Budgeting A student earns $2,500 a month while working in college. She sets aside $\frac{1}{20}$ of this money for gas to travel to and from campus, $\frac{1}{16}$ for food, and $\frac{1}{25}$ for savings. What fraction of her income does she plan to spend on these three items?

82. Popular Majors Enrollment figures show that the most popular programs at a local college are liberal art studies and business programs. The liberal arts studies program accounts for $\frac{1}{5}$ of the student enrollment while business programs account for $\frac{1}{10}$ of the enrollment. What fraction of student enrollment chooses one of these two areas of study?

83. Exercising According to national studies, childhood obesity is on the rise. Doctors recommend a minimum of 30 minutes of exercise three times a week to help keep us fit. Suppose during a given week you walk for $\frac{1}{4}$ hour one day, $\frac{2}{3}$ of an hour a second day and $\frac{3}{4}$ of an hour on a third day. Find the total number of hours walked as a fraction.

84. Cooking You are making pancakes for breakfast and need $\frac{3}{4}$ of a cup of milk for your batter. You discover that you only have $\frac{1}{2}$ cup of milk in the refrigerator. How much more milk do you need?

85. Conference Attendees At a recent mathematics conference $\frac{1}{3}$ of the attendees were teachers, $\frac{1}{4}$ were software salespersons, and $\frac{1}{12}$ were representatives from various book publishing companies. The remainder of the people in the conference center were employees of the center. What fraction represents the employees of the conference center?

86. Painting Recently you purchased $\frac{1}{2}$ gallon of paint to paint your dorm room. Once the job was finished you realized that you only used $\frac{1}{3}$ of the gallon. What fractional amount of the paint is left in the can?

87. Subdivision A 6-acre piece of land is subdivided into $\frac{3}{5}$-acre lots. How many lots are there?

88. Cutting Wood A 12-foot piece of wood is cut into shelves. If each is $\frac{3}{4}$ foot in length, how many shelves are there?

Find the perimeter of each figure.

89.

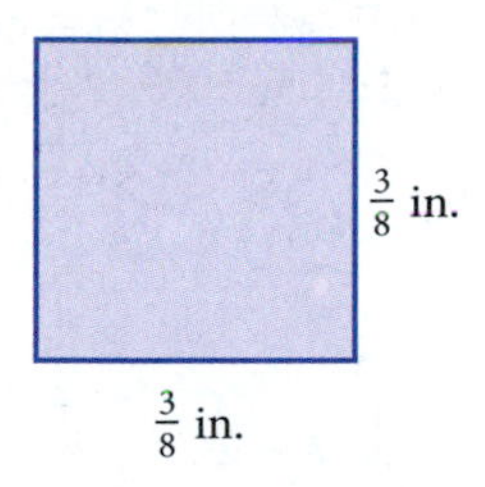

90.

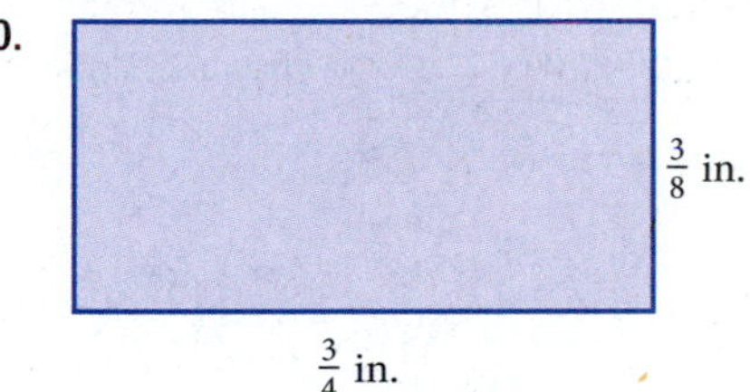

91.

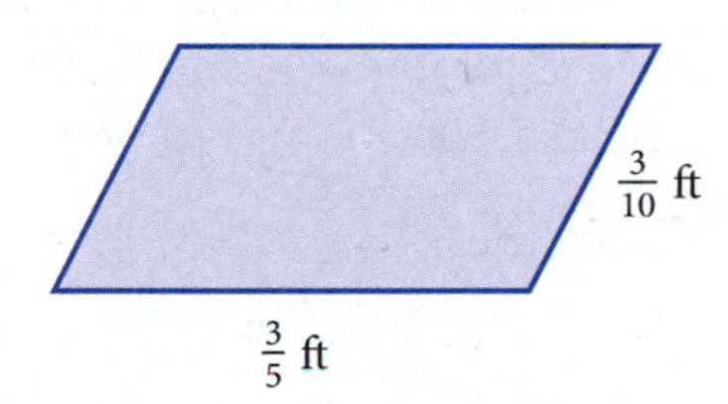

92.

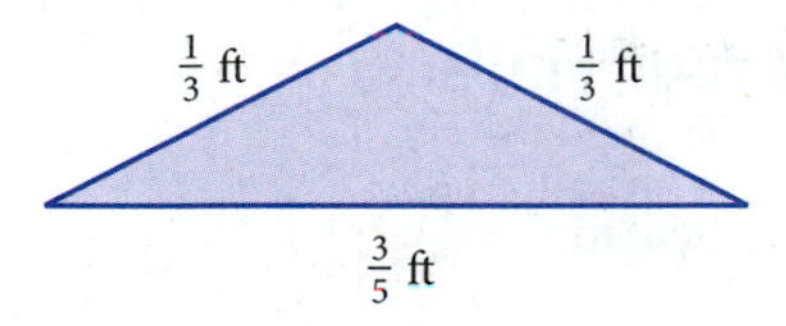

Arithmetic Sequences Recall that an arithmetic sequence is a sequence in which each term comes from the previous term by adding the same number each time. For example, the sequence $1, \frac{3}{2}, 2, \frac{5}{2}, \ldots$ is an arithmetic sequence that starts with the number 1. Then each term after that is found by adding $\frac{1}{2}$ to the previous term. By observing this fact, we know that the next term in the sequence will be $\frac{5}{2} + \frac{1}{2} = \frac{6}{2} = 3$.

Find the next number in each arithmetic sequence below.

93. $1, \frac{4}{3}, \frac{5}{3}, 2, \ldots$

94. $1, \frac{5}{4}, \frac{3}{2}, \frac{7}{4}, \ldots$

95. $\frac{3}{2}, 2, \frac{5}{2}, \ldots$

96. $\frac{2}{3}, 1, \frac{4}{3}, \ldots$

Getting Ready for the Next Section

Simplify.

97. $9 \cdot 6 + 5$

98. $4 \cdot 6 + 3$

99. Write 2 as a fraction with denominator 8.

100. Write 2 as a fraction with denominator 4.

101. Write 1 as a fraction with denominator 8.

102. Write 5 as a fraction with denominator 4.

Add.

103. $\frac{8}{4} + \frac{3}{4}$

104. $\frac{16}{8} + \frac{1}{8}$

105. $2 + \frac{1}{8}$

106. $2 + \frac{3}{4}$

107. $1 + \frac{1}{8}$

108. $5 + \frac{3}{4}$

Divide.

109. $11 \div 4$

110. $10 \div 3$

111. $208 \div 24$

112. $207 \div 26$

Maintaining Your Skills

Multiply or divide as indicated.

113. $\frac{3}{4} \div \frac{5}{6}$

114. $12 \div \frac{1}{2}$

115. $12 \cdot \frac{2}{3}$

116. $12 \cdot \frac{3}{4}$

117. $4 \cdot \frac{3}{4}$

118. $4 \cdot \frac{1}{2}$

119. $\frac{7}{6} \div \frac{7}{12}$

120. $\frac{9}{10} \div \frac{7}{10}$

121. $\frac{2}{3} \cdot \frac{3}{4} \cdot \frac{4}{5} \cdot \frac{5}{6} \cdot \frac{6}{7}$

122. $\frac{11}{12} \cdot \frac{10}{11} \cdot \frac{9}{10} \cdot \frac{8}{9} \cdot \frac{7}{8}$

123. $\frac{35}{110} \cdot \frac{80}{63} \div \frac{16}{27}$

124. $\frac{20}{72} \cdot \frac{42}{18} \div \frac{20}{16}$

Mixed-Number Notation

Objectives

A Change mixed numbers to improper fractions.

B Change improper fractions to mixed numbers.

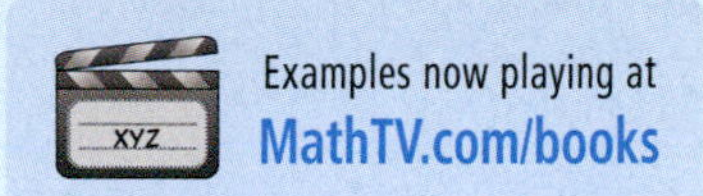

Introduction . . .

If you are interested in the stock market, you know that, prior to the year 2000, stock prices were given in eighths. For example, at one point in 1990, one share of Intel Corporation was selling at $\$73\frac{5}{8}$, or seventy-three and five-eighths dollars. The number $73\frac{5}{8}$ is called a *mixed number.* It is the sum of a whole number and a proper fraction. With mixed-number notation, we leave out the addition sign.

Notation

A number such as $5\frac{3}{4}$ is called a *mixed number* and is equal to $5 + \frac{3}{4}$. It is simply the sum of the whole number 5 and the proper fraction $\frac{3}{4}$, written without a + sign. Here are some further examples:

$$2\frac{1}{8} = 2 + \frac{1}{8}, \quad 6\frac{5}{9} = 6 + \frac{5}{9}, \quad 11\frac{2}{3} = 11 + \frac{2}{3}$$

The notation used in writing mixed numbers (writing the whole number and the proper fraction next to each other) must always be interpreted as addition. It is a mistake to read $5\frac{3}{4}$ as meaning 5 times $\frac{3}{4}$. If we want to indicate multiplication, we must use parentheses or a multiplication symbol. That is:

$5\frac{3}{4}$ is not the same as $5\left(\frac{3}{4}\right)$

This implies addition — This implies multiplication

$5\frac{3}{4}$ is not the same as $5 \cdot \frac{3}{4}$

A Changing Mixed Numbers to Improper Fractions

To change a mixed number to an improper fraction, we write the mixed number with the + sign showing and then add the two numbers, as we did earlier.

EXAMPLE 1 Change $2\frac{3}{4}$ to an improper fraction.

SOLUTION

$2\frac{3}{4} = 2 + \frac{3}{4}$ — Write the mixed number as a sum

$= \frac{2}{1} + \frac{3}{4}$ — Show that the denominator of 2 is 1

$= \frac{\mathbf{4} \cdot 2}{\mathbf{4} \cdot 1} + \frac{3}{4}$ — Multiply the numerator and the denominator of $\frac{2}{1}$ by 4 so both fractions will have the same denominator

$= \frac{8}{4} + \frac{3}{4}$

$= \frac{11}{4}$ — Add the numerators; keep the common denominator

The mixed number $2\frac{3}{4}$ is equal to the improper fraction $\frac{11}{4}$. The diagram that follows further illustrates the equivalence of $2\frac{3}{4}$ and $\frac{11}{4}$.

PRACTICE PROBLEMS

1. Change $5\frac{2}{3}$ to an improper fraction.

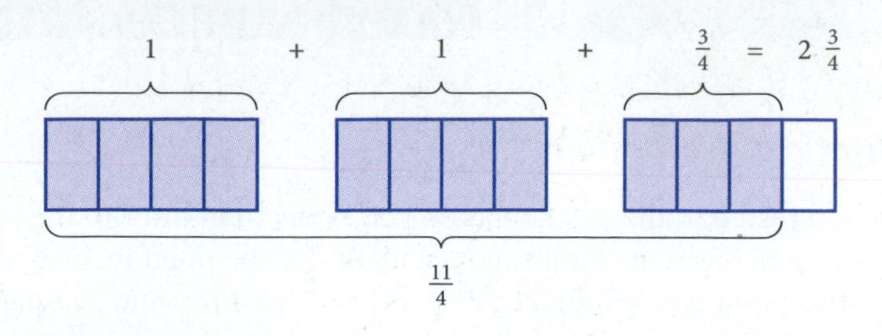

2. Change $3\frac{1}{6}$ to an improper fraction.

EXAMPLE 2 Change $2\frac{1}{8}$ to an improper fraction.

SOLUTION

$$2\frac{1}{8} = 2 + \frac{1}{8} \quad \text{Write as addition}$$

$$= \frac{2}{1} + \frac{1}{8} \quad \text{Write the whole number 2 as a fraction}$$

$$= \frac{\mathbf{8} \cdot 2}{\mathbf{8} \cdot 1} + \frac{1}{8} \quad \text{Change } \frac{2}{1} \text{ to a fraction with denominator 8}$$

$$= \frac{16}{8} + \frac{1}{8}$$

$$= \frac{17}{8} \quad \text{Add the numerators}$$

If we look closely at Examples 1 and 2, we can see a shortcut that will let us change a mixed number to an improper fraction without so many steps.

Strategy Changing a Mixed Number to an Improper Fraction (Shortcut)

Step 1: Multiply the whole number part of the mixed number by the denominator.

Step 2: Add your answer to the numerator of the fraction.

Step 3: Put your new number over the original denominator.

3. Use the shortcut to change $5\frac{2}{3}$ to an improper fraction.

EXAMPLE 3 Use the shortcut to change $5\frac{3}{4}$ to an improper fraction.

SOLUTION

1. First, we multiply 4×5 to get 20.
2. Next, we add 20 to 3 to get 23.
3. The improper fraction equal to $5\frac{3}{4}$ is $\frac{23}{4}$.

Here is a diagram showing what we have done:

Step 1 Multiply $4 \times 5 = 20$.

$$5\frac{3}{4}$$

Step 2 Add $20 + 3 = 23$.

Mathematically, our shortcut is written like this:

$$5\frac{3}{4} = \frac{(4 \cdot 5) + 3}{4} = \frac{20 + 3}{4} = \frac{23}{4}$$

The result will always have the same denominator as the original mixed number

The shortcut shown in Example 3 works because the whole-number part of a mixed number can always be written with a denominator of 1. Therefore, the LCD for a whole number and fraction will always be the denominator of the fraction. That is why we multiply the whole number by the denominator of the fraction:

$$5\frac{3}{4} = 5 + \frac{3}{4} = \frac{5}{1} + \frac{3}{4} = \frac{\mathbf{4} \cdot 5}{\mathbf{4} \cdot 1} + \frac{3}{4} = \frac{4 \cdot 5 + 3}{4} = \frac{23}{4}$$

Answers

1. $\frac{17}{3}$ 2. $\frac{19}{6}$ 3. $\frac{17}{3}$

EXAMPLE 4 Change $6\frac{5}{9}$ to an improper fraction.

SOLUTION Using the first method, we have

$$6\frac{5}{9} = 6 + \frac{5}{9} = \frac{6}{1} + \frac{5}{9} = \frac{\mathbf{9 \cdot 6}}{\mathbf{9 \cdot 1}} + \frac{5}{9} = \frac{54}{9} + \frac{5}{9} = \frac{59}{9}$$

Using the shortcut method, we have

$$6\frac{5}{9} = \frac{(9 \cdot 6) + 5}{9} = \frac{54 + 5}{9} = \frac{59}{9}$$

4. Change $6\frac{4}{9}$ to an improper fraction.

CALCULATOR NOTE

The sequence of keys to press on a calculator to obtain the numerator in Example 4 looks like this:

9 [×] 6 [+] 5 [=]

B Changing Improper Fractions to Mixed Numbers

To change an improper fraction to a mixed number, we divide the numerator by the denominator. The result is used to write the mixed number.

EXAMPLE 5 Change $\frac{11}{4}$ to a mixed number.

SOLUTION Dividing 11 by 4 gives us

$$\begin{array}{r} 2 \\ 4\overline{)11} \\ \underline{8} \\ 3 \end{array}$$

We see that 4 goes into 11 two times with 3 for a remainder. We write this as

$$\frac{11}{4} = 2 + \frac{3}{4} = 2\frac{3}{4}$$

The improper fraction $\frac{11}{4}$ is equivalent to the mixed number $2\frac{3}{4}$.

5. Change $\frac{11}{3}$ to a mixed number.

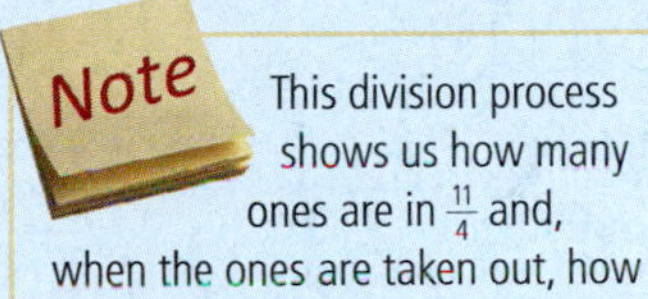

This division process shows us how many ones are in $\frac{11}{4}$ and, when the ones are taken out, how many fourths are left.

An easy way to visualize the results in Example 5 is to imagine having 11 quarters. Your 11 quarters are equivalent to $\frac{11}{4}$ dollars. In dollars, your quarters are worth 2 dollars plus 3 quarters, or $2\frac{3}{4}$ dollars.

EXAMPLE 6 Change $\frac{10}{3}$ to a mixed number.

SOLUTION $\frac{10}{3}$: $\begin{array}{r} \mathbf{3} \\ 3\overline{)10} \\ \underline{9} \\ \mathbf{1} \end{array}$ so $\frac{10}{3} = \mathbf{3} + \frac{\mathbf{1}}{3} = 3\frac{1}{3}$

Change each improper fraction to a mixed number.

6. $\frac{14}{5}$

EXAMPLE 7 Change $\frac{208}{24}$ to a mixed number.

SOLUTION $\frac{208}{24}$: $\begin{array}{r} \mathbf{8} \\ 24\overline{)208} \\ \underline{192} \\ \mathbf{16} \end{array}$ so $\frac{208}{24} = \mathbf{8} + \frac{\mathbf{16}}{24} = 8 + \frac{2}{3} = 8\frac{2}{3}$

Reduce to lowest terms

7. $\frac{207}{26}$

Answers

4. $\frac{58}{9}$ 5. $3\frac{2}{3}$ 6. $2\frac{4}{5}$ 7. $7\frac{25}{26}$

In the first part of this section, we changed mixed numbers to improper fractions. For example we changed $2\frac{3}{4}$ to an improper fraction by adding 2 and $\frac{3}{4}$ (see Example 1). An extension of this concept to algebra would be to add 2 and $\frac{3}{x}$.

8. Add: $3 + \frac{2}{x}$

EXAMPLE 8 Add: $2 + \frac{3}{x}$

SOLUTION We can write 2 as $\frac{2}{1}$:

$$2 + \frac{3}{x} = \frac{2}{1} + \frac{3}{x}$$

Now, the LCD for 1 and x is $1x$ or just x. To change to equivalent fractions, we multiply the numerator and the denominator of $\frac{2}{1}$ by x.

$$\begin{aligned} \frac{2}{1} + \frac{3}{x} &= \frac{2 \cdot \mathbf{x}}{1 \cdot \mathbf{x}} + \frac{3}{x} && \text{The LCD is } x \\ &= \frac{2x}{x} + \frac{3}{x} \\ &= \frac{2x + 3}{x} && \text{Add numerators} \end{aligned}$$

9. Subtract: $x - \frac{2}{3}$

EXAMPLE 9 Subtract: $x - \frac{3}{4}$

SOLUTION This time we write x as $\frac{x}{1}$. The LCD for 1 and for 4 is 4.

$$\begin{aligned} x - \frac{3}{4} &= \frac{x}{1} - \frac{3}{4} && \text{Write } x \text{ as } \frac{x}{1} \\ &= \frac{\mathbf{4} \cdot x}{\mathbf{4} \cdot 1} - \frac{3}{4} && \text{The LCD is 4} \\ &= \frac{4x}{4} - \frac{3}{4} \\ &= \frac{4x - 3}{4} && \text{Subtract numerators} \end{aligned}$$

As a final note in this section, we should mention that negative mixed numbers are thought of this way:

$$-3\frac{2}{5} = -3 - \frac{2}{5}$$

Both the whole-number part and the fraction part of the mixed number are negative when the mixed number is negative.

Getting Ready for Class

After reading through the preceding section, respond in your own words and in complete sentences.

1. What is a mixed number?
2. The expression $5\frac{3}{4}$ is equivalent to what addition problem?
3. The improper fraction $\frac{11}{4}$ is equivalent to what mixed number?
4. Why is $\frac{13}{5}$ an improper fraction, but $\frac{3}{5}$ is not an improper fraction?

Answers

8. $\frac{3x + 2}{x}$ 9. $\frac{3x - 2}{3}$

Problem Set 3.6

A Change each mixed number to an improper fraction. [Examples 1–4]

1. $4\frac{2}{3}$ **2.** $3\frac{5}{8}$ **3.** $5\frac{1}{4}$ **4.** $7\frac{1}{2}$ **5.** $1\frac{5}{8}$ **6.** $1\frac{6}{7}$

7. $15\frac{2}{3}$ **8.** $17\frac{3}{4}$ **9.** $4\frac{20}{21}$ **10.** $5\frac{18}{19}$ **11.** $12\frac{31}{33}$ **12.** $14\frac{29}{31}$

B Change each improper fraction to a mixed number. [Examples 5–7]

13. $\frac{9}{8}$ **14.** $\frac{10}{9}$ **15.** $\frac{19}{4}$ **16.** $\frac{23}{5}$ **17.** $\frac{29}{6}$ **18.** $\frac{7}{2}$

19. $\frac{13}{4}$ **20.** $\frac{41}{15}$ **21.** $\frac{109}{27}$ **22.** $\frac{319}{23}$ **23.** $\frac{428}{15}$ **24.** $\frac{769}{27}$

Add or subtract as indicated. [Examples 8, 9]

25. $8 + \frac{3}{x}$ **26.** $7 + \frac{9}{x}$ **27.** $2 - \frac{5}{y}$ **28.** $3 - \frac{4}{y}$ **29.** $x + \frac{5}{6}$ **30.** $x + \frac{2}{7}$

31. $a - \frac{1}{3}$ **32.** $a - \frac{1}{4}$ **33.** $5 + \frac{3}{2x}$ **34.** $7 - \frac{1}{2x}$ **35.** $3 - \frac{2}{3x}$ **36.** $4 + \frac{5}{3x}$

Getting Ready for the Next Section

Change to improper fractions.

37. $2\frac{3}{4}$ **38.** $3\frac{1}{5}$ **39.** $4\frac{5}{8}$ **40.** $1\frac{3}{5}$ **41.** $2\frac{4}{5}$ **42.** $5\frac{9}{10}$

Find the following products. (Multiply.)

43. $\frac{3}{8} \cdot \frac{3}{5}$ **44.** $\frac{11}{4} \cdot \frac{16}{5}$ **45.** $\frac{2}{3}\left(\frac{9}{16}\right)$ **46.** $\frac{7}{10}\left(\frac{5}{21}\right)$

Find the quotients. (Divide.)

47. $\frac{4}{5} \div \frac{7}{8}$ **48.** $\frac{3}{4} \div \frac{1}{2}$ **49.** $\frac{8}{5} \div \frac{14}{5}$ **50.** $\frac{59}{10} \div 2$

Maintaining Your Skills

Perform the indicated operations.

51. $4(3x - 2)$ **52.** $-5(2x + 3)$ **53.** $7(4x - 7)$ **54.** $-6(5x + 4)$

Perform the indicated operations.

55. $\frac{2}{3} \cdot \frac{3}{4} \div \frac{5}{8}$ **56.** $4 \cdot \frac{7}{6} \div 7$

57. $\frac{3}{4} \div \frac{1}{2} \cdot 6$ **58.** $\frac{2}{3} \div \frac{1}{6} \cdot 12$

59. $12 \div \frac{6}{7} \cdot 7$ **60.** $15 \div \frac{5}{8} \cdot 16$

Multiplication and Division with Mixed Numbers

3.7

Objectives

A Multiply mixed numbers.

B Divide mixed numbers.

C Solve applications involving multiplying and dividing mixed numbers.

Examples now playing at MathTV.com/books

Introduction . . .

The figure here shows one of the nutrition labels we worked with in Chapter 1. It is from a can of Italian tomatoes. Notice toward the top of the label, the number of servings in the can is $3\frac{1}{2}$. The number $3\frac{1}{2}$ is called a *mixed number*. If we want to know how many calories are in the whole can of tomatoes, we must be able to multiply $3\frac{1}{2}$ by 25 (the number of calories per serving). Multiplication with mixed numbers is one of the topics we will cover in this section.

The procedures for multiplying and dividing mixed numbers are the same as those we used in Sections 3.3 and 3.4 to multiply and divide fractions. The only additional work involved is in changing the mixed numbers to improper fractions before we actually multiply or divide.

CANNED ITALIAN TOMATOES

Nutrition Facts
Serving Size 1/2 cup (121g)
Servings Per Container: about 3 1/2

Amount Per Serving	
Calories 25	Calories from fat 0
	% Daily Value*
Total Fat 0g	0%
Saturated Fat 0g	0%
Cholesterol 0mg	0%
Sodium 300mg	12%
Potassium 145mg	4%
Total Carbohydrate 4g	2%
Dietary Fiber 1g	4%
Sugars 4g	
Protein 1g	
Vitamin A 20% •	Vitamin C 15%
Calcium 4% •	Iron 15%

*Percent Daily Values are based on a 2,000 calorie diet. Your daily values may be higher or lower depending on your calorie needs.

A Multiplying Mixed Numbers

EXAMPLE 1 Multiply: $2\frac{3}{4} \cdot 3\frac{1}{5}$

SOLUTION We begin by changing each mixed number to an improper fraction:

$$2\frac{3}{4} = \frac{11}{4} \quad \text{and} \quad 3\frac{1}{5} = \frac{16}{5}$$

Using the resulting improper fractions, we multiply as usual. (That is, we multiply numerators and multiply denominators.)

$$\frac{11}{4} \cdot \frac{16}{5} = \frac{11 \cdot 16}{4 \cdot 5}$$
$$= \frac{11 \cdot \cancel{4} \cdot 4}{\cancel{4} \cdot 5}$$
$$= \frac{44}{5} \quad \text{or} \quad 8\frac{4}{5}$$

EXAMPLE 2 Multiply: $3 \cdot 4\frac{5}{8}$

SOLUTION Writing each number as an improper fraction, we have

$$3 = \frac{3}{1} \quad \text{and} \quad 4\frac{5}{8} = \frac{37}{8}$$

The complete problem looks like this:

$$3 \cdot 4\frac{5}{8} = \frac{3}{1} \cdot \frac{37}{8}$$ Change to improper fractions

$$= \frac{111}{8}$$ Multiply numerators and multiply denominators

$$= 13\frac{7}{8}$$ Write the answer as a mixed number

PRACTICE PROBLEMS

1. Multiply: $2\frac{3}{4} \cdot 4\frac{1}{3}$

2. Multiply: $2 \cdot 3\frac{5}{8}$

As you can see, once you have changed each mixed number to an improper fraction, you multiply the resulting fractions the same way you did in Section 3.3.

Answers

1. $11\frac{11}{12}$ 2. $7\frac{1}{4}$

3. Divide: $1\frac{3}{5} \div 3\frac{2}{5}$

4. Divide: $4\frac{5}{8} \div 2$

B Dividing Mixed Numbers

Dividing mixed numbers also requires that we change all mixed numbers to improper fractions before we actually do the division.

EXAMPLE 3 Divide: $1\frac{3}{5} \div 2\frac{4}{5}$

SOLUTION We begin by rewriting each mixed number as an improper fraction:

$$1\frac{3}{5} = \frac{8}{5} \quad \text{and} \quad 2\frac{4}{5} = \frac{14}{5}$$

We then divide using the same method we used in Section 3.4. Remember? We multiply by the reciprocal of the divisor. Here is the complete problem:

$$1\frac{3}{5} \div 2\frac{4}{5} = \frac{8}{5} \div \frac{14}{5}$$ Change to improper fractions

$$= \frac{8}{5} \cdot \frac{5}{14}$$ To divide by $\frac{14}{5}$, multiply by $\frac{5}{14}$

$$= \frac{8 \cdot 5}{5 \cdot 14}$$ Multiply numerators and multiply denominators

$$= \frac{4 \cdot \cancel{2} \cdot \cancel{5}}{\cancel{5} \cdot \cancel{2} \cdot 7}$$ Divide out factors common to the numerator and denominator

$$= \frac{4}{7}$$ Answer in lowest terms

EXAMPLE 4 Divide: $5\frac{9}{10} \div 2$

SOLUTION We change to improper fractions and proceed as usual:

$$5\frac{9}{10} \div 2 = \frac{59}{10} \div \frac{2}{1}$$ Write each number as an improper fraction

$$= \frac{59}{10} \cdot \frac{1}{2}$$ Write division as multiplication by the reciprocal

$$= \frac{59}{20}$$ Multiply numerators and multiply denominators

$$= 2\frac{19}{20}$$ Change to a mixed number

Getting Ready for Class

After reading through the preceding section, respond in your own words and in complete sentences.

1. What is the first step when multiplying or dividing mixed numbers?
2. What is the reciprocal of $2\frac{4}{5}$?
3. Dividing $5\frac{9}{10}$ by 2 is equivalent to multiplying $5\frac{9}{10}$ by what number?
4. Find $4\frac{5}{8}$ of 3.

Answers

3. $\frac{8}{17}$ 4. $2\frac{5}{16}$

Problem Set 3.7

A Write your answers as proper fractions or mixed numbers, not as improper fractions. Find the following products. (Multiply.) [Examples 1, 2]

1. $3\frac{2}{5} \cdot 1\frac{1}{2}$

2. $2\frac{1}{3} \cdot 6\frac{3}{4}$

3. $5\frac{1}{8} \cdot 2\frac{2}{3}$

4. $1\frac{5}{6} \cdot 1\frac{4}{5}$

5. $2\frac{1}{10} \cdot 3\frac{3}{10}$

6. $4\frac{7}{10} \cdot 3\frac{1}{10}$

7. $1\frac{1}{4} \cdot 4\frac{2}{3}$

8. $3\frac{1}{2} \cdot 2\frac{1}{6}$

9. $2 \cdot 4\frac{7}{8}$

10. $10 \cdot 1\frac{1}{4}$

11. $\frac{3}{5} \cdot 5\frac{1}{3}$

12. $\frac{2}{3} \cdot 4\frac{9}{10}$

13. $2\frac{1}{2} \cdot 3\frac{1}{3} \cdot 1\frac{1}{2}$

14. $3\frac{1}{5} \cdot 5\frac{1}{6} \cdot 1\frac{1}{8}$

15. $\frac{3}{4} \cdot 7 \cdot 1\frac{4}{5}$

16. $\frac{7}{8} \cdot 6 \cdot 1\frac{5}{6}$

B Find the following quotients. (Divide.) [Examples 3, 4]

17. $3\frac{1}{5} \div 4\frac{1}{2}$

18. $1\frac{4}{5} \div 2\frac{5}{6}$

19. $6\frac{1}{4} \div 3\frac{3}{4}$

20. $8\frac{2}{3} \div 4\frac{1}{3}$

21. $10 \div 2\frac{1}{2}$

22. $12 \div 3\frac{1}{6}$

23. $8\frac{3}{5} \div 2$

24. $12\frac{6}{7} \div 3$

25. $\left(\frac{3}{4} \div 2\frac{1}{2}\right) \div 3$

26. $\frac{7}{8} \div \left(1\frac{1}{4} \div 4\right)$

27. $\left(8 \div 1\frac{1}{4}\right) \div 2$

28. $8 \div \left(1\frac{1}{4} \div 2\right)$

29. $2\frac{1}{2} \cdot \left(3\frac{2}{5} \div 4\right)$

30. $4\frac{3}{5} \cdot \left(2\frac{1}{4} \div 5\right)$

31. Find the product of $2\frac{1}{2}$ and 3.

32. Find the product of $\frac{1}{5}$ and $3\frac{2}{3}$.

33. What is the quotient of $2\frac{3}{4}$ and $3\frac{1}{4}$?

34. What is the quotient of $1\frac{1}{5}$ and $2\frac{2}{5}$?

C Applying the Concepts

35. Cooking A certain recipe calls for $2\frac{3}{4}$ cups of sugar. If the recipe is to be doubled, how much sugar should be used?

36. Cooking A recipe calls for $3\frac{1}{4}$ cups of flour. If Diane is using only half the recipe, how much flour should she use?

37. Number Problem Find $\frac{3}{4}$ of $1\frac{7}{9}$. (Remember that *of* means multiply.)

38. Number Problem Find $\frac{5}{6}$ of $2\frac{4}{15}$.

39. Cost of Gasoline If a gallon of gas costs $335\frac{9}{10}$¢, how much do 8 gallons cost?

40. Cost of Gasoline If a gallon of gas costs $353\frac{9}{10}$¢, how much does $\frac{1}{2}$ gallon cost?

41. Distance Traveled If a car can travel $32\frac{3}{4}$ miles on a gallon of gas, how far will it travel on 5 gallons of gas?

42. Sewing If it takes $1\frac{1}{2}$ yards of material to make a pillow cover, how much material will it take to make 3 pillow covers?

43. Buying Stocks Assume that you have $1000 to invest in the stock market. Because you own an iPod™ and an iPhone™, you decide to buy Apple stock. It is currently selling at a cost of $\$150\frac{7}{8}$ per share. At this price how many shares can you buy?

44. Subdividing Land A local developer owns $145\frac{3}{4}$ acres of land that he hopes to subdivide into $2\frac{1}{2}$ acre home site lots to sell. How many home sites can be developed from this tract of land?

45. Selling Stocks You inherit 100 shares of Cisco stock that has a current value of $\$25\frac{1}{6}$ per share. How much will you receive when you sell the stock?

46. Gas Mileage You won a new car and are anxious to see what kind of gas mileage you get. You travel $427\frac{1}{5}$ miles before needing to fill your tank. You purchase $13\frac{3}{4}$ gallons of gas. How many miles were you able to travel on a single gallon of gas?

47. Area Find the area of a bedroom that measures $11\frac{1}{2}$ ft by $15\frac{7}{8}$ ft.

48. Building Shelves You are building a small bookcase. You need three shelves, each with a length of $4\frac{7}{8}$ ft. You bought a piece of wood that is 15 ft long. Will this board be long enough?

49. The Google Earth image shows some fields in the midwestern part of the United States. The rectangle outlines a corn field, and gives the dimensions in miles. Find the area of the corn field written as a mixed number.

50. The Google Earth map shows Crater Lake National Park in Oregon. If Crater Lake is roughly the shape of a circle with a radius of $2\frac{1}{2}$ miles, how long is the shoreline? Use $\frac{22}{7}$ for π.

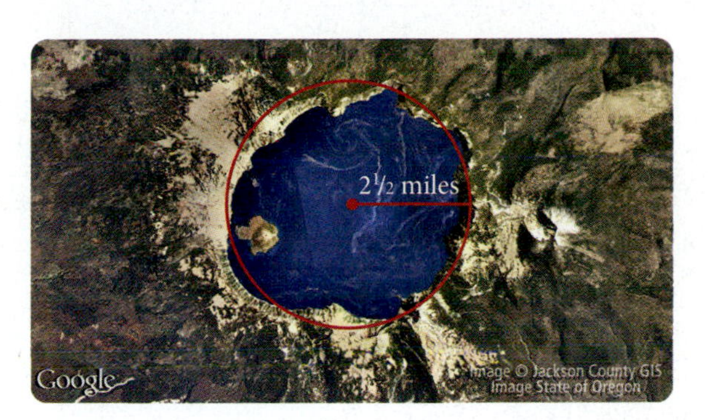

Nutrition The figure below shows nutrition labels for two different cans of Italian tomatoes.

CANNED TOMATOES 1

Nutrition Facts
Serving Size 1/2 cup (121g)
Servings Per Container: about 3 1/2

Amount Per Serving	
Calories 45	Calories from fat 0
	% Daily Value*
Total Fat 0g	**0%**
Saturated Fat 0g	**0%**
Cholesterol 0mg	**0%**
Sodium 560mg	**23%**
Total Carbohydrate 11g	**4%**
Dietary Fiber 2g	**8%**
Sugars 9g	
Protein 1g	
Vitamin A 10% •	Vitamin C 25%
Calcium 2% •	Iron 2%

*Percent Daily Values are based on a 2,000 calorie diet.

CANNED TOMATOES 2

Nutrition Facts
Serving Size 1/2 cup (121g)
Servings Per Container: about 3 1/2

Amount Per Serving	
Calories 25	Calories from fat 0
	% Daily Value*
Total Fat 0g	**0%**
Saturated Fat 0g	**0%**
Cholesterol 0mg	**0%**
Sodium 300mg	**12%**
Potassium 145mg	**4%**
Total Carbohydrate 4g	**2%**
Dietary Fiber 1g	**4%**
Sugars 4g	
Protein 1g	
Vitamin A 20% •	Vitamin C 15%
Calcium 4% •	Iron 15%

*Percent Daily Values are based on a 2,000 calorie diet. Your daily values may be higher or lower depending on your calorie needs.

51. Compare the total number of calories in the two cans of tomatoes.

52. Compare the total amount of sugar in the two cans of tomatoes.

53. Compare the total amount of sodium in the two cans of tomatoes.

54. Compare the total amount of protein in the two cans of tomatoes.

Getting Ready for the Next Section

55. Write as equivalent fractions with denominator 15.

a. $\frac{2}{3}$ **b.** $\frac{1}{5}$ **c.** $\frac{3}{5}$ **d.** $\frac{1}{3}$

56. Write as equivalent fractions with denominator 12.

a. $\frac{3}{4}$ **b.** $\frac{1}{3}$ **c.** $\frac{5}{6}$ **d.** $\frac{1}{4}$

57. Write as equivalent fractions with denominator 20.

a. $\frac{1}{4}$ **b.** $\frac{3}{5}$ **c.** $\frac{9}{10}$ **d.** $\frac{1}{10}$

58. Write as equivalent fractions with denominator 24.

a. $\frac{3}{4}$ **b.** $\frac{7}{8}$ **c.** $\frac{5}{8}$ **d.** $\frac{3}{8}$

Maintaining Your Skills

Add or subtract the following fractions, as indicated.

59. $\frac{2}{3} + \frac{1}{5}$

60. $\frac{3}{4} + \frac{5}{6}$

61. $\frac{2}{3} + \frac{8}{9}$

62. $\frac{1}{4} + \frac{3}{5} + \frac{9}{10}$

63. $\frac{9}{10} - \frac{3}{10}$

64. $\frac{7}{10} - \frac{3}{5}$

65. $\frac{1}{14} + \frac{3}{21}$

66. $\frac{5}{12} - \frac{1}{3}$

Extending the Concepts

To find the square of a mixed number, we first change the mixed number to an improper fraction, and then we square the result. For example:

$$\left(2\frac{1}{2}\right)^2 = \left(\frac{5}{2}\right)^2 = \frac{25}{4}$$

If we are asked to write our answer as a mixed number, we write it as $6\frac{1}{4}$.

Find each of the following squares, and write your answers as mixed numbers.

67. $\left(1\frac{1}{2}\right)^2$

68. $\left(3\frac{1}{2}\right)^2$

69. $\left(1\frac{3}{4}\right)^2$

70. $\left(2\frac{3}{4}\right)^2$

Addition and Subtraction with Mixed Numbers

3.8

Objectives

A Perform addition and subtraction with mixed numbers.

B Perform subtraction involving borrowing with mixed numbers.

C Solve application problems involving addition and subtraction with mixed numbers.

Examples now playing at MathTV.com/books

Introduction . . .

In March 1995, rumors that Michael Jordan would return to basketball sent stock prices for the companies whose products he endorsed higher. The price of one share of General Mills, the maker of Wheaties, which Michael Jordan endorses, went from $\$60\frac{1}{2}$ to $\$63\frac{3}{8}$. To find the increase in the price of this stock, we must be able to subtract mixed numbers.

The notation we use for mixed numbers is especially useful for addition and subtraction. When adding and subtracting mixed numbers, we will assume you recall how to go about finding a least common denominator (LCD). (If you don't remember, then review Section 3.5.)

A Combining Mixed Numbers

Add: $3\frac{2}{3} + 4\frac{1}{5}$

SOLUTION **Method 1:** We begin by writing each mixed number showing the + sign. We then apply the commutative and associative properties to rearrange the order and grouping:

$$3\frac{2}{3} + 4\frac{1}{5} = 3 + \frac{2}{3} + 4 + \frac{1}{5} \quad \text{Expand each number to show the + sign}$$

$$= 3 + 4 + \frac{2}{3} + \frac{1}{5} \quad \text{Commutative property}$$

$$= (3 + 4) + \left(\frac{2}{3} + \frac{1}{5}\right) \quad \text{Associative property}$$

$$= 7 + \left(\frac{\mathbf{5} \cdot 2}{\mathbf{5} \cdot 3} + \frac{\mathbf{3} \cdot 1}{\mathbf{3} \cdot 5}\right) \quad \text{Add } 3 + 4 = 7\text{; then multiply to get the LCD}$$

$$= 7 + \left(\frac{10}{15} + \frac{3}{15}\right) \quad \text{Write each fraction with the LCD}$$

$$= 7 + \frac{13}{15} \quad \text{Add the numerators}$$

$$= 7\frac{13}{15} \quad \text{Write the answer in mixed-number notation}$$

Method 2: As you can see, we obtain our result by adding the whole-number parts $(3 + 4 = 7)$ and the fraction parts $\left(\frac{2}{3} + \frac{1}{5} = \frac{13}{15}\right)$ of each mixed number. Knowing this, we can save ourselves some writing by doing the same problem in columns:

$$\begin{array}{rl} 3\frac{2}{3} = 3\frac{2 \cdot \mathbf{5}}{3 \cdot \mathbf{5}} = 3\frac{10}{15} & \text{Add whole numbers} \\ + 4\frac{1}{5} = 4\frac{1 \cdot \mathbf{3}}{5 \cdot \mathbf{3}} = 4\frac{3}{15} & \text{Then add fractions} \\ \hline 7\frac{13}{15} & \end{array}$$

Write each fraction with LCD 15

The second method shown above requires less writing and lends itself to mixed-number notation. We will use this method for the rest of this section.

PRACTICE PROBLEMS

1. Add: $3\frac{2}{3} + 2\frac{1}{4}$

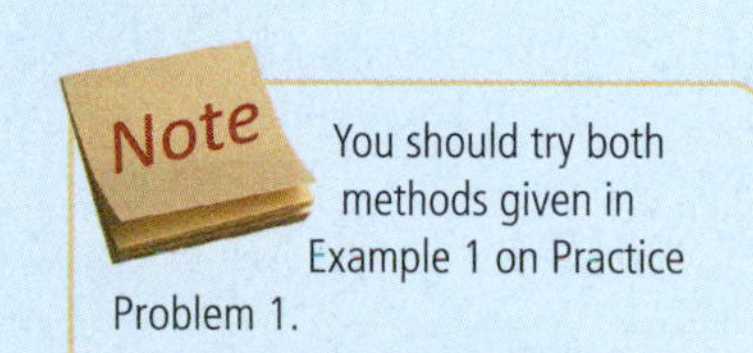

You should try both methods given in Example 1 on Practice Problem 1.

Answer

1. $5\frac{11}{12}$

2. Add: $5\frac{3}{4} + 6\frac{4}{5}$

EXAMPLE 2 Add: $5\frac{3}{4} + 9\frac{5}{6}$

SOLUTION The LCD for 4 and 6 is 12. Writing the mixed numbers in a column and then adding looks like this:

$$\begin{array}{rcl} 5\frac{3}{4} = 5\frac{3\cdot 3}{4\cdot 3} & = & 5\frac{9}{12} \\ +\,9\frac{5}{6} = 9\frac{5\cdot 2}{6\cdot 2} & = & 9\frac{10}{12} \\ \hline & & 14\frac{19}{12} \end{array}$$

The fraction part of the answer is an improper fraction. We rewrite it as a whole number and a proper fraction:

$$14\frac{19}{12} = 14 + \frac{19}{12} \quad \text{Write the mixed number with a + sign}$$

$$= 14 + 1\frac{7}{12} \quad \text{Write } \frac{19}{12} \text{ as a mixed number}$$

$$= 15\frac{7}{12} \quad \text{Add 14 and 1}$$

Note Once you see how to change from a whole number and an improper fraction to a whole number and a proper fraction, you will be able to do this step without showing any work.

3. Add: $6\frac{3}{4} + 2\frac{7}{8}$

EXAMPLE 3 Add: $5\frac{2}{3} + 6\frac{8}{9}$

SOLUTION

$$\begin{array}{rcl} 5\frac{2}{3} = 5\frac{2\cdot 3}{3\cdot 3} & = & 5\frac{6}{9} \\ +\,6\frac{8}{9} = 6\frac{8}{9} & = & 6\frac{8}{9} \\ \hline & & 11\frac{14}{9} = 12\frac{5}{9} \end{array}$$

The last step involves writing $\frac{14}{9}$ as $1\frac{5}{9}$ and then adding 11 and 1 to get 12.

4. Add: $2\frac{1}{3} + 1\frac{1}{4} + 3\frac{11}{12}$

EXAMPLE 4 Add: $3\frac{1}{4} + 2\frac{3}{5} + 1\frac{9}{10}$

SOLUTION The LCD is 20. We rewrite each fraction as an equivalent fraction with denominator 20 and add:

$$\begin{array}{rcl} 3\frac{1}{4} = 3\frac{1\cdot 5}{4\cdot 5} & = & 3\frac{5}{20} \\ 2\frac{3}{5} = 2\frac{3\cdot 4}{5\cdot 4} & = & 2\frac{12}{20} \\ +\,1\frac{9}{10} = 1\frac{9\cdot 2}{10\cdot 2} & = & 1\frac{18}{20} \\ \hline & & 6\frac{35}{20} = 7\frac{15}{20} = 7\frac{3}{4} \end{array}$$

Reduce to lowest terms

$$\frac{35}{20} = 1\frac{15}{20}$$

Change to a mixed number

We should note here that we could have worked each of the first four examples in this section by first changing each mixed number to an improper fraction and

Answers

2. $12\frac{11}{20}$ 3. $9\frac{5}{8}$ 4. $7\frac{1}{2}$

then adding as we did in Section 3.5. To illustrate, if we were to work Example 4 this way, it would look like this:

$$3\frac{1}{4} + 2\frac{3}{5} + 1\frac{9}{10} = \frac{13}{4} + \frac{13}{5} + \frac{19}{10} \quad \text{Change to improper fractions}$$

$$= \frac{13 \cdot \mathbf{5}}{4 \cdot \mathbf{5}} + \frac{13 \cdot \mathbf{4}}{5 \cdot \mathbf{4}} + \frac{19 \cdot \mathbf{2}}{10 \cdot \mathbf{2}} \quad \text{LCD is 20}$$

$$= \frac{65}{20} + \frac{52}{20} + \frac{38}{20} \quad \text{Equivalent fractions}$$

$$= \frac{155}{20} \quad \text{Add numerators}$$

$$= 7\frac{15}{20} = 7\frac{3}{4} \quad \text{Change to a mixed number, and reduce}$$

As you can see, the result is the same as the result we obtained in Example 4.

There are advantages to both methods. The method just shown works well when the whole-number parts of the mixed numbers are small. The vertical method shown in Examples 1–4 works well when the whole-number parts of the mixed numbers are large.

Subtraction with mixed numbers is very similar to addition with mixed numbers.

EXAMPLE 5 Subtract: $3\frac{9}{10} - 1\frac{3}{10}$

SOLUTION Because the denominators are the same, we simply subtract the whole numbers and subtract the fractions:

$$\begin{array}{r} 3\frac{9}{10} \\ -\,1\frac{3}{10} \\ \hline 2\frac{6}{10} \end{array} = 2\frac{3}{5}$$

Reduce to lowest terms

5. Subtract: $4\frac{7}{8} - 1\frac{5}{8}$

An easy way to visualize the results in Example 5 is to imagine 3 dollar bills and 9 dimes in your pocket. If you spend 1 dollar and 3 dimes, you will have 2 dollars and 6 dimes left.

EXAMPLE 6 Subtract: $12\frac{7}{10} - 8\frac{3}{5}$

SOLUTION The common denominator is 10. We must rewrite $\frac{3}{5}$ as an equivalent fraction with denominator 10:

$$\begin{array}{rcrcr} 12\frac{7}{10} & = & 12\frac{7}{10} & = & 12\frac{7}{10} \\ -\,8\frac{3}{5} & = & -\,8\frac{3 \cdot 2}{5 \cdot 2} & = & -\,8\frac{6}{10} \\ \hline & & & & 4\frac{1}{10} \end{array}$$

6. Subtract: $12\frac{7}{10} - 7\frac{2}{5}$

Answers

5. $3\frac{1}{4}$ 6. $5\frac{3}{10}$

B Borrowing with Mixed Numbers

7. Subtract: $10 - 5\frac{4}{7}$

Note Convince yourself that 10 is the same as $9\frac{7}{7}$. The reason we choose to write the 1 we borrowed as $\frac{7}{7}$ is that the fraction we eventually subtracted from $\frac{7}{7}$ was $\frac{2}{7}$. Both fractions must have the same denominator, 7, so that we can subtract.

EXAMPLE 7 Subtract: $10 - 5\frac{2}{7}$

SOLUTION In order to have a fraction from which to subtract $\frac{2}{7}$, we borrow 1 from 10 and rewrite the 1 we borrow as $\frac{7}{7}$. The process looks like this:

$$10 = 9\frac{7}{7} \longleftarrow \text{We rewrite 10 as } 9 + 1\text{, which is } 9 + \frac{7}{7} = 9\frac{7}{7}$$

$$-\ 5\frac{2}{7} = -5\frac{2}{7} \quad \text{Then we can subtract as usual}$$

$$4\frac{5}{7}$$

8. Subtract: $6\frac{1}{3} - 2\frac{2}{3}$

EXAMPLE 8 Subtract: $8\frac{1}{4} - 3\frac{3}{4}$

SOLUTION Because $\frac{3}{4}$ is larger than $\frac{1}{4}$, we again need to borrow 1 from the whole number. The 1 that we borrow from the 8 is rewritten as $\frac{4}{4}$, because 4 is the denominator of both fractions:

$$8\frac{1}{4} = 7\frac{5}{4} \longleftarrow \text{Borrow 1 in the form } \frac{4}{4}\text{; then } \frac{4}{4} + \frac{1}{4} = \frac{5}{4}$$

$$-\ 3\frac{3}{4} = -3\frac{3}{4}$$

$$4\frac{2}{4} = 4\frac{1}{2} \quad \text{Reduce to lowest terms}$$

9. Subtract: $6\frac{3}{4} - 2\frac{5}{6}$

EXAMPLE 9 Subtract: $4\frac{3}{4} - 1\frac{5}{6}$

SOLUTION This is about as complicated as it gets with subtraction of mixed numbers. We begin by rewriting each fraction with the common denominator 12:

$$4\frac{3}{4} = 4\frac{3 \cdot 3}{4 \cdot 3} = 4\frac{9}{12}$$

$$-\ 1\frac{5}{6} = -1\frac{5 \cdot 2}{6 \cdot 2} = -1\frac{10}{12}$$

Because $\frac{10}{12}$ is larger than $\frac{9}{12}$, we must borrow 1 from 4 in the form $\frac{12}{12}$ before we subtract:

$$4\frac{9}{12} = 3\frac{21}{12} \longleftarrow 4 = 3 + 1 = 3 + \frac{12}{12}\text{, so } 4\frac{9}{12} = \left(3 + \frac{12}{12}\right) + \frac{9}{12}$$

$$-\ 1\frac{10}{12} = -1\frac{10}{12} \qquad = 3 + \left(\frac{12}{12} + \frac{9}{12}\right)$$

$$2\frac{11}{12} \qquad = 3 + \frac{21}{12}$$

$$= 3\frac{21}{12}$$

Getting Ready for Class

After reading through the preceding section, respond in your own words and in complete sentences.

1. Is it necessary to "borrow" when subtracting $1\frac{3}{10}$ from $3\frac{9}{10}$?
2. To subtract $1\frac{2}{7}$ from 10, it is necessary to rewrite 10 as what mixed number?
3. To subtract $11\frac{20}{30}$ from $15\frac{3}{30}$, it is necessary to rewrite $15\frac{3}{30}$ as what mixed number?
4. Rewrite $14\frac{19}{12}$ so that the fraction part is a proper fraction instead of an improper fraction.

Answers

7. $4\frac{3}{7}$ **8.** $3\frac{2}{3}$ **9.** $3\frac{11}{12}$

Problem Set 3.8

A Add and subtract the following mixed numbers as indicated. [Examples 1–6]

1. $2\frac{1}{5} + 3\frac{3}{5}$

2. $8\frac{2}{9} + 1\frac{5}{9}$

3. $4\frac{3}{10} + 8\frac{1}{10}$

4. $5\frac{2}{7} + 3\frac{3}{7}$

5. $6\frac{8}{9} - 3\frac{4}{9}$

6. $12\frac{5}{12} - 7\frac{1}{12}$

7. $9\frac{1}{6} + 2\frac{5}{6}$

8. $9\frac{1}{4} + 5\frac{3}{4}$

9. $3\frac{5}{8} - 2\frac{1}{4}$

10. $7\frac{9}{10} - 6\frac{3}{5}$

11. $11\frac{1}{3} + 2\frac{5}{6}$

12. $1\frac{5}{8} + 2\frac{1}{2}$

13. $7\frac{5}{12} - 3\frac{1}{3}$

14. $7\frac{3}{4} - 3\frac{5}{12}$

15. $6\frac{1}{3} - 4\frac{1}{4}$

16. $5\frac{4}{5} - 3\frac{1}{3}$

17. $10\frac{5}{6} + 15\frac{3}{4}$

18. $11\frac{7}{8} + 9\frac{1}{6}$

19. $5\frac{2}{3} + 6\frac{1}{3}$

20. $8\frac{5}{6} + 9\frac{5}{6}$

21. $\begin{array}{r} 10\frac{13}{16} \\ -\ 8\frac{5}{16} \\ \hline \end{array}$

22. $\begin{array}{r} 17\frac{7}{12} \\ -\ 9\frac{5}{12} \\ \hline \end{array}$

23. $\begin{array}{r} 6\frac{1}{2} \\ +\ 2\frac{5}{14} \\ \hline \end{array}$

24. $\begin{array}{r} 9\frac{11}{12} \\ +\ 4\frac{1}{6} \\ \hline \end{array}$

25. $\begin{array}{r} 1\frac{5}{8} \\ +\ 1\frac{3}{4} \\ \hline \end{array}$

26. $\begin{array}{r} 7\frac{6}{7} \\ +\ 2\frac{3}{14} \\ \hline \end{array}$

27. $\begin{array}{r} 4\frac{2}{3} \\ +\ 5\frac{3}{5} \\ \hline \end{array}$

28. $\begin{array}{r} 9\frac{4}{9} \\ +\ 1\frac{1}{6} \\ \hline \end{array}$

29. $\begin{array}{r} 5\frac{4}{10} \\ -\ 3\frac{1}{3} \\ \hline \end{array}$

30. $\begin{array}{r} 12\frac{7}{8} \\ -\ 3\frac{5}{6} \\ \hline \end{array}$

31. $\begin{array}{r} 10\frac{1}{20} \\ +\ 11\frac{4}{5} \\ \hline \end{array}$

32. $\begin{array}{r} 18\frac{7}{8} \\ +\ 19\frac{1}{12} \\ \hline \end{array}$

A Find the following sums. (Add.) [Examples 1–4]

33. $1\frac{1}{4} + 2\frac{3}{4} + 5$

34. $6 + 5\frac{3}{5} + 8\frac{2}{5}$

35. $7\frac{1}{10} + 8\frac{3}{10} + 2\frac{7}{10}$

36. $5\frac{2}{7} + 8\frac{1}{7} + 3\frac{5}{7}$

37. $\frac{3}{4} + 8\frac{1}{4} + 5$

38. $\frac{5}{8} + 1\frac{1}{8} + 7$

39. $3\frac{1}{2} + 8\frac{1}{3} + 5\frac{1}{6}$

40. $4\frac{1}{5} + 7\frac{1}{3} + 8\frac{1}{15}$

41. $\begin{array}{r} 8\frac{2}{3} \\ 9\frac{1}{8} \\ +\ 6\frac{1}{4} \\ \hline \end{array}$

42. $\begin{array}{r} 7\frac{3}{5} \\ 8\frac{2}{3} \\ +\ 1\frac{1}{5} \\ \hline \end{array}$

43. $\begin{array}{r} 6\frac{1}{7} \\ 9\frac{3}{14} \\ +\ 12\frac{1}{2} \\ \hline \end{array}$

44. $\begin{array}{r} 1\frac{5}{6} \\ 2\frac{3}{4} \\ +\ 5\frac{1}{2} \\ \hline \end{array}$

B The following problems all involve the concept of borrowing. Subtract in each case. [Examples 7–9]

45. $8 - 1\frac{3}{4}$

46. $5 - 3\frac{1}{3}$

47. $15 - 5\frac{3}{10}$

48. $24 - 10\frac{5}{12}$

49. $8\frac{1}{4} - 2\frac{3}{4}$

50. $12\frac{3}{10} - 5\frac{7}{10}$

51. $9\frac{1}{3} - 8\frac{2}{3}$

52. $7\frac{1}{6} - 6\frac{5}{6}$

53. $4\frac{1}{4} - 2\frac{1}{3}$

54. $6\frac{1}{5} - 1\frac{2}{3}$

55. $9\frac{2}{3} - 5\frac{3}{4}$

56. $12\frac{5}{6} - 8\frac{7}{8}$

57. $16\frac{3}{4} - 10\frac{4}{5}$

58. $18\frac{5}{12} - 9\frac{3}{4}$

59. $10\frac{3}{10} - 4\frac{4}{5}$

60. $9\frac{4}{7} - 7\frac{2}{3}$

61. $13\frac{1}{6} - 12\frac{5}{8}$

62. $21\frac{2}{5} - 20\frac{5}{6}$

63. $15\frac{3}{10} - 11\frac{4}{5}$

64. $19\frac{3}{15} - 10\frac{2}{3}$

C Applying the Concepts

Stock Prices As we mentioned in the introduction to this section, in March 1995, rumors that Michael Jordan would return to basketball sent stock prices for the companies whose products he endorses higher. The table at the right gives some of the details of those increases. Use the table to work Problems 65–70.

Stock Prices for Companies with Michael Jordan Endorsements

Company	Product Endorsed	Stock Price (Dollars) 3/8/95	3/13/95
Nike	Air Jordans	$74\frac{7}{8}$	$77\frac{3}{8}$
Quaker Oats	Gatorade	$32\frac{1}{4}$	$32\frac{5}{8}$
General Mills	Wheaties	$60\frac{1}{2}$	$63\frac{3}{8}$
McDonald's		$32\frac{7}{8}$	$34\frac{3}{8}$

65. Find the difference in the price of Nike stock between March 13 and March 8.

66. Find the difference in price of General Mills stock between March 13 and March 8.

67. If you owned 100 shares of Nike stock, how much more are the 100 shares worth on March 13 than on March 8?

68. If you owned 1,000 shares of General Mills stock on March 8, how much more would they be worth on March 13?

69. If you owned 200 shares of McDonald's stock on March 8, how much more would they be worth on March 13?

70. If you owned 100 shares of McDonald's stock on March 8, how much more would they be worth on March 13?

71. Area and Perimeter The diagrams below show the dimensions of playing fields for the National Football League (NFL), the Canadian Football League, and Arena Football.

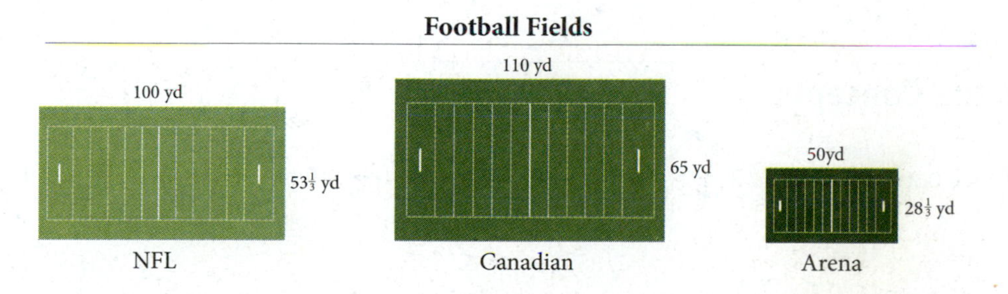

a. Find the perimeter of each football field.

b. Find the area of each football field.

72. Triple Crown The three races that constitute the Triple Crown in horse racing are shown in the table. The information comes from the ESPN website.

a. Write the distances in order from smallest to largest.

b. How much longer is the Belmont Stakes race than the Preakness Stakes?

Race	Distance (miles)
Kentucky Derby	$1\frac{1}{4}$
Preakness Stakes	$1\frac{3}{16}$
Belmont Stakes	$1\frac{1}{2}$

73. Length of Jeans A pair of jeans is $32\frac{1}{2}$ inches long. How long are the jeans after they have been washed if they shrink $1\frac{1}{3}$ inches?

74. Manufacturing A clothing manufacturer has two rolls of cloth. One roll is $35\frac{1}{2}$ yards, and the other is $62\frac{5}{8}$ yards. What is the total number of yards in the two rolls?

Getting Ready for the Next Section

Multiply or divide as indicated.

75. $\frac{11}{8} \cdot \frac{29}{8}$

76. $\frac{3}{4} \div \frac{5}{6}$

77. $\frac{7}{6} \cdot \frac{12}{7}$

78. $10\frac{1}{3} \div 8\frac{2}{3}$

Combine.

79. $\frac{3}{4} + \frac{5}{8}$

80. $\frac{1}{2} + \frac{2}{3}$

81. $2\frac{3}{8} + 1\frac{1}{4}$

82. $3\frac{2}{3} + 4\frac{1}{3}$

Maintaining Your Skills

Use the rule for order of operations to combine the following.

83. $3 + 2 \cdot 7$

84. $8 \cdot 3 - 2$

85. $4 \cdot 5 - 3 \cdot 2$

86. $9 \cdot 7 + 6 \cdot 5$

87. $3 \cdot 2^3 + 5 \cdot 4^2$

88. $6 \cdot 5^2 + 2 \cdot 3^3$

89. $3[2 + 5(6)]$

90. $4[2(3) + 3(5)]$

91. $(7 - 3)(8 + 2)$

92. $(9 - 5)(9 + 5)$

Extending the Concepts

93. Find the difference between $6\frac{1}{5}$ and $2\frac{7}{10}$.

94. Give the difference between $5\frac{1}{3}$ and $1\frac{5}{6}$.

95. Find the sum of $3\frac{1}{8}$ and $2\frac{3}{5}$.

96. Find the sum of $1\frac{5}{6}$ and $3\frac{4}{9}$.

97. Improving Your Quantitative Literacy A column on horse racing in the *Daily News* in Los Angeles reported that the horse Action This Day ran 3 furlongs in $35\frac{1}{5}$ seconds and another horse, Halfbridled, went two-fifths of a second faster. How many seconds did it take Halfbridled to run 3 furlongs?

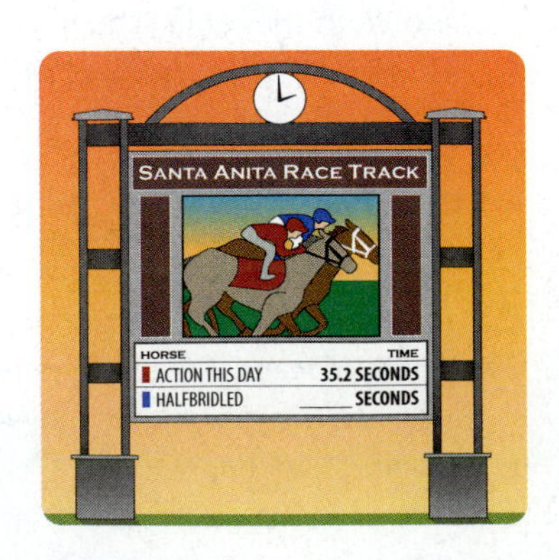

Combinations of Operations and Complex Fractions

3.9

Objectives

A Simplify expressions involving fractions and mixed numbers.

B Simplify complex fractions.

C Solve application problems involving mixed numbers.

Examples now playing at **MathTV.com/books**

Introduction . . .

Now that we have developed skills with both fractions and mixed numbers, we can simplify expressions that contain both types of numbers.

A Simplifying Expressions Involving Fractions and Mixed Numbers

EXAMPLE 1 Simplify the expression: $5 + \left(2\frac{1}{2}\right)\left(3\frac{2}{3}\right)$

SOLUTION The rule for order of operations indicates that we should multiply $2\frac{1}{2}$ times $3\frac{2}{3}$ and then add 5 to the result:

$$5 + \left(2\frac{1}{2}\right)\left(3\frac{2}{3}\right) = 5 + \left(\frac{5}{2}\right)\left(\frac{11}{3}\right)$$ Change the mixed numbers to improper fractions

$$= 5 + \frac{55}{6}$$ Multiply the improper fractions

$$= \frac{30}{6} + \frac{55}{6}$$ Write 5 as $\frac{30}{6}$ so both numbers have the same denominator

$$= \frac{85}{6}$$ Add fractions by adding their numerators

$$= 14\frac{1}{6}$$ Write the answer as a mixed number

PRACTICE PROBLEMS

1. Simplify the expression:

$$4 + \left(1\frac{1}{2}\right)\left(2\frac{3}{4}\right)$$

EXAMPLE 2 Simplify: $\left(\frac{3}{4} + \frac{5}{8}\right)\left(2\frac{3}{8} + 1\frac{1}{4}\right)$

SOLUTION We begin by combining the numbers inside the parentheses:

$$\frac{3}{4} + \frac{5}{8} = \frac{3 \cdot \mathbf{2}}{4 \cdot \mathbf{2}} + \frac{5}{8} = \frac{6}{8} + \frac{5}{8} = \frac{11}{8}$$

and

$$\begin{array}{rcrcr} 2\frac{3}{8} & = & 2\frac{3}{8} & = & 2\frac{3}{8} \\ +\,1\frac{1}{4} & = & +\,1\frac{1 \cdot 2}{4 \cdot 2} & = & +\,1\frac{2}{8} \\ \hline & & & & 3\frac{5}{8} \end{array}$$

2. Simplify:

$$\left(\frac{2}{3} + \frac{1}{6}\right)\left(2\frac{5}{6} + 1\frac{1}{3}\right)$$

Now that we have combined the expressions inside the parentheses, we can complete the problem by multiplying the results:

$$\left(\frac{3}{4} + \frac{5}{8}\right)\left(2\frac{3}{8} + 1\frac{1}{4}\right) = \left(\frac{11}{8}\right)\left(3\frac{5}{8}\right)$$

$$= \frac{11}{8} \cdot \frac{29}{8}$$ Change $3\frac{5}{8}$ to an improper fraction

$$= \frac{319}{64}$$ Multiply fractions

$$= 4\frac{63}{64}$$ Write the answer as a mixed number

Answers

1. $8\frac{1}{8}$ 2. $3\frac{17}{36}$

3. Simplify:

$\frac{3}{7} + \frac{1}{3}\left(1\frac{1}{2} + 4\frac{1}{2}\right)^2$

EXAMPLE 3 Simplify: $\frac{3}{5} + \frac{1}{2}\left(3\frac{2}{3} + 4\frac{1}{3}\right)^2$

SOLUTION We begin by combining the expressions inside the parentheses:

$$\frac{3}{5} + \frac{1}{2}\left(3\frac{2}{3} + 4\frac{1}{3}\right)^2 = \frac{3}{5} + \frac{1}{2}(8)^2 \quad \text{The sum inside the parentheses is 8}$$

$$= \frac{3}{5} + \frac{1}{2}(64) \quad \text{The square of 8 is 64}$$

$$= \frac{3}{5} + 32 \quad \tfrac{1}{2} \text{ of 64 is 32}$$

$$= 32\frac{3}{5} \quad \text{The result is a mixed number}$$

B Complex Fractions

Definition

A **complex fraction** is a fraction in which the numerator and/or the denominator are themselves fractions or combinations of fractions.

Each of the following is a complex fraction:

$$\frac{\frac{3}{4}}{\frac{5}{6}}, \quad \frac{3 + \frac{1}{2}}{2 - \frac{3}{4}}, \quad \frac{\frac{1}{2} + \frac{2}{3}}{\frac{3}{4} - \frac{1}{6}}$$

4. Simplify:

$\frac{\frac{2}{3}}{\frac{5}{9}}$

EXAMPLE 4 Simplify: $\frac{\frac{3}{4}}{\frac{5}{6}}$

SOLUTION This is actually the same as the problem $\frac{3}{4} \div \frac{5}{6}$, because the bar between $\frac{3}{4}$ and $\frac{5}{6}$ indicates division. Therefore, it must be true that

$$\frac{\frac{3}{4}}{\frac{5}{6}} = \frac{3}{4} \div \frac{5}{6}$$

$$= \frac{3}{4} \cdot \frac{6}{5}$$

$$= \frac{18}{20}$$

$$= \frac{9}{10}$$

As you can see, we continue to use properties we have developed previously when we encounter new situations. In Example 4 we use the fact that division by a number and multiplication by its reciprocal produce the same result. We are taking a new problem, simplifying a complex fraction, and thinking of it in terms of a problem we have done previously, division by a fraction.

Answers

3. $12\frac{3}{7}$ 4. $1\frac{1}{5}$

EXAMPLE 5 Simplify: $\dfrac{\frac{1}{2}+\frac{2}{3}}{\frac{3}{4}-\frac{1}{6}}$

SOLUTION Let's decide to call the numerator of this complex fraction the *top* of the fraction and its denominator the *bottom* of the complex fraction. It will be less confusing if we name them this way. The LCD for all the denominators on the top and bottom is 12, so we can multiply the top and bottom of this complex fraction by 12 and be sure all the denominators will divide it exactly. This will leave us with only whole numbers on the top and bottom:

$$\frac{\frac{1}{2}+\frac{2}{3}}{\frac{3}{4}-\frac{1}{6}} = \frac{\mathbf{12}\left(\frac{1}{2}+\frac{2}{3}\right)}{\mathbf{12}\left(\frac{3}{4}-\frac{1}{6}\right)} \quad \text{Multiply the top and bottom by the LCD}$$

$$= \frac{\mathbf{12}\cdot\frac{1}{2}+\mathbf{12}\cdot\frac{2}{3}}{\mathbf{12}\cdot\frac{3}{4}-\mathbf{12}\cdot\frac{1}{6}} \quad \text{Distributive property}$$

$$= \frac{6+8}{9-2} \quad \text{Multiply each fraction by 12}$$

$$= \frac{14}{7} \quad \text{Add on top and subtract on bottom}$$

$$= 2 \quad \text{Reduce to lowest terms}$$

The problem can be worked in another way also. We can simplify the top and bottom of the complex fraction separately. Simplifying the top, we have

$$\frac{1}{2}+\frac{2}{3} = \frac{1\cdot\mathbf{3}}{2\cdot\mathbf{3}}+\frac{2\cdot\mathbf{2}}{3\cdot\mathbf{2}} = \frac{3}{6}+\frac{4}{6} = \frac{7}{6}$$

Simplifying the bottom, we have

$$\frac{3}{4}-\frac{1}{6} = \frac{3\cdot\mathbf{3}}{4\cdot\mathbf{3}}-\frac{1\cdot\mathbf{2}}{6\cdot\mathbf{2}} = \frac{9}{12}-\frac{2}{12} = \frac{7}{12}$$

We now write the original complex fraction again using the simplified expressions for the top and bottom. Then we proceed as we did in Example 4.

$$\frac{\frac{1}{2}+\frac{2}{3}}{\frac{3}{4}-\frac{1}{6}} = \frac{\frac{7}{6}}{\frac{7}{12}}$$

$$= \frac{7}{6}\div\frac{7}{12} \quad \text{The divisor is } \frac{7}{12}$$

$$= \frac{7}{6}\cdot\frac{12}{7} \quad \text{Replace } \frac{7}{12} \text{ by its reciprocal and multiply}$$

$$= \frac{\not{7}\cdot 2\cdot\not{6}}{\not{6}\cdot\not{7}} \quad \text{Divide out common factors}$$

$$= 2$$

STUDY SKILLS

Review with the Exam in Mind

Each day you should review material that will be covered on the next exam. Your review should consist of working problems. Preferably, the problems you work should be problems from your list of difficult problems.

5. Simplify: $\dfrac{\frac{1}{2}+\frac{3}{4}}{\frac{2}{3}-\frac{1}{4}}$

Note We are going to simplify this complex fraction by two different methods. This is the first method.

Note The fraction bar that separates the numerator of the complex fraction from its denominator works like parentheses. If we were to rewrite this problem without it, we would write it like this:

$$\left(\frac{1}{2}+\frac{2}{3}\right)\div\left(\frac{3}{4}-\frac{1}{6}\right)$$

That is why we simplify the top and bottom of the complex fraction separately and then divide.

Answer

5. 3

6. Simplify: $\dfrac{4+\frac{2}{3}}{3-\frac{1}{4}}$

EXAMPLE 6 Simplify: $\dfrac{3+\frac{1}{2}}{2-\frac{3}{4}}$

SOLUTION The simplest approach here is to multiply both the top and bottom by the LCD for all fractions, which is 4:

$$\frac{3+\frac{1}{2}}{2-\frac{3}{4}} = \frac{4\left(3+\frac{1}{2}\right)}{4\left(2-\frac{3}{4}\right)} \quad \text{Multiply the top and bottom by 4}$$

$$= \frac{4 \cdot 3 + 4 \cdot \frac{1}{2}}{4 \cdot 2 - 4 \cdot \frac{3}{4}} \quad \text{Distributive property}$$

$$= \frac{12+2}{8-3} \quad \text{Multiply each number by 4}$$

$$= \frac{14}{5} \quad \text{Add on top and subtract on bottom}$$

$$= 2\frac{4}{5}$$

7. Simplify: $\dfrac{12\frac{1}{3}}{6\frac{2}{3}}$

EXAMPLE 7 Simplify: $\dfrac{10\frac{1}{3}}{8\frac{2}{3}}$

SOLUTION The simplest way to simplify this complex fraction is to think of it as a division problem.

$$\frac{10\frac{1}{3}}{8\frac{2}{3}} = 10\frac{1}{3} \div 8\frac{2}{3} \quad \text{Write with a} \div \text{symbol}$$

$$= \frac{31}{3} \div \frac{26}{3} \quad \text{Change to improper fractions}$$

$$= \frac{31}{3} \cdot \frac{3}{26} \quad \text{Write in terms of multiplication}$$

$$= \frac{31 \cdot \cancel{3}}{\cancel{3} \cdot 26} \quad \text{Divide out the common factor 3}$$

$$= \frac{31}{26} = 1\frac{5}{26} \quad \text{Answer as a mixed number}$$

Getting Ready for Class

After reading through the preceding section, respond in your own words and in complete sentences.

1. What is a complex fraction?
2. Rewrite $\dfrac{\frac{5}{6}}{\frac{1}{3}}$ as a multiplication problem.
3. True or false? The rules for order of operations tell us to work inside parentheses first.
4. True or false? We find the LCD when we add or subtract fractions, but not when we multiply them.

Answers

6. $1\frac{23}{33}$ 7. $1\frac{17}{20}$

Problem Set 3.9

A Use the rule for order of operations to simplify each of the following. [Examples 1–3]

1. $3 + \left(1\frac{1}{2}\right)\left(2\frac{2}{3}\right)$

2. $7 - \left(1\frac{3}{5}\right)\left(2\frac{1}{2}\right)$

3. $8 - \left(\frac{6}{11}\right)\left(1\frac{5}{6}\right)$

4. $10 + \left(2\frac{4}{5}\right)\left(\frac{5}{7}\right)$

5. $\frac{2}{3}\left(1\frac{1}{2}\right) + \frac{3}{4}\left(1\frac{1}{3}\right)$

6. $\frac{2}{5}\left(2\frac{1}{2}\right) + \frac{5}{8}\left(3\frac{1}{5}\right)$

7. $2\left(1\frac{1}{2}\right) + 5\left(6\frac{2}{5}\right)$

8. $4\left(5\frac{3}{4}\right) + 6\left(3\frac{5}{6}\right)$

9. $\left(\frac{3}{5} + \frac{1}{10}\right)\left(\frac{1}{2} + \frac{3}{4}\right)$

10. $\left(\frac{2}{9} + \frac{1}{3}\right)\left(\frac{1}{5} + \frac{1}{10}\right)$

11. $\left(2 + \frac{2}{3}\right)\left(3 + \frac{1}{8}\right)$

12. $\left(3 - \frac{3}{4}\right)\left(3 + \frac{1}{3}\right)$

13. $\left(1 + \frac{5}{6}\right)\left(1 - \frac{5}{6}\right)$

14. $\left(2 - \frac{1}{4}\right)\left(2 + \frac{1}{4}\right)$

15. $\frac{2}{3} + \frac{1}{3}\left(2\frac{1}{2} + \frac{1}{2}\right)^2$

16. $\frac{3}{5} + \frac{1}{4}\left(2\frac{1}{2} - \frac{1}{2}\right)^3$

17. $2\frac{3}{8} + \frac{1}{2}\left(\frac{1}{3} + \frac{5}{3}\right)^3$

18. $8\frac{2}{3} + \frac{1}{3}\left(\frac{8}{5} + \frac{7}{5}\right)^2$

19. $2\left(\frac{1}{2} + \frac{1}{3}\right) + 3\left(\frac{2}{3} + \frac{1}{4}\right)$

20. $5\left(\frac{1}{5} + \frac{3}{10}\right) + 2\left(\frac{1}{10} + \frac{1}{2}\right)$

B Simplify each complex fraction as much as possible. [Examples 4–7]

21. $\dfrac{\frac{2}{3}}{\frac{3}{4}}$

22. $\dfrac{\frac{5}{6}}{\frac{3}{12}}$

23. $\dfrac{\frac{2}{3}}{\frac{4}{3}}$

24. $\dfrac{\frac{7}{9}}{\frac{5}{9}}$

25. $\dfrac{\frac{11}{20}}{\frac{5}{10}}$

26. $\dfrac{\frac{9}{16}}{\frac{3}{4}}$

27. $\dfrac{\frac{1}{2}+\frac{1}{3}}{\frac{1}{2}-\frac{1}{3}}$

28. $\dfrac{\frac{1}{4}+\frac{1}{5}}{\frac{1}{4}-\frac{1}{5}}$

29. $\dfrac{\frac{5}{8}-\frac{1}{4}}{\frac{1}{8}+\frac{1}{2}}$

30. $\dfrac{\frac{3}{4}+\frac{1}{3}}{\frac{2}{3}+\frac{1}{6}}$

31. $\dfrac{\frac{9}{20}-\frac{1}{10}}{\frac{1}{10}+\frac{9}{20}}$

32. $\dfrac{\frac{1}{2}+\frac{2}{3}}{\frac{3}{4}+\frac{5}{6}}$

33. $\dfrac{1+\frac{2}{3}}{1-\frac{2}{3}}$

34. $\dfrac{5-\frac{3}{4}}{2+\frac{3}{4}}$

35. $\dfrac{2+\frac{5}{6}}{5-\frac{1}{3}}$

36. $\dfrac{9-\frac{11}{5}}{3+\frac{13}{10}}$

37. $\dfrac{3+\frac{5}{6}}{1+\frac{5}{3}}$

38. $\dfrac{10+\frac{9}{10}}{5+\frac{4}{5}}$

39. $\dfrac{\frac{1}{3}+\frac{3}{4}}{2-\frac{1}{6}}$

40. $\dfrac{3+\frac{5}{2}}{\frac{5}{6}+\frac{1}{4}}$

41. $\dfrac{\frac{5}{6}}{3+\frac{2}{3}}$

42. $\dfrac{9-\frac{3}{2}}{\frac{7}{4}}$

B Simplify each of the following complex fractions. [Examples 5–7]

43. $\dfrac{2\frac{1}{2} + \frac{1}{2}}{3\frac{3}{5} - \frac{2}{5}}$

44. $\dfrac{5\frac{3}{8} + \frac{5}{8}}{4\frac{1}{4} + 1\frac{3}{4}}$

45. $\dfrac{2 + 1\frac{2}{3}}{3\frac{5}{6} - 1}$

46. $\dfrac{5 + 8\frac{3}{5}}{2\frac{3}{10} + 4}$

47. $\dfrac{3\frac{1}{4} - 2\frac{1}{2}}{5\frac{3}{4} + 1\frac{1}{2}}$

48. $\dfrac{9\frac{3}{8} + 2\frac{5}{8}}{6\frac{1}{2} + 7\frac{1}{2}}$

49. $\dfrac{3\frac{1}{4} + 5\frac{1}{6}}{2\frac{1}{3} + 3\frac{1}{4}}$

50. $\dfrac{8\frac{5}{6} + 1\frac{2}{3}}{7\frac{1}{3} + 2\frac{1}{4}}$

51. $\dfrac{6\frac{2}{3} + 7\frac{3}{4}}{8\frac{1}{2} + 9\frac{7}{8}}$

52. $\dfrac{3\frac{4}{5} - 1\frac{9}{10}}{6\frac{5}{6} - 2\frac{3}{4}}$

53. What is twice the sum of $2\frac{1}{5}$ and $\frac{3}{6}$?

54. Find 3 times the difference of $1\frac{7}{9}$ and $\frac{2}{9}$.

55. Add $5\frac{1}{4}$ to the sum of $\frac{3}{4}$ and 2.

56. Subtract $\frac{7}{8}$ from the product of 2 and $3\frac{1}{2}$.

C Applying the Concepts

57. Tri-cities The Google Earth image shows a right triangle between three cities in Colorado. If the distance between Edgewater and Denver is 4 miles, and the distance between Denver and North Washington is $2\frac{1}{2}$ miles, what is the area of the triangle created by the three cities?

58. Tri-cities The Google Earth image shows a right triangle between three cities in California. If the distance between Pomona and Ontario is $5\frac{7}{10}$ miles, and the distance between Ontario and Upland is $3\frac{3}{5}$ miles, what is the area of the triangle created by the three cities?

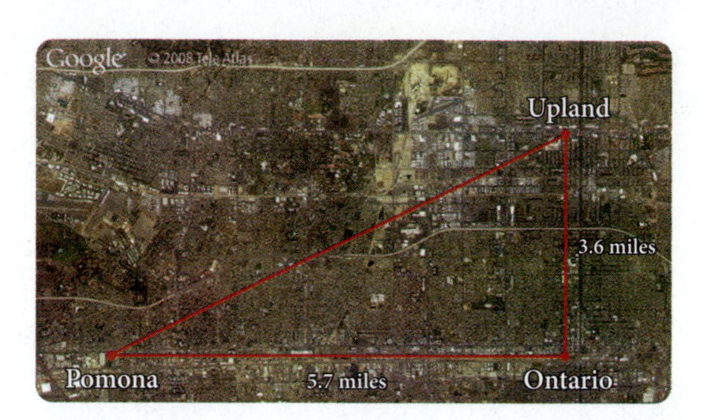

59. Manufacturing A dress manufacturer usually buys two rolls of cloth, one of $32\frac{1}{2}$ yards and the other of $25\frac{1}{3}$ yards, to fill his weekly orders. If his orders double one week, how much of the cloth should he order? (Give the total yardage.)

60. Body Temperature Suppose your normal body temperature is $98\frac{3}{5}°$ Fahrenheit. If your temperature goes up $3\frac{1}{5}°$ on Monday and then down $1\frac{4}{5}°$ on Tuesday, what is your temperature on Tuesday?

Maintaining Your Skills

These problems review the four basic operations with fractions from this chapter.

Perform the indicated operations.

61. $\frac{3}{4} \cdot \frac{8}{9}$

62. $8 \cdot \frac{5}{6}$

63. $\frac{2}{3} \div 4$

64. $\frac{7}{8} \div \frac{14}{24}$

65. $\frac{3}{7} - \frac{2}{7}$

66. $\frac{6}{7} + \frac{9}{14}$

67. $10 - \frac{2}{9}$

68. $\frac{2}{3} - \frac{3}{5}$

Chapter 3 Summary

Definition of Fractions [3.1]

A fraction is any number that can be written in the form $\frac{a}{b}$, where a and b are numbers and b is not 0. The number a is called the *numerator*, and the number b is called the *denominator*.

EXAMPLES

1. Each of the following is a fraction:

$\frac{1}{2}, \frac{3}{4}, \frac{8}{1}, \frac{7}{3}$

Properties of Fractions [3.1]

Multiplying the numerator and the denominator of a fraction by the same nonzero number will produce an equivalent fraction. The same is true for dividing the numerator and denominator by the same nonzero number. In symbols the properties look like this: If a, b, and c are numbers and b and c are not 0, then

Property 1 $\frac{a}{b} = \frac{a \cdot c}{b \cdot c}$ **Property 2** $\frac{a}{b} = \frac{a \div c}{b \div c}$

2. Change $\frac{3}{4}$ to an equivalent fraction with denominator 12.

$\frac{3}{4} = \frac{3 \cdot 3}{4 \cdot 3} = \frac{9}{12}$

Fractions and the Number 1 [3.1]

If a represents any number, then

$$\frac{a}{1} = a \quad \text{and} \quad \frac{a}{a} = 1 \quad \text{(where } a \text{ is not 0)}$$

3. $\frac{5}{1} = 5, \frac{5}{5} = 1$

Reducing Fractions to Lowest Terms [3.2]

To reduce a fraction to lowest terms, factor the numerator and the denominator, and then divide both the numerator and denominator by any factors they have in common.

4.
$$\frac{90}{588} = \frac{\not{2} \cdot \not{3} \cdot 3 \cdot 5}{\not{2} \cdot 2 \cdot \not{3} \cdot 7 \cdot 7} = \frac{3 \cdot 5}{2 \cdot 7 \cdot 7} = \frac{15}{98}$$

Multiplying Fractions [3.3]

To multiply fractions, multiply numerators and multiply denominators.

5. $\frac{3}{5} \cdot \frac{4}{7} = \frac{3 \cdot 4}{5 \cdot 7} = \frac{12}{35}$

The Area of a Triangle [3.3]

The formula for the area of a triangle with base b and height h is

$$A = \frac{1}{2}bh$$

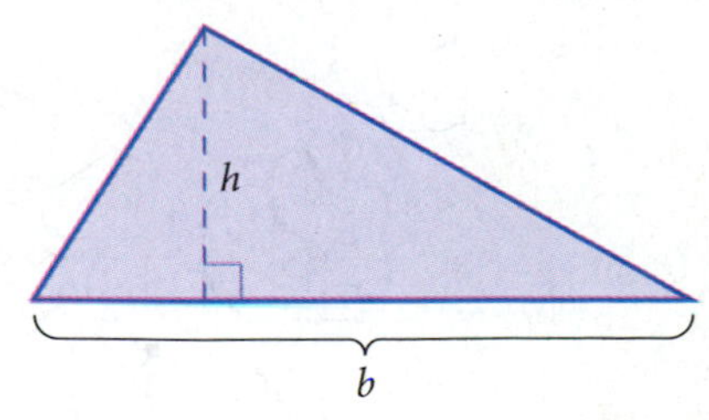

6. If the base of a triangle is 10 inches and the height is 7 inches, then the area is

$$A = \frac{1}{2}bh = \frac{1}{2} \cdot 10 \cdot 7 = 5 \cdot 7 = 35 \text{ square inches}$$

Reciprocals [3.4]

Any two numbers whose product is 1 are called *reciprocals*. The numbers $\frac{2}{3}$ and $\frac{3}{2}$ are reciprocals, because their product is 1.

Division with Fractions [3.4]

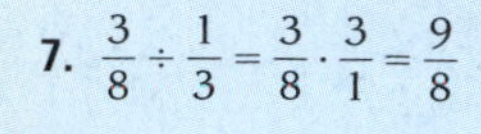

7. $\frac{3}{8} \div \frac{1}{3} = \frac{3}{8} \cdot \frac{3}{1} = \frac{9}{8}$

To divide by a fraction, multiply by its reciprocal. That is, the quotient of two fractions is defined to be the product of the first fraction with the reciprocal of the second fraction (the divisor).

Least Common Denominator (LCD) [3.5]

The *least common denominator* (LCD) for a set of denominators is the smallest number that is exactly divisible by each denominator.

Addition and Subtraction of Fractions [3.5]

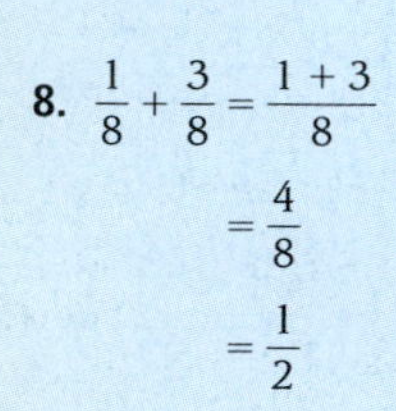

8. $\frac{1}{8} + \frac{3}{8} = \frac{1+3}{8} = \frac{4}{8} = \frac{1}{2}$

To add (or subtract) two fractions with a common denominator, add (or subtract) numerators and use the common denominator. In symbols: If a, b, and c are numbers with c not equal to 0, then

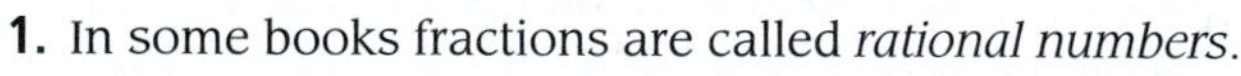

$$\frac{a}{c} + \frac{b}{c} = \frac{a+b}{c} \quad \text{and} \quad \frac{a}{c} - \frac{b}{c} = \frac{a-b}{c}$$

Additional Facts about Fractions

1. In some books fractions are called *rational numbers*.
2. Every whole number can be written as a fraction with a denominator of 1.
3. The commutative, associative, and distributive properties are true for fractions.
4. The word *of* as used in the expression "$\frac{2}{3}$ *of* 12" indicates that we are to multiply $\frac{2}{3}$ and 12.
5. Two fractions with the same value are called *equivalent fractions.*

Mixed-Number Notation [3.6]

A mixed number is the sum of a whole number and a fraction. The + sign is not shown when we write mixed numbers; it is implied. The mixed number $4\frac{2}{3}$ is actually the sum $4 + \frac{2}{3}$.

Changing Mixed Numbers to Improper Fractions [3.6]

To change a mixed number to an improper fraction, we write the mixed number showing the + sign and add as usual. The result is the same if we multiply the denominator of the fraction by the whole number and add what we get to the numerator of the fraction, putting this result over the denominator of the fraction.

9. $4\frac{2}{3} = \frac{3 \cdot 4 + 2}{3} = \frac{14}{3}$

Mixed number ↑ Improper fraction ↑

Changing an Improper Fraction to a Mixed Number [3.6]

To change an improper fraction to a mixed number, divide the denominator into the numerator. The quotient is the whole-number part of the mixed number. The fraction part is the remainder over the divisor.

10. Change $\frac{14}{3}$ to a mixed number.

$$\begin{array}{r} 4 \\ 3\overline{)14} \\ \underline{12} \\ 2 \end{array} \qquad \frac{14}{3} = 4\frac{2}{3}$$

Quotient: 4; Divisor: 3; Remainder: 2

Multiplication and Division with Mixed Numbers [3.7]

To multiply or divide two mixed numbers, change each to an improper fraction and multiply or divide as usual.

11. $2\frac{1}{3} \cdot 1\frac{3}{4} = \frac{7}{3} \cdot \frac{7}{4} = \frac{49}{12} = 4\frac{1}{12}$

Addition and Subtraction with Mixed Numbers [3.8]

To add or subtract two mixed numbers, add or subtract the whole-number parts and the fraction parts separately. This is best done with the numbers written in columns.

12.

$$\begin{array}{rcrcr} 3\frac{4}{9} & = & 3\frac{4}{9} & = & 3\frac{4}{9} \\ +\,2\frac{2}{3} & = & 2\frac{2 \cdot 3}{3 \cdot 3} & = & 2\frac{6}{9} \\ \hline & & & & 5\frac{10}{9} = 6\frac{1}{9} \end{array}$$

Common denominator; Add fractions; Add whole numbers

Borrowing in Subtraction with Mixed Numbers [3.8]

It is sometimes necessary to borrow when doing subtraction with mixed numbers. We always change to a common denominator before we actually borrow.

13.

$$\begin{array}{rcrcr} 4\frac{1}{3} & = & 4\frac{2}{6} & = & 3\frac{8}{6} \\ -\,1\frac{5}{6} & = & -\,1\frac{5}{6} & = & -\,1\frac{5}{6} \\ \hline & & & & 2\frac{3}{6} = 2\frac{1}{2} \end{array}$$

Complex Fractions [3.9]

A fraction that contains a fraction in its numerator or denominator is called a *complex fraction.*

14.

$$\frac{4 + \frac{1}{3}}{2 - \frac{5}{6}} = \frac{6\left(4 + \frac{1}{3}\right)}{6\left(2 - \frac{5}{6}\right)} = \frac{6 \cdot 4 + 6 \cdot \frac{1}{3}}{6 \cdot 2 - 6 \cdot \frac{5}{6}} = \frac{24 + 2}{12 - 5} = \frac{26}{7} = 3\frac{5}{7}$$

COMMON MISTAKES

1. The most common mistake when working with fractions occurs when we try to add two fractions without using a common denominator. For example,

$$\frac{2}{3} + \frac{4}{5} \neq \frac{2+4}{3+5}$$

If the two fractions we are trying to add don't have the same denominators, then we *must* rewrite each one as an equivalent fraction with a common denominator. *We never add denominators when adding fractions.* **Note:** We do *not* need a common denominator when multiplying fractions.

2. A common mistake made with division of fractions occurs when we multiply by the reciprocal of the first fraction instead of the reciprocal of the divisor. For example,

$$\frac{2}{3} \div \frac{5}{6} \neq \frac{3}{2} \cdot \frac{5}{6}$$

Remember, we perform division by multiplying by the reciprocal of the divisor (the fraction to the right of the division symbol).

3. If the answer to a problem turns out to be a fraction, that fraction should always be written in lowest terms. It is a mistake not to reduce to lowest terms.

4. A common mistake when working with mixed numbers is to confuse mixed-number notation for multiplication of fractions. The notation $3\frac{2}{5}$ does *not* mean 3 *times* $\frac{2}{5}$. It means 3 *plus* $\frac{2}{5}$.

5. Another mistake occurs when multiplying mixed numbers. The mistake occurs when we don't change the mixed number to an improper fraction before multiplying and instead try to multiply the whole numbers and fractions separately. Like this:

$$\begin{aligned} 2\frac{1}{2} \cdot 3\frac{1}{3} &= (2 \cdot 3) + \left(\frac{1}{2} \cdot \frac{1}{3}\right) \quad \text{Mistake} \\ &= 6 + \frac{1}{6} \\ &= 6\frac{1}{6} \end{aligned}$$

Remember, the correct way to multiply mixed numbers is to first change to improper fractions and then multiply numerators and multiply denominators. This is correct:

$$2\frac{1}{2} \cdot 3\frac{1}{3} = \frac{5}{2} \cdot \frac{10}{3} = \frac{50}{6} = 8\frac{2}{6} = 8\frac{1}{3} \quad \text{Correct}$$

Chapter 3 Review

Reduce each of the following fractions to lowest terms. [3.2]

1. $\frac{6}{8}$ **2.** $\frac{12}{36}$ **3.** $\frac{110a^3}{70a}$ **4.** $\frac{45xy}{75y}$

Multiply the following fractions. (That is, find the product in each case, and reduce to lowest terms.) [3.3]

5. $\frac{1}{5}(5x)$ **6.** $-\frac{80}{27}\left(\frac{3}{20}\right)$ **7.** $\frac{96}{25} \cdot \frac{15}{98} \cdot \frac{35}{54}$ **8.** $\frac{3}{5} \cdot 75 \cdot \frac{2}{3}$

Find the following quotients. (That is, divide and reduce to lowest terms.) [3.4]

9. $\frac{8}{9} \div \frac{4}{3}$ **10.** $\frac{9}{10} \div (-3)$ **11.** $\frac{a^3}{b} \div \frac{a}{b^2}$ **12.** $-\frac{18}{49} \div \left(-\frac{36}{28}\right)$

Perform the indicated operations. Reduce all answers to lowest terms. [3.5]

13. $\frac{6}{8} - \frac{2}{8}$ **14.** $\frac{9}{x} + \frac{11}{x}$ **15.** $-3 - \frac{1}{2}$ **16.** $\frac{3}{x} - \frac{5}{6}$

17. $\frac{11}{126} - \frac{5}{84}$ **18.** $\frac{3}{10} + \frac{7}{25} + \frac{3}{4}$

19. Change $3\frac{5}{8}$ to an improper fraction. [3.6]

20. Add: $3 + \frac{2}{x}$ [3.6]

21. Subtract: $x - \frac{3}{4}$ [3.6]

22. Change $\frac{110}{8}$ to a mixed number. [3.6]

Perform the indicated operations. [3.7, 3.8]

23. $2 \div 3\frac{1}{4}$ **24.** $4\frac{7}{8} \div 2\frac{3}{5}$ **25.** $6 \cdot 2\frac{1}{2} \cdot \frac{4}{5}$ **26.** $3\frac{1}{5} + 4\frac{2}{5}$ **27.** $8\frac{2}{3} + 9\frac{1}{4}$ **28.** $5\frac{1}{3} - 2\frac{8}{9}$

Simplify each of the following as much as possible. [3.9]

29. $3 + 2\left(4\frac{1}{3}\right)$

30. $\left(2\frac{1}{2} + \frac{3}{4}\right)\left(2\frac{1}{2} - \frac{3}{4}\right)$

Simplify each complex fraction as much as possible. [3.9]

31. $\dfrac{1 + \frac{2}{3}}{1 - \frac{2}{3}}$

32. $\dfrac{3 - \frac{3}{4}}{3 + \frac{3}{4}}$

33. $\dfrac{\frac{7}{8} - \frac{1}{2}}{\frac{1}{4} + \frac{1}{2}}$

34. $\dfrac{2\frac{1}{8} + 3\frac{1}{3}}{1\frac{1}{6} + 5\frac{1}{4}}$

35. Defective Parts If $\frac{1}{10}$ of the items in a shipment of 200 items are defective, how many are defective? [3.3]

36. Number of Students If 80 students took a math test and $\frac{3}{4}$ of them passed, then how many students passed the test? [3.3]

37. Translating What is 3 times the sum of $2\frac{1}{4}$ and $\frac{3}{4}$? [3.9]

38. Translating Subtract $\frac{5}{6}$ from the product of $1\frac{1}{2}$ and $\frac{2}{3}$. [3.9]

39. Cooking If a recipe that calls for $2\frac{1}{2}$ cups of flour will make 48 cookies, how much flour is needed to make 36 cookies? [3.7]

40. Length of Wood A piece of wood $10\frac{3}{4}$ inches long is divided into 6 equal pieces. How long is each piece? [3.7]

41. Cooking A recipe that calls for $3\frac{1}{2}$ tablespoons of oil is tripled. How much oil must be used in the tripled recipe? [3.7]

42. Sheep Feed A rancher fed his sheep $10\frac{1}{2}$ pounds of feed on Monday, $9\frac{3}{4}$ pounds on Tuesday, and $12\frac{1}{4}$ pounds on Wednesday. How many pounds of feed did he use on these 3 days? [3.8]

43. Find the area and the perimeter of the triangle below. [3.7, 3.8]

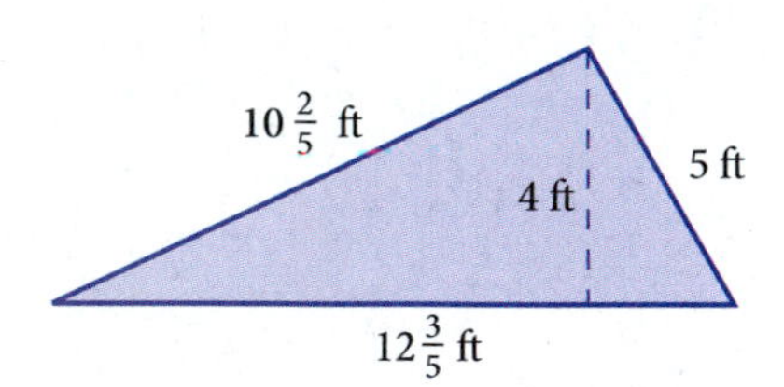

44. Comparing Area On April 3, 2000, *USA TODAY* changed the size of its paper. Previous to this date, each page of the paper was $13\frac{1}{2}$ inches wide and $22\frac{1}{4}$ inches long. The new paper size is $1\frac{1}{4}$ inches narrower and $\frac{1}{2}$ inch longer. [3.7, 3.8]

a. What was the area of a page previous to April 3, 2000?

b. What is the area of a page after April 3, 2000?

c. What is the difference in the areas of the two page sizes?

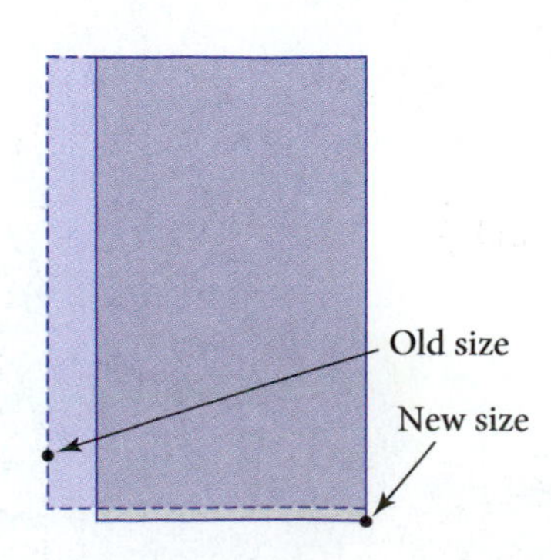

Chapter 3 Cumulative Review

Simplify.

1. $(6 + 2) + (3 + 6)$

2. $\begin{array}{r} 99 \\ 144 \\ 81 \\ +\ 49 \\ \hline \end{array}$

3. $1985 - 141$

4. $13 - (9 - 4)$

5. $13x - 19x$

6. $17\frac{13}{16} - 9\frac{5}{12}$

7. $\frac{3}{4}\left(\frac{8}{3}x\right)$

8. $11(2n)$

9. $\begin{array}{r} 5280 \\ \times\ 26 \\ \hline \end{array}$

10. $9050(373x)$

11. $2(-2)2 - 5(-3)$

12. $\frac{2}{5} \cdot \frac{1}{3}$

13. $\frac{2}{3} \cdot 6$

14. $\frac{3}{5} \cdot \frac{2}{3}$

15. $\frac{3}{4} \cdot 4$

16. $6 \div (2 \cdot 3) + 5$

17. $8 \div 4(2 + 6)$

18. $\left(\frac{3}{4}\right)^2 \cdot \left(-\frac{1}{2}\right)^3$

19. $\frac{11}{77}$

20. $\frac{104}{33}$

21. $121 \div 11 \div 11$

22. $\dfrac{4\frac{3}{8}}{5\frac{2}{3}}$

23. $\left(\frac{3}{5} + \frac{2}{25}\right) - \frac{4}{125}$

24. $6 + \frac{3}{11} \div \frac{1}{2}$

25. $2\frac{1}{3}\left(1\frac{1}{4} \div \frac{1}{5}\right)$

26. Round the following numbers to the nearest ten, then add.

$\begin{array}{r} 747 \\ 116 \\ +\ 222 \\ \hline \end{array}$

27. Find the sum of $\frac{2}{3}$, $\frac{1}{9}$, and $\frac{3}{4}$.

28. Write the fraction $\frac{3}{13}$ as an equivalent fraction with a denominator of $39x$.

29. Find $\frac{7}{9}$ of $3\frac{1}{4}$.

30. Find the sum of 12 times 2 and 19 times 4.

31. Find the area of the figure:

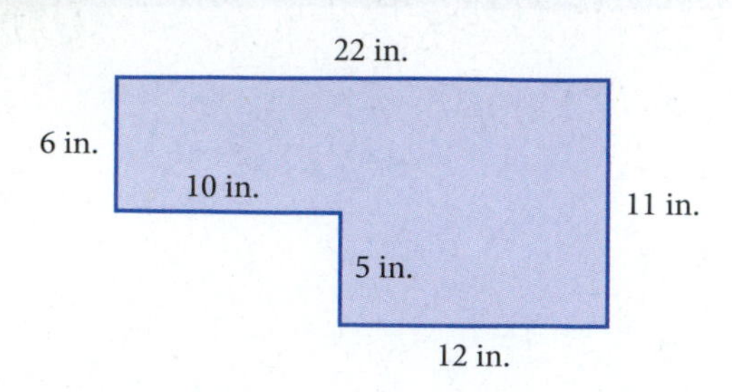

32. Medical Costs The table below shows the average yearly cost of visits to the doctor. Fill in the last column of the table by rounding each cost to the nearest hundred.

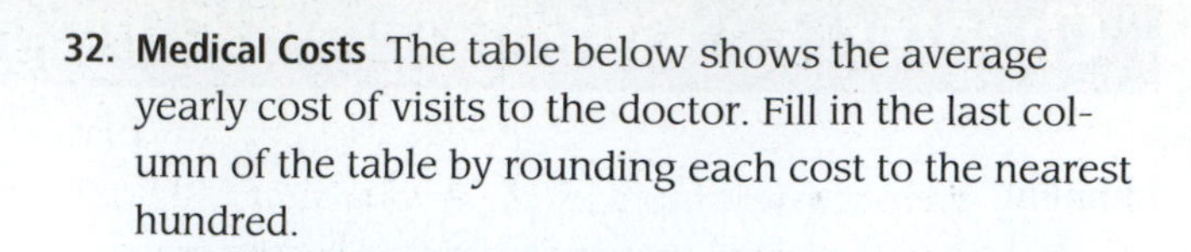

MEDICAL COSTS

Year	Average Annual Cost	Cost to the Nearest Hundred
1990	$583	
1995	$739	
2000	$906	
2005	$1,172	

33. Reduce to lowest terms.

$$\frac{14x^2y}{49x^3y^4}$$

34. Find the area of the figure below:

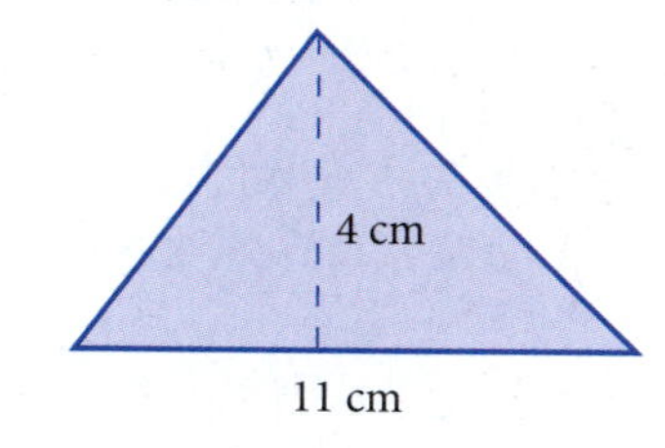

35. Add $-\frac{4}{5}$ to half of $\frac{4}{9}$.

36. Neptune's Diameter The planet Neptune has an equatorial diameter of about 30,760 miles. Write out Neptune's diameter in words and expanded form.

37. Photography A photographer buys one roll of 36-exposure 400-speed film and two rolls of 24-exposure 100-speed film. She uses $\frac{5}{6}$ of each type of film. How many pictures did she take?

38. Stopping Distances The table below shows how many feet it takes to stop a car traveling at different rates of speed, once the brakes are applied. Use the template to construct a line graph of the information in the table.

Speed (Mi/Hr)	Distance (Ft)
0	0
20	22
30	49
40	88
50	137
60	198
70	269
80	352

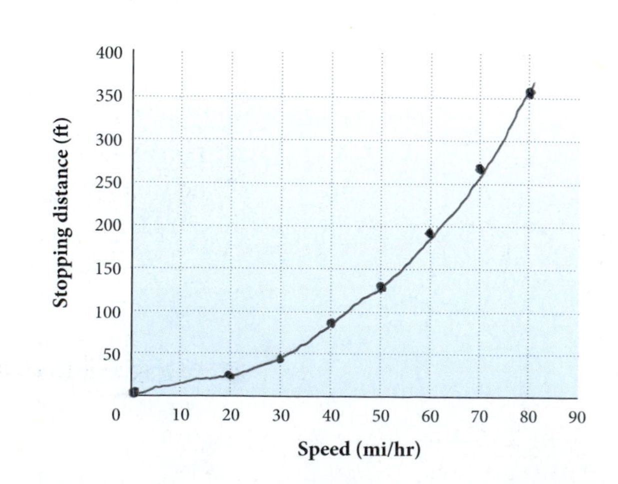

Chapter 3 Test

1. Each circle below is divided into 8 equal parts. Shade each circle to represent the fraction below the circle.

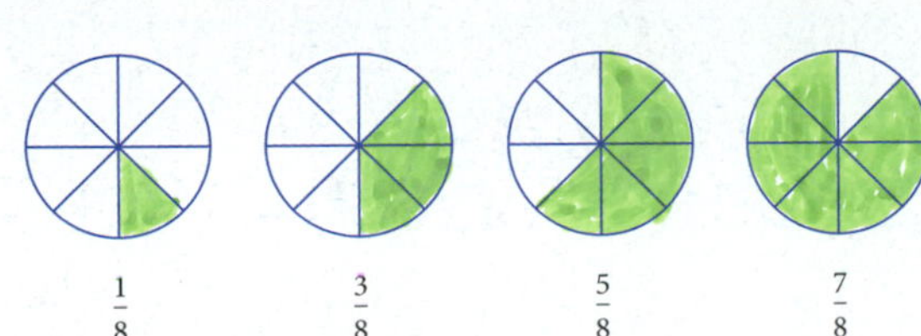

2. Reduce each fraction to lowest terms.

   a. $\frac{10}{15}$

   b. $\frac{130xy}{50x}$

Find each product, and reduce your answer to lowest terms.

3. $\frac{3}{5}(30x)$

4. $\frac{48}{49} \cdot \frac{35}{50} \cdot \frac{6}{18}$

Find each quotient, and reduce your answer to lowest terms.

5. $\frac{5}{18} \div \frac{15}{16}$

6. $\frac{4}{5} \div (-8)$

Perform the indicated operations. Reduce all answers to lowest terms.

7. $\frac{3}{10} + \frac{1}{10}$

8. $\frac{5}{x} - \frac{2}{x}$

9. $-4 - \frac{3}{5}$

10. $\frac{3}{x} + \frac{2}{5}$

11. $\frac{5}{6} + \frac{2}{9} + \frac{1}{4}$

12. Change $5\frac{2}{7}$ to an improper fraction.

13. Change $\frac{43}{5}$ to a mixed number.

14. Add: $5 + \frac{4}{x}$

Perform the indicated operations.

15. $6 \div 1\frac{1}{3}$

16. $7\frac{1}{3} + 2\frac{3}{8}$

17. $5\frac{1}{6} - 1\frac{1}{2}$

Simplify each of the following as much as possible.

18. $4 + 3\left(4\frac{1}{4}\right)$

19. $\left(2\frac{1}{3} + \frac{1}{2}\right)\left(3\frac{2}{3} - \frac{1}{6}\right)$

20. $\dfrac{\frac{11}{12} - \frac{2}{3}}{\frac{1}{6} + \frac{1}{3}}$

21. Number of Grapefruit If $\frac{1}{3}$ of a shipment of 120 grapefruit is spoiled, how many grapefruit are spoiled?

22. Sewing A dress that is $31\frac{1}{6}$ inches long is shortened by $3\frac{2}{3}$ inches. What is the new length of the dress?

23. Cooking A recipe that calls for $4\frac{2}{3}$ cups of sugar is doubled. How much sugar must be used in the doubled recipe?

24. Length of Rope A piece of rope $15\frac{2}{3}$ feet long is divided into 5 equal pieces. How long is each piece?

25. Find the area and the perimeter of the triangle below.

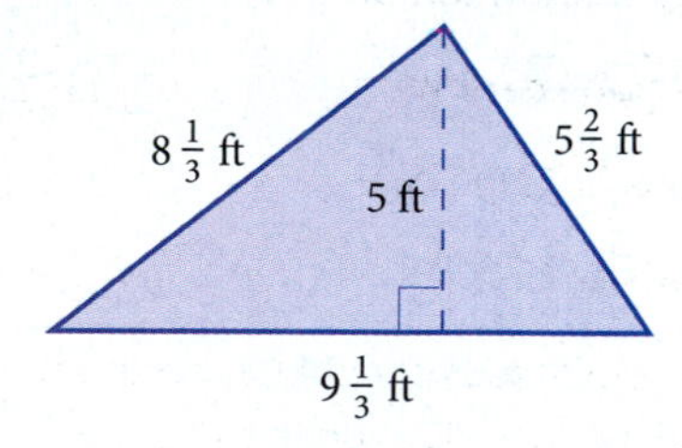

Chapter 3 Projects

FRACTIONS AND MIXED NUMBERS

GROUP PROJECT

Recipe

Number of People 2

Time Needed 5 minutes

Equipment Pencil and paper

Background Here is Martha Stewart's recipe for chocolate chip cookies.

Chocolate Chip Cookies

Makes 2 dozen
You can substitute bittersweet chocolate for half of the semisweet chocolate chips.

1 cup (2 sticks) unsalted butter, room temperature
1 1/2 cups packed light-brown sugar
1/2 cup granulated sugar
1 teaspoon pure vanilla extract
1 large egg, room temperature
2 cups all-purpose flour
1/2 teaspoon baking soda
1/2 teaspoon salt
12 ounces semisweet chocolate, coarsely chopped, or one 12-ounce bag semisweet chocolate chips

1. Heat oven to 375°. Line several baking sheets with parchment paper, and set aside.
2. Combine butter and both sugars in the bowl of an electric mixer fitted with the paddle attachment, and beat until light and fluffy. Add vanilla, and mix to combine. Add egg, and continue beating until well combined.
3. In a medium bowl, whisk together the flour, baking soda, and salt. Slowly add the dry ingredients to the butter mixture. Mix on low speed until just combined. Stir in chocolate chips.
4. Scoop out 2 tablespoons of dough, and place on a prepared baking sheet. Repeat with remaining dough, placing scoops 3 inches apart. Bake until just brown around the edges, 16 to 18 minutes, rotating the pans between the oven shelves halfway through baking. Remove from the oven, and let cool slightly before removing cookies from the baking sheets. Store in an airtight container at room temperature for up to 1 week.

Procedure Rewrite the recipe to make 3 dozen cookies by multiplying the quantities by $1\frac{1}{2}$.

____ cups (____ sticks) unsalted butter, room temperature

____ cups packed light-brown sugar

____ cups granulated sugar

____ teaspoons pure vanilla extract

____ large eggs, room temperature

____ cups all-purpose flour

____ teaspoons baking soda

____ teaspoons salt

____ ounces semisweet chocolate, coarsely chopped, or ____ 12-ounce bags semisweet chocolate chips

RESEARCH PROJECT

Sophie Germain

The photograph at the right shows the street sign in Paris named for the French mathematician Sophie Germain (1776-1831). Among her contributions to mathematics is her work with prime numbers. In this chapter we had an introductory look at some of the classifications for numbers, including the prime numbers. Within the prime numbers themselves, there are still further classifications. In fact, a Sophie Germain prime is a prime number P, for which both P and $2P + 1$ are primes. For example, the prime number 2 is the first Sophie Germain prime because both 2 and $2 \cdot 2 + 1 = 5$ are prime numbers. The next Germain prime is 3 because both 3 and $2 \cdot 3 + 1 = 7$ are primes.

Sophie Germain was born on April 1, 1776, in Paris, France. She taught herself mathematics by reading the books in her father's library at home. Today she is recognized most for her work in number theory, which includes her work with prime numbers. Research the life of Sophie Germain. Write a short essay that includes information on her work with prime numbers and how her results contributed to solving Fermat's Theorem almost 200 years later.

Cheryl Slaughter

Solving Equations

4

Chapter Outline

Introduction

Central Park in New York City was the first landscaped public park in the United States. More than 25 million people visit the park each year. Central Park is $\frac{1}{2}$ mile wide and covers an area of 1.4 square miles. A person who jogs around the perimeter of the park will cover approximately 6.6 miles. Because the park can be modeled with a rectangle, we can use these numbers to find the length of the park. In fact, solving either of the two equations below will give us the length.

$$\frac{1}{2}x = 1.4 \qquad 2x + 2 \cdot \frac{1}{2} = 6.6$$

In this chapter, we will learn how to take the numbers and relationships given in the paragraph above and translate them into equations like the ones above. Before we do that, we will learn how to solve these equations, and many others as well.

The illustration here shows the area of Central Park compared to other prominent parks in large cities.

Chapter Pretest

The pretest below contains problems that are representative of the problems you will find in the chapter.

Simplify.

1. $4a - 1 - 5a + 8$
2. $2(5y - 6) + 4y$

Solve each equation.

3. $a + 4 = -2$
4. $5y + 9 - 4y = -7 + 11$
5. $\frac{1}{5}x = 3$
6. $-2x + 9 = 11$
7. $2a + 1 = 5(a - 2) - 1$
8. $\frac{x}{3} - \frac{x}{4} = 5$
9. Find the value of $3x - 4$ when $x = -2$.
10. Is $x = 5$ a solution to the equation $6x - 28 = 1$?
11. The sum of a number and 6 is -17. Find the number.
12. If four times a number is decreased by 7, the result is 25. Find the number.
13. Plot the following points:
    $(3, 1), (-3, 0), (2, -1), (-2, -2)$.

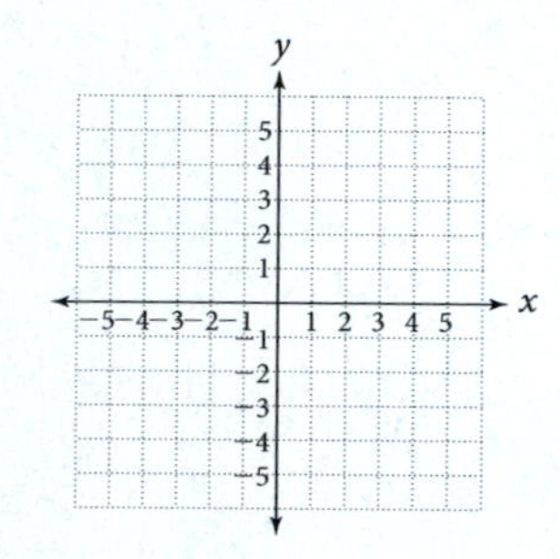

14. Graph $y = 3x - 1$.

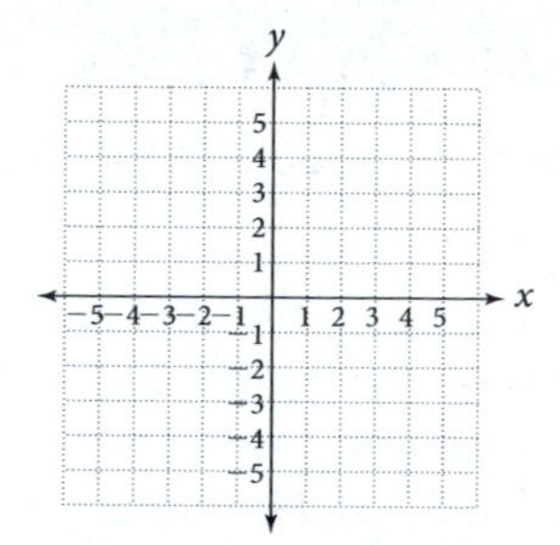

Getting Ready for Chapter 4

The problems below review material covered previously that you need to know in order to be successful in Chapter 4. If you have any difficulty with the problems here, you need to go back and review before going on to Chapter 4.

Simplify.

1. $-2 + 7$
2. $180 - 45$
3. $-2 + (-4)$
4. $\frac{5}{8} + \frac{3}{4}$
5. $(-4)(5)$
6. $\frac{6}{-3}$
7. $\frac{3}{2}(12)$
8. $\left(-\frac{5}{4}\right)\left(-\frac{4}{5}\right)$
9. $\left(-\frac{5}{4}\right)\left(\frac{8}{15}\right)$
10. $-4(-1) + 9$
11. $\frac{1}{3}(15) + 2$
12. $\frac{5}{9}(95 - 32)$
13. $3x + 7x$
14. $-4(3x)$
15. $4(x - 5)$
16. $2\left(\frac{1}{2}x\right)$
17. Write in symbols: the sum of x and 2.
18. Find the perimeter.

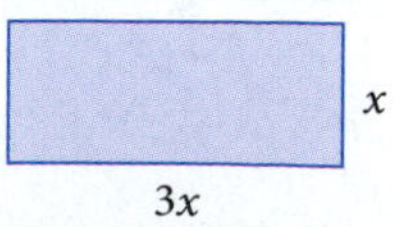

The Distributive Property and Algebraic Expressions

4.1

We recall that the distributive property from Section 1.5 can be used to find the area of a rectangle using two different methods.

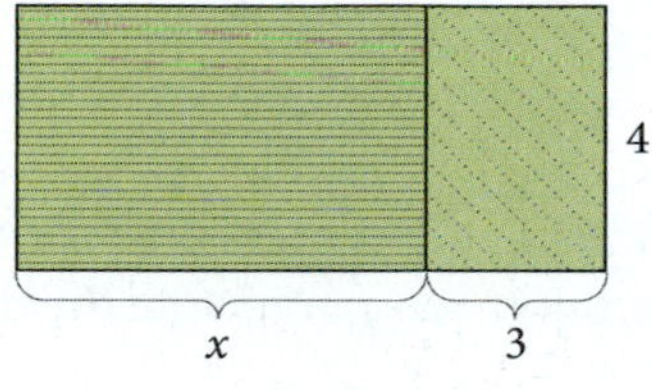

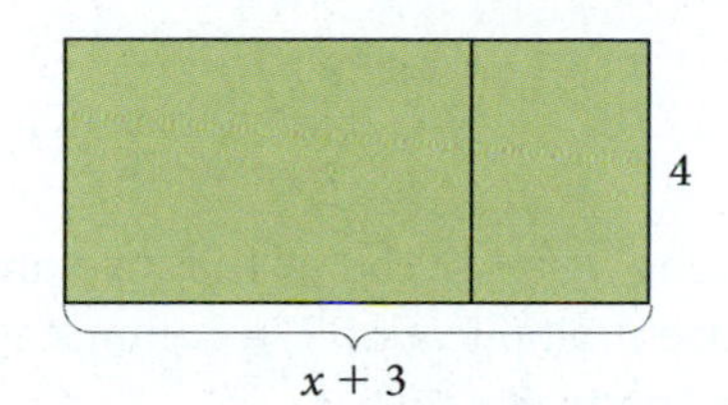

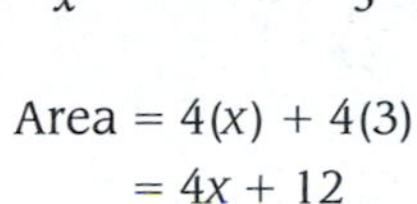

$$\text{Area} = 4(x) + 4(3) = 4x + 12 \qquad \text{Area} = 4(x + 3) = 4x + 12$$

Since the areas are equal, the equation $4(x + 3) = 4(x) + 4(3)$ is true.

Objectives

A Apply the distributive property to an expression.

B Combine similar terms.

C Find the value of an algebraic expression.

D Solve applications involving complementary and supplementary angles.

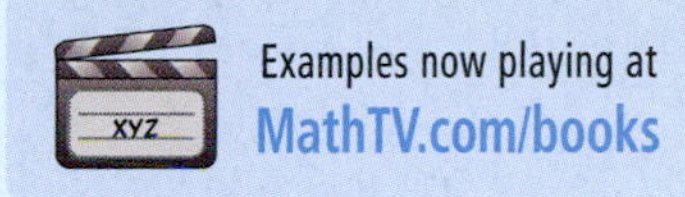

A The Distributive Property

EXAMPLE 1 Apply the distributive property to the expression:

$5(x + 3)$

SOLUTION Distributing the 5 over x and 3, we have

$$\begin{aligned} 5(x + 3) &= 5(x) + 5(3) && \text{Distributive property} \\ &= 5x + 15 && \text{Multiplication} \end{aligned}$$

Remember, $5x$ means "5 times x."

The distributive property can be applied to more complicated expressions involving negative numbers.

EXAMPLE 2 Multiply: $-4(3x + 5)$

SOLUTION Multiplying both the $3x$ and the 5 by -4, we have

$$\begin{aligned} -4(3x + 5) &= -4(3x) + (-4)5 && \text{Distributive property} \\ &= -12x + (-20) && \text{Multiplication} \\ &= -12x - 20 && \text{Definition of subtraction} \end{aligned}$$

Notice, first of all, that when we apply the distributive property here, we multiply through by -4. It is important to include the sign with the number when we use the distributive property. Second, when we multiply -4 and $3x$, the result is $-12x$ because

$$\begin{aligned} -4(3x) &= (-4 \cdot 3)x && \text{Associative property} \\ &= -12x && \text{Multiplication} \end{aligned}$$

PRACTICE PROBLEMS

1. Apply the distributive property to the expression $6(x + 4)$.

2. Multiply: $-3(2x + 4)$

Answers

1. $6x + 24$ **2.** $-6x - 12$

3. Multiply: $\frac{1}{2}(2x - 4)$

EXAMPLE 3 Multiply: $\frac{1}{3}(3x - 12)$

SOLUTION

$$\begin{aligned}\frac{1}{3}(3x - 12) &= \frac{1}{3}(3x) - \frac{1}{3}(12) && \text{Distributive property}\\ &= 1x - \frac{12}{3} && \text{Simplify}\\ &= x - 4 && \text{Divide}\end{aligned}$$

We can also use the distributive property to simplify expressions like $4x + 3x$. Because multiplication is a commutative operation, we can rewrite the distributive property like this:

$$b \cdot a + c \cdot a = (b + c)a$$

Applying the distributive property in this form to the expression $4x + 3x$, we have:

$$\begin{aligned}4x + 3x &= (4 + 3)x && \text{Distributive property}\\ &= 7x && \text{Addition}\end{aligned}$$

B Similar Terms

Note We are using the word *term* in a different sense here than we did with fractions. (The terms of a fraction are the numerator and the denominator.)

Recall that expressions like $4x$ and $3x$ are called *similar terms* because the variable parts are the same. Some other examples of similar terms are $5y$ and $-6y$, and the terms $7a$, $-13a$, $\frac{3}{4}a$. To simplify an algebraic expression (an expression that involves both numbers and variables), we combine similar terms by applying the distributive property. Table 1 reviews how we combine similar terms using the distributive property.

TABLE 1

Original Expression		Apply Distribution Property		Simplified Expression
$4x + 3x$	=	$(4 + 3)x$	=	$7x$
$7a + a$	=	$(7 + 1)a$	=	$8a$
$-5x + 7x$	=	$(-5 + 7)x$	=	$2x$
$8y - y$	=	$(8 - 1)y$	=	$7y$
$-4a - 2a$	=	$(-4 - 2)a$	=	$-6a$
$3x - 7x$	=	$(3 - 7)x$	=	$-4x$

As you can see from the table, the distributive property can be applied to any combination of positive and negative terms so long as they are similar terms.

4. Simplify: $6x - 2 + 3x + 8$

EXAMPLE 4 Simplify: $5x - 2 + 3x + 7$

SOLUTION We begin by changing subtraction to addition of the opposite and applying the commutative property to rearrange the order of the terms. We want similar terms to be written next to each other.

$$\begin{aligned}5x - 2 + 3x + 7 &= 5x + 3x + (-2) + 7 && \text{Commutative property}\\ &= (5 + 3)x + (-2) + 7 && \text{Distributive property}\\ &= 8x + 5 && \text{Addition}\end{aligned}$$

Notice that we take the negative sign in front of the 2 with the 2 when we rearrange terms. How do we justify doing this?

Answers
3. $x - 2$ 4. $9x + 6$

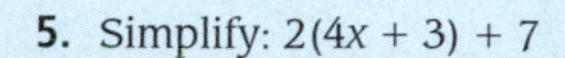

EXAMPLE 5 Simplify: $3(4x + 5) + 6$

SOLUTION We begin by distributing the 3 across the sum of $4x$ and 5. Then we combine similar terms.

$$
\begin{aligned}
3(4x + 5) + 6 &= 12x + 15 + 6 && \text{Distributive property} \\
&= 12x + 21 && \text{Add 15 and 6}
\end{aligned}
$$

EXAMPLE 6 Simplify: $2(3x + 1) + 4(2x - 5)$

SOLUTION Again, we apply the distributive property first; then we combine similar terms. Here is the solution showing only the essential steps:

$$
\begin{aligned}
2(3x + 1) + 4(2x - 5) &= 6x + 2 + 8x - 20 && \text{Distributive property} \\
&= 14x - 18 && \text{Combine similar terms}
\end{aligned}
$$

C The Value of an Algebraic Expression

An expression such as $3x + 5$ will take on different values depending on what x is. If we were to let x equal 2, the expression $3x + 5$ would become 11. On the other hand, if x is 10, the same expression has a value of 35:

When	$x = 2$	When	$x = 10$
the expression	$3x + 5$	the expression	$3x + 5$
becomes	$3(2) + 5$	becomes	$3(10) + 5$
	$= 6 + 5$		$= 30 + 5$
	$= 11$		$= 35$

EXAMPLES Find the value of each of the following expressions by replacing the variable with the given number.

Original Expression	Value of the Variable	Value of the Expression
7. $3x - 1$	$x = 2$	$3(2) - 1 = 6 - 1 = 5$
8. $2x - 3 + 4x$	$x = -1$	$2(-1) - 3 + 4(-1) = -2 - 3 + (-4) = -9$
9. $y^2 - 6y + 9$	$y = 4$	$4^2 - 6(4) + 9 = 16 - 24 + 9 = 1$

EXAMPLE 10 Find the area of a 30-W solar panel shown here with a length of 15 inches and a width of $10 + 3x$ inches.

15"

$10 + 3x$

SOLUTION Previously we worked with area, so we know that Area = (length) (width). Using the values for length and width, we have:

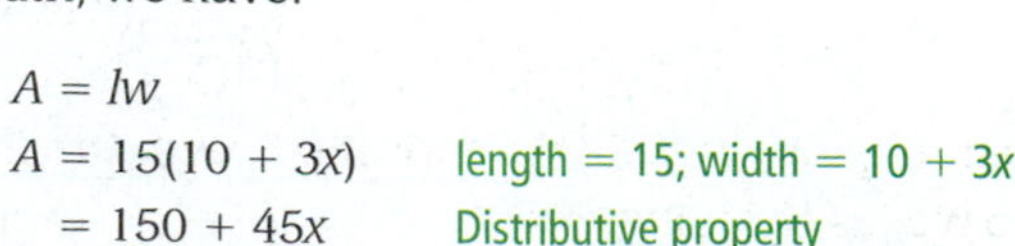

$$
\begin{aligned}
A &= lw \\
A &= 15(10 + 3x) && \text{length} = 15;\ \text{width} = 10 + 3x \\
&= 150 + 45x && \text{Distributive property}
\end{aligned}
$$

The area of this solar panel is $150 + 45x$ square inches.

5. Simplify: $2(4x + 3) + 7$

6. Simplify: $3(2x + 1) + 5(4x - 3)$

7. Find the value of $4x - 7$ when $x = 3$.

8. Find the value of $2x - 5 + 6x$ when $x = -2$.

9. Find the value of $y^2 - 10y + 25$ when $y = -2$.

10. Find the area of a 30-W solar panel with a length of 25 cm and a width of $8 + 2x$ cm.

Answers

5. $8x + 13$ **6.** $26x - 12$ **7.** 5 **8.** -21 **9.** 49 **10.** $200 + 50x$ cm^2

FACTS FROM GEOMETRY Angles

An angle is formed by two rays with the same endpoint. The common endpoint is called the *vertex* of the angle, and the rays are called the *sides* of the angle.

In Figure 1, angle θ (theta) is formed by the two rays OA and OB. The vertex of θ is O. Angle θ is also denoted as angle AOB, where the letter associated with the vertex is always the middle letter in the three letters used to denote the angle.

Degree Measure The angle formed by rotating a ray through one complete revolution about its endpoint (Figure 2) has a measure of 360 degrees, which we write as 360°.

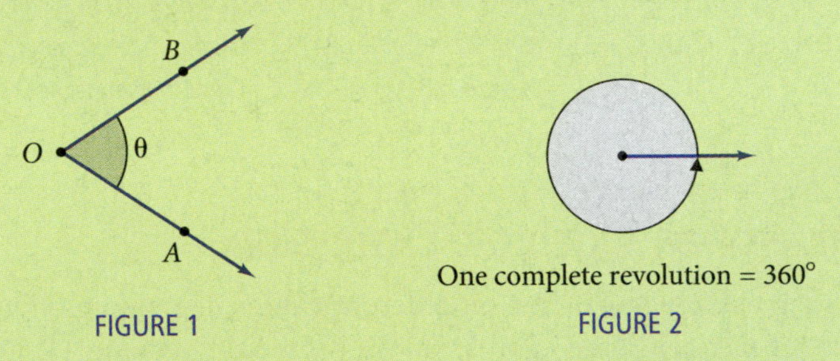

FIGURE 1 FIGURE 2

One degree of angle measure, written 1°, is $\frac{1}{360}$ of a complete rotation of a ray about its endpoint; there are 360° in one full rotation. (The number 360 was decided upon by early civilizations because it was believed that the Earth was at the center of the universe and the Sun would rotate once around the Earth every 360 days.) Similarly, 180° is half of a complete rotation, and 90° is a quarter of a full rotation. Angles that measure 90° are called *right angles*, and angles that measure 180° are called *straight angles*. If an angle measures between 0° and 90° it is called an *acute angle*, and an angle that measures between 90° and 180° is an *obtuse angle*. Figure 3 illustrates further.

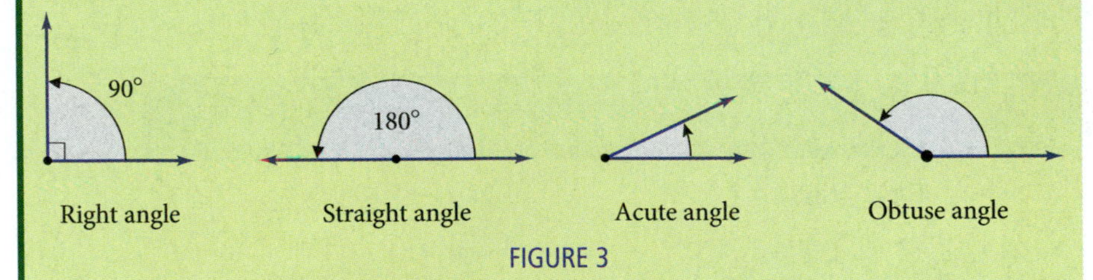

FIGURE 3

D **Complementary Angles and Supplementary Angles** If two angles add up to 90°, we call them *complementary angles*, and each is called the *complement* of the other. If two angles have a sum of 180°, we call them *supplementary angles*, and each is called the supplement of the other. Figure 4 illustrates the relationship between angles that are complementary and angles that are supplementary.

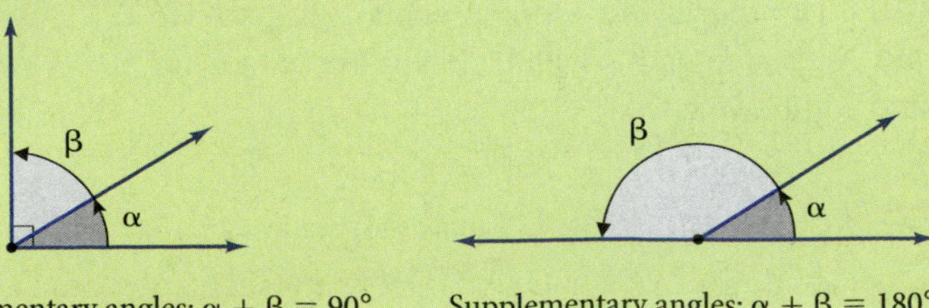

Complementary angles: $\alpha + \beta = 90°$ Supplementary angles: $\alpha + \beta = 180°$

FIGURE 4

EXAMPLE 11 Find x in each of the following diagrams.

a.

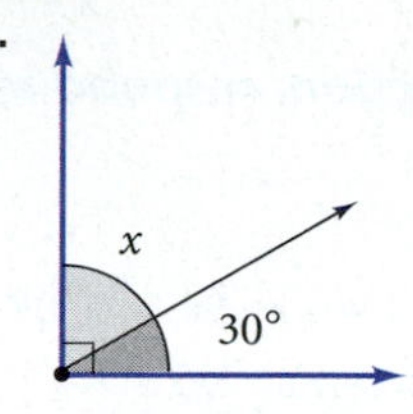

Complementary angles

b.

x
45°

Supplementary angles

SOLUTION We use subtraction to find each angle.

a. Because the two angles are complementary, we can find x by subtracting 30° from 90°:

$x = 90° - 30° = 60°$

We say 30° and 60° are complementary angles. The complement of 30° is 60°.

b. The two angles in the diagram are supplementary. To find x, we subtract 45° from 180°:

$x = 180° - 45° = 135°$

We say 45° and 135° are supplementary angles. The supplement of 45° is 135°.

11. Find x in each of the following diagrams.

a.

x
45°

Complementary angles

b.

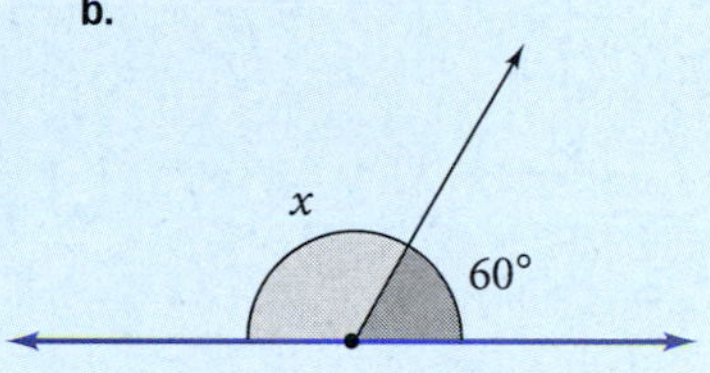

Supplementary angles

USING TECHNOLOGY

Protractors

When we think of technology, we think of computers and calculators. However, some simpler devices are also in the category of technology, because they help us do things that would be difficult to do without them. The protractor below can be used to draw and measure angles. In the diagram below, the protractor is being used to measure an angle of 120°. It can also be used to draw angles of any size.

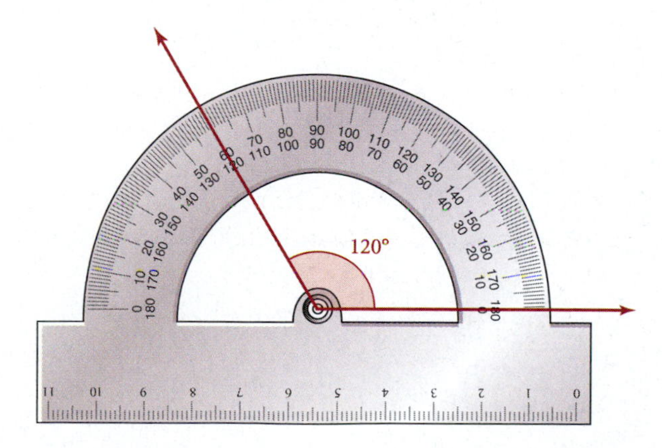

If you have a protractor, use it to draw the following angles: 30°, 45°, 60°, 120°, 135°, and 150°. Then imagine how you would draw these angles without a protractor.

Answer

11. a. 45° **b.** 120°

Getting Ready for Class

After reading through the preceding section, respond in your own words and in complete sentences.

1. What is the distributive property?
2. What property allows $5(x + 3)$ to be rewritten as $5x + 5(3)$?
3. What property allows $3x + 4x$ to be rewritten as $7x$?
4. True or false? The expression $3x$ means 3 multiplied by x.

Problem Set 4.1

A For review, use the distributive property to combine each of the following pairs of similar terms. [Examples 1–3]

1. $2x + 8x$
2. $3x + 7x$
3. $-4y + 5y$
4. $-3y + 10y$
5. $4a - a$
6. $9a - a$
7. $8(x + 2)$
8. $8(x - 2)$
9. $2(3a + 7)$
10. $5(3a + 2)$
11. $\frac{1}{3}(3x + 6)$
12. $\frac{1}{2}(2x + 4)$

B Simplify the following expressions by combining similar terms. In some cases the order of the terms must be rearranged first by using the commutative property. [Examples 4–6]

13. $4x + 2x + 3 + 8$
14. $7x + 5x + 2 + 9$
15. $7x - 5x + 6 - 4$
16. $10x - 7x + 9 - 6$
17. $-2a + a + 7 + 5$
18. $-8a + 3a + 12 + 1$
19. $6y - 2y - 5 + 1$
20. $4y - 3y - 7 + 2$
21. $4x + 2x - 8x + 4$
22. $6x + 5x - 12x + 6$
23. $9x - x - 5 - 1$
24. $2x - x - 3 - 8$
25. $2a + 4 + 3a + 5$
26. $9a + 1 + 2a + 6$
27. $3x + 2 - 4x + 1$
28. $7x + 5 - 2x + 6$
29. $12y + 3 + 5y$
30. $8y + 1 + 6y$
31. $4a - 3 - 5a + 2a$
32. $6a - 4 - 2a + 6a$

Apply the distributive property to each expression and then simplify.

33. $2(3x + 4) + 8$

34. $2(5x + 1) + 10$

35. $5(2x - 3) + 4$

36. $6(4x - 2) + 7$

37. $8(2y + 4) + 3y$

38. $2(5y + 1) + 2y$

39. $6(4y - 3) + 6y$

40. $5(2y - 6) + 4y$

41. $2(x + 3) + 4(x + 2)$

42. $3(x + 1) + 2(x + 5)$

43. $3(2a + 4) + 7(3a - 1)$

44. $7(2a + 2) + 4(5a - 1)$

C Find the value of each of the following expressions when $x = 5$. [Examples 7–9]

45. $2x + 4$

46. $3x + 2$

47. $7x - 8$

48. $8x - 9$

49. $-4x + 1$

50. $-3x + 7$

51. $-8 + 3x$

52. $-7 + 2x$

Find the value of each of the following expressions when $a = -2$.

53. $2a + 5$

54. $3a + 4$

55. $-7a + 4$

56. $-9a + 3$

57. $-a + 10$

58. $-a + 8$

59. $-4 + 3a$

60. $-6 + 5a$

Find the value of each of the following expressions when $x = 3$. You may substitute 3 for x in each expression the way it is written, or you may simplify each expression first and then substitute 3 for x.

61. $3x + 5x + 4$

62. $6x + 8x + 7$

63. $9x + x + 3 + 7$

64. $5x + 3x + 2 + 4$

65. $4x + 3 + 2x + 5$

66. $7x + 6 + 2x + 9$

67. $3x - 8 + 2x - 3$

68. $7x - 2 + 4x - 1$

Find the value of each of $12x - 3$ for each of the following values of x.

69. $\frac{1}{2}$

70. $\frac{1}{3}$

71. $\frac{1}{4}$

72. $\frac{1}{6}$

73. $\frac{3}{2}$

74. $\frac{2}{3}$

75. $\frac{3}{4}$

76. $\frac{5}{6}$

Use the distributive property to write two equivalent expressions for the area of each figure.

77.

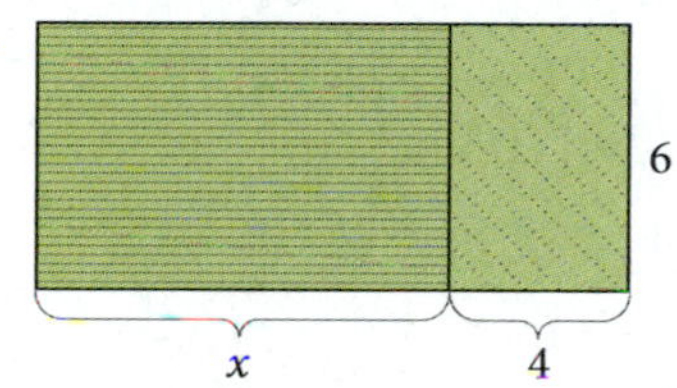

78.

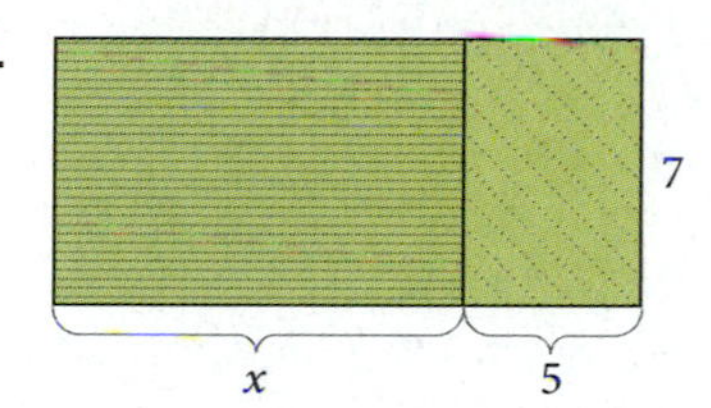

Write an expression for the perimeter of each figure.

79.

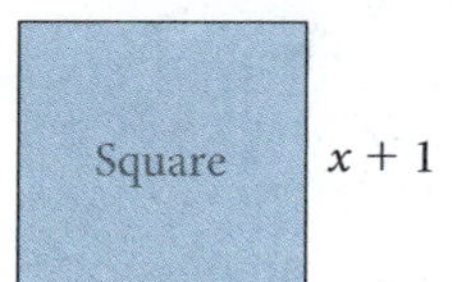

80.

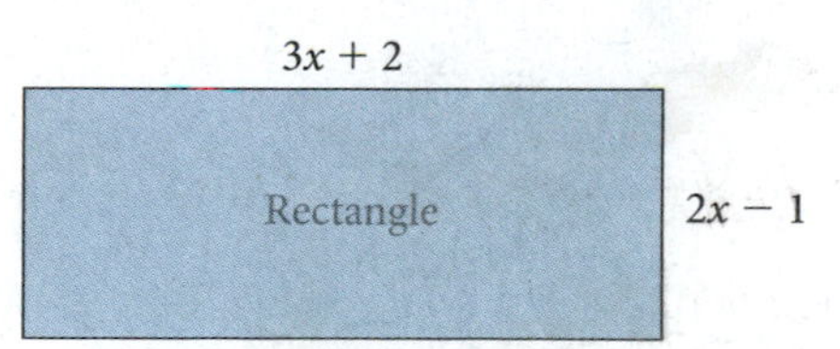

81.

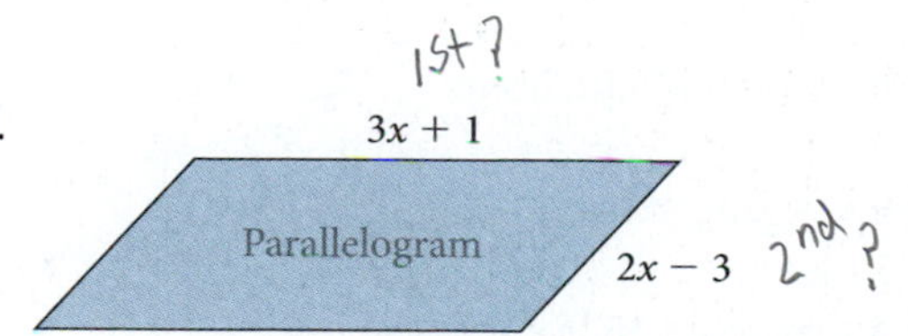

82.

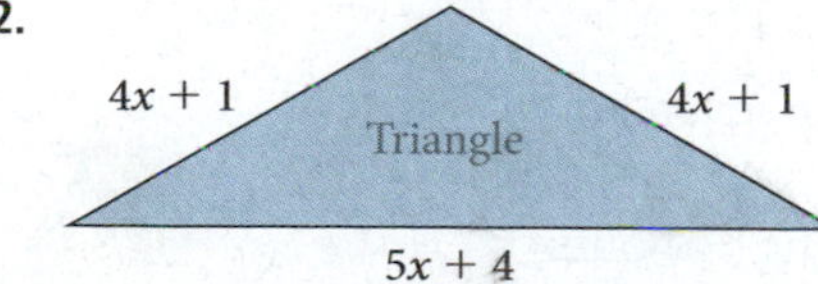

D Applying the Concepts

83. Buildings This Google Earth image shows the Leaning Tower of Pisa. Most buildings stand at a right angle, but the tower is sinking on one side. The angle of inclination is the angle between the vertical and the tower. If the angle between the tower and the ground is 85° what is the angle of inclination?

5°

84. Geometry This Google Earth image shows the Pentagon. The interior angles of a regular pentagon are all the same and sum to 540°. Find the size of each angle.

108°

Find x in each figure and decide if the two angles are complementary or supplementary. [Example 11]

85.

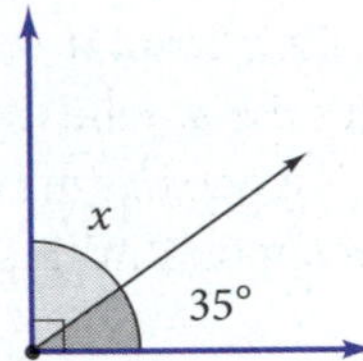

86.

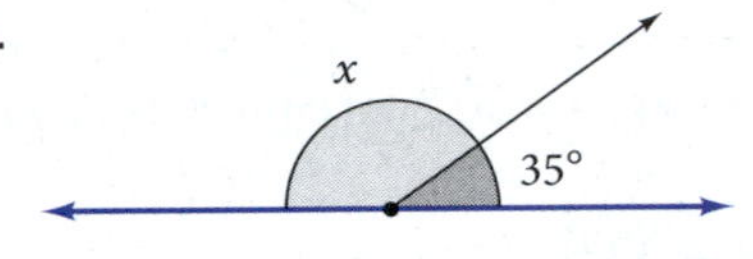

87.

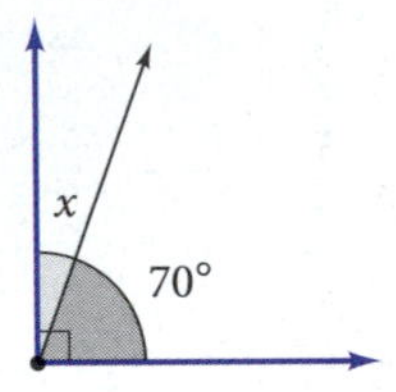

88.

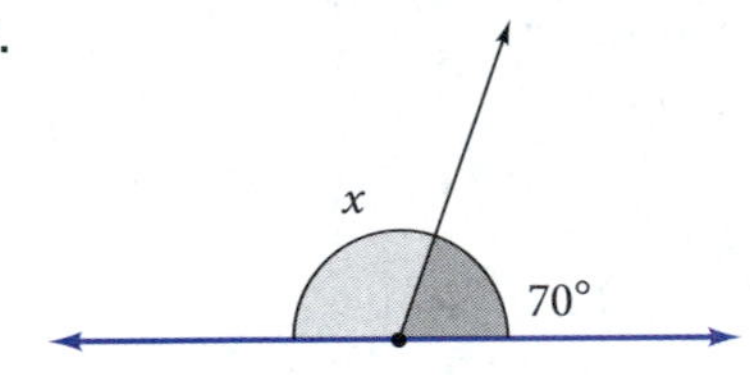

89. Luke earns \$12 per hour working as a math tutor. We can express the amount he earns each week for working x hours with the expression $12x$. Indicate with a yes or no, which of the following could be one of Luke's paychecks. If you answer no, explain your answer.

a. \$60 for working five hours

b. \$100 for working nine hours

c. \$80 for working seven hours

d. \$168 for working 14 hours

90. Kelly earns \$15 per hour working as a graphic designer. We can express the amount she earns each week for working x hours with the expression $15x$. Indicate with a yes or no which of the following could be one of Kelly's paychecks. If you answer no, explain your answer.

a. \$75 for working five hours

b. \$125 for working nine hours

c. \$90 for working six hours

d. \$500 for working 35 hours

91. Temperature and Altitude On a certain day, the temperature on the ground is 72 degrees Fahrenheit, and the temperature at an altitude of A feet above the ground is found from the expression $72 - \frac{A}{300}$. Find the temperature at the following altitudes.

a. 12,000 feet **b.** 15,000 feet **c.** 27,000 feet

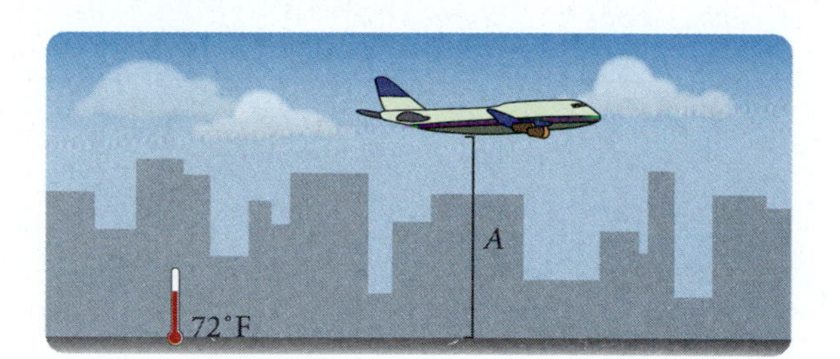

92. Perimeter of a Rectangle As you know, the expression $2l + 2w$ gives the perimeter of a rectangle with length l and width w. The garden below has a width of $3\frac{1}{2}$ feet and a length of 8 feet. What is the length of the fence that surrounds the garden?

93. Cost of Bottled Water A water bottling company charges \$7 per month for their water dispenser and \$2 for each gallon of water delivered. If you have g gallons of water delivered in a month, then the expression $7 + 2g$ gives the amount of your bill for that month. Find the monthly bill for each of the following deliveries.

a. 10 gallons **b.** 20 gallons

WBC **WATER BOTTLE CO.**

MONTHLY BILL

234 5th Street Glendora, CA 91740		DUE 07/23/08
Water dispenser	1	\$7.00
Gallons of water	8	\$2.00
		\$23.00

94. Cellular Phone Rates A cellular phone company charges \$35 per month plus 25 cents for each minute, or fraction of a minute, that you use one of their cellular phones. The expression $\frac{3500 + 25t}{100}$ gives the amount of money, in dollars, you will pay for using one of their phones for t minutes a month. Find the monthly bill for using one of their phones:

a. 20 minutes in a month **b.** 40 minutes in a month

Cell Phone Company
Grover Beach, CA

August 2008		DUE 08/15/08
Monthly Access per Phone:	1	\$35.00
Charges: \$0.25/minute	50	\$12.50
		\$47.50

Getting Ready for the Next Section

Add.

95. $4 + (-4)$

96. $2 + (-2)$

97. $-2 + (-4)$

98. $-2 + (-5)$

99. $-5 + 2$

100. $-3 + 12$

101. $\frac{5}{8} + \frac{3}{4}$

102. $\frac{5}{6} + \frac{2}{3}$

103. $-\frac{3}{4} + \frac{3}{4}$

104. $-\frac{2}{3} + \frac{2}{3}$

Simplify.

105. $x + 0$ **106.** $y + 0$ **107.** $y + 4 - 6$ **108.** $y + 6 - 2$

Maintaining Your Skills

Give the opposite of each number.

109. 9 **110.** 12 **111.** −6 **112.** −5

Problems 113–118 review material we covered in Chapter 1. Match each statement on the left with the property that justifies it on the right.

113. $2(6 + 5) = 2(6) + 2(5)$

114. $3 + (4 + 1) = (3 + 4) + 1$

115. $x + 5 = 5 + x$

116. $(a + 3) + 2 = a + (3 + 2)$

117. $(x + 5) + 1 = 1 + (x + 5)$

118. $(a + 4) + 2 = (4 + 2) + a$

a. Distributive property
b. Associative property
c. Commutative property
d. Commutative and associative properties

Perform the indicated operation.

119. $-\frac{5}{4}\left(\frac{8}{15}\right)$ **120.** $-\frac{4}{3}\left(\frac{6}{5}\right)$ **121.** $12 \div \frac{2}{3}$

122. $6 \div \frac{3}{5}$ **123.** $\frac{2}{3} - \frac{3}{4}$ **124.** $\frac{3}{5} - \frac{5}{8}$

The Addition Property of Equality

4.2

Objectives

A Identify a solution to an equation.

B Use the addition property of equality to solve linear equations.

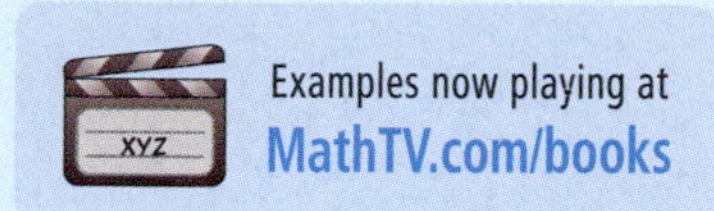

Introduction . . .

Previously we defined complementary angles as two angles whose sum is 90°. If A and B are complementary angles, then

$$A + B = 90°$$

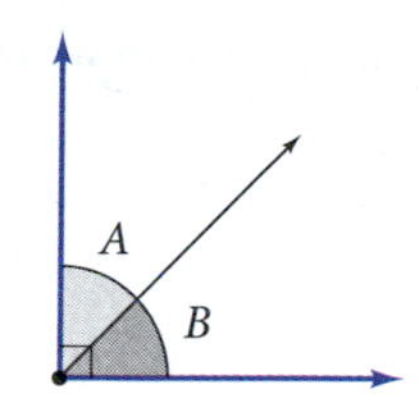

Complementary angles

If we know that $A = 30°$, then we can substitute 30° for A in the formula above to obtain the equation

$$30° + B = 90°$$

In this section we will learn how to solve equations like this one that involve addition and subtraction with one variable.

In general, solving an equation involves finding all replacements for the variable that make the equation a true statement.

A Solutions to Equations

Definition

A **solution** for an equation is a number that when used in place of the variable makes the equation a true statement.

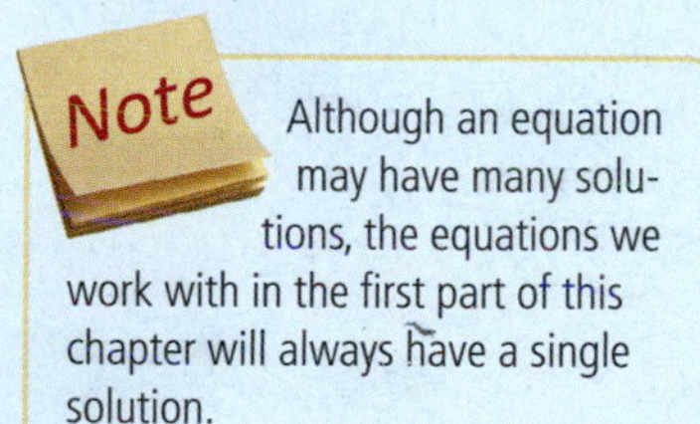

Although an equation may have many solutions, the equations we work with in the first part of this chapter will always have a single solution.

For example, the equation $x + 3 = 7$ has as its solution the number 4, because replacing x with 4 in the equation gives a true statement:

When	$x = 4$	
the equation	$x + 3 = 7$	
becomes	$4 + 3 = 7$	
or	$7 = 7$	A true statement

Is $x = 5$ the solution to the equation $3x + 2 = 17$?

SOLUTION To see if it is, we replace x with 5 in the equation and find out if the result is a true statement:

When	$x = 5$	
the equation	$3x + 2 = 17$	
becomes	$3(5) + 2 = 17$	
	$15 + 2 = 17$	
	$17 = 17$	A true statement

Because the result is a true statement, we can conclude that $x = 5$ is the solution to $3x + 2 = 17$.

PRACTICE PROBLEMS

1. Show that $x = 3$ is the solution to the equation $5x - 4 = 11$.

Answer

1. See solutions section.

2. Is $a = -3$ the solution to the equation $6a - 3 = 2a + 4$?

EXAMPLE 2 Is $a = -2$ the solution to the equation $7a + 4 = 3a - 2$?

SOLUTION

When $a = -2$
the equation $7a + 4 = 3a - 2$
becomes $7(-2) + 4 = 3(-2) - 2$
$-14 + 4 = -6 - 2$
$-10 = -8$ A false statement

Because the result is a false statement, we must conclude that $a = -2$ is *not* the solution to the equation $7a + 4 = 3a - 2$.

B Addition Property of Equality

We want to develop a process for solving equations with one variable. The most important property needed for solving the equations in this section is called the *addition property of equality.* The formal definition looks like this:

Addition Property of Equality

Let A, B, and C represent algebraic expressions.

If $A = B$
then $A + C = B + C$

In words: Adding the same quantity to both sides of an equation never changes the solution to the equation.

This property is extremely useful in solving equations. Our goal in solving equations is to isolate the variable on one side of the equation. We want to end up with an equation of the form

$x =$ a number

To do so we use the addition property of equality. Remember to follow this basic rule of algebra: *Whatever is done to one side of an equation must be done to the other side in order to preserve the equality.*

3. Solve for x: $x + 5 = -2$

EXAMPLE 3 Solve for x: $x + 4 = -2$

SOLUTION We want to isolate x on one side of the equation. If we add -4 to both sides, the left side will be $x + 4 + (-4)$, which is $x + 0$ or just x.

$$x + 4 = -2$$
$$x + 4 + \mathbf{(-4)} = -2 + \mathbf{(-4)} \quad \text{Add } -4 \text{ to both sides}$$
$$x + 0 = -6 \quad \text{Addition}$$
$$x = -6 \quad x + 0 = x$$

The solution is -6. We can check it if we want to by replacing x with -6 in the original equation:

When $x = -6$
the equation $x + 4 = -2$
becomes $-6 + 4 = -2$
$-2 = -2$ A true statement

Note With some of the equations in this section, you will be able to see the solution just by looking at the equation. But it is important that you show all the steps used to solve the equations anyway. The equations you come across in the future will not be as easy to solve, so you should learn the steps involved very well.

Answers
2. No 3. -7

EXAMPLE 4 Solve for a: $a - 3 = 5$

SOLUTION

$$\begin{aligned} a - 3 &= 5 \\ a - 3 + \mathbf{3} &= 5 + \mathbf{3} && \text{Add 3 to both sides} \\ a + 0 &= 8 && \text{Addition} \\ a &= 8 && a + 0 = a \end{aligned}$$

The solution to $a - 3 = 5$ is $a = 8$.

EXAMPLE 5 Solve for y: $y + 4 - 6 = 7 - 1$

SOLUTION Before we apply the addition property of equality, we must simplify each side of the equation as much as possible:

$$\begin{aligned} y + 4 - 6 &= 7 - 1 \\ y - 2 &= 6 && \text{Simplify each side} \\ y - 2 + \mathbf{2} &= 6 + \mathbf{2} && \text{Add 2 to both sides} \\ y + 0 &= 8 && \text{Addition} \\ y &= 8 && y + 0 = y \end{aligned}$$

EXAMPLE 6 Solve for x: $3x - 2 - 2x = 4 - 9$

SOLUTION Simplifying each side as much as possible, we have

$$\begin{aligned} 3x - 2 - 2x &= 4 - 9 \\ x - 2 &= -5 && 3x - 2x = x \\ x - 2 + \mathbf{2} &= -5 + \mathbf{2} && \text{Add 2 to both sides} \\ x + 0 &= -3 && \text{Addition} \\ x &= -3 && x + 0 = x \end{aligned}$$

EXAMPLE 7 Solve for x: $-3 - 6 = x + 4$

SOLUTION The variable appears on the right side of the equation in this problem. This makes no difference; we can isolate x on either side of the equation. We can leave it on the right side if we like:

$$\begin{aligned} -3 - 6 &= x + 4 \\ -9 &= x + 4 && \text{Simplify the left side} \\ -9 + \mathbf{(-4)} &= x + 4 + \mathbf{(-4)} && \text{Add } -4 \text{ to both sides} \\ -13 &= x + 0 && \text{Addition} \\ -13 &= x && x + 0 = x \end{aligned}$$

The statement $-13 = x$ is equivalent to the statement $x = -13$. In either case the solution to our equation is -13.

EXAMPLE 8 Solve: $a - \frac{3}{4} = \frac{5}{8}$

SOLUTION To isolate a we add $\frac{3}{4}$ to each side:

$$a - \frac{3}{4} = \frac{5}{8}$$

$$a - \frac{3}{4} + \mathbf{\frac{3}{4}} = \frac{5}{8} + \mathbf{\frac{3}{4}}$$

$$a = \frac{11}{8}$$

When solving equations we will leave answers like $\frac{11}{8}$ as improper fractions, rather than change them to mixed numbers.

4. Solve for a: $a - 2 = 7$

5. Solve for y: $y + 6 - 2 = 8 - 9$

6. Solve for x: $5x - 3 - 4x = 4 - 7$

7. Solve for x: $-5 - 7 = x + 2$

8. Solve: $a - \frac{2}{3} = \frac{5}{6}$

Answers

4. 9 **5.** -5 **6.** 0 **7.** -14
8. $\frac{3}{2}$

9. Solve: $5(3a - 4) - 14a = 25$

EXAMPLE 9 Solve: $4(2a - 3) - 7a = 2 - 5$.

SOLUTION We must begin by applying the distributive property to separate terms on the left side of the equation. Following that, we combine similar terms and then apply the addition property of equality.

$4(2a - 3) - 7a = 2 - 5$	Original equation
$8a - 12 - 7a = 2 - 5$	Distributive property
$a - 12 = -3$	Simplify each side
$a - 12 + \mathbf{12} = -3 + \mathbf{12}$	Add 12 to each side
$a = 9$	Addition

A Note on Subtraction

Although the addition property of equality is stated for addition only, we can subtract the same number from both sides of an equation as well. Because subtraction is defined as addition of the opposite, subtracting the same quantity from both sides of an equation will not change the solution. If we were to solve the equation in Example 3 using subtraction instead of addition, the steps would look like this:

$x + 4 = -2$	Original equation
$x + 4 - \mathbf{4} = -2 - \mathbf{4}$	Subtract 4 from each side
$x = -6$	Subtraction

In my experience teaching algebra, I find that students make fewer mistakes if they think in terms of addition rather than subtraction. So, you are probably better off if you continue to use the addition property just the way we have used it in the examples in this section. But, if you are curious as to whether you can subtract the same number from both sides of an equation, the answer is yes.

Getting Ready for Class

After reading through the preceding section, respond in your own words and in complete sentences. An answer of true or false should be accompanied by a sentence explaining why the answer is true or false.

1. What is a solution to an equation?
2. True or false? According to the addition property of equality, adding the same value to both sides of an equation will never change the solution to the equation.
3. Show that $x = 5$ is a solution to the equation $3x + 2 = 17$ without solving the equation.
4. True or false? The equations below have the same solution.

    Equation 1: $7x + 5 = 19$

    Equation 2: $7x + 5 + 3 = 19 + 3$

Answer

9. 45

Problem Set 4.2

A Check to see if the number to the right of each of the following equations is the solution to the equation. [Examples 1, 2]

1. $2x + 1 = 5; 2$

2. $4x + 3 = 7; 1$

3. $3x + 4 = 19; 5$

4. $3x + 8 = 14; 2$

5. $2x - 4 = 2; 4$

6. $5x - 6 = 9; 3$

7. $2x + 1 = 3x + 3; -2$

8. $4x + 5 = 2x - 1; -6$

9. $x - 4 = 2x + 1; -4$

10. $x - 8 = 3x + 2; -5$

B Solve each equation. [Examples 3, 4, 8]

11. $x + 2 = 8$

12. $x + 3 = 5$

13. $x - 4 = 7$

14. $x - 6 = 2$

15. $a + 9 = -6$

16. $a + 3 = -1$

17. $x - 5 = -4$

18. $x - 8 = -3$

19. $y - 3 = -6$

20. $y - 5 = -1$

21. $a + \frac{1}{3} = -\frac{2}{3}$

22. $a + \frac{1}{4} = -\frac{3}{4}$

23. $x - \frac{3}{5} = \frac{4}{5}$

24. $x - \frac{7}{8} = \frac{3}{8}$

25. $y + 73 = -27$

26. $y + 82 = -28$

B Simplify each side of the following equations before applying the addition property. [Examples 5–7]

27. $x + 4 - 7 = 3 - 10$

28. $x + 6 - 2 = 5 - 12$

29. $x - 6 + 4 = -3 - 2$

30. $x - 8 + 2 = -7 - 1$

31. $3 - 5 = a - 4$

32. $2 - 6 = a - 1$

33. $3a + 7 - 2a = 1$

34. $5a + 6 - 4a = 4$

35. $6a - 2 - 5a = -9 + 1$

36. $7a - 6 - 6a = -3 + 1$

37. $8 - 5 = 3x - 2x + 4$

38. $10 - 6 = 8x - 7x + 6$

B The following equations contain parentheses. Apply the distributive property to remove the parentheses, then simplify each side before using the addition property of equality. [Example 9]

39. $2(x + 3) - x = 4$

40. $5(x + 1) - 4x = 2$

41. $-3(x - 4) + 4x = 3 - 7$

42. $-2(x - 5) + 3x = 4 - 9$

43. $5(2a + 1) - 9a = 8 - 6$

44. $4(2a - 1) - 7a = 9 - 5$

45. $-(x + 3) + 2x - 1 = 6$

46. $-(x - 7) + 2x - 8 = 4$

Find the value of x for each of the figures, given the perimeter.

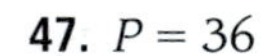

47. $P = 36$

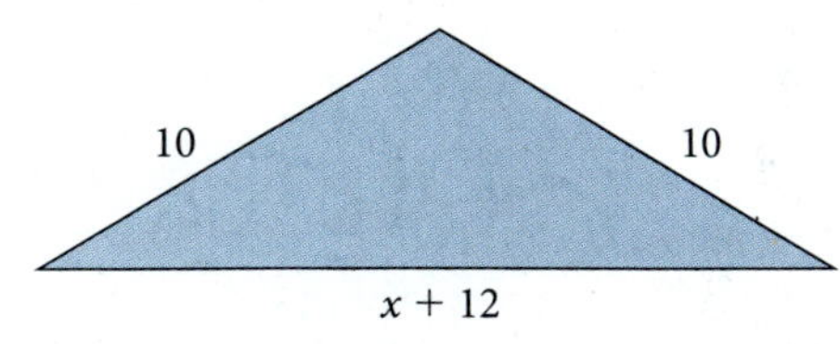

48. $P = 30$

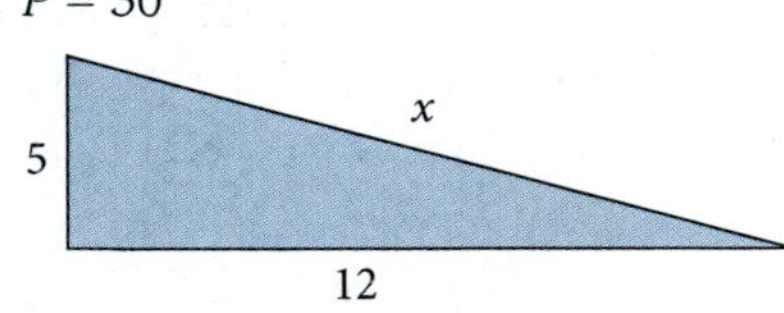

49. $P = 16$

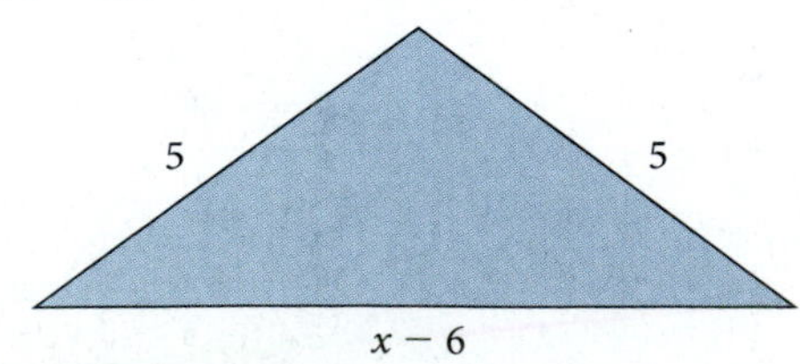

50. $P = 60$

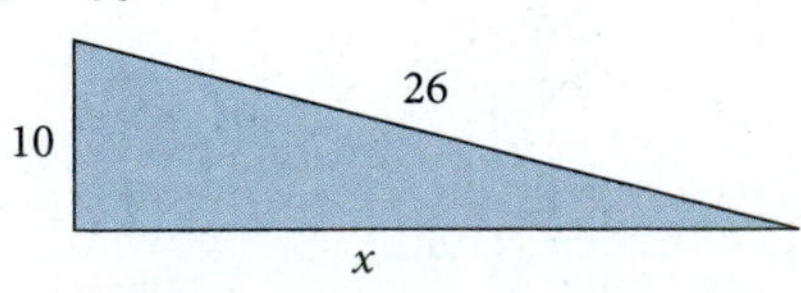

Applying the Concepts

Temperature The chart shows the temperatures for some of the world's hottest places. To convert from Celsius to Kelvin we use the formula $y = x + 273$, where y is the temperature in Kelvin and x is the temperature in Celsius. Use the formula to answer Questions 51 and 52.

Heating Up

137°F Al'Aziziyah, Libya
134°F Greenland Ranch, Death Valley, United States
131°F Ghudamis, Libya
131°F Kebili, Tunisia
130°F Tombouctou, Mali

Source: Aneki.com

51. The hottest temperature in Al'Aziziyah was 331 Kelvin. Convert this to Celsius.

52. The hottest temperature in Kebili, Tunisia, was 328 Kelvin. Convert this to Celsius.

53. **Geometry** Two angles are complementary angles. If one of the angles is 23°, then solving the equation $x + 23° = 90°$ will give you the other angle. Solve the equation.

Complementary angles

54. **Geometry** Two angles are supplementary angles. If one of the angles is 23°, then solving the equation $x + 23° = 180°$ will give you the other angle. Solve the equation.

55. **Theater Tickets** The El Portal Center for the Arts in North Hollywood, California, holds a maximum of 400 people. The two balconies hold 86 and 89 people each; the rest of the seats are at the stage level. Solving the equation $x + 86 + 89 = 400$ will give you the number of seats on the stage level.
    a. Solve the equation for x.
    b. If tickets on the stage level are \$30 each, and tickets in either balcony are \$25 each, what is the maximum amount of money the theater can bring in for a show?

56. **Geometry** The sum of the angles in the triangle on the swing set is 180°. Use this fact to write an equation containing x. Then solve the equation.

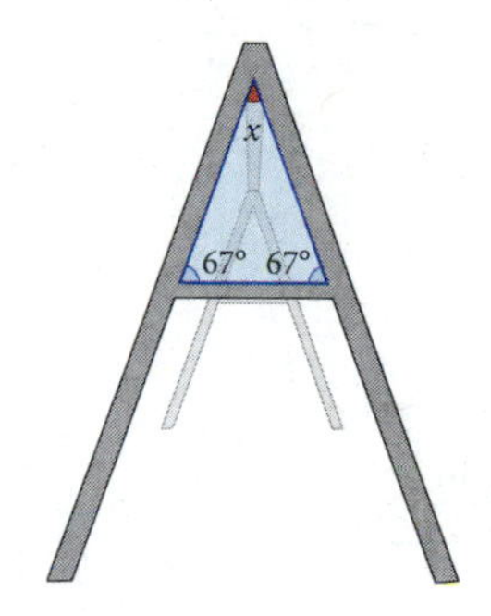

Getting Ready for the Next Section

Find the reciprocal of each number.

57. 4

58. 3

59. $\frac{1}{2}$

60. $\frac{1}{3}$

61. $\frac{2}{3}$

62. $\frac{3}{5}$

Multiply.

63. $2 \cdot \frac{1}{2}$

64. $\frac{1}{4} \cdot 4$

65. $-\frac{1}{3}(-3)$

66. $-\frac{1}{4}(-4)$

67. $\frac{3}{2}\left(\frac{2}{3}\right)$

68. $\frac{5}{3}\left(\frac{3}{5}\right)$

69. $\left(-\frac{5}{4}\right)\left(-\frac{4}{5}\right)$

70. $\left(-\frac{4}{3}\right)\left(-\frac{3}{4}\right)$

Simplify.

71. $1 \cdot x$

72. $1 \cdot a$

73. $4x - 11 + 3x$

74. $2x - 11 + 3x$

Maintaining Your Skills

Add or subtract as indicated.

75. $\frac{3}{2} + \frac{5}{10}$

76. $\frac{1}{3} + \frac{4}{12}$

77. $\frac{2}{7} + \frac{1}{14}$

78. $\frac{3}{8} + \frac{1}{16}$

79. $\frac{1}{3} - \frac{2}{5}$

80. $\frac{3}{4} - \frac{3}{7}$

81. $\frac{1}{6} - \frac{4}{3}$

82. $\frac{2}{5} - \frac{5}{10}$

Translating Translate each of the following into an equation, and then solve the equation.

83. The sum of x and 12 is 30.

84. The difference of x and 12 is 30.

85. The difference of 8 and 5 is equal to the sum of x and 7.

86. The sum of 8 and 5 is equal to the difference of x and 7.

4.3 The Multiplication Property of Equality

Objectives

A Use the multiplication property of equality to solve equations.

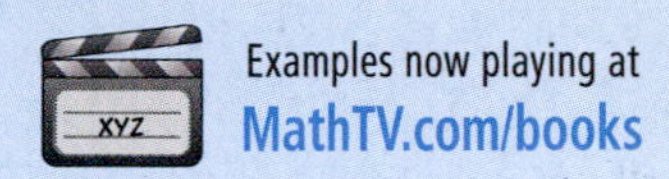

Examples now playing at MathTV.com/books

In this section we will continue to solve equations in one variable. We will again use the addition property of equality, but we will also use another property—the *multiplication property of equality*—to solve the equations in this section. We will state the multiplication property of equality and then see how it is used by looking at some examples.

The most popular Internet video download of all time was a *Star Wars* movie trailer. The video was compressed so it would be small enough for people to download over the Internet. In movie theaters, a film plays at 24 frames per second. Over the Internet, that number is sometimes cut in half, to 12 frames per second, to make the file size smaller.

We can use the equation $240 = \frac{x}{12}$ to find the number of total frames, x, in a 240-second movie clip that plays at 12 frames per second.

A Multiplication Property of Equality

Multiplication Property of Equality

Let A, B, and C represent algebraic expressions, with C not equal to 0.

$$\text{If} \quad A = B$$
$$\text{then} \quad AC = BC$$

In words: Multiplying both sides of an equation by the same nonzero quantity never changes the solution to the equation.

Now, because division is defined as multiplication by the reciprocal, we are also free to divide both sides of an equation by the same nonzero quantity and always be sure we have not changed the solution to the equation.

EXAMPLE 1 Solve for x: $\frac{1}{2}x = 3$

SOLUTION Our goal here is the same as it was in Section 4.2. We want to isolate x (that is, $1x$) on one side of the equation. We have $\frac{1}{2}x$ on the left side. If we multiply both sides by 2, we will have $1x$ on the left side. Here is how it looks:

$$\frac{1}{2}x = 3$$

$$\mathbf{2}\left(\frac{1}{2}x\right) = \mathbf{2}(3) \qquad \text{Multiply both sides by 2}$$

$$x = 6 \qquad \text{Multiplication}$$

To see why $2(\frac{1}{2}x)$ is equivalent to x, we use the associative property:

$$2\left(\frac{1}{2}x\right) = \left(2 \cdot \frac{1}{2}\right)x \qquad \text{Associative property}$$

$$= 1 \cdot x \qquad 2 \cdot \frac{1}{2} = 1$$

$$= x \qquad 1 \cdot x = x$$

Although we will not show this step when solving problems, it is implied.

PRACTICE PROBLEMS

1. Solve for x: $\frac{1}{3}x = 5$

Answer

1. 15

2. Solve for a: $\frac{1}{5}a + 3 = 7$

EXAMPLE 2 Solve for a: $\frac{1}{3}a + 2 = 7$

SOLUTION We begin by adding -2 to both sides to get $\frac{1}{3}a$ by itself. We then multiply by 3 to solve for a.

$$\frac{1}{3}a + 2 = 7$$

$$\frac{1}{3}a + 2 + (\mathbf{-2}) = 7 + (\mathbf{-2}) \quad \text{Add } -2 \text{ to both sides}$$

$$\frac{1}{3}a = 5 \quad \text{Addition}$$

$$\mathbf{3} \cdot \frac{1}{3}a = \mathbf{3} \cdot 5 \quad \text{Multiply both sides by 3}$$

$$a = 15 \quad \text{Multiplication}$$

We can check our solution to see that it is correct:

When $a = 15$

the equation $\frac{1}{3}a + 2 = 7$

becomes $\frac{1}{3}(15) + 2 = 7$

$$5 + 2 = 7$$

$$7 = 7 \quad \text{A true statement}$$

3. Solve for y: $\frac{3}{5}y = 6$

EXAMPLE 3 Solve for y: $\frac{2}{3}y = 12$

SOLUTION In this case we multiply each side of the equation by the reciprocal of $\frac{2}{3}$, which is $\frac{3}{2}$.

$$\frac{2}{3}y = 12$$

$$\mathbf{\frac{3}{2}}\left(\frac{2}{3}y\right) = \mathbf{\frac{3}{2}}(12)$$

$$y = 18$$

The solution checks because $\frac{2}{3}$ of 18 is 12.

Note The reciprocal of a negative number is also a negative number. Remember, reciprocals are two numbers that have a product of 1. Since 1 is a positive number, any two numbers we multiply to get 1 must both have the same sign. Here are some negative numbers and their reciprocals:

The reciprocal of -2 is $-\frac{1}{2}$.

The reciprocal of -7 is $-\frac{1}{7}$.

The reciprocal of $-\frac{1}{3}$ is -3.

The reciprocal of $-\frac{3}{4}$ is $-\frac{4}{3}$.

The reciprocal of $-\frac{9}{5}$ is $-\frac{5}{9}$.

Answers
2. 20 3. 10

EXAMPLE 4 Solve for x: $-\frac{4}{5}x = \frac{8}{15}$

SOLUTION The reciprocal of $-\frac{4}{5}$ is $-\frac{5}{4}$.

$$-\frac{4}{5}x = \frac{8}{15}$$

$$-\frac{\mathbf{5}}{\mathbf{4}}\left(-\frac{4}{5}x\right) = -\frac{\mathbf{5}}{\mathbf{4}}\left(\frac{8}{15}\right)$$

$$x = -\frac{2}{3}$$

Many times, it is convenient to divide both sides by a nonzero number to solve an equation, as the next example shows.

EXAMPLE 5 Solve for x: $4x = -20$

SOLUTION If we divide both sides by 4, the left side will be just x, which is what we want. It is okay to divide both sides by 4 because division by 4 is equivalent to multiplication by $\frac{1}{4}$, and the multiplication property of equality states that we can multiply both sides by any number so long as it isn't 0.

$$4x = -20$$

$$\frac{4x}{\mathbf{4}} = \frac{-20}{\mathbf{4}} \quad \text{Divide both sides by 4}$$

$$x = -5 \quad \text{Division}$$

Because $4x$ means "4 times x," the factors in the numerator of $\frac{4x}{4}$ are 4 and x. Because the factor 4 is common to the numerator and the denominator, we divide it out to get just x.

EXAMPLE 6 Solve for x: $-3x + 7 = -5$

SOLUTION We begin by adding -7 to both sides to reduce the left side to $-3x$.

$$-3x + 7 = -5$$

$$-3x + 7 + \mathbf{(-7)} = -5 + \mathbf{(-7)} \quad \text{Add } -7 \text{ to both sides}$$

$$-3x = -12 \quad \text{Addition}$$

$$\frac{-3x}{\mathbf{-3}} = \frac{-12}{\mathbf{-3}} \quad \text{Divide both sides by } -3$$

$$x = 4 \quad \text{Division}$$

With more complicated equations we simplify each side separately before applying the addition or multiplication properties of equality. The examples below illustrate.

EXAMPLE 7 Solve for x: $5x - 8x + 3 = 4 - 10$

SOLUTION We combine similar terms to simplify each side and then solve as usual.

$$5x - 8x + 3 = 4 - 10$$

$$-3x + 3 = -6 \quad \text{Simplify each side}$$

$$-3x + 3 + \mathbf{(-3)} = -6 + \mathbf{(-3)} \quad \text{Add } -3 \text{ to both sides}$$

$$-3x = -9 \quad \text{Addition}$$

$$\frac{-3x}{\mathbf{-3}} = \frac{-9}{\mathbf{-3}} \quad \text{Divide both sides by } -3$$

$$x = 3 \quad \text{Division}$$

4. Solve for x: $-\frac{3}{4}x = \frac{6}{5}$

5. Solve for x: $6x = -42$

Note If we multiply each side by $\frac{1}{4}$, the solution looks like this:

$$\frac{\mathbf{1}}{\mathbf{4}}(4x) = \frac{\mathbf{1}}{\mathbf{4}}(-20)$$

$$\left(\frac{1}{4} \cdot 4\right)x = -5$$

$$1x = -5$$

$$x = -5$$

6. Solve for x: $-5x + 6 = -14$

7. Solve for x: $3x - 7x + 5 = 3 - 18$

Answers

4. $-\frac{8}{5}$ **5.** -7 **6.** 4 **7.** 5

8. Solve for x:
$-5 + 4 = 2x - 11 + 3x$

EXAMPLE 8 Solve for x: $-8 + 11 = 4x - 11 + 3x$

SOLUTION We begin by simplifying each side separately.

$$-8 + 11 = 4x - 11 + 3x$$

$$3 = 7x - 11 \quad \text{Simplify both sides}$$

$$3 + \mathbf{11} = 7x - 11 + \mathbf{11} \quad \text{Add 11 to both sides}$$

$$14 = 7x \quad \text{Addition}$$

$$\frac{14}{\mathbf{7}} = \frac{7x}{\mathbf{7}} \quad \text{Divide both sides by 7}$$

$$2 = x \text{ or } x = 2$$

Again, it makes no difference which side of the equation x ends up on, so long as it is just one x.

COMMON MISTAKES

Before we end this section, we should mention a very common mistake made by students when they first begin to solve equations. It involves trying to subtract away the number in front of the variable—like this:

$$7x = 21$$

$$7x - \mathbf{7} = 21 - \mathbf{7} \quad \text{Add } -7 \text{ to both sides}$$

$$x = 14 \longleftarrow \text{Mistake}$$

The mistake is not in trying to subtract 7 from both sides of the equation. The mistake occurs when we say $7x - 7 = x$. It just isn't true. We can add and subtract only similar terms. The terms $7x$ and 7 are not similar, because one contains x and the other doesn't. The correct way to do the problem is like this:

$$7x = 21$$

$$\frac{7x}{\mathbf{7}} = \frac{21}{\mathbf{7}} \quad \text{Divide both sides by 7}$$

$$x = 3 \quad \text{Division}$$

Getting Ready for Class

After reading through the preceding section, respond in your own words and in complete sentences.

1. True or false? Multiplying both sides of an equation by the same nonzero quantity will never change the solution to the equation.
2. If we were to multiply the right side of an equation by 2, then the left side should be multiplied by __________.
3. Dividing both sides of the equation $4x = -20$ by 4 is the same as multiplying both sides by what number?

Answer
8. 2

Problem Set 4.3

A Use the multiplication property of equality to solve each of the following equations. In each case, show all the steps. [Examples 1, 3–5]

1. $\frac{1}{4}x = 2$

2. $\frac{1}{3}x = 7$

3. $\frac{1}{2}x = -3$

4. $\frac{1}{5}x = -6$

5. $-\frac{1}{3}x = 2$

6. $-\frac{1}{3}x = 5$

7. $-\frac{1}{6}x = -1$

8. $-\frac{1}{2}x = -4$

9. $\frac{3}{4}y = 12$

10. $\frac{2}{3}y = 18$

11. $3a = 48$

12. $2a = 28$

13. $-\frac{3}{5}x = \frac{9}{10}$

14. $-\frac{4}{5}x = -\frac{8}{15}$

15. $5x = -35$

16. $7x = -35$

17. $-8y = 64$

18. $-9y = 27$

19. $-7x = -42$

20. $-6x = -42$

A Using the addition property of equality first, solve each of the following equations. [Examples 2, 6]

21. $3x - 1 = 5$

22. $2x + 4 = 6$

23. $-4a + 3 = -9$

24. $-5a + 10 = 50$

25. $6x - 5 = 19$

26. $7x - 5 = 30$

27. $\frac{1}{3}a + 3 = -5$

28. $\frac{1}{2}a + 2 = -7$

29. $-\frac{1}{4}a + 5 = 2$

30. $-\frac{1}{5}a + 3 = 7$

31. $2x - 4 = -20$

32. $3x - 5 = -26$

33. $\frac{2}{3}x - 4 = 6$

34. $\frac{3}{4}x - 2 = 7$

35. $-11a + 4 = -29$

36. $-12a + 1 = -47$

37. $-3y - 2 = 1$

38. $-2y - 8 = 2$

39. $-2x - 5 = -7$

40. $-3x - 6 = -36$

A Simplify each side of the following equations first, then solve. [Examples 7, 8]

41. $2x + 3x - 5 = 7 + 3$

42. $4x + 5x - 8 = 6 + 4$

43. $4x - 7 + 2x = 9 - 10$

44. $5x - 6 + 3x = -6 - 8$

45. $3a + 2a + a = 7 - 13$

46. $8a - 6a + a = 8 - 14$

47. $5x + 4x + 3x = 4 - 8$

48. $4x + 8x - 2x = 15 - 10$

49. $5 - 18 = 3y - 2y + 1$

50. $7 - 16 = 4y - 3y + 2$

Find the value of x for each of the figures, given the perimeter.

51. $P = 72$

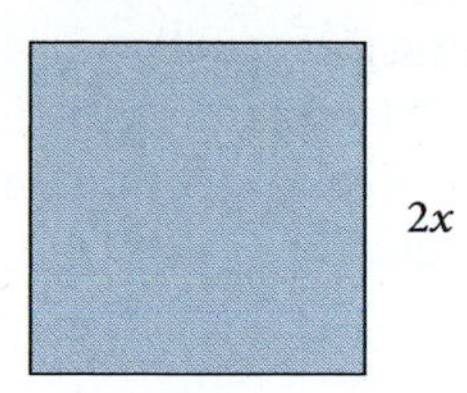

52. $P = 96$

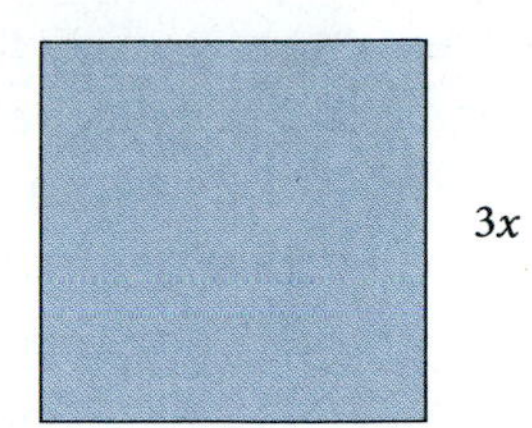

53. $P = 80$

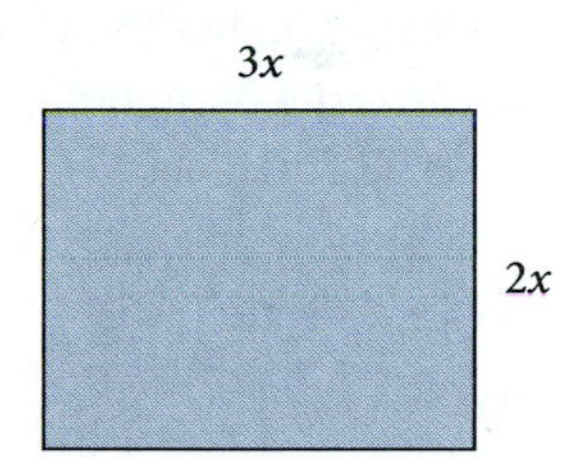

54. $P = 64$

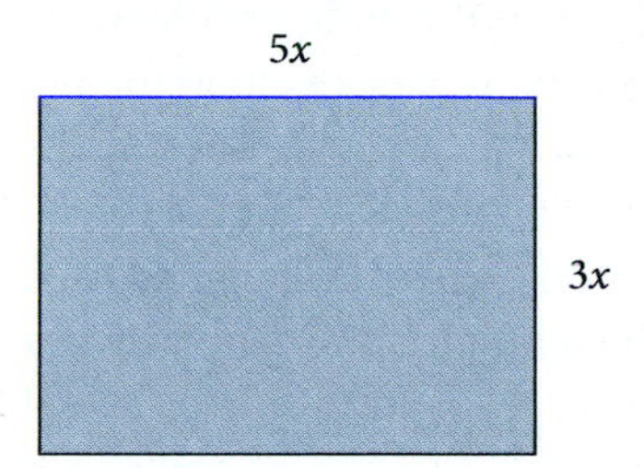

Applying the Concepts

55. **Cars** The chart shows the fastest cars in America. To convert miles per hour to feet per second we use the formula $y = \frac{15}{22}x$ where x is the car's speed in feet per second and y is the speed in miles per hour. Find the speed of the Ford GT in feet per second. Round to the nearest tenth.

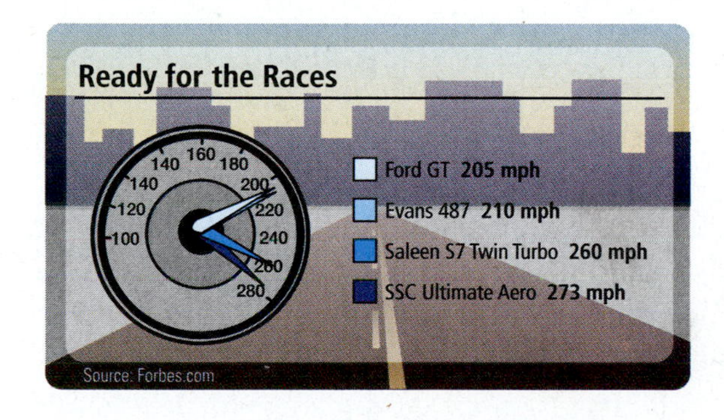

56. **Mountains** The map shows the heights of the tallest mountains in the world. To convert the heights of the mountains into miles, we use the formula $y = 5{,}280x$, where y is in feet and x is in miles. Find the height of K2 in miles. Round to the nearest tenth of a mile.

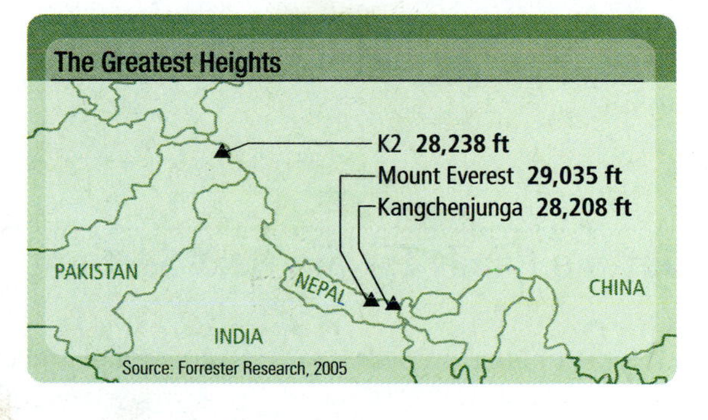

57. MP3 Players Southwest Electronics tracked the number of MP3 players it sold each month for a year. The store manager found that when he raised the price of the MP3 players just slightly, sales went down. He used the equation $60 = -2x + 130$ to determine the price x he needs to charge if he wants to sell 60 MP3 players a month. Solve this equation.

58. Part-time Tuition Costs Many two-year colleges have a large number of students who take courses on a part-time basis. Students pay a charge for each credit hour taken plus an activity fee. Suppose the equation $\$1960 = \$175x + \$35$ can be used to determine the number of credit hours a student is taking during the upcoming semester. Solve this equation.

59. Super Bowl XLII According to Nielsen Media Research, the New York Giants' victory over the New England Patriots in Super Bowl XLII was the most watched Super Bowl ever, with 3 million more viewers than the previous record for Super Bowl XXX in 1996. The equation $192{,}000{,}000 = 2x - 3{,}000{,}000$ shows that the total number of viewers for both Super Bowl games was 192 million. Solve for x to determine how many viewers watched Super Bowl XLII.

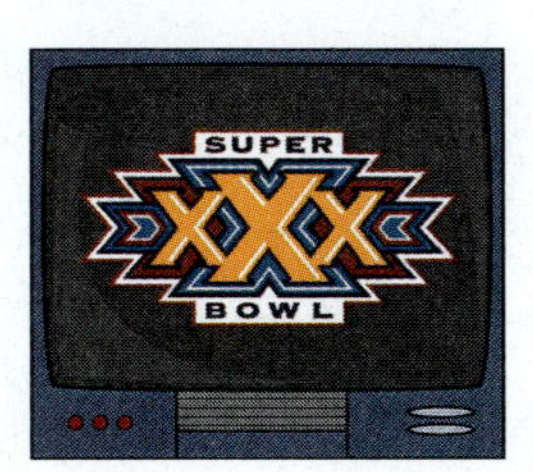

192,000,000 viewers total

60. Blending Gasoline In an attempt to save money at the gas pump, customers will combine two different octane gasolines to get a blend that is slightly higher in octane than regular gas but not as expensive as premium gas. The equation $14x + 120 - 6x = 200$ can be used to find out how many gallons of one octane are needed. Solve this equation.

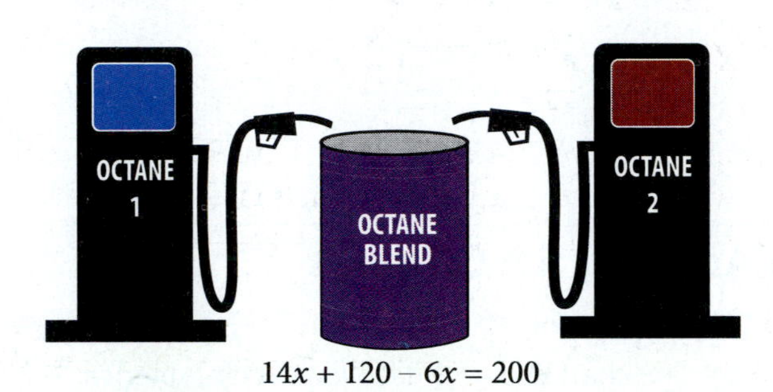

$14x + 120 - 6x = 200$

Translations Translate each sentence below into an equation, then solve the equation.

61. The sum of $2x$ and 5 is 19.

62. The sum of 8 and $3x$ is 2.

63. The difference of $5x$ and 6 is -9.

64. The difference of 9 and $6x$ is 21.

Getting Ready for the Next Section

Apply the distributive property to each of the following expressions.

65. $2(3a - 8)$

66. $4(2a - 5)$

67. $-3(5x - 1)$

68. $-2(7x - 3)$

Simplify each of the following expressions as much as possible.

69. $3(y - 5) + 6$

70. $5(y + 3) + 7$

71. $6(2x - 1) + 4x$

72. $8(3x - 2) + 4x$

Linear Equations in One Variable

4.4

Objectives

A Solve linear equations with one variable.

B Solve linear equations involving fractions.

C Solve application problems using linear equations in one variable.

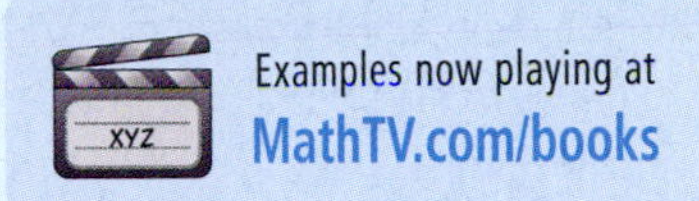
Examples now playing at MathTV.com/books

Introduction . . .

The Rhind Papyrus is an ancient Egyptian document, created around 1650 BC, that contains some mathematical riddles. One problem on the Rhind Papyrus asked the reader to find a quantity such that when it is added to one-fourth of itself the sum is 15. The equation that describes this situation is

$$x + \frac{1}{4}x = 15$$

Bridgeman Art Library/Getty Images

As you can see, this equation contains a fraction. One of the topics we will discuss in this section is how to solve equations that contain fractions.

In this chapter we have been solving what are called *linear equations in one variable*. They are equations that contain only one variable, and that variable is always raised to the first power and never appears in a denominator. Here are some examples of linear equations in one variable:

$$3x + 2 = 17, \quad 7a + 4 = 3a - 2, \quad 2(3y - 5) = 6$$

Because of the work we have done in the first three sections of this chapter, we are now able to solve any linear equation in one variable. The steps outlined below can be used as a guide to solving these equations.

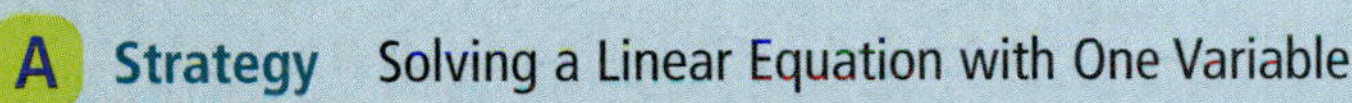

A Strategy Solving a Linear Equation with One Variable

Step 1: Simplify each side of the equation as much as possible. This step is done using the commutative, associative, and distributive properties.

Step 2: Use the addition property of equality to get all *variable terms* on one side of the equation and all *constant terms* on the other, and then combine like terms. A *variable term* is any term that contains the variable. A *constant term* is any term that contains only a number.

Step 3: Use the multiplication property of equality to get the variable by itself on one side of the equation.

Step 4: Check your solution in the original equation if you think it is necessary.

Note Once you have some practice at solving equations, these steps will seem almost automatic. Until that time, it is a good idea to pay close attention to these steps.

EXAMPLE 1

Solve: $3(x + 2) = -9$

SOLUTION We begin by applying the distributive property to the left side:

Step 1

$$3(x + 2) = -9$$

$$3x + 6 = -9 \qquad \text{Distributive property}$$

Step 2

$$3x + 6 + (-6) = -9 + (-6) \qquad \text{Add } -6 \text{ to both sides}$$

$$3x = -15 \qquad \text{Addition}$$

Step 3

$$\frac{3x}{3} = \frac{-15}{3} \qquad \text{Divide both sides by 3}$$

$$x = -5 \qquad \text{Division}$$

PRACTICE PROBLEMS

1. Solve: $4(x + 3) = -8$

Answer

1. -5

2. Solve: $6a + 7 = 4a - 3$

3. Solve: $5(x - 2) + 3 = -12$

4. Solve: $3(4x - 5) + 6 = 3x + 9$

Answers
2. -5 3. -1 4. 2

This general method of solving linear equations involves using the two properties developed in Sections 4.2 and 4.3. We can add any number to both sides of an equation or multiply (or divide) both sides by the same nonzero number and always be sure we have not changed the solution to the equation. The equations may change in form, but the solution to the equation stays the same. Looking back to Example 1, we can see that each equation looks a little different from the preceding one. What is interesting, and useful, is that each of the equations says the same thing about x. They all say that x is -5. The last equation, of course, is the easiest to read. That is why our goal is to end up with x isolated on one side of the equation.

EXAMPLE 2 Solve: $4a + 5 = 2a - 7$

SOLUTION Neither side can be simplified any further. What we have to do is get the variable terms ($4a$ and $2a$) on the same side of the equation. We can eliminate the variable term from the right side by adding $-2a$ to both sides:

Step 2

$$4a + 5 = 2a - 7$$

$$4a + \mathbf{(-2a)} + 5 = 2a + \mathbf{(-2a)} - 7 \quad \text{Add } -2a \text{ to both sides}$$

$$2a + 5 = -7 \quad \text{Addition}$$

$$2a + 5 + \mathbf{(-5)} = -7 + \mathbf{(-5)} \quad \text{Add } -5 \text{ to both sides}$$

$$2a = -12 \quad \text{Addition}$$

Step 3

$$\frac{2a}{\mathbf{2}} = \frac{-12}{\mathbf{2}} \quad \text{Divide by 2}$$

$$a = -6 \quad \text{Division}$$

EXAMPLE 3 Solve: $2(x - 4) + 5 = -11$

SOLUTION We begin by applying the distributive property to multiply 2 and $x - 4$:

Step 1

$$2(x - 4) + 5 = -11$$

$$2x - 8 + 5 = -11 \quad \text{Distributive property}$$

$$2x - 3 = -11 \quad \text{Addition}$$

Step 2

$$2x - 3 + \mathbf{3} = -11 + \mathbf{3} \quad \text{Add 3 to both sides}$$

$$2x = -8 \quad \text{Addition}$$

Step 3

$$\frac{2x}{\mathbf{2}} = \frac{-8}{\mathbf{2}} \quad \text{Divide by 2}$$

$$x = -4 \quad \text{Division}$$

EXAMPLE 4 Solve: $5(2x - 4) + 3 = 4x - 5$

SOLUTION We apply the distributive property to multiply 5 and $2x - 4$. We then combine similar terms and solve as usual:

Step 1

$$5(2x - 4) + 3 = 4x - 5$$

$$10x - 20 + 3 = 4x - 5 \quad \text{Distributive property}$$

$$10x - 17 = 4x - 5 \quad \text{Simplify the left side}$$

Step 2

$$10x + \mathbf{(-4x)} - 17 = 4x + \mathbf{(-4x)} - 5 \quad \text{Add } -4x \text{ to both sides}$$

$$6x - 17 = -5 \quad \text{Addition}$$

$$6x - 17 + \mathbf{17} = -5 + \mathbf{17} \quad \text{Add 17 to both sides}$$

$$6x = 12 \quad \text{Addition}$$

Step 3

$$\frac{6x}{\mathbf{6}} = \frac{12}{\mathbf{6}} \quad \text{Divide by 6}$$

$$x = 2 \quad \text{Division}$$

4.4 Linear Equations in One Variable

Objectives

A Solve linear equations with one variable.

B Solve linear equations involving fractions.

C Solve application problems using linear equations in one variable.

Examples now playing at MathTV.com/books

Introduction . . .

The Rhind Papyrus is an ancient Egyptian document, created around 1650 BC, that contains some mathematical riddles. One problem on the Rhind Papyrus asked the reader to find a quantity such that when it is added to one-fourth of itself the sum is 15. The equation that describes this situation is

$$x + \frac{1}{4}x = 15$$

Bridgeman Art Library/Getty Images

As you can see, this equation contains a fraction. One of the topics we will discuss in this section is how to solve equations that contain fractions.

In this chapter we have been solving what are called *linear equations in one variable.* They are equations that contain only one variable, and that variable is always raised to the first power and never appears in a denominator. Here are some examples of linear equations in one variable:

$$3x + 2 = 17, \quad 7a + 4 = 3a - 2, \quad 2(3y - 5) = 6$$

Because of the work we have done in the first three sections of this chapter, we are now able to solve any linear equation in one variable. The steps outlined below can be used as a guide to solving these equations.

Strategy Solving a Linear Equation with One Variable

Step 1: Simplify each side of the equation as much as possible. This step is done using the commutative, associative, and distributive properties.

Step 2: Use the addition property of equality to get all *variable terms* on one side of the equation and all *constant terms* on the other, and then combine like terms. A *variable term* is any term that contains the variable. A *constant term* is any term that contains only a number.

Step 3: Use the multiplication property of equality to get the variable by itself on one side of the equation.

Step 4: Check your solution in the original equation if you think it is necessary.

Note Once you have some practice at solving equations, these steps will seem almost automatic. Until that time, it is a good idea to pay close attention to these steps.

EXAMPLE 1 Solve: $3(x + 2) = -9$

SOLUTION We begin by applying the distributive property to the left side:

Step 1

$$3(x + 2) = -9$$

$$3x + 6 = -9 \qquad \text{Distributive property}$$

Step 2

$$3x + 6 + (-6) = -9 + (-6) \qquad \text{Add } -6 \text{ to both sides}$$

$$3x = -15 \qquad \text{Addition}$$

Step 3

$$\frac{3x}{3} = \frac{-15}{3} \qquad \text{Divide both sides by 3}$$

$$x = -5 \qquad \text{Division}$$

PRACTICE PROBLEMS

1. Solve: $4(x + 3) = -8$

Answer

1. -5

This general method of solving linear equations involves using the two properties developed in Sections 4.2 and 4.3. We can add any number to both sides of an equation or multiply (or divide) both sides by the same nonzero number and always be sure we have not changed the solution to the equation. The equations may change in form, but the solution to the equation stays the same. Looking back to Example 1, we can see that each equation looks a little different from the preceding one. What is interesting, and useful, is that each of the equations says the same thing about x. They all say that x is -5. The last equation, of course, is the easiest to read. That is why our goal is to end up with x isolated on one side of the equation.

2. Solve: $6a + 7 = 4a - 3$

EXAMPLE 2 Solve: $4a + 5 = 2a - 7$

SOLUTION Neither side can be simplified any further. What we have to do is get the variable terms ($4a$ and $2a$) on the same side of the equation. We can eliminate the variable term from the right side by adding $-2a$ to both sides:

Step 2:

$$4a + 5 = 2a - 7$$

$$4a + \mathbf{(-2a)} + 5 = 2a + \mathbf{(-2a)} - 7 \quad \text{Add } -2a \text{ to both sides}$$

$$2a + 5 = -7 \quad \text{Addition}$$

$$2a + 5 + \mathbf{(-5)} = -7 + \mathbf{(-5)} \quad \text{Add } -5 \text{ to both sides}$$

$$2a = -12 \quad \text{Addition}$$

Step 3:

$$\frac{2a}{\mathbf{2}} = \frac{-12}{\mathbf{2}} \quad \text{Divide by 2}$$

$$a = -6 \quad \text{Division}$$

3. Solve: $5(x - 2) + 3 = -12$

EXAMPLE 3 Solve: $2(x - 4) + 5 = -11$

SOLUTION We begin by applying the distributive property to multiply 2 and $x - 4$:

Step 1:

$$2(x - 4) + 5 = -11$$

$$2x - 8 + 5 = -11 \quad \text{Distributive property}$$

$$2x - 3 = -11 \quad \text{Addition}$$

Step 2:

$$2x - 3 + \mathbf{3} = -11 + \mathbf{3} \quad \text{Add 3 to both sides}$$

$$2x = -8 \quad \text{Addition}$$

Step 3:

$$\frac{2x}{\mathbf{2}} = \frac{-8}{\mathbf{2}} \quad \text{Divide by 2}$$

$$x = -4 \quad \text{Division}$$

4. Solve: $3(4x - 5) + 6 = 3x + 9$

EXAMPLE 4 Solve: $5(2x - 4) + 3 = 4x - 5$

SOLUTION We apply the distributive property to multiply 5 and $2x - 4$. We then combine similar terms and solve as usual:

Step 1:

$$5(2x - 4) + 3 = 4x - 5$$

$$10x - 20 + 3 = 4x - 5 \quad \text{Distributive property}$$

$$10x - 17 = 4x - 5 \quad \text{Simplify the left side}$$

Step 2:

$$10x + \mathbf{(-4x)} - 17 = 4x + \mathbf{(-4x)} - 5 \quad \text{Add } -4x \text{ to both sides}$$

$$6x - 17 = -5 \quad \text{Addition}$$

$$6x - 17 + \mathbf{17} = -5 + \mathbf{17} \quad \text{Add 17 to both sides}$$

$$6x = 12 \quad \text{Addition}$$

Step 3:

$$\frac{6x}{\mathbf{6}} = \frac{12}{\mathbf{6}} \quad \text{Divide by 6}$$

$$x = 2 \quad \text{Division}$$

Answers
2. -5 3. -1 4. 2

B Equations Involving Fractions

We will now solve some equations that involve fractions. Because integers are usually easier to work with than fractions, we will begin each problem by clearing the equation we are trying to solve of all fractions. To do this, we will use the multiplication property of equality to multiply each side of the equation by the LCD for all fractions appearing in the equation. Here is an example.

EXAMPLE 5 Solve the equation $\frac{x}{2} + \frac{x}{6} = 8$.

SOLUTION The LCD for the fractions $\frac{x}{2}$ and $\frac{x}{6}$ is 6. It has the property that both 2 and 6 divide it evenly. Therefore, if we multiply both sides of the equation by 6, we will be left with an equation that does not involve fractions.

$$\mathbf{6}\left(\frac{x}{2} + \frac{x}{6}\right) = \mathbf{6}(8) \quad \text{Multiply each side by 6}$$
$$6\left(\frac{x}{2}\right) + 6\left(\frac{x}{6}\right) = 6(8) \quad \text{Apply the distributive property}$$
$$3x + x = 48 \quad \text{Multiplication}$$
$$4x = 48 \quad \text{Combine similar terms}$$
$$x = 12 \quad \text{Divide each side by 4}$$

We could check our solution by substituting 12 for x in the original equation. If we do so, the result is a true statement. The solution is 12.

5. Solve: $\frac{x}{3} + \frac{x}{6} = 9$

As you can see from Example 5, the most important step in solving an equation that involves fractions is the first step. In that first step we multiply both sides of the equation by the LCD for all the fractions in the equation. After we have done so, the equation is clear of fractions because the LCD has the property that all the denominators divide it evenly.

EXAMPLE 6 Solve the equation $2x + \frac{1}{2} = \frac{3}{4}$.

SOLUTION This time the LCD is 4. We begin by multiplying both sides of the equation by 4 to clear the equation of fractions.

$$\mathbf{4}\left(2x + \frac{1}{2}\right) = \mathbf{4}\left(\frac{3}{4}\right) \quad \text{Multiply each side by the LCD, 4}$$
$$4(2x) + 4\left(\frac{1}{2}\right) = 4\left(\frac{3}{4}\right) \quad \text{Apply the distributive property}$$
$$8x + 2 = 3 \quad \text{Multiplication}$$
$$8x = 1 \quad \text{Add } -2 \text{ to each side}$$
$$x = \frac{1}{8} \quad \text{Divide each side by 8}$$

6. Solve: $3x + \frac{1}{4} = \frac{5}{8}$

EXAMPLE 7 Solve for x: $\frac{3}{x} + 2 = \frac{1}{2}$. (Assume x is not 0.)

SOLUTION This time the LCD is $2x$. Following the steps we used in Examples 5 and 6, we have

$$\mathbf{2x}\left(\frac{3}{x} + 2\right) = \mathbf{2x}\left(\frac{1}{2}\right) \quad \text{Multiply through by the LCD, } 2x$$
$$2x\left(\frac{3}{x}\right) + 2x(2) = 2x\left(\frac{1}{2}\right) \quad \text{Distributive property}$$
$$6 + 4x = x \quad \text{Multiplication}$$
$$6 = -3x \quad \text{Add } -4x \text{ to each side}$$
$$-2 = x \quad \text{Divide each side by } -3$$

7. Solve: $\frac{4}{x} + 3 = \frac{11}{5}$

Answers
5. 18 6. $\frac{1}{8}$ 7. -5

STUDY SKILLS

Begin to Develop Confidence with Word Problems

The main difference between people who are good at working word problems and those who are not seems to be confidence. People with confidence know that no matter how long it takes them, they will eventually be able to solve the problem they are working on. Those without confidence begin by saying to themselves, "I'll never be able to work this problem." If you are in this second group, then instead of telling yourself that you can't do word problems, that you don't like them, or that they're not good for anything anyway, decide to do whatever it takes to master them.

Getting Ready for Class

After reading through the preceding section, respond in your own words and in complete sentences.

1. Apply the distributive property to the expression $3(x + 4)$.
2. Write the equation that results when $-4a$ is added to both sides of the equation below.

$$6a + 9 = 4a - 3$$

3. Solve the equation $2x + \frac{1}{2} = \frac{3}{4}$ by first adding $-\frac{1}{2}$ to each side. Compare your answer with the solution to the equation shown in Example 6.

Problem Set 4.4

A Solve each equation using the methods shown in this section. [Examples 1–4]

1. $5(x + 1) = 20$

2. $4(x + 2) = 24$

3. $6(x - 3) = -6$

4. $7(x - 2) = -7$

5. $2x + 4 = 3x + 7$

6. $5x + 3 = 2x + (-3)$

7. $7y - 3 = 4y - 15$

8. $3y + 5 = 9y + 8$

9. $12x + 3 = -2x + 17$

10. $15x + 1 = -4x + 20$

11. $6x - 8 = -x - 8$

12. $7x - 5 = -x - 5$

13. $7(a - 1) + 4 = 11$

14. $3(a - 2) + 1 = 4$

15. $8(x + 5) - 6 = 18$

16. $7(x + 8) - 4 = 10$

17. $2(3x - 6) + 1 = 7$

18. $5(2x - 4) + 8 = 38$

19. $10(y + 1) + 4 = 3y + 7$

20. $12(y + 2) + 5 = 2y - 1$

21. $4(x - 6) + 1 = 2x - 9$

22. $7(x - 4) + 3 = 5x - 9$

23. $2(3x + 1) = 4(x - 1)$

24. $7(x - 8) = 2(x - 13)$

25. $3a + 4 = 2(a - 5) + 15$

26. $10a + 3 = 4(a - 1) + 1$

27. $9x - 6 = -3(x + 2) - 24$

28. $8x - 10 = -4(x + 3) + 2$

29. $3x - 5 = 11 + 2(x - 6)$

30. $5x - 7 = -7 + 2(x + 3)$

B Solve each equation by first finding the LCD for the fractions in the equation and then multiplying both sides of the equation by it. (Assume x is not 0 in Problems 39–46.) [Examples 5–7]

31. $\frac{x}{3} + \frac{x}{6} = 5$

32. $\frac{x}{2} - \frac{x}{4} = 3$

33. $\frac{x}{5} - x = 4$

34. $\frac{x}{3} + x = 8$

35. $3x + \frac{1}{2} = \frac{1}{4}$

36. $3x - \frac{1}{3} = \frac{1}{6}$

37. $\frac{x}{3} + \frac{1}{2} = -\frac{1}{2}$

38. $\frac{x}{2} + \frac{4}{3} = -\frac{2}{3}$

39. $\frac{4}{x} = \frac{1}{5}$

40. $\frac{2}{3} = \frac{6}{x}$

41. $\frac{3}{x} + 1 = \frac{2}{x}$

42. $\frac{4}{x} + 3 = \frac{1}{x}$

43. $\frac{3}{x} - \frac{2}{x} = \frac{1}{5}$

44. $\frac{7}{x} + \frac{1}{x} = 2$

45. $\frac{1}{x} - \frac{1}{2} = -\frac{1}{4}$

46. $\frac{3}{x} - \frac{4}{5} = -\frac{1}{5}$

Find the value of x for each of the figures, given the perimeter.

47. $P = 36$

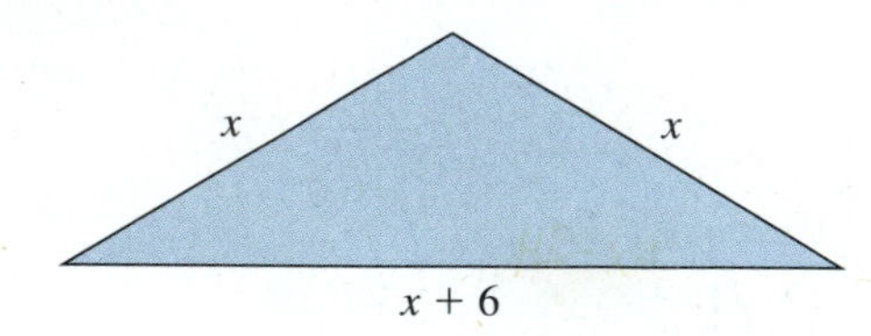

48. $P = 30$

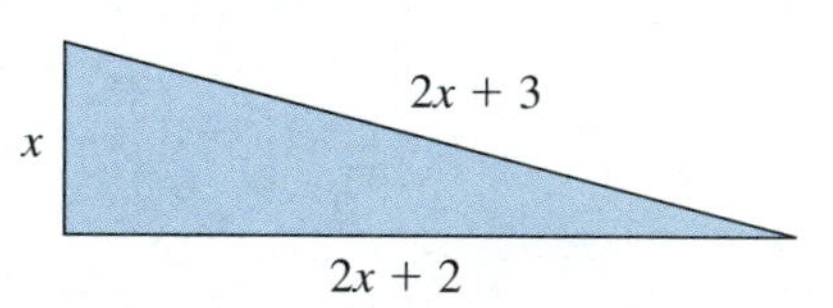

49. $P = 16$

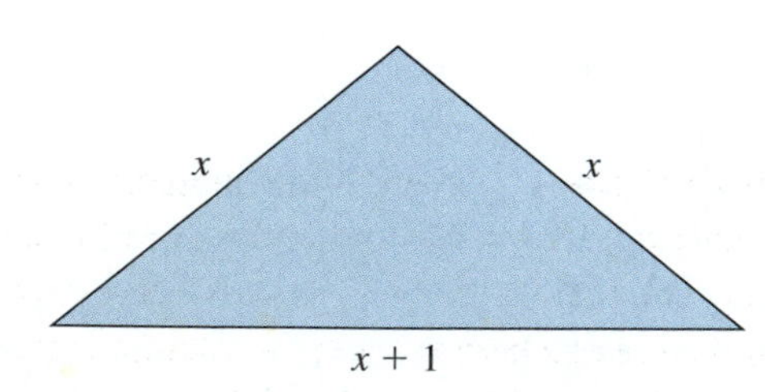

50. $P = 60$

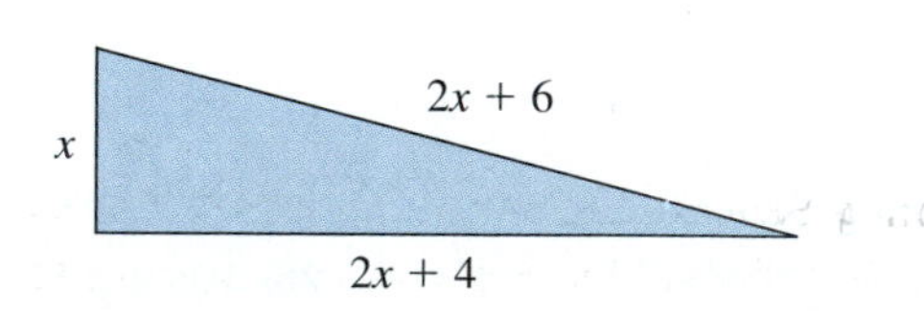

C Applying the Concepts

51. Skyscrapers The chart shows the heights of the three tallest buildings in the world. The height of the Empire State Building relative to the Petronas Towers can be given by the equation $1483 = 233 + x$. What is the height of the Empire State Building?

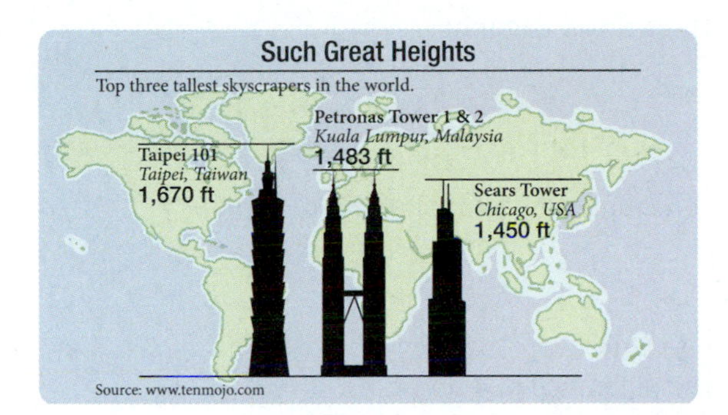

52. Sound The chart shows the decibel level o sounds. The human threshold of pain relative decibel level at a football stadium is given by the tion $117 = x - 3$. What is the human threshold of pa

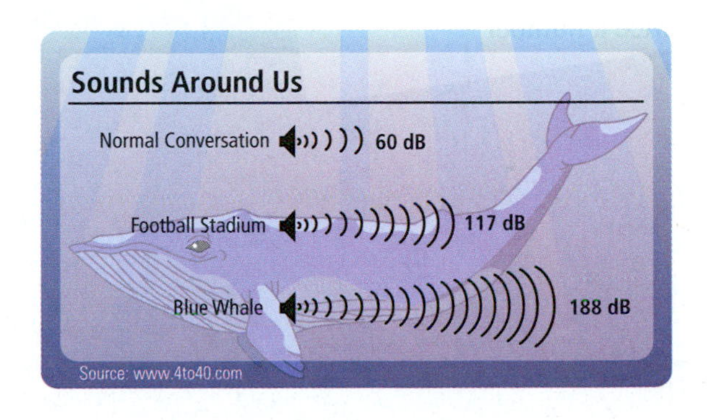

53. Geometry The figure shows part of a room. From a point on the floor, the angle of elevation to the top of the window is 45°, while the angle of elevation to the ceiling above the window is 58°. Solving either of the equations $58 - x = 45$ or $45 + x = 58$ will give us the number of degrees in the angle labeled $x°$. Solve both equations.

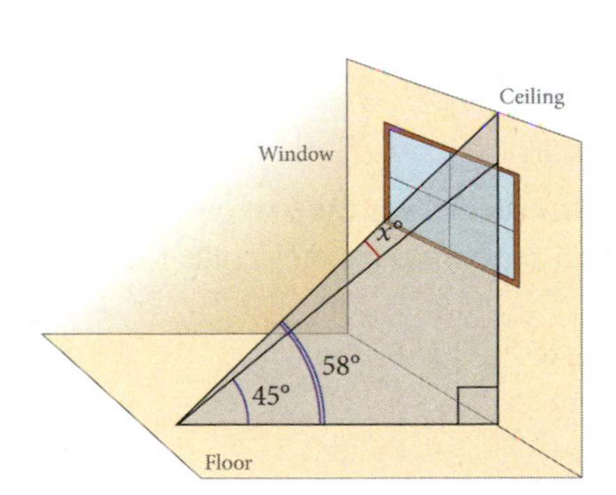

54. Rhind Papyrus As we mentioned in the introduction to this section, the Rhind Papyrus was created around 1650 BC and contains the riddle, "What quantity when added to one-fourth of itself becomes 15?" This riddle can be solved by finding x in the equation below. Solve this equation.

$$x + \frac{1}{4}x = 15$$

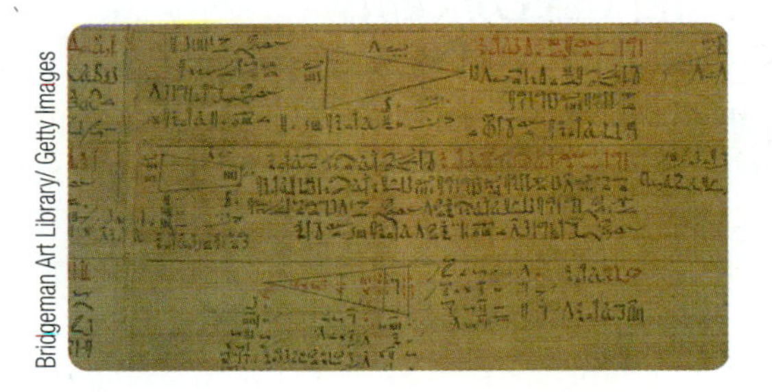
Bridgeman Art Library/ Getty Images

55. Math Tutoring Several students on campus decide to start a small business that offers tutoring services to students enrolled in mathematics courses. The following equation shows the amount of money they collected at the end of the month, assuming expenses of \$400 and a charge of \$30 an hour per student. Solve the equation $\$500 = \$30x - 400$ to determine the number of hours students came in for tutoring in one month.

56. Shopping for a Calculator You find that you need to purchase a specific calculator for your college mathematics class. The equation $\$18.75 = p + \frac{1}{4}p$ shows the price charged by the college bookstore for a calculator after it has been marked up. How much did the bookstore pay the manufacturer for the calculator?

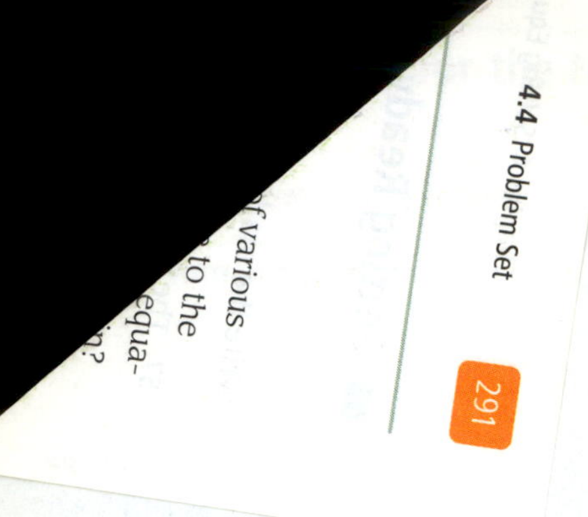

ext Section

hat are equivalent to each of the following English phrases.

58. The sum of a number and 5

60. Three times a number

61. Twice the sum of a number and 6

62. Three times the sum of a number and 8

63. The difference of x and 4

64. The difference of 4 and x

65. The sum of twice a number and 5

66. The sum of three times a number and 4

Extending the Concepts

67. Admission to the school basketball game is $4 for students and $6 for general admission. For the first game of the season, 100 more student tickets than general admission tickets were sold. The total amount of money collected was $2,400.
    - a. Write an equation that will help us find the number of students in attendance.
    - b. Solve this equation for x.
    - c. What was the total attendance for the game?

Applications

Objectives

A Set up and solve number problems using linear equations.

B Set up and solve geometry problems using linear equations.

C Set up and solve age problems using linear equations.

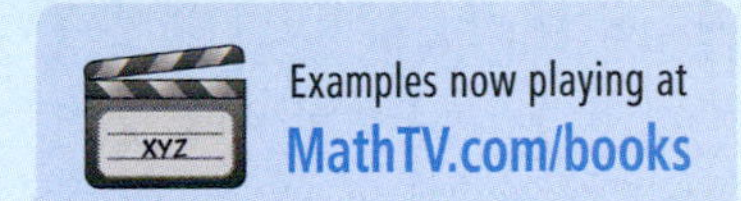

Introduction . . .

As you begin reading through the examples in this section, you may find yourself asking why some of these problems seem so contrived. The title of the section is "Applications," but many of the problems here don't seem to have much to do with real life. You are right about that. Example 5 is what we refer to as an "age problem." Realistically, it is not the kind of problem you would expect to find if you choose a career in which you use algebra. However, solving age problems is good practice for someone with little experience with application problems, because the solution process has a form that can be applied to all similar age problems.

To begin this section we list the steps used in solving application problems. We call this strategy the *Blueprint for Problem Solving*. It is an outline that will overlay the solution process we use on all application problems.

Blueprint for Problem Solving

Step 1: **Read** the problem, and then mentally **list** the items that are known and the items that are unknown.

Step 2: **Assign a variable** to one of the unknown items. (In most cases this will amount to letting x equal the item that is asked for in the problem.) Then **translate** the other **information** in the problem to expressions involving the variable.

Step 3: **Reread** the problem, and then **write an equation**, using the items and variables listed in Steps 1 and 2, that describes the situation.

Step 4: **Solve the equation** found in Step 3.

Step 5: **Write** your **answer** using a complete sentence.

Step 6: **Reread** the problem, and **check** your solution with the original words in the problem.

There are a number of substeps within each of the steps in our blueprint. For instance, with Steps 1 and 2 it is always a good idea to draw a diagram or picture if it helps you to visualize the relationship between the items in the problem.

It is important for you to remember that solving application problems is more of an art than a science. Be flexible. No one strategy works all of the time. Try to stay away from looking for the "one way" to set up and solve a problem. Think of the blueprint for problem solving as guidelines that will help you organize your approach to these problems, rather than as a set of rules.

A Number Problems

The sum of a number and 2 is 8. Find the number.

SOLUTION Using our blueprint for problem solving as an outline, we solve the problem as follows:

Step 1 *Read* the problem, and then mentally *list* the items that are known and the items that are unknown.

Known items: The numbers 2 and 8

Unknown item: The number in question

PRACTICE PROBLEMS

1. The sum of a number and 3 is 10. Find the number.

Step 2 *Assign a variable* to one of the unknown items. Then *translate* the other *information* in the problem to expressions involving the variable.

Let x = the number asked for in the problem
Then "The sum of a number and 2" translates to $x + 2$.

Step 3 *Reread* the problem, and then *write an equation,* using the items and variables listed in Steps 1 and 2, that describes the situation.

With all word problems, the word "is" translates to = .

The sum of x and 2 is 8.
$x + 2 = 8$

Step 4 *Solve the equation* found in Step 3.

$$x + 2 = 8$$
$$x + 2 + \mathbf{(-2)} = 8 + \mathbf{(-2)} \quad \text{Add } -2 \text{ to each side}$$
$$x = 6$$

Step 5 *Write* your *answer* using a complete sentence.

The number is 6.

Step 6 *Reread* the problem, and *check* your solution with the original words in the problem.

The sum of **6** and 2 is 8. A true statement

To help with other problems of the type shown in Example 1, here are some common English words and phrases and their mathematical translations.

English	Algebra	
The sum of a and b	$a + b$	
The difference of a and b	$a - b$	
The product of a and b	$a \cdot b$	
The quotient of a and b	$\frac{a}{b}$	
Of	$\cdot$	(multiply)
Is	$=$	(equals)
A number	x	
4 more than x	$x + 4$	
4 times x	$4x$	
4 less than x	$x - 4$	

You may find some examples and problems in this section and the problem set that follows that you can solve without using algebra or our blueprint. It is very important that you solve those problems using the methods we are showing here. The purpose behind these problems is to give you experience using the blueprint as a guide to solving problems written in words. Your answers are much less important than the work that you show in obtaining your answer.

EXAMPLE 2 If 5 is added to the sum of twice a number and three times the number, the result is 25. Find the number.

SOLUTION

Step 1 *Read and list.*

Known items: The numbers 5 and 25, twice a number, and three times a number

Unknown item: The number in question

2. If 4 is added to the sum of twice a number and three times the number, the result is 34. Find the number.

Answer

1. The number is 7.

Step 2 *Assign a variable and translate the information.*

Let x = the number asked for in the problem.
Then "The sum of twice a number and three times the number" translates to $2x + 3x$.

Step 3 *Reread and write an equation.*

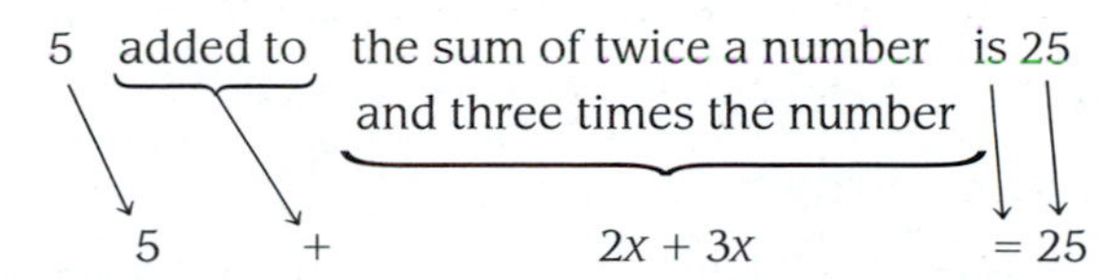

Step 4 *Solve the equation.*

$$5 + 2x + 3x = 25$$

$5x + 5 = 25$	Simplify the left side
$5x + 5 + (-5) = 25 + (-5)$	Add −5 to both sides
$5x = 20$	Addition
$\frac{5x}{5} = \frac{20}{5}$	Divide by 5
$x = 4$	

Step 5 *Write your answer.*

The number is 4.

Step 6 *Reread and check.*

Twice **4** is 8, and three times **4** is 12. Their sum is $8 + 12 = 20$. Five added to this is 25. Therefore, 5 added to the sum of twice **4** and three times **4** is 25.

B Geometry Problems

EXAMPLE 3 The length of a rectangle is three times the width. The perimeter is 72 centimeters. Find the width and the length.

SOLUTION

Step 1 *Read and list.*

Known items: The length is three times the width.
The perimeter is 72 centimeters.
Unknown items: The length and the width

Step 2 *Assign a variable, and translate the information.* We let x = the width. Because the length is three times the width, the length must be $3x$. A picture will help.

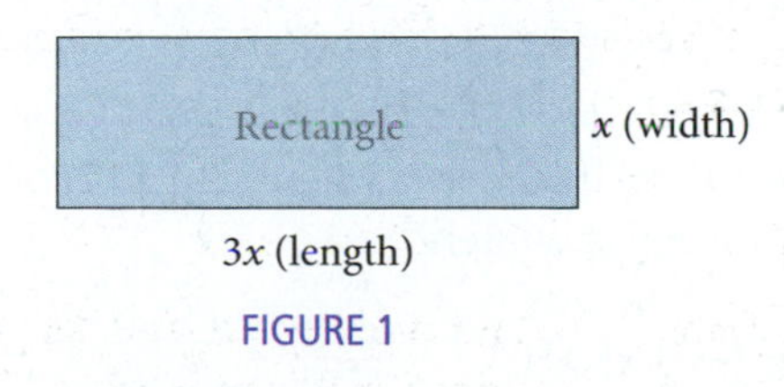

FIGURE 1

3. The length of a rectangle is twice the width. The perimeter is 42 centimeters. Find the length and the width.

Answer
2. The number is 6.

Step 3 *Reread and write an equation.* Because the perimeter is the sum of the sides, it must be $x + x + 3x + 3x$ (the sum of the four sides). But the perimeter is also given as 72 centimeters. Hence,

$$x + x + 3x + 3x = 72$$

Step 4 *Solve the equation.*

$$\begin{aligned} x + x + 3x + 3x &= 72 \\ 8x &= 72 \\ x &= 9 \end{aligned}$$

Step 5 *Write your answer.* The width, x, is 9 centimeters. The length, $3x$, must be 27 centimeters.

Step 6 *Reread and check.* From the diagram below, we see that these solutions check:

Perimeter is 72	Length = 3 × Width
$9 + 9 + 27 + 27 = 72$	$27 = 3 \cdot 9$

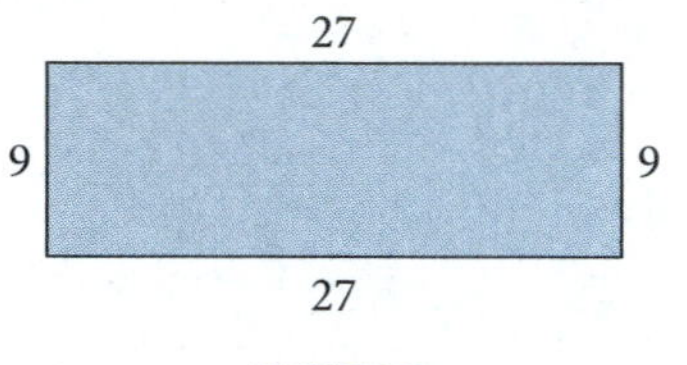

FIGURE 2

Next we review some facts about triangles that we introduced in a previous chapter.

FACTS FROM GEOMETRY Labeling Triangles and the Sum of the Angles in a Triangle

One way to label the important parts of a triangle is to label the vertices with capital letters and the sides with small letters, as shown in Figure 3.

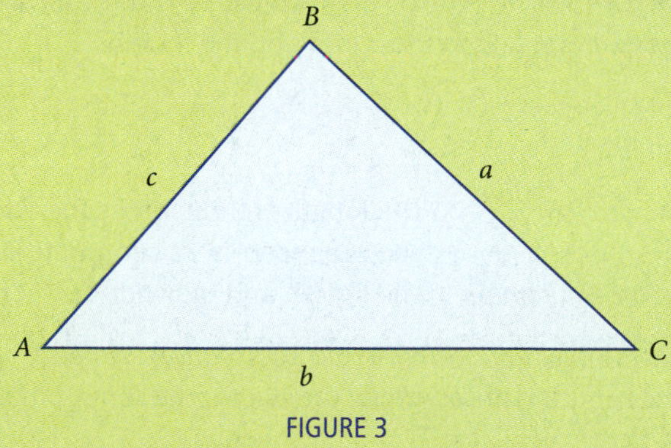

FIGURE 3

In Figure 3, notice that side a is opposite vertex A, side b is opposite vertex B, and side c is opposite vertex C. Also, because each vertex is the vertex of one of the angles of the triangle, we refer to the three interior angles as A, B, and C.

In any triangle, the sum of the interior angles is 180°. For the triangle shown in Figure 3, the relationship is written

$$A + B + C = 180°$$

Answer

3. The width is 7 cm, and the length is 14 cm.

EXAMPLE 4 The angles in a triangle are such that one angle is twice the smallest angle, while the third angle is three times as large as the smallest angle. Find the measure of all three angles.

SOLUTION

Step 1 *Read and list.*

Known items: The sum of all three angles is 180°; one angle is twice the smallest angle; and the largest angle is three times the smallest angle.

Unknown items: The measure of each angle

Step 2 *Assign a variable and translate information.* Let x be the smallest angle, then $2x$ will be the measure of another angle, and $3x$ will be the measure of the largest angle.

Step 3 *Reread and write an equation.* When working with geometric objects, drawing a generic diagram will sometimes help us visualize what it is that we are asked to find. In Figure 4, we draw a triangle with angles A, B, and C.

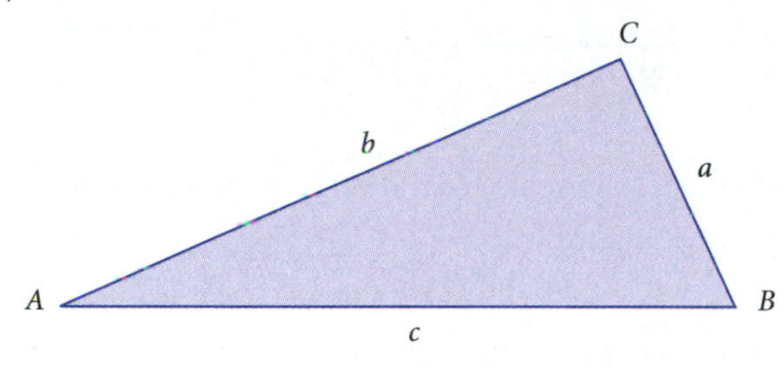

FIGURE 4

We can let the value of $A = x$, the value of $B = 2x$, and the value of $C = 3x$. We know that the sum of angles A, B, and C will be 180°, so our equation becomes

$$x + 2x + 3x = 180°$$

Step 4 *Solve the equation.*

$$\begin{aligned} x + 2x + 3x &= 180° \\ 6x &= 180° \\ x &= 30° \end{aligned}$$

Step 5 *Write the answer.*

The smallest angle A measures 30°
Angle B measures $2x$, or $2(30°) = 60°$
Angle C measures $3x$, or $3(30°) = 90°$

Step 6 *Reread and check.* The angles must add to 180°:

$$\begin{aligned} A + B + C &= 180° \\ 30° + 60° + 90° &= 180° \\ 180° &= 180° \quad \text{Our answers check} \end{aligned}$$

STUDY SKILLS

Continue to List Difficult problems

You should continue to list and rework the problems that give you the most difficulty. It is this list that you will use to study for the next exam. Your goal is to go into the next exam knowing that you can successfully work any problem from your list of difficult problems.

4. The angles in a triangle are such that one angle is three times the smallest angle, while the largest angle is five times the smallest angle. Find the measure of all three angles.

Answer

4. The angles are 20°, 60°, and 100°.

5. Joyce is 21 years older than her son Travis. In six years the sum of their ages will be 49. How old are they now?

C Age Problem

EXAMPLE 5 Jo Ann is 22 years older than her daughter Stacey. In six years the sum of their ages will be 42. How old are they now?

SOLUTION

Step 1 *Read and list:*

Known items: Jo Ann is 22 years older than Stacey. Six years from now their ages will add to 42.

Unknown items: Their ages now

Step 2 *Assign a variable and translate the information.* Let x = Stacey's age now. Because Jo Ann is 22 years older than Stacey, her age is $x + 22$.

Step 3 *Reread and write an equation.* As an aid in writing the equation we use the following table:

	Now	In Six years
Stacey	x	$x + 6$
Jo Ann	$x + 22$	$x + 28$

Their ages in six years will be their ages now plus 6

Because the sum of their ages six years from now is 42, we write the equation as

$$(x + 6) + (x + 28) = 42$$

↑ Stacey's age in 6 years ↑ Jo Ann's age in 6 years

Step 4 *Solve the equation.*

$$x + 6 + x + 28 = 42$$
$$2x + 34 = 42$$
$$2x = 8$$
$$x = 4$$

Step 5 *Write your answer.* Stacey is now 4 years old, and Jo Ann is $4 + 22 = 26$ years old.

Step 6 *Reread and check.* To check, we see that in six years, Stacey will be 10, and Jo Ann will be 32. The sum of 10 and 32 is 42, which checks.

Getting Ready for Class

After reading through the preceding section, respond in your own words and in complete sentences.

1. What is the first step in solving a word problem?
2. Write a mathematical expression equivalent to the phrase "the sum of x and ten."
3. Write a mathematical expression equivalent to the phrase "twice the sum of a number and ten."
4. Suppose the length of a rectangle is three times the width. If we let x represent the width of the rectangle, what expression do we use to represent the length?

Answer

5. Travis is 8; Joyce is 29.

Problem Set 4.5

Write each of the following English phrases in symbols using the variable x.

1. The sum of x and 3
2. The difference of x and 2
3. The sum of twice x and 1
4. The sum of three times x and 4
5. Five x decreased by 6
6. Twice the sum of x and 5
7. Three times the sum of x and 1
8. Four times the sum of twice x and 1
9. Five times the sum of three x and 4
10. Three x added to the sum of twice x and 1

Use the six steps in the "Blueprint for Problem Solving" to solve the following word problems. You may recognize the solution to some of them by just reading the problem. In all cases, be sure to assign a variable and write the equation used to describe the problem. Write your answer using a complete sentence.

A Number Problems [Examples 1, 2]

11. The sum of a number and 3 is 5. Find the number.
12. If 2 is subtracted from a number, the result is 4. Find the number.
13. The sum of twice a number and 1 is -3. Find the number.
14. If three times a number is increased by 4, the result is -8. Find the number.
15. When 6 is subtracted from five times a number, the result is 9. Find the number.
16. Twice the sum of a number and 5 is 4. Find the number.
17. Three times the sum of a number and 1 is 18. Find the number.
18. Four times the sum of twice a number and 6 is -8. Find the number.
19. Five times the sum of three times a number and 4 is -10. Find the number.
20. If the sum of three times a number and two times the same number is increased by 1, the result is 16. Find the number.

B Geometry Problems [Examples 3, 4]

21. The length of a rectangle is twice its width. The perimeter is 30 meters. Find the length and the width.

22. The width of a rectangle is 3 feet less than its length. If the perimeter is 22 feet, what is the width?

23. The perimeter of a square is 32 centimeters. What is the length of one side?

24. Two sides of a triangle are equal in length, and the third side is 10 inches. If the perimeter is 26 inches, how long are the two equal sides?

25. Two angles in a triangle are equal, and their sum is equal to the third angle in the triangle. What are the measures of each of the three interior angles?

26. One angle in a triangle measures twice the smallest angle, while the largest angle is six times the smallest angle. Find the measures of all three angles.

27. The smallest angle in a triangle is $\frac{1}{3}$ as large as the largest angle. The third angle is twice the smallest angle. Find the three angles.

28. One angle in a triangle is half the largest angle, but three times the smallest. Find all three angles.

C Age Problems [Example 5]

29. Pat is 20 years older than his son Patrick. In 2 years, the sum of their ages will be 90. How old are they now?

	Now	In 2 Years
Patrick	x	
Pat		

30. Diane is 23 years older than her daughter Amy. In 5 years, the sum of their ages will be 91. How old are they now?

	Now	In 5 Years
Amy	x	
Diane		

31. Dale is 4 years older than Sue. Five years ago the sum of their ages was 64. How old are they now?

32. Pat is 2 years younger than his wife, Wynn. Ten years ago the sum of their ages was 48. How old are they now?

Miscellaneous Problems

33. Magic Square The sum of the numbers in each row, each column, and each diagonal of the square below is 15. Use this fact, along with the information in the first column of the square, to write an equation containing the variable x, then solve the equation to find x. Next, write and solve equations that will give you y and z.

x	1	y
3	5	7
4	z	2

34. Magic Square The sum of the numbers in each row, each column, and each diagonal of the square below is 3. Use this fact, along with the information in the second row of the square, to write an equation containing the variable a, then solve the equation to find a. Next, write and solve an equation that will allow you to find the value of b. Next, write and solve equations that will give you c and d.

4	d	b
a	1	3
0	c	−2

35. Wages JoAnn works in the publicity office at the state university. She is paid $14 an hour for the first 35 hours she works each week and $21 an hour for every hour after that. If she makes $574 one week, how many hours did she work?

36. Ticket Sales Stacey is selling tickets to the school play. The tickets are $6 for adults and $4 for children. She sells twice as many adult tickets as children's tickets and brings in a total of $112. How many of each kind of ticket did she sell?

37. Cars The chart shows the fastest cars in America. The maximum speed of an Evans 487 is twice the sum of the speed of a trucker and 45 miles per hour. What is the speed of the trucker?

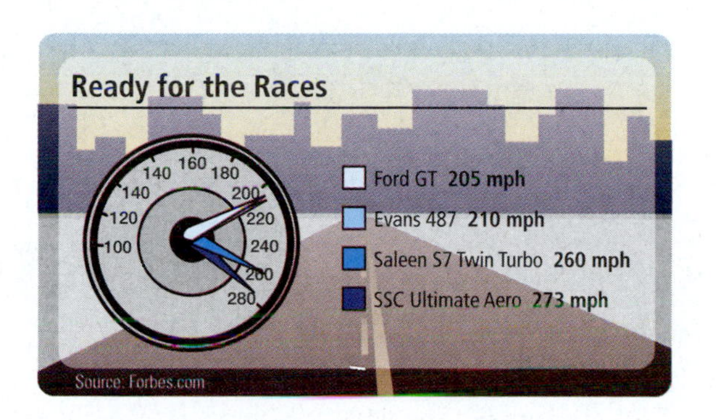

38. Skyscrapers The chart shows the heights of the three tallest buildings in the world. The Sears Tower is 80 feet less than 5 times the height of the Statue of Liberty. What is the height of the Statue of Liberty?

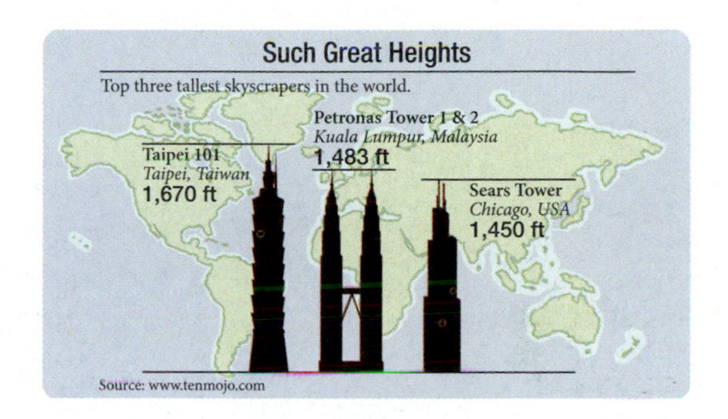

Getting Ready for the Next Section

Simplify.

39. $\frac{5}{9}(95 - 32)$

40. $\frac{5}{9}(77 - 32)$

41. Find the value of $90 - x$ when $x = 25$.

42. Find the value of $180 - x$ when $x = 25$.

43. Find the value of $2x + 6$ when $x = -2$

44. Find the value of $2x + 6$ when $x = 0$.

Solve.

45. $40 = 2l + 12$

46. $80 = 2l + 12$

47. $6 + 3y = 4$

48. $8 + 3y = 4$

Evaluating Formulas

Objectives

A Solve a formula for a given variable.

B Solve problems using the rate equation.

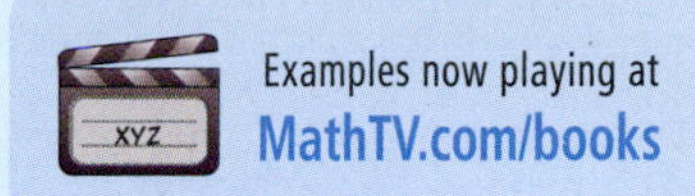

Introduction . . .

In mathematics a formula is an equation that contains more than one variable. The equation $P = 2w + 2l$ is an example of a formula. This formula tells us the relationship between the perimeter P of a rectangle, its length l, and its width w.

There are many formulas with which you may be familiar already. Perhaps you have used the formula $d = r \cdot t$ to find out how far you would go if you traveled at 50 miles an hour for 3 hours. If you take a chemistry class while you are in college, you will certainly use the formula that gives the relationship between the two temperature scales, Fahrenheit and Celsius:

$$F = \frac{9}{5}C + 32$$

Although there are many kinds of problems we can work using formulas, we will limit ourselves to those that require only substitutions. The examples that follow illustrate this type of problem.

A Formulas

EXAMPLE 1 The perimeter P of a rectangular livestock pen is 40 feet. If the width w is 6 feet, find the length.

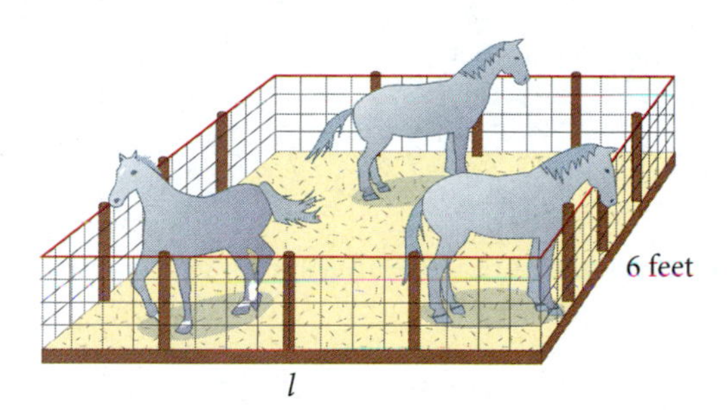

SOLUTION First we substitute 40 for P and 6 for w in the formula $P = 2l + 2w$. Then we solve for l:

When	$P = 40$ and $w = 6$	
the formula	$P = 2l + 2w$	
becomes	$40 = 2l + 2(6)$	
or	$40 = 2l + 12$	Multiply 2 and 6
	$28 = 2l$	Add -12 to each side
	$14 = l$	Multiply each side by $\frac{1}{2}$

To summarize our results, if a rectangular pen has a perimeter of 40 feet and a width of 6 feet, then the length must be 14 feet.

PRACTICE PROBLEMS

1. Suppose the livestock pen in Example 1 has a perimeter of 80 feet. If the width is still 6 feet, what is the new length?

Answer

1. 34 feet

2. Use the formula in Example 2 to find C when F is 77 degrees.

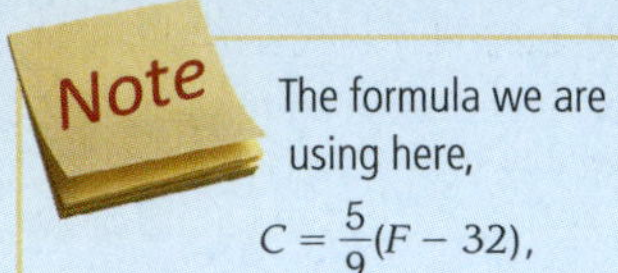

The formula we are using here,

$$C = \frac{5}{9}(F - 32),$$

is an alternative form of the formula we mentioned in the introduction to this section:

$$F = \frac{9}{5}C + 32$$

Both formulas describe the same relationship between the two temperature scales. If you go on to take an algebra class, you will learn how to convert one formula into the other.

3. Use the formula in Example 3 to find y when x is 0.

4. Use the formula in Example 4 to find y when x is -3.

Answers

2. 25 degrees Celsius 3. 6

4. $\frac{10}{3}$

EXAMPLE 2 Use the formula $C = \frac{5}{9}(F - 32)$ to find C when F is 95 degrees.

SOLUTION Substituting 95 for F in the formula gives us the following:

When $F = 95$

the formula $C = \frac{5}{9}(F - 32)$

becomes

$$\begin{aligned} C &= \frac{5}{9}(95 - 32) \\ &= \frac{5}{9}(63) \\ &= \frac{5}{9} \cdot \frac{63}{1} \\ &= \frac{315}{9} \\ &= 35 \end{aligned}$$

A temperature of 95 degrees Fahrenheit is the same as a temperature of 35 degrees Celsius.

EXAMPLE 3 Use the formula $y = 2x + 6$ to find y when x is -2.

SOLUTION Proceeding as we have in the previous examples, we have:

When $x = -2$

the formula $y = 2x + 6$

becomes

$$\begin{aligned} y &= 2(-2) + 6 \\ &= -4 + 6 \\ &= 2 \end{aligned}$$

In some cases evaluating a formula also involves solving an equation, as the next example illustrates.

EXAMPLE 4 Find y when x is 3 in the formula $2x + 3y = 4$.

SOLUTION First we substitute 3 for x; then we solve the resulting equation for y.

When $x = 3$

the equation $2x + 3y = 4$

becomes

$$\begin{aligned} 2(3) + 3y &= 4 \\ 6 + 3y &= 4 \\ 3y &= -2 && \text{Add } -6 \text{ to each side} \\ y &= -\frac{2}{3} && \text{Divide each side by 3} \end{aligned}$$

B Rate Equation

Now we will look at some problems that use what is called the *rate equation*. You use this equation on an intuitive level when you are estimating how long it will take you to drive long distances. For example, if you drive at 50 miles per hour for 2 hours, you will travel 100 miles. Here is the rate equation:

$$\text{Distance} = \text{rate} \cdot \text{time, or } d = r \cdot t$$

The rate equation has two equivalent forms, one of which is obtained by solving for r, while the other is obtained by solving for t. Here they are:

$$r = \frac{d}{t} \text{ and } t = \frac{d}{r}$$

The rate in this equation is also referred to as *average speed.*

EXAMPLE 5 At 1 P.M., Jordan leaves her house and drives at an average speed of 50 miles per hour to her sister's house. She arrives at 4 P.M.

a. How many hours was the drive to her sister's house?
b. How many miles from her sister does Jordan live?

SOLUTION

a. If she left at 1:00 P.M. and arrived at 4:00 P.M., we simply subtract 1 from 4 for an answer of 3 hours.

b. We are asked to find a distance in miles given a rate of 50 miles per hour and a time of 3 hours. We will use the rate equation, $d = r \cdot t$, to solve this. We have:

$$d = 50 \text{ miles per hour} \cdot 3 \text{ hours}$$
$$d = 50(3)$$
$$d = 150 \text{ miles}$$

Notice that we were asked to find a distance in miles, so our answer has a unit of miles. When we are asked to find a time, our answer will include a unit of time, like days, hours, minutes, or seconds.

When we are asked to find a rate, our answer will include units of rate, like miles per hour, feet per second, problems per minute, and so on.

FACTS FROM GEOMETRY

Earlier we defined complementary angles as angles that add to 90°. That is, if x and y are complementary angles, then

$$x + y = 90°$$

If we solve this formula for y, we obtain a formula equivalent to our original formula:

$$y = 90° - x$$

Because y is the complement of x, we can generalize by saying that the complement of angle x is the angle $90° - x$. By a similar reasoning process, we can say that the supplement of angle x is the angle $180° - x$. To summarize, if x is an angle, then

the complement of x is $90° - x$, and
the supplement of x is $180° - x$

If you go on to take a trigonometry class, you will see these formulas again.

5. At 9 A.M. Maggie leaves her house and drives at an average speed of 60 miles per hour to her sister's house. She arrives at 11 A.M.

b. How many hours was the drive to her sister's house?
c. How many miles from her sister does Maggie live?

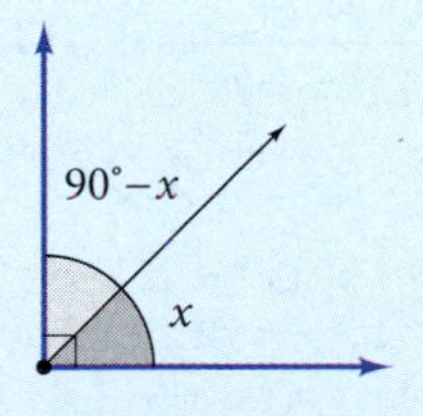

Complementary angles

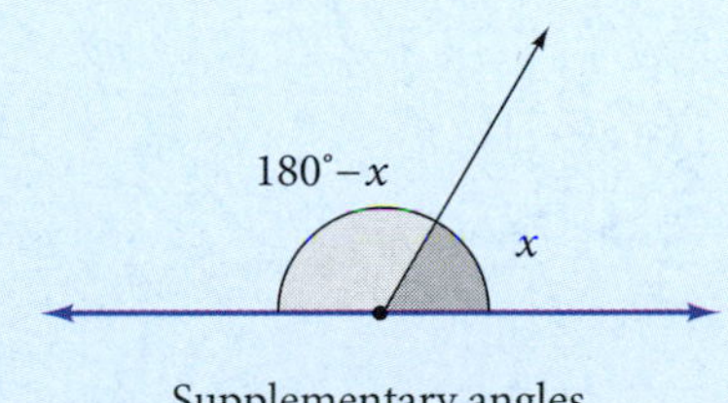

Supplementary angles

Answer
5. a. 2 hours **b.** 120 miles

6. Find the complement and the supplement of 35°.

EXAMPLE 6 Find the complement and the supplement of 25°.

SOLUTION We can use the formulas above with $x = 25°$.

The complement of 25° is $90° - 25° = 65°$.
The supplement of 25° is $180° - 25° = 155°$.

Getting Ready for Class

After reading through the preceding section, respond in your own words and in complete sentences.

1. What is a formula?
2. How do you solve a formula for one of its variables?
3. What are complementary angles?
4. What is the formula that converts temperature on the Celsius scale to temperature on the Fahrenheit scale?

Answer

6. Complement = 55°; Supplement = 145°

Problem Set 4.6

A The formula for the area A of a rectangle with length l and width w is $A = l \cdot w$. Find A if: [Examples 1–4]

1. $l = 32$ feet and $w = 22$ feet

2. $l = 22$ feet and $w = 12$ feet

3. $l = \frac{3}{2}$ inch and $w = \frac{3}{4}$ inch

4. $l = \frac{3}{5}$ inch and $w = \frac{3}{10}$ inch

The formula $G = H \cdot R$ tells us how much gross pay G a person receives for working H hours at an hourly rate of pay R. In Problems 5-8, find G.

5. $H = 40$ hours and $R = \$6$

6. $H = 36$ hours and $R = \$8$

7. $H = 30$ hours and $R = \$9\frac{1}{2}$

8. $H = 20$ hours and $R = \$6\frac{3}{4}$

Because there are 3 feet in every yard, the formula $F = 3 \cdot Y$ will convert Y yards into F feet. In Problems 9-12, find F.

9. $Y = 4$ yards

10. $Y = 8$ yards

11. $Y = 2\frac{2}{3}$ yards

12. $Y = 6\frac{1}{3}$ yards

If you invest P dollars (P is for *principal*) at simple interest rate R for T years, the amount of interest you will earn is given by the formula $I = P \cdot R \cdot T$. In Problems 13 and 14, find I.

13. $P = \$1{,}000$, $R = \frac{7}{100}$, and $T = 2$ years

14. $P = \$2{,}000$, $R = \frac{6}{100}$, and $T = 2\frac{1}{2}$ years

In Problems 15-18, use the formula $P = 2w + 2l$ to find P.

15. $w = 10$ inches and $l = 19$ inches

16. $w = 12$ inches and $l = 22$ inches

17. $w = \frac{3}{4}$ foot and $l = \frac{7}{8}$ foot

18. $w = \frac{1}{2}$ foot and $l = \frac{3}{2}$ feet

We have mentioned the two temperature scales, Fahrenheit and Celsius. Table 1 is intended to give you a more intuitive idea of the relationship between the two temperatures scales.

TABLE 1
COMPARING TWO TEMPERATURE SCALES

Situation	Temperature (Fahrenheit)	Temperature (Celsius)
Water freezes	32°F	0°C
Room temperature	68°F	20°C
Normal body temperature	$98\frac{3}{5}$°F	37°C
Water boils	212°F	100°C
Bake cookies	365°F	185°C

Table 2 gives the formulas, in both symbols and words, that are used to convert between the two scales.

TABLE 2
FORMULAS FOR CONVERTING BETWEEN TEMPERATURE SCALES

To Convert From	Formula In Symbols	Formula In Words
Fahrenheit to Celsius	$C = \frac{5}{9}(F - 32)$	Subtract 32, then multiply by $\frac{5}{9}$.
Celsius to Fahrenheit	$F = \frac{9}{5}C + 32$	Multiply by $\frac{9}{5}$, then add 32.

19. Let $F = 212$ in the formula $C = \frac{5}{9}(F - 32)$, and solve for C. Does the value of C agree with the information in Table 1?

20. Let $C = 100$ in the formula $F = \frac{9}{5}C + 32$, and solve for F. Does the value of F agree with the information in Table 1?

21. Let $F = 68$ in the formula $C = \frac{5}{9}(F - 32)$, and solve for C. Does the value of C agree with the information in Table 1?

22. Let $C = 37$ in the formula $F = \frac{9}{5}C + 32$, and solve for F. Does the value of F agree with the information in Table 1?

23. Find C when F is 32°.

24. Find C when F is -4°.

25. Find F when C is -15°.

26. Find F when C is 35°.

B **Maximum Heart Rate** In exercise physiology, a person's maximum heart rate, in beats per minute, is found by subtracting his age in years from 220. So, if A represents your age in years, then your maximum heart rate is

$$M = 220 - A$$

Use this formula to complete the following tables.

27.

Age (years)	Maximum Heart Rate (beats per minute)
18	202
19	201
20	200
21	199
22	198
23	197

28.

Age (years)	Maximum Heart Rate (beats per minute)
15	
20	
25	
30	
35	
40	

Training Heart Rate A person's training heart rate, in beats per minute, is the person's resting heart rate plus 60% of the difference between maximum heart rate and his resting heart rate. If resting heart rate is R and maximum heart rate is M, then the formula that gives training heart rate is

$$T = R + \frac{3}{5}(M - R)$$

Use this formula along with the results of Problems 29 and 30 to fill in the following two tables.

29. For a 20-year-old person

Resting Heart Rate (beats per minute)	Training Heart Rate (beats per minute)
60	144
65	146
70	148
75	150
80	152
85	154

30. For a 40-year-old person

Resting Heart Rate (beats per minute)	Training Heart Rate (beats per minute)
60	
65	
70	
75	
80	
85	

B Use the rate equation $d = r \cdot t$ to solve Problems 31 and 32. [Example 5]

31. At 2:30 P.M. Shelly leaves her house and drives at an average speed of 55 miles per hour to her sister's house. She arrives at 6:30 P.M.

a. How many hours was the drive to her sister's house?

b. How many miles from her sister does Shelly live?

32. At 1:30 P.M. Cary leaves his house and drives at an average speed of 65 miles per hour to his brother's house. He arrives at 5:30 P.M.

a. How many hours was the drive to his brother's house?

b. How many miles from his brother's house does Cary live?

Use the rate equation $r = \frac{d}{t}$ to solve Problems 33 and 34.

33. At 2:30 P.M. Brittney leaves her house and drives 260 miles to her sister's house. She arrives at 6:30 P.M.

a. How many hours was the drive to her sister's house?

b. What was Brittney's average speed?

34. At 8:30 A.M. Ethan leaves his house and drives 220 miles to his brother's house. He arrives at 12:30 P.M.

a. How many hours was the drive to his brother's house?

b. What was Ethan's average speed?

As you know, the volume V enclosed by a rectangular solid with length l, width w, and height h is $V = l \cdot w \cdot h$. In Problems 35-38, find V if:

35. $l = 6$ inches, $w = 12$ inches, and $h = 5$ inches

36. $l = 16$ inches, $w = 22$ inches, and $h = 15$ inches

37. $l = 6$ yards, $w = \frac{1}{2}$ yard, and $h = \frac{1}{3}$ yard

38. $l = 30$ yards, $w = \frac{5}{2}$ yards, and $h = \frac{5}{3}$ yards

Suppose $y = 3x - 2$. In Problems 39–44, find y if:

39. $x = 3$

40. $x = -5$

41. $x = -\frac{1}{3}$

42. $x = \frac{2}{3}$

43. $x = 0$

44. $x = 5$

Suppose $x + y = 5$. In Problems 45–50, find x if:

45. $y = 2$

46. $y = -2$

47. $y = 0$

48. $y = 5$

49. $y = -3$

50. $y = 3$

Suppose $x + y = 3$. In Problems 51–56, find y if:

51. $x = 2$

52. $x = -2$

53. $x = 0$

54. $x = 3$

55. $x = \frac{1}{2}$

56. $x = -\frac{1}{2}$

Suppose $4x + 3y = 12$. In Problems 57–62, find y if:

57. $x = 3$

58. $x = -5$

59. $x = -\frac{1}{4}$

60. $x = \frac{3}{2}$

61. $x = 0$

62. $x = -3$

Suppose $4x + 3y = 12$. In Problems 63-68, find x if:

63. $y = 4$

64. $y = -4$

65. $y = -\frac{1}{3}$

66. $y = \frac{5}{3}$

67. $y = 0$

68. $y = -3$

Find the complement and supplement of each angle. [Example 6]

69. 45°

70. 75°

71. 31°

72. 59°

Applying the Concepts

73. Digital Video The most popular video download of all time was a *Star Wars* movie trailer. The video was compressed so it would be small enough for people to download over the Internet. A formula for estimating the size, in kilobytes, of a compressed video is

$$S = \frac{height \cdot width \cdot fps \cdot time}{35{,}000}$$

where *height* and *width* are in pixels, *fps* is the number of frames per second the video is to play (television plays at 30 fps), and *time* is given in seconds.

a. Estimate the size in kilobytes of the *Star Wars* trailer that has a height of 480 pixels, has a width of 216 pixels, plays at 30 fps, and runs for 150 seconds.

b. Estimate the size in kilobytes of the *Star Wars* trailer that has a height of 320 pixels, has a width of 144 pixels, plays at 15 fps, and runs for 150 seconds.

74. Vehicle Weight If you can measure the area that the tires on your car contact the ground, and you know the air pressure in the tires, then you can estimate the weight of your car, in pounds, with the following formula:

$$W = APN$$

where W is the vehicle's weight in pounds, A is the average tire contact area with a hard surface in square inches, P is the air pressure in the tires in pounds per square inch (psi, or lb/in²), and N is the number of tires.

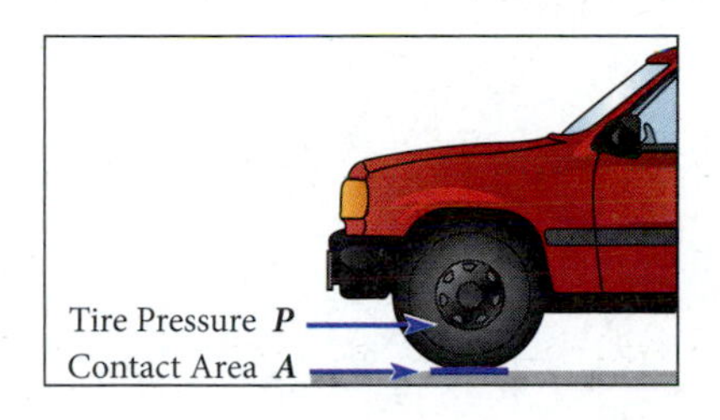

a. What is the approximate weight of a car if the average tire contact area is a rectangle 6 inches by 5 inches and if the air pressure in the tires is 30 psi?

b. What is the approximate weight of a car if the average tire contact area is a rectangle 5 inches by 4 inches, and the tire pressure is 30 psi?

75. Temperature The chart shows the temperatures for some of the world's hottest places.

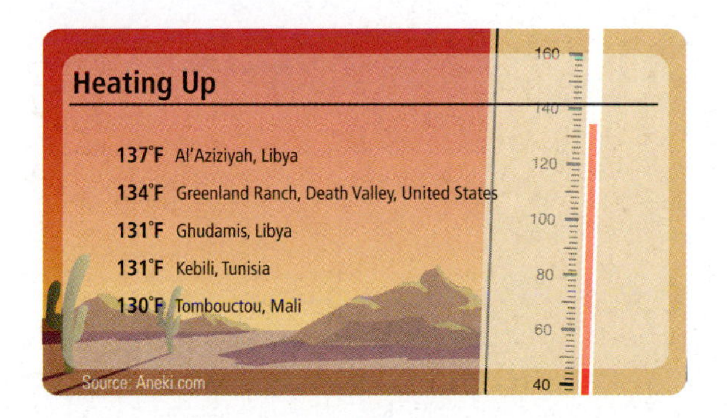

a. Use the formula $C = \frac{5}{9}(F - 32)$ to find the temperature in Celsius for Al'Aziziyah.

b. Use the formula $K = \frac{5}{9}(F - 32) + 273$ to find the temperature in Ghudamis, Libya, in Kelvin.

76. Fermat's Last Theorem The postage stamp shows Fermat's last theorem, which states that if n is an integer greater than 2, then there are no positive integers x, y, and z that will make the formula $x^n + y^n = z^n$ true.

Use the formula $x^n + y^n = z^n$ to

a. Find x if $n = 1$, $y = 7$, and $z = 15$.

b. Find y if $n = 1$, $x = 23$, and $z = 37$.

Maintaining Your Skills

Simplify.

77. $\dfrac{\frac{3}{5}}{\frac{4}{5}}$

78. $\dfrac{\frac{5}{7}}{\frac{6}{7}}$

79. $\dfrac{1 + \frac{1}{2}}{1 - \frac{1}{2}}$

80. $\dfrac{1 + \frac{1}{3}}{1 - \frac{1}{3}}$

81. $\dfrac{\frac{1}{2} + \frac{1}{4}}{\frac{1}{4} + \frac{1}{8}}$

82. $\dfrac{\frac{1}{2} - \frac{1}{4}}{\frac{1}{4} - \frac{1}{8}}$

83. $\dfrac{\frac{3}{5} + \frac{3}{7}}{\frac{3}{5} - \frac{3}{7}}$

84. $\dfrac{\frac{5}{7} + \frac{5}{8}}{\frac{5}{7} - \frac{5}{8}}$

4.7 Paired Data and Equations in Two Variables

Objectives

A Write solutions to equations in two variables as ordered pairs.

B Fill in a table from solutions to equations in two variables.

C Evaluate ordered pairs as possible solutions to equations in two variables.

D Solve application problems involving equations in two variables.

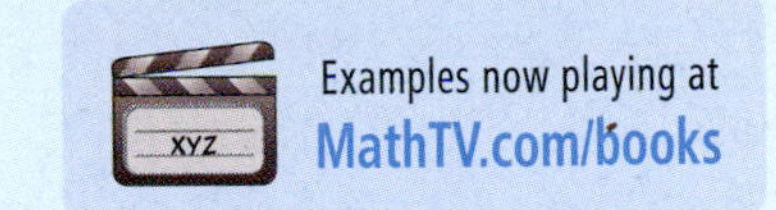

Introduction . . .

The table and a line graph show the relationship between the Fahrenheit and Celsius temperature scales.

TABLE 1

COMPARING TEMPERATURES ON TWO SCALES (NUMERIC)

Temperature in Degrees Celsius	Temperature in Degrees Fahrenheit
0°C	32°F
25°C	77°F
50°C	122°F
75°C	167°F
100°C	212°F

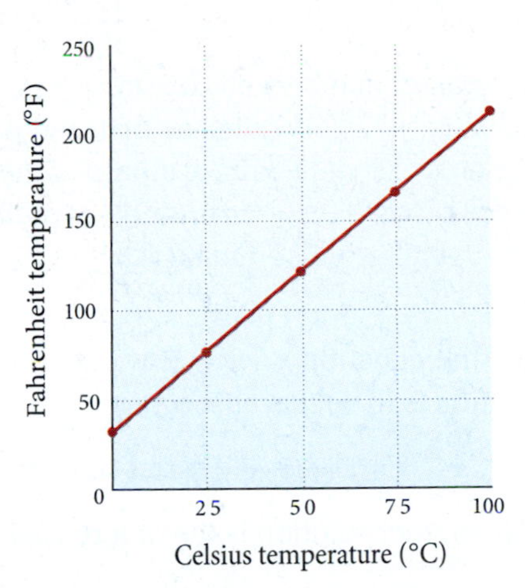

FIGURE 1 A line graph of the data in the table (geometric)

The data in Table 1 are called *paired data* because the table is organized so that each number in the first column is paired with a specific number in the second column: A Celsius temperature of 100°C (first column) corresponds to a Fahrenheit temperature of 212°F (second column). The information in Figure 1 is also paired data because each dot on the line graph comes from one of the pairs of numbers in the table: The upper rightmost dot on the line graph corresponds to 100°C and 212°F.

We considered paired data in Section 4.6 also, when we used formulas to produce pairs of numbers: When we substitute $C = 100$ into the formula below, we obtain $F = 212$, the same pair of numbers shown in Table 1 and again in the line graph in Figure 1.

$$F = \frac{9}{5}C + 32$$

The concept of paired data is very important in mathematics, so we want to continue our study of it. We start with a more detailed look at equations that contain two variables.

Recall that a solution to an equation in one variable is a single number that, when substituted for the variable in the equation, turns the equation into a true statement. For example, $x = 5$ is a solution to the equation $2x - 4 = 6$, because replacing x with 5 turns the equation into $6 = 6$, a true statement.

Next, consider the equation $x + y = 5$. It has two variables instead of one. Therefore, a solution to $x + y = 5$ will have to consist of two numbers, one for x and one for y, that together make the equation a true statement. One pair of numbers is $x = 2$ and $y = 3$, because when we substitute 2 for x and 3 for y into the equation $x + y = 5$, the result is a true statement. That is,

When	$x = 2$ and $y = 3$
the equation	$x + y = 5$
becomes	$2 + 3 = 5$
	$5 = 5$ A true statement

To simplify our work, we write the pair of numbers $x = 2, y = 3$ in the shorthand form (2, 3). The expression (2, 3) is called an *ordered pair* of numbers. Here is the formal definition:

A Solutions as Ordered Pairs

Definition

A pair of numbers enclosed in parentheses and separated by a comma, such as (2, 3), is called an **ordered pair** of numbers. The first number in the pair is called the ***x*-coordinate** of the ordered pair, while the second number is called the ***y*-coordinate.** For the ordered pair (2, 3), the x-coordinate is 2 and the y-coordinate is 3.

In the equation $x + y = 5$, we find that (2, 3) is not the only solution. Another solution is (0, 5), because when $x = 0$ and $y = 5$, then

$$0 + 5 = 5 \quad \text{A true statement}$$

Still another solution is the ordered pair $(-7, 12)$, because

$$\begin{aligned} \text{When} \quad & x = -7 \text{ and } y = 12 \\ \text{the equation} \quad & x + y = 5 \\ \text{becomes} \quad & -7 + 12 = 5 \\ & 5 = 5 \quad \text{A true statement} \end{aligned}$$

As you can imagine, there are many more ordered pairs that are solutions to the equation $x + y = 5$. As a matter of fact, for any number we choose for x, there is another number we can use for y that will make the equation a true statement. There are an infinite number of ordered pairs that are solutions to the equation $x + y = 5$.

PRACTICE PROBLEMS

1. Fill in the ordered pairs (0,), (, 0), and $(-5,\ )$ so that they are solutions to the equation $3x + 5y = 15$.

EXAMPLE 1 Fill in the ordered pairs (0,), (, −2), and (3,) so that they are solutions to the equation $2x + 3y = 6$.

SOLUTION To complete the ordered pair (0,), we substitute 0 for x in the equation and then solve for y:

$$\begin{aligned} \text{When} \quad & x = 0 \\ \text{the equation} \quad & 2x + 3y = 6 \\ \text{becomes} \quad & 2 \cdot 0 + 3y = 6 \\ & 3y = 6 \\ & y = 2 \end{aligned}$$

Therefore, the ordered pair (0, 2) is a solution to $2x + 3y = 6$.

To complete the ordered pair (, −2), we substitute −2 for y and then solve for x.

$$\begin{aligned} \text{When} \quad & y = -2 \\ \text{the equation} \quad & 2x + 3y = 6 \\ \text{becomes} \quad & 2x + 3(-2) = 6 \\ & 2x - 6 = 6 \\ & 2x = 12 \\ & x = 6 \end{aligned}$$

Therefore, the ordered pair $(6, -2)$ is another solution to our equation.

Finally, to complete the ordered pair (3,), we substitute 3 for x and then solve for y. The result is $y = 0$. The ordered pair (3, 0) is a third solution to our equation.

Answer

1. (0, 3), (5, 0), and (−5, 6)

B Tables

EXAMPLE 2 Use the equation $5x - 2y = 20$ to complete the table below.

x	y
2	
0	
	5
	0

SOLUTION Filling in the table is equivalent to completing the following ordered pairs: (2,), (0,), (, 5), and (, 0). We proceed as in Example 1.

When $x = 2$, we have

$$5 \cdot 2 - 2y = 20$$
$$10 - 2y = 20$$
$$-2y = 10$$
$$y = -5$$

When $x = 0$, we have

$$5 \cdot 0 - 2y = 20$$
$$0 - 2y = 20$$
$$-2y = 20$$
$$y = -10$$

When $y = 5$, we have

$$5x - 2 \cdot 5 = 20$$
$$5x - 10 = 20$$
$$5x = 30$$
$$x = 6$$

When $y = 0$, we have

$$5x - 2 \cdot 0 = 20$$
$$5x - 0 = 20$$
$$5x = 20$$
$$x = 4$$

Using these results, we complete our table.

x	y
2	−5
0	−10
6	5
4	0

EXAMPLE 3 Complete the table below for the equation $y = 2x + 1$.

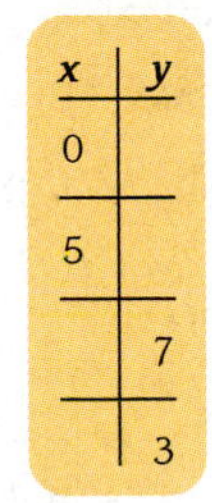

x	y
0	
5	
	7
	3

SOLUTION When $x = 0$, we have

$$y = 2 \cdot 0 + 1$$
$$y = 0 + 1$$
$$y = 1$$

When $x = 5$, we have

$$y = 2 \cdot 5 + 1$$
$$y = 10 + 1$$
$$y = 11$$

When $y = 7$, we have

$$7 = 2x + 1$$
$$6 = 2x$$
$$3 = x$$

When $y = 3$, we have

$$3 = 2x + 1$$
$$2 = 2x$$
$$1 = x$$

2. Use the equation $5x + 2y = 20$ to complete the table below.

x	y
2	
0	
	5
	0

3. Complete the table below for the equation $y = \frac{1}{2}x + 1$.

x	y
0	
4	
	7
	−3

Answer

2.

x	y
2	5
0	10
2	5
4	0

The completed table looks like this:

x	y
0	1
5	11
3	7
1	3

C Evaluating Ordered Pairs

EXAMPLE 4 Which of the ordered pairs (1, 5) and (2, 4) are solutions to the equation $y = 3x + 2$?

SOLUTION If an ordered pair is a solution to an equation, then it must yield a true statement when the coordinates of the ordered pair are substituted for x and y in the equation.

First, we try (1, 5) in the equation $y = 3x + 2$:

$$5 = 3 \cdot 1 + 2$$
$$5 = 3 + 2$$
$$5 = 5 \quad \text{A true statement}$$

Next, we try (2, 4) in the equation:

$$4 = 3 \cdot 2 + 2$$
$$4 = 6 + 2$$
$$4 = 8 \quad \text{A false statement}$$

The ordered pair (1, 5) is a solution to the equation $y = 3x + 2$, but (2, 4) is not a solution to the equation.

4. Which of the ordered pairs (1, 5) and (2, 4) are solutions to the equation $y = 5x - 6$?

Getting Ready for Class

After reading the preceding section and working the practice problems in the margin, answer the following questions.

1. How can you tell if an ordered pair is a solution to an equation?
2. How would you find a solution to $y = 3x + 2$?
3. Why is (3, 2) not a solution to $y = 3x + 2$?
4. How many solutions are there to an equation that contains two variables?

Answers

3.

x	y
0	1
4	3
12	7
−8	−3

4. (2, 4)

Problem Set 4.7

A For each equation, complete the given ordered pairs. [Example 1]

1. $x + y = 4$ (0,), (3,), (−2,)
2. $x + y = 6$ (0,), (2,), (, 2)
3. $x + 2y = 6$ (0,), (2,), (, −6)
4. $3x + y = 5$ (0,), (1,), (, 5)
5. $4x + 3y = 12$ (0,), (, 0), (−3,)
6. $5x + 5y = 20$ (0,), (, −2), (1,)
7. $y = 4x - 3$ (1,), (, −3), (5,)
8. $y = 3x - 5$ (, 13), (0,), (−2,)
9. $y = 2x + 3$ (0,), (2,), (−2,)
10. $y = 3x + 2$ (0,), (2,), (−2,)
11. $y = 7x$ (2,), (, 6), (0,)
12. $y = 8x$ (3,), (, 0), (, −6)
13. $y = -2x$ (0,), (−2,), (2,)
14. $y = -5x$ (0,), (−1,), (1,)
15. $y = \frac{1}{2}x$ (0,), (2,), (4,)
16. $y = \frac{1}{3}x$ (−3,), (0,), (3,)
17. $y = -\frac{1}{2}x + 2$ (−2,), (0,), (2,)
18. $y = -\frac{1}{3}x + 1$ (−3,), (0,), (3,)

B For each of the following equations, complete the given table. [Example 2, 3]

19. $x + y = 5$

x	y
2	
3	
	4
	5

20. $x - y = 2$

x	y
3	
−2	
	2
	6

21. $y = -4x$

x	y
0	
	4
	8
1	

22. $y = 5x$

x	y
	5
	0
−1	
2	

23. $3x - 2y = 6$

x	y
−2	
	0
4	
	−3

24. $2x - y = 6$

x	y
1	
	6
−6	
	−6

25. $y = 6x - 1$

x	y
−1	
	5
	−13
0	

26. $y = 3x + 1$

x	y
−2	
	−2
	4
0	

27. Comparing Tables Which of the tables below could be produced from the equation $y = 2x - 6$?

a.

x	y
0	6
1	4
2	2
3	0

b.

x	y
0	−6
1	−4
2	−2
3	0

c.

x	y
0	−6
1	−5
2	−4
3	−3

28. Comparing Tables Which of the tables below could be produced from the equation $y = 3x + 5$?

a.

x	y
−3	−4
−2	−7
−1	−10
0	−10

b.

x	y
−3	14
−2	11
−1	8
0	5

c.

x	y
−3	−4
−2	−1
−1	2
0	5

C In Problems 29–36, indicate which of the given ordered pairs are solutions for each equation. [Example 4]

29. $2x - 5y = 10$ $(2, 3), (0, -2), \left(\frac{5}{2}, 1\right)$

30. $3x + 7y = 21$ $(0, 3), (7, 0), (1, 2)$

31. $y = 2x + 3$ $(0, 3), (5, 4), (2, 0)$

32. $y = -3x + 2$ $(0, -3), (0, 2), (-3, 0)$

33. $x + y = 0$ $(0, 0), (5, -5), (-3, 3)$

34. $x - y = 0$ $(0, 0), (5, -5), (3, 3)$

35. $y = 5x$ $(0, 5), (1, 5), \left(2, \frac{5}{2}\right)$

36. $y = -3x$ $(0, 0), (-3, 0), (-1, 3)$

D Applying the Concepts

37. Light Bulbs The chart shows a comparison of power usage between incandescent and energy-efficient light bulbs. The line segment from 1,100 to 1,600 lumens for an incandescent bulb is given by the equation $y = \frac{1}{25}x + 36$. How many watts are used in a bulb that puts out 1,500 lumens?

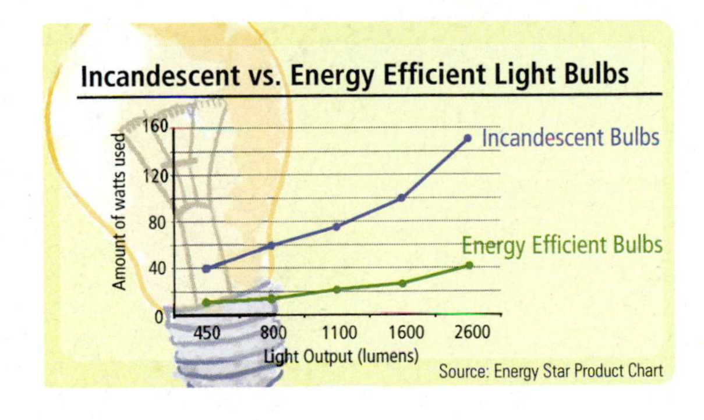

38. Mothers The graph shows the average age of first-time mothers from 1970 to 2005. The line segment from 1990 to 2000 has the equation $y = \frac{x}{10} - 175$. Find the average age of first-time mothers in 1995.

39. Complementary Angles This diagram shows sunlight hitting the ground. Angle α (alpha) is called the angle of inclination and angle θ (theta) is called the angle of incidence. As the sun moves across the sky, the values of these angles change, but the two angles are always complementary, meaning that $\alpha + \theta = 90°$. Use this information to fill in the table.

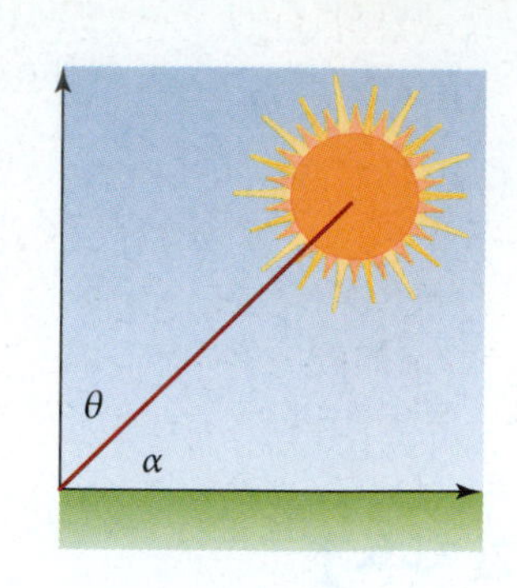

α	θ
0°	
30°	
	45°
60°	
	15°
90°	

40. Cost of Bottled Water A water bottling company charges $7 per month for their water dispenser and $2 for each gallon of water delivered. If you have g gallons of water delivered in a month, then the equation $y = 7 + 2g$ gives the amount of your bill for that month. Use this equation to fill in the following table.

WATER (GALLONS) g	MONTHLY BILL (DOLLARS) y
0	
1	
2	
	13
	15
5	

WBC **WATER BOTTLE CO.**

MONTHLY BILL

234 5th Street
Glendora, CA 91740

DUE 07/23/08

Water dispenser	1	$7.00
Gallons of water	8	$2.00
		$23.00

Maintaining Your Skills

Find the area and the perimeter of each triangle.

41.

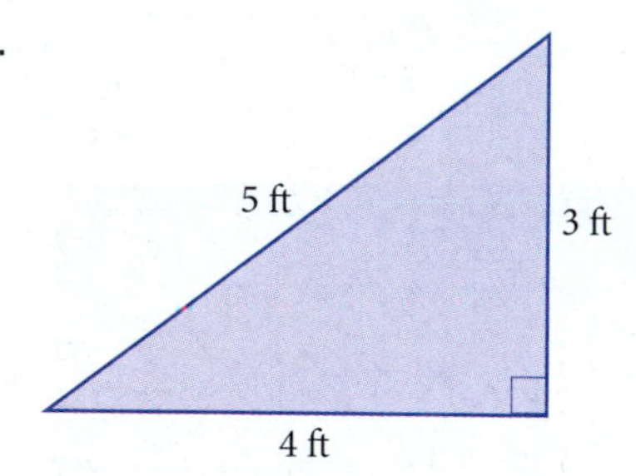

42.

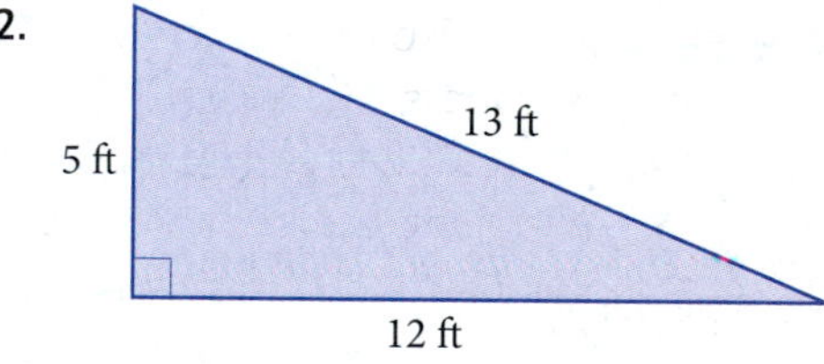

43.

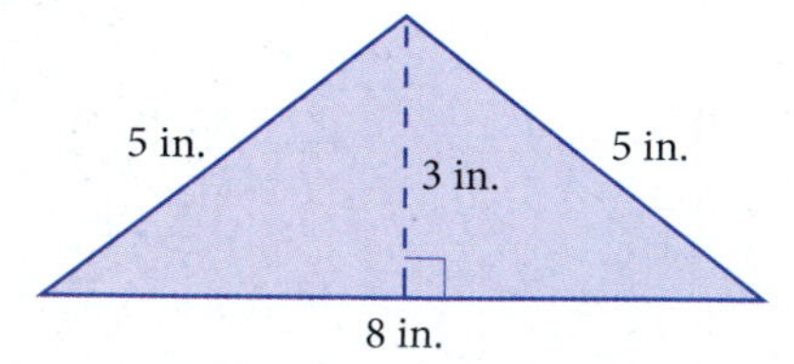

44.

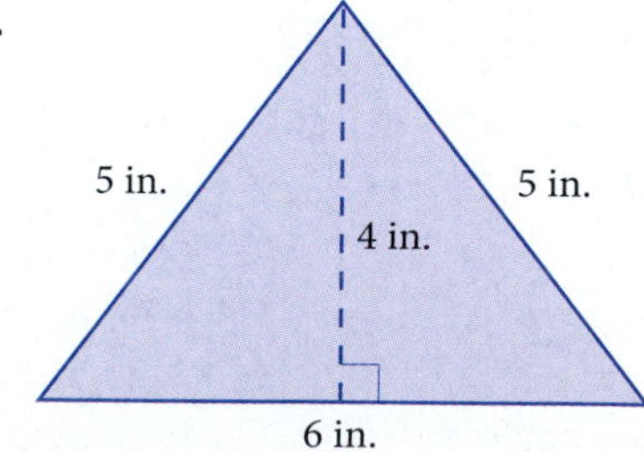

The Rectangular Coordinate System

4.8

Objectives

A Plot ordered pairs on a coordinate system.

B Name the coordinates of a point on the rectangular coordinate system.

C Graph a line given two points.

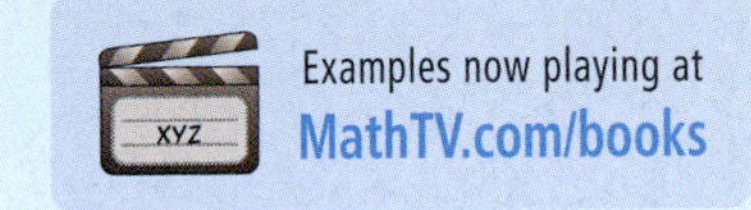

Introduction . . .

The table and line graph below are similar to those that we have used previously in this chapter to discuss temperature. Note that we have extended both the table and the line graph to show some temperatures below zero on both scales.

TABLE 1

COMPARING TEMPERATURES ON TWO SCALES

Temperature In Degrees Celsius	Temperature In Degrees Fahrenheit
−100°C	−148°F
−75°C	−103°F
−50°C	−58°F
−25°C	−13°F
0°C	32°F
25°C	77°F
50°C	122°F
75°C	167°F
100°C	212°F

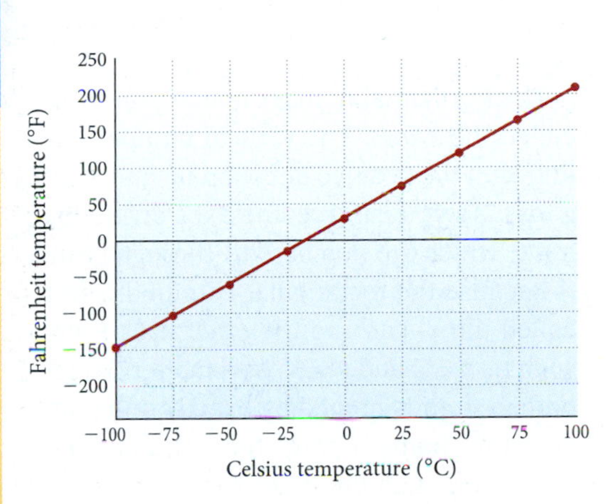

FIGURE 1 A line graph of the data in Table 1

We know from the previous section that the information in both the table and the line graph is paired data. We also know that solutions to equations in two variables consist of pairs of numbers that together satisfy the equations. What we want to do next is look at solutions to equations in two variables from a visual perspective. In order to do so, we need to standardize the way in which we present paired data visually. This is accomplished with the *rectangular coordinate system*.

A The Rectangular Coordinate System

The rectangular coordinate system can be used to plot (or graph) pairs of numbers (see Figure 2 on the following page). It consists of two number lines, called *axes*, which intersect at right angles. (A right angle is a 90° angle.) The point at which the axes intersect is called the *origin*.

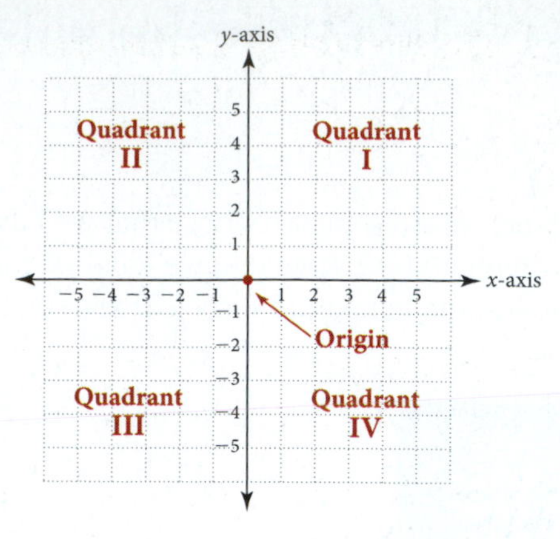

FIGURE 2

The horizontal number line is exactly the same as the real number line and is called the *x-axis.* The vertical number line is also the same as the real number line with the positive direction up and the negative direction down. It is called the *y-axis.* As you can see, the axes divide the plane into four regions, called *quadrants,* which are numbered I through IV in a counterclockwise direction.

Because the rectangular coordinate system consists of two number lines, one called the x-axis and the other called the y-axis, we can plot pairs of numbers such as $x = 2$ and $y = 3$. As a matter of fact, each point in the rectangular coordinate system is named by exactly one pair of numbers. We call the pair of numbers that name a point the *coordinates* of that point. To find the point that is associated with the pair of numbers $x = 2$ and $y = 3$, we start at the origin and move 2 units horizontally to the right and then 3 units vertically up (see Figure 3). The place where we end up is the point named by the pair of numbers $x = 2, y = 3$, which we write in shorthand form as the ordered pair (2, 3).

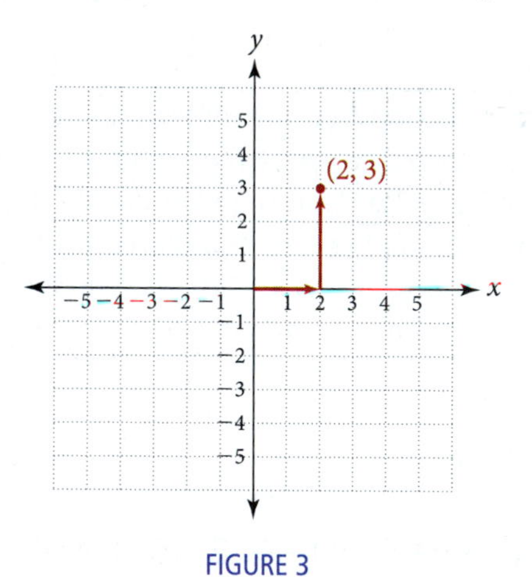

FIGURE 3

In general, to graph an ordered pair (a, b) on the rectangular coordinate system, we start at the origin and move a units right or left (right if a is positive, left if a is negative). From there we move b units up or down (up if b is positive, down if b is negative). The point where we end up is the graph of the ordered pair (a, b).

PRACTICE PROBLEMS

1. Plot the ordered pairs (3, 2), (3, −2), (−3, −2), and (−3, 2) on the coordinate system used in Example 1.

EXAMPLE 1 Plot (graph) the ordered pairs (2, 3), (−2, 3), (−2, −3), and (2, −3).

SOLUTION To graph the ordered pair (2, 3), we start at the origin and move 2 units to the right, then 3 units up. We are now at the point whose coordinates are

(2, 3). We plot the other three ordered pairs in the same manner (Figure 4).

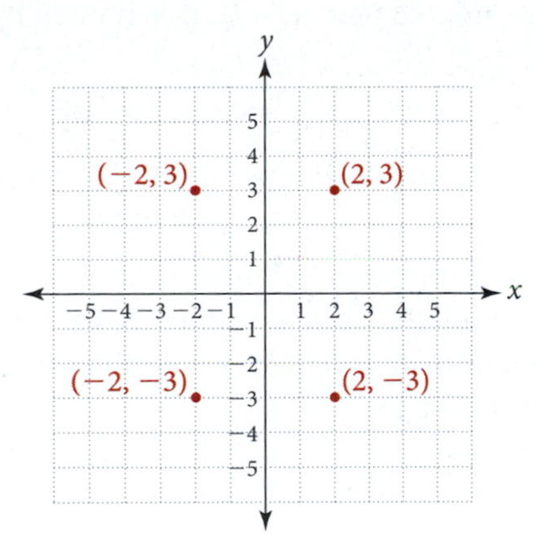

FIGURE 4

Note Looking at Example 1, we see that any point in quadrant I must have positive x- and y-coordinates (+, +). In quadrant II, x-coordinates are negative and y-coordinates are positive, (−, +). In quadrant III, both coordinates are negative (−, −). Finally, in quadrant IV, all ordered pairs must have the form (+, −).

EXAMPLE 2 Plot the ordered pairs (1, −4), $(\frac{1}{2}, 3)$, (2, 0), (0, −2), and (−3, 0).

SOLUTION

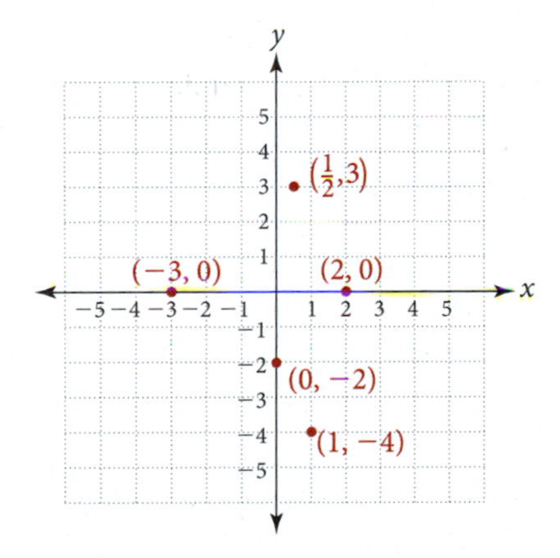

2. Plot the ordered pairs (−1, −4), $(-\frac{1}{2}, 3)$, (0, 2), (5, 0), (0, −5), and (−1, 0) on the coordinate system used in Example 2.

B Points on a Rectangular Coordinate System

EXAMPLE 3 Give the coordinates of each point in Figure 5.

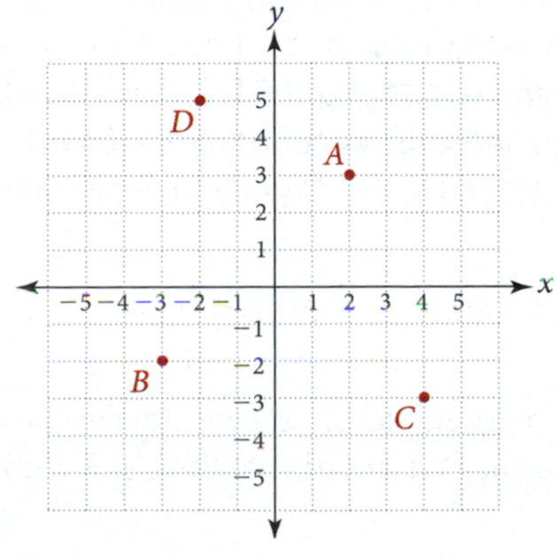

FIGURE 5

3. Give the coordinates of each point in the figure below.

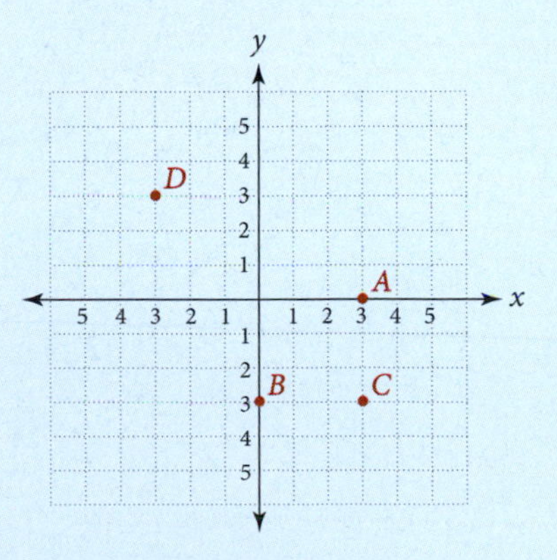

Answers

1. See solutions section.
2. See solutions section.

SOLUTION A is named by the ordered pair (2, 3). B is named by the ordered pair (−3, −2). C is named by the ordered pair (4, −3). D is named by the ordered pair (−2, 5).

EXAMPLE 4 Where are all the points that have coordinates of the form $(x, 0)$?

SOLUTION Because the y-coordinate is 0, these points must lie on the x-axis. Remember, the y-coordinate tells us how far up or down we move to find the point in question. If the y-coordinate is 0, then we don't move up or down at all. Therefore, we must stay on the x-axis.

C Graphing Lines

EXAMPLE 5 Graph the points (1, 2) and (3, 4), and draw a line through them. Then use your result to answer these questions.

a. Does the graph of (2, 3) lie on this line?

b. Does the graph of (−3, −5) lie on this line?

SOLUTION Figure 6 shows the graphs of (1, 2) and (3, 4) and the line that connects them. The line does not pass through the point (−3, −5) but does pass through (2, 3).

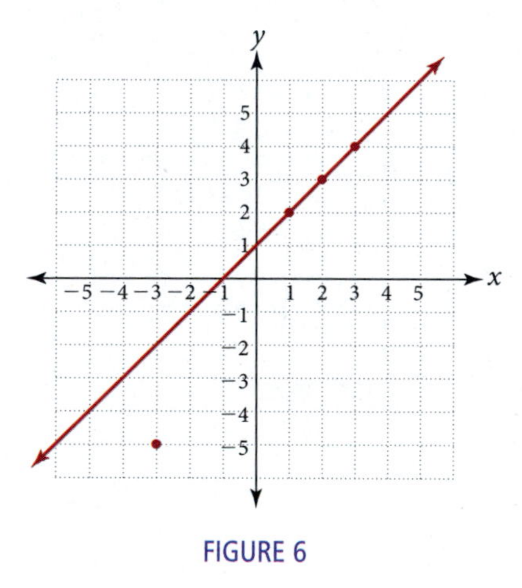

FIGURE 6

Getting Ready for Class

After reading the preceding section and working the practice problems in the margin, answer the following questions.

1. What is an ordered pair of numbers?
2. Explain in words how you would graph the ordered pair (2, 3).
3. How do you construct a rectangular coordinate system?
4. Where is the origin on a rectangular coordinate system?

---

4. The points (2, 0), (−3, 0), $(\frac{5}{2}, 0)$, and (−8, 0) all lie on which axis?

5. a. Does the point (−3, −2) lie on the line shown in Example 5?

b. Does the point (3, 2) lie on the line shown in Example 5?

Answers

3. A is (3, 0); B is (0, −3); C is (3, −3); D is (−3, 3).

4. The x-axis **5. a.** Yes **b.** No

Problem Set 4.8

A Graph each of the following ordered pairs. [Examples 1, 2]

1. $(4, 2)$ **2.** $(4, -2)$ **3.** $(-4, 2)$ **4.** $(-4, -2)$ **5.** $(3, 4)$ **6.** $(-3, 4)$

7. $(-3, -4)$ **8.** $(3, -4)$ **9.** $(4, 3)$ **10.** $(-4, 3)$ **11.** $\left(5, \frac{1}{2}\right)$ **12.** $\left(-5, -\frac{1}{2}\right)$

13. $\left(1, -\frac{3}{2}\right)$ **14.** $\left(-\frac{3}{2}, 1\right)$ **15.** $(2, 0)$ **16.** $(0, -5)$ **17.** $(-2, 0)$ **18.** $(0, 5)$

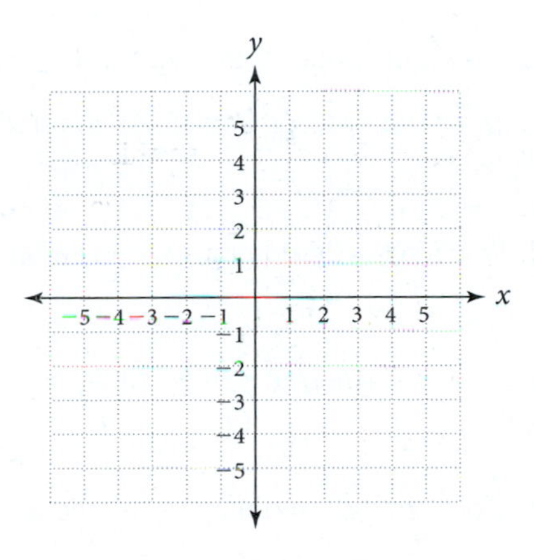

USE FOR ODD-NUMBERED PROBLEMS

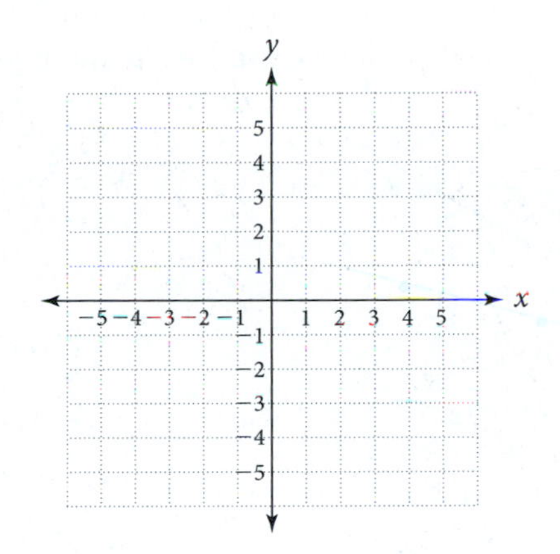

USE FOR EVEN-NUMBERED PROBLEMS

B **19.–25.** Give the coordinates of each point in the figure below. [Examples 3, 4]

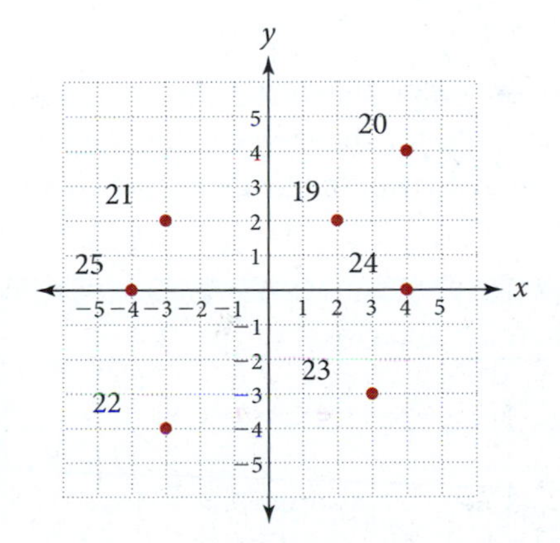

26. Where will you find all the ordered pairs of the form $(0, y)$?

C Graph the points (1, 3) and (2, 4), and draw a line through them. Use that graph to answer Problems 27–30. [Example 5]

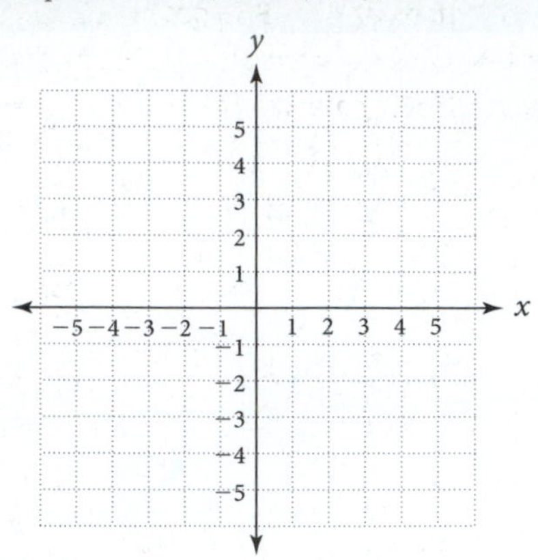

27. Does the ordered pair (−1, 1) lie on this line?

28. Does the ordered pair (3, 4) lie on this line?

29. Does the ordered pair (−3, −5) lie on this line?

30. Does the ordered pair (3, 5) lie on this line?

Graph the points (−3, 2) and (1, 3), and draw a line through them. Use that graph to answer Problems 31–34.

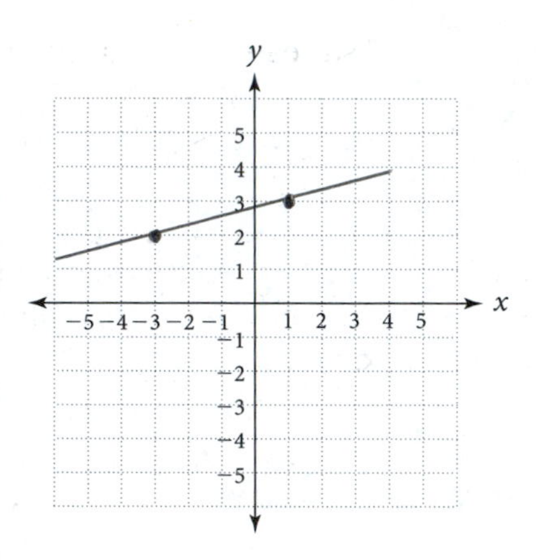

31. Does the ordered pair (5, 4) lie on this line?

32. Does the ordered pair (3, −1) lie on this line?

33. Does the ordered pair (4, 5) lie on this line?

34. Does the ordered pair (−7, 1) lie on this line?

Applying the Concepts

35. Horse Racing The graph shows the total amount of money wagered on the Kentucky Derby over the years. List three ordered pairs that lie on the graph.

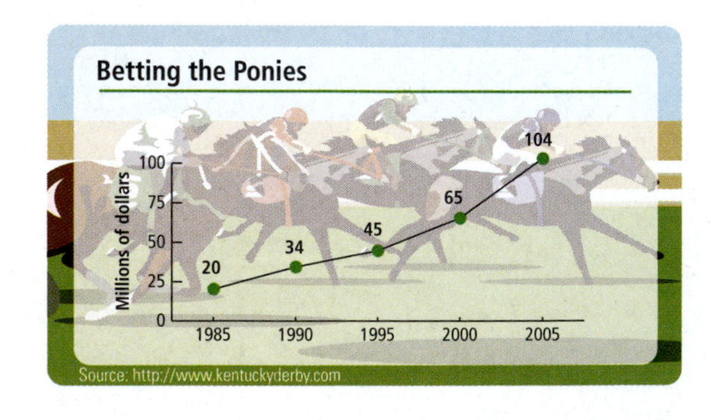

36. Solar Energy The graph shows the annual number of solar thermal collector shipments in the United States. Use the graph to answer the following questions.

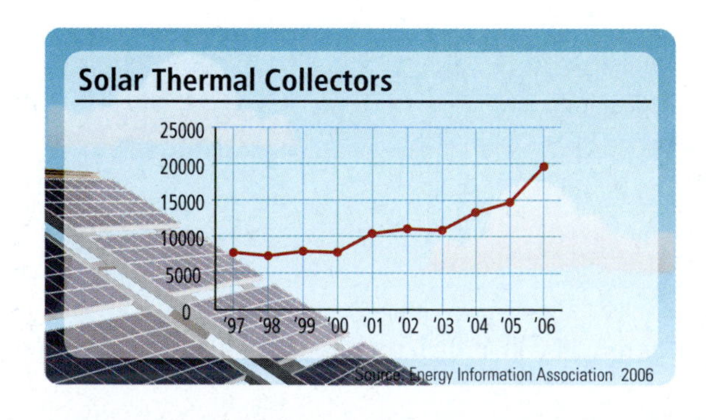

a. Does the graph contain the point (2000, 7,500)?
b. Does the graph contain the point (2004, 15,000)?
c. Does the graph contain the point (2005, 15,000)?

37. Hourly Wages Jane takes a job at the local Marcy's department store. Her job pays \$8.00 per hour. The graph shows how much Jane earns for working from 0 to 40 hours in a week.

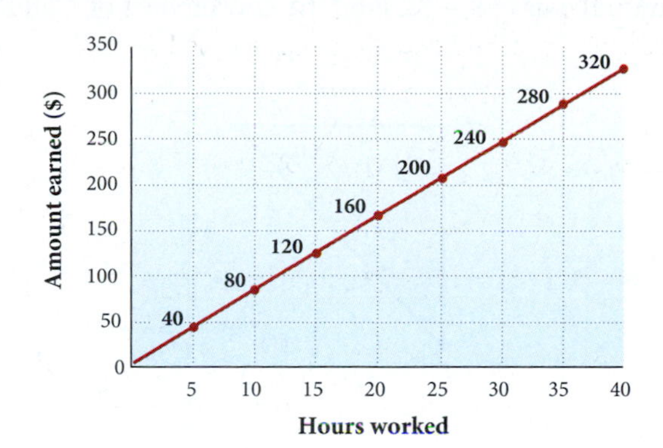

a. List three ordered pairs that lie on the line graph.

b. How much will she earn for working 40 hours?

c. If her check for one week is \$240, how many hours did she work?

d. She works 35 hours one week, but her paycheck before deductions are subtracted out is for \$260. Is this correct? Explain

38. Hourly Wages Judy takes a job at Gigi's boutique. Her job pays \$6.00 per hour plus \$50 per week in commission. The graph shows how much Judy earns for working from 0 to 40 hours in a week.

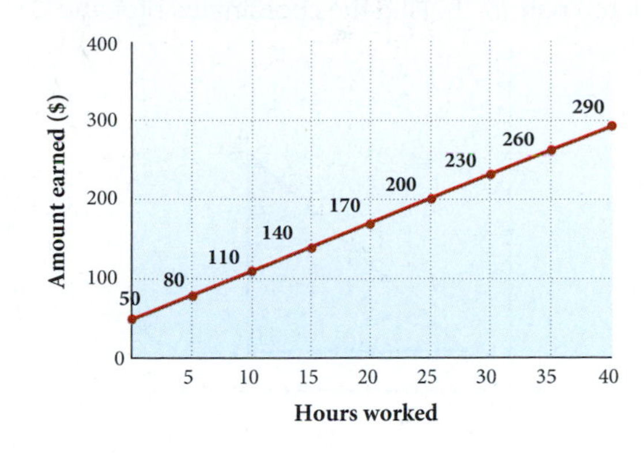

a. List three ordered pairs that lie on the line graph.

b. How much will she earn for working 40 hours?

c. If her check for one week is \$230, how many hours did she work?

d. She works 35 hours one week, but her paycheck before deductions are subtracted out is for \$260. Is this correct? Explain.

Maintaining Your Skills

Multiply or divide as indicated.

39. $\frac{x^2}{y} \cdot \frac{y^3}{x}$

40. $\frac{a^3}{15} \cdot \frac{12}{a^2}$

41. $\frac{x^2}{y^3} \div \frac{x^3}{y^2}$

42. $\frac{a^2b}{c^3} \div \frac{ab^2}{c^3}$

Add or subtract as indicated.

43. $\frac{x}{5} + \frac{3}{4}$

44. $5 + \frac{2}{x}$

45. $3 - \frac{1}{x}$

46. $\frac{x}{8} - \frac{5}{6}$

Extending the Concepts

47. Right triangle ABC has legs of length 5. Point C is the ordered pair (6, 2). Find the coordinates of A and B.

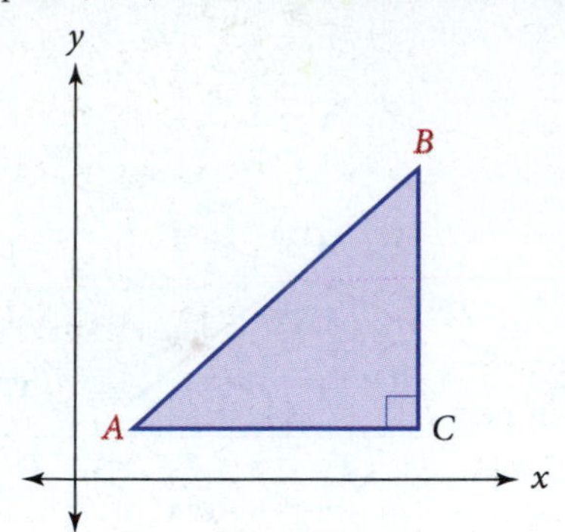

48. Right triangle ABC has legs of length 7. Point C is the ordered pair (−8, −3). Find the coordinates of A and B.

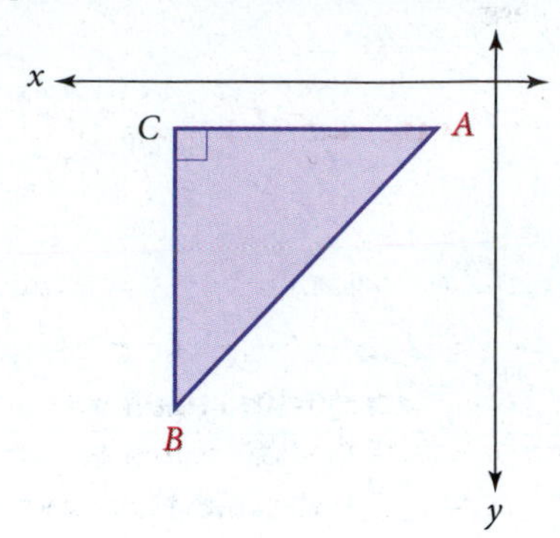

49. Rectangle $ABCD$ has a length of 5 and a width of 3. Point D is the ordered pair (7, 2). Find points A, B, and C.

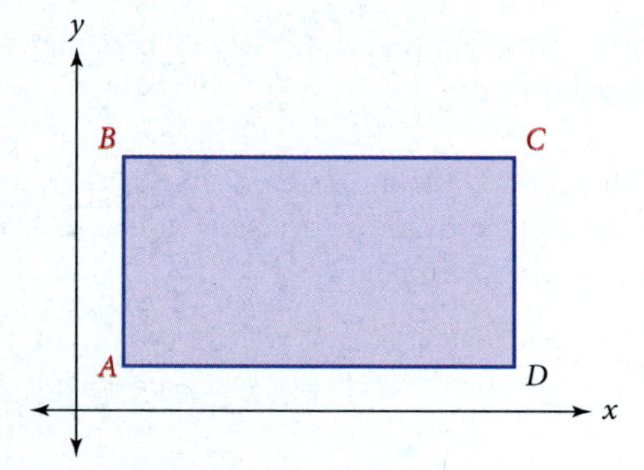

50. Rectangle $ABCD$ has a length of 5 and a width of 3. Point D is the ordered pair (−1, 1). Find points A, B, and C.

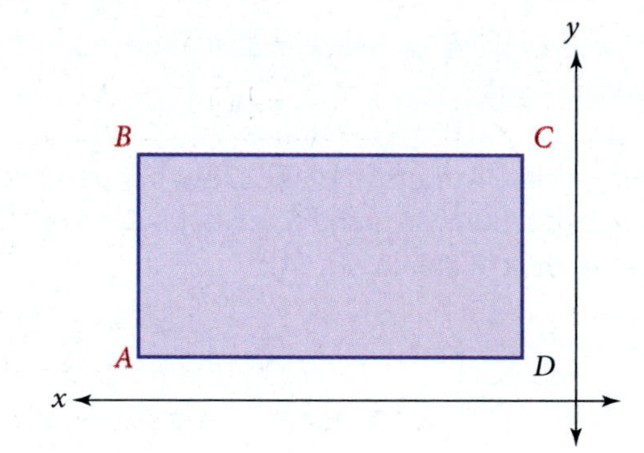

4.9 Graphing Straight Lines

Objectives

A Graph a line given a linear equation in two variables.

B Graph a vertical or horizontal line.

In this section we use what we have learned in the previous two sections to graph straight lines.

A Graphing a Linear Equation in Two Variables

Examples now playing at MathTV.com/books

EXAMPLE 1 Graph the solution set for $x + y = 4$.

SOLUTION We know from our work in the previous section that there are an infinite number of solutions to this equation, and that each solution is an ordered pair of numbers. Some of the ordered pairs that satisfy the equation $x + y = 4$ are $(0, 4)$, $(2, 2)$, $(4, 0)$, and $(5, -1)$. If we plot these ordered pairs, we have the points shown in Figure 1.

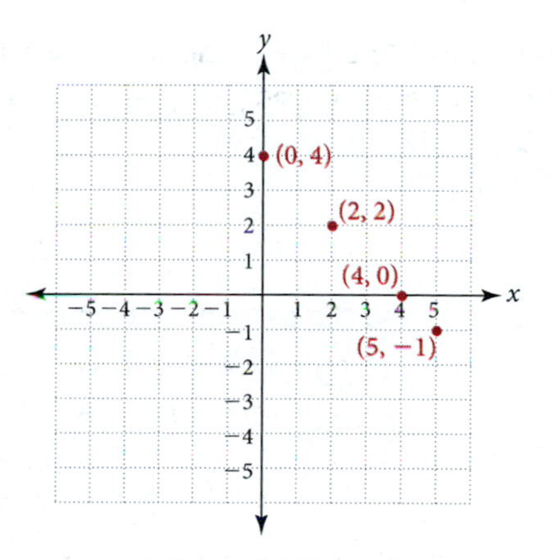

FIGURE 1 The graph of four solutions to $x + y = 4$

Notice that all four points lie in a straight line. If we were to find other solutions to the equation $x + y = 4$, we would find that they too would line up with the points shown in Figure 1. In fact, every solution to $x + y = 4$ is a point that lies in line with the points shown in Figure 1. Therefore, to graph the solution set to $x + y = 4$, we simply draw a line through the points in Figure 1, to obtain the graph shown in Figure 2.

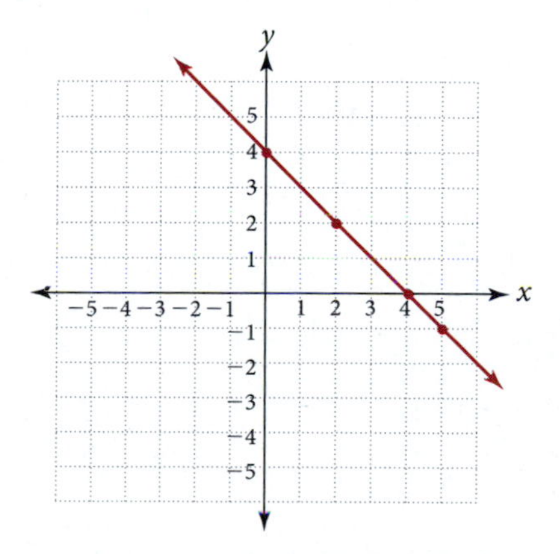

FIGURE 2 The graph of the line $x + y = 4$

PRACTICE PROBLEMS

1. Graph the solution set for $x + y = 3$.

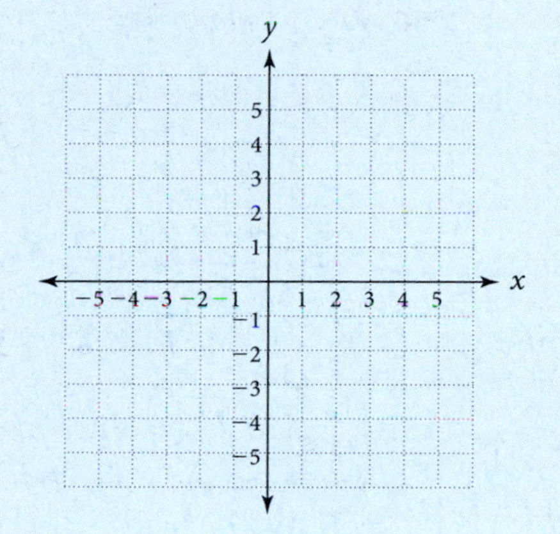

Note In Figure 2, every ordered pair that satisfies $x + y = 4$ has its graph on the line, and any point on the line has coordinates that satisfy the equation $x + y = 4$.

Answers

For answers to the Practice Problems in this section, see the solutions section.

Strategy Graphing a Straight Line

Step 1: Find any three ordered pairs that satisfy the equation. This is usually accomplished by substituting a number for one of the variables in the equation, and then solving for the other variable.

Step 2: Plot the three ordered pairs found in Step 1. (Actually, we need only two points to graph a straight line. The third point serves as a check. If all three points don't line up, then we have made a mistake in our work.)

Step 3: Draw a line through the three points you plotted in Step 2.

2. Graph the equation $y = -2x + 1$ by completing the ordered pairs $(1, \ )$, $(0, \ )$, and $(-1, \ )$.

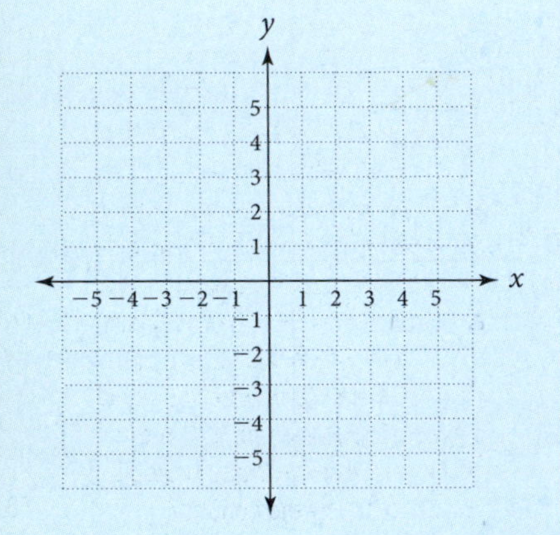

EXAMPLE 2 Graph the equation $y = 2x + 1$ by completing the ordered pairs $(1, \ )$, $(0, \ )$, and $(-1, \ )$.

SOLUTION To complete the ordered pair $(1, \ )$, we let $x = 1$ in the equation $y = 2x + 1$:

$$y = 2 \cdot 1 + 1$$
$$y = 2 + 1$$
$$y = 3$$

To complete the ordered pair $(0, \ )$, we let $x = 0$ in the equation:

$$y = 2 \cdot 0 + 1$$
$$y = 0 + 1$$
$$y = 1$$

To complete the ordered pair $(-1, \ )$, we let $x = -1$ in the equation:

$$y = 2(-1) + 1$$
$$y = -2 + 1$$
$$y = -1$$

The ordered pairs $(1, 3)$, $(0, 1)$, and $(-1, -1)$ each satisfy the equation $y = 2x + 1$. Graphing each ordered pair and then drawing a line through them, we have the graph shown in Figure 3.

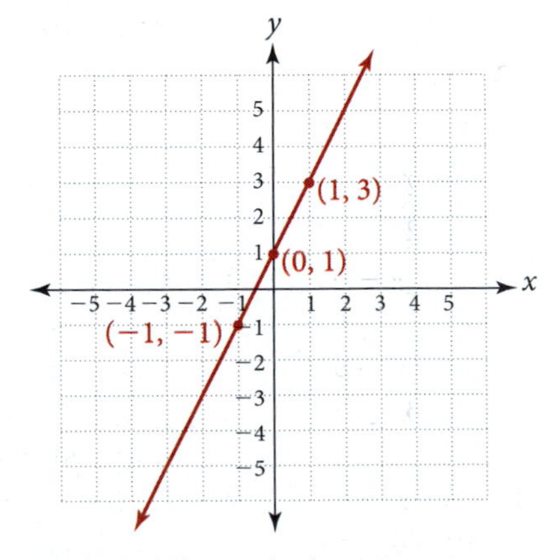

FIGURE 3 The graph of $y = 2x + 1$

EXAMPLE 3

Graph the equation $y = 3x - 2$.

SOLUTION We need to find three solutions to the equation. We can do so by choosing some numbers to use for x and then seeing what values of y they yield. Since the choice of the numbers we will use for x is up to us, let's make it easy on ourselves and use numbers like $x = -1$, 0, and 2, that are easy to work with.

$$\begin{aligned}
\text{Let } x = -1: \quad & y = 3(-1) - 2 \\
& y = -3 - 2 \\
& y = -5 \qquad (-1, -5) \text{ is one solution.} \\
\text{Let } x = 0: \quad & y = 3 \cdot 0 - 2 \\
& y = 0 - 2 \\
& y = -2 \qquad (0, -2) \text{ is another solution.} \\
\text{Let } x = 2: \quad & y = 3 \cdot 2 - 2 \\
& y = 6 - 2 \\
& y = 4 \qquad (2, 4) \text{ is a third solution.}
\end{aligned}$$

Now we plot the solutions we found above and then draw a line through them.

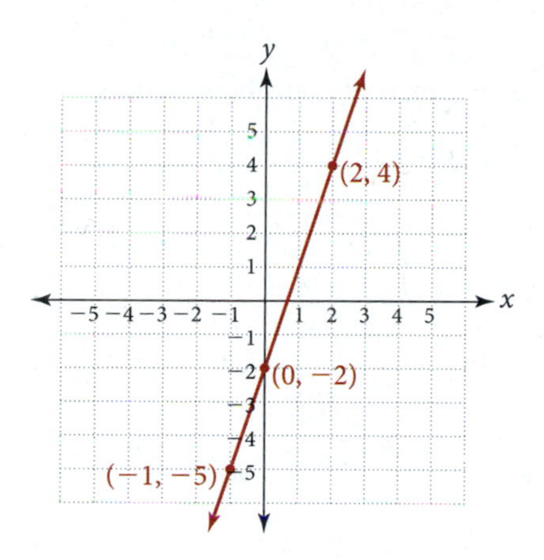

FIGURE 4 The graph of $y = 3x - 2$

3. Graph the equation $y = 2x - 3$.

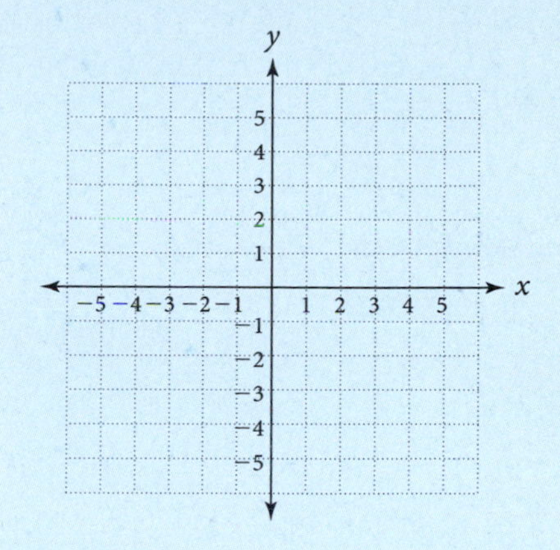

Note Every point on the line shown in Figure 4 has coordinates that satisfy the equation $y = 3x - 2$. Likewise, every ordered pair that satisfies the equation $y = 3x - 2$ has its graph on the line shown in Figure 4. We say that there is a one-to-one correspondence between points on the graph of $y = 3x - 2$ and ordered pairs that satisfy the equation $y = 3x - 2$.

EXAMPLE 4

Graph the line $3x + 2y = 6$.

SOLUTION This time, let's substitute values for both x and y to find the three solutions we need to draw the graph.

$$\begin{aligned}
\text{Let } x = 0: \quad & 3 \cdot 0 + 2y = 6 \\
& 0 + 2y = 6 \\
& 2y = 6 \\
& y = 3 \qquad (0, 3) \text{ is one solution.} \\
\text{Let } y = 0: \quad & 3x + 2 \cdot 0 = 6 \\
& 3x + 0 = 6 \\
& 3x = 6 \\
& x = 2 \qquad (2, 0) \text{ is a second solution.} \\
\text{Let } y = -3: \quad & 3x + 2(-3) = 6 \\
& 3x - 6 = 6 \\
& 3x = 12 \\
& x = 4 \qquad (4, -3) \text{ is a third solution.}
\end{aligned}$$

Plotting these three solutions and drawing a line through them, we have the graph shown in Figure 5.

4. Graph the line $3x - 2y = 6$.

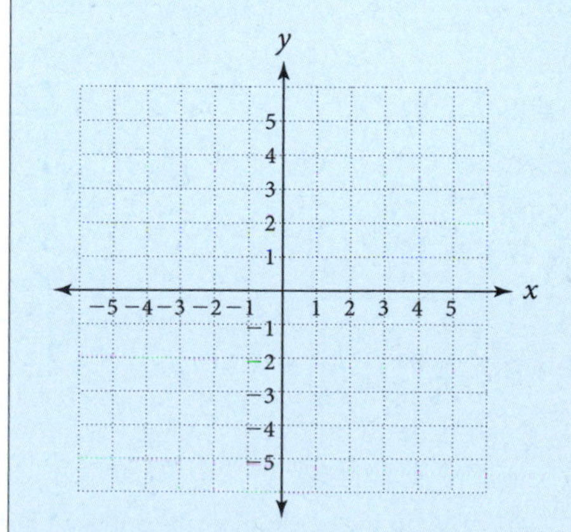

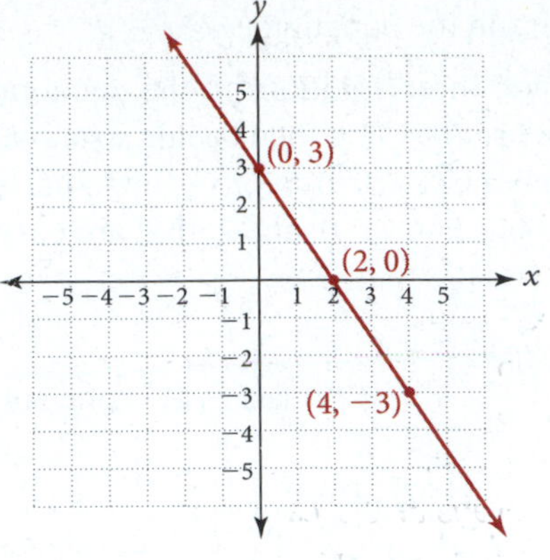

FIGURE 5 The graph of $3x + 2y = 6$

B Horizontal and Vertical Lines

EXAMPLE 5 Graph each of the following lines.

a. $y = \frac{1}{2}x$ **b.** $x = 3$ **c.** $y = -2$

SOLUTION **a.** The line $y = \frac{1}{2}x$ passes through the origin because (0, 0) satisfies the equation. To sketch the graph we need at least one more point on the line. When x is 2, we obtain the point (2, 1), and when x is -4, we obtain the point $(-4, -2)$. The graph of $y = \frac{1}{2}x$ is shown in Figure 6A.

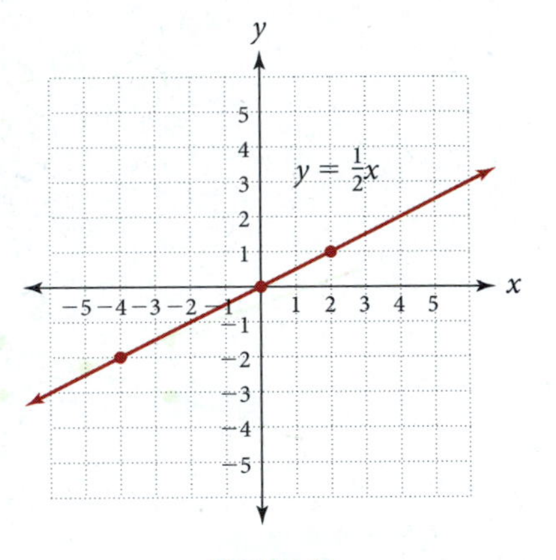

FIGURE 6A

b. The line $x = 3$ is the set of all points whose x-coordinate is 3. The variable y does not appear in the equation, so the y-coordinate can be any number. Note that we can write our equation as a linear equation in two variables by writing it as $x + 0y = 3$. Because the product of 0 and y will always be 0, y can be any number. The graph of $x = 3$ is the vertical line shown in Figure 6B.

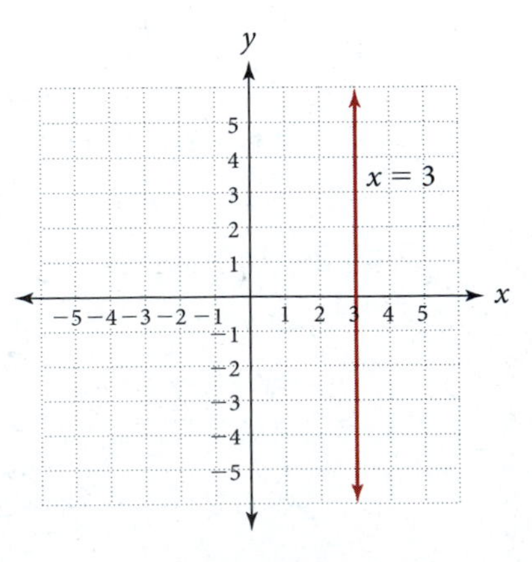

FIGURE 6B

5. **a.** Graph the line $y = 2x$.

b. Graph the line $x = -3$.

c. The line $y = -2$ is the set of all points whose y-coordinate is -2. The variable x does not appear in the equation, so the x-coordinate can be any number. Again, we can write our equation as a linear equation in two variables by writing it as $0x + y = -2$. Because the product of 0 and x will always be 0, x can be any number. The graph of $y = -2$ is the horizontal line shown in Figure 6C.

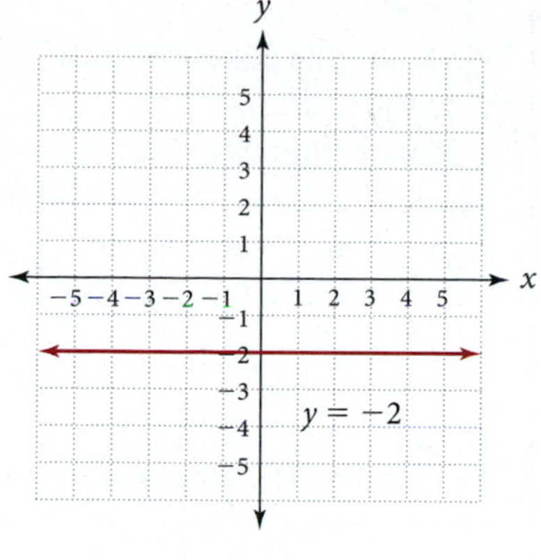

FIGURE 6C

c. Graph the line $y = 2$.

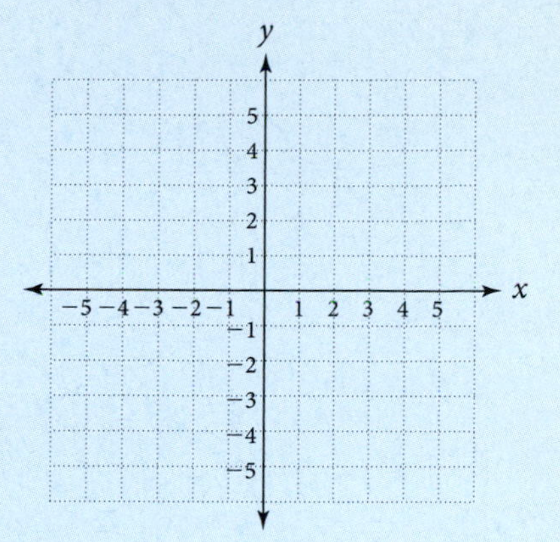

FACTS FROM GEOMETRY Special Equations and Their Graphs

For the equations below, m, a, and b are real numbers.

Through the Origin

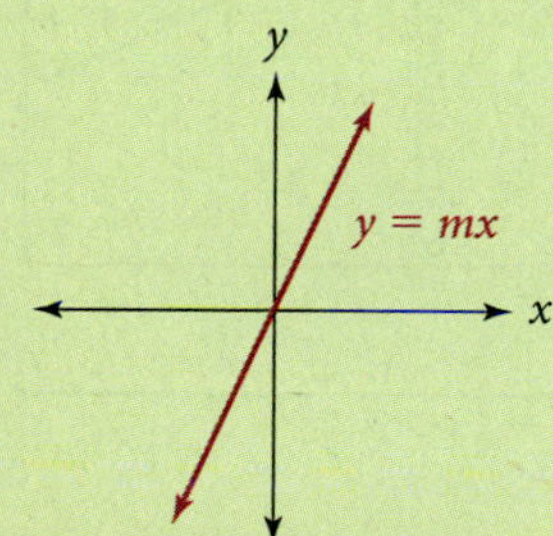

FIGURE 7A Any equation of the form $y = mx$ has a graph that passes through the origin.

Vertical Line

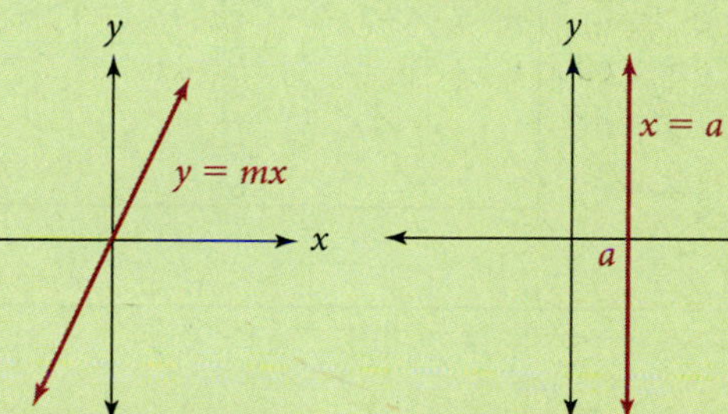

FIGURE 7B Any equation of the form $x = a$ has a vertical line for its graph.

Horizontal Line

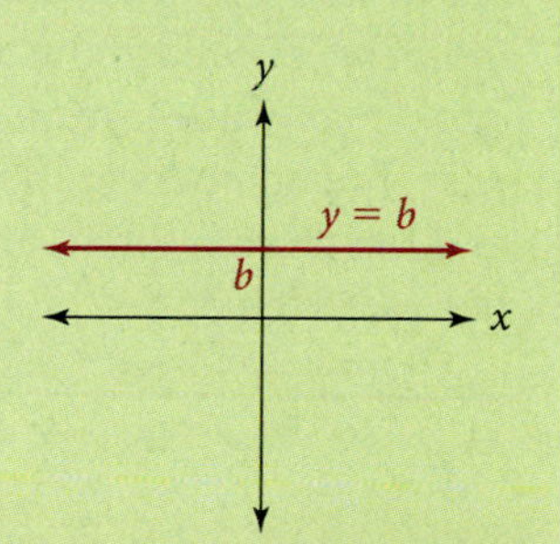

FIGURE 7C Any equation of the form $y = b$ has a horizontal line for its graph.

USING TECHNOLOGY

Graphing a Line

To graph the equation $2x + 3y = 6$ on a graphing calculator, we must first solve for y:

$$2x + 3y = 6$$

$$3y = -2x + 6 \quad \text{Add } -2x \text{ to both sides}$$

$$y = -\frac{2}{3}x + 2 \quad \text{Divide each term by 3}$$

Continued

We enter this equation in the calculator as Y_1=-(2/3)X+2

Notice that we use the negative sign (-), rather than the subtraction sign, −, and that we use the parentheses around the fraction, as $-\frac{2}{3} = -(2/3)$.

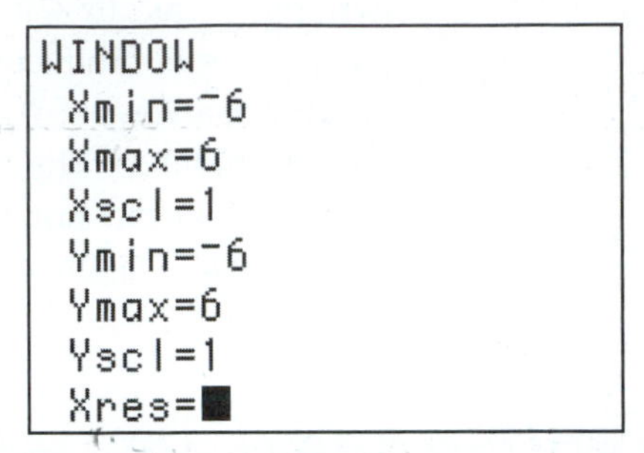

To match the other graphs we have shown in this section, we can set the window to show the x-axis from −6 to 6 and the y-axis from −6 to 6. On a graphing calculator, this means Xmin = −6, Xmax = 6, Ymin = −6 and Ymax = 6. (It is generally OK to leave the Xscl and Yscl set to 1.)

WINDOW
Xmin=⁻6
Xmax=6
Xscl=1
Ymin=⁻6
Ymax=6
Yscl=1
Xres=■

Even though your calculator will not show the numbers along the axis we do so to illustrate a point about our window. Notice the numbers at the very left and right are the Xmin and Xmax values, respectively. Likewise, the numbers at the very bottom and top are the Ymin, and Ymax values we entered.

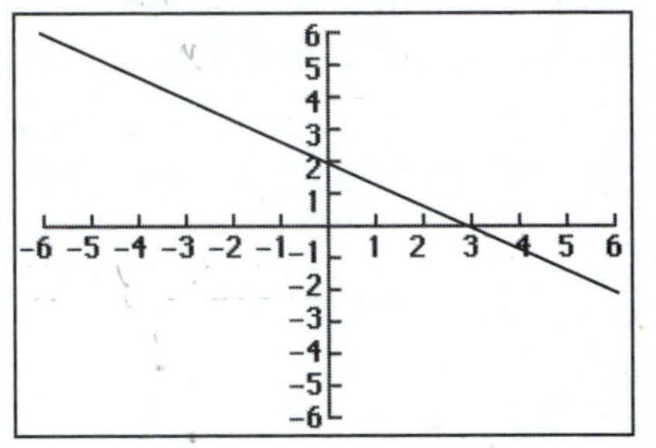

STUDY SKILLS

Pay Attention to Instructions

Each of the following is a valid instruction with respect to the equation $y = 3x - 2$, and the result of applying the instructions will be different in each case:

- Find x when y is 10.
- Solve for x.
- Graph the equation.

There are many things to do with the equation $y = 3x - 2$. If you train yourself to pay attention to the instructions that accompany a problem as you work through the assigned problems, you will not find yourself confused about what to do with a problem when you see it on a test.

Getting Ready for Class

After reading the preceding section and working the practice problems in the margin, answer the following questions.

1. Explain how you would go about graphing the line $x + y = 4$.
2. When graphing straight lines, why is it a good idea to find three points, when every straight line is determined by only two points?
3. What kind of equations have vertical lines for graphs?
4. What kind of equations have horizontal lines for graphs?

Problem Set 4.9

A For the equations in Problems 1-16, complete the given ordered pairs, and use the results to graph the equation. [Examples 1–5]

1. $x + y = 2$ $(0, 2)$, $(2, 0)$, $(-1, 3)$

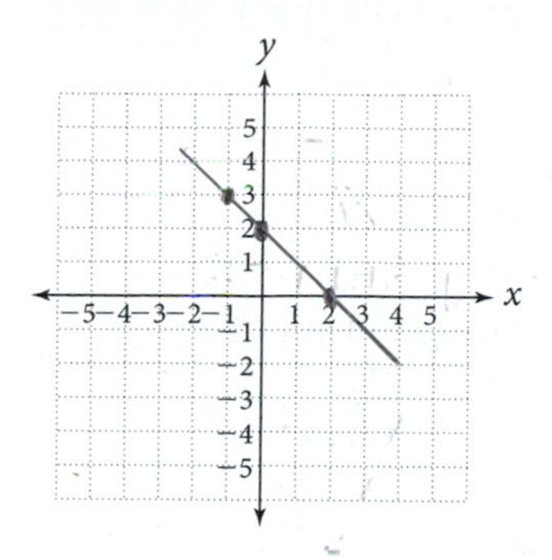

2. $x - y = 2$ $(1, -1)$, $(2, 0)$, $(4, 2)$

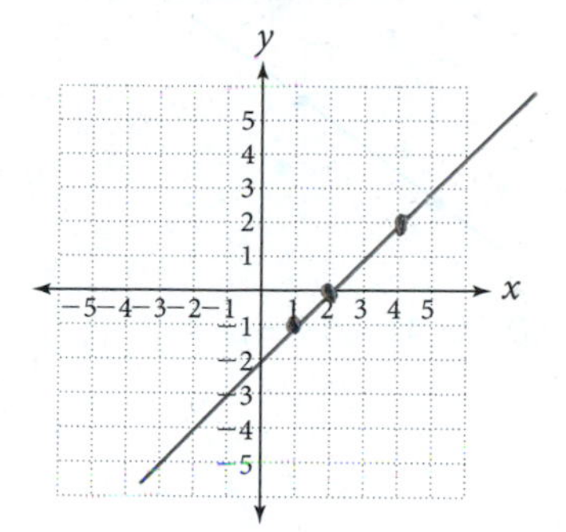

3. $y = 2x - 4$ $(0, -4)$, $(1, -2)$, $(2, 0)$

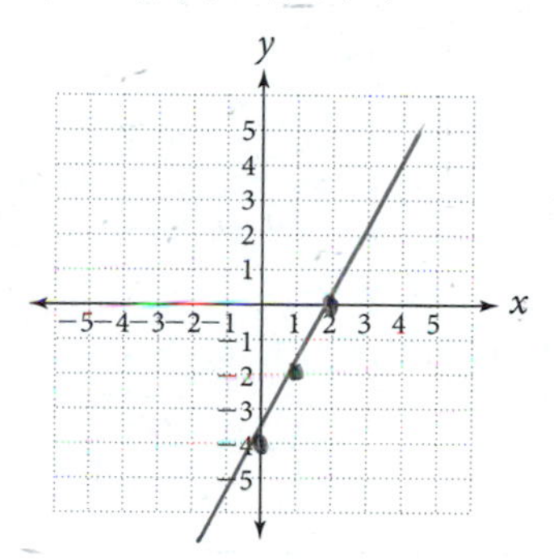

4. $y = -2x + 4$ $(0, 4)$, $(1, 2)$, $(2, 0)$

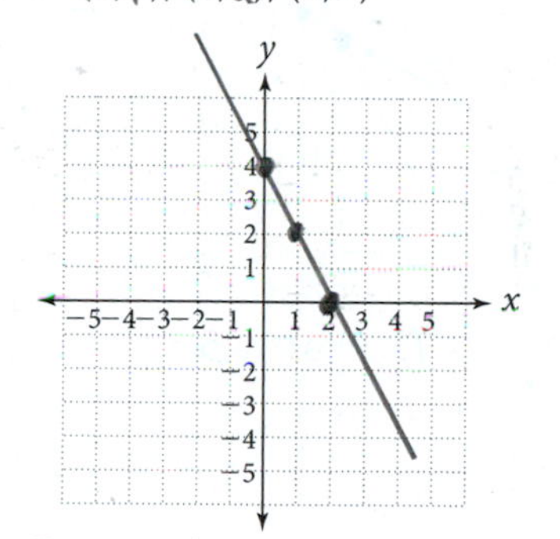

5. $3x + y = 3$ $(0, 3)$, $(1, 0)$, $(3, -6)$

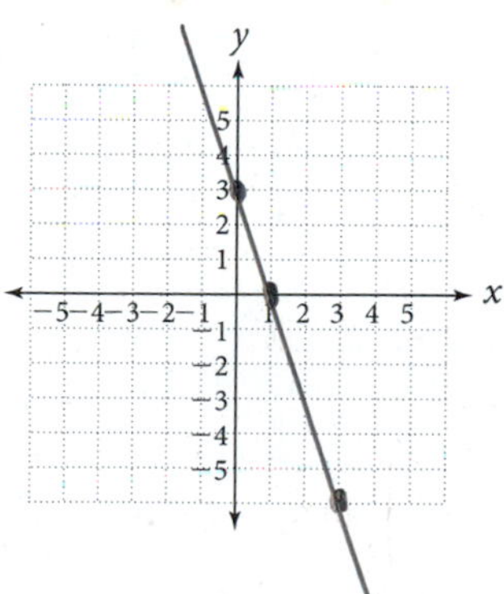

6. $x + 3y = 3$ $(0, 1)$, $(3, 0)$, $(-3, 2)$

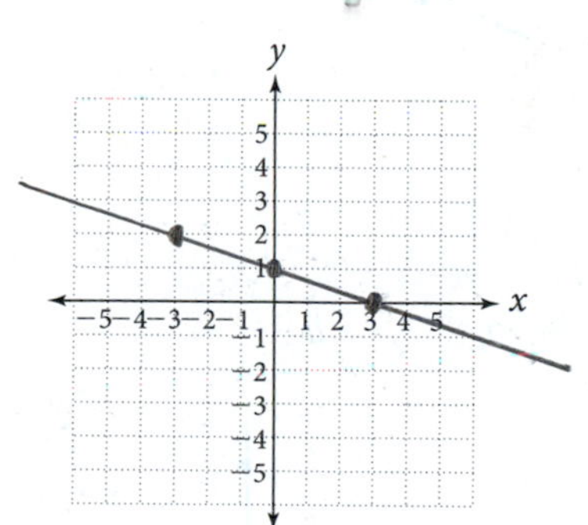

7. $3x + 4y = 12$ $(0, 3)$, $(4, 0)$, $(-4, 6)$

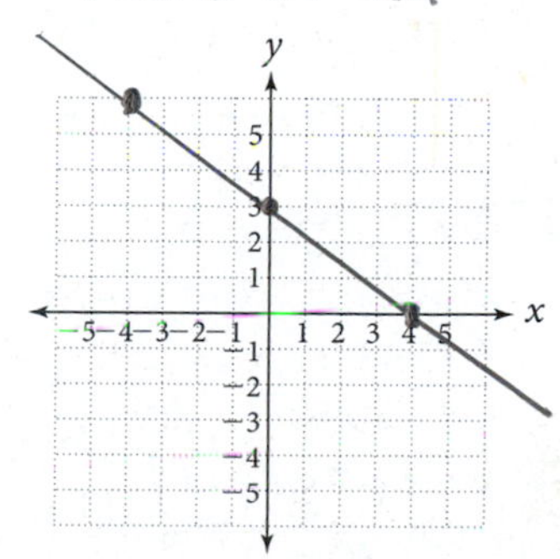

8. $4x + 3y = 12$ $(0, 4)$, $(3, 0)$, $(6, -4)$

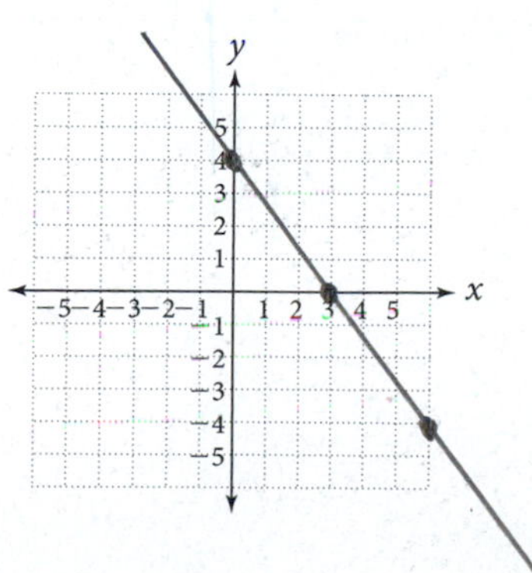

9. $3x - 4y = 12$ (0, -3), (4, 0), (−4, -6)

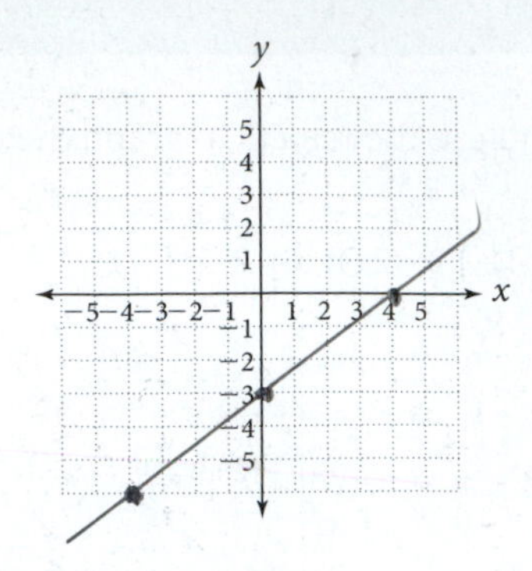

10. $4x - 3y = 12$ (0, -4), (3, 0), (6, 4)

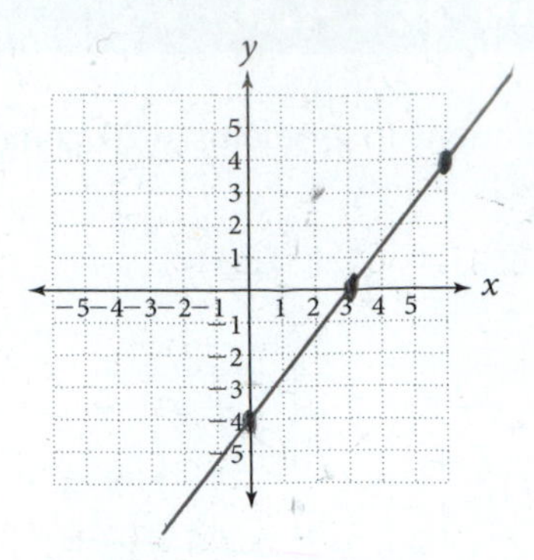

11. $y = \frac{1}{2}x$ (0, 0), (2, 1), (−2, -1)

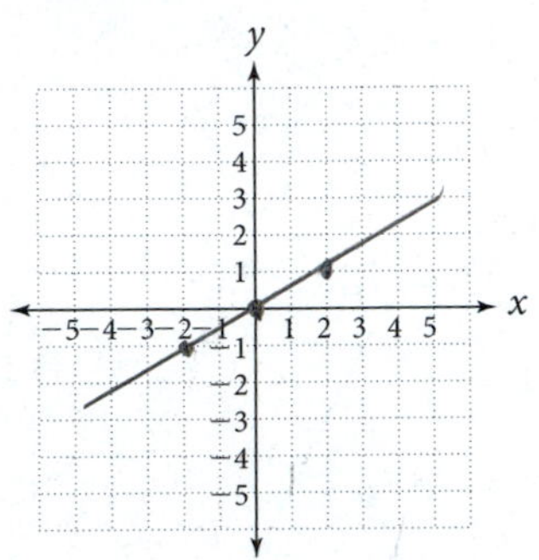

12. $y = -\frac{1}{2}x$ (0, 0), (2, -1), (−2, 1)

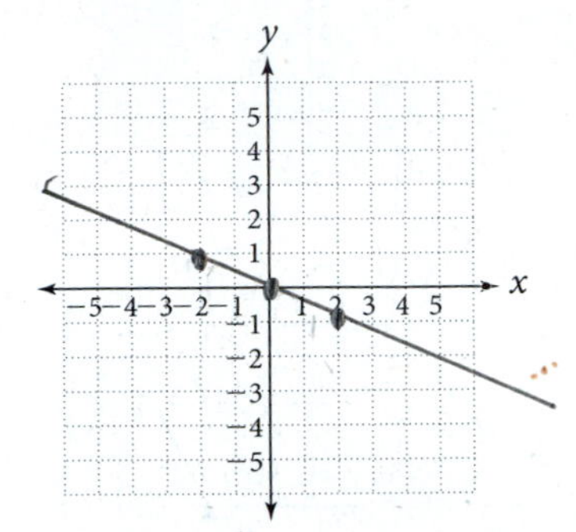

13. $y = \frac{1}{2}x + 2$ (−2, 1), (0, 2), (2, 3)

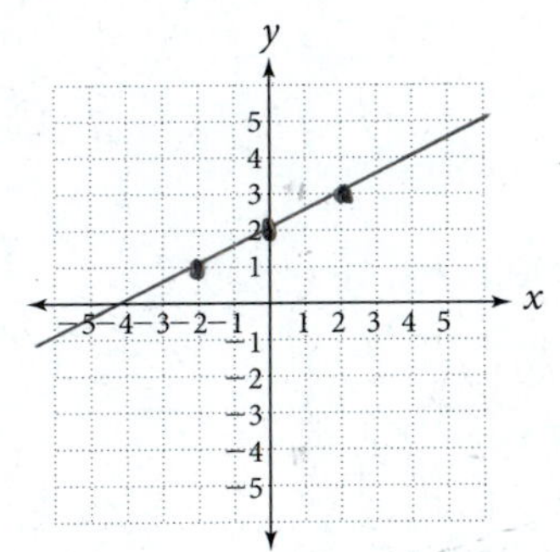

14. $y = \frac{1}{3}x + 1$ (−3, 0), (0, 1), (3, 2)

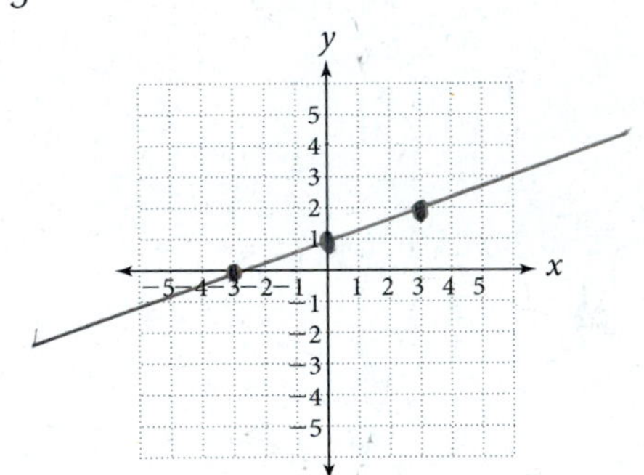

B [Example 5]

15. $x = 2$ (2, 0), (2, 3), (2, 6)

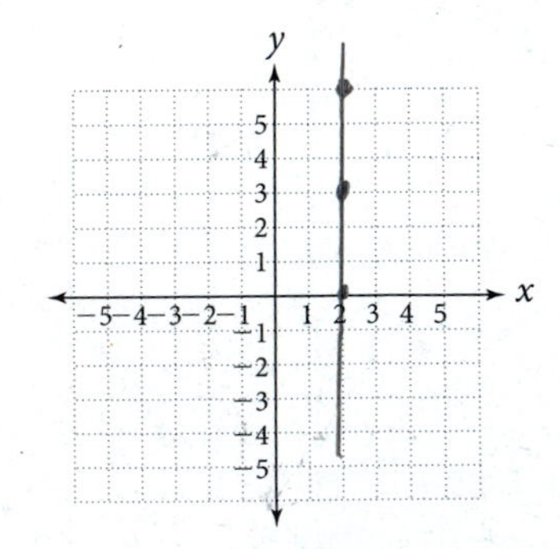

16. $y = 2$ (0, 2), (3, 2), (6, 2)

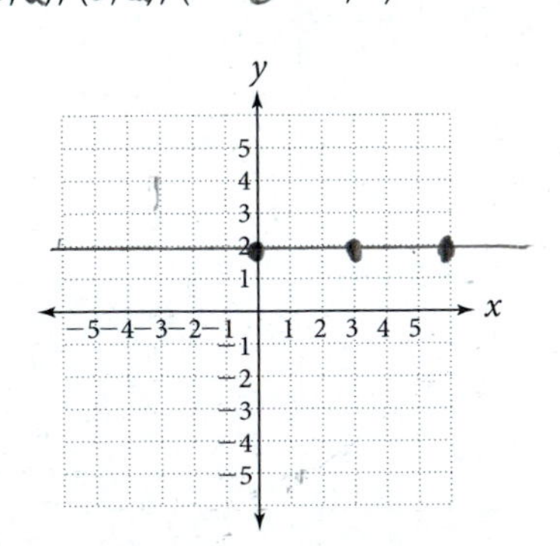

A Find three solutions to each of the equations in Problems 17–34, and use them to draw the graph. [Examples 1–5]

17. $x + y = 5$

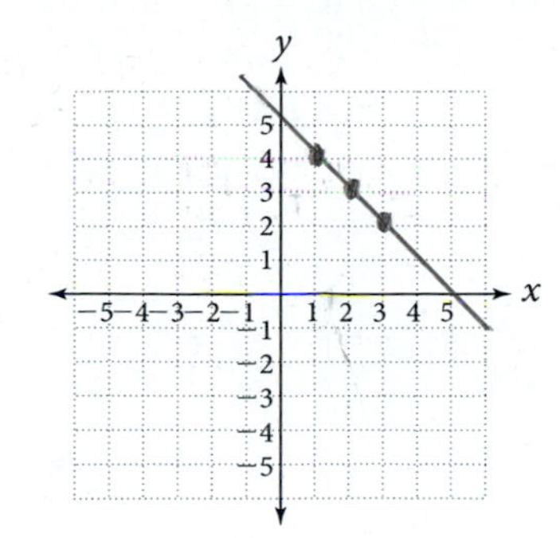

18. $x - y = 4$

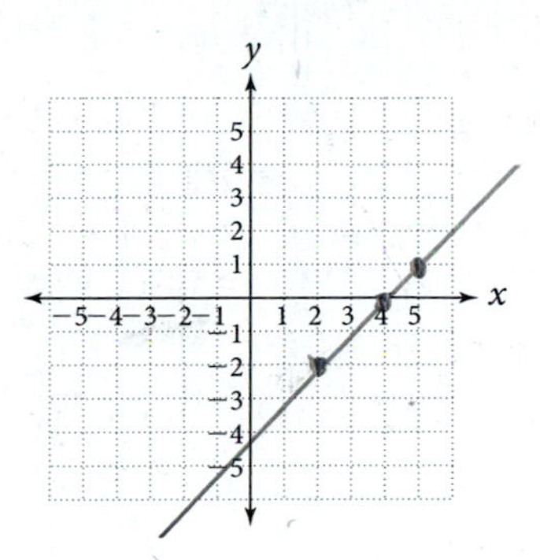

19. $x - y = 5$

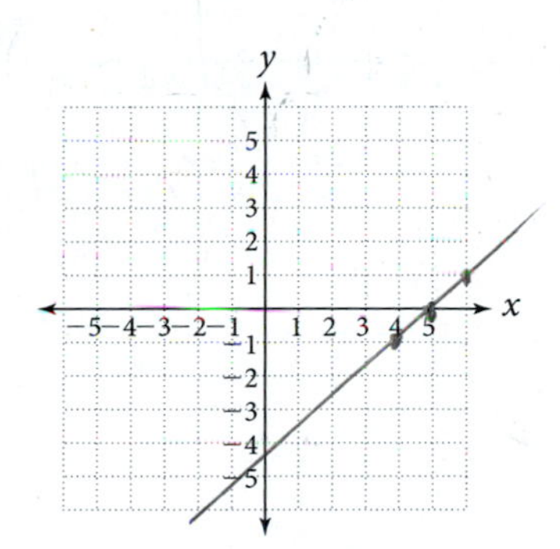

20. $x + y = 3$

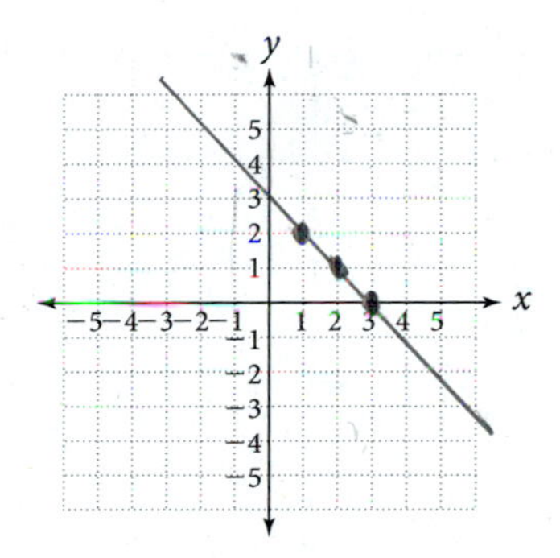

21. $2x - 4y = 4$

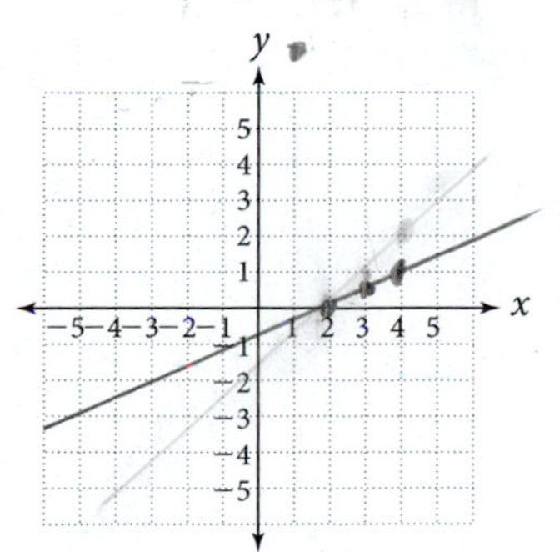

22. $4x + 2y = 4$

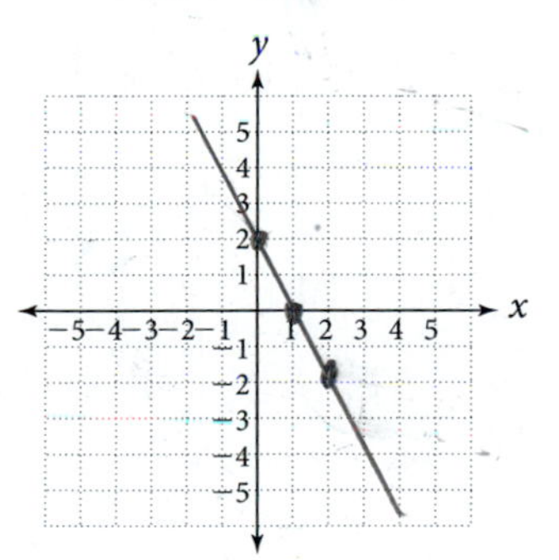

23. $2x + 4y = 4$

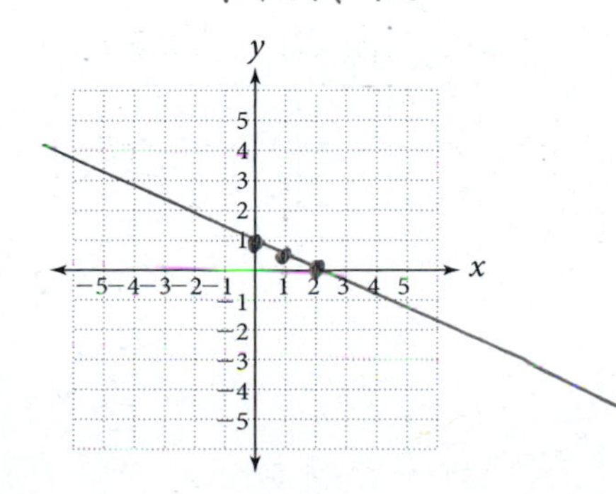

24. $4x - 2y = 4$

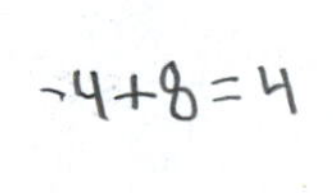

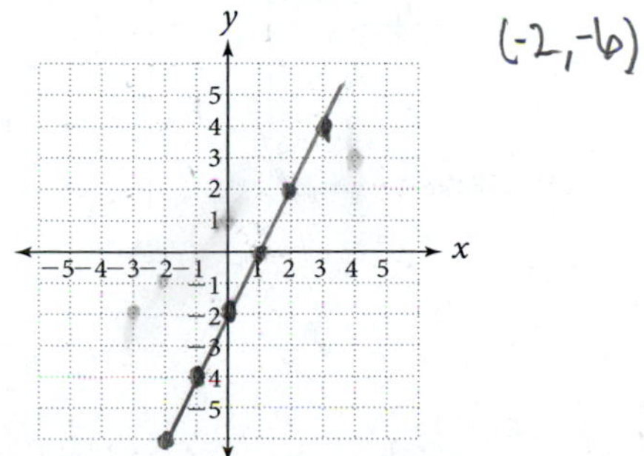

25. $y = 2x + 3$

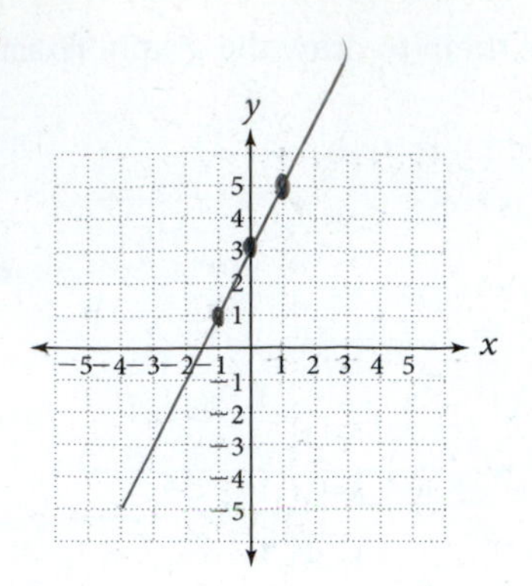

26. $y = 3x - 4$

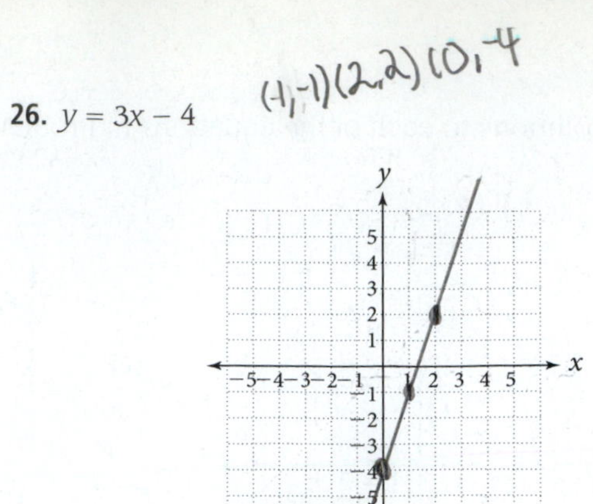

27. $y = 2x$

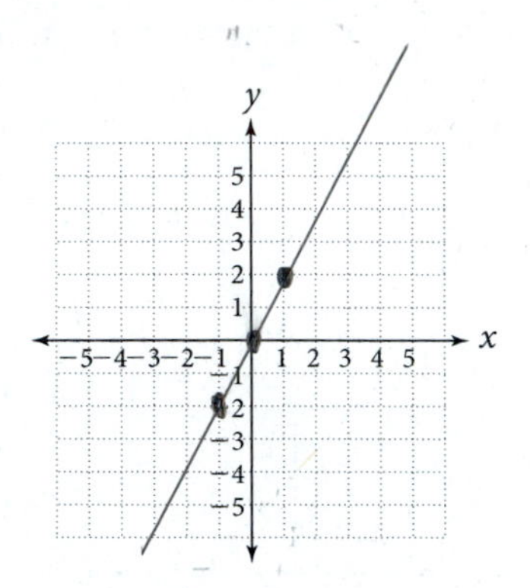

28. $y = -2x$ (-1,2)(0,0)(1,-2)

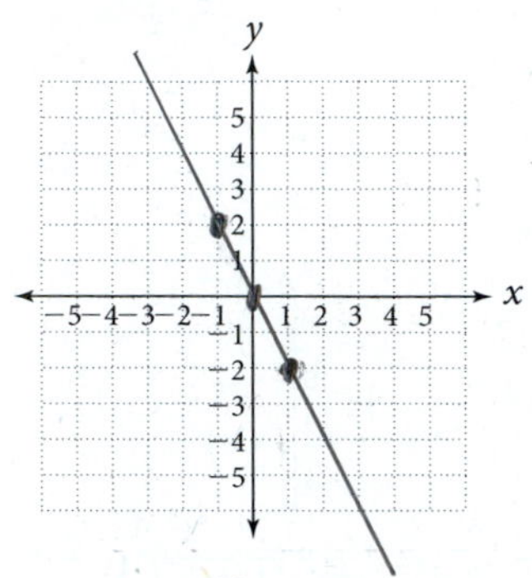

29. $y = \frac{1}{3}x$

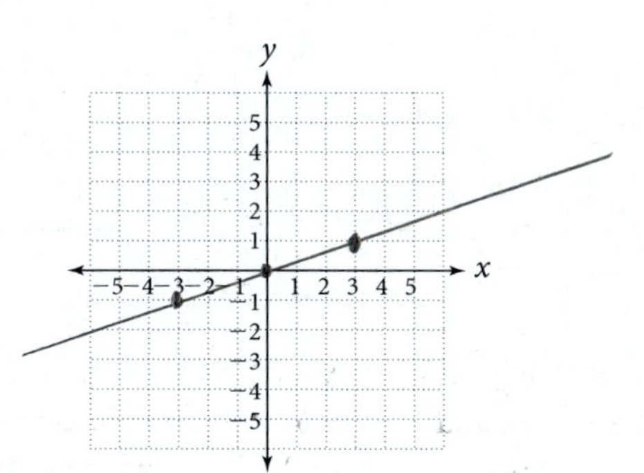

30. $y = -\frac{1}{3}x$ (3,-1)(0,0)(-3,1)

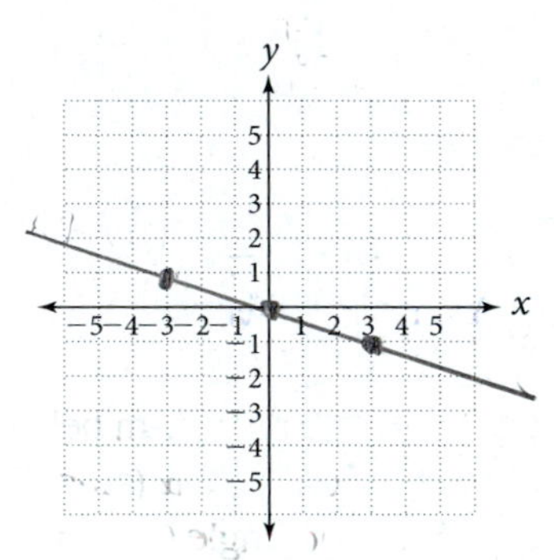

B [Example 5]

31. $x = 4$

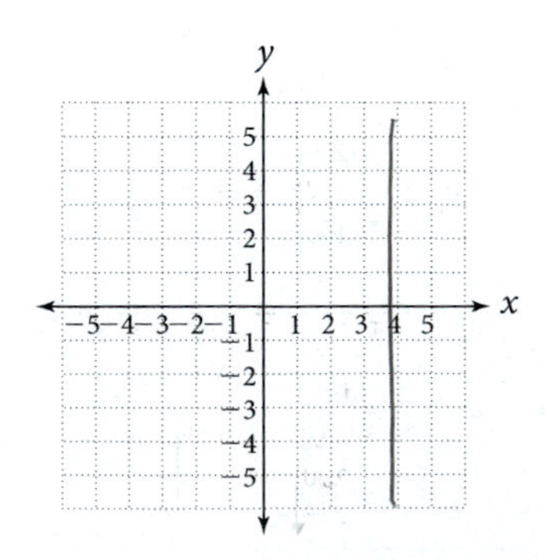

32. $x = -3$

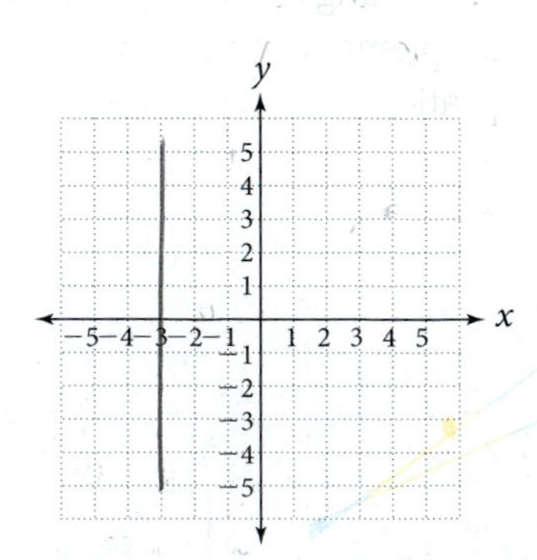

33. $y = -2$

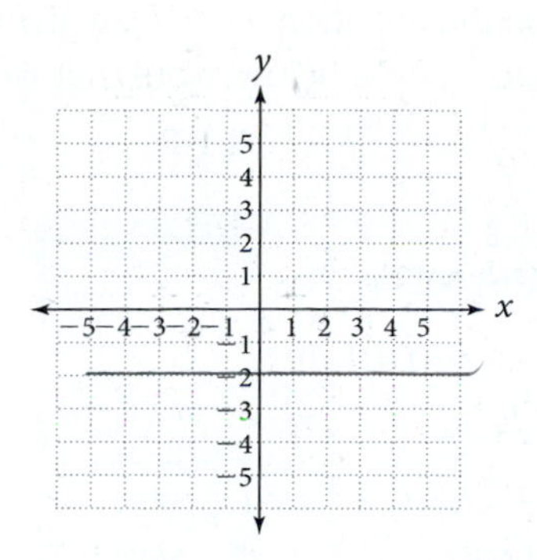

34. $y = 3$

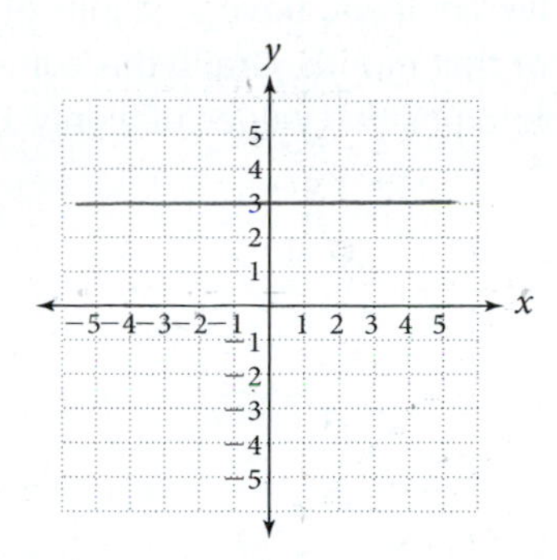

35. The graph shown here is the graph of which of the following equations?

a. $3x - 2y = 6$

b. $2x - 3y = 6$

c. $2x + 3y = 6$

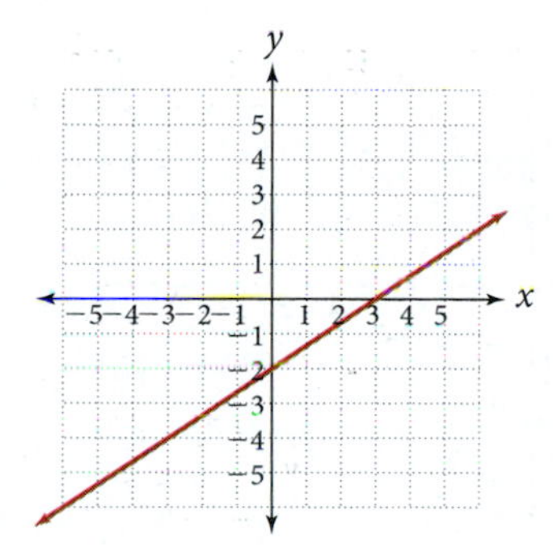

36. The graph shown here is the graph of which of the following equations?

a. $3x - 2y = 8$

b. $2x - 3y = 8$

c. $2x + 3y = 8$

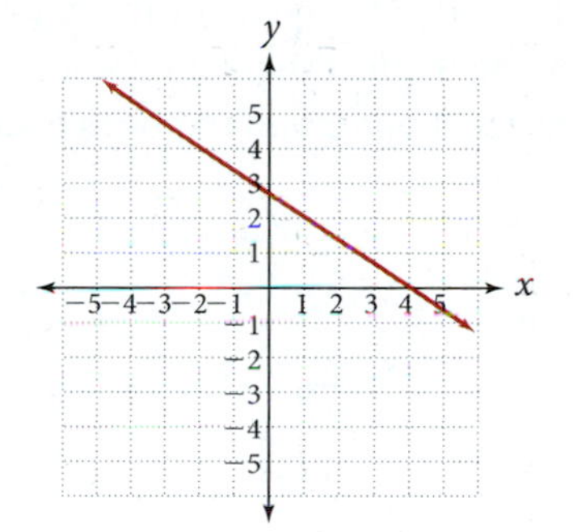

Applying the Concepts

37. Complementary Angles The diagram below shows sunlight hitting the ground. Angle α (alpha) is called the angle of inclination, and angle θ (theta) is called the angle of incidence. As the sun moves across the sky, the values of these angles change, but the two angles are always complementary, meaning that $\alpha + \theta = 90°$. Graph this equation on the template below. (The graph is restricted to the first quadrant so that we don't have to define negative angles.)

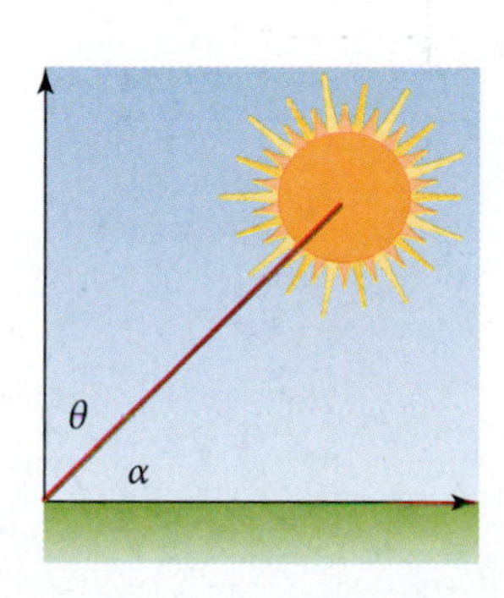

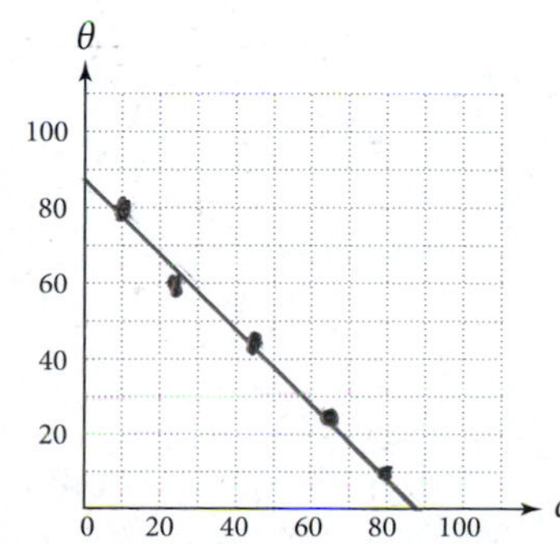

38. Temperature Scales The relationship between the Fahrenheit and Celsius temperature scales is given by the equation $F = \frac{9}{5}C + 32$. Graph this equation on the template below.

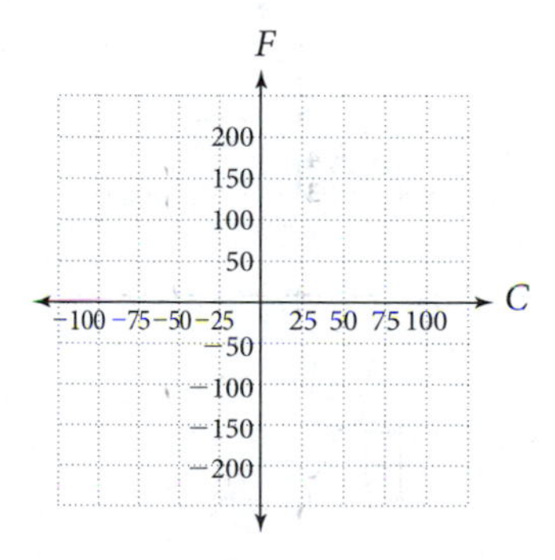

39. Cost of Bottled Water A water bottling company charges \$7 per month for their water dispenser and \$2 for each gallon of water delivered. If you have g gallons of water delivered in a month, then the equation $y = 7 + 2g$ gives the amount of your bill for that month. Graph this equation on the rectangular coordinate system below. Note that we are restricting our graph to positive values of g only. Do you know why?

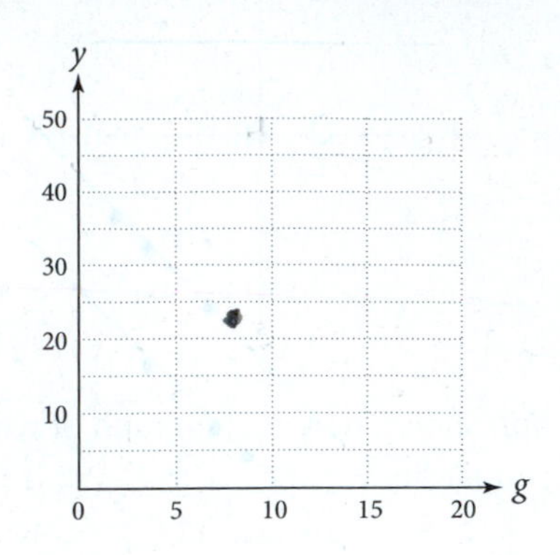

WATER BOTTLE CO. (WBC)

MONTHLY BILL

234 5th Street
Glendora, CA 91740

DUE 07/23/08

Water dispenser	1	\$7.00
Gallons of water	8	\$2.00
		\$23.00

40. Temperature and Altitude On a certain day, the temperature on the ground is 72° Fahrenheit, and the temperature T at an altitude of A feet above the ground is given by the equation $T = 72 - \frac{1}{300}A$. Graph this equation on the rectangular coordinate system here. Note that we have restricted our graph to positive values of A only.

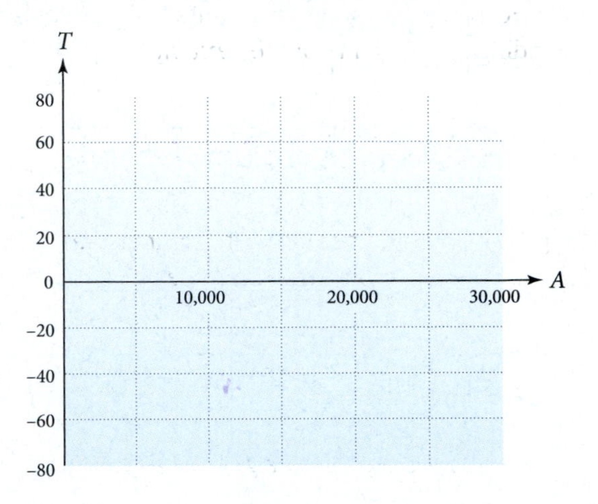

Maintaining Your Skills

Simplify each expression.

41. $\frac{3}{5}\left(2\frac{1}{5} - 1\frac{1}{10}\right)$

42. $\frac{7}{8}\left(5\frac{3}{4} - 2\frac{1}{2}\right)$

43. $\left(\frac{3}{5} + \frac{1}{3}\right)\left(\frac{3}{5} - \frac{1}{3}\right)$

44. $\left(1 + \frac{3}{8}\right)\left(1 - \frac{3}{8}\right)$

45. $\left(\frac{7}{15} - \frac{11}{30}\right)^2$

46. $\left(\frac{13}{21} - \frac{13}{35}\right)^2$

Chapter 4 Summary

Combining Similar Terms [4.1]

Two terms are similar terms if they have the same variable part. The expressions $7x$ and $2x$ are similar because the variable part in each is the same. Similar terms are combined by using the distributive property.

EXAMPLES

1. $$7x + 2x = (7 + 2)x$$
$$= 9x$$

Finding the Value of an Algebraic Expression [4.1]

An algebraic expression is a mathematical expression that contains numbers and variables. Expressions that contain a variable will take on different values depending on the value of the variable.

2. When $x = 5$, the expression $2x + 7$ becomes $2(5) + 7 = 10 + 7 = 17$

The Solution to an Equation [4.2]

A solution to an equation is a number that, when used in place of the variable, makes the equation a true statement.

The Addition Property of Equality [4.2]

Let A, B, and C represent algebraic expressions.

If $\quad A = B$
then $\quad A + C = B + C$

In words: Adding the same quantity to both sides of an equation will not change the solution.

3. We solve $x - 4 = 9$ by adding 4 to each side.
$$x - 4 = 9$$
$$x - 4 + \mathbf{4} = 9 + \mathbf{4}$$
$$x + 0 = 13$$
$$x = 13$$

The Multiplication Property of Equality [4.3]

Let A, B, and C represent algebraic expressions, with C not equal to 0.

If $\quad A = B$
then $\quad AC = BC$

In words: Multiplying both sides of an equation by the same nonzero number will not change the solution to the equation. This property holds for division as well.

4. Solve $\frac{1}{3}x = 5$.
$$\frac{1}{3}x = 5$$
$$\mathbf{3} \cdot \frac{1}{3}x = \mathbf{3} \cdot 5$$
$$x = 15$$

Steps Used to Solve a Linear Equation in One Variable [4.4]

Step 1 Simplify each side of the equation.

Step 2 Use the addition property of equality to get all variable terms on one side and all constant terms on the other side.

5. $$2(x - 4) + 5 = -11$$
$$2x - 8 + 5 = -11$$
$$2x - 3 = -11$$
$$2x - 3 + \mathbf{3} = -11 + \mathbf{3}$$
$$2x = -8$$
$$\frac{2x}{\mathbf{2}} = \frac{-8}{\mathbf{2}}$$
$$x = -4$$

Step 3 Use the multiplication property of equality to get just one x isolated on either side of the equation.

Step 4 Check the solution in the original equation if necessary.

If the original equation contains fractions, you can begin by multiplying each side by the LCD for all fractions in the equation.

Evaluating Formulas [4.6]

6. When $w = 8$ and $l = 13$ the formula $P = 2w + 2l$ becomes

$$P = 2 \cdot 8 + 2 \cdot 13 = 16 + 26 = 42$$

In mathematics, a formula is an equation that contains more than one variable. For example, the formula for the perimeter of a rectangle is $P = 2l + 2w$. We evaluate a formula by substituting values for all but one of the variables and then solving the resulting equation for that variable.

The Rectangular Coordinate System [4.8]

The rectangular coordinate system consists of two number lines, called axes, which intersect at right angles. The point at which the axes intersect is called the origin.

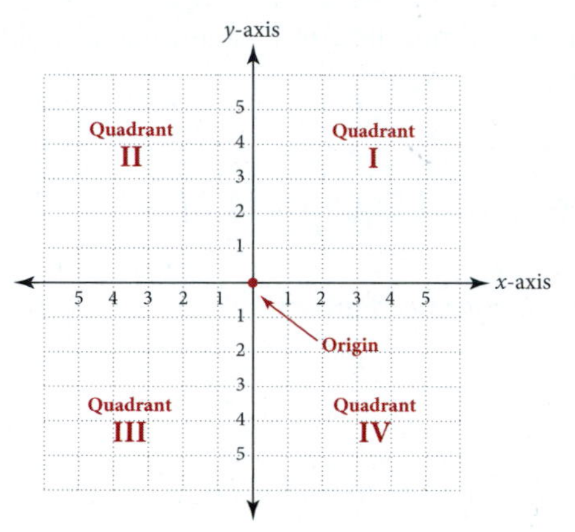

Graphing Ordered Pairs [4.8]

To graph an ordered pair (a, b) on the rectangular coordinate system, we start at the origin and move a units right or left (right if a is positive, left if a is negative). From there we move b units up or down (up if b is positive, down if b is negative). The point where we end up is the graph of the ordered pair (a, b).

To Graph a Straight Line [4.9]

7. The graph of $3x + 2y = 6$

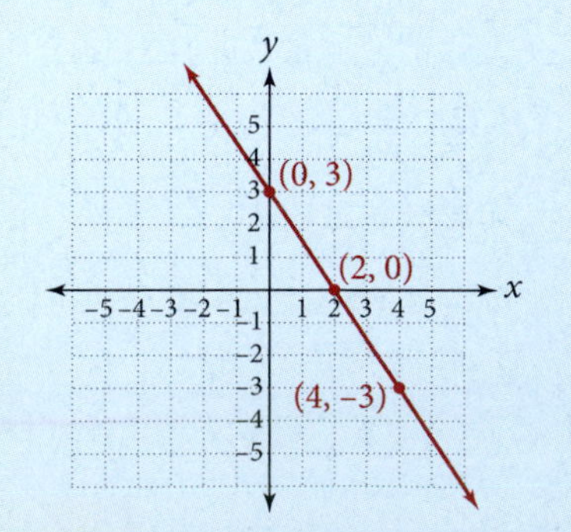

Step 1 Find any three ordered pairs that satisfy the equation. This is usually accomplished by substituting a number for one of the variables in the equation and then solving for the other variable.

Step 2 Plot the three ordered pairs found in Step 1. (Actually, we need only two points to graph a straight line. The third point serves as a check. If all three points don't line up, then we have made a mistake in our work.)

Step 3 Draw a line through the three points you plotted in Step 2.

Chapter 4 Review

Simplify the expressions by combining similar terms. [4.1]

1. $10x + 7x$

2. $8x - 12x$

3. $2a + 9a + 3 - 6$

4. $4y - 7y + 8 - 10$

5. $6x - x + 4$

6. $-5a + a + 4 - 3$

7. $2a - 6 + 8a + 2$

8. $12y - 4 + 3y - 9$

Find the value of each expression when x is 4. [4.1]

9. $10x + 2$

10. $5x - 12$

11. $-2x + 9$

12. $-x + 8$

13. Is $x = -3$ a solution to $5x - 2 = -17$? [4.2]

14. Is $x = 4$ a solution to $3x - 2 = 2x + 1$? [4.2]

Solve the equations. [4.2, 4.3, 4.4]

15. $x - 5 = 4$

16. $-x + 3 + 2x = 6 - 7$

17. $2x + 1 = 7$

18. $3x - 5 = 1$

19. $2x + 4 = 3x - 5$

20. $4x + 8 = 2x - 10$

21. $3(x - 2) = 9$

22. $4(x - 3) = -20$

23. $3(2x + 1) - 4 = -7$

24. $4(3x + 1) = -2(5x - 2)$

25. $5x + \frac{3}{8} = -\frac{1}{4}$

26. $\frac{7}{x} - \frac{2}{5} = 1$

27. Number Problem The sum of a number and -3 is -5. Find the number. [4.5]

28. Number Problem If twice a number is added to 3, the result is 7. Find the number. [4.5]

29. Number Problem Three times the sum of a number and 2 is -6. Find the number. [4.5]

30. Number Problem If 7 is subtracted from twice a number, the result is 5. Find the number. [4.5]

31. Geometry The length of a rectangle is twice its width. If the perimeter is 42 meters, find the length and the width. [4.5]

32. Age Problem Patrick is 3 years older than Amy. In 5 years the sum of their ages will be 31. How old are they now? [4.5]

In Problems 33-36, use the equation $3x + 2y = 6$ to find y. [4.6, 4.7]

33. $x = -2$

34. $x = 6$

35. $x = 0$

36. $x = \frac{1}{3}$

In Problems 37–39, use the equation $3x + 2y = 6$ to find x when y has the given value. [4.6, 4.7]

37. $y = 3$

38. $y = -3$

39. $y = 0$

40. Medical Costs The table below shows the average yearly cost of visits to the doctor as reported in 1999.

MEDICAL COSTS	
YEAR	AVERAGE ANNUAL COST
1990	$583
1995	$739
2000	$906
2005	$1,172

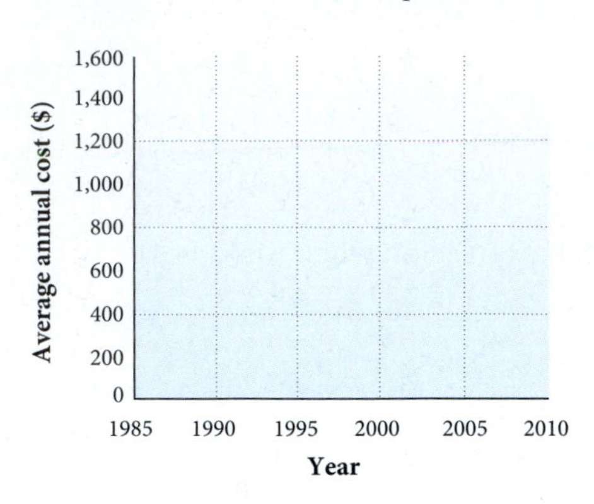

a. Use the template to construct a line graph of the information in the table.

b. From the line graph, estimate the annual cost of seeing a doctor in 2010.

In Problems 41–48, plot the points. [4.8]

41. (4, 2)

42. (4, −2)

43. (−4, 2)

44. (−4, −2)

45. (4, 0)

46. (0, 4)

47. (0, −4)

48. (−4, 0)

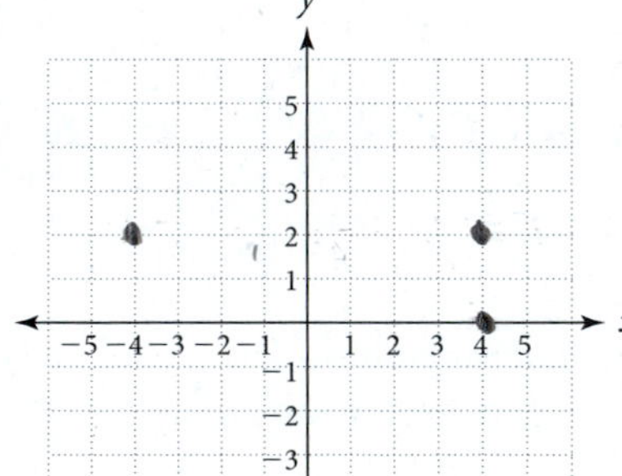
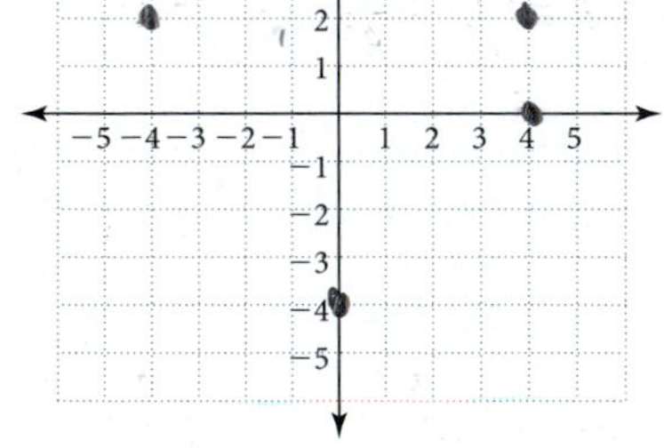

In Problems 49-58, graph each line. [4.9]

49. $x + y = 2$

50. $x - y = 2$

51. $3x + 2y = 6$

52. $2x + 3y = 6$

53. $y = \frac{1}{2}x$

54. $y = 2x$

55. $y = 2x - 3$

56. $y = \frac{1}{3}x + 1$

57. $x = -2$

58. $y = 3$

Chapter 4 Cumulative Review

Simplify.

1. $\begin{array}{r} 7520 \\ 599 \\ +\,8640 \\ \hline \end{array}$

2. $\begin{array}{r} 6000 \\ -3999 \\ \hline \end{array}$

3. $156 \div 13$

4. $9(7 \cdot 2)$

5. $64\overline{)31{,}362}$

6. 2^8

7. $12 + 81 \div 3^2$

8. $\dfrac{329}{47}$

9. $-25 + (-13)$

10. $(-10 + 4) + (212 - 100)$

11. $\dfrac{-39}{-3}$

12. $\dfrac{-4(3-8) - 2(7-9)}{-6(2)}$

13. $\dfrac{3 \cdot 2^3 - 3 \cdot 4^2}{-2^3}$

14. $\dfrac{-3 \cdot 5^2 - 5(3-8)}{-5^2}$

15. $\dfrac{-6(2)^3 - 4^2}{2(2)^2}$

16. $\left(\dfrac{1}{4}\right)^3\left(-\dfrac{1}{2}\right)^2$

17. $17 \div \left(\dfrac{1}{3}\right)^2$

18. $\dfrac{t}{25} + \left(-\dfrac{7}{50}\right)$

19. $\left(16 \div 1\dfrac{1}{4}\right) \div 2$

20. $15 - 3\dfrac{1}{2}$

21. $3(z + 2) - 2(3z - 1)$

22. $-2a - 3 + a + 7$

23. $13 + \dfrac{3}{14} \div \dfrac{5}{42}$

Solve.

24. $4b + 3 + 2b = 12b - 6$

25. $5a - 3a + 6a = -8 + (-8)$

26. Graph the equation $2x + y = 6$, and name three ordered pairs on the graph.

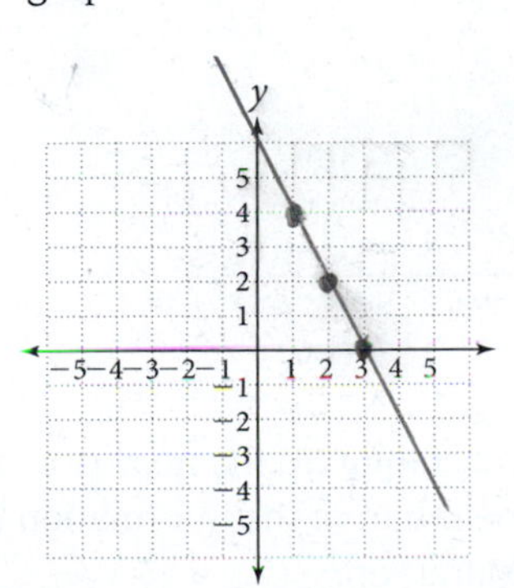

27. Find the perimeter and area of the figure below.

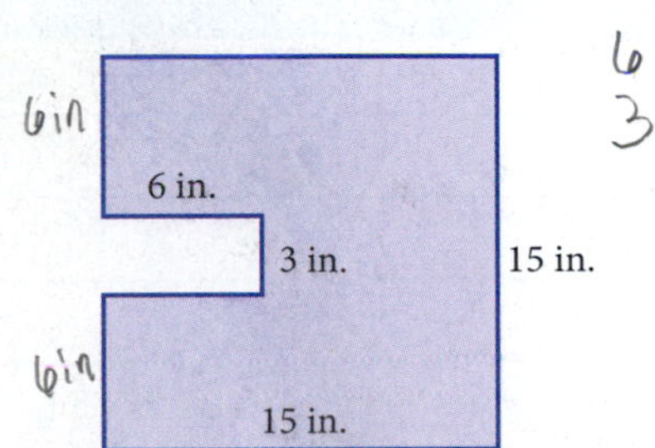

28. Find the perimeter of the figure below.

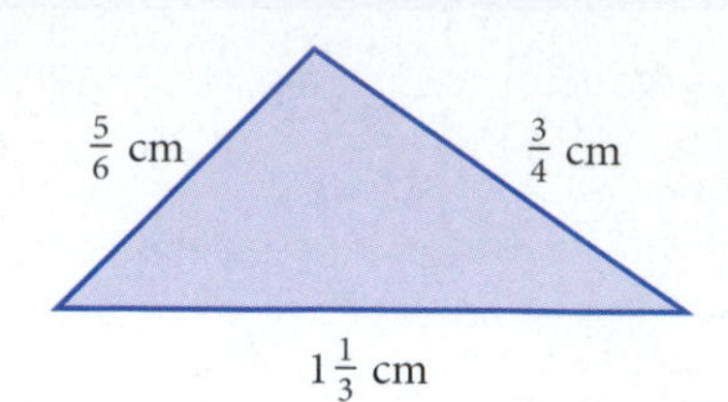

29. Find the difference between -15 and -62.

30. Find the value of $3x + 5y$ when $x = 2$ and $y = 1$.

31. Find the volume of a cube with sides 21 cm.

32. Factor 126 into a product of prime factors.

33. Find $\frac{2}{3}$ of the product of 7 and 9.

34. Use the equation $4x + y = 26$ to find y when $x = 3$.

35. Use the equation $2x - 2y = 12$ to complete the following ordered pairs. (10,), (, 2), (4,)

36. On the graph of $x = -2$, when $y = 4$, what is the value of x?

37. Hammerhead Shark The largest hammerhead shark ever recorded was $18\frac{1}{3}$ feet long. Multiply the shark's length by 12 to find its length in inches.

38. Running Over two days, James ran 18 miles. He ran three times farther on the first day than he did on the second day. What was the distance of James's shorter run?

39. Volume What is the volume of a rectangular solid that is 36 inches long, 24 inches wide, and 12 inches deep?

40. NYC Marathon The table below shows the number of people who finished the New York City Marathon in certain years.

NEW YORK CITY MARATHON

Year	Number of Finishers	Rounded to the Nearest Hundred
1975	339	
1980	12,512	
1985	15,881	
1990	23,774	
1995	26,754	

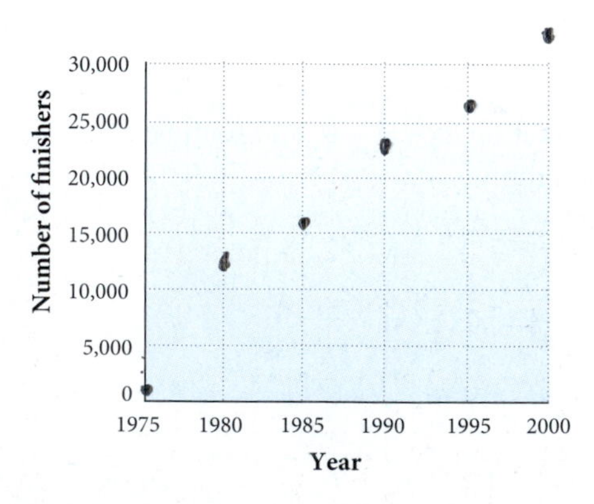

a. Fill in the last column of the table by rounding each number in the second column to the nearest hundred.
b. Use the template and the rounded numbers from Part a to construct a line graph of the information in the table.
c. From the line graph, estimate the number of finishers for the 2000 NYC Marathon.

Chapter 4 Test

Simplify each expression by combining similar terms.

1. $9x - 3x + 7 - 12$

2. $4b - 1 - b - 3$

Find the value of each expression when $x = 3$.

3. $3x - 12$

4. $-x + 12$

5. Is $x = -1$ a solution to $4x - 3 = -7$?

Solve each equation.

6. $x - 7 = -3$

7. $\frac{2}{3}y = 18$

8. $3x - 7 = 5x + 1$

9. $2(x - 5) = -8$

10. $3(2x + 3) = -3(x - 5)$

11. $6(3x - 2) - 8 = 4x - 6$

12. Number Problem Twice the sum of a number and 3 is -10. Find the number.

13. Hot Air Balloon The first successful crossing of the Atlantic in a hot air balloon was made in August 1978 by Maxie Anderson, Ben Abruzzo, and Larry Newman of the United States. The 3,100 mile trip took approximately 140 hours. Use the formula $r = \frac{d}{t}$ to find their average speed to the nearest whole number.

14. Geometry The length of a rectangle is 4 centimeters longer than its width. If the perimeter is 28 centimeters, find the length and the width.

15. Age problem Karen is 5 years younger than Susan. Three years ago, the sum of their ages was 11. How old are they now?

16. Use the equation $4x + 3y = 12$ to find y when $x = -3$.

17. Plot the following points: $(3, 2)$, $(-3, 2)$, $(3, 0)$, $(0, -2)$.

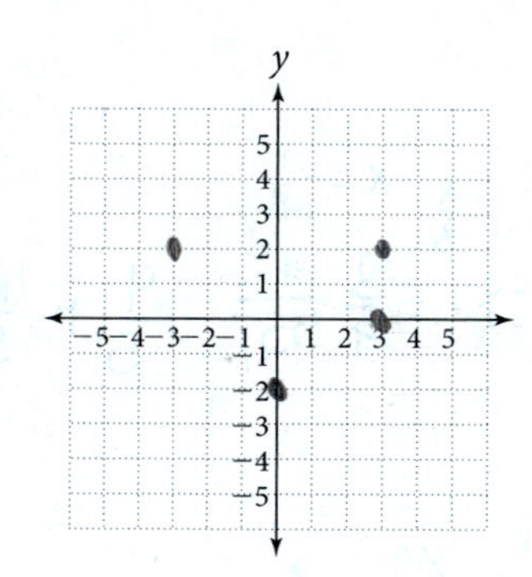

18. Graph $2x + y = 4$. (-1,6)(0,4)(1,2)

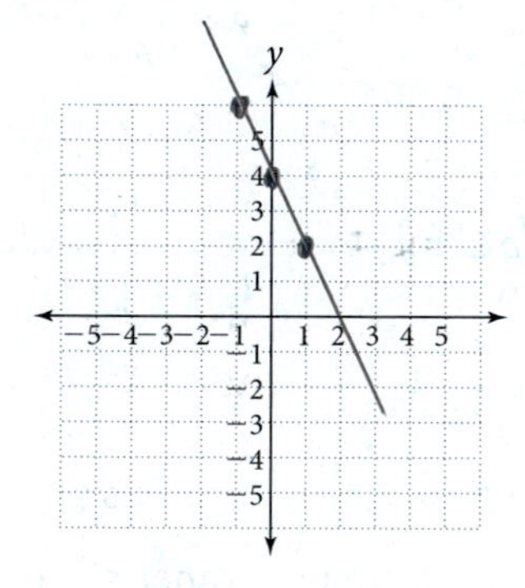

19. Graph $y = 2x - 1$. (-1,-3)(0,-1)(1,1)

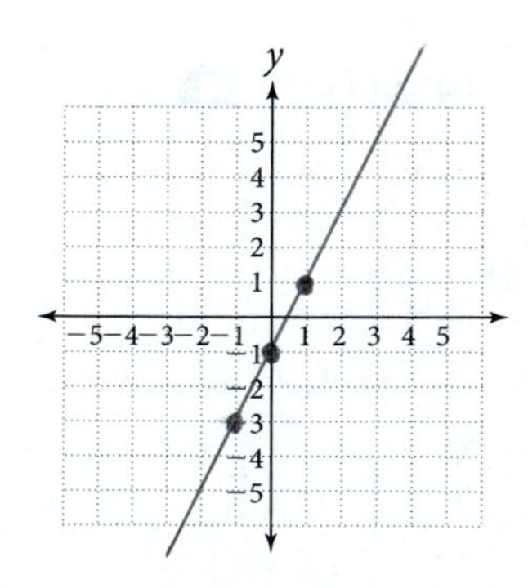

20. Graph $x = 2$.

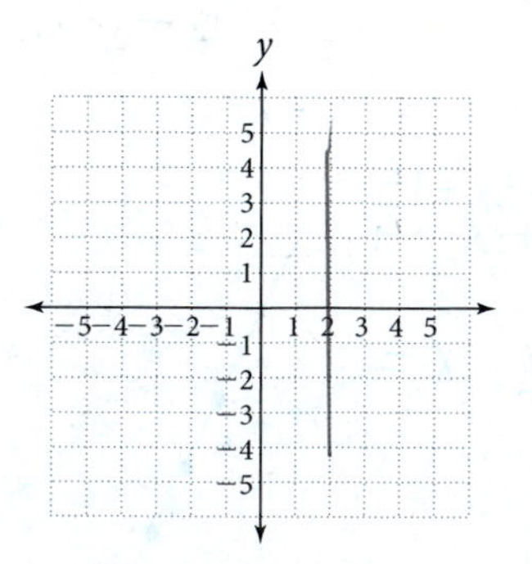

Chapter 4 Projects

SOLVING EQUATIONS

GROUP PROJECT

The Equation Game

Number of People 2–5

Time Needed 30 minutes

Equipment Per group: deck of cards, timer or clock, pencil and paper, copy of rules.

Background The Equation Game is a fun way to practice working with equations.

Procedure Remove all the face cards from the deck. Aces will be 1's. The dealer deals four cards face up, a fifth card face down. Each player writes down the four numbers that are face up. Set the timer for 5 minutes, then flip the fifth card. Each player writes down equations that use the numbers on the first four cards to equal the number on the fifth card. When the five minutes are up, figure out the scores. An equation that uses

1 of the four cards scores 0 points
2 of the four cards scores 4 points
3 of the four cards scores 9 points
4 of the four cards scores 16 points

Check the other players' equations. If you find an error, you get 7 points. The person with the mistake gets no points for that equation.

Example The first four cards are a four, a nine, an ace, and a two. The fifth card is a seven. Here are some equations you could make:

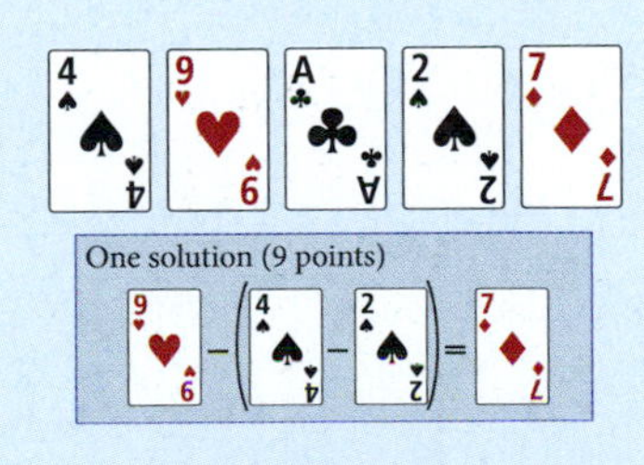

$9 - 2 = 7$

$4 + 2 + 1 = 7$

$\dfrac{9 - (4 - 2)}{1} = 7$

*This project was adapted from www.exploratorium.edu/math_explorer/fantasticFour.html.

RESEARCH PROJECT

Algebraic Symbolism

Algebra made a beginning as early as 1850 B.C. in Egypt. However, the symbols we use in algebra today took some time to develop. For example, the algebraic use of letters for numbers began much later with Diophantus, a mathematician famous for studying *Diophantine equations.* In the early centuries, the full words *plus, minus, multiplied by, divided by,* and *equals* were written out. Imagine how much more difficult your homework would be if you had to write out all these words instead of using symbols. Algebraists began to come up with a system of symbols to make writing algebra easier. At first, not everyone agreed on the symbols to be used. For example, the present division sign $\div$ was often used for subtraction. People in different countries used different symbols: the Italians preferred to use p and m for plus and minus, while the less traditional Germans were starting to use $+$ and $-$.

Research the history of algebraic symbolism. Find out when the algebraic symbols we use today (such as letters to represent variables, $+$, $-$, $\div$, $\cdot$, a/b and $\frac{a}{b}$) came into common use. Summarize your results in an essay.

Decimals

5

Chapter Outline

Introduction

The 2000 Summer Olympic Games in Sydney, Australia, featured more than 10,000 athletes competing from 199 countries. The image above shows many of the venues as they appear in Google Earth in 2008, almost eight years after the 2000 Olympics. The chart below shows the top four finishes in the 400-meter Freestyle swimming event in Sydney.

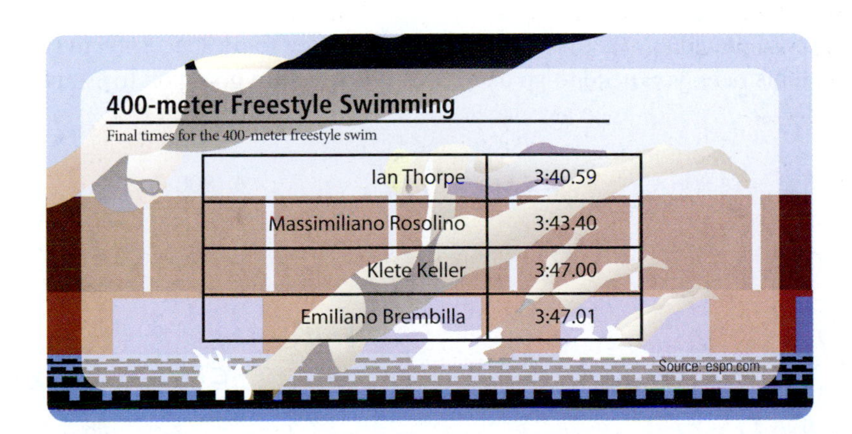

400-meter Freestyle Swimming

Final times for the 400-meter freestyle swim

Ian Thorpe	3:40.59
Massimiliano Rosolino	3:43.40
Klete Keller	3:47.00
Emiliano Brembilla	3:47.01

Source: espn.com

The times in the chart are in minutes and seconds, accurate to the nearest hundredth of a second. In this chapter we work with numbers like these to obtain a good working knowledge of the decimal numbers we see everywhere around us.

Chapter Pretest

The pretest below contains problems that are representative of the problems you will find in the chapter.

1. Write the number 4.013 in words.
2. Write 12.09 as a mixed number.
3. Write *thirty-four hundredths* as a decimal number.
4. Write $\frac{21}{50}$ as a decimal
5. Write the following numbers in order from smallest to largest.
   0.04, 0.4, 0.51, 0.5, 0.45, 0.41
6. Change 0.85 to a fraction, and then reduce to lowest terms.

Perform the indicated operations.

7. $7.36 + 8.05$
8. $20.3 - 15.09$
9. 3.6×2.7
10. $56.78(10)$
11. $32\overline{)131.84}$
12. $1.04 \div 0.12$

Simplify each expression as much as possible.

13. $\sqrt{36} + 2\sqrt{100}$
14. $\sqrt{\frac{25}{144}}$
15. $\sqrt{180}$

Simplify each expression as much as possible. Assume all variables represent positive numbers.

16. $\sqrt{18x^3y^2}$
17. $3\sqrt{y} - 2\sqrt{y} + \sqrt{y}$
18. $3\sqrt{12} + 5\sqrt{27}$

Getting Ready for Chapter 5

The problems below review material covered previously that you need to know in order to be successful in Chapter 5. If you have any difficulty with the problems here, you need to go back and review before going on to Chapter 5.

Simplify.

1. $25{,}430 + 2{,}897 + 379{,}600$
2. $39{,}812 - 14{,}236$
3. $2{,}000 - 1{,}564$
4. $800 - 137$
5. 305×436
6. $13(56)$
7. $480 + 12(32)^2$
8. $384 \div 4$
9. $49{,}896 \div 27$
10. $5{,}974 \div 20$
11. $5^2 + 7^2$
12. $\left(\frac{1}{4}\right)^2$
13. $\frac{3}{4}\left(\frac{2}{5}\right)$
14. $\frac{1}{2}\left(2\frac{5}{8}\right)\left(1\frac{1}{4}\right)$
15. Round 9,235 to the nearest hundred.
16. Reduce: $\frac{38}{100}$

Factor into a product of prime factors.

17. 48
18. 180

5.1 Decimal Notation and Place Value

Objectives

A Understand place value for decimal numbers.

B Write decimal numbers in words and with digits.

C Convert decimals to fractions and fractions to decimals.

D Round a decimal number.

E Solve applications involving decimals.

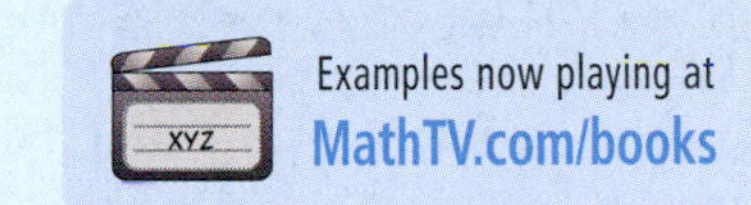

Introduction . . .

In this chapter we will focus our attention on *decimals*. Anyone who has used money in the United States has worked with decimals already. For example, if you have been paid an hourly wage, such as

$6.25 per hour
↑ Decimal point

you have had experience with decimals. What is interesting and useful about decimals is their relationship to fractions and to powers of ten. The work we have done up to now—especially our work with fractions—can be used to develop the properties of decimal numbers.

A Place Value

In Chapter 1 we developed the idea of place value for the digits in a whole number. At that time we gave the name and the place value of each of the first seven columns in our number system, as follows:

Millions Column	Hundred Thousands Column	Ten Thousands Column	Thousands Column	Hundreds Column	Tens Column	Ones Column
1,000,000	100,000	10,000	1,000	100	10	1

As we move from right to left, we multiply by 10 each time. The value of each column is 10 times the value of the column on its right, with the rightmost column being 1. Up until now we have always looked at place value as increasing by a factor of 10 each time we move one column to the left:

Ten Thousands		Thousands		Hundreds		Tens		Ones
10,000	← Multiply by 10	1,000	← Multiply by 10	100	← Multiply by 10	10	← Multiply by 10	1

To understand the idea behind decimal numbers, we notice that moving in the opposite direction, from left to right, we *divide* by 10 each time:

Ten Thousands		Thousands		Hundreds		Tens		Ones
10,000	Divide by 10 →	1,000	Divide by 10 →	100	Divide by 10 →	10	Divide by 10 →	1

If we keep going to the right, the next column will have to be

$$1 \div 10 = \frac{1}{10} \qquad \text{Tenths}$$

The next one after that will be

$$\frac{1}{10} \div 10 = \frac{1}{10} \cdot \frac{1}{10} = \frac{1}{100} \qquad \text{Hundredths}$$

After that, we have

$$\frac{1}{100} \div 10 = \frac{1}{100} \cdot \frac{1}{10} = \frac{1}{1{,}000} \qquad \text{Thousandths}$$

We could continue this pattern as long as we wanted. We simply divide by 10 to move one column to the right. (And remember, dividing by 10 gives the same result as multiplying by $\frac{1}{10}$.)

To show where the ones column is, we use a *decimal point* between the ones column and the tenths column.

> **Note** Because the digits to the right of the decimal point have fractional place values, numbers with digits to the right of the decimal point are called decimal fractions. In this book we will also call them decimal numbers, or simply decimals for short.

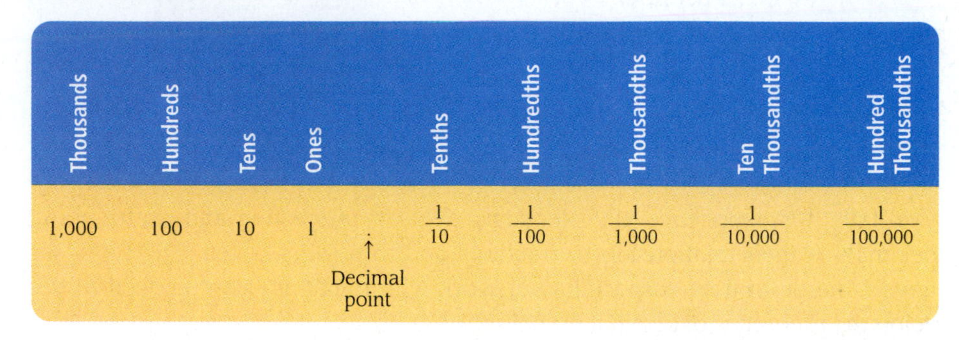

The ones column can be thought of as the middle column, with columns larger than 1 to the left and columns smaller than 1 to the right. The first column to the right of the ones column is the tenths column, the next column to the right is the hundredths column, the next is the thousandths column, and so on. The decimal point is always written between the ones column and the tenths column.

We can use the place value of decimal fractions to write them in expanded form.

EXAMPLE 1 Write 423.576 in expanded form.

SOLUTION $423.576 = 400 + 20 + 3 + \frac{5}{10} + \frac{7}{100} + \frac{6}{1{,}000}$

B Writing Decimals with Words

EXAMPLE 2 Write each number in words.

a. 0.4
b. 0.04
c. 0.004

SOLUTION **a.** 0.4 is "four tenths."
b. 0.04 is "four hundredths."
c. 0.004 is "four thousandths."

When a decimal fraction contains digits to the left of the decimal point, we use the word "and" to indicate where the decimal point is when writing the number in words.

EXAMPLE 3 Write each number in words.

a. 5.4
b. 5.04
c. 5.004

SOLUTION **a.** 5.4 is "five and four tenths."
b. 5.04 is "five and four hundredths."
c. 5.004 is "five and four thousandths."

PRACTICE PROBLEMS

1. Write 785.462 in expanded form.
2. Write in words.
   a. 0.06
   b. 0.7
   c. 0.008
3. Write in words.
   a. 5.06
   b. 4.7
   c. 3.008

> **Note** Sometimes we name decimal fractions by simply reading the digits from left to right and using the word "point" to indicate where the decimal point is. For example, using this method the number 5.04 is read "five point zero four."

Answers
1–3. See solutions section.

EXAMPLE 4 Write 3.64 in words.

SOLUTION The number 3.64 is read "three and sixty-four hundredths." The place values of the digits are as follows:

3 . 6 4
↑ ↑ ↖
3 ones 6 tenths 4 hundredths

We read the decimal part as "sixty-four hundredths" because

$$6 \text{ tenths} + 4 \text{ hundredths} = \frac{6}{10} + \frac{4}{100} = \frac{60}{100} + \frac{4}{100} = \frac{64}{100}$$

EXAMPLE 5 Write 25.4936 in words.

SOLUTION Using the idea given in Example 4, we write 25.4936 in words as "twenty-five and four thousand, nine hundred thirty-six ten thousandths."

C Converting Between Fractions and Decimals

In order to understand addition and subtraction of decimals in the next section, we need to be able to convert decimal numbers to fractions or mixed numbers.

EXAMPLE 6 Write each number as a fraction or a mixed number. Do not reduce to lowest terms.

a. 0.004 **b.** 3.64 **c.** 25.4936

SOLUTION **a.** Because 0.004 is 4 thousandths, we write

$$0.004 = \frac{4}{1{,}000}$$

Three digits after the decimal point — Three zeros

b. Looking over the work in Example 4, we can write

$$3.64 = 3\frac{64}{100}$$

Two digits after the decimal point — Two zeros

c. From the way in which we wrote 25.4936 in words in Example 5, we have

$$25.4936 = 25\frac{4936}{10{,}000}$$

Four digits after the decimal point — Four zeros

D Rounding Decimal Numbers

The rule for rounding decimal numbers is similar to the rule for rounding whole numbers. If the digit in the column to the right of the one we are rounding to is 5 or more, we add 1 to the digit in the column we are rounding to; otherwise, we leave it alone. We then replace all digits to the right of the column we are rounding to with zeros if they are to the left of the decimal point; otherwise, we simply delete them. Table 1 illustrates the procedure.

4. Write in words.
a. 5.98
b. 5.098

5. Write 305.406 in words.

6. Write each number as a fraction or a mixed number. Do not reduce to lowest terms.
a. 0.06
b. 5.98
c. 305.406

Answers
4–5. See solutions section.
6. a. $\frac{6}{100}$ **b.** $5\frac{98}{100}$
c. $305\frac{406}{1{,}000}$

TABLE 1

	Rounded to the Nearest		
Number	Whole Number	Tenth	Hundredth
24.785	25	24.8	24.79
2.3914	2	2.4	2.39
0.98243	1	1.0	0.98
14.0942	14	14.1	14.09
0.545	1	0.5	0.55

7. Round 8,935.042 to the nearest:
   a. hundred
   b. hundredth

EXAMPLE 7 Round 9,235.492 to the nearest hundred.

SOLUTION The number next to the hundreds column is 3, which is less than 5. We change all digits to the right to 0, and we can drop all digits to the right of the decimal point, so we write

9,200

8. Round 0.05067 to the nearest:
   a. ten thousandth
   b. tenth

EXAMPLE 8 Round 0.00346 to the nearest ten thousandth.

SOLUTION Because the number to the right of the ten thousandths column is more than 5, we add 1 to the 4 and get

0.0035

E Applications with Decimals

9. Round each number in the bar chart to the nearest tenth of a dollar

EXAMPLE 9 The bar chart below shows some ticket prices for a recent major league baseball season. Round each ticket price to the nearest dollar.

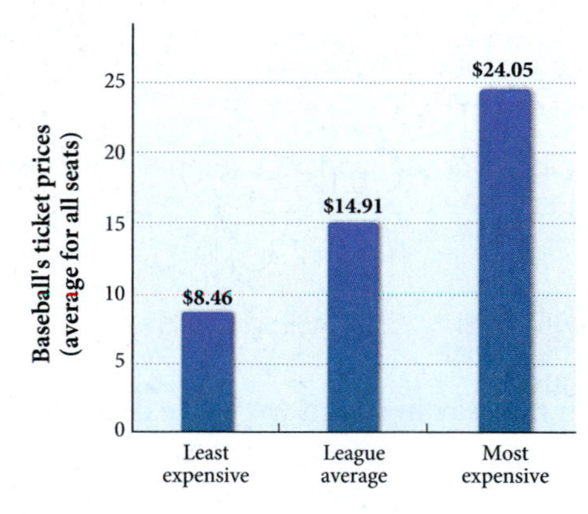

SOLUTION Using our rule for rounding decimal numbers, we have the following results:

Least expensive: $8.46 rounds to $8
League average: $14.91 rounds to $15
Most expensive: $24.05 rounds to $24

Getting Ready for Class

After reading through the preceding section, respond in your own words and in complete sentences.

1. Write 754.326 in expanded form.
2. Write $400 + 70 + 5 + \frac{1}{10} + \frac{3}{100} + \frac{7}{1,000}$ in decimal form.
3. Write seventy-two and three tenths in decimal form.

Answers
7. a. 8,900 b. 8,935.04
8. a. 0.0507 b. 0.1
9. Least expensive, $8.50; league average, $14.90; most expensive, $24.10

Problem Set 5.1

B Write out the name of each number in words. [Examples 2–5]

1. 0.3
2. 0.03
3. 0.015
4. 0.0015
5. 3.4
6. 2.04
7. 52.7
8. 46.8

C Write each number as a fraction or a mixed number. Do not reduce your answers. [Example 6]

9. 405.36
10. 362.78
11. 9.009
12. 60.06
13. 1.234
14. 12.045
15. 0.00305
16. 2.00106

A Give the place value of the 5 in each of the following numbers. [Example 1]

17. 458.327
18. 327.458
19. 29.52
20. 25.92
21. 0.00375
22. 0.00532
23. 275.01
24. 0.356
25. 539.76
26. 0.123456

B Write each of the following as a decimal number.

27. Fifty-five hundredths
28. Two hundred thirty-five ten thousandths
29. Six and nine tenths
30. Forty-five thousand and six hundred twenty-one thousandths
31. Eleven and eleven hundredths
32. Twenty-six thousand, two hundred forty-five and sixteen hundredths
33. One hundred and two hundredths
34. Seventy-five and seventy-five hundred thousandths
35. Three thousand and three thousandths
36. One thousand, one hundred eleven and one hundred eleven thousandths

For each pair of numbers, place the correct symbol, $<$ or $>$, between the numbers.

37. **a.** 0.02 0.2
b. 0.3 0.032

38. **a.** 0.45 0.5
b. 0.5 0.56

39. Write the following numbers in order from smallest to largest.
0.02 0.05 0.025 0.052 0.005 0.002
3 5 4 6 2 1

40. Write the following numbers in order from smallest to largest.
0.2 0.02 0.4 0.04 0.42 0.24

41. Which of the following numbers will round to 7.5?
7.451 7.449 7.54 7.56

42. Which of the following numbers will round to 3.2?
3.14999 3.24999 3.279 3.16111

C Change each decimal to a fraction, and then reduce to lowest terms.

43. 0.25 **44.** 0.75 **45.** 0.125 **46.** 0.375

47. 0.625 **48.** 0.0625 **49.** 0.875 **50.** 0.1875

Estimating For each pair of numbers, choose the number that is closest to 10.

51. 9.9 and 9.99 **52.** 8.5 and 8.05 **53.** 10.5 and 10.05 **54.** 10.9 and 10.99

Estimating For each pair of numbers, choose the number that is closest to 0.

55. 0.5 and 0.05 **56.** 0.10 and 0.05 **57.** 0.01 and 0.02 **58.** 0.1 and 0.01

D Complete the following table. [Examples 7, 8]

		Rounded to the Nearest			
	Number	**Whole Number**	**Tenth**	**Hundredth**	**Thousandth**
59.	47.5479	48	47.5	47.55	47.548
60.	100.9256	101	100.9	100.93	100.926
61.	0.8175	1	0.8	0.82	0.818
62.	29.9876	30	30	29.99	29.988
63.	0.1562	0	0.2	0.16	0.156
64.	128.9115	129	128.9	128.91	128.912
65.	2,789.3241	2789	2789.3	2789.32	2789.324
66.	0.8743	1	0.9	0.87	0.874
67.	99.9999	100	100	100	100
68.	71.7634	72	71.8	71.76	71.763

E Applying the Concepts [Example 9]

100 Meters At the 1928 Olympic Games in Amsterdam, the winning time for the women's 100 meters was 12.2 seconds. Since then, the time has continued to get faster. The chart shows the fastest times for the women's 100 meters in the Olympics. Use the chart to answer Problems 69 and 70.

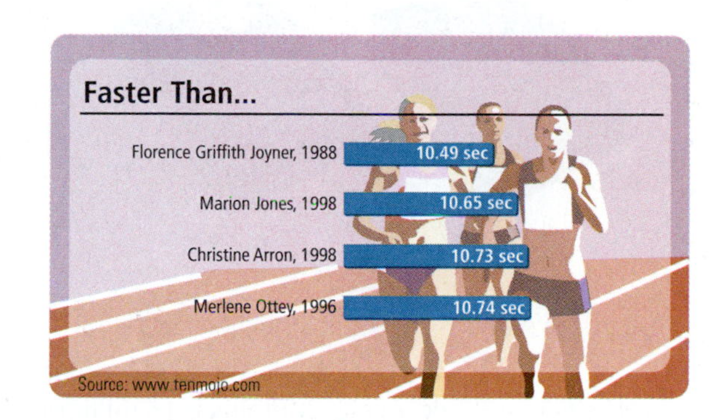

69. What is the place value of the 3 in Christine Arron's time in 1998?

70. Write Christine Arron's time using words.

71. Gasoline Prices The bar chart below was created from a survey by the U.S. Department of Energy's Energy Information Administration during the month of May 2008. It gives the average price of regular gasoline for the state of California on each Monday of the month. Use the information in the chart to fill in the table.

PRICE OF 1 GALLON OF REGULAR GASOLINE	
Date	Price (Dollars)
5/5/08	3.903
5/12/08	3.919
5/19/08	3.952
5/26/08	4.099

Price (dollars)
4.100
4.000
3.900
3.800
3.903
3.919
3.952
4.099
5/5/08
5/12/08
5/19/08
5/26/08
Date

72. Speed and Time The bar chart below was created from data given by *Car and Driver* magazine. It gives the minimum time in seconds for a Toyota Echo to reach various speeds from a complete stop. Use the information in the chart to fill in the table.

Speed (Miles per Hour)	Time (Seconds)
30	
40	
50	
60	
70	
80	
90	

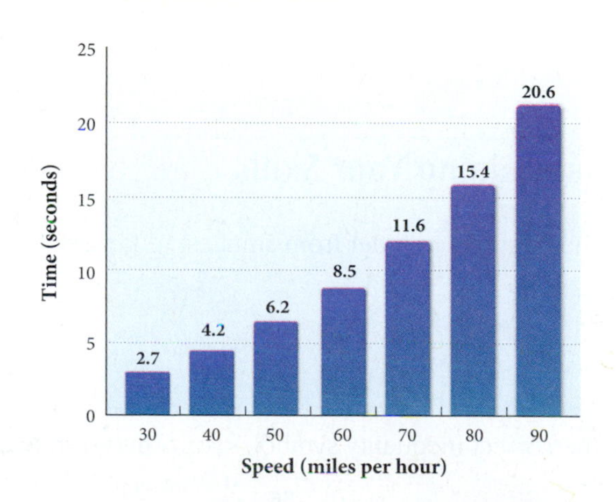

73. Penny Weight If you have a penny dated anytime from 1959 through 1982, its original weight was 3.11 grams. If the penny has a date of 1983 or later, the original weight was 2.5 grams. Write the two weights in words.

74. Halley's Comet Halley's comet was seen from the earth during 1986. It will be another 76.1 years before it returns. Write 76.1 in words.

NASA

75. Nutrition A 50-gram egg contains 0.15 milligram of riboflavin. Write 0.15 in words.

76. Nutrition One medium banana contains 0.64 milligram of B_6. Write 0.64 in words.

Getting Ready for the Next Section

In the next section we will do addition and subtraction with decimals. To understand the process of addition and subtraction, we need to understand the process of addition and subtraction with mixed numbers.

Find each of the following sums and differences. (Add or subtract.)

77. $4\frac{3}{10} + 2\frac{1}{100}$

78. $5\frac{35}{100} + 2\frac{3}{10}$

79. $8\frac{5}{10} - 2\frac{4}{100}$

80. $6\frac{3}{100} - 2\frac{125}{1,000}$

81. $5\frac{1}{10} + 6\frac{2}{100} + 7\frac{3}{1,000}$

82. $4\frac{27}{100} + 6\frac{3}{10} + 7\frac{123}{1,000}$

Maintaining Your Skills

Write the fractions in order from smallest to largest.

83. $\frac{3}{8}$ $\frac{3}{16}$ $\frac{3}{4}$ $\frac{3}{10}$

84. $\frac{3}{4}$ $\frac{1}{4}$ $\frac{5}{4}$ $\frac{1}{2}$

Place the correct inequality symbol, < or > between each pair of numbers.

85. $\frac{3}{8}$ $\frac{5}{6}$

86. $\frac{9}{10}$ $\frac{10}{11}$

87. $\frac{1}{12}$ $\frac{1}{13}$

88. $\frac{3}{4}$ $\frac{5}{8}$

Addition and Subtraction with Decimals

5.2

Objectives

A Add and subtract decimals.

B Solve applications involving addition and subtraction of decimals.

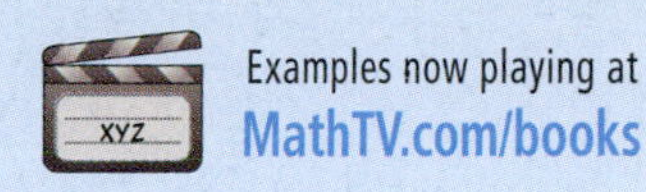

Introduction . . .

The chart shows the top finishing times for the women's 400-meter race during the Sydney Olympics in 2000. In order to analyze the different finishing times, it is important that you are able to add and subtract decimals, and that is what we will cover in this section.

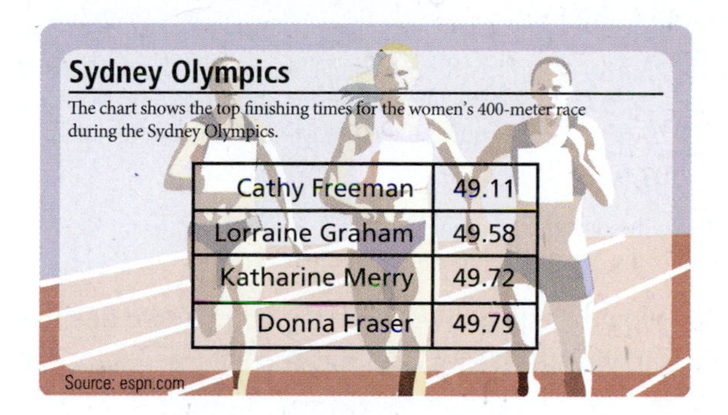

Cathy Freeman	49.11
Lorraine Graham	49.58
Katharine Merry	49.72
Donna Fraser	49.79

A Combining Decimals

Suppose you are earning \$8.50 an hour and you receive a raise of \$1.25 an hour. Your new hourly rate of pay is

$$\begin{array}{r} \$8.50 \\ +\ \$1.25 \\ \hline \$9.75 \end{array}$$

To add the two rates of pay, we align the decimal points, and then add in columns.

To see why this is true in general, we can use mixed-number notation:

$$\begin{aligned} 8.50 &= 8\frac{50}{100} \\ +\ 1.25 &= 1\frac{25}{100} \\ \hline &\ \ \ 9\frac{75}{100} = 9.75 \end{aligned}$$

We can visualize the mathematics above by thinking in terms of money:

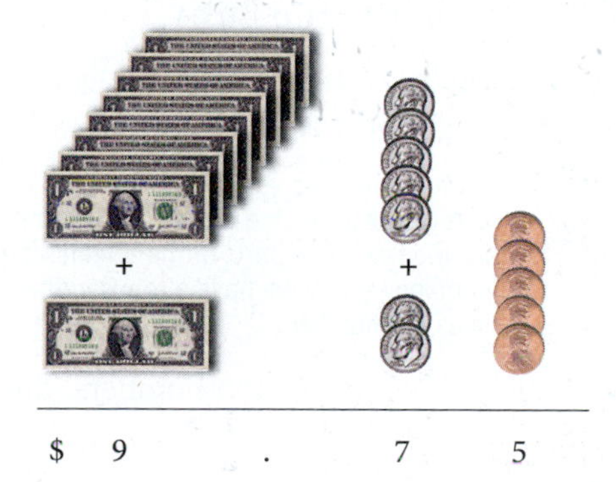

EXAMPLE 1 Add by first changing to fractions: 25.43 + 2.897 + 379.6

SOLUTION We first change each decimal to a mixed number. We then write each fraction using the least common denominator and add as usual:

PRACTICE PROBLEMS

1. Change each decimal to a fraction, and then add. Write your answer as a decimal.
   a. 38.45 + 456.073
   b. 38.045 + 456.73

$$
\begin{aligned}
25.43 &= 25\frac{43}{100} = 25\frac{430}{1,000} \\
2.897 &= 2\frac{897}{1,000} = 2\frac{897}{1,000} \\
+\ 379.6 &= 379\frac{6}{10} = 379\frac{600}{1,000} \\
& \qquad 406\frac{1,927}{1,000} = 407\frac{927}{1,000} = 407.927
\end{aligned}
$$

Again, the result is the same if we just line up the decimal points and add as if we were adding whole numbers:

$$
\begin{array}{r}
25.430 \\
2.897 \\
+\ 379.600 \\
\hline
407.927
\end{array}
$$

Notice that we can fill in zeros on the right to help keep the numbers in the correct columns. Doing this does not change the value of any of the numbers.

Note: The decimal point in the answer is directly below the decimal points in the problem

The same thing would happen if we were to subtract two decimal numbers. We can use these facts to write a rule for addition and subtraction of decimal numbers.

Rule

To add (or subtract) decimal numbers, we line up the decimal points and add (or subtract) as usual. The decimal point in the result is written directly below the decimal points in the problem.

We will use this rule for the rest of the examples in this section.

2. Subtract: 78.674 − 23.431

EXAMPLE 2 Subtract: 39.812 − 14.236

SOLUTION We write the numbers vertically, with the decimal points lined up, and subtract as usual.

$$
\begin{array}{r}
39.812 \\
-\ 14.236 \\
\hline
25.576
\end{array}
$$

3. Add: 16 + 0.033 + 4.6 + 0.08

EXAMPLE 3 Add: 8 + 0.002 + 3.1 + 0.04

SOLUTION To make sure we keep the digits in the correct columns, we can write zeros to the right of the rightmost digits.

$$
\begin{aligned}
8 &= 8.000 \\
3.1 &= 3.100 \\
0.04 &= 0.040
\end{aligned}
$$

Writing the extra zeros here is really equivalent to finding a common denominator for the fractional parts of the original four numbers—now we have a thousandths column in all the numbers

This doesn't change the value of any of the numbers, and it makes our task easier. Now we have

$$
\begin{array}{r}
8.000 \\
0.002 \\
3.100 \\
+\ 0.040 \\
\hline
11.142
\end{array}
$$

Answers

1. a. 494.523 **b.** 494.775
2. 55.243 **3.** 20.713

EXAMPLE 4 Subtract: 5.9 − 3.0814

SOLUTION In this case it is very helpful to write 5.9 as 5.9000, since we will have to borrow in order to subtract.

$$\begin{array}{r} 5.9000 \\ -\ 3.0814 \\ \hline 2.8186 \end{array}$$

EXAMPLE 5 Subtract 3.09 from the sum of 9 and 5.472.

SOLUTION Writing the problem in symbols, we have

$$\begin{aligned}(9 + 5.472) - 3.09 &= 14.472 - 3.09 \\ &= 11.382\end{aligned}$$

EXAMPLE 6 Add: 2.89 + (−5.93)

SOLUTION Recall from Chapter 2 that to add two numbers with different signs, we subtract the smaller absolute value from the larger. The sign of the answer is the same as the sign of the number with the larger absolute value.

$$2.89 + (-5.93) = \underbrace{-3.04}$$

The answer has the sign of the number with the larger absolute value

EXAMPLE 7 Subtract: −8 − 1.37

SOLUTION Because subtraction can be thought of as addition of the opposite, instead of subtracting 1.37 we can add −1.37.

$$-8 - 1.37 = -8 + (-1.37)$$

Now, to add two numbers with the same sign, we add their absolute values. The answer has the same sign as the original numbers.

$$-8 + (-1.37) = \underbrace{-9.37}$$

The answer has the same sign as the original two numbers

$$\begin{array}{r} 8.00 \\ +1.37 \\ \hline 9.37 \end{array}$$

Add the absolute values of the two numbers

B Applications

EXAMPLE 8 While I was writing this section of the book, I stopped to have lunch with a friend at a coffee shop near my office. The bill for lunch was $15.64. I gave the person at the cash register a $20 bill. For change, I received four $1 bills, a quarter, a nickel, and a penny. Was my change correct?

SOLUTION To find the total amount of money I received in change, we add:

Four $1 bills	=	$4.00
One quarter	=	0.25
One nickel	=	0.05
One penny	=	0.01
Total	=	$4.31

4. Subtract:
   a. 6.7 − 2.05
   b. 6.7 − 2.0563

5. Subtract 5.89 from the sum of 7 and 3.567.

6. Add: 4.93 + (−7.85)

7. Subtract: −4.09 − 3

8. If you pay for a purchase of $9.56 with a $10 bill, how much money should you receive in change? What will you do if the change that is given to you is one quarter, two dimes, and four pennies?

Answers
4. a. 4.65 b. 4.6437
5. 4.677 6. −2.92 7. −7.09

To find out if this is the correct amount, we subtract the amount of the bill from \$20.00.

$$\begin{array}{r} \$20.00 \\ -\ 15.64 \\ \hline \$\ 4.36 \end{array}$$

The change was not correct. It is off by 5 cents. Instead of the nickel, I should have been given a dime.

EXAMPLE 9 Find the perimeter of each of the following stamps. Write your answer as a decimal, rounded to the nearest tenth, if necessary.

a.

Each side is 3.5 centimeters

b.

Base = 2.625 inches
Other two sides = 1.875 inches

SOLUTION To find the perimeter, we add the lengths of all the sides together.

a. $P = 3.5 + 3.5 + 3.5 + 3.5 = 14.0$ cm

b. $P = 2.625 + 1.875 + 1.875 = 6.4$ in.

9. Find the perimeter of each stamp in Example 9 from the dimensions given below.

a. Each side is 1.38 inches

b. Base = 6.6 centimeters, other two sides = 4.7 centimeters

Getting Ready for Class

After reading through the preceding section, respond in your own words and in complete sentences.

1. When adding numbers with decimals, why is it important to line up the decimal points?
2. Write 379.6 in mixed-number notation.
3. Look at Example 8 in this section of your book. If I had given the person at the cash register a \$20 bill and four pennies, how much change should I then have received?
4. How many quarters does the decimal 0.75 represent?

Answers

8. \$0.44; Tell the clerk that you have been given too much change. Instead of two dimes, you should have received one dime and one nickel.

9. a. 5.52 in. **b.** 16.0 cm

Problem Set 5.2

A Find each of the following sums. (Add.) [Examples 1, 3]

1. 2.91 + 3.28

2. 8.97 + 2.04

3. 0.04 + 0.31 + 0.78

4. 0.06 + 0.92 + 0.65

5. 3.89 + 2.4

6. 7.65 + 3.8

7. 4.532 + 1.81 + 2.7

8. 9.679 + 3.49 + 6.5

9. 0.081 + 5 + 2.94

10. 0.396 + 7 + 3.96

11. 5.0003 + 6.78 + 0.004

12. 27.0179 + 7.89 + 0.009

13. 7.123
8.12
9.1

14. 5.432
4.32
3.2

15. 9.001
8.01
7.1

16. 6.003
5.02
4.1

17. 89.7854
3.4
65.35
100.006

18. 57.4698
9.89
32.032
572.0079

19. 543.21
123.45

20. 987.654
456.789

A Find each of the following differences. (Subtract.) [Examples 2, 4]

21. 99.34 − 88.23

22. 47.69 − 36.58

23. 5.97 − 2.4

24. 9.87 − 1.04

25. 6.3 − 2.08

26. 7.5 − 3.04

27. 149.37 − 28.96

28. 796.45 − 32.68

29. 45 − 0.067

30. 48 − 0.075

31. 8 − 0.327

32. 12 − 0.962

33. 765.432 − 234.567

34. 654.321 − 123.456

A Subtract. [Example 4]

35. $34.07 - 6.18$

36. $25.008 - 3.119$

37. $40.04 - 4.4$

38. $50.05 - 5.5$

39. $768.436 - 356.998$

40. $495.237 - 247.668$

A Add and subtract as indicated. [Examples 1–7]

41. $(7.8 - 4.3) + 2.5$

42. $(8.3 - 1.2) + 3.4$

43. $7.8 - (4.3 + 2.5)$

44. $8.3 - (1.2 + 3.4)$

45. $(9.7 - 5.2) - 1.4$

46. $(7.8 - 3.2) - 1.5$

47. $9.7 - (5.2 - 1.4)$

48. $7.8 - (3.2 - 1.5)$

49. Subtract 5 from the sum of 8.2 and 0.072.

50. Subtract 8 from the sum of 9.37 and 2.5.

51. What number is added to 0.035 to obtain 4.036?

52. What number is added to 0.043 to obtain 6.054?

B Applying the Concepts [Examples 8, 9]

53. 100 Meters The chart shows the fastest times for the women's 100 meters in the Olympics. How much faster was Christine Arron's time than the first time recorded in 1928?

Faster Than...

Florence Griffith Joyner, 1988	10.49 sec
Marion Jones, 1998	10.65 sec
Christine Arron, 1998	10.73 sec
Merlene Ottey, 1996	10.74 sec

Source: www.tenmojo.com

54. Computers The chart shows how many computers can be found in the countries containing the most computers. What is the total number of computers that can be found in these three countries?

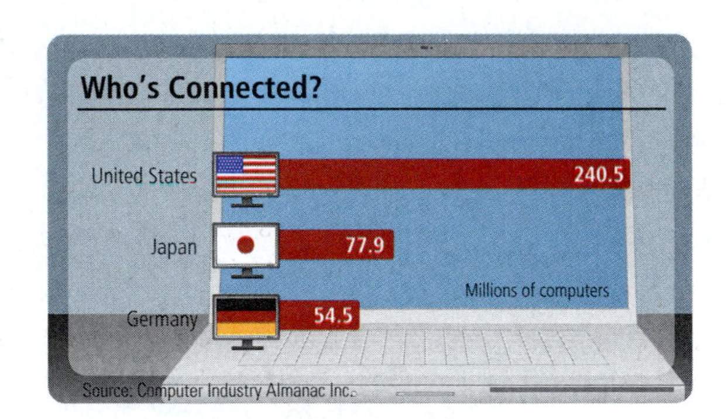

55. Take-Home Pay A college professor making $2,105.96 per month has deducted from her check $311.93 for federal income tax, $158.21 for retirement, and $64.72 for state income tax. How much does the professor take home after the deductions have been taken from her monthly income?

56. Take-Home Pay A cook making $1,504.75 a month has deductions of $157.32 for federal income tax, $58.52 for Social Security, and $45.12 for state income tax. How much does the cook take home after the deductions have been taken from his check?

57. Perimeter of a Stamp This stamp shows the Mexican artist Frida Kahlo. The stamp was issued in 2001 and is the first U.S. stamp to honor a Hispanic woman. The image area of the stamp has a width of 0.84 inches and a length of 1.41 inches. Find the perimeter of the image.

58. Perimeter of a Stamp This stamp was issued in 2001 to honor the Italian scientist Enrico Fermi. The stamp caused some discussion because some of the mathematics in the upper left corner of the stamp is incorrect. The image area of the stamp has a width of 21.4 millimeters and a length of 35.8 millimeters. Find the perimeter of the image.

59. Change A person buys $4.57 worth of candy. If he pays for the candy with a $10 bill, how much change should he receive?

60. Checking Account A checking account contains $342.38. If checks are written for $25.04, $36.71, and $210, how much money is left in the account?

RECORD ALL CHARGES OR CREDITS THAT AFFECT YOUR ACCOUNT

NUMBER	DATE	DESCRIPTION OF TRANSACTION	PAYMENT/DEBIT (-)	DEPOSIT/CREDIT (+)	BALANCE
	2/8	Deposit		$342 38	$342 38
1457	2/8	Woolworths	$25 04		
1458	2/9	Walgreens	$36 71		
1459	2/11	Electric Company	$210 00		?

61. Sydney Olympics The chart show the top finishing times for the mens' 400-meter freestyle swim during Sydney's Olympics. How much faster was Ian Thorpe than Emiliano Brembilla?

400-meter Freestyle Swimming

Final times for the 400-meter freestyle swim.

Ian Thorpe	3:40.59
Massimiliano Rosolino	3:43.40
Klete Keller	3:47.00
Emiliano Brembilla	3:47.01

Source: espn.com

62. Sydney Olympics The chart shows the top finishing times for the women's 400-meter race during the Sydney Olympics. How much faster was Lorraine Graham than Katharine Merry?

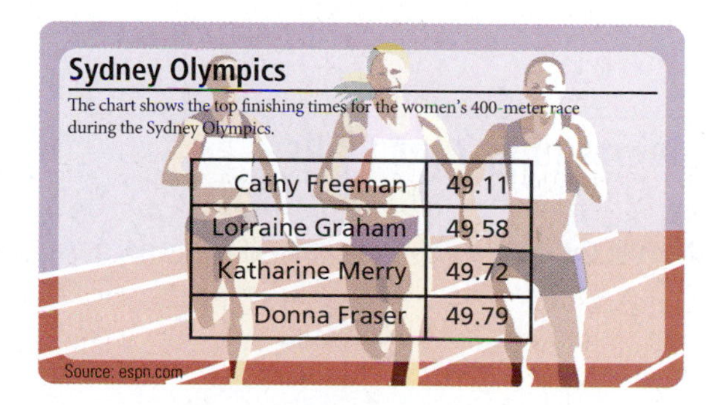

Sydney Olympics

The chart shows the top finishing times for the women's 400-meter race during the Sydney Olympics.

Cathy Freeman	49.11
Lorraine Graham	49.58
Katharine Merry	49.72
Donna Fraser	49.79

Source: espn.com

63. Geometry A rectangle has a perimeter of 9.5 inches. If the length is 2.75 inches, find the width.

64. Geometry A rectangle has a perimeter of 11 inches. If the width is 2.5 inches, find the length.

65. Change Suppose you eat dinner in a restaurant and the bill comes to \$16.76. If you give the cashier a \$20 bill and a penny, how much change should you receive? List the bills and coins you should receive for change.

66. Change Suppose you buy some tools at the hardware store and the bill comes to \$37.87. If you give the cashier two \$20 bills and 2 pennies, how much change should you receive? List the bills and coins you should receive for change.

Sequences Find the next number in each sequence.

67. 2.5, 2.75, 3, . . .

68. 3.125, 3.375, 3.625, . . .

Getting Ready for the Next Section

To understand how to multiply decimals, we need to understand multiplication with whole numbers, fractions, and mixed numbers. The following problems review these concepts.

69. $\frac{1}{10} \cdot \frac{3}{10}$

70. $\frac{5}{10} \cdot \frac{6}{10}$

71. $\frac{3}{100} \cdot \frac{17}{100}$

72. $\frac{7}{100} \cdot \frac{31}{100}$

73. $5\left(\frac{3}{10}\right)$

74. $7 \cdot \frac{7}{10}$

75. $56 \cdot 25$

76. $39(48)$

77. $\frac{5}{10} \times \frac{3}{10}$

78. $\frac{5}{100} \times \frac{3}{1{,}000}$

79. $2\frac{1}{10} \times \frac{7}{100}$

80. $3\frac{5}{10} \times \frac{4}{100}$

81. $305(436)$

82. $403(522)$

83. $5(420 + 3)$

84. $3(550 + 2)$

Maintaining Your Skills

Use the rule for order of operations to simplify each expression.

85. $30 \div 5 \cdot 2$

86. $60 \div 3 \cdot 10$

87. $22 - 2 \cdot 3$

88. $37 - 7 \cdot 2$

89. $12 + 18 \div 2 - 1$

90. $15 + 10 \div 5 - 4$

91. $3 \cdot 5^2 - 75 \div 5 + 2^3$

92. $2 \cdot 3^2 - 18 \div 3 + 2^4$

Multiplication with Decimals

5.3

Objectives

A Multiply decimal numbers.

B Solve application problems involving decimals.

C Find the circumference of a circle.

Introduction . . .

The distance around a circle is called the circumference. If you know the circumference of a bicycle wheel, and you ride the bicycle for one mile, you can calculate how many times the wheel has turned through one complete revolution. In this section we learn how to multiply decimal numbers, and this gives us the information we need to work with circles and their circumferences.

A Multiplying with Decimals

Before we introduce circumference, we need to back up and discuss multiplication with decimals. Suppose that during a half-price sale a calendar that usually sells for \$6.42 is priced at \$3.21. Therefore it must be true that

$$\frac{1}{2} \text{ of } 6.42 \text{ is } 3.21$$

But, because $\frac{1}{2}$ can be written as 0.5 and *of* translates to *multiply*, we can write this problem again as

$$0.5 \times 6.42 = 3.21$$

If we were to ignore the decimal points in this problem and simply multiply 5 and 642, the result would be 3,210. So, multiplication with decimal numbers is similar to multiplication with whole numbers. The difference lies in deciding where to place the decimal point in the answer. To find out how this is done, we can use fraction notation.

EXAMPLE 1 Change each decimal to a fraction and multiply:

0.5×0.3 — To indicate multiplication we are using a × sign here instead of a dot so we won't confuse the decimal points with the multiplication symbol.

SOLUTION Changing each decimal to a fraction and multiplying, we have

$$\begin{aligned} 0.5 \times 0.3 &= \frac{5}{10} \times \frac{3}{10} && \text{Change to fractions} \\ &= \frac{15}{100} && \text{Multiply numerators and multiply denominators} \\ &= 0.15 && \text{Write the answer in decimal form} \end{aligned}$$

The result is 0.15, which has two digits to the right of the decimal point.

What we want to do now is find a shortcut that will allow us to multiply decimals without first having to change each decimal number to a fraction. Let's look at another example.

PRACTICE PROBLEMS

1. Change each decimal to a fraction and multiply. Write your answer as a decimal.
   a. 0.4×0.6
   b. 0.04×0.06

Answer
1. **a.** 0.24 **b.** 0.0024

2. Change each decimal to a fraction and multiply. Write your answer as a decimal.
   a. 0.5×0.007
   b. 0.05×0.07

3. Change to fractions and multiply:
   a. 3.5×0.04
   b. 0.35×0.4

4. How many digits will be to the right of the decimal point in the following products?
   a. 3.706×55.88
   b. 37.06×0.5588

Answers
2. Both are 0.0035
3. Both are 0.14 4. a. 5 b. 6

EXAMPLE 2 Change each decimal to a fraction and multiply: 0.05×0.003

SOLUTION

$$0.05 \times 0.003 = \frac{5}{100} \times \frac{3}{1,000} \quad \text{Change to fractions}$$

$$= \frac{15}{100,000} \quad \text{Multiply numerators and multiply denominators}$$

$$= 0.00015 \quad \text{Write the answer in decimal form}$$

The result is 0.00015, which has a total of five digits to the right of the decimal point.

Looking over these first two examples, we can see that the digits in the result are just what we would get if we simply forgot about the decimal points and multiplied; that is, $3 \times 5 = 15$. The decimal point in the result is placed so that the total number of digits to its right is the same as the total number of digits to the right of both decimal points in the original two numbers. The reason this is true becomes clear when we look at the denominators after we have changed from decimals to fractions.

EXAMPLE 3 Multiply: 2.1×0.07

SOLUTION

$$2.1 \times 0.07 = 2\frac{1}{10} \times \frac{7}{100} \quad \text{Change to fractions}$$

$$= \frac{21}{10} \times \frac{7}{100}$$

$$= \frac{147}{1,000} \quad \text{Multiply numerators and multiply denominators}$$

$$= 0.147 \quad \text{Write the answer as a decimal}$$

Again, the digits in the answer come from multiplying $21 \times 7 = 147$. The decimal point is placed so that there are three digits to its right, because that is the total number of digits to the right of the decimal points in 2.1 and 0.07.

We summarize this discussion with a rule.

Rule
To multiply two decimal numbers:
1. Multiply as you would if the decimal points were not there.
2. Place the decimal point in the answer so that the number of digits to its right is equal to the total number of digits to the right of the decimal points in the original two numbers in the problem.

EXAMPLE 4 How many digits will be to the right of the decimal point in the following product?

$$2.987 \times 24.82$$

SOLUTION There are three digits to the right of the decimal point in 2.987 and two digits to the right in 24.82. Therefore, there will be $3 + 2 = 5$ digits to the right of the decimal point in their product.

EXAMPLE 5 Multiply: 3.05×4.36

SOLUTION We can set this up as if it were a multiplication problem with whole numbers. We multiply and then place the decimal point in the correct position in the answer.

$$\begin{array}{r} 3.05 \\ \times\, 4.36 \\ \hline 1830 \\ 915 \\ 12\,20 \\ \hline 13.2980 \end{array}$$

3.05 ← 2 digits to the right of decimal point

× 4.36 ← 2 digits to the right of decimal point

13.2980 ← The decimal point is placed so that there are $2 + 2 = 4$ digits to its right

As you can see, multiplying decimal numbers is just like multiplying whole numbers, except that we must place the decimal point in the result in the correct position.

EXAMPLE 6 Multiply: $-1.3(-5.6)$

SOLUTION In this case we are multiplying two negative numbers. To do so we simply multiply their absolute values. The answer is positive, because the original two numbers have the same sign.

$$-1.3(-5.6) = 7.28$$

EXAMPLE 7 Multiply: $4.56(-100)$

SOLUTION The product of two numbers with different signs is negative.

$$4.56(-100) = -456$$

Estimating

Look back to Example 5. We could have placed the decimal point in the answer by rounding the two numbers to the nearest whole number and then multiplying them. Because 3.05 rounds to 3 and 4.36 rounds to 4, and the product of 3 and 4 is 12, we estimate that the answer to 3.05×4.36 will be close to 12. We then place the decimal point in the product 132980 between the 3 and the 2 in order to make it into a number close to 12.

EXAMPLE 8 Estimate the answer to each of the following products.

a. 29.4×8.2 **b.** 68.5×172 **c.** $(6.32)^2$

SOLUTION

a. Because 29.4 is approximately 30 and 8.2 is approximately 8, we estimate this product to be about $30 \times 8 = 240$. (If we were to multiply 29.4 and 8.2, we would find the product to be exactly 241.08.)

b. Rounding 68.5 to 70 and 172 to 170, we estimate this product to be $70 \times 170 = 11{,}900$. (The exact answer is 11,782.) Note here that we do not always round the numbers to the nearest whole number when making estimates. The idea is to round to numbers that will be easy to multiply.

c. Because 6.32 is approximately 6 and $6^2 = 36$, we estimate our answer to be close to 36. (The actual answer is 39.9424.)

---

5. Multiply.
a. 4.03×5.22
b. 40.3×0.522

6. Multiply: $1.3(-5.6)$

7. Multiply: $-4.56(-100)$

8. Estimate the answer to each product.
a. 82.3×5.8
b. 37.5×178
c. $(8.21)^2$

Answers
5. Both are 21.0366 **6.** -7.28
7. 456 **8. a.** 480 **b.** 7,200 **c.** 64

Combined Operations

We can use the rule for order of operations to simplify expressions involving decimal numbers and addition, subtraction, and multiplication.

EXAMPLE 9 Perform the indicated operations: $0.05(4.2 + 0.03)$

SOLUTION We begin by adding inside the parentheses:

$$\begin{aligned} 0.05(4.2 + 0.03) &= 0.05(4.23) && \text{Add} \\ &= 0.2115 && \text{Multiply} \end{aligned}$$

Notice that we could also have used the distributive property first, and the result would be unchanged:

$$\begin{aligned} 0.05(4.2 + 0.03) &= 0.05(4.2) + 0.05(0.03) && \text{Distributive property} \\ &= 0.210 + 0.0015 && \text{Multiply} \\ &= 0.2115 && \text{Add} \end{aligned}$$

9. Perform the indicated operations.
a. $0.03(5.5 + 0.02)$
b. $0.03(0.55 + 0.002)$

EXAMPLE 10 Simplify: $4.8 + 12(3.2)^2$

SOLUTION According to the rule for order of operations, we must first evaluate the number with an exponent, then multiply, and finally add.

$$\begin{aligned} 4.8 + 12(3.2)^2 &= 4.8 + 12(10.24) && (3.2)^2 = 10.24 \\ &= 4.8 + 122.88 && \text{Multiply} \\ &= 127.68 && \text{Add} \end{aligned}$$

10. Simplify.
a. $5.7 + 14(2.4)^2$
b. $0.57 + 1.4(2.4)^2$

B Applications

EXAMPLE 11 Find the area of each of the following stamps.

a.

Each side is 35.0 millimeters

b. Round to the nearest hundredth.

Length = 1.56 inches
Width = 0.99 inches

SOLUTION Applying our formulas for area we have

a. $A = s^2 = (35 \text{ mm})^2 = 1{,}225 \text{ mm}^2$

b. $A = lw = (1.56 \text{ in.})(0.99 \text{ in.}) = 1.54 \text{ in}^2$

11. Find the area of each stamp in Example 11 from the dimensions given below. Round answers to the nearest hundredth.
a. Each side is 1.38 inches
b. Length = 39.6 millimeters, width = 25.1 millimeters

Answers
9. a. 0.1656 **b.** 0.01656
10. a. 86.34 **b.** 8.634
11. a. 1.90 in. **b.** 993.96 mm

EXAMPLE 12 Sally earns \$6.82 for each of the first 36 hours she works in one week and \$10.23 in overtime pay for each additional hour she works in the same week. How much money will she make if she works 42 hours in one week?

SOLUTION The difference between 42 and 36 is 6 hours of overtime pay. The total amount of money she will make is

$$\underbrace{6.82(36)}_{\text{Pay for the first 36 hours}} + \underbrace{10.23(6)}_{\text{Pay for the next 6 hours}} = 245.52 + 61.38$$
$$= 306.90$$

She will make \$306.90 for working 42 hours in one week.

12. How much will Sally make if she works 50 hours in one week?

Note To estimate the answer to Example 12 before doing the actual calculations, we would do the following:
$6(40) + 10(6) = 240 + 60 = 300$

C Circumference

FACTS FROM GEOMETRY The Circumference of a Circle

The *circumference* of a circle is the distance around the outside, just as the perimeter of a polygon is the distance around the outside. The circumference of a circle can be found by measuring its radius or diameter and then using the appropriate formula. The *radius* of a circle is the distance from the center of the circle to the circle itself. The radius is denoted by the letter r. The *diameter* of a circle is the distance from one side to the other, through the center. The diameter is denoted by the letter d. In Figure 1 we can see that the diameter is twice the radius, or

$$d = 2r$$

The relationship between the circumference and the diameter or radius is not as obvious. As a matter of fact, it takes some fairly complicated mathematics to show just what the relationship between the circumference and the diameter is.

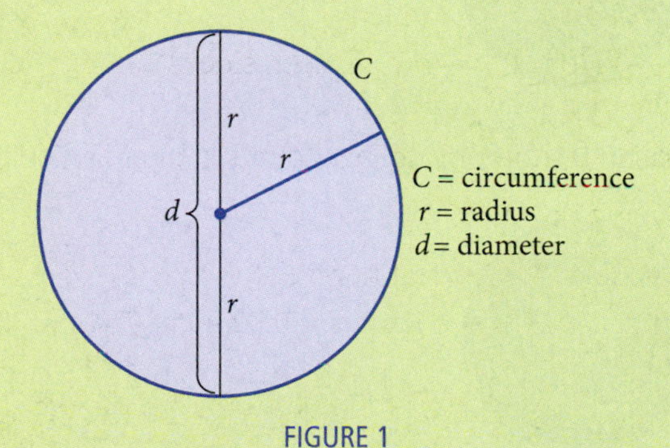

FIGURE 1

If you took a string and actually measured the circumference of a circle by wrapping the string around the circle and then measured the diameter of the same circle, you would find that the ratio of the circumference to the diameter, C/d, would be approximately equal to 3.14. The actual ratio of C to d in any circle is an irrational number. It can't be written in decimal form. We use the symbol π (Greek pi) to represent this ratio. In symbols the relationship between the circumference and the diameter in any circle is

$$\frac{C}{d} = \pi$$

Answer
12. \$388.74

Knowing what we do about the relationship between division and multiplication, we can rewrite this formula as

$$C = \pi d$$

This is the formula for the circumference of a circle. When we do the actual calculations, we will use the approximation 3.14 for π.

Because $d = 2r$, the same formula written in terms of the radius is

$$C = 2\pi r$$

Here are some examples that show how we use the formulas given above to find the circumference of a circle.

13. Find the circumference of a circle with a diameter of 3 centimeters.

EXAMPLE 13 Find the circumference of a circle with a diameter of 5 feet.

SOLUTION Substituting 5 for d in the formula $C = \pi d$, and using 3.14 for π, we have

$$C = 3.14(5)$$
$$= 15.7 \text{ feet}$$

14. Find the circumference for each coin in Example 14 from the dimensions given below. Round answers to the nearest hundredth.
a. Diameter = 0.92 inches
b. Radius = 13.20 millimeters

EXAMPLE 14 Find the circumference of each coin.

a. 1 Euro coin (Round to the nearest whole number.)

Diameter = 23.25 millimeters

b. Susan B. Anthony dollar (Round to the nearest hundredth.)

Radius = 0.52 inch

SOLUTION Applying our formulas for circumference we have:

a. $C = \pi d \approx (3.14)(23.25) \approx 73$ mm

b. $C = 2\pi r \approx 2(3.14)(0.52) \approx 3.27$ in.

Answers
13. 9.42 cm
14. a. 2.89 in. **b.** 82.90 mm

FACTS FROM GEOMETRY Other Formulas Involving π

Two figures are presented here, along with some important formulas that are associated with each figure. As you can see, each of the formulas contains the number π. When we do the actual calculations, we will use the approximation 3.14 for π.

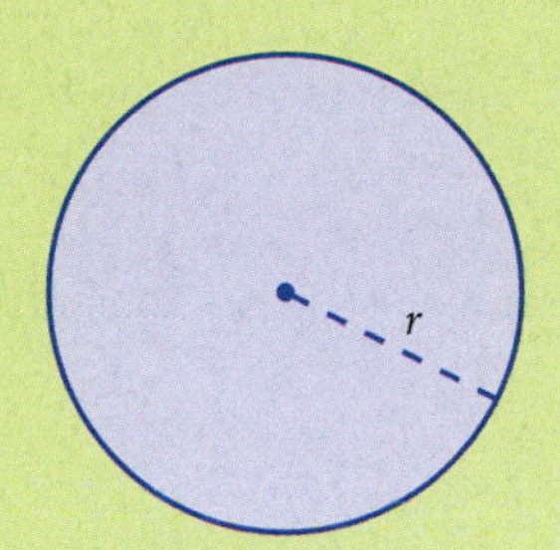

Area = π(radius)^2
$A = \pi r^2$

FIGURE 2 Circle

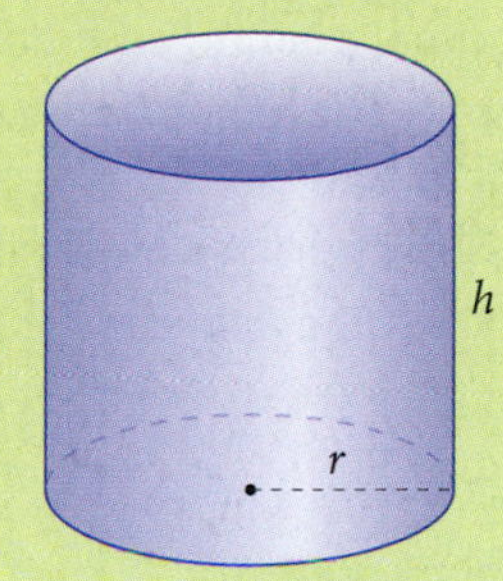

Volume = π(radius)^2(height)
$V = \pi r^2 h$

FIGURE 3 Right circular cylinder

EXAMPLE 15 Find the area of a circle with a diameter of 10 feet.

SOLUTION The formula for the area of a circle is $A = \pi r^2$. Because the radius r is half the diameter and the diameter is 10 feet, the radius is 5 feet. Therefore,

$$A = \pi r^2 = (3.14)(5)^2 = (3.14)(25) = 78.5 \text{ ft}^2$$

EXAMPLE 16 The drinking straw shown in Figure 4 has a radius of 0.125 inch and a length of 6 inches. To the nearest thousandth, find the volume of liquid that it will hold.

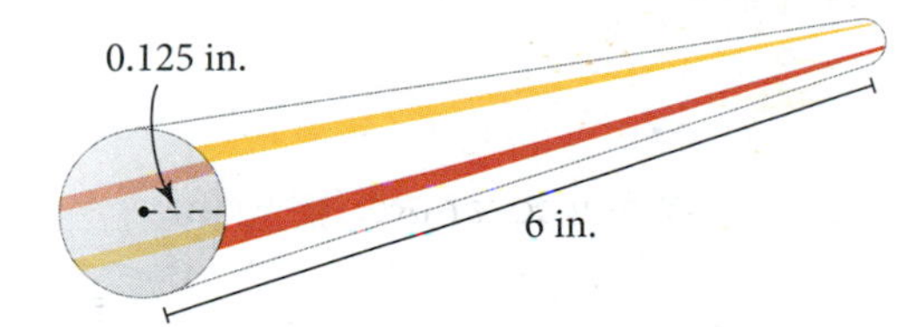

FIGURE 4

SOLUTION The total volume is found from the formula for the volume of a right circular cylinder. In this case, the radius is $r = 0.125$, and the height is $h = 6$. We approximate π with 3.14.

$$\begin{aligned} V &= \pi r^2 h \\ &= (3.14)(0.125)^2(6) \\ &= (3.14)(0.015625)(6) \\ &= 0.294 \text{ in}^3 \text{ to the nearest thousandth} \end{aligned}$$

15. Find the area of a circle with a diameter of 20 feet.

16. Find the volume of the straw in Example 16, if the radius is doubled. Round your answer to the nearest thousandth.

Answers
15. 314 ft^2 **16.** 1.178 in^3

Getting Ready for Class

After reading through the preceding section, respond in your own words and in complete sentences.

1. If you multiply 34.76 and 0.072, how many digits will be to the right of the decimal point in your answer?

2. To simplify the expression 0.053(9) + 67.42, what would be the first step according to the rule for order of operations?

3. What is the purpose of estimating?

4. What are some applications of decimals that we use in our everyday lives?

Problem Set 5.3

A Find each of the following products. (Multiply.) [Examples 1–3, 5]

1. $\begin{array}{r} 0.7 \\ \times\ 0.4 \\ \hline \end{array}$

2. $\begin{array}{r} 0.8 \\ \times\ 0.3 \\ \hline \end{array}$

3. $\begin{array}{r} 0.07 \\ \times\ 0.4 \\ \hline \end{array}$

4. $\begin{array}{r} 0.8 \\ \times\ 0.03 \\ \hline \end{array}$

5. $\begin{array}{r} 0.03 \\ \times\ 0.09 \\ \hline \end{array}$

6. $\begin{array}{r} 0.07 \\ \times\ 0.002 \\ \hline \end{array}$

7. $2.6(0.3)$

8. $8.9(0.2)$

9. $\begin{array}{r} 0.9 \\ \times\ 0.88 \\ \hline \end{array}$

10. $\begin{array}{r} 0.8 \\ \times\ 0.99 \\ \hline \end{array}$

11. $\begin{array}{r} 3.12 \\ \times\ 0.005 \\ \hline \end{array}$

12. $\begin{array}{r} 4.69 \\ \times\ 0.006 \\ \hline \end{array}$

13. $\begin{array}{r} 4.003 \\ \times\ 6.07 \\ \hline \end{array}$

14. $\begin{array}{r} 7.0001 \\ \times\ \ \ 3.04 \\ \hline \end{array}$

15. $5(0.006)$

16. $7(0.005)$

17. $\begin{array}{r} 75.14 \\ \times\ 2.5 \\ \hline \end{array}$

18. $\begin{array}{r} 963.8 \\ \times\ 0.24 \\ \hline \end{array}$

19. $\begin{array}{r} 0.1 \\ \times\ 0.02 \\ \hline \end{array}$

20. $\begin{array}{r} 0.3 \\ \times\ 0.02 \\ \hline \end{array}$

21. $2.796(10)$

22. $97.531(100)$

23. $\begin{array}{r} 0.0043 \\ \times\ 100 \\ \hline \end{array}$

24. $\begin{array}{r} 12.345 \\ \times\ 1{,}000 \\ \hline \end{array}$

25. $\begin{array}{r} 49.94 \\ \times\ 1{,}000 \\ \hline \end{array}$

26. $\begin{array}{r} 157.02 \\ \times\ 10{,}000 \\ \hline \end{array}$

27. $\begin{array}{r} 987.654 \\ \times\ 10{,}000 \\ \hline \end{array}$

28. $\begin{array}{r} 1.23 \\ \times\ 100{,}000 \\ \hline \end{array}$

A Perform the following operations according to the rule for order of operations. [Examples 6, 7, 9, 10]

29. $2.1(3.5 - 2.6)$

30. $5.4(9.9 - 6.6)$

31. $0.05(0.02 + 0.03)$

32. $0.04(0.07 + 0.09)$

33. $2.02(0.03 + 2.5)$

34. $4.04(0.05 + 6.6)$

35. $(2.1 + 0.03)(3.4 + 0.05)$

36. $(9.2 + 0.01)(3.5 + 0.03)$

37. $(2.1 - 0.1)(2.1 + 0.1)$

38. $(9.6 - 0.5)(9.6 + 0.5)$

39. $3.08 - 0.2(5 + 0.03)$

40. $4.09 + 0.5(6 + 0.02)$

41. $4.23 - 5(0.04 + 0.09)$

42. $7.89 - 2(0.31 + 0.76)$

43. $2.5 + 10(4.3)^2$

44. $3.6 + 15(2.1)^2$

45. $100(1 + 0.08)^2$

46. $500(1 + 0.12)^2$

47. $(1.5)^2 + (2.5)^2 + (3.5)^2$

48. $(1.1)^2 + (2.1)^2 + (3.1)^2$

B Applying the Concepts [Examples 11–16]

Solve each of the following word problems. Note that not all of the problems are solved by simply multiplying the numbers in the problems. Many of the problems involve addition and subtraction as well as multiplication.

49. Google Earth This Google Earth image shows an aerial view of a crop circle found near Wroughton, England. If the crop circle has a radius of 59.13 meters, what is its circumference? Use the approximation 3.14 for π. Round to the nearest hundredth.

50. Google Earth This is a 3D model of the Louvre Museum in Paris, France. The pyramid that dominates the Napoleon Courtyard has a height of 21.65 meters and a square base with sides of 35.50 meters. What is the volume of the pyramid to the nearest whole number? Hint: The volume of a pyramid can be found by the equation $V = \left(\frac{1}{3}\right)$(area of the base)(height).

51. Number Problem What is the product of 6 and the sum of 0.001 and 0.02?

52. Number Problem Find the product of 8 and the sum of 0.03 and 0.002.

53. Number Problem What does multiplying a decimal number by 100 do to the decimal point?

54. Number Problem What does multiplying a decimal number by 1,000 do to the decimal point?

55. Home Mortgage On a certain home mortgage, there is a monthly payment of \$9.66 for every \$1,000 that is borrowed. What is the monthly payment on this type of loan if \$143,000 is borrowed?

56. Caffeine Content If 1 cup of regular coffee contains 105 milligrams of caffeine, how much caffeine is contained in 3.5 cups of coffee?

57. Geometry of a Coin The \$1 coin shown here depicts Sacagawea and her infant son. The diameter of the coin is 26.5 mm, and the thickness is 2.00 mm. Find the following, rounding your answers to the nearest hundredth. Use 3.14 for π.

a. The circumference of the coin.

b. The area of one face of the coin.

c. The volume of the coin.

58. Geometry of a Coin The Susan B. Anthony dollar shown here has a radius of 0.52 inches and a thickness of 0.0079 inches. Find the following, rounding your answers to the nearest ten thousandth, if necessary. Use 3.14 for π.

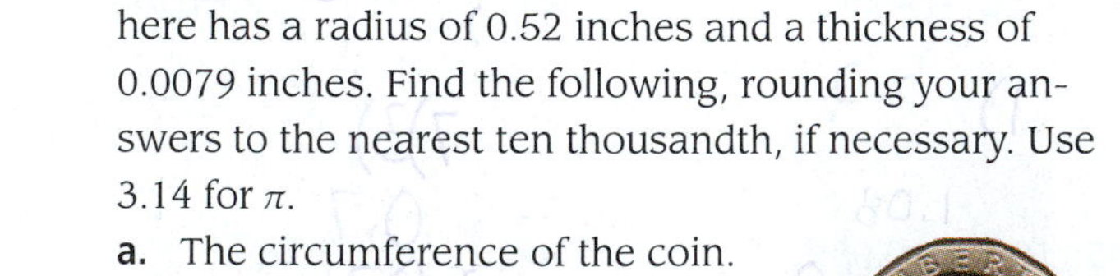

a. The circumference of the coin.

b. The area of one face of the coin.

c. The volume of the coin.

59. Area of a Stamp This stamp shows the Mexican artist Frida Kahlo. The image area of the stamp has a width of 0.84 inches and a length of 1.41 inches. Find the area of the image. Round to the nearest hundredth.

60. Area of a Stamp This stamp was issued in 2001 to honor the Italian scientist Enrico Fermi. The image area of the stamp has a width of 21.4 millimeters and a length of 35.8 millimeters. Find the area of the image. Round to the nearest whole number.

C Circumference Find the circumference and the area of each circle. Use 3.14 for π. [Examples 13–16]

61.

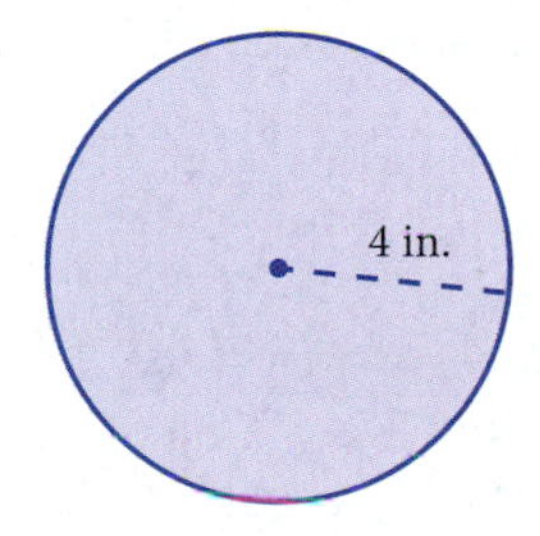

62.

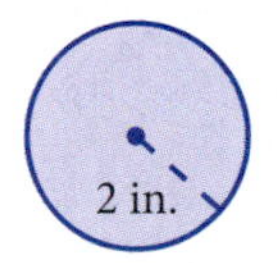

63. Circumference The radius of the earth is approximately 3,900 miles. Find the circumference of the earth at the equator. (The equator is a circle around the earth that divides the earth into two equal halves.)

64. Circumference The radius of the moon is approximately 1,100 miles. Find the circumference of the moon around its equator.

65. Bicycle Wheel The wheel on a 26-inch bicycle is such that the distance from the center of the wheel to the outside of the tire is 26.75 inches. If you walk the bicycle so that the wheel turns through one complete revolution, how many inches did you walk? Round to the nearest inch.

66. Model Plane A model plane is flying in a circle with a radius of 40 feet. To the nearest foot, how far does it fly in one complete trip around the circle?

Find the volume of each right circular cylinder.

67.

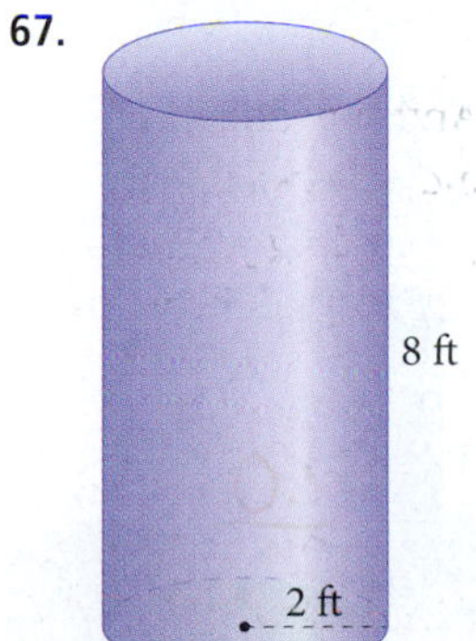

68.

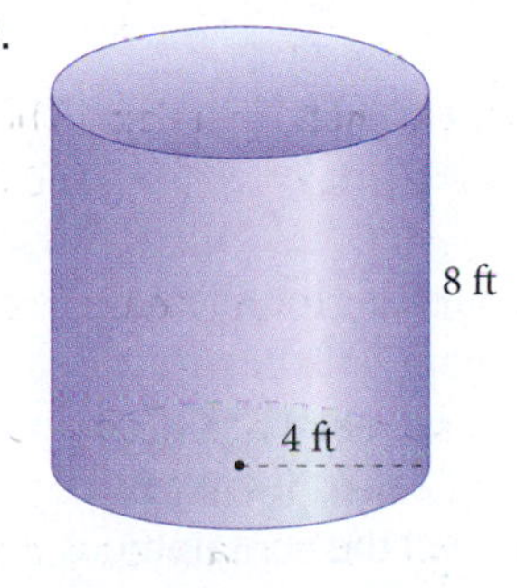

69.

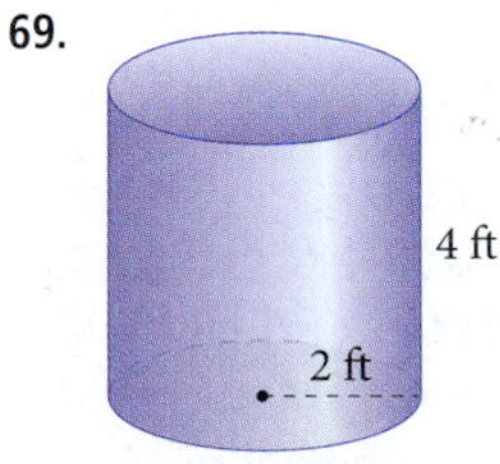

70.

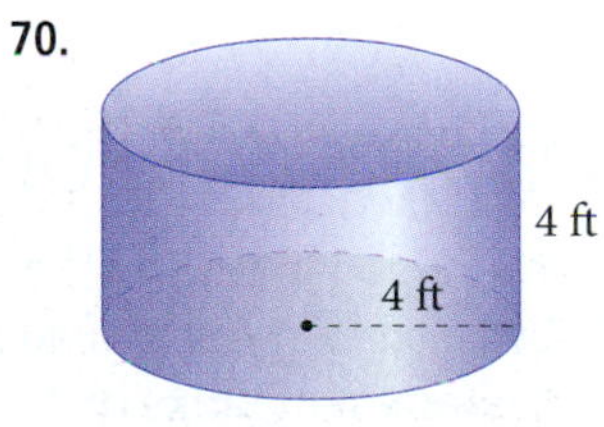

Getting Ready for the Next Section

To get ready for the next section, which covers division with decimals, we will review division with whole numbers and fractions.

Perform each of the following divisions. (Find the quotients.)

71. $3{,}758 \div 2$

72. $9{,}900 \div 22$

73. $50{,}032 \div 33$

74. $90{,}902 \div 5$

75. $20\overline{)5{,}960}$

76. $30\overline{)4{,}620}$

77. 4×8.7

78. 5×6.7

79. 27×1.848

80. 35×32.54

81. $38\overline{)31{,}350}$

82. $25\overline{)377{,}800}$

Maintaining Your Skills

83. Write the fractions in order from smallest to largest.

$\frac{2}{5} \quad \frac{4}{5} \quad \frac{3}{10} \quad \frac{1}{2}$

$\frac{8}{20} \quad \frac{16}{20} \quad \frac{6}{20} \quad \frac{10}{20}$

84. Write the fractions in order from smallest to largest.

$\frac{4}{5} \quad \frac{1}{4} \quad \frac{1}{10} \quad \frac{17}{100}$

85. Write the numbers in order from smallest to largest.

$1\frac{5}{6} \quad \frac{3}{2} \quad 1\frac{2}{3} \quad \frac{25}{12}$

$\frac{11}{6} \quad \frac{18}{12} \quad \frac{5}{3}$

$\frac{22}{12} \quad \frac{20}{12}$

86. Write the numbers in order from smallest to largest.

$1\frac{11}{12} \quad \frac{19}{12} \quad \frac{4}{3} \quad 1\frac{1}{6}$

Extending the Concepts

87. Containment System Holding tanks for hazardous liquids are often surrounded by containment tanks that will hold the hazardous liquid if the main tank begins to leak. We see that the center tank has a height of 16 feet and a radius of 6 feet. The outside containment tank has a height of 4 feet and a radius of 8 feet. If the center tank is full of heating fuel and develops a leak at the bottom, will the containment tank be able to hold all the heating fuel that leaks out?

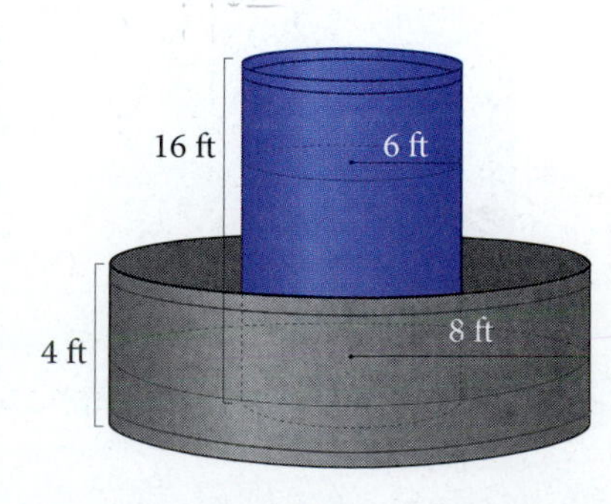

Division with Decimals

5.4

Objectives

A Divide decimal numbers.

B Solve application problems involving decimals.

Examples now playing at
MathTV.com/books

Introduction . . .

The chart shows the top finishing times for the men's 400-meter freestyle swim during Sydney's Olympics. An Olympic pool is 50 meters long, so each swimmer will have to complete 8 lengths during a 400-meter race.

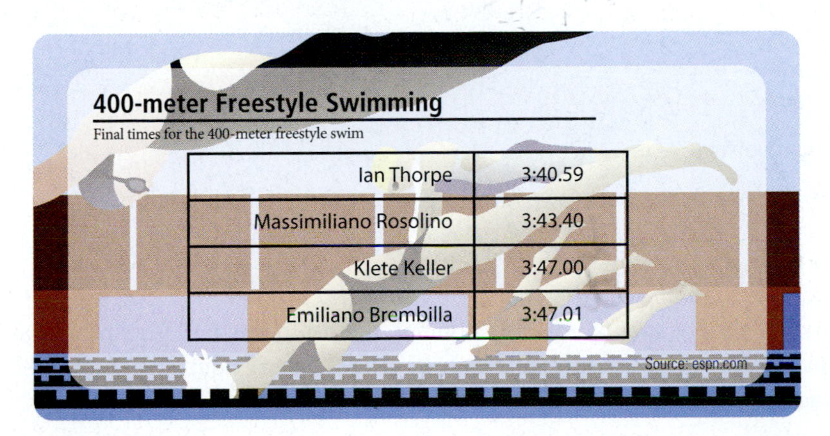

Ian Thorpe	3:40.59
Massimiliano Rosolino	3:43.40
Klete Keller	3:47.00
Emiliano Brembilla	3:47.01

During the race, each swimmer keeps track of how long it takes him to complete each length. To find the time of a swimmer's average lap, we need to be able to divide with decimal numbers, which we will learn in this section.

A Dividing with Decimals

EXAMPLE 1 Divide: 5,974 ÷ 20

SOLUTION

```
       298
  20)5,974
     4 0
     1 97
     1 80
       174
       160
        14
```

In the past we have written this answer as $298\frac{14}{20}$ or, after reducing the fraction, $298\frac{7}{10}$. Because $\frac{7}{10}$ can be written as 0.7, we could also write our answer as 298.7. This last form of our answer is exactly the same result we obtain if we write 5,974 as 5,974.0 and continue the division until we have no remainder. Here is how it looks:

```
       298.7
  20)5,974.0
     4 0↓
     1 97
     1 80↓
       174
       160↓
        14 0
        14 0
           0
```

Notice that we place the decimal point in the answer directly above the decimal point in the problem

Let's try another division problem. This time one of the numbers in the problem will be a decimal.

PRACTICE PROBLEMS

1. Divide: 4,626 ÷ 30

We can estimate the answer to Example 1 by rounding 5,974 to 6,000 and dividing by 20:

$$\frac{6{,}000}{20} = 300$$

Note We never need to make a mistake with division, because we can always check our results with multiplication.

Answer

1. 154.2

2. Divide.
   - a. $33.5 \div 5$
   - b. $34.5 \div 5$
   - c. $35.5 \div 5$

EXAMPLE 2 Divide: $34.8 \div 4$

SOLUTION We can use the ideas from Example 1 and divide as usual. The decimal point in the answer will be placed directly above the decimal point in the problem.

```
   8.7         Check:   8.7
4)34.8                × 4
  32↓                  34.8
   2 8
   2 8
     0
```

The answer is 8.7.

We can use these facts to write a rule for dividing decimal numbers.

Rule

To divide a decimal by a whole number, we do the usual long division as if there were no decimal point involved. The decimal point in the answer is placed directly above the decimal point in the problem.

Here are some more examples to illustrate the procedure.

3. Divide.
   - a. $47.448 \div 18$
   - b. $474.48 \div 18$

EXAMPLE 3 Divide: $49.896 \div 27$

SOLUTION

```
     1.848
27)49.896      Check this result by multiplication:
   27↓
   22 8              1.848
   21 6↓           ×    27
    1 29            12 936
    1 08↓           36 96
      216           49.896
      216
        0
```

We can write as many zeros as we choose after the rightmost digit in a decimal number without changing the value of the number. For example,

$$6.91 = 6.910 = 6.9100 = 6.91000$$

There are times when this can be very useful, as Example 4 shows.

4. Divide.
   - a. $1,138.5 \div 25$
   - b. $113.85 \div 25$

EXAMPLE 4 Divide: $1,138.9 \div 35$

SOLUTION

```
      32.54
35)1,138.90     Write 0 after the 9. It doesn't
   1 05↓        change the original number,
     88         but it gives us another digit
     70↓        to bring down.
     18 9       Check:    32.54
     17 5↓              ×    35
      1 40              162 70
      1 40              976 2
         0            1,138.90
```

Answers

2. a. 6.7 b. 6.9 c. 7.1
3. a. 2.636 b. 26.36
4. a. 45.54 b. 4.554

Until now we have considered only division by whole numbers. Extending division to include division by decimal numbers is a matter of knowing what to do about the decimal point in the divisor.

EXAMPLE 5 Divide: $31.35 \div 3.8$

SOLUTION In fraction form, this problem is equivalent to

$$\frac{31.35}{3.8}$$

If we want to write the divisor as a whole number, we can multiply the numerator and the denominator of this fraction by 10:

$$\frac{31.35 \times \mathbf{10}}{3.8 \times \mathbf{10}} = \frac{313.5}{38}$$

So, since this fraction is equivalent to the original fraction, our original division problem is equivalent to

```
      8.25
38)313.50    Put 0 after the last digit
   304 ↓
     9 5
     7 6↓
     1 90
     1 90
        0
```

5. Divide.
   a. $13.23 \div 4.2$
   b. $13.23 \div 0.42$

Note We do not always use the rules for rounding numbers to make estimates. For example, to estimate the answer to Example 5, $31.35 \div 3.8$, we can get a rough estimate of the answer by reasoning that 3.8 is close to 4 and 31.35 is close to 32. Therefore, our answer will be approximately $32 \div 4 = 8$.

We can summarize division with decimal numbers by listing the following points, as illustrated by the first five examples.

Summary of Division with Decimals

1. We divide decimal numbers by the same process used in Chapter 1 to divide whole numbers. The decimal point in the answer is placed directly above the decimal point in the dividend.
2. We are free to write as many zeros after the last digit in a decimal number as we need.
3. If the divisor is a decimal, we can change it to a whole number by moving the decimal point to the right as many places as necessary so long as we move the decimal point in the dividend the same number of places.

EXAMPLE 6 Divide, and round the answer to the nearest hundredth:

$0.3778 \div 0.25$

SOLUTION First, we move the decimal point two places to the right:

$0.25.\overline{)\,.37.78}$

6. Divide, and round your answer to the nearest hundredth: $0.4553 \div 0.32$

Note Moving the decimal point two places in both the divisor and the dividend is justified like this:

$$\frac{0.3778 \times \mathbf{100}}{0.25 \times \mathbf{100}} = \frac{37.78}{25}$$

Answers
5. a. 3.15 b. 31.5

Then we divide, using long division:

```
     1.5112
25)37.7800
   25
   12 7
   12 5
      28
      25
       30
       25
        50
        50
         0
```

Rounding to the nearest hundredth, we have 1.51. We actually did not need to have this many digits to round to the hundredths column. We could have stopped at the thousandths column and rounded off.

7. Divide, and round to the nearest tenth.
   a. 19 ÷ 0.06
   b. 1.9 ÷ 0.06

EXAMPLE 7 Divide, and round to the nearest tenth: 17 ÷ 0.03

SOLUTION Because we are rounding to the nearest tenth, we will continue dividing until we have a digit in the hundredths column. We don't have to go any further to round to the tenths column.

```
           5 66.66
0.03.)17.00.00
      15
       2 0
       1 8
         20
         18
          2 0
          1 8
            20
            18
             2
```

Rounding to the nearest tenth, we have 566.7.

B Applications

8. A woman earning $6.54 an hour receives a paycheck for $186.39. How many hours did the woman work?

EXAMPLE 8 If a man earning $7.26 an hour receives a paycheck for $235.95, how many hours did he work?

SOLUTION To find the number of hours the man worked, we divide $235.95 by $7.26.

```
              32.5
7.26.)235.95.0
      217 8
       18 15
       14 52
        3 63 0
        3 63 0
             0
```

The man worked 32.5 hours.

Answers
6. 1.42 7. a. 316.7 b. 31.7
8. 28.5 hours

EXAMPLE 9 A telephone company charges $0.43 for the first minute and then $0.33 for each additional minute for a long-distance call. If a long-distance call costs $3.07, how many minutes was the call?

SOLUTION To solve this problem we need to find the number of additional minutes for the call. To do so, we first subtract the cost of the first minute from the total cost, and then we divide the result by the cost of each additional minute. Without showing the actual arithmetic involved, the solution looks like this:

$$\text{The number of additional minutes} = \frac{\overset{\text{Total cost of the call}}{3.07} - \overset{\text{Cost of the first minute}}{0.43}}{\underset{\text{Cost of each additional minute}}{0.33}} = \frac{2.64}{0.33} = 8$$

The call was 9 minutes long. (The number 8 is the number of additional minutes past the first minute.)

9. If the phone company in Example 9 charged $4.39 for a call, how long was the call?

DESCRIPTIVE STATISTICS

Grade Point Average

I have always been surprised by the number of my students who have difficulty calculating their grade point average (GPA). During her first semester in college, my daughter, Amy, earned the following grades:

Class	Units	Grade
Algebra	5	B
Chemistry	4	C
English	3	A
History	3	B

When her grades arrived in the mail, she told me she had a 3.0 grade point average, because the A and C grades averaged to a B. I told her that her GPA was a little less than a 3.0. What do you think? Can you calculate her GPA? If not, you will be able to after you finish this section.

When you calculate your grade point average (GPA), you are calculating what is called a *weighted average*. To calculate your grade point average, you must first calculate the number of grade points you have earned in each class that you have completed. The number of grade points for a class is the product of the number of units the class is worth times the value of the grade received. The table below shows the value that is assigned to each grade.

Grade	Value
A	4
B	3
C	2
D	1
F	0

If you earn a B in a 4-unit class, you earn $4 \times 3 = 12$ grade points. A grade of C in the same class gives you $4 \times 2 = 8$ grade points. To find your grade point average for one term (a semester or quarter), you must add your grade points and divide that total by the number of units. Round your answer to the nearest hundredth.

Answer
9. 13 minutes

10. If Amy had earned a B in chemistry, instead of a C, what grade point average would she have?

EXAMPLE 10 Calculate Amy's grade point average using the information above.

SOLUTION We begin by writing in two more columns, one for the value of each grade (4 for an A, 3 for a B, 2 for a C, 1 for a D, and 0 for an F), and another for the grade points earned for each class. To fill in the grade points column, we multiply the number of units by the value of the grade:

Class	Units	Grade	Value	Grade Points
Algebra	5	B	3	$5 \times 3 = 15$
Chemistry	4	C	2	$4 \times 2 = 8$
English	3	A	4	$3 \times 4 = 12$
History	3	B	3	$3 \times 3 = 9$
Total Units	15			Total Grade Points: 44

To find her grade point average, we divide 44 by 15 and round (if necessary) to the nearest hundredth:

$$\text{Grade point average} = \frac{44}{15} = 2.93$$

STUDY SKILLS
Pay Attention to Instructions

Taking a test is not like doing homework. On a test, the problems will be varied. When you do your homework, you usually work a number of similar problems. I have some students who do very well on their homework but become confused when they see the same problems on a test. The reason for their confusion is that they have not paid attention to the instructions on their homework. If a test problem asks for the *mean* of some numbers, then you must know the definition of the word *mean*. Likewise, if a test problem asks you to find a *sum* and then to *round* your answer to the nearest hundred, then you must know that the word *sum* indicates addition, and after you have added, you must round your answer as indicated.

Getting Ready for Class

After reading through the preceding section, respond in your own words and in complete sentences.

1. The answer to the division problem in Example 1 is $298\frac{14}{20}$. Write this number in decimal notation.
2. In Example 4 we place a 0 at the end of a number without changing the value of the number. Why is the placement of this 0 helpful?
3. The expression $0.3778 \div 0.25$ is equivalent to the expression $37.78 \div 25$ because each number was multiplied by what?
4. Round 372.1675 to the nearest tenth.

Answer
10. 3.20

Problem Set 5.4

A Perform each of the following divisions. [Examples 1–5]

1. $394 \div 20$
2. $486 \div 30$
3. $248 \div 40$
4. $372 \div 80$
5. $5\overline{)26}$
6. $8\overline{)36}$
7. $25\overline{)276}$
8. $50\overline{)276}$
9. $28.8 \div 6$
10. $15.5 \div 5$
11. $77.6 \div 8$
12. $31.48 \div 4$
13. $35\overline{)92.05}$
14. $26\overline{)146.38}$
15. $45\overline{)190.8}$
16. $55\overline{)342.1}$

17. $86.7 \div 34$

18. $411.4 \div 44$

19. $29.7 \div 22$

20. $488.4 \div 88$

21. $4.5\overline{)29.25}$

22. $3.3\overline{)21.978}$

23. $0.11\overline{)1.089}$

24. $0.75\overline{)2.40}$

25. $2.3\overline{)0.115}$

26. $6.6\overline{)0.198}$

27. $0.012\overline{)1.068}$

28. $0.052\overline{)0.23712}$

29. $1.1\overline{)2.42}$

30. $2.2\overline{)7.26}$

Carry out each of the following divisions only so far as needed to round the results to the nearest hundredth. [Examples 6, 7]

31. $26\overline{)35}$

32. $18\overline{)47}$

33. $3.3\overline{)56}$

34. $4.4\overline{)75}$

35. $0.1234 \div 0.5$

36. $0.543 \div 2.1$

37. $19 \div 7$

38. $16 \div 6$

39. $0.059\overline{)0.69}$

40. $0.048\overline{)0.49}$

41. $1.99 \div 0.5$

42. $0.99 \div 0.5$

43. $2.99 \div 0.5$

44. $3.99 \div 0.5$

Calculator Problems Work each of the following problems on your calculator. If rounding is necessary, round to the nearest hundred thousandth.

45. $7 \div 9$

46. $11 \div 13$

47. $243 \div 0.791$

48. $67.8 \div 37.92$

49. $0.0503 \div 0.0709$

50. $429.87 \div 16.925$

B Applying the Concepts [Examples 8–10]

51. Google Earth The Google Earth map shows Yellowstone National Park. There is an average of 2.3 moose per square mile. If there are about 7,986 moose in Yellowstone, how many square miles does Yellowstone cover? Round to the nearest square mile.

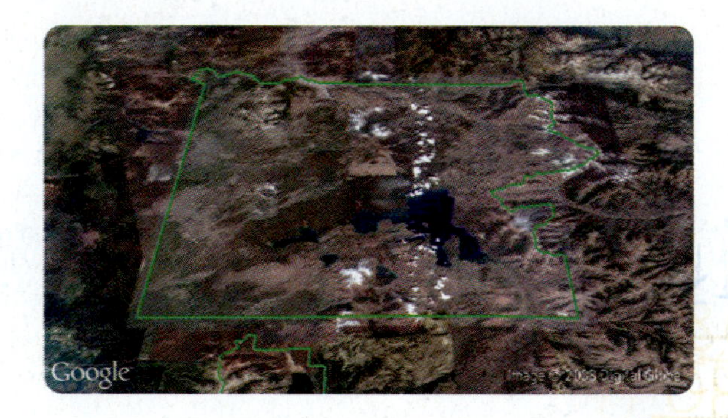

52. Google Earth The Google Earth image shows a corn field. A farmer harvests 29,952 bushels of corn. If the farmer harvested 130 bushels per acre, how many acres does the field cover?

53. Hot Air Balloon Since the pilot of a hot air balloon can only control the balloon's altitude, he relies on the winds for travel. To ride on the jet streams, a hot air balloon must rise as high as 12 kilometers. Convert this to miles by dividing by 1.61. Round your answer to the nearest tenth of a mile.

54. Hot Air Balloon December and January are the best times for traveling in a hot-air balloon because the jet streams in the Northern Hemisphere are the strongest. They reach speeds of 400 kilometers per hour. Convert this to miles per hour by dividing by 1.61. Round to the nearest whole number.

55. Wages If a woman earns $39.90 for working 6 hours, how much does she earn per hour?

56. Wages How many hours does a person making $6.78 per hour have to work in order to earn $257.64?

57. Gas Mileage If a car travels 336 miles on 15 gallons of gas, how far will the car travel on 1 gallon of gas?

58. Gas Mileage If a car travels 392 miles on 16 gallons of gas, how far will the car travel on 1 gallon of gas?

59. Wages Suppose a woman earns $6.78 an hour for the first 36 hours she works in a week and then $10.17 an hour in overtime pay for each additional hour she works in the same week. If she makes $294.93 in one week, how many hours did she work overtime?

60. Wages Suppose a woman makes $286.08 in one week. If she is paid $5.96 an hour for the first 36 hours she works and then $8.94 an hour in overtime pay for each additional hour she works in the same week, how many hours did she work overtime that week?

61. Phone Bill Suppose a telephone company charges $0.41 for the first minute and then $0.32 for each additional minute for a long-distance call. If a long-distance call costs $2.33, how many minutes was the call?

62. Phone Bill Suppose a telephone company charges $0.45 for the first three minutes and then $0.29 for each additional minute for a long-distance call. If a long-distance call costs $2.77, how many minutes was the call?

63. Women's Golf The table gives the top five money earners for the Ladies' Professional Golf Association (LPGA) in 2008, through June 1. Fill in the last column of the table by finding the average earnings per event for each golfer. Round your answers to the nearest dollar.

Rank	Name	Number of Events	Total Earnings	Average per Event
1.	Lorena Ochoa	25	$1,838,616	73545
2.	Annika Sorenstam	13	$1,295,585	99660
3.	Paula Creamer	24	$891,804	37159
4.	Seon Hwa Lee	28	$656,313	23430
5.	Jeong Jang	27	$642,320	23790

64. Men's Golf The table gives the top five money earners for the men's Professional Golf Association (PGA) in 2008, through June 1. Fill in the last column of the table by finding the average earnings per event for each golfer. Round your answers to the nearest dollar.

Rank	Name	Number of Events	Total Earnings	Average per Event
1.	Tiger Woods	5	$4,425,000	
2.	Phil Mickelson	13	$3,872,270	
3.	Geoff Ogilvy	13	$2,584,685	
4.	Stewart Cink	13	$2,516,512	
5.	Kenny Perry	15	$2,437,655	

Grade Point Average The following grades were earned by Steve during his first term in college. Use these data to answer Problems 65–68.

Class	Units	Grade
Basic mathematics	3	A
Health	2	B
History	3	B
English	3	C
Chemistry	4	C

65. Calculate Steve's GPA.

66. If his grade in chemistry had been a B instead of a C, by how much would his GPA have increased?

67. If his grade in health had been a C instead of a B, by how much would his grade point average have dropped?

68. If his grades in both English and chemistry had been B's, what would his GPA have been?

Getting Ready for the Next Section

In the next section we will consider the relationship between fractions and decimals in more detail. The problems below review some of the material that is necessary to make a successful start in the next section.

Reduce to lowest terms.

69. $\frac{75}{100}$

70. $\frac{220}{1,000}$

71. $\frac{12x}{18xy}$

72. $\frac{15xy}{30x}$

73. $\frac{75x^2y^3}{100xy^2}$

74. $\frac{220x^3y^2}{1,000x^2y}$

75. $\frac{38}{100}$

76. $\frac{75}{1,000}$

Write each fraction as an equivalent fraction with denominator 10.

77. $\frac{3}{5}$

78. $\frac{1}{2}$

Write each fraction as an equivalent fraction with denominator 100.

79. $\frac{3}{5}$

80. $\frac{17}{20}$

Write each fraction as an equivalent fraction with denominator $15x$.

81. $\frac{4}{5}$

82. $\frac{2}{3}$

83. $\frac{4}{x}$

84. $\frac{2}{x}$

85. $\frac{6}{5x}$

86. $\frac{7}{3x}$

Divide.

87. $3 \div 4$

88. $3 \div 5$

89. $7 \div 8$

90. $3 \div 8$

Maintaining Your Skills

Simplify.

91. $15\left(\frac{2}{3} + \frac{3}{5}\right)$

92. $15\left(\frac{4}{5} - \frac{1}{3}\right)$

93. $4\left(\frac{1}{2} + \frac{1}{4}\right)$

94. $6\left(\frac{1}{3} + \frac{1}{2}\right)$

Fractions and Decimals

5.5

Objectives

A Convert fractions to decimals.

B Convert decimals to fractions.

C Simplify expressions containing fractions and decimals.

D Solve applications involving fractions and decimals.

Introduction . . .

If you are shopping for clothes and a store has a sale advertising $\frac{1}{3}$ off the regular price, how much can you expect to pay for a pair of pants that normally sells for $31.95? If the sale price of the pants is $22.30, have they really been marked down by $\frac{1}{3}$? To answer questions like these, we need to know how to solve problems that involve fractions and decimals together.

We begin this section by showing how to convert back and forth between fractions and decimals.

Examples now playing at MathTV.com/books

A Converting Fractions to Decimals

You may recall that the notation we use for fractions can be interpreted as implying division. That is, the fraction $\frac{3}{4}$ can be thought of as meaning "3 divided by 4." We can use this idea to convert fractions to decimals.

EXAMPLE 1 Write $\frac{3}{4}$ as a decimal.

SOLUTION Dividing 3 by 4, we have

$$\begin{array}{r} .75 \\ 4\overline{)3.00} \\ \underline{2\,8} \\ 20 \\ \underline{20} \\ 0 \end{array}$$

The fraction $\frac{3}{4}$ is equal to the decimal 0.75.

EXAMPLE 2 Write $\frac{7}{12}$ as a decimal correct to the thousandths column.

SOLUTION Because we want the decimal to be rounded to the thousandths column, we divide to the ten thousandths column and round off to the thousandths column:

$$\begin{array}{r} .5833 \\ 12\overline{)7.0000} \\ \underline{6\,0} \\ 1\,00 \\ \underline{96} \\ 40 \\ \underline{36} \\ 40 \\ \underline{36} \\ 4 \end{array}$$

Rounding off to the thousandths column, we have 0.583. Because $\frac{7}{12}$ is not exactly the same as 0.583, we write

$$\frac{7}{12} \approx 0.583$$

where the symbol $\approx$ is read "is approximately."

PRACTICE PROBLEMS

1. Write as a decimal.
   a. $\frac{2}{5}$
   b. $\frac{3}{5}$
   c. $\frac{4}{5}$

2. Write as a decimal correct to the thousandths column.
   a. $\frac{11}{12}$
   b. $\frac{12}{13}$

Answers

1. a. 0.4 **b.** 0.6 **c.** 0.8

2. a. 0.917 **b.** 0.923

If we wrote more zeros after 7.0000 in Example 2, the pattern of 3's would continue for as many places as we could want. When we get a sequence of digits that repeat like this, 0.58333 . . . , we can indicate the repetition by writing

$0.58\overline{3}$ The bar over the 3 indicates that the 3 repeats from there on

3. Write $\frac{5}{11}$ as a decimal.

EXAMPLE 3 Write $\frac{3}{11}$ as a decimal.

SOLUTION Dividing 3 by 11, we have

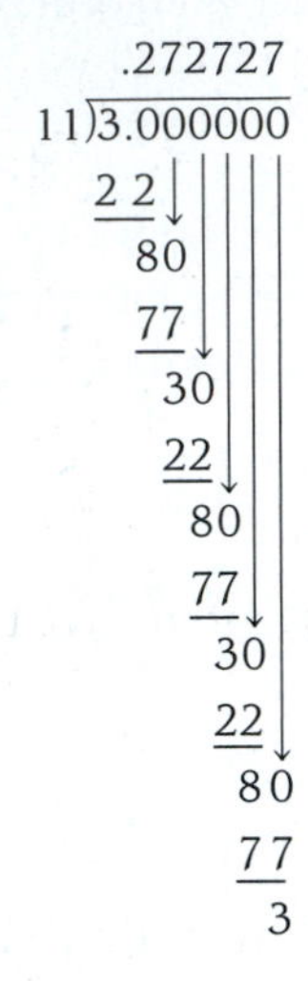

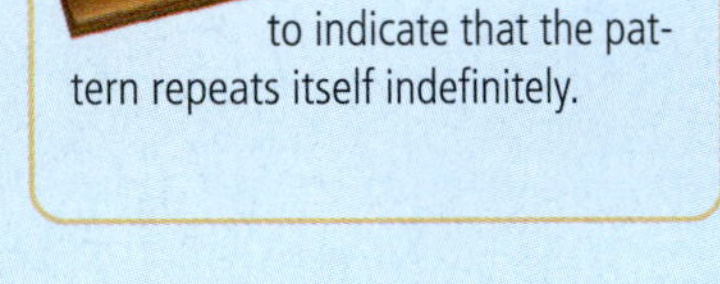

The bar over the 2 and the 7 in $0.\overline{27}$ is used to indicate that the pattern repeats itself indefinitely.

No matter how long we continue the division, the remainder will never be 0, and the pattern will continue. We write the decimal form of $\frac{3}{11}$ as $0.\overline{27}$, where

$0.\overline{27} = 0.272727\ldots$ The dots mean "and so on"

B Converting Decimals to Fractions

To convert decimals to fractions, we take advantage of the place values we assigned to the digits to the right of the decimal point.

4. Write as a fraction in lowest terms.
   a. 0.48
   b. 0.048

EXAMPLE 4 Write 0.38 as a fraction in lowest terms.

SOLUTION 0.38 is 38 hundredths, or

$$0.38 = \frac{38}{100}$$

$$= \frac{19}{50}$$ Divide the numerator and the denominator by 2 to reduce to lowest terms

The decimal 0.38 is equal to the fraction $\frac{19}{50}$.

We could check our work here by converting $\frac{19}{50}$ back to a decimal. We do this by dividing 19 by 50. That is,

```
     .38
50)19.00
   15 0
    4 00
    4 00
       0
```

Answers
3. $0.\overline{45}$ 4. a. $\frac{12}{25}$ b. $\frac{6}{125}$

EXAMPLE 5 Convert 0.075 to a fraction.

SOLUTION We have 75 thousandths, or

$$0.075 = \frac{75}{1,000}$$

$$= \frac{3}{40} \quad \text{Divide the numerator and the denominator by 25 to reduce to lowest terms}$$

EXAMPLE 6 Write 15.6 as a mixed number.

SOLUTION Converting 0.6 to a fraction, we have

$$0.6 = \frac{6}{10} = \frac{3}{5} \quad \text{Reduce to lowest terms}$$

Since $0.6 = \frac{3}{5}$, we have $15.6 = 15\frac{3}{5}$.

C Problems Containing Both Fractions and Decimals

We continue this section by working some problems that involve both fractions and decimals.

EXAMPLE 7 Simplify: $\frac{19}{50}(1.32 + 0.48)$

SOLUTION In Example 4, we found that $0.38 = \frac{19}{50}$. Therefore we can rewrite the problem as

$$\frac{19}{50}(1.32 + 0.48) = 0.38(1.32 + 0.48) \quad \text{Convert all numbers to decimals}$$

$$= 0.38(1.80) \quad \text{Add: } 1.32 + 0.48$$

$$= 0.684 \quad \text{Multiply: } 0.38 \times 1.80$$

EXAMPLE 8 Simplify: $\frac{1}{2} + (0.75)\left(\frac{2}{5}\right)$

SOLUTION We could do this problem one of two different ways. First, we could convert all fractions to decimals and then simplify:

$$\frac{1}{2} + (0.75)\left(\frac{2}{5}\right) = 0.5 + 0.75(0.4) \quad \text{Convert to decimals}$$

$$= 0.5 + 0.300 \quad \text{Multiply: } 0.75 \times 0.4$$

$$= 0.8 \quad \text{Add}$$

Or, we could convert 0.75 to $\frac{3}{4}$ and then simplify:

$$\frac{1}{2} + 0.75\left(\frac{2}{5}\right) = \frac{1}{2} + \frac{3}{4}\left(\frac{2}{5}\right) \quad \text{Convert decimals to fractions}$$

$$= \frac{1}{2} + \frac{3}{10} \quad \text{Multiply: } \frac{3}{4} \times \frac{2}{5}$$

$$= \frac{5}{10} + \frac{3}{10} \quad \text{The common denominator is 10}$$

$$= \frac{8}{10} \quad \text{Add numerators}$$

$$= \frac{4}{5} \quad \text{Reduce to lowest terms}$$

The answers are equivalent. That is, $0.8 = \frac{8}{10} = \frac{4}{5}$. Either method can be used with problems of this type.

5. Convert 0.025 to a fraction.

6. Write 12.8 as a mixed number.

7. Simplify: $\frac{14}{25}(2.43 + 0.27)$

8. Simplify: $\frac{1}{4} + 0.25\left(\frac{3}{5}\right)$

Answers

5. $\frac{1}{40}$ 6. $12\frac{4}{5}$ 7. 1.512

8. $\frac{2}{5}$, or 0.4

9. Simplify: $\left(\frac{1}{3}\right)^3(5.4) + \left(\frac{1}{5}\right)^2(2.5)$

EXAMPLE 9 Simplify: $\left(\frac{1}{2}\right)^3(2.4) + \left(\frac{1}{4}\right)^2(3.2)$

SOLUTION This expression can be simplified without any conversions between fractions and decimals. To begin, we evaluate all numbers that contain exponents. Then we multiply. After that, we add.

$$\left(\frac{1}{2}\right)^3(2.4) + \left(\frac{1}{4}\right)^2(3.2) = \frac{1}{8}(2.4) + \frac{1}{16}(3.2) \quad \text{Evaluate exponents}$$
$$= 0.3 + 0.2 \quad \text{Multiply by } \tfrac{1}{8} \text{ and } \tfrac{1}{16}$$
$$= 0.5 \quad \text{Add}$$

A Applications

10. A shirt that normally sells for \$35.50 is on sale for $\frac{1}{4}$ off. What is the sale price of the shirt? (Round to the nearest cent.)

EXAMPLE 10 If a shirt that normally sells for \$27.99 is on sale for $\frac{1}{3}$ off, what is the sale price of the shirt?

SOLUTION To find out how much the shirt is marked down, we must find $\frac{1}{3}$ of 27.99. That is, we multiply $\frac{1}{3}$ and 27.99, which is the same as dividing 27.99 by 3.

$$\frac{1}{3}(27.99) = \frac{27.99}{3} = 9.33$$

The shirt is marked down \$9.33. The sale price is the original price less the amount it is marked down:

$$\text{Sale price} = 27.99 - 9.33 = 18.66$$

The sale price is \$18.66. We also could have solved this problem by simply multiplying the original price by $\frac{2}{3}$, since, if the shirt is marked $\frac{1}{3}$ off, then the sale price must be $\frac{2}{3}$ of the original price. Multiplying by $\frac{2}{3}$ is the same as dividing by 3 and then multiplying by 2. The answer would be the same.

11. Find the area of the stamp in Example 11 if

Base = 6.6 centimeters, height = 3.3 centimeters

EXAMPLE 11 Find the area of the stamp.

Write your answer as a decimal, rounded to the nearest hundredth.

Base = $2\frac{5}{8}$ inches

Height = $1\frac{1}{4}$ inches

SOLUTION We can work the problem using fractions and then convert the answer to a decimal.

$$A = \frac{1}{2}bh = \frac{1}{2}\left(2\frac{5}{8}\right)\left(1\frac{1}{4}\right) = \frac{1}{2} \cdot \frac{21}{8} \cdot \frac{5}{4} = \frac{105}{64} \approx 1.64 \text{ in}^2$$

Or, we can convert the fractions to decimals and then work the problem.

$$A = \frac{1}{2}bh = \frac{1}{2}(2.625)(1.25) \approx 1.64 \text{ in}^2$$

Answers
9. 0.3 **10.** \$26.63 **11.** 10.89 cm^2

FACTS FROM GEOMETRY The Volume of a Sphere

Figure 1 shows a sphere and the formula for its volume. Because the formula contains both the fraction $\frac{4}{3}$ and the number π, and we have been using 3.14 for π, we can think of the formula as containing both a fraction and a decimal.

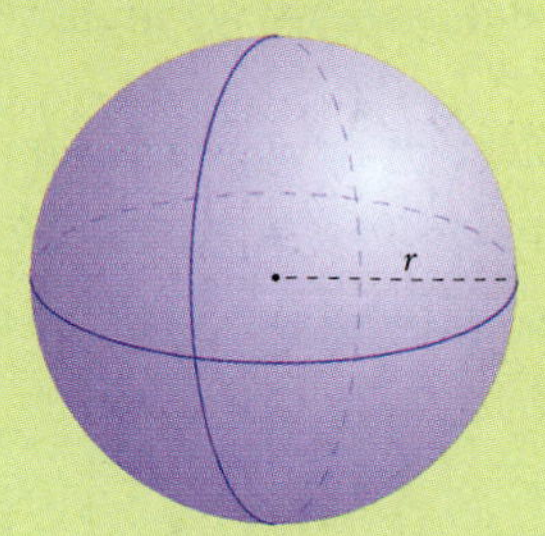

$$\text{Volume} = \frac{4}{3}\pi(\text{radius})^3$$
$$= \frac{4}{3}\pi r^3$$

FIGURE 1 Sphere

EXAMPLE 12 Figure 2 is composed of a right circular cylinder with half a sphere on top. (A half-sphere is called a *hemisphere*.) To the nearest tenth, find the total volume enclosed by the figure.

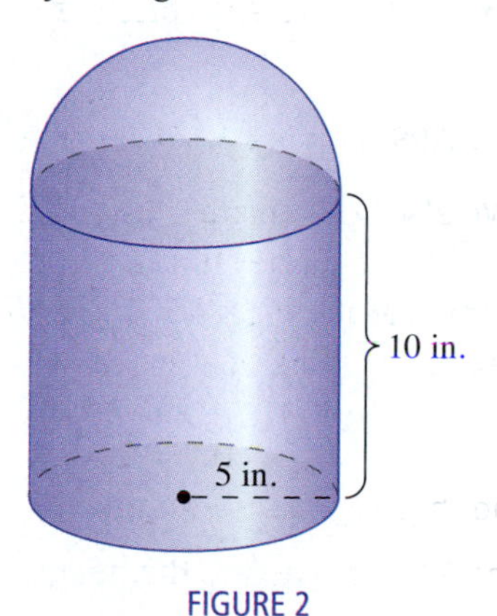

FIGURE 2

SOLUTION The total volume is found by adding the volume of the cylinder to the volume of the hemisphere.

$$V = \text{volume of cylinder} + \text{volume of hemisphere}$$
$$= \pi r^2 h + \frac{1}{2} \cdot \frac{4}{3}\pi r^3$$
$$= (3.14)(5)^2(10) + \frac{1}{2} \cdot \frac{4}{3}(3.14)(5)^3$$
$$= (3.14)(25)(10) + \frac{1}{2} \cdot \frac{4}{3}(3.14)(125)$$
$$= 785 + \frac{2}{3}(392.5) \qquad \text{Multiply: } \frac{1}{2} \cdot \frac{4}{3} = \frac{4}{6} = \frac{2}{3}$$
$$= 785 + \frac{785}{3} \qquad \text{Multiply: } 2(392.5) = 785$$
$$\approx 785 + 261.7 \qquad \text{Divide 785 by 3, and round to the nearest tenth}$$
$$= 1{,}046.7 \text{ in}^3$$

12. If the radius in Figure 2 is doubled so that it becomes 10 inches instead of 5 inches, what is the new volume of the figure? Round your answer to the nearest tenth.

Answer
12. 5,233.3 in^3

Getting Ready for Class

After reading through the preceding section, respond in your own words and in complete sentences.

1. To convert fractions to decimals, do we multiply or divide the numerator by the denominator?

2. The decimal 0.13 is equivalent to what fraction?

3. Write 36 thousandths in decimal form and in fraction form.

4. Explain how to write the fraction $\frac{84}{1,000}$ in lowest terms.

Problem Set 5.5

A Each circle below is divided into 8 equal parts. The number below each circle indicates what fraction of the circle is shaded. Convert each fraction to a decimal. [Examples 1–3]

1. 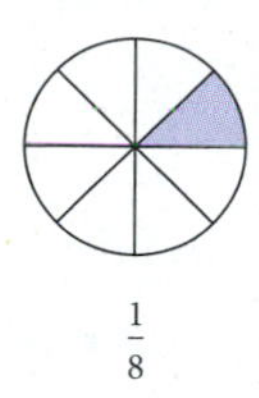

$\frac{1}{8}$

2. 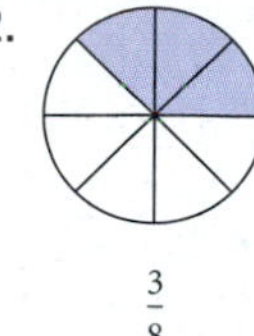

$\frac{3}{8}$

3. 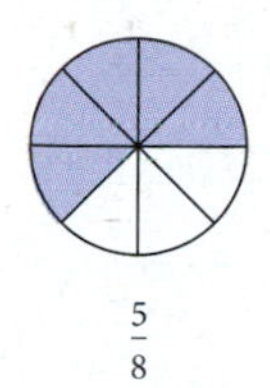

$\frac{5}{8}$

4. 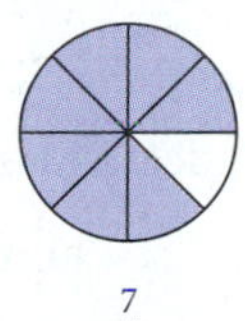

$\frac{7}{8}$

A Complete the following tables by converting each fraction to a decimal. [Examples 1–3]

5.

Fraction	$\frac{1}{4}$	$\frac{2}{4}$	$\frac{3}{4}$	$\frac{4}{4}$
Decimal				

6.

Fraction	$\frac{1}{5}$	$\frac{2}{5}$	$\frac{3}{5}$	$\frac{4}{5}$	$\frac{5}{5}$
Decimal					

7.

Fraction	$\frac{1}{6}$	$\frac{2}{6}$	$\frac{3}{6}$	$\frac{4}{6}$	$\frac{5}{6}$	$\frac{6}{6}$
Decimal						

A Convert each of the following fractions to a decimal. [Examples 1–3]

8. $\frac{1}{2}$

9. $\frac{12}{25}$

10. $\frac{14}{25}$

11. $\frac{14}{32}$

12. $\frac{18}{32}$

A Write each fraction as a decimal correct to the hundredths column. [Examples 1–3]

13. $\frac{12}{13}$

14. $\frac{17}{19}$

15. $\frac{3}{11}$

16. $\frac{5}{11}$

17. $\frac{2}{23}$

18. $\frac{3}{28}$

19. $\frac{12}{43}$

20. $\frac{15}{51}$

B Complete the following table by converting each decimal to a fraction.

21.

Decimal	0.125	0.250	0.375	0.500	0.625	0.750	0.875
Fraction							

22.

Decimal	0.1	0.2	0.3	0.4	0.5	0.6	0.7	0.8	0.9
Fraction									

B Write each decimal as a fraction in lowest terms. [Examples 4–6]

23. 0.15 **24.** 0.45 **25.** 0.08 **26.** 0.06 **27.** 0.375 **28.** 0.475

B Write each decimal as a mixed number. [Examples 6]

29. 5.6 **30.** 8.4 **31.** 5.06 **32.** 8.04 **33.** 1.22 **34.** 2.11

C Simplify each of the following as much as possible, and write all answers as decimals. [Examples 7–9]

35. $\frac{1}{2}(2.3 + 2.5)$

36. $\frac{3}{4}(1.8 + 7.6)$

37. $\dfrac{1.99}{\frac{1}{2}}$

38. $\dfrac{2.99}{\frac{1}{2}}$

39. $3.4 - \frac{1}{2}(0.76)$

40. $6.7 - \frac{1}{5}(0.45)$

41. $\frac{2}{5}(0.3) + \frac{3}{5}(0.3)$

42. $\frac{1}{8}(0.7) + \frac{3}{8}(0.7)$

43. $6\left(\frac{3}{5}\right)(0.02)$

44. $8\left(\frac{4}{5}\right)(0.03)$

45. $\frac{5}{8} + 0.35\left(\frac{1}{2}\right)$

46. $\frac{7}{8} + 0.45\left(\frac{3}{4}\right)$

47. $\left(\frac{1}{3}\right)^2(5.4) + \left(\frac{1}{2}\right)^3(3.2)$

48. $\left(\frac{1}{5}\right)^2(7.5) + \left(\frac{1}{4}\right)^2(6.4)$

49. $(0.25)^2 + \left(\frac{1}{4}\right)^2(3)$

50. $(0.75)^2 + \left(\frac{1}{4}\right)^2(7)$

D Applying the Concepts [Examples 10–12]

51. Commuting The map shows the average number of days spent commuting per year in the United States' largest cities. Change the data for Houston to a mixed number.

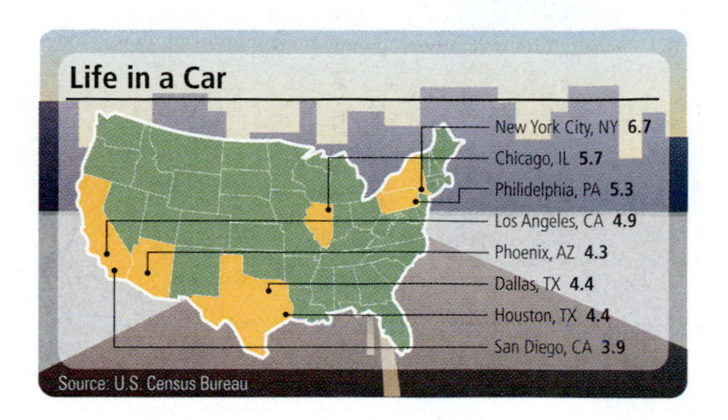

52. Pitchers The chart shows the active major league pitchers with the most career strikeouts. To compute the number of strikeouts per nine-inning game, divide by the total innings pitched and then multiply by 9. If Pedro Martinez had pitched 2,783 innings, write his strikeouts per game as a mixed number.

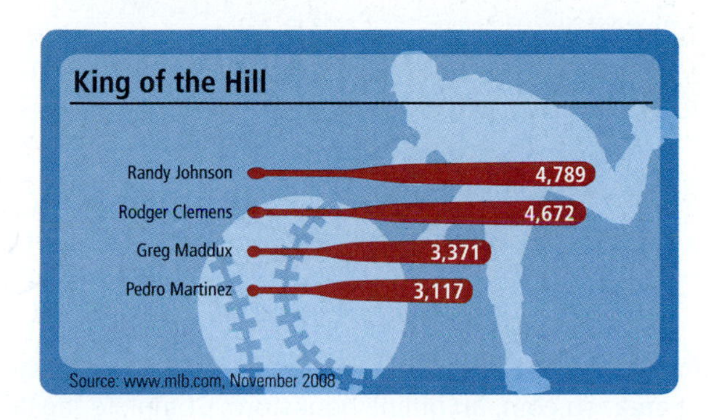

53. Price of Beef If each pound of beef costs $4.99, how much does $3\frac{1}{4}$ pounds cost?

54. Price of Gasoline What does it cost to fill a $15\frac{1}{2}$-gallon gas tank if the gasoline is priced at 429.9¢ per gallon?

55. Sale Price A dress that costs $78.99 is on sale for $\frac{1}{3}$ off. What is the sale price of the dress?

56. Sale Price A suit that normally sells for $221 is on sale for $\frac{1}{4}$ off. What is the sale price of the suit?

57. Perimeter of the Sierpinski Triangle The diagram below shows one stage of what is known as the Sierpinski triangle. Each triangle in the diagram has three equal sides. The large triangle is made up of 4 smaller triangles. If each side of the large triangle is 2 inches, and each side of the smaller triangles is 1 inch, what is the perimeter of the shaded region?

58. Perimeter of the Sierpinski Triangle The diagram below shows another stage of the Sierpinski triangle. Each triangle in the diagram has three equal sides. The largest triangle is made up of a number of smaller triangles. If each side of the large triangle is 2 inches, and each side of the smallest triangles is 0.5 inch, what is the perimeter of the shaded region?

59. Average Gain in Stock Price The table below shows the amount of gain each day of one week in 2000 for the price of an Internet company specializing in distance learning for college students. Complete the table by converting each fraction to a decimal, rounding to the nearest hundredth if necessary.

CHANGE IN STOCK PRICE		
Date	Gain ($)	As a Decimal ($) (To the Nearest hundredth)
Monday, March 6, 2000	$\frac{3}{4}$	
Tuesday, March 7, 2000	$\frac{9}{16}$	
Wednesday, March 8, 2000	$\frac{3}{32}$	
Thursday, March 9, 2000	$\frac{7}{32}$	
Friday, March 10, 2000	$\frac{1}{16}$	

60. Average Gain in Stock Price The table below shows the amount of gain each day of one week in 2000 for the stock price of amazon.com, an online bookstore. Complete the table by converting each fraction to a decimal, rounding to the nearest hundredth, if necessary.

CHANGE IN STOCK PRICE		
Date	Gain	As a Decimal ($) (To the Nearest Hundredth)
Monday, March 6, 2000	$\frac{1}{16}$	
Tuesday, March 7, 2000	$1\frac{3}{8}$	
Wednesday, March 8, 2000	$\frac{3}{8}$	
Thursday, March 9, 2000	$5\frac{13}{16}$	
Friday, March 10, 2000	$\frac{3}{8}$	

61. Nutrition If 1 ounce of ground beef contains 50.75 calories and 1 ounce of halibut contains 27.5 calories, what is the difference in calories between a $4\frac{1}{2}$-ounce serving of ground beef and a $4\frac{1}{2}$-ounce serving of halibut?

62. Nutrition If a 1-ounce serving of baked potato contains 48.3 calories and a 1-ounce serving of chicken contains 24.6 calories, how many calories are in a meal of $5\frac{1}{4}$ ounces of chicken and a $3\frac{1}{3}$-ounce baked potato?

Taxi Ride Recently, the Texas Junior College Teachers Association annual conference was held in Austin. At that time a taxi ride in Austin was $1.25 for the first $\frac{1}{5}$ of a mile and $0.25 for each additional $\frac{1}{5}$ of a mile. The charge for a taxi to wait is $12.00 per hour. Use this information for Problems 63 through 66.

63. If the distance from one of the convention hotels to the airport is 7.5 miles, how much will it cost to take a taxi from that hotel to the airport?

64. If you were to tip the driver of the taxi in Problem 63 $1.50, how much would it cost to take a taxi from the hotel to the airport?

65. Suppose the distance from one of the hotels to one of the western dance clubs in Austin is 12.4 miles. If the fare meter in the taxi gives the charge for that trip as \$16.50, is the meter working correctly?

66. Suppose t
8.2 miles,
pensive tc
taxi?

Volume Find the volume of each sphere. Round to the nearest hundredth. Use

67.

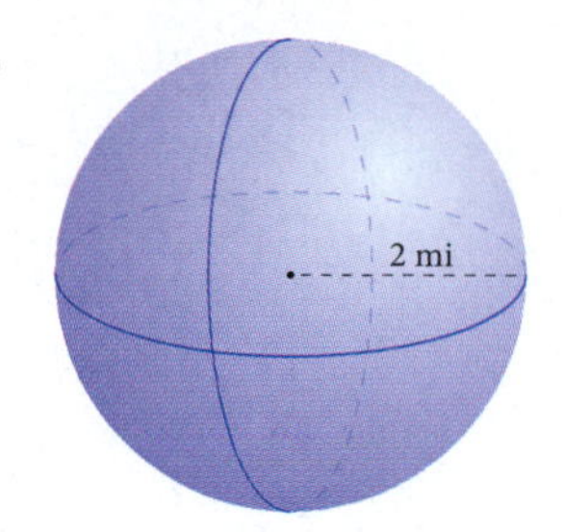

68.

69. Volume The radius of a sphere is 3.9 inches. Find the volume to the nearest hundredth.

70. Volume The radius of a sphere is 1.1 inches. Find the volume to the nearest hundredth.

Area Find the total area enclosed by each figure below. Use 3.14 for π.

71.

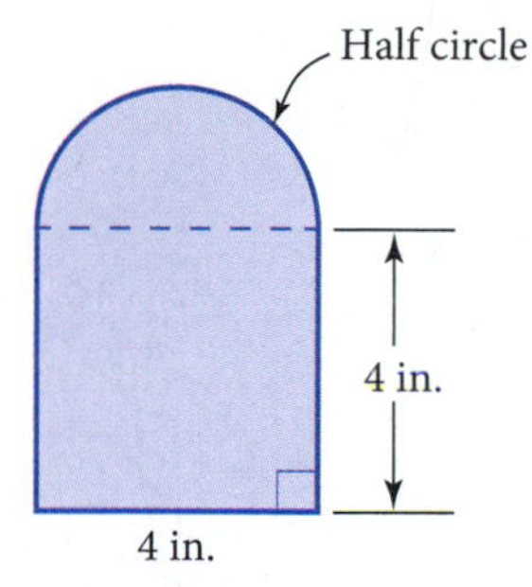

72.

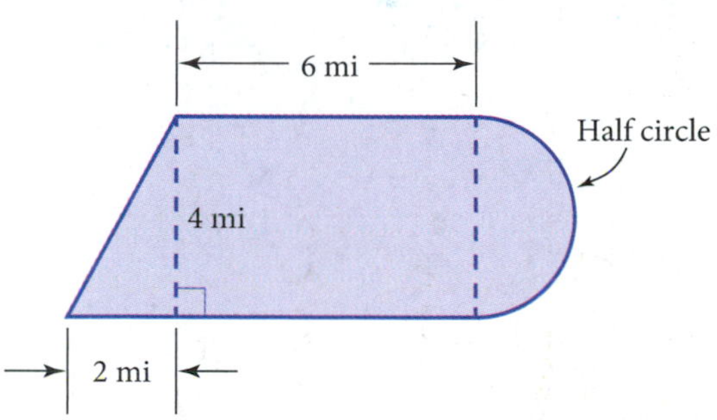

Maintaining Your Skills

The problems below review some of the material on solving equations. Reviewing these problems will help you with the next section.

Solve.

73. $x + 3 = -6$

74. $-4y = 28$

75. $\frac{1}{7}a = -7$

76. $\frac{1}{5}a = -3$

77. $5n + 4 = -26$

78. $6n - 2 = 40$

79. $5x + 8 = 3x + 2$

80. $7x - 3 = 5x + 9$

81. $2(x + 3) = 10$

82. $3(x - 2) = 6$

83. $3(y - 4) + 5 = -4$

84. $5(y - 1) + 6 = -9$

86. $\left(\frac{3}{4}\right)^3$

87. $\left(\frac{5}{6}\right)^2$

88. $\left(\frac{3}{5}\right)^3$

90. $(0.1)^3$

91. $(1.2)^2$

92. $(2.1)^2$

93. $3^2 + 4^2$

94. $5^2 + 12^2$

95. $6^2 + 8^2$

96. $2^2 + 3^2$

97. Find the sum of 827 and 25.

98. Find the difference of 827 and 25.

99. Find the product of 827 and 25.

100. Find the quotient of 827 and 25.

Equations Containing Decimals

5.6

Objectives

A Solve equations containing decimals.

B Solve applications involving equations with decimals.

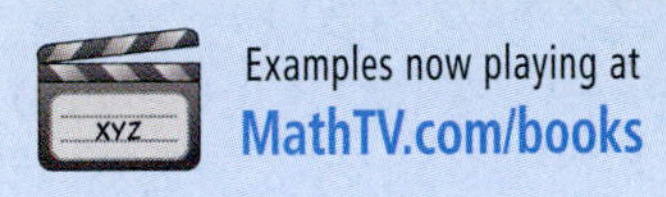

In this section we will continue our work with equations by considering some equations that involve decimals. We will also look at some application problems whose solutions come from equations with decimals.

 A Solving Equations

EXAMPLE 1 Solve the equation $x + 8.2 = 5.7$.

SOLUTION We use the addition property of equality to add -8.2 to each side of the equation.

$$x + 8.2 = 5.7$$
$$x + 8.2 + \mathbf{(-8.2)} = 5.7 + \mathbf{(-8.2)} \quad \text{Add } -8.2 \text{ to each side}$$
$$x + 0 = -2.5 \quad \text{Simplify each side}$$
$$x = -2.5$$

EXAMPLE 2 Solve: $3y = 2.73$

SOLUTION To isolate y on the left side, we divide each side by 3.

$$3y = 2.73$$
$$\frac{3y}{\mathbf{3}} = \frac{2.73}{\mathbf{3}} \quad \text{Divide each side by 3}$$
$$y = 0.91 \quad \text{Division}$$

EXAMPLE 3 Solve: $\frac{1}{2}x - 3.78 = 2.52$

SOLUTION We begin by adding 3.78 to each side of the equation. Then we multiply each side by 2.

$$\frac{1}{2}x - 3.78 = 2.52$$
$$\frac{1}{2}x - 3.78 + \mathbf{3.78} = 2.52 + \mathbf{3.78} \quad \text{Add 3.78 to each side}$$
$$\frac{1}{2}x = 6.30$$
$$\mathbf{2}\left(\frac{1}{2}x\right) = \mathbf{2}(6.30) \quad \text{Multiply each side by 2}$$
$$x = 12.6$$

EXAMPLE 4 Solve: $5a - 0.42 = -3a + 0.98$

SOLUTION We begin to isolate a on the left side of the equation by adding $3a$ to each side.

$$5a + \mathbf{3a} - 0.42 = -3a + \mathbf{3a} + 0.98 \quad \text{Add } 3a \text{ to each side}$$
$$8a - 0.42 = 0.98$$
$$8a - 0.42 + \mathbf{0.42} = 0.98 + \mathbf{0.42} \quad \text{Add 0.42 to each side}$$
$$8a = 1.40$$
$$\frac{8a}{\mathbf{8}} = \frac{1.40}{\mathbf{8}} \quad \text{Divide each side by 8}$$
$$a = 0.175$$

PRACTICE PROBLEMS

1. Solve: $x - 3.4 = 6.7$
2. Solve: $4y = 3.48$
3. Solve: $\frac{1}{5}x - 2.4 = 8.3$
4. Solve: $7a - 0.18 = 2a + 0.77$

Answers

1. 10.1 2. 0.87 3. 53.5 4. 0.19

5. If a car was rented from the company in Example 5 for 2 days and the total charge was \$40.88, how many miles was the car driven?

B Applications

EXAMPLE 5 A car rental company charges \$11 per day and 16 cents per mile for their cars. If a car was rented for 1 day and the charge was \$25.40, how many miles was it driven?

SOLUTION We use our Blueprint for Problem Solving as a guide to solving this application problem.

Step 1 *Read and list*

Known items: Charges are \$11 per day and 16 cents per mile; car was rented for 1 day for a total cost of \$25.40.

Unknown item: Number of miles it was driven

Step 2 *Assign a variable and translate information*

If we let x = the number of miles driven, then the charge for the miles driven will be $0.16x$, the cost per mile times the number of miles.

Step 3 *Reread and write an equation*

$$\underbrace{\text{\$11 per day}}_{11} + \underbrace{\text{16 cents per mile}}_{0.16x} = \underbrace{\text{Total cost}}_{25.40}$$

Step 4 *Solve the equation*

To solve the equation, we add -11 to each side and then divide each side by 0.16.

$$11 + \mathbf{(-11)} + 0.16x = 25.40 + \mathbf{(-11)} \quad \text{Add } -11 \text{ to each side}$$

$$0.16x = 14.40$$

$$\frac{0.16x}{\mathbf{0.16}} = \frac{14.40}{\mathbf{0.16}} \quad \text{Divide each side by 0.16}$$

$$x = 90 \quad 14.40 \div 0.16 = 90$$

Step 5 *Write the answer*

The car was driven 90 miles.

Step 6 *Reread and check*

$$\begin{aligned} \text{The 1-day charge} &= \$11.00 \\ \text{The mileage charge is } \$.16(90) &= \$14.40 \\ \hline \text{Total charge} &= \$25.40 \end{aligned}$$

6. Amy has \$1.75 in dimes and quarters. If she has 7 more dimes than quarters, how many of each coin does she have?

EXAMPLE 6 Diane has \$1.60 in dimes and nickels. If she has 7 more dimes than nickels, how many of each coin does she have?

SOLUTION We use our Blueprint for Problem Solving as a guide to solving this application problem.

Step 1 *Read and list*

Known items: We have dimes and nickels, seven more dimes than nickels.

Unknown items: Number of dimes and the number of nickels.

Step 2 *Assign a variable and translate information*

If we let x = the number of nickels, then the number of dimes must be $x + 7$, because Diane has 7 more dimes than nickels. Since each nickel is worth 5 cents, the amount of money she has in nickels is $0.05x$. Similarly,

Answer

5. 118 miles

since each dime is worth 10 cents, the amount of money she has in dimes is $0.10(x + 7)$. Here is a table that summarizes what we have so far:

	Nickels	Dimes
Number of	x	$x + 7$
Value of	$0.05x$	$0.10(x + 7)$

Step 3 *Reread and write an equation*

Because the total value of all the coins is \$1.60, the equation that describes this situation is

$$\underbrace{\text{Amount of money in nickels}}_{0.05x} + \underbrace{\text{Amount of money in dimes}}_{0.10(x+7)} = \underbrace{\text{Total amount of money}}_{1.60}$$

Step 4 *Solve the equation*

This time, let's show only the essential steps in the solution.

$$0.05x + 0.10x + 0.70 = 1.60 \quad \text{Distributive property}$$
$$0.15x + 0.70 = 1.60 \quad \text{Add } 0.05x \text{ and } 0.10x \text{ to get } 0.15x$$
$$0.15x = 0.90 \quad \text{Add } -0.70 \text{ to each side}$$
$$x = 6 \quad \text{Divide each side by } 0.15$$

Step 5 *Write the answer*

Because $x = 6$, Diane has 6 nickels. To find the number of dimes, we add 7 to the number of nickels (she has 7 more dimes than nickels). The number of dimes is $6 + 7 = 13$.

Step 6 *Reread and check*

6 nickels are worth 6(\$0.05) = \$0.30
13 dimes are worth 13(\$0.10) = \$1.30
The total value is \$1.60

Getting Ready for Class

After reading through the preceding section, respond in your own words and in complete sentences.

1. In your opinion, is solving an equation that contains decimals more difficult than solving an equation that does not contain decimals? Explain.
2. Multiply both sides of the equation in Example 1 by 10, then solve the resulting equation.
3. Repeat Example 3, but multiply both sides of the original equation by 100 for your first step. Does your answer match the answer shown in Example 3?
4. If x is the number of nickels in a cash register, how can we represent the *value* of those nickels?

Answer
6. 3 quarters, 10 dimes

Problem Set 5.6

A Solve each equation. [Examples 1–4]

1. $x + 3.7 = 2.2$
2. $x + 4.8 = 9.1$
3. $x - 0.45 = 0.32$
4. $x - 23.3 = -4.5$
5. $8a = 1.2$
6. $6a = 18.6$
7. $-4y = 1.4$
8. $-7y = -0.63$
9. $0.5n = -0.4$
10. $0.6n = -0.12$
11. $4x - 4.7 = 3.5$
12. $2x + 3.8 = -7.7$
13. $0.02 + 5y = -0.3$
14. $0.8 + 10y = -0.7$
15. $\frac{1}{3}x - 2.99 = 1.02$
16. $\frac{1}{7}x + 2.87 = -3.01$
17. $7n - 0.32 = 5n + 0.56$
18. $6n + 0.88 = 2n - 0.77$
19. $3a + 4.6 = 7a + 5.3$
20. $2a - 3.3 = 7a - 5.2$
21. $0.5x + 0.1(x + 20) = 3.2$
22. $0.1x + 0.5(x + 8) = 7$
23. $0.08x + 0.09(x + 2000) = 690$
24. $0.11x + 0.12(x + 4000) = 940$

B Applying the Concepts [Examples 5, 6]

25. Google Earth The Google Earth image shows a corn field with a length of 0.5 mile and a width of 0.25 mile. The farmer harvested 80,000 bushels of corn per square mile and made $19,000 from the harvest. How much did the farmer receive for each bushel?

26. Picture Messaging The chart shows the number of picture messages (in millions) sent in the first nine months of the year. If the revenue from picture messages for April was $2.5 million, how much was charged for each picture message sent?

27. Car Rental A car rental company charges $18 a day and 16 cents per mile to rent their cars. If a car was rented for 1 day for a total charge of $35.60, how many miles was it driven?

28. Car Rental A car rental company charges $27 a day and 18 cents per mile to rent their cars. If the total charge for a 1-day rental was $48.78, how many miles was the car driven?

29. Car Rental A car rental company charges $24.95 per day and 15 cents a mile to rent their cars. If a car was rented for 2 days for a total charge of $71.95, how many miles was it driven?

30. Car Rental A car rental company charges $19.95 a day and 18 cents per mile to rent their cars. If the total charge for a 2-day rental was $79.50, how many miles was it driven?

31. Coin Problem Mary has $2.20 in dimes and nickels. If she has 10 more dimes than nickels, how many of each coin does she have?

32. Coin Problem Bob has $1.65 in dimes and nickels. If he has 9 more nickels than dimes, how many of each coin does he have?

33. Coin Problem Suppose you have $9.60 in dimes and quarters. How many of each coin do you have if you have twice as many quarters as dimes?

34. Coin Problem A collection of dimes and quarters has a total value of $2.75. If there are three times as many dimes as quarters, how many of each coin is in the collection?

35. **Long-Distance Charges** The cost of a long-distance phone call is \$0.41 for the first minute and \$0.32 for each additional minute. If the total charge for a long-distance call is \$5.21, how many minutes was the call?

36. **Long-Distan** ing with th one of the her friend realizes Da trying to g for the firs that. If the did it take phone?

Maintaining Your S

Find the value of each

51. $3(x - 4)$

54.

37. **Coin Problem** Katie has a collection of nickels, dimes, and quarters with a total value of \$4.35. There are 3 more dimes than nickels and 5 more quarters than nickels. How many of each coin is in her collection? (*Hint:* Let $x =$ the number of nickels.)

38. **Coin Problem** Mary Jo has \$3.90 worth of nickels, dimes, and quarters. The number of nickels is 3 more than the number of dimes. The number of quarters is 7 more than the number of dimes. How many of each coin does she have? (*Hint:* Let $x =$ the number of dimes.)

Getting Ready for the Next Section

The problems below review the material on exponents we have covered previously. Expand and simplify.

39. 5^3

40. 2^5

41. $(-3)^2$

42. $(-2)^3$

43. $\left(\frac{1}{3}\right)^4$

44. $\left(\frac{3}{4}\right)^3$

45. $\left(-\frac{5}{6}\right)^2$

46. $\left(-\frac{3}{5}\right)^3$

47. $(0.5)^2$

48. $(0.1)^3$

49. $(-1.2)^2$

50. $(-2.1)^2$

kills

expression when $x = -4$.

52. $-3(x - 4)$

53. $-5x + 8$

$x + 8$

55. $\frac{x - 14}{36}$

56. $\frac{x - 12}{36}$

57. $\frac{16}{x} + 3x$

58. $\frac{16}{x} - 3x$

59. $7x - \frac{12}{x}$

60. $7x + \frac{12}{x}$

61. $8\left(\frac{x}{2} + 5\right)$

62. $-8\left(\frac{x}{2} + 5\right)$

Square Roots and the Pythagorean Theorem

5.7

Objectives

A Find square roots of numbers.

B Use decimals to approximate square roots.

C Solve problems with the Pythagorean Theorem.

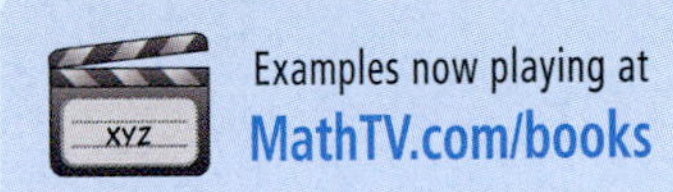

Examples now playing at **MathTV.com/books**

Introduction . . .

Figure 1 shows the front view of the roof of a tool shed. How do we find the length d of the diagonal part of the roof? (Imagine that you are drawing the plans for the shed. Since the shed hasn't been built yet, you can't just measure the diagonal, but you need to know how long it will be so you can buy the correct amount of material to build the shed.)

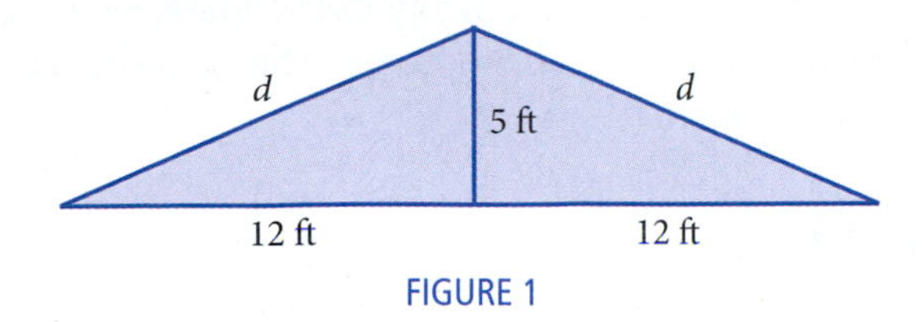

FIGURE 1

There is a formula from geometry that gives the length d:

$$d = \sqrt{12^2 + 5^2}$$

where $\sqrt{}$ is called the *square root symbol.* If we simplify what is under the square root symbol, we have this:

$$d = \sqrt{144 + 25}$$
$$= \sqrt{169}$$

The expression $\sqrt{169}$ stands for the number we *square* to get 169. Because $13 \cdot 13 = 169$, that number is 13. Therefore the length d in our original diagram is 13 feet.

A Square Roots

Here is a more detailed discussion of square roots. In Chapter 1, we did some work with exponents. In particular, we spent some time finding squares of numbers. For example, we considered expressions like this:

$$5^2 = 5 \cdot 5 = 25$$
$$7^2 = 7 \cdot 7 = 49$$
$$x^2 = x \cdot x$$

We say that "the square of 5 is 25" and "the square of 7 is 49." To square a number, we multiply it by itself. When we ask for the *square root* of a given number, we want to know what number we *square* in order to obtain the given number. We say that the square root of 49 is 7, because 7 is the number we square to get 49. Likewise, the square root of 25 is 5, because $5^2 = 25$. The symbol we use to denote square root is $\sqrt{}$, which is also called a *radical sign.* Here is the precise definition of square root.

The square root we are describing here is actually the principal square root. There is another square root that is a negative number. We won't see it in this book, but, if you go on to take an algebra course, you will see it there.

Definition

The **square root** of a positive number a, written $\sqrt{a}$, is the number we square to get a. In symbols:

$$\text{If} \quad \sqrt{a} = b \quad \text{then} \quad b^2 = a.$$

We list some common square roots in Table 1.

TABLE 1

Statement	In Words	Reason
$\sqrt{0} = 0$	The square root of 0 is 0	Because $0^2 = 0$
$\sqrt{1} = 1$	The square root of 1 is 1	Because $1^2 = 1$
$\sqrt{4} = 2$	The square root of 4 is 2	Because $2^2 = 4$
$\sqrt{9} = 3$	The square root of 9 is 3	Because $3^2 = 9$
$\sqrt{16} = 4$	The square root of 16 is 4	Because $4^2 = 16$
$\sqrt{25} = 5$	The square root of 25 is 5	Because $5^2 = 25$

Numbers like 1, 9, and 25, whose square roots are whole numbers, are called *perfect squares.* To find the square root of a perfect square, we look for the whole number that is squared to get the perfect square. The following examples involve square roots of perfect squares.

PRACTICE PROBLEMS

1. Simplify: $4\sqrt{25}$

2. Simplify: $\sqrt{36} + \sqrt{4}$

3. Simplify: $\sqrt{\frac{36}{100}}$

Simplify each expression as much as possible.

4. $14\sqrt{36}$

5. $\sqrt{81} - \sqrt{25}$

6. $\sqrt{\frac{64}{121}}$

Answers

1. 20 2. 8 3. $\frac{3}{5}$ 4. 84 5. 4 6. $\frac{8}{11}$

EXAMPLE 1 Simplify: $7\sqrt{64}$

SOLUTION The expression $7\sqrt{64}$ means 7 times $\sqrt{64}$. To simplify this expression, we write $\sqrt{64}$ as 8 and multiply:

$$7\sqrt{64} = 7 \cdot 8 = 56$$

We know $\sqrt{64} = 8$, because $8^2 = 64$.

EXAMPLE 2 Simplify: $\sqrt{9} + \sqrt{16}$

SOLUTION We write $\sqrt{9}$ as 3 and $\sqrt{16}$ as 4. Then we add:

$$\sqrt{9} + \sqrt{16} = 3 + 4 = 7$$

EXAMPLE 3 Simplify: $\sqrt{\frac{25}{81}}$

SOLUTION We are looking for the number we square (multiply times itself) to get $\frac{25}{81}$. We know that when we multiply two fractions, we multiply the numerators and multiply the denominators. Because $5 \cdot 5 = 25$ and $9 \cdot 9 = 81$, the square root of $\frac{25}{81}$ must be $\frac{5}{9}$.

$$\sqrt{\frac{25}{81}} = \frac{5}{9} \quad \text{because} \quad \left(\frac{5}{9}\right)^2 = \frac{5}{9} \cdot \frac{5}{9} = \frac{25}{81}$$

In Examples 4–6, we simplify each expression as much as possible.

EXAMPLE 4 Simplify: $12\sqrt{25} = 12 \cdot 5 = 60$

EXAMPLE 5 Simplify: $\sqrt{100} - \sqrt{36} = 10 - 6 = 4$

EXAMPLE 6 Simplify: $\sqrt{\frac{49}{121}} = \frac{7}{11}$ because $\left(\frac{7}{11}\right)^2 = \frac{7}{11} \cdot \frac{7}{11} = \frac{49}{121}$

B Approximating Square Roots

So far in this section we have been concerned only with square roots of perfect squares. The next question is, "What about square roots of numbers that are not perfect squares, like $\sqrt{7}$, for example?" We know that

$$\sqrt{4} = 2 \quad \text{and} \quad \sqrt{9} = 3$$

And because 7 is between 4 and 9, $\sqrt{7}$ should be between $\sqrt{4}$ and $\sqrt{9}$. That is, $\sqrt{7}$ should be between 2 and 3. But what is it exactly? The answer is, we cannot write it exactly in decimal or fraction form. Because of this, it is called an *irrational number*. We can approximate it with a decimal, but we can never write it exactly with a decimal. Table 2 gives some decimal approximations for $\sqrt{7}$. The decimal approximations were obtained by using a calculator. We could continue the list to any accuracy we desired. However, we would never reach a number in decimal form whose square was exactly 7.

TABLE 2

APPROXIMATIONS FOR THE SQUARE ROOT OF 7

Accurate to the Nearest	The Square Root of 7 is	Check by Squaring
Tenth	$\sqrt{7} = 2.6$	$(2.6)^2 = 6.76$
Hundredth	$\sqrt{7} = 2.65$	$(2.65)^2 = 7.0225$
Thousandth	$\sqrt{7} = 2.646$	$(2.646)^2 = 7.001316$
Ten thousandth	$\sqrt{7} = 2.6458$	$(2.6458)^2 = 7.00025764$

EXAMPLE 7 Give a decimal approximation for the expression $5\sqrt{12}$ that is accurate to the nearest ten thousandth.

SOLUTION Let's agree not to round to the nearest ten thousandth until we have first done all the calculations. Using a calculator, we find $\sqrt{12} \approx 3.4641016$. Therefore,

$$
\begin{aligned}
5\sqrt{12} &\approx 5(3.4641016) && \sqrt{12} \text{ on calculator} \\
&= 17.320508 && \text{Multiplication} \\
&= 17.3205 && \text{To the nearest ten thousandth}
\end{aligned}
$$

EXAMPLE 8 Approximate $\sqrt{301} + \sqrt{137}$ to the nearest hundredth.

SOLUTION Using a calculator to approximate the square roots, we have

$$\sqrt{301} + \sqrt{137} \approx 17.349352 + 11.704700 = 29.054052$$

To the nearest hundredth, the answer is 29.05.

EXAMPLE 9 Approximate $\sqrt{\frac{7}{11}}$ to the nearest thousandth.

SOLUTION Because we are using calculators, we first change $\frac{7}{11}$ to a decimal and then find the square root:

$$\sqrt{\frac{7}{11}} \approx \sqrt{0.6363636} \approx 0.7977240$$

To the nearest thousandth, the answer is 0.798.

C The Pythagorean Theorem

FACTS FROM GEOMETRY Perimeter

A *right triangle* is a triangle that contains a 90° (or right) angle. The longest side in a right triangle is called the *hypotenuse*, and we use the letter c to denote it. The two shorter sides are denoted by the letters a and b. The Pythagorean theorem states that the hypotenuse is the square root of the sum of the squares of the two shorter sides. In symbols:

$$c = \sqrt{a^2 + b^2}$$

7. Give a decimal approximation for the expression $5\sqrt{14}$ that is accurate to the nearest ten thousandth.

8. Approximate $\sqrt{405} + \sqrt{147}$ to the nearest hundredth.

9. Approximate $\sqrt{\frac{7}{12}}$ to the nearest thousandth.

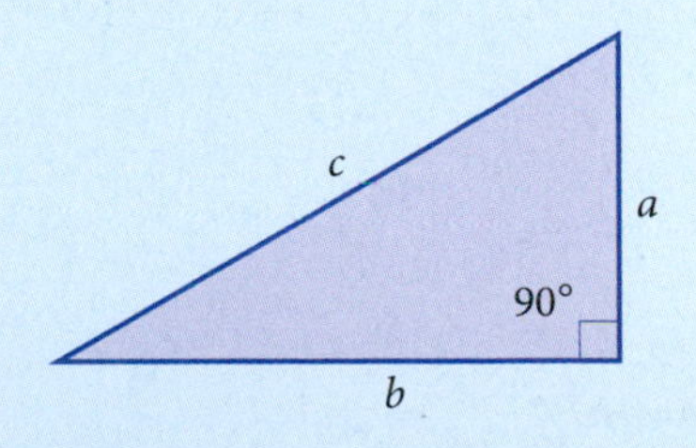

Answers

7. 18.7083 **8.** 32.25 **9.** 0.764

10. Find the length of the hypotenuse in each right triangle.

a.

c

5 ft

5 ft

b.

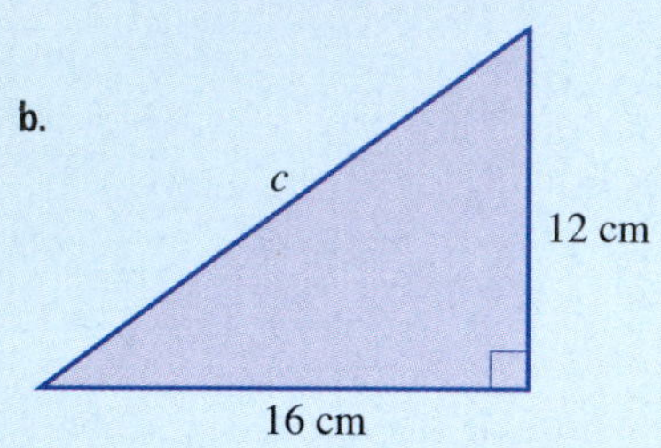

11. A wire from the top of a 12-foot pole is fastened to the ground by a stake that is 5 feet from the bottom of the pole. What is the length of the wire?

Answers

10. a. approx. 7.07 ft **b.** 20 cm

11. 13 ft

EXAMPLE 10

Find the length of the hypotenuse in each right triangle.

a.

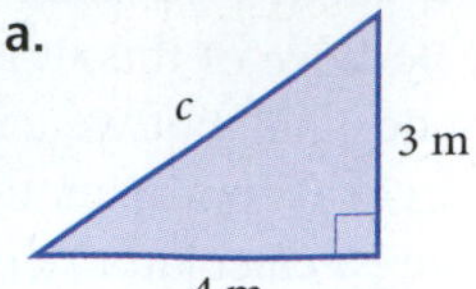

b.

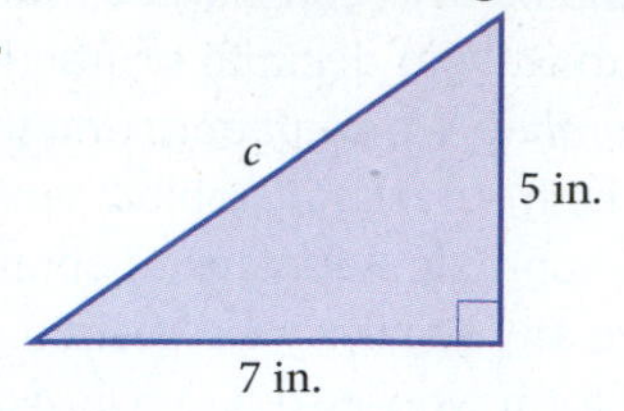

SOLUTION We apply the formula given above.

a.

When $a = 3$ and $b = 4$:

$$\begin{aligned} c &= \sqrt{3^2 + 4^2} \\ &= \sqrt{9 + 16} \\ &= \sqrt{25} \\ c &= 5 \text{ meters} \end{aligned}$$

b.

When $a = 5$ and $b = 7$:

$$\begin{aligned} c &= \sqrt{5^2 + 7^2} \\ &= \sqrt{25 + 49} \\ &= \sqrt{74} \\ c &\approx 8.60 \text{ inches} \end{aligned}$$

In part a, the solution is a whole number, whereas in part b, we must use a calculator to get 8.60 as an approximation to $\sqrt{74}$.

EXAMPLE 11

A ladder is leaning against the top of a 6-foot wall. If the bottom of the ladder is 8 feet from the wall, how long is the ladder?

SOLUTION A picture of the situation is shown in Figure 2. We let c denote the length of the ladder. Applying the Pythagorean theorem, we have

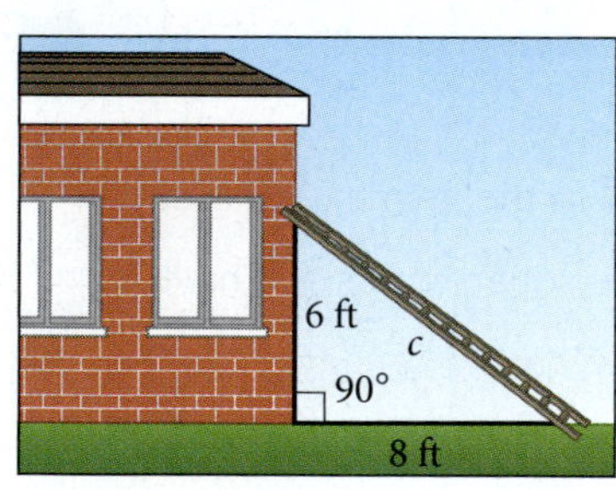

FIGURE 2

$$\begin{aligned} c &= \sqrt{6^2 + 8^2} \\ &= \sqrt{36 + 64} \\ &= \sqrt{100} \\ &= 10 \text{ feet} \end{aligned}$$

The ladder is 10 feet long.

Getting Ready for Class

After reading through the preceding section, respond in your own words and in complete sentences.

1. Which number is larger, the square of 10 or the square root of 10?
2. Give a definition for the square root of a number.
3. What two numbers will the square root of 20 fall between?
4. What is the Pythagorean theorem?

Problem Set 5.7

A Find each of the following square roots without using a calculator. [Example 1]

1. $\sqrt{64}$
2. $\sqrt{100}$
3. $\sqrt{81}$
4. $\sqrt{49}$
5. $\sqrt{36}$
6. $\sqrt{144}$
7. $\sqrt{25}$
8. $\sqrt{169}$

A Simplify each of the following expressions without using a calculator. [Examples 1–6]

9. $3\sqrt{25}$
10. $9\sqrt{49}$
11. $6\sqrt{64}$
12. $11\sqrt{100}$
13. $15\sqrt{9}$
14. $8\sqrt{36}$
15. $16\sqrt{9}$
16. $9\sqrt{16}$
17. $\sqrt{49} + \sqrt{64}$
18. $\sqrt{1} + \sqrt{0}$
19. $\sqrt{16} - \sqrt{9}$
20. $\sqrt{25} - \sqrt{4}$
21. $3\sqrt{25} + 9\sqrt{49}$
22. $6\sqrt{64} + 11\sqrt{100}$
23. $15\sqrt{9} - 9\sqrt{16}$
24. $7\sqrt{49} - 2\sqrt{4}$
25. $\sqrt{\frac{16}{49}}$
26. $\sqrt{\frac{100}{121}}$
27. $\sqrt{\frac{36}{64}}$
28. $\sqrt{\frac{81}{144}}$

Indicate whether each of the statements in Problems 29–32 is *True* or *False*.

29. $\sqrt{4} + \sqrt{9} = \sqrt{4 + 9}$
30. $\sqrt{\frac{16}{25}} = \frac{\sqrt{16}}{\sqrt{25}}$
31. $\sqrt{25 \cdot 9} = \sqrt{25} \cdot \sqrt{9}$
32. $\sqrt{100} - \sqrt{36} = \sqrt{100 - 36}$

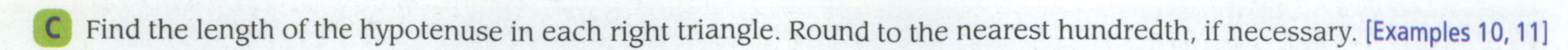
C Find the length of the hypotenuse in each right triangle. Round to the nearest hundredth, if necessary. [Examples 10, 11]

33.

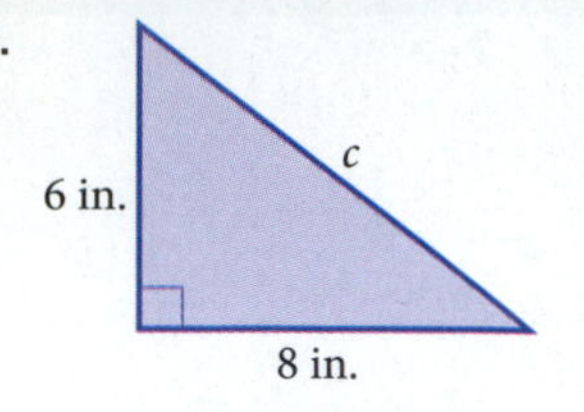

34.

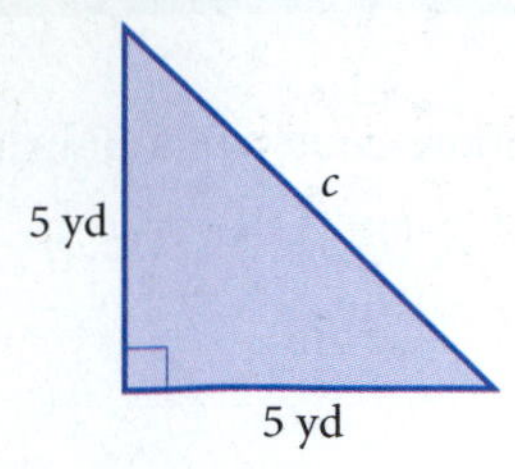

35.

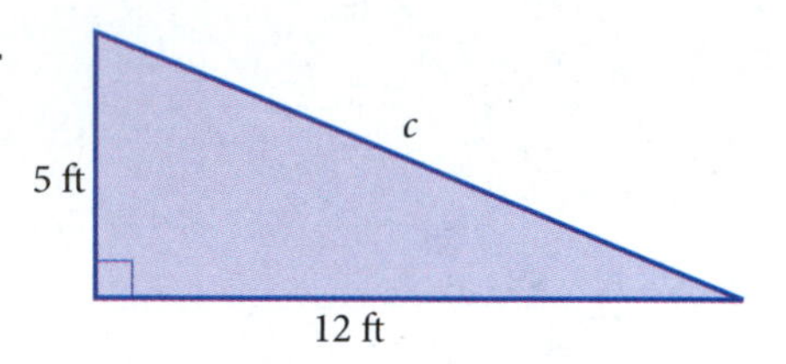

36.

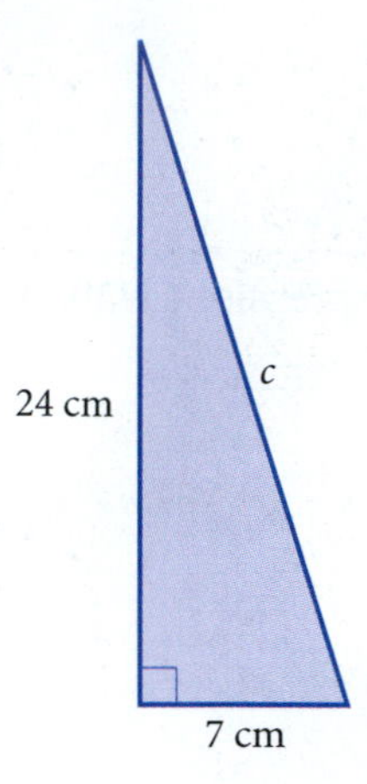

37.

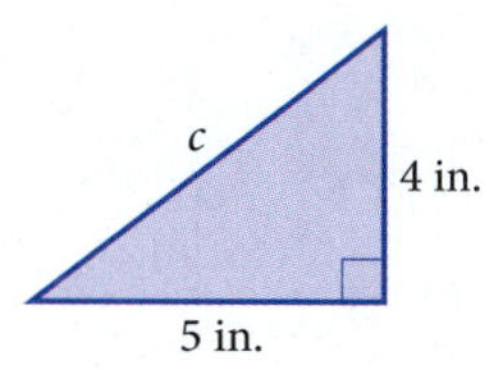

38.

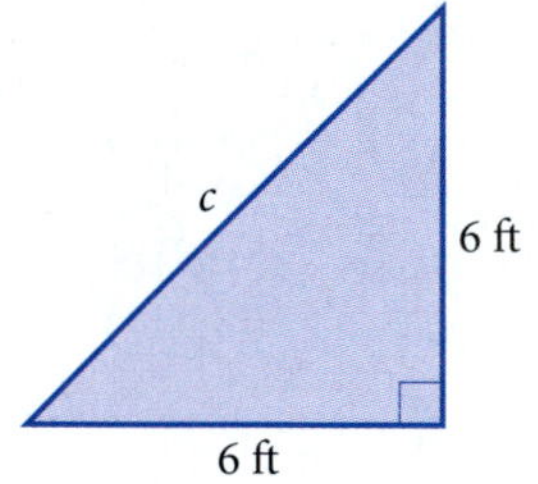

39.

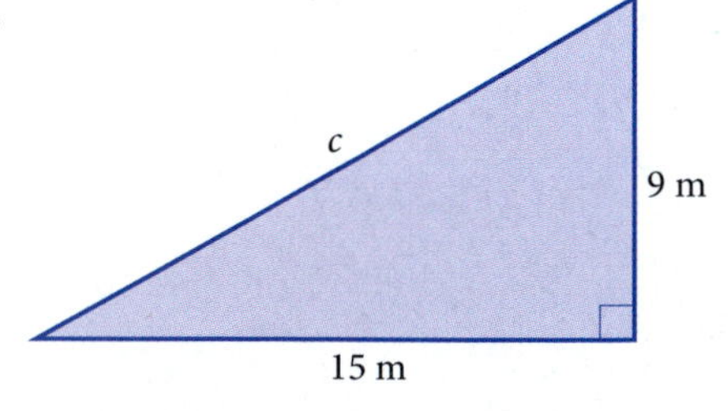

40.

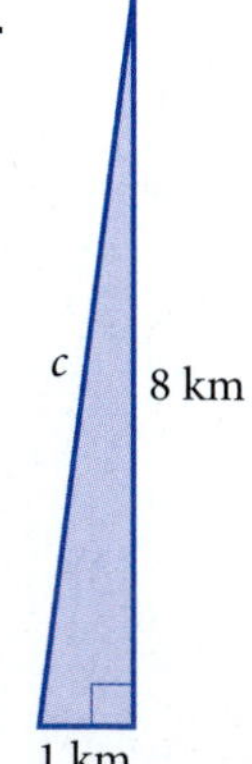

B Calculator Problems [Examples 7–9]

Use a calculator to work problems 41 through 60.

Approximate each of the following square roots to the nearest ten thousandth.

41. $\sqrt{1.25}$ **42.** $\sqrt{12.5}$ **43.** $\sqrt{125}$ **44.** $\sqrt{1250}$

Approximate each of the following expressions to the nearest hundredth.

45. $2\sqrt{3}$ **46.** $3\sqrt{2}$ **47.** $5\sqrt{5}$ **48.** $5\sqrt{3}$

49. $\frac{\sqrt{3}}{3}$ **50.** $\frac{\sqrt{2}}{2}$ **51.** $\sqrt{\frac{1}{3}}$ **52.** $\sqrt{\frac{1}{2}}$

Approximate each of the following expressions to the nearest thousandth.

53. $\sqrt{12} + \sqrt{75}$ **54.** $\sqrt{18} + \sqrt{50}$ **55.** $\sqrt{87}$ **56.** $\sqrt{68}$

57. $2\sqrt{3} + 5\sqrt{3}$ **58.** $3\sqrt{2} + 5\sqrt{2}$ **59.** $7\sqrt{3}$ **60.** $8\sqrt{2}$

Applying the Concepts

61. Google Earth The Google Earth image shows a right triangle between three cities in the Los Angeles area. If the distance between Pomona and Ontario is 5.7 miles, and the distance between Ontario and Upland is 3.6 miles, what is the distance between Pomona and Upland? Round to the nearest tenth of a mile.

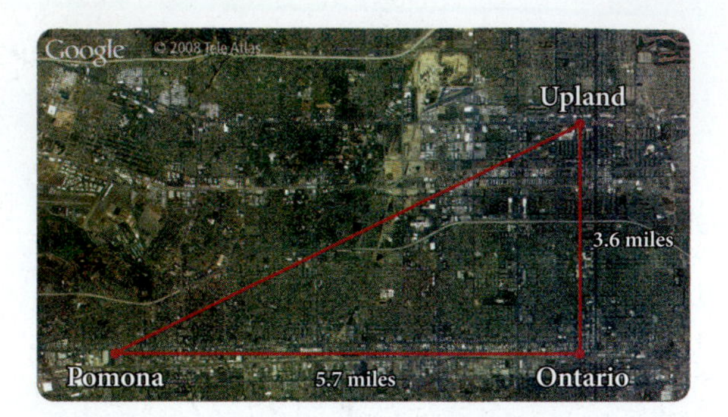

62. Google Earth The Google Earth image shows three cities in Colorado. If the distance between Denver and North Washington is 2.5 miles, and the distance between Edgewater and Denver is 4 miles, what is the distance between North Washington and Edgewater? Round to the nearest tenth.

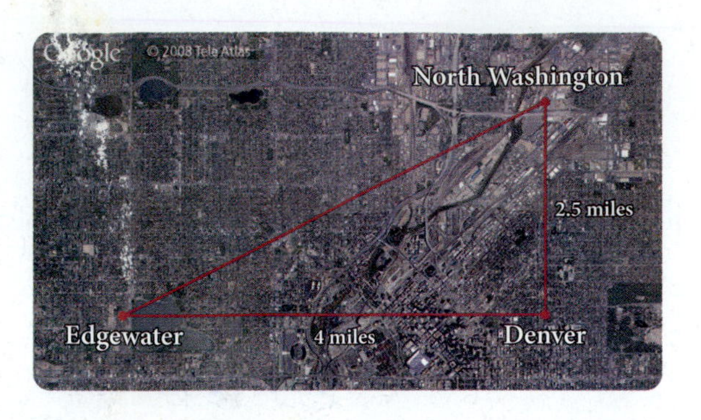

63. Geometry One end of a wire is attached to the top of a 24-foot pole; the other end of the wire is anchored to the ground 18 feet from the bottom of the pole. If the pole makes an angle of 90° with the ground, find the length of the wire.

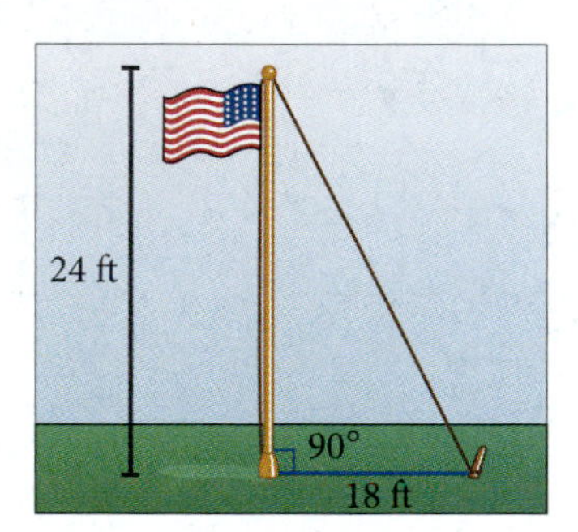

64. Geometry Two children are trying to cross a stream. They want to use a log that goes from one bank to the other. If the left bank is 5 feet higher than the right bank and the stream is 12 feet wide, how long must a log be to just barely reach?

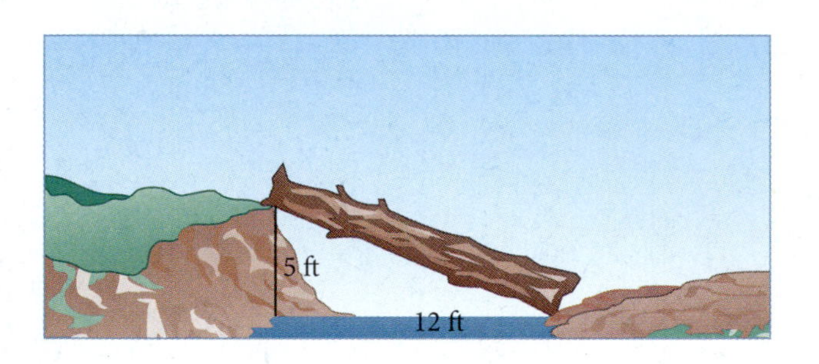

65. Geometry A ladder is leaning against the top of a 15-foot wall. If the bottom of the ladder is 20 feet from the wall, how long is the ladder?

66. Geometry A wire from the top of a 24-foot pole is fastened to the ground by a stake that is 10 feet from the bottom of the pole. How long is the wire?

67. Surveying A surveying team wants to calculate the length of a straight tunnel through a mountain. They form a right angle by connecting lines from each end of the proposed tunnel. One of the connecting lines is 3 miles, and the other is 4 miles. What is the length of the proposed tunnel?

68. Surveying A surveying team wants to calculate the length of a straight tunnel through a mountain. They form a right angle by connecting lines from each end of the proposed tunnel. One of the connecting lines is 6 miles, and the other is 8 miles. What is the length of the proposed tunnel?

69. Lighthouse Problem The higher you are above the ground, the farther you can see. If your view is unobstructed, then the distance in miles that you can see from h feet above the ground is given by the formula

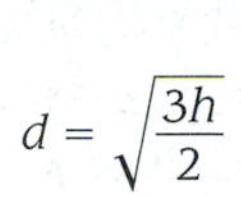

$$d = \sqrt{\frac{3h}{2}}$$

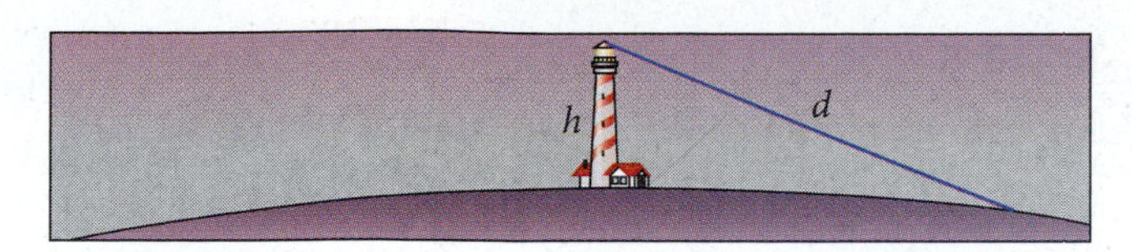

The following figure shows a lighthouse with a door and windows at various heights. The preceding formula can be used to find the distance to the ocean horizon from these heights. Use the formula and a calculator to complete the following table. Round your answers to the nearest whole number.

Height h(feet)	Distance d(miles)
10	4
50	9
90	12
130	14
170	16
190	17

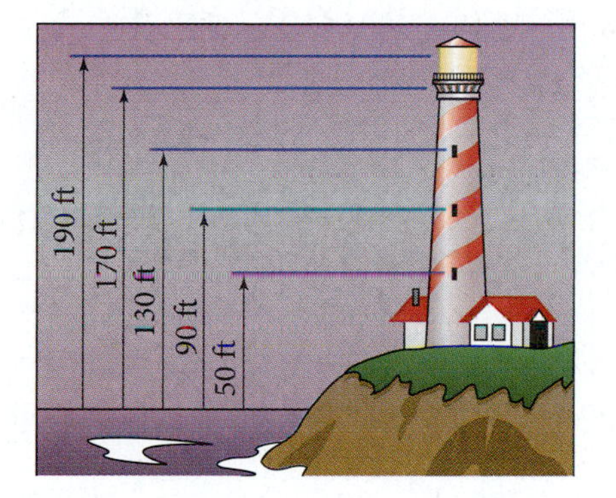

70. Pendulum Problem The time (in seconds) it takes for the pendulum on a clock to swing through one complete cycle is given by the formula

$$T = \frac{11}{7}\sqrt{\frac{L}{2}}$$

where L is the length (in feet) of the pendulum. Use this formula and a calculator to complete the following table. Round your answers to the nearest hundredth.

Length L (feet)	Time T (seconds)
1	
2	
3	
4	
5	
6	

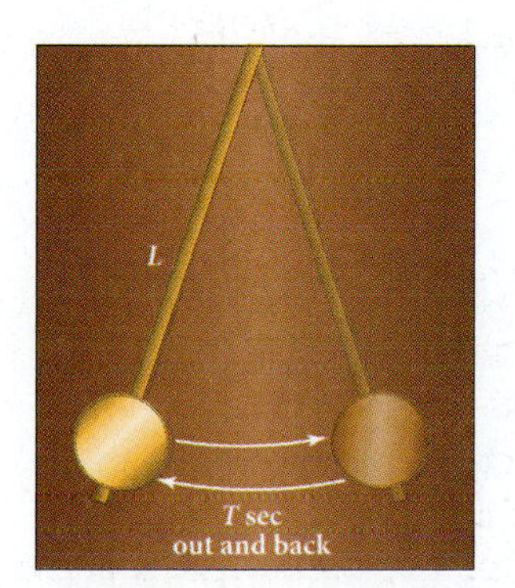

Getting Ready for the Next Section

Factor each of the following numbers into the product of two numbers, one of which is a perfect square. (Remember from Chapter 1, a perfect square is 1, 4, 9, 16, 25, 36, . . ., etc.)

71. 75

72. 12

73. 50

74. 20

75. 40

76. 18

77. 32

78. 27

79. 98

80. 72

81. 48

82. 121

Maintaining Your Skills

The problems below review material involving fractions and mixed numbers. Perform the indicated operations. Write your answers as whole numbers, proper fractions, or mixed numbers.

83. $\frac{5}{7} \cdot \frac{14}{25}$

84. $1\frac{1}{4} \div 2\frac{1}{8}$

85. $4\frac{3}{10} + 5\frac{2}{100}$

86. $8\frac{1}{5} + 1\frac{1}{10}$

87. $3\frac{2}{10} \cdot 2\frac{5}{10}$

88. $6\frac{9}{10} \div 2\frac{3}{10}$

89. $7\frac{1}{10} - 4\frac{3}{10}$

90. $3\frac{7}{10} - 1\frac{97}{100}$

91. $\dfrac{\frac{3}{8}}{\frac{6}{7}}$

92. $\dfrac{\frac{3}{4}}{\frac{1}{2} + \frac{1}{4}}$

93. $\dfrac{\frac{2}{3} + \frac{3}{5}}{\frac{2}{3} - \frac{3}{5}}$

94. $\dfrac{\frac{4}{5} - \frac{1}{3}}{\frac{4}{5} + \frac{1}{3}}$

Simplifying Square Roots

5.8

Objectives

A Use the multiplication property to simplify square roots.

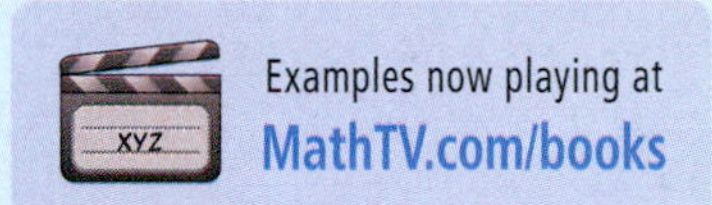

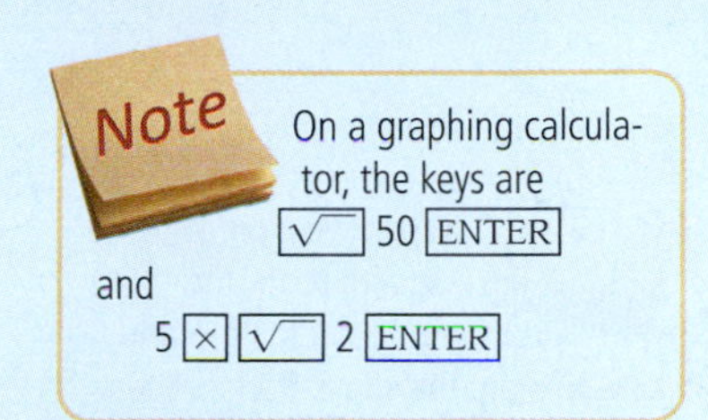

On a graphing calculator, the keys are

[$\sqrt{}$] 50 [ENTER]

and

5 [$\times$] [$\sqrt{}$] 2 [ENTER]

Do you know that $\sqrt{50}$ and $5 \cdot \sqrt{2}$ are the same number? One way to convince yourself that this is true is with a calculator. On a scientific calculator, to find a decimal approximation to the first expression, we enter 50 and then press the $\sqrt{}$ key:

50 [$\sqrt{}$] The calculator shows 7.0710678

To find a decimal approximation to the second expression, we multiply 5 and $\sqrt{2}$:

5 [$\times$] 2 [$\sqrt{}$] [=] The calculator shows 7.0710678

Although a calculator will give the same result for both $\sqrt{50}$ and $5 \cdot \sqrt{2}$, it does not tell us *why* the answers are the same. The discussion below shows why the results are the same.

First, notice that the expressions $\sqrt{4 \cdot 9}$ and $\sqrt{4} \cdot \sqrt{9}$ have the same value:

$$\sqrt{4 \cdot 9} = \sqrt{36} = 6 \quad \text{and} \quad \sqrt{4} \cdot \sqrt{9} = 2 \cdot 3 = 6$$

Both are equal to 6. When we are multiplying and taking square roots, we can either multiply first and then take the square root of what we get, or we can take square roots first and then multiply. In symbols, we write it this way:

A Simplifying Square Roots

Multiplication Property for Square Roots

If a and b are positive numbers, then

$$\sqrt{a \cdot b} = \sqrt{a} \cdot \sqrt{b}$$

In words: The square root of a product is the product of the square roots.

Second, when a number occurs twice as a factor in a square root, then the square root simplifies to just that number. For example, $\sqrt{5 \cdot 5}$ is really $\sqrt{25}$, which is the same as just 5. Therefore, $\sqrt{5 \cdot 5} = 5$.

Repeated Factor Property for Square Roots

If a is a positive number, then

$$\sqrt{a \cdot a} = a$$

But how do these two properties help us simplify expressions such as $\sqrt{50}$? To see the answer to this question, we must factor 50 into the product of its prime factors:

$$\sqrt{50} = \sqrt{5 \cdot 5 \cdot 2}$$

The factor 5 occurs twice, meaning that we have a perfect square ($5 \cdot 5 = 25$) under the radical. Writing this as two separate square roots, we have

$$\begin{aligned}\sqrt{5 \cdot 5 \cdot 2} &= \sqrt{5 \cdot 5} \cdot \sqrt{2} \\ &= 5 \cdot \sqrt{2}\end{aligned}$$

Rule

When the number under a square root is factored completely, any factor that occurs twice can be taken out from under the square root symbol.

PRACTICE PROBLEMS

1. Simplify: $\sqrt{63}$

EXAMPLE 1

Simplify: $\sqrt{45}$

SOLUTION To begin we factor 45 into the product of prime factors:

$$\begin{aligned} \sqrt{45} &= \sqrt{3 \cdot 3 \cdot 5} && \text{Factor} \\ &= \sqrt{3 \cdot 3} \cdot \sqrt{5} && \text{Multiplication property} \\ &= 3 \cdot \sqrt{5} && \text{Repeated factor property} \end{aligned}$$

The expressions $\sqrt{45}$ and $3 \cdot \sqrt{5}$ are equivalent. The expression $3 \cdot \sqrt{5}$ is said to be in *simplified form* because the number under the radical is as small as possible.

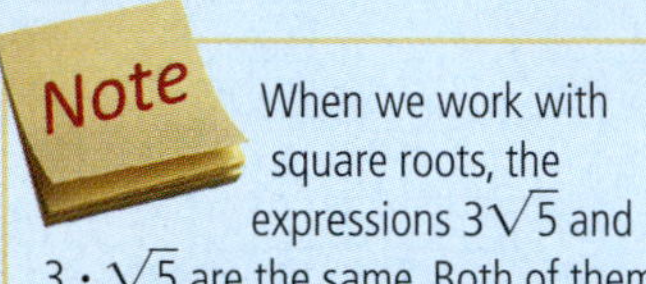

When we work with square roots, the expressions $3\sqrt{5}$ and $3 \cdot \sqrt{5}$ are the same. Both of them represent the product of 3 and $\sqrt{5}$. For simplicity, we usually omit the multiplication dot.

In our next example, a variable appears under the square root symbol. We will assume that all variables that appear under a radical represent positive numbers.

2. Simplify: $\sqrt{45x^2}$

EXAMPLE 2

Simplify: $\sqrt{18x^2}$

SOLUTION We factor $18x^2$ into $3 \cdot 3 \cdot 2 \cdot x \cdot x$. Because the factor 3 occurs twice, it can be taken out from under the radical. Likewise, because the factor x appears twice, it also can be taken out from under the radical. Therefore,

$$\begin{aligned} \sqrt{18x^2} &= \sqrt{3 \cdot 3 \cdot 2 \cdot x \cdot x} \\ &= 3 \cdot x \cdot \sqrt{2} \\ &= 3x\sqrt{2} \end{aligned}$$

You may be wondering if we can check this answer on a calculator. The answer is yes, but we need to substitute a value for x first. Suppose x is 5. Then

$$\sqrt{18x^2} = \sqrt{18 \cdot 25} = \sqrt{450} \approx 21.213203$$

and

$$3x\sqrt{2} = 3 \cdot 5\sqrt{2} = 15\sqrt{2} \approx 15(1.4142136) = 21.213203$$

3. Simplify: $\sqrt{300}$

EXAMPLE 3

Simplify: $\sqrt{180}$

SOLUTION We factor and then look for factors occurring twice.

$$\begin{aligned} \sqrt{180} &= \sqrt{2 \cdot 2 \cdot 3 \cdot 3 \cdot 5} \\ &= 2 \cdot 3 \cdot \sqrt{5} \\ &= 6\sqrt{5} \end{aligned}$$

4. Simplify: $\sqrt{50x^3}$

EXAMPLE 4

Simplify: $\sqrt{48x^3}$

SOLUTION

$$\begin{aligned} \sqrt{48x^3} &= \sqrt{2 \cdot 2 \cdot 2 \cdot 2 \cdot 3 \cdot x \cdot x \cdot x} \\ &= 2 \cdot 2 \cdot x \cdot \sqrt{3 \cdot x} \\ &= 4x\sqrt{3x} \end{aligned}$$

Answers

1. $3\sqrt{7}$ 2. $3x\sqrt{5}$ 3. $10\sqrt{3}$
4. $5x\sqrt{2x}$

FACTS FROM GEOMETRY The Spiral of Roots

To visualize the square roots of the positive integers, we can construct the spiral of roots. To begin, we draw two line segments, each of length 1, at right angles to each other. Then we use the Pythagorean theorem to find the length of the diagonal. Figure 1 illustrates:

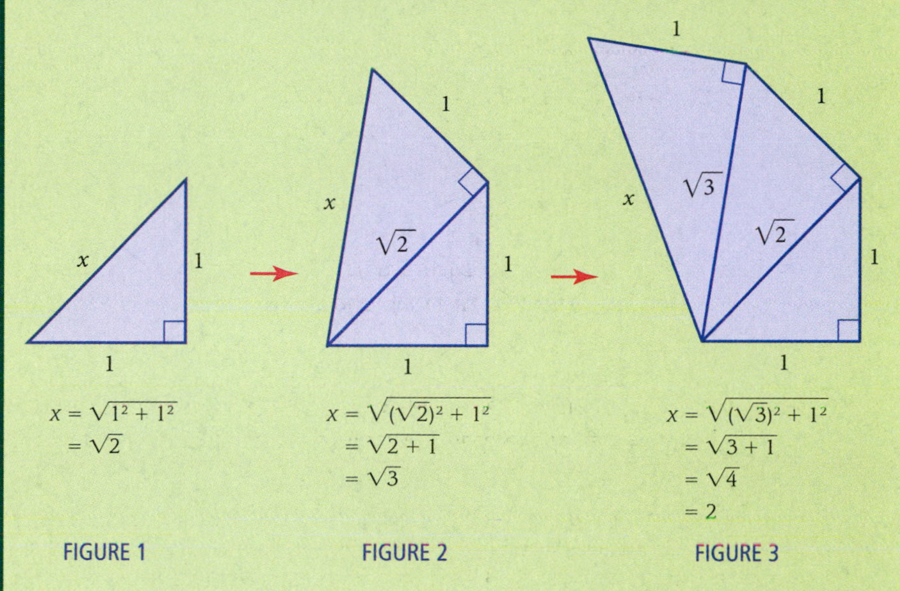

FIGURE 1 FIGURE 2 FIGURE 3

Next, we construct a second triangle by connecting a line segment of length 1 to the end of the first diagonal so that the angle formed is a right angle. We find the length of the second diagonal using the Pythagorean theorem. Figure 2 illustrates this procedure. As we continue to draw new triangles by connecting line segments of length 1 to the end of each previous diagonal so that the angle formed is a right angle, the spiral of roots begins to appear (see Figure 3).

Getting Ready for Class

After reading through the preceding section, respond in your own words and in complete sentences.

1. Why is factoring important when simplifying square roots?
2. Two sides of a right triangle are 3 inches and 4 inches. How long is the hypotenuse?
3. Two sides of a right triangle are each 1 inch. How do we represent the hypotenuse?
4. Why are the expressions $\sqrt{45}$ and $3\sqrt{5}$ equal?

Problem Set 5.8

A Simplify each expression by taking as much out from under the radical as possible. You may assume that all variables represent positive numbers. [Examples 1–4]

1. $\sqrt{12}$

2. $\sqrt{18}$

3. $\sqrt{20}$

4. $\sqrt{27}$

5. $\sqrt{72}$

6. $\sqrt{48}$

7. $\sqrt{98}$

8. $\sqrt{75}$

9. $\sqrt{28}$

10. $\sqrt{44}$

11. $\sqrt{200}$

12. $\sqrt{300}$

13. $\sqrt{12x^2}$

14. $\sqrt{18x^2}$

15. $\sqrt{50x^2}$

16. $\sqrt{45x^2}$

17. $\sqrt{75x^3}$

18. $\sqrt{8x^3}$

19. $\sqrt{50x^3}$

20. $\sqrt{45x^3}$

21. $\sqrt{32x^2y^3}$

22. $\sqrt{90x^2y^3}$

23. $\sqrt{243x^4}$

24. $\sqrt{288x^4}$

25. $\sqrt{72x^2y^4}$

26. $\sqrt{72x^4y^2}$

27. $\sqrt{12x^3y^3}$

28. $\sqrt{20x^3y^3}$

The triangles below are called *isosceles* right triangles because the two shorter sides are the same length. In each case, use the Pythagorean theorem to find the length of the hypotenuse. Simplify your answers, but do not use a calculator to approximate them.

29.

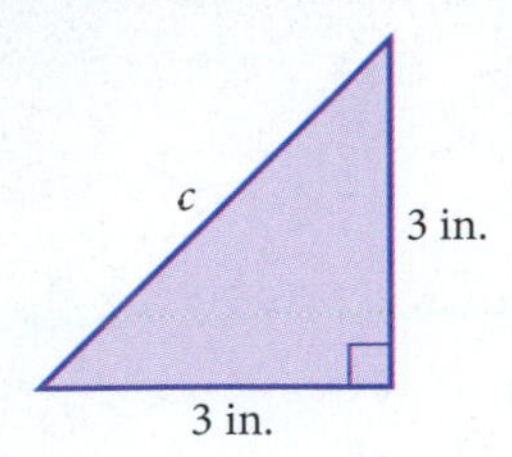

30.

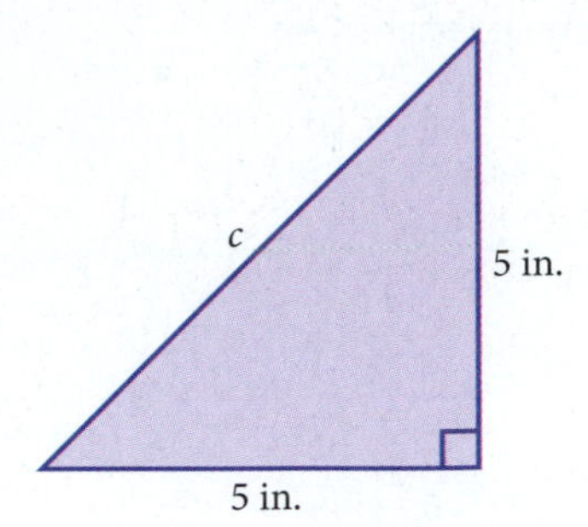

31.

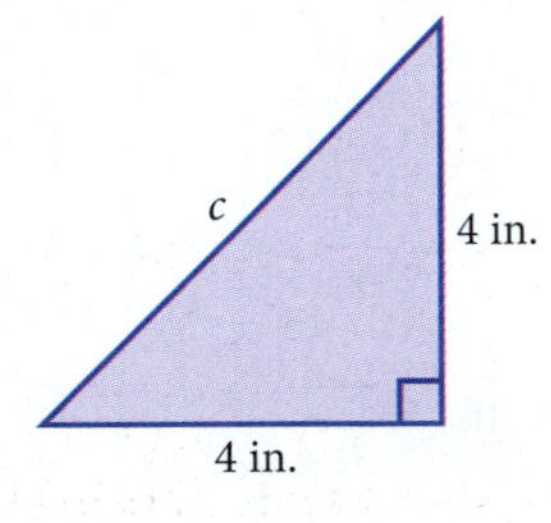

32.

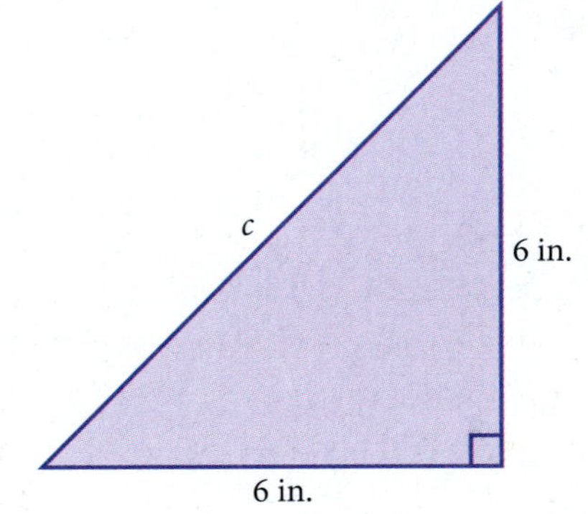

33.

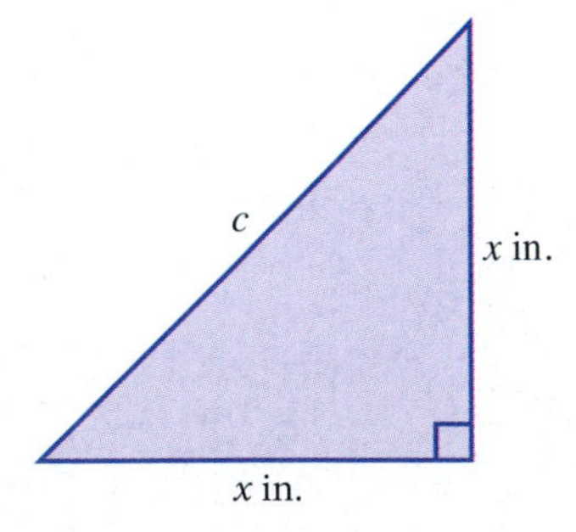

34.

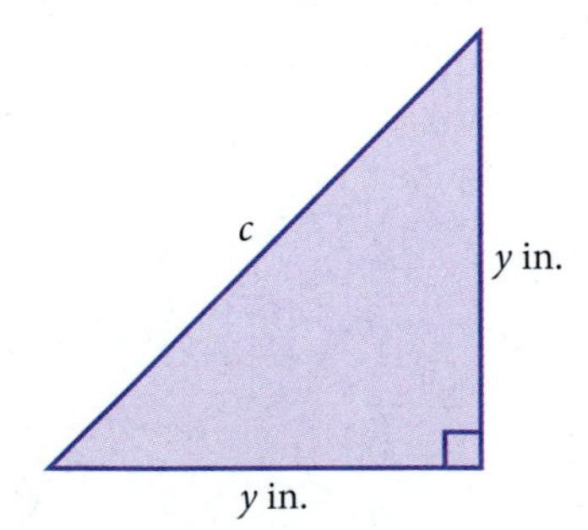

Applying the Concepts

35. Falling Time The formula that gives the number of seconds t it takes for an object to reach the ground when dropped from a height of h feet is

$$t = \sqrt{\frac{h}{16}}$$

If a rock is dropped from the top of a building 25 feet high, how long will it take for the rock to hit the ground?

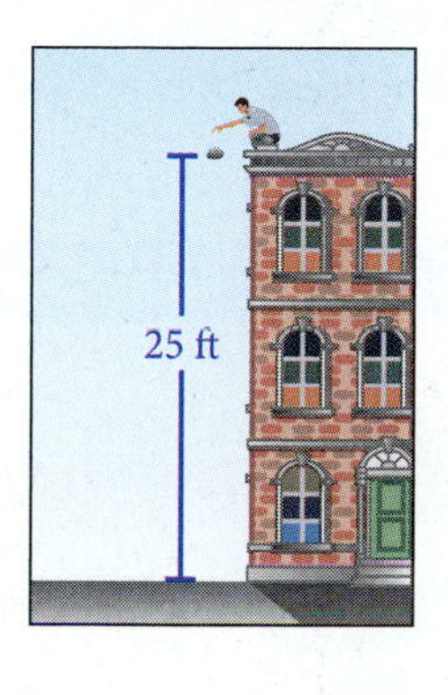

36. Falling Time Using the formula given in Problem 35, how long will it take for an object dropped from a building 100 feet high to reach the ground?

37. Spiral of Roots Construct your own spiral of roots by using a ruler. Draw the first triangle by using two 1-inch lines. The first diagonal will have a length of $\sqrt{2}$ inches. Each new triangle will be formed by drawing a 1-inch line segment at the end of the previous diagonal so that the angle formed is 90°. Draw your spiral until you have at least six right triangles.

38. Spiral of Roots Construct a spiral of roots by using line segments of length 2 inches. The length of the first diagonal will be $2\sqrt{2}$ inches. The length of the second diagonal will be $2\sqrt{3}$ inches. What will be the length of the third diagonal?

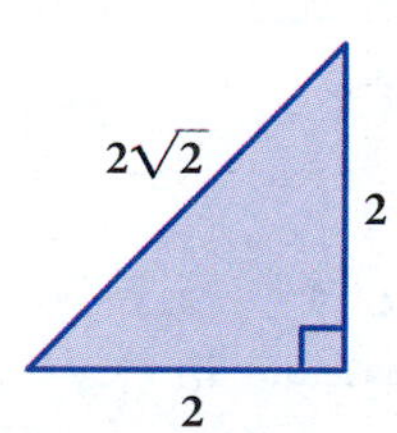

Calculator Problems

Use a calculator to find decimal approximations for each of the following numbers.

39. $\sqrt{72}$ and $6\sqrt{2}$

40. $\sqrt{75}$ and $5\sqrt{3}$

Substitute $x = 5$ into each of the following expressions, and then use a calculator to obtain a decimal approximation to each.

41. $x\sqrt{x}$

42. $\sqrt{x^3}$

43. $x^2\sqrt{x}$

44. $\sqrt{x^5}$

Use a calculator to help complete the following tables. If an answer needs rounding, round to the nearest thousandth.

45.

x	$\sqrt{x}$	$2\sqrt{x}$	$\sqrt{4x}$
1	1	2	2
2	1.414	2.828	2.828
3	1.732	3.464	3.464
4	2	4	4

46.

x	$\sqrt{x}$	$2\sqrt{x}$	$\sqrt{4x}$
1			
4			
9			
16			

47.

x	$\sqrt{x}$	$3\sqrt{x}$	$\sqrt{9x}$
1	1	3	3
2	1.414	4.242	4.242
3	1.732	5.196	5.196
4	2	6	6

48.

x	$\sqrt{x}$	$3\sqrt{x}$	$\sqrt{9x}$
1			
4			
9			
16			

Getting Ready for the Next Section

Combine like terms.

49. $15x + 8x$

50. $6x + 20x$

51. $25y + 3y - y$

52. $12y + 4y - y$

53. $2ab + 5ab$

54. $3ab + 7ab$

55. $2xy - 9xy + 50x$

56. $2xy - 18xy + 3x$

Maintaining Your Skills

In Problems 49–54, use the formula $y = \frac{1}{2}x - 3$ to find y if:

57. $x = 0$

58. $x = 1$

59. $x = -4$

60. $x = 4$

61. $x = 2$

62. $x = -2$

In Problems 55–58, use the formula $2x + 5y = 10$ to find y if:

63. $x = 0$

64. $x = -5$

65. $x = 5$

66. $x = \frac{5}{2}$

In Problems 59–62, use the formula $2x + 5y = 10$ to find x if:

67. $y = 0$

68. $y = 2$

69. $y = -2$

70. $y = \frac{2}{5}$

Adding and Subtracting Roots

5.9

Objectives

A Add and subtract expressions containing square roots.

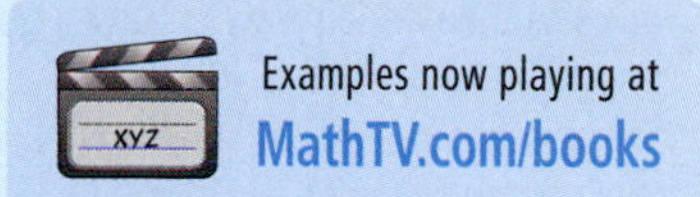

In the past we have combined similar terms by applying the distributive property. Here are some examples that will remind you of that procedure. The middle step in each example shows the distributive property.

$$7x + 3x = (7 + 3)x = 10x$$
$$5a^2 - 2a^2 = (5 - 2)a^2 = 3a^2$$
$$6xy + 3xy = (6 + 3)xy = 9xy$$
$$4x + 3x - 2x = (4 + 3 - 2)x = 5x$$

The distributive property is the only property that allows us to combine similar terms as we have done above. In order for the distributive property to be applied, the variable parts in each expression must be the same.

A Simplifying Square Roots

We add radical expressions in the same way we add similar terms, that is, by applying the distributive property. Here is how we use the distributive property to add $7\sqrt{2}$ and $3\sqrt{2}$:

$$7\sqrt{2} + 3\sqrt{2} = (7 + 3)\sqrt{2} \quad \text{Distributive property}$$
$$= 10\sqrt{2} \quad \text{Add 7 and 3}$$

To understand the steps shown here, you must remember that $7\sqrt{2}$ means 7 times $\sqrt{2}$; the 7 and the $\sqrt{2}$ are not "stuck together." Here are some additional examples. Compare them with the problems we did at the beginning of this section. Combine using the distributive property. (Assume all variables represent positive numbers.)

EXAMPLES Combine using the distributive property. (Assume all variables represent positive numbers.)

1. $5\sqrt{6} - 2\sqrt{6} = (5 - 2)\sqrt{6} = 3\sqrt{6}$
2. $6\sqrt{x} + 3\sqrt{x} = (6 + 3)\sqrt{x} = 9\sqrt{x}$
3. $4\sqrt{3} + 3\sqrt{3} - 2\sqrt{3} = (4 + 3 - 2)\sqrt{3} = 5\sqrt{3}$

PRACTICE PROBLEMS

Use the distributive property to combine each of the following.

1. $5\sqrt{3} - 2\sqrt{3}$
2. $7\sqrt{y} + 3\sqrt{y}$
3. $8\sqrt{5} - 2\sqrt{5} + 9\sqrt{5}$

As you can see, it is easy to combine radical expressions when each term contains the same square root.

Next, suppose we try to add $\sqrt{12}$ and $\sqrt{75}$. How should we go about it? You may think we should add 12 and 75 to get $\sqrt{87}$, but notice that we have not done that in any of the examples above. In fact, in Examples 1–3, the square root in the answer is the same square root we started with; we never added the number under the square roots!

A calculator can help us decide if $\sqrt{12} + \sqrt{75}$ is the same as $\sqrt{87}$. Here are the decimal approximations a calculator will give us:

$$\sqrt{12} + \sqrt{75} \approx 3.4641016 + 8.6602540 = 12.1243556$$
$$\sqrt{12 + 75} = \sqrt{87} \approx 9.3273791$$

As you can see, the two results are quite different, so we can assume that it would be a mistake to add the numbers under the square roots. That is:

$$\sqrt{12} + \sqrt{75} \neq \sqrt{12 + 75}$$

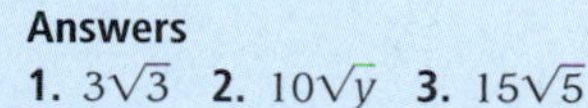

Answers
1. $3\sqrt{3}$ 2. $10\sqrt{y}$ 3. $15\sqrt{5}$

Note You may be thinking "Why did he show us the wrong way to do the problem first?" The reason is simple: Many people will try to add $\sqrt{12}$ and $\sqrt{75}$ by adding 12 and 75—it is a natural thing to want to do. The reason we don't is that it gives us the wrong answer every time! One of the things you need to know about learning algebra is that your intuition may lead you to a mistake.

The correct way to add $\sqrt{12}$ and $\sqrt{75}$ is to simplify each expression by taking as much out from under each square root as possible. Then, if the square roots in the resulting expressions are the same, we can add using the distributive property. Here is the way the problem is done correctly:

$$\begin{aligned} \sqrt{12} + \sqrt{75} &= \sqrt{2 \cdot 2 \cdot 3} + \sqrt{5 \cdot 5 \cdot 3} && \text{Simplify each square root} \\ &= 2\sqrt{3} + 5\sqrt{3} \\ &= (2 + 5)\sqrt{3} && \text{Distributive property} \\ &= 7\sqrt{3} && \text{Add 2 and 5} \end{aligned}$$

On a calculator, $7\sqrt{3} \approx 7(1.7320508) = 12.1243556$, which matches the approximation a calculator gives for $\sqrt{12} + \sqrt{75}$.

EXAMPLE 4 Combine, if possible: $\sqrt{18} + \sqrt{50} - \sqrt{8}$

SOLUTION First we simplify each term by taking as much out from under the square root as possible. Then we use the distributive property to combine terms if they contain the same square root.

$$\begin{aligned} \sqrt{18} + \sqrt{50} - \sqrt{8} &= \sqrt{3 \cdot 3 \cdot 2} + \sqrt{5 \cdot 5 \cdot 2} - \sqrt{2 \cdot 2 \cdot 2} \\ &= 3\sqrt{2} + 5\sqrt{2} - 2\sqrt{2} \\ &= (3 + 5 - 2)\sqrt{2} \\ &= 6\sqrt{2} \end{aligned}$$

EXAMPLE 5 Combine, if possible: $5\sqrt{54} - 3\sqrt{24}$

SOLUTION Proceeding as we did in the previous example, we simplify each term first; then we subtract by applying the distributive property.

$$\begin{aligned} 5\sqrt{54} - 3\sqrt{24} &= 5\sqrt{3 \cdot 3 \cdot 6} - 3\sqrt{2 \cdot 2 \cdot 6} \\ &= 5 \cdot 3\sqrt{6} - 3 \cdot 2\sqrt{6} \\ &= 15\sqrt{6} - 6\sqrt{6} \\ &= (15 - 6)\sqrt{6} \\ &= 9\sqrt{6} \end{aligned}$$

EXAMPLE 6 Assume x is a positive number and combine, if possible:

$$5\sqrt{12x^3} - 3\sqrt{75x^3}$$

SOLUTION We simplify each square root, and then we subtract.

$$\begin{aligned} 5\sqrt{12x^3} - 3\sqrt{75x^3} &= 5\sqrt{2 \cdot 2 \cdot 3 \cdot x \cdot x \cdot x} - 3\sqrt{5 \cdot 5 \cdot 3 \cdot x \cdot x \cdot x} \\ &= 5 \cdot 2 \cdot x\sqrt{3x} - 3 \cdot 5 \cdot x\sqrt{3x} \\ &= 10x\sqrt{3x} - 15x\sqrt{3x} \\ &= -5x\sqrt{3x} \end{aligned}$$

4. Combine, if possible: $\sqrt{27} + \sqrt{75} - \sqrt{12}$

5. Combine, if possible: $5\sqrt{45} - 3\sqrt{20}$

6. Combine, if possible: $7\sqrt{18x^3} - 3\sqrt{50x^3}$

Getting Ready for Class

After reading through the preceding section, respond in your own words and in complete sentences.

1. What property is used to add $7x$ and $3x$?
2. How is the distributive property used to add $6\sqrt{x}$ and $3\sqrt{x}$?
3. Why is it wrong to add $\sqrt{12}$ and $\sqrt{75}$ to get $\sqrt{87}$?
4. What is the sum of $\sqrt{12}$ and $\sqrt{75}$?

Problem Set 5.9

A Combine by applying the distributive property. Assume all variables represent positive numbers. [Examples 1–6]

1. $2\sqrt{3} + 8\sqrt{3}$

2. $2\sqrt{3} - 8\sqrt{3}$

3. $7\sqrt{5} - 3\sqrt{5}$

4. $7\sqrt{5} + 3\sqrt{5}$

5. $9\sqrt{x} + 3\sqrt{x} - 5\sqrt{x}$

6. $6\sqrt{x} + 10\sqrt{x} - 3\sqrt{x}$

7. $8\sqrt{7} + \sqrt{7}$

8. $9\sqrt{7} + \sqrt{7}$

9. $2\sqrt{y} + \sqrt{y} + 3\sqrt{y}$

10. $7\sqrt{y} + \sqrt{y} + 2\sqrt{y}$

A Simplify each square root, then combine if possible. Assume all variables represent positive numbers. [Examples 1–6]

11. $\sqrt{18} + \sqrt{32}$

12. $\sqrt{12} + \sqrt{27}$

13. $\sqrt{75} + \sqrt{27}$

14. $\sqrt{50} + \sqrt{8}$

15. $2\sqrt{75} - 4\sqrt{27}$

16. $4\sqrt{50} - 5\sqrt{8}$

17. $2\sqrt{90} + 3\sqrt{40} - 4\sqrt{10}$

18. $5\sqrt{40} - 2\sqrt{90} + 3\sqrt{10}$

19. $\sqrt{72x^2} - \sqrt{50x^2}$

20. $\sqrt{98x^2} - \sqrt{72x^2}$

21. $4\sqrt{20x^3} + 3\sqrt{45x^3}$

22. $8\sqrt{48x^3} + 2\sqrt{12x^3}$

In each diagram below, find the distance from A to B. Simplify your answers, but do not use a calculator.

23.

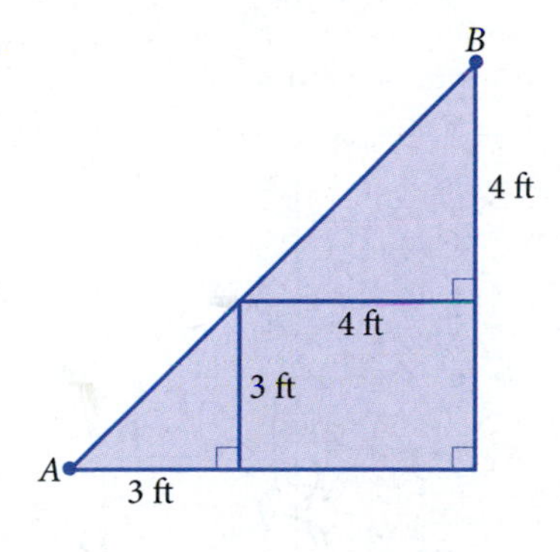

24.

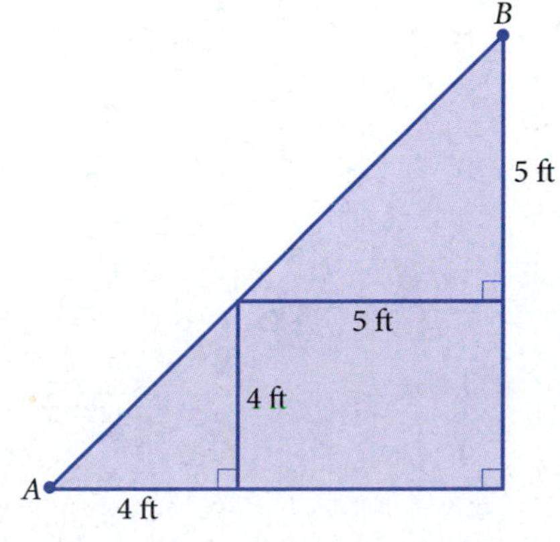

Calculator Problems

25. Use a calculator to show that $\sqrt{2} + \sqrt{3}$ is not the same as $\sqrt{5}$.

26. Use a calculator to show that $\sqrt{5} - \sqrt{2}$ is not the same as $\sqrt{3}$.

Use a calculator to help complete the following tables. If an answer needs rounding, round to the nearest thousandth.

27.

x	$\sqrt{x^2 + 9}$	$x + 3$
1	3.162	4
2	3.606	5
3	4.243	6
4	5	7
5	5.831	8
6	6.708	9

28.

x	$\sqrt{x^2 + 16}$	$x + 4$
1		
2		
3		
4		
5		
6		

29.

x	$\sqrt{x + 3}$	$\sqrt{x} + \sqrt{3}$
1	2	3.732
2	2.236	3.146
3	2.449	3.464
4	2.646	3.732
5	2.828	3.968
6	3	4.732

30.

x	$\sqrt{x + 4}$	$\sqrt{x} + 2$
1		
2		
3		
4		
5		
6		

Maintaining Your Skills

The problems below review material we covered in Section 4.9.
Graph each equation.

31. $x + y = 3$

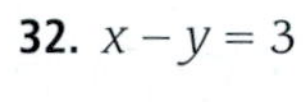

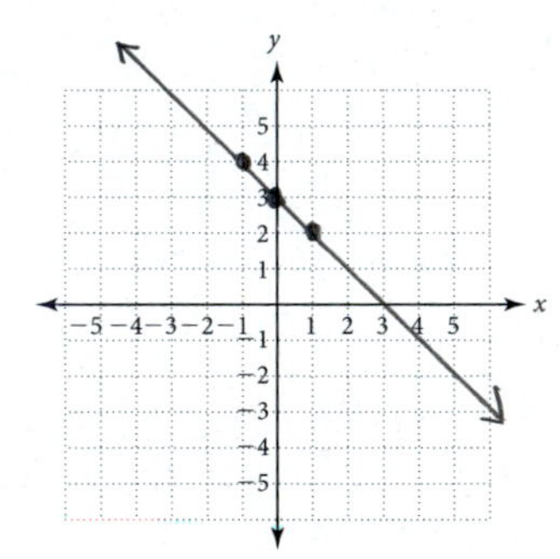

32. $x - y = 3$

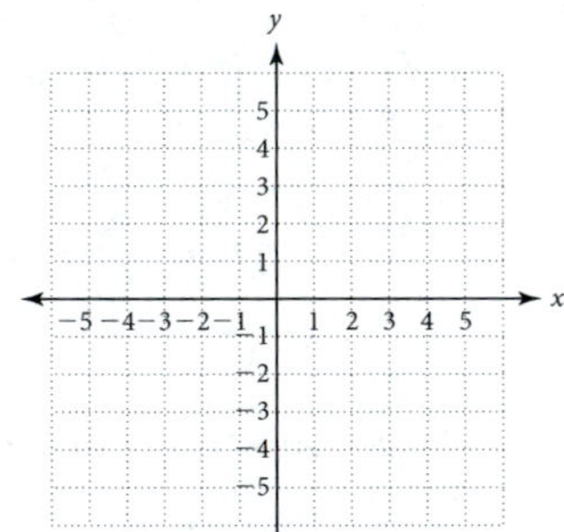

33. $y = \quad x - 3$

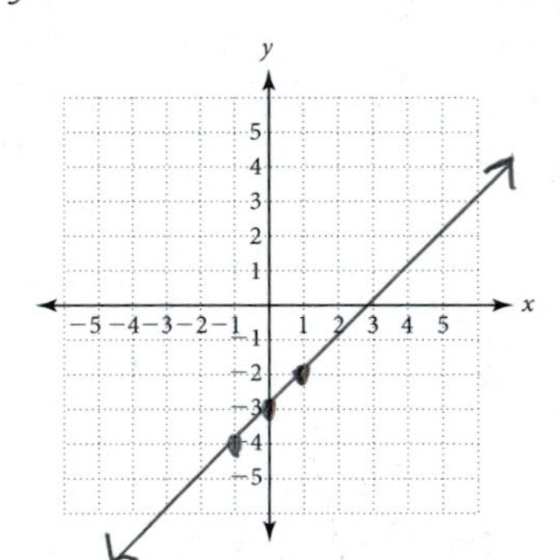

34. $y = \quad x + 3$

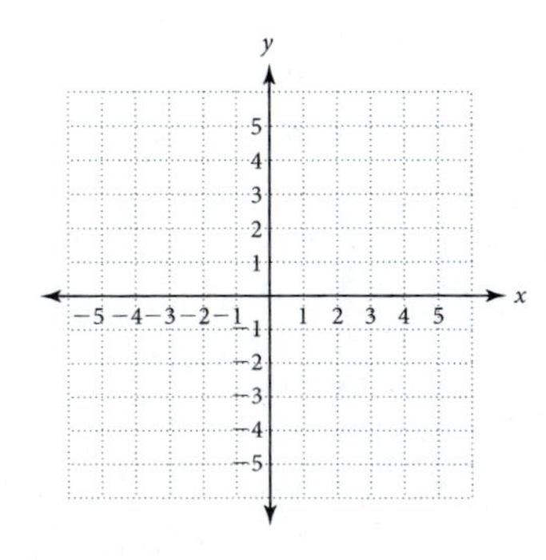

35. $2x + 5y = 10$

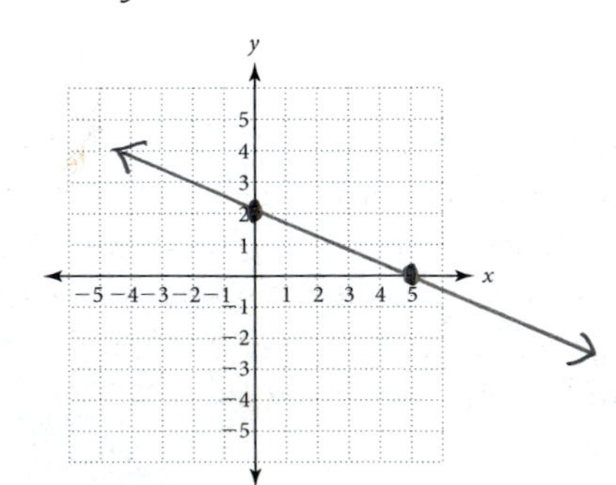

36. $5x + 2y = 10$

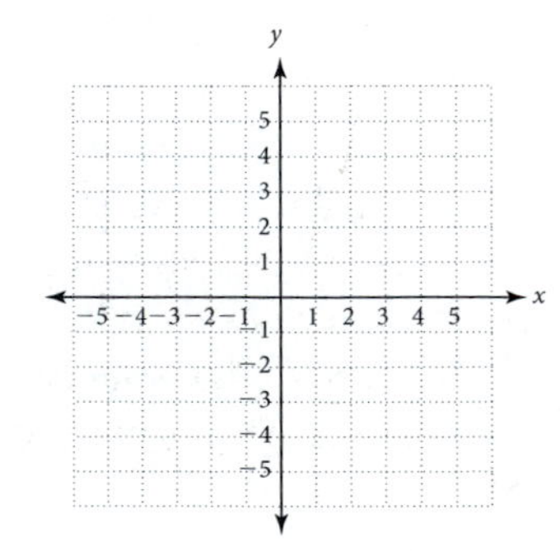

Chapter 5 Summary

Place Value [5.1]

The place values for the first five places to the right of the decimal point are

Decimal Point	Tenths	Hundredths	Thousandths	Ten Thousandths	Hundred Thousandths
.	$\frac{1}{10}$	$\frac{1}{100}$	$\frac{1}{1,000}$	$\frac{1}{10,000}$	$\frac{1}{100,000}$

EXAMPLES

1. The number 4.123 in words is "four and one hundred twenty-three thousandths."

Rounding Decimals [5.1]

If the digit in the column to the right of the one we are rounding to is 5 or more, we add 1 to the digit in the column we are rounding to; otherwise, we leave it alone. We then replace all digits to the right of the column we are rounding to with zeros if they are to the left of the decimal point; otherwise, we simply delete them.

2. 357.753 rounded to the nearest
   Tenth: 357.8
   Ten: 360

Addition and Subtraction with Decimals [5.2]

To add (or subtract) decimal numbers, we align the decimal points and add (or subtract) as if we were adding (or subtracting) whole numbers. The decimal point in the answer goes directly below the decimal points in the problem.

3. $$\begin{array}{r} 3.400 \\ 25.060 \\ +\ 0.347 \\ \hline 28.807 \end{array}$$

Multiplication with Decimals [5.3]

To multiply two decimal numbers, we multiply as if the decimal points were not there. The decimal point in the product has as many digits to the right as there are total digits to the right of the decimal points in the two original numbers.

4. If we multiply 3.49×5.863, there will be a total of $2 + 3 = 5$ digits to the right of the decimal point in the answer.

Division with Decimals [5.4]

To begin a division problem with decimals, we make sure that the divisor is a whole number. If it is not, we move the decimal point in the divisor to the right as many places as it takes to make it a whole number. We must then be sure to move the decimal point in the dividend the same number of places to the right. Once the divisor is a whole number, we divide as usual. The decimal point in the answer is placed directly above the decimal point in the dividend.

5. $$\begin{array}{r} 1.39 \\ 2.5.\overline{)3.4.75} \\ \underline{2\ 5} \\ 9\ 7 \\ \underline{7\ 5} \\ 2\ 25 \\ \underline{2\ 25} \\ 0 \end{array}$$

Changing Fractions to Decimals [5.5]

6. $\frac{4}{15} = 0.2\overline{6}$ because

$$\begin{array}{r} .266 \\ 15\overline{)4.000} \\ \underline{3\,0} \\ 1\,00 \\ \underline{90} \\ 100 \\ \underline{90} \\ 10 \end{array}$$

To change a fraction to a decimal, we divide the numerator by the denominator.

Changing Decimals to Fractions [5.5]

7. $0.781 = \frac{781}{1{,}000}$

To change a decimal to a fraction, we write the digits to the right of the decimal point over the appropriate power of 10.

Equations Containing Decimals [5.6]

8.

$$\frac{1}{2}x - 3.78 = 2.52$$
$$\frac{1}{2}x - 3.78 + \mathbf{3.78} = 2.52 + \mathbf{3.78}$$
$$\frac{1}{2}x = 6.30$$
$$\mathbf{2}\left(\frac{1}{2}x\right) = \mathbf{2}(6.30)$$
$$x = 12.6$$

We solve equations that contain decimals by applying the addition property of equality and the multiplication property of equality.

Square Roots [5.7]

9. $\sqrt{49} = 7$ because $7^2 = 7 \cdot 7 = 49$

The square root of a positive number a, written $\sqrt{a}$, is the number we square to get a.

Pythagorean Theorem [5.7]

In any right triangle, the length of the longest side (the hypotenuse) is equal to the square root of the sum of the squares of the two shorter sides.

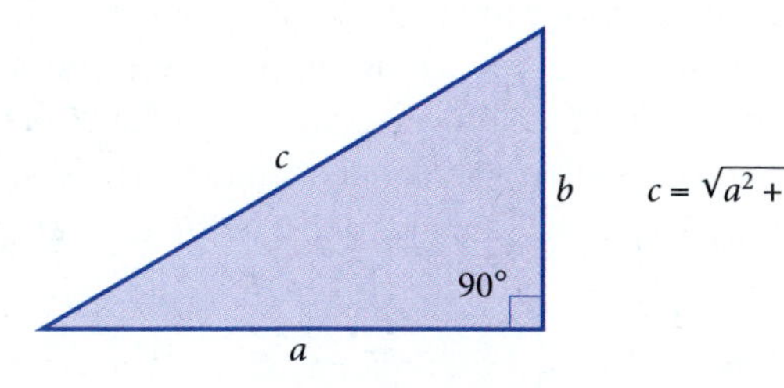

Multiplication Property for Square Roots [5.8]

If a and b are positive numbers, then

$$\sqrt{a \cdot b} = \sqrt{a} \cdot \sqrt{b}$$

In words: The square root of a product is the product of the square roots.

Repeated Factor Property for Square Roots [5.8]

If a is a positive number, then

$$\sqrt{a \cdot a} = a$$

Simplifying Square Roots [5.8]

To simplify a square root, we factor the number under the square root symbol into the product of prime factors. Then we use the two properties of square roots shown above to take as much out from under the square root symbol as possible.

10. Simplify $\sqrt{50}$

$\sqrt{50} = \sqrt{5 \cdot 5 \cdot 2}$ Factor 50

$= \sqrt{5 \cdot 5} \cdot \sqrt{2}$ Multiplication property

$= 5 \cdot \sqrt{2}$ Repeated factor property

Adding and Subtracting Roots [5.9]

We add expressions containing square roots by first simplifying each square root and then, if the square roots in the resulting terms are the same, applying the distributive property.

11. $\sqrt{12} + \sqrt{75}$

$= \sqrt{2 \cdot 2 \cdot 3} + \sqrt{5 \cdot 5 \cdot 3}$

$= 2\sqrt{3} + 5\sqrt{3}$

$= (2 + 5)\sqrt{3}$

$= 7\sqrt{3}$

Chapter 5 Review

Give the place value of the 7 in each of the following numbers. [5.1]

1. 36.007

2. 121.379

Write each of the following as a decimal number. [5.1]

3. Thirty-seven and forty-two ten thousandths

4. One hundred and two hundred two hundred thousandths

5. Round 98.7654 to the nearest hundredth. [5.1]

Perform the following operations. [5.2, 5.3, 5.4]

6. $3.78 + 2.036$

7. $11.076 - 3.297$

8. 6.7×5.43

9. $-0.89(24.24)$

10. $-29.07 \div (-3.8)$

11. $0.7134 \div 0.58$

12. Write $\frac{7}{8}$ as a decimal. [5.5]

13. Write 0.705 as a fraction in lowest terms. [5.5]

14. Write 14.125 as a mixed number. [5.5]

Simplify each of the following expressions as much as possible. [5.5]

15. $3.3 - 4(0.22)$

16. $54.987 - 2(3.05 + 0.151)$

17. $125\left(\frac{3}{5}\right) + 4$

18. $\frac{3}{5}(0.9) + \frac{2}{5}(0.4)$

Solve each equation. [5.6]

19. $x + 9.8 = 3.9$

20. $5x = 23.4$

21. $0.5y - 0.2 = 3$

22. $5x - 7.2 = 3x + 3.8$

Simplify each expression as much as possible. [5.7]

23. $3\sqrt{25}$

24. $\sqrt{64} - \sqrt{36}$

25. $4\sqrt{25} + 3\sqrt{81}$

26. $\sqrt{\frac{16}{49}}$

27. **Perimeter of a Dollar** A $10 bill has a width of 2.56 inches and a length of 6.14 inches. Find the perimeter.

28. A person purchases $7.23 worth of goods at a drugstore. If a $10 bill is used to pay for the purchases, how much change is received? [5.2]

Find the length of the hypotenuse in each right triangle. Round your answers to the nearest tenth. [5.7]

29.

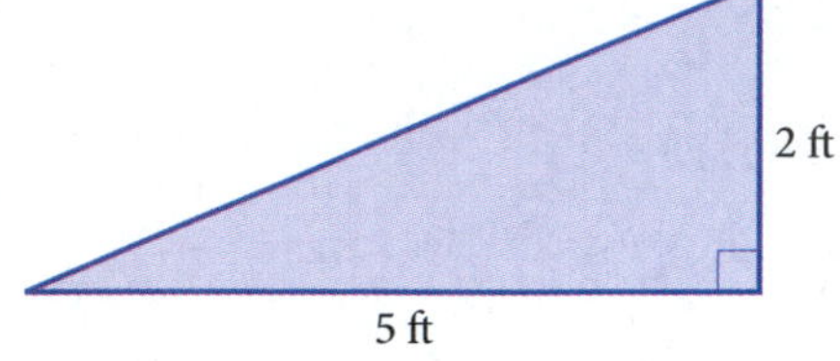

30.

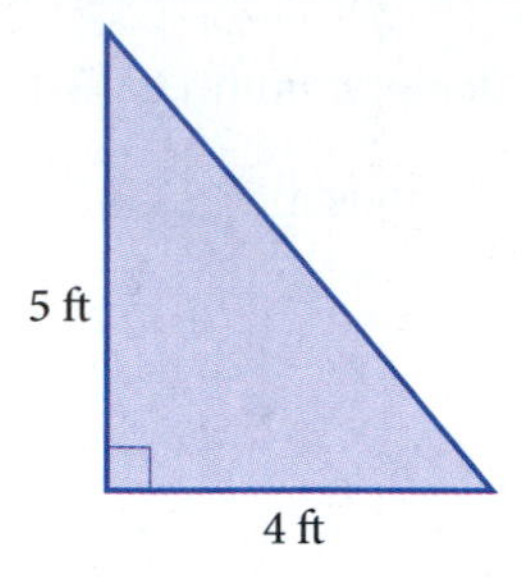

31. Find the perimeter (to the nearest tenth) and the area of the triangle in Problem 29. [5.2, 5.3]

32. Find the perimeter (to the nearest tenth) and the area of the triangle in Problem 30. [5.2, 5.3]

33. **Hula Hoop** A youngster wants to make a hula hoop from pipe purchased at the hardware store. If the hula hoop is to have a diameter of 32 inches, what length pipe should be purchased?

Ariel Skelley/Corbis

34. **Spaceship Earth** Spaceship Earth is a geodesic sphere at Walt Disney World's Epcot Center in Florida. The inner sphere has a diameter of 165 feet, while the outer sphere is 180 feet in diameter. Find the difference in volume between the outer sphere and the inner sphere.

Douglas Peebles/Corbis

Simplify each expression by taking as much out from under the radical as possible. [5.8]

35. $\sqrt{12}$

36. $\sqrt{50}$

37. $\sqrt{20x^2}$

38. $\sqrt{20x^3}$

39. $\sqrt{18x^2y^3}$

40. $\sqrt{75x^3y^2}$

Combine. [5.9]

41. $8\sqrt{3} + 2\sqrt{3}$

42. $7\sqrt{6} + 10\sqrt{6}$

43. $4\sqrt{3} + \sqrt{3}$

44. $9\sqrt{2} + \sqrt{2}$

Simplify each square root, and then combine if possible. [5.9]

45. $\sqrt{24} + \sqrt{54}$

46. $\sqrt{18} + \sqrt{8}$

47. $3\sqrt{75} - 8\sqrt{27}$

48. $7\sqrt{20} - 2\sqrt{45}$

Chapter 5 Cumulative Review

Simplify.

1. $6 + \frac{5}{x}$

2. $\frac{0}{-8}$

3. $56(287)$

4. $-6(7 - 12)$

5. $-\frac{2}{5}(5x + 10) - 4$

6. $-\frac{3}{4}(8x + 16) + 7$

7. $-4.3(-12.96)$

8. $1{,}292 \div 17$

9. $\frac{5}{14} \div \frac{15}{21}$

10. Round 463,612 to the nearest thousand.

11. Change $\frac{63}{4}$ to a mixed number.

12. Give the opposite and absolute value of 11.

13. Is $x = -2$ a solution to $3x - 5 = 1$?

$\frac{25}{200}$ $\frac{5}{40}$ $\frac{50}{200}$ $\frac{10}{40}$ $\frac{75}{200}$ $\frac{15}{40}$ $\frac{3}{8}$ $\frac{100}{200}$ $\frac{20}{40}$ $\frac{125}{200}$ $\frac{25}{40}$ $\frac{150}{200}$ $\frac{30}{40}$ $\frac{6}{8}$ $\frac{3}{4}$

$\frac{175}{200}$ $\frac{35}{40}$ $\frac{7}{8}$

14. Change each decimal into a fraction.

Decimal	0.125	0.250	0.375	0.500	0.625	0.750	0.875	1
Fraction	$\frac{1}{8}$	$\frac{1}{4}$	$\frac{3}{8}$	$\frac{1}{2}$	$\frac{5}{8}$	$\frac{3}{4}$	$\frac{7}{8}$	$\frac{8}{8}$

15. Give the quotient of 72 and -8.

16. Identify the property or properties used in the following: $2 \cdot (x \cdot 3) = (2 \cdot 3) \cdot x$

17. Translate into symbols, then simplify: Three times the sum of 13 and 4 is 51.

18. Reduce $\frac{120ab}{70b}$

19. Solve the equation $5x - 10 = 15$.

$+10 \quad +10$ $\quad \frac{5x}{5} = \frac{25}{5}$ $\quad x = 5$

Simplify.

20. $6(4)^2 - 8(2)^3$

21. $\dfrac{-6 + 2(-4)}{8 - 10}$

22. $\sqrt{72x^3y^2}$

23. $\dfrac{2}{3}(0.45) - \dfrac{4}{5}(0.8)$

24. $2b + 6 - 4b - 5$

25. $\left(3\dfrac{1}{3} - \dfrac{1}{2}\right)\left(4\dfrac{1}{2} + \dfrac{3}{4}\right)$

26. $3\sqrt{25} - 5\sqrt{49}$

Solve.

27. $\dfrac{3}{4}x = 21$

28. $8b - 0.22 = 2b + 0.68$

29. $2(x - 6) = -10$

30. $5(3x + 4) - 9 = -2(x - 8)$

31. $a - \dfrac{3}{4} = \dfrac{3}{8}$

32. Gambling A gambler loses $15 playing poker one night, wins $25 the next night, and then loses $30 the last night. How much did the gambler win or lose overall?

33. Average Score Lorena has scores of 83, 85, 79, 93, and 80 on her first five math tests. What is her average score for these five tests?

34. Geometry Find the length of the hypotenuse of a right triangle with shorter sides of 6 in. and 8 in.

35. Perimeter of a Banknote The 10 euro banknote shown here has a width of 6.7 centimeters and a length of 12.7 centimeters. Find the perimeter.

36. Recipe A muffin recipe calls for $2\frac{3}{4}$ cups of flour. If the recipe is tripled, how many cups of flour will be needed?

37. Volume Find the volume of a rectangular solid with length 5 in., width 3 in., and height 7 in.

38. Hourly Wage If you earn $384 for working 40 hours, what is your hourly wage?

39. Test Taking If 40 students take a Spanish test and $\frac{4}{5}$ of them pass, how many students passed?

Chapter 5 Test

1. Write the decimal number 5.053 in words.

2. Give the place value of the 4 in the number 53.0543.

3. Write seventeen and four hundred six ten thousandths as a decimal number.

4. Round 46.7549 to the nearest hundredth.

Perform the following operations.

5. $7 + 0.6 + 0.58$

6. $12.032 - 5.976$

7. $5.7(6.24)$

8. $-22.672 \div (-2.6)$

9. Write $\frac{23}{25}$ as a decimal.

10. Write 0.56 as a fraction in lowest terms.

Simplify each expression as much as possible.

11. $5.2(2.8 + 0.02)$

12. $5.2 - 3(0.17)$

13. $23.852 - 3(2.01 + 0.231)$

14. $\frac{3}{5}(0.6) - \frac{2}{3}(0.15)$

15. Solve: $6a - 0.18 = a + 0.77$

$-5 \quad +0.18 \; -a \quad +0.18$

$\frac{5a}{5} = \frac{0.95}{5} \qquad a = 0.19$

Simplify each expression as much as possible.

16. $2\sqrt{36} + 3\sqrt{64}$

17. $\sqrt{\frac{25}{81}}$

18. A person purchases \$8.47 worth of goods at a drugstore. If a \$20 bill is used to pay for the purchases, how much change is received?

19. If coffee sells for \$6.99 per pound, how much will 3.5 pounds of coffee cost?

20. If a person earns \$262 for working 40 hours, what is the person's hourly wage?

21. Find the length of the hypotenuse of the right triangle below.

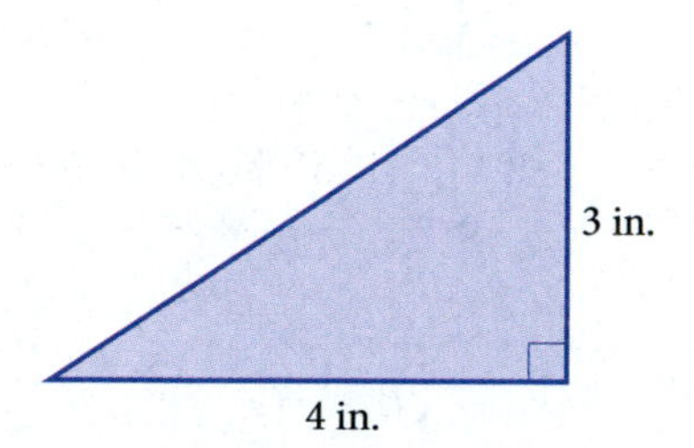

Simplify by taking as much out from under the radical as possible.

22. $\sqrt{72}$

23. $\sqrt{48x^2y}$

24. Combine $5\sqrt{7} + 3\sqrt{7}$.

25. Simplify each square root, and then combine, if possible: $6\sqrt{12} - 5\sqrt{48}$

Chapter 5 Projects

DECIMALS

GROUP PROJECT

Unwinding the Spiral of Roots

Number of People 2–3

Time Needed 8–12 minutes

Equipment Pencil, ruler, graph paper, scissors, and tape

Background In this chapter, we used the Spiral of Roots to visualize square roots of whole numbers. If we "unwind" the spiral of roots, we can produce the graph of a simple equation on a rectangular coordinate system.

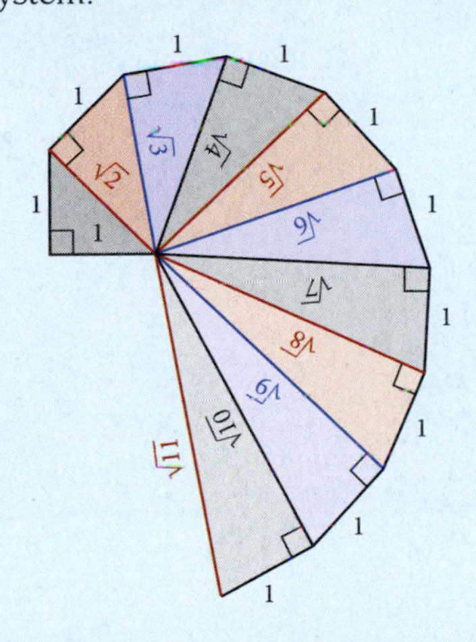

Procedure

1. Carefully cut out each triangle from the Spiral of Roots above.
2. Line up the triangles horizontally on the coordinate system shown here so that the side of length 1 is on the x-axis and the hypotenuse is on the left. Note that the first triangle is shown in place, and the outline of the second triangle is next to it. The 1-unit side of each triangle should fit in each of the 1-unit spaces on the x-axis.
3. On the coordinate system, plot a point at the tip of each triangle. Then, connect these points with a smooth curve.
4. What is the equation of the curve you have just drawn?

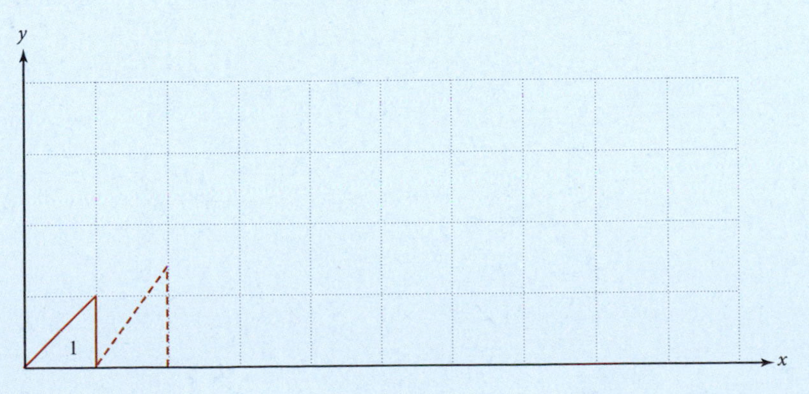

RESEARCH PROJECT

The Wizard of Oz

In the 1939 movie *The Wizard of Oz,* the Scarecrow (played by Ray Bolger) sings "If I only had a brain." Upon receiving a diploma from the great Oz, he rapidly recites a math theorem in an attempt to display his new knowledge. Unfortunately, the Scarecrow's inability to recite the Pythagorean Theorem might lead one to doubt the effectiveness of his diploma. Watch this scene in the movie. Write down the Scarecrow's speech and explain the errors.

The Kobal Collection

Ratio and Proportion

6

Chapter Outline

Introduction

The Eiffel Tower in Paris, France, is one of the most recognizable structures in the world. It was built in 1889 on the Champ de Mars beside the Seine River and named after its designer, engineer Gustave Eiffel. At that time it was the world's tallest tower, measuring 300 meters. Today it remains the tallest building in Paris.

This structure has inspired many reproductions, that is, towers built as replicas of the original Eiffel tower. The chart below shows the location and heights of some of these towers:

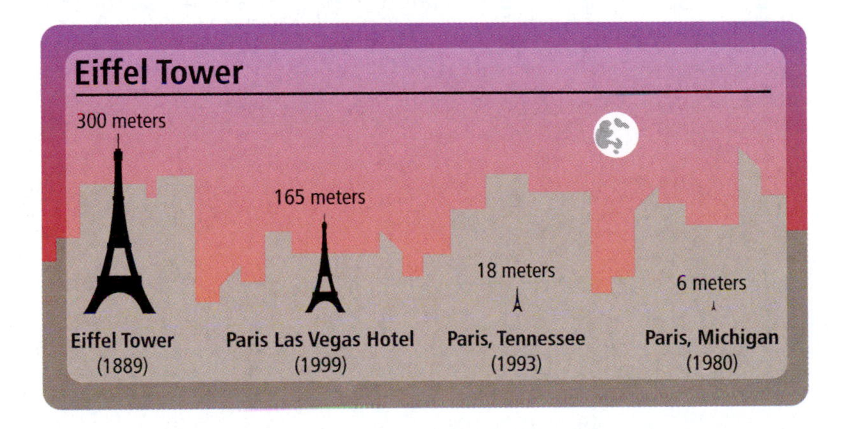

As you can see, these replicas are all different sizes. In mathematics, we can use ratios to compare those different size towers. For instance, we say the largest height and smallest height in the chart are in a ratio of 50 to 1. In this chapter, we study ratios like this one. As you will see, ratios are very closely related to fractions and decimals, which we have already studied.

Chapter Pretest

The Pretest below contains problems that are representative of the problems you will find in the chapter.

Express each ratio as a fraction in lowest terms.

1. 15 to 25

2. 400 to 150

3. $\frac{5}{4}$ to $\frac{7}{4}$

4. 3.2 to 4.6

5. A car travels 434 miles in 7 hours. What is the rate of the car in miles per hour?

6. A 16-ounce container of heavy whipping cream costs \$2.40. Find the price per ounce.

Solve each proportion.

7. $\frac{5}{6} = \frac{x}{12}$

8. $\frac{2}{y} = \frac{4}{10}$

9. $\frac{9}{7} = \frac{6}{x}$

10. $\frac{n}{8} = \frac{\frac{1}{6}}{\frac{1}{9}}$

11. A trucker drives his rig 480 miles in 8 hours. At this rate, how far will he travel in 12 hours?

12. A company reimburses its employees 36.5¢ for every mile of business travel. If an employee drives 150 miles, how much will she be reimbursed?

Assume the figures presented are similar.

13. Find length x.

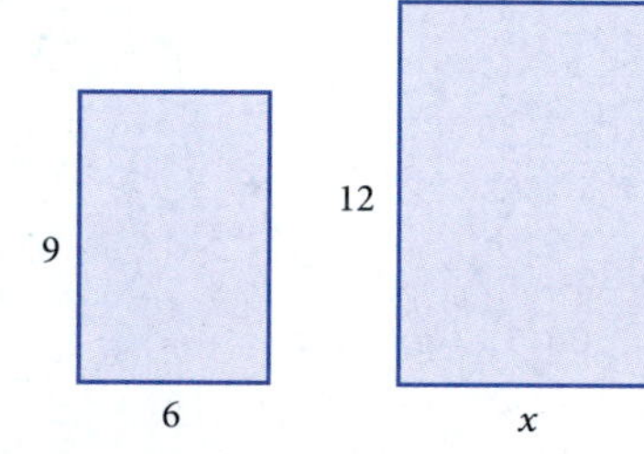

14. Find length BC.

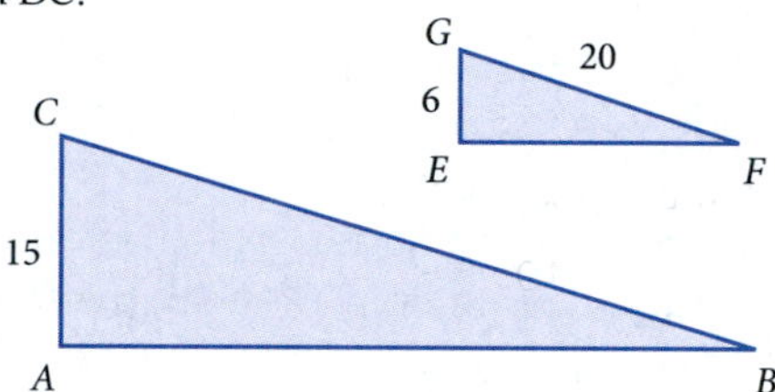

Getting Ready for Chapter 6

The problems below review material covered previously that you need to know in order to be successful in Chapter 6.

Reduce to lowest terms.

1. $\frac{16}{48}$

2. $\frac{320}{160}$

Write as a decimal.

3. $\frac{1}{4}$

4. $\frac{1}{8}$

Multiply or divide as indicated.

5. $5 \cdot 13$

6. 3(0.4)

7. 3.5(85)

8. $\frac{2}{3} \cdot 6$

9. 0.08×100

10. 0.12×100

11. $125 \div 2$

12. $1.39 \div 2$

13. $\frac{1.99}{\frac{1}{2}}$

14. $\frac{\frac{2}{3}}{\frac{4}{9}}$

Divide. Round answers to the nearest tenth.

15. $48 \div 5.5$

16. $75 \div 11.5$

6.1 Ratios

Objectives

A Express ratios as fractions in lowest terms.

B Use ratios to solve application problems.

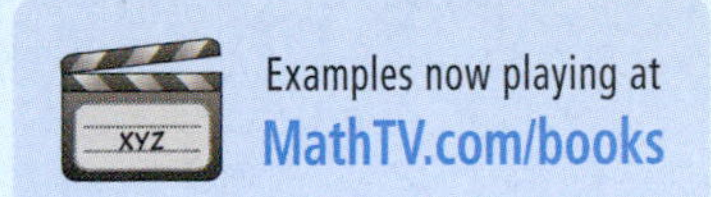

Introduction

The *ratio* of two numbers is a way of comparing them. If we say that the ratio of two numbers is 2 to 1, then the first number is twice as large as the second number. For example, if there are 10 men and 5 women enrolled in a math class, then the ratio of men to women is 10 to 5. Because 10 is twice as large as 5, we can also say that the ratio of men to women is 2 to 1.

We can define the ratio of two numbers in terms of fractions.

A Express Ratios as Fractions in Lowest Terms

> **Definition**
>
> A **ratio** is a comparison between two numbers and is represented as a fraction, where the first number in the ratio is the numerator and the second number in the ratio is the denominator. *In symbols:*
>
> If a and b are any two numbers,
>
> then the ratio of a to b is $\frac{a}{b}$. $(b \neq 0)$

We handle ratios the same way we handle fractions. For example, when we said that the ratio of 10 men to 5 women was the same as the ratio 2 to 1, we were actually saying

$$\frac{10}{5} = \frac{2}{1} \quad \text{Reducing to lowest terms}$$

Because we have already studied fractions in detail, much of the introductory material on ratios will seem like review.

EXAMPLE 1 Express the ratio of 16 to 48 as a fraction in lowest terms.

SOLUTION Because the ratio is 16 to 48, the numerator of the fraction is 16 and the denominator is 48:

$$\frac{16}{48} = \frac{1}{3} \quad \text{In lowest terms}$$

Notice that the first number in the ratio becomes the numerator of the fraction, and the second number in the ratio becomes the denominator.

PRACTICE PROBLEMS

1. a. Express the ratio of 32 to 48 as a fraction in lowest terms.
   b. Express the ratio of 3.2 to 4.8 as a fraction in lowest terms.
   c. Express the ratio of 0.32 to 0.48 as a fraction in lowest terms.

Answer

1. All are $\frac{2}{3}$

2. a. Give the ratio of $\frac{3}{5}$ to $\frac{9}{10}$ as a fraction in lowest terms.
   b. Give the ratio of 0.6 to 0.9 as a fraction in lowest terms.

3. a. Write the ratio of 0.06 to 0.12 as a fraction in lowest terms.
   b. Write the ratio of 600 to 1200 as a fraction in lowest terms.

Note Another symbol used to denote ratio is the colon (:). The ratio of, say, 5 to 4 can be written as 5:4. Although we will not use it here, this notation is fairly common.

4. Suppose the basketball player in Example 4 makes 12 out of 16 free throws. Write the ratio again using these new numbers.

Answers

2. Both are $\frac{2}{3}$ 3. Both are $\frac{1}{2}$
4. $\frac{3}{4}$

EXAMPLE 2 Give the ratio of $\frac{2}{3}$ to $\frac{4}{9}$ as a fraction in lowest terms.

SOLUTION We begin by writing the ratio of $\frac{2}{3}$ to $\frac{4}{9}$ as a complex fraction. The numerator is $\frac{2}{3}$, and the denominator is $\frac{4}{9}$. Then we simplify.

$$\frac{\frac{2}{3}}{\frac{4}{9}} = \frac{2}{3} \cdot \frac{9}{4} \quad \text{Division by } \tfrac{4}{9} \text{ is the same as multiplication by } \tfrac{9}{4}$$

$$= \frac{18}{12} \quad \text{Multiply}$$

$$= \frac{3}{2} \quad \text{Reduce to lowest terms}$$

EXAMPLE 3 Write the ratio of 0.08 to 0.12 as a fraction in lowest terms.

SOLUTION When the ratio is in reduced form, it is customary to write it with whole numbers and not decimals. For this reason we multiply the numerator and the denominator of the ratio by 100 to clear it of decimals. Then we reduce to lowest terms.

$$\frac{0.08}{0.12} = \frac{0.08 \times 100}{0.12 \times 100} \quad \text{Multiply the numerator and the denominator by 100 to clear the ratio of decimals}$$

$$= \frac{8}{12} \quad \text{Multiply}$$

$$= \frac{2}{3} \quad \text{Reduce to lowest terms}$$

Table 1 shows several more ratios and their fractional equivalents. Notice that in each case the fraction has been reduced to lowest terms. Also, the ratio that contains decimals has been rewritten as a fraction that does not contain decimals.

TABLE 1

Ratio	Fraction	Fraction In Lowest Terms
25 to 35	$\frac{25}{35}$	$\frac{5}{7}$
35 to 25	$\frac{35}{25}$	$\frac{7}{5}$
8 to 2	$\frac{8}{2}$	$\frac{4}{1}$ We can also write this as just 4.
$\frac{1}{4}$ to $\frac{3}{4}$	$\frac{\frac{1}{4}}{\frac{3}{4}}$	$\frac{1}{3}$ because $\frac{\frac{1}{4}}{\frac{3}{4}} = \frac{1}{4} \cdot \frac{4}{3} = \frac{1}{3}$
0.6 to 1.7	$\frac{0.6}{1.7}$	$\frac{6}{17}$ because $\frac{0.6 \times 10}{1.7 \times 10} = \frac{6}{17}$

B Applications of Ratios

EXAMPLE 4 During a game, a basketball player makes 12 out of the 18 free throws he attempts. Write the ratio of the number of free throws he makes to the number of free throws he attempts as a fraction in lowest terms.

SOLUTION Because he makes 12 out of 18, we want the ratio 12 to 18, or

$$\frac{12}{18} = \frac{2}{3}$$

Because the ratio is 2 to 3, we can say that, in this particular game, he made 2 out of every 3 free throws he attempted.

EXAMPLE 5 A solution of alcohol and water contains 15 milliliters of water and 5 milliliters of alcohol. Find the ratio of alcohol to water, water to alcohol, water to total solution, and alcohol to total solution. Write each ratio as a fraction and reduce to lowest terms.

SOLUTION There are 5 milliliters of alcohol and 15 milliliters of water, so there are 20 milliliters of solution (alcohol + water). The ratios are as follows:

The ratio of alcohol to water is 5 to 15, or

$$\frac{5}{15} = \frac{1}{3} \quad \text{In lowest terms}$$

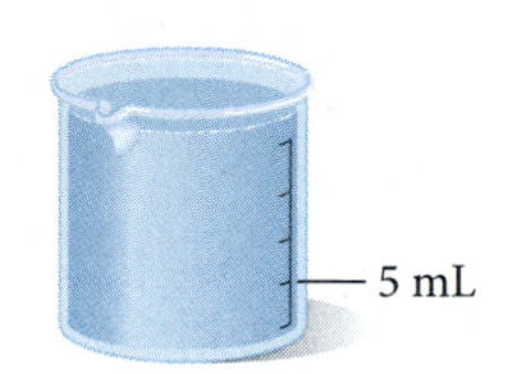

The ratio of water to alcohol is 15 to 5, or

$$\frac{15}{5} = \frac{3}{1} \quad \text{In lowest terms}$$

The ratio of water to total solution is 15 to 20, or

$$\frac{15}{20} = \frac{3}{4} \quad \text{In lowest terms}$$

The ratio of alcohol to total solution is 5 to 20, or

$$\frac{5}{20} = \frac{1}{4} \quad \text{In lowest terms}$$

5. A solution of alcohol and water contains 12 milliliters of water and 4 milliliters of alcohol. Find the ratio of alcohol to water, water to alcohol, and water to total solution. Write each ratio as a fraction and reduce to lowest terms.

Getting Ready for Class

After reading through the preceding section, respond in your own words and in complete sentences.

1. In your own words, write a definition for the ratio of two numbers.
2. What does a ratio compare?
3. What are some different ways of using mathematics to write the ratio of *a* to *b*?
4. When will the ratio of two numbers be a complex fraction?

Answer

5. $\frac{1}{3}, \frac{3}{1}, \frac{3}{4}$

Problem Set 6.1

A Write each of the following ratios as a fraction in lowest terms. None of the answers should contain decimals. [Examples 1–3]

1. 8 to 6

2. 6 to 8

3. 64 to 12

4. 12 to 64

5. 100 to 250

6. 250 to 100

7. 13 to 26

8. 36 to 18

9. $\frac{3}{4}$ to $\frac{1}{4}$

10. $\frac{5}{8}$ to $\frac{3}{8}$

11. $\frac{7}{3}$ to $\frac{6}{3}$

12. $\frac{9}{5}$ to $\frac{11}{5}$

13. $\frac{6}{5}$ to $\frac{6}{7}$

14. $\frac{5}{3}$ to $\frac{1}{3}$

15. $2\frac{1}{2}$ to $3\frac{1}{2}$

16. $5\frac{1}{4}$ to $1\frac{3}{4}$

17. $2\frac{2}{3}$ to $\frac{5}{3}$

18. $\frac{1}{2}$ to $3\frac{1}{2}$

19. 0.05 to 0.15

20. 0.21 to 0.03

21. 0.3 to 3

22. 0.5 to 10

23. 1.2 to 10

24. 6.4 to 0.8

25. a. What is the ratio of shaded squares to nonshaded squares?

b. What is the ratio of shaded squares to total squares?

c. What is the ratio of nonshaded squares to total squares?

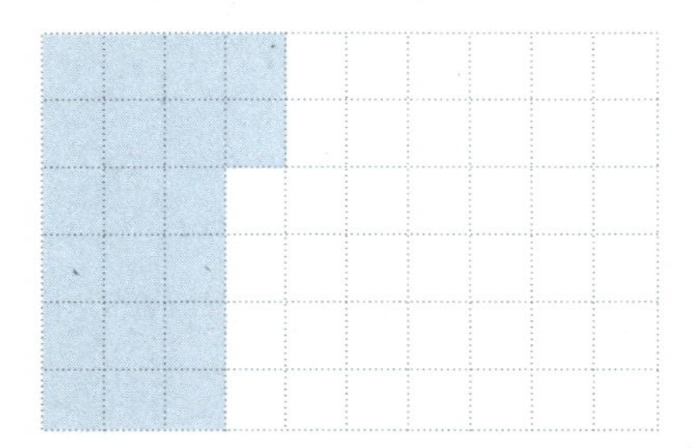

26. a. What is the ratio of shaded squares to nonshaded squares?

b. What is the ratio of shaded squares to total squares?

c. What is the ratio of nonshaded squares to total squares?

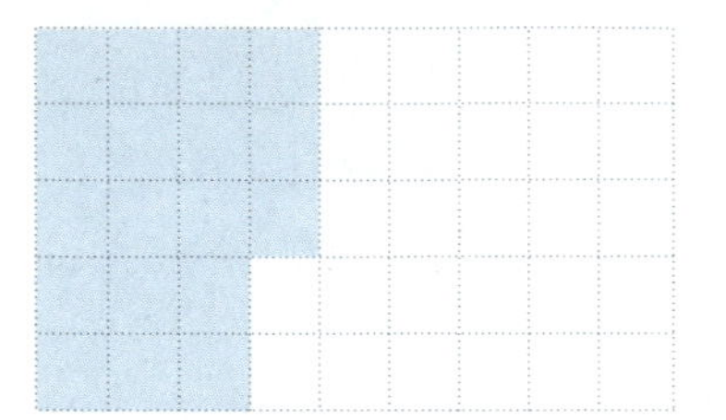

B Applying the Concepts [Examples 4, 5]

27. Biggest Hits The chart shows the number of hits for the three best charting artists in the United States. Use the information to find the ratio of hits the Beach Boys had to hits the Beatles had.

28. Google Earth The Google Earth image shows Crater Lake National Park in Oregon. The park covers 266 square miles and the lake covers 20 square miles. What is the ratio of the park's area to the lake's area? Write your answer as a decimal.

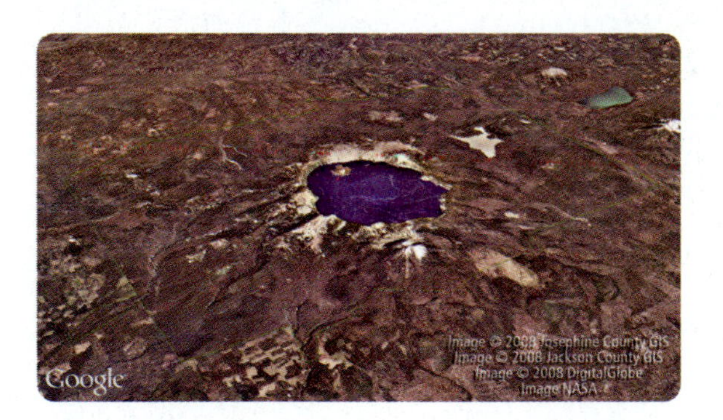

Write each of the following ratios as a fraction in lowest terms. None of the answers should contain decimals.

29. 100 mg to 5 mL

30. 25 g to 1 L

31. 375 mg to 10 mL

32. 450 mg to 20 mL

33. Family Budget A family of four budgeted the following amounts for some of their monthly bills:

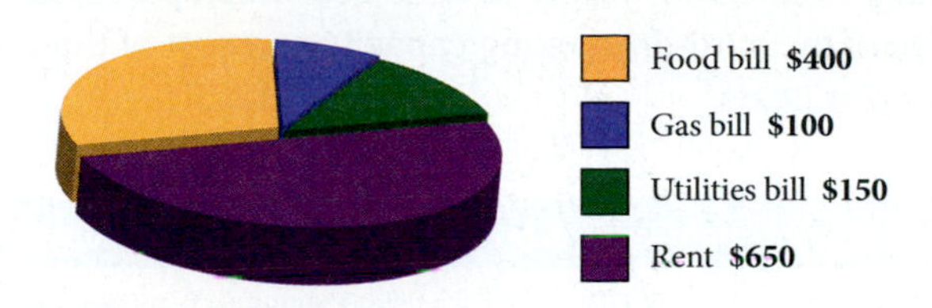

a. What is the ratio of the rent to the food bill?

b. What is the ratio of the gas bill to the food bill?

c. What is the ratio of the utilities bill to the food bill?

d. What is the ratio of the rent to the utilities bill?

34. Nutrition One cup of breakfast cereal was found to contain the following nutrients:

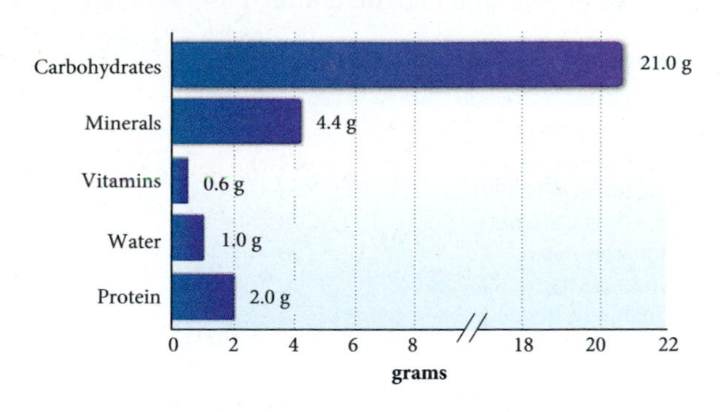

a. Find the ratio of water to protein.

b. Find the ratio of carbohydrates to protein.

c. Find the ratio of vitamins to minerals.

d. Find the ratio of protein to vitamins and minerals.

35. Profit and Revenue The following bar chart shows the profit and revenue of the Baby Steps Shoe Company each quarter for one year.

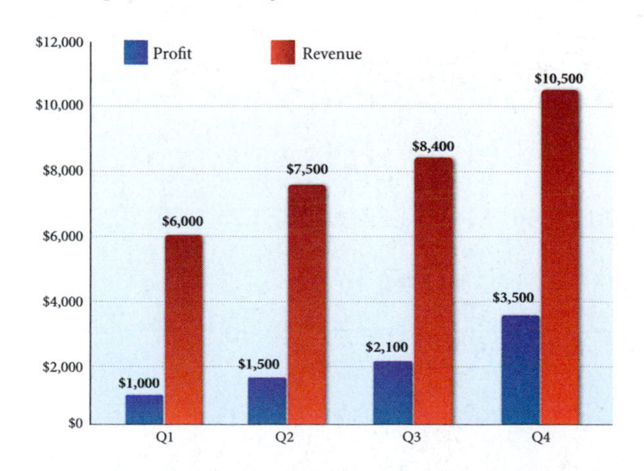

Find the ratio of revenue to profit for each of the following quarters. Write your answer in lowest terms.

a. Q1 b. Q2 c. Q3 d. Q4

e. Find the ratio of revenue to profit for the entire year.

36. Geometry Regarding the diagram below, AC represents the length of the line segment that starts at A and ends at C. From the diagram we see that $AC = 8$.

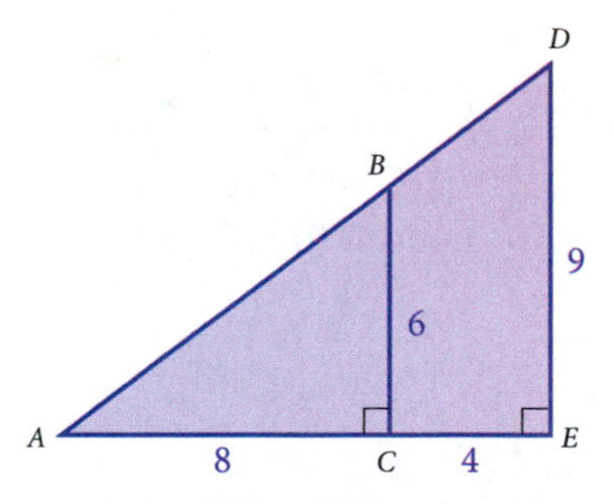

a. Find the ratio of BC to AC.

b. What is the length AE?

c. Find the ratio of DE to AE.

37. Major League Baseball The following table shows the number of games won during the 2007 baseball season by several National League teams.

Team	Number of Wins
New York Mets	88
Atlanta Braves	84
Washington Nationals	73
St. Louis Cardinals	78
Houston Astros	73
Arizona Diamondbacks	90
Cincinnati Reds	72

a. What is the ratio of wins of the New York Mets to the St. Louis Cardinals?

b. What is the ratio of wins of the Washington Nationals to the Houston Astros?

c. What is the ratio of wins of the Cincinnati Reds to the Atlanta Braves?

38. Buying an iPod™ The hard drive of an Apple iPod determines how many songs you will be able to store and carry around with you. The table below compares the size of the hard-drive, song capacity and cost of three popular iPods.

iPod Type	Hard-Drive Size	Number of Songs	Cost
Shuffle	2 GB	500 songs	$69.00
Nano	4 GB	1000 songs	$149.00
Classic	80 GB	20,000 songs	$249.00

a. What is the ratio of hard-drive size between the Shuffle and the Nano? Between the Shuffle and the Classic?

b. What is the ratio of number of songs between the Shuffle and the Nano? Between the Shuffle and the Classic?

Getting Ready for the Next Section

The following problems review material from a previous section. Reviewing these problems will help you with the next section.

Write as a decimal.

39. $\frac{90}{5}$ **40.** $\frac{120}{3}$ **41.** $\frac{125}{2}$ **42.** $\frac{2}{10}$

43. $\frac{1.23}{2}$ **44.** $\frac{1.39}{2}$ **45.** $\frac{88}{0.5}$ **46.** $\frac{1.99}{0.5}$ **47.** $\frac{46}{0.25}$ **48.** $\frac{92}{0.25}$

Divide. Round answers to the nearest thousandth.

49. $0.48 \div 5.5$ **50.** $0.75 \div 11.5$ **51.** $2.19 \div 46$ **52.** $1.25 \div 50$

Maintaining Your Skills

Multiply and divide as indicated.

53. $\frac{3}{4} \cdot \frac{5}{6}$ **54.** $\frac{7}{8} \cdot 32$ **55.** $\frac{11}{16} \div \frac{1}{8}$ **56.** $13 \div \frac{1}{3}$

57. $\frac{65}{72} \cdot \frac{108}{273}$ **58.** $\frac{165}{84} \cdot \frac{24}{195}$ **59.** $\frac{3}{4} \div \frac{1}{8} \cdot 16$ **60.** $\frac{1}{4} \div \frac{1}{12} \cdot 6$

Rates and Unit Pricing

Objectives

A Express rates as ratios.

B Use ratios to write a unit price.

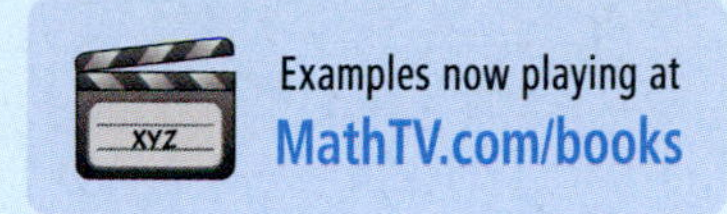

Here is the first paragraph of an article that appeared in *USA Today* in 2003.

Culture Clash

Dannon recently shrank its 8-ounce cup of yogurt by 25% to 6 ounces—but cut its suggested retail price by only 20% from 89 cents to 71 cents, which would raise the unit price a penny an ounce—9%—to 12 cents. At the Hoboken store, which charges more than Dannon's suggested prices, the unit price went from 12 cents to 13 cents with the size change.

DANNON YOGURT

	Old	New
Size	8 ounces	6 ounces
Container cost	88 cents	72 cents
Price per ounce	11 cents	12 cents
Price difference per ounce: 9%		

In this section we cover material that will give you a better understanding of the information in this article. We start this section with a discussion of rates, then we move on to unit pricing.

A Rates

Whenever a ratio compares two quantities that have different units (and neither unit can be converted to the other), then the ratio is called a *rate*. For example, if we were to travel 120 miles in 3 hours, then our average rate of speed expressed as the ratio of miles to hours would be

$$\frac{120 \text{ miles}}{3 \text{ hours}} = \frac{40 \text{ miles}}{1 \text{ hour}}$$ Divide the numerator and the denominator by 3 to reduce to lowest terms

The ratio $\frac{40 \text{ miles}}{1 \text{ hour}}$ can be expressed as

$$40 \frac{\text{miles}}{\text{hour}} \quad \text{or} \quad 40 \text{ miles/hour} \quad \text{or} \quad 40 \text{ miles per hour}$$

A rate is expressed in simplest form when the numerical part of the denominator is 1. To accomplish this we use division.

EXAMPLE 1 A train travels 125 miles in 2 hours. What is the train's rate in miles per hour?

SOLUTION The ratio of miles to hours is

$$\frac{125 \text{ miles}}{2 \text{ hours}} = 62.5 \frac{\text{miles}}{\text{hour}}$$ Divide 125 by 2

$$= 62.5 \text{ miles per hour}$$

If the train travels 125 miles in 2 hours, then its average rate of speed is 62.5 miles per hour.

EXAMPLE 2 A car travels 90 miles on 5 gallons of gas. Give the ratio of miles to gallons as a rate in miles per gallon.

SOLUTION The ratio of miles to gallons is

$$\frac{90 \text{ miles}}{5 \text{ gallons}} = 18 \frac{\text{miles}}{\text{gallon}}$$ Divide 90 by 5

$$= 18 \text{ miles/gallon}$$

The gas mileage of the car is 18 miles per gallon.

PRACTICE PROBLEMS

1. A car travels 107 miles in 2 hours. What is the car's rate in miles per hour?

2. A car travels 192 miles on 6 gallons of gas. Give the ratio of miles to gallons as a rate in miles per gallon.

Answers

1. 53.5 miles/hour
2. 32 miles/gallon

B Unit Pricing

One kind of rate that is very common is *unit pricing*. Unit pricing is the ratio of price to quantity when the quantity is one unit. Suppose a 1-liter bottle of a certain soft drink costs \$1.19, whereas a 2-liter bottle of the same drink costs \$1.39. Which is the better buy? That is, which has the lower price per liter?

$$\frac{\$1.19}{1 \text{ liter}} = \$1.19 \text{ per liter}$$

$$\frac{\$1.39}{2 \text{ liters}} = \$0.695 \text{ per liter}$$

The unit price for the 1-liter bottle is \$1.19 per liter, whereas the unit price for the 2-liter bottle is 69.5¢ per liter. The 2-liter bottle is a better buy.

EXAMPLE 3 A supermarket sells low-fat milk in three different containers at the following prices:

1 gallon	\$3.59
$\frac{1}{2}$ gallon	\$1.99
1 quart	\$1.29

(1 quart $= \frac{1}{4}$ gallon)

Give the unit price in dollars per gallon for each one.

SOLUTION Because 1 quart $= \frac{1}{4}$ gallon, we have

1-gallon container $\frac{\$3.59}{1 \text{ gallon}} = \frac{\$3.59}{1 \text{ gallon}} = \$3.59$ per gallon

$\frac{1}{2}$-gallon container $\frac{\$1.99}{\frac{1}{2} \text{ gallon}} = \frac{\$1.99}{0.5 \text{ gallon}} = \$3.98$ per gallon

1-quart container $\frac{\$1.29}{1 \text{ quart}} = \frac{\$1.29}{0.25 \text{ gallon}} = \$5.16$ per gallon

The 1-gallon container has the lowest unit price, whereas the 1-quart container has the highest unit price.

3. A supermarket sells vegetable juice in three different containers at the following prices:
   5.5 ounces, 48¢
   11.5 ounces, 75¢
   46 ounces, \$2.19
   Give the unit price in cents per ounce for each one. Round to the nearest tenth of a cent, if necessary.

Answer
3. 8.7¢/ounce, 6.5¢/ounce, 4.8¢/ounce

Getting Ready for Class

After reading through the preceding section, respond in your own words and in complete sentences.

1. A rate is a special type of ratio. In your own words, explain what a rate is.
2. When is a rate written in simplest terms?
3. What is *unit pricing*?
4. Give some examples of rates *not* found in your textbook.

Problem Set 6.2

A Express each of the following rates as a ratio with the given units. [Examples 1, 2]

1. **Miles/Hour** A car travels 220 miles in 4 hours. What is the rate of the car in miles per hour?

2. **Miles/Hour** A train travels 360 miles in 5 hours. What is the rate of the train in miles per hour?

3. **Kilometers/Hour** It takes a car 3 hours to travel 252 kilometers. What is the rate in kilometers per hour?

4. **Kilometers/Hour** In 6 hours an airplane travels 4,200 kilometers. What is the rate of the airplane in kilometers per hour?

5. **Gallons/Second** The flow of water from a water faucet can fill a 3-gallon container in 15 seconds. Give the ratio of gallons to seconds as a rate in gallons per second.

6. **Gallons/Minute** A 225-gallon drum is filled in 3 minutes. What is the rate in gallons per minute?

7. **Liters/Minute** It takes 4 minutes to fill a 56-liter gas tank. What is the rate in liters per minute?

8. **Liters/Hour** The gas tank on a car holds 60 liters of gas. At the beginning of a 6-hour trip, the tank is full. At the end of the trip, it contains only 12 liters. What is the rate at which the car uses gas in liters per hour?

9. **Miles/Gallon** A car travels 95 miles on 5 gallons of gas. Give the ratio of miles to gallons as a rate in miles per gallon.

10. **Miles/Gallon** On a 384-mile trip, an economy car uses 8 gallons of gas. Give this as a rate in miles per gallon.

11. **Miles/Liter** The gas tank on a car has a capacity of 75 liters. On a full tank of gas, the car travels 325 miles. What is the gas mileage in miles per liter?

12. **Miles/Liter** A car pulling a trailer can travel 105 miles on 70 liters of gas. What is the gas mileage in miles per liter?

13. Gas Prices The snapshot shows the gas prices for the different regions of the United States. If a man bought 12 gallons of gas for $48.72, where might he live?

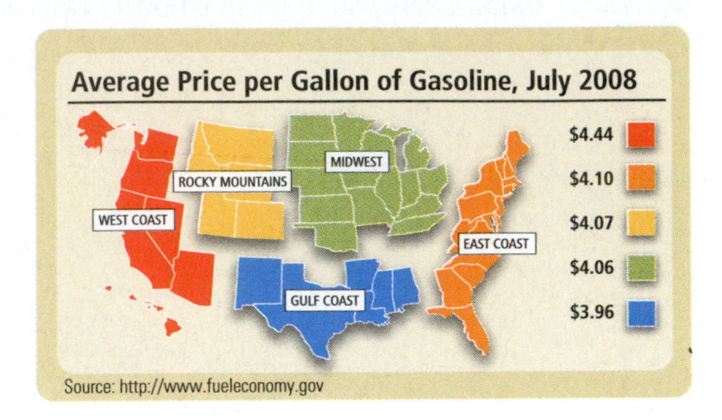

14. Pitchers The chart shows the active major league pitchers with the most career strikeouts. If Pedro Martinez pitched 2,783 innings, how many strikeouts does he throw per inning? Round to the nearest hundredth.

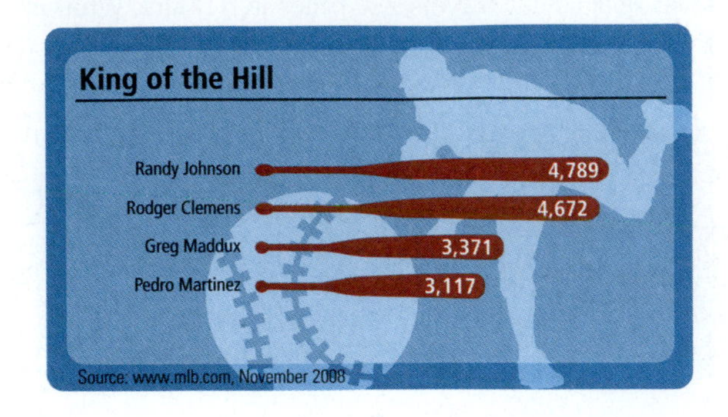

Nursing Intravenous (IV) infusions are often ordered in either milliliters per hour or milliliters per minute.

15. What was the infusion rate in milliliters per hour if it took 5 hours to administer 2,400 mL?

16. What was the infusion rate in milliliters per minute if 42 milliliters were administered in 6 minutes?

B Unit Pricing [Example 3]

17. Cents/Ounce A 20-ounce package of frozen peas is priced at 99¢. Give the unit price in cents per ounce.

18. Dollars/Pound A 4-pound bag of cat food costs $8.12. Give the unit price in dollars per pound.

19. Best Buy Find the unit price in cents per diaper for each of the packages shown here. Which is the better buy? Round to the nearest tenth of a cent.

36 disposable diapers
$12.49

38 disposable diapers
$11.99

20. Best Buy Find the unit price in cents per pill for each of the packages shown here. Which is the better buy? Round to the nearest tenth of a cent.

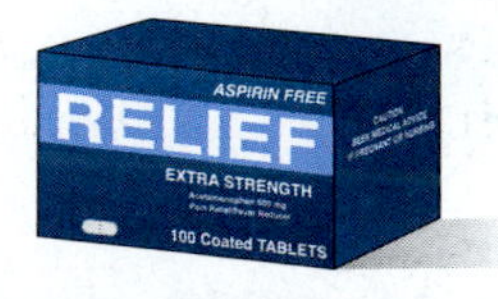

100 pills
$5.99

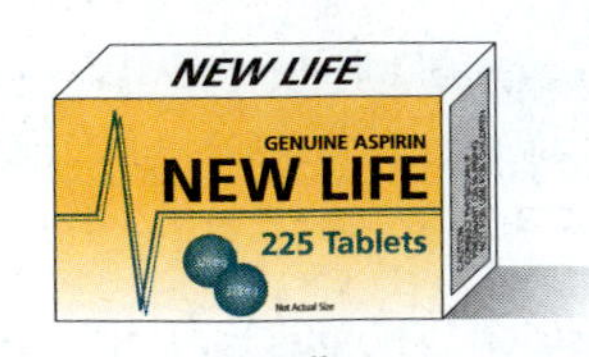

225 pills
$13.96

Currency Conversions There are a number of online calculators that will show what the money in one country is worth in another country. One such converter, the XE Universal Currency Converter®, uses live, up-to-the-minute currency rates. Use the information shown here to determine what the equivalent to one U.S. dollar for each of the following denominations. Round to the nearest thousandth, wherre necessary.

21. $100.00 U.S. dollars are equivalent to 64.582 euros

22. $50.00 U.S. dollars are equivalent to $51.0775 Canadian dollars

23. $40.00 U.S. dollars are equivalent to 20.3765 British pounds

24. $25.00 U.S. dollars are equivalent to 2704.0125 Japanese yen

25. Food Prices Using unit rates is a way to compare prices of different sized packages to see which price is really the best deal. Suppose we compare the cost of a box of Cheerios sold at three different stores for the following prices:

Store	Size	Cost
A	11.3 ounce box	$4.00
B	18 ounce box	$4.99
C	180 ounce case	$52.90

Which size is the best buy? Give the cost per ounce for that size.

26. Cell Phone Plans All cell phone plans are not created equal. The number of minutes and the monthly charges can vary greatly. The table shows four plans presented by four different cell phone providers.

Carrier	AT&T	Sprint	T Mobile	Verizon
Plan name	Nation 450	Sprint Basic	Individual Value	Nationwide Basic
Monthly minutes	450	200	600	450
Monthly cost	$39.99	$29.99	$39.99	$39.99
Plan cost per minute				

Find the cost per minute for each plan. Based on your results, which plan should you go with?

27. Miles/Hour A car travels 675.4 miles in $12\frac{1}{2}$ hours. Give the rate in miles per hour to the nearest hundredth.

28. Miles/Hour At the beginning of a trip, the odometer on a car read 32,567.2 miles. At the end of the trip, it read 32,741.8 miles. If the trip took $4\frac{1}{4}$ hours, what was the rate of the car in miles per hour to the nearest tenth?

29. Miles/Gallon If a truck travels 128.4 miles on 13.8 gallons of gas, what is the gas mileage in miles per gallon? (Round to the nearest tenth.)

30. Cents/Day If a 15-day supply of vitamins costs $1.62, what is the price in cents per day?

Hourly Wages Jane has a job at the local Marcy's department store. The graph shows how much Jane earns for working 8 hours per day for 5 days.

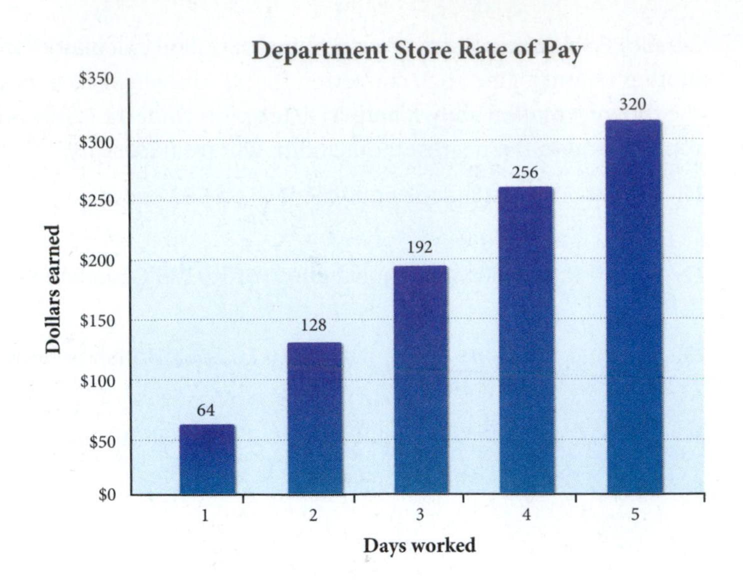

31. What is her daily rate of pay? (Assume she works 8 hours per day.)

32. What is her weekly rate of pay? (Assume she works 5 days per week.)

33. What is her annual rate of pay? (Assume she works 50 weeks per year.)

34. What is her hourly rate of pay? (Assume she works 8 hours per day.)

Getting Ready for the Next Section

Solve each equation by finding a number to replace n that will make the equation a true statement.

35. $2 \cdot n = 12$

36. $3 \cdot n = 27$

37. $6 \cdot n = 24$

38. $8 \cdot n = 16$

39. $20 = 5 \cdot n$

40. $35 = 7 \cdot n$

41. $650 = 10 \cdot n$

42. $630 = 7 \cdot n$

Maintaining Your Skills

Add and subtract as indicated.

43. $\frac{1}{2} + \frac{3}{8}$

44. $\frac{7}{6} - \frac{1}{3}$

45. $\frac{2}{5} - \frac{3}{8}$

46. $\frac{5}{8} + \frac{3}{4}$

47. $\frac{11}{12} - \frac{9}{10}$

48. $\frac{13}{15} + \frac{1}{10}$

49. $\frac{5}{6} - \frac{1}{3} + \frac{4}{3}$

50. $\frac{7}{8} + \frac{1}{8} - \frac{1}{16}$

Extending the Concepts

51. **Unit Pricing** The makers of Wisk liquid detergent cut the size of its popular midsize jug from 100 ounces (3.125 quarts) to 80 ounces (2.5 quarts). At the same time it lowered the price from $6.99 to $5.75. Fill in the table below and use your results to decide which of the two sizes is the better buy.

WISK LAUNDRY DETERGENT

	Old	New
Size	100 ounces	80 ounces
Container cost	$6.99	$5.75
Price per quart		

Proportions

6.3

Objectives

A Name the terms in a proportion.
B Use the fundamental property of proportions to solve a proportion.

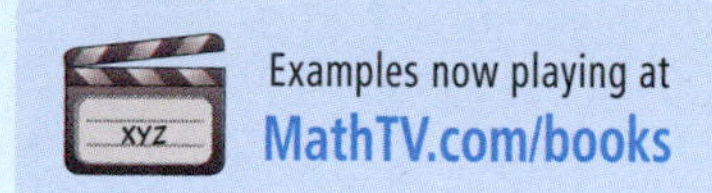

Millions of people are turning to the Internet to view music videos of their favorite musician. Many Web sites offer different sizes of video based on the speed of a user's Internet connection. Even though the figures below are not the same size, their sides are proportional. Later in this chapter we will use proportions to find the unknown height in the larger figure.

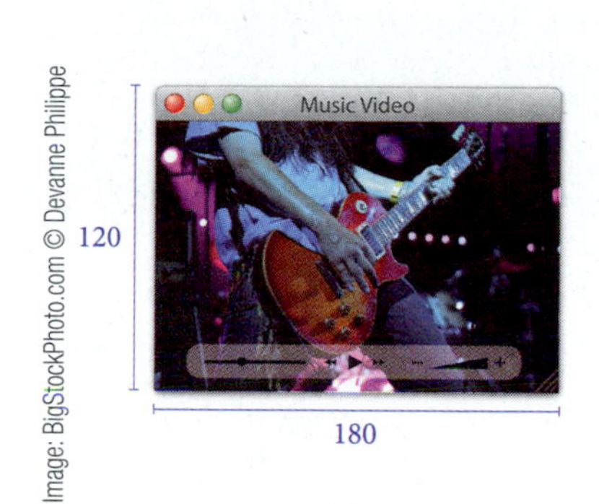

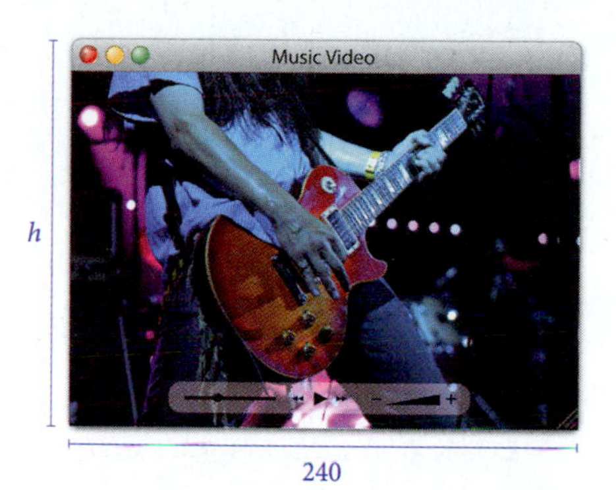

Image: BigStockPhoto.com © Devanne Philippe

In this section we will solve problems using proportions. As you will see later in this chapter, proportions can model a number of everyday applications.

> **Definition**
>
> A statement that two ratios are equal is called a **proportion**. If $\frac{a}{b}$ and $\frac{c}{d}$ are two equal ratios, then the statement
>
> $$\frac{a}{b} = \frac{c}{d}$$
>
> is called a proportion.

A Terms of a Proportion

Each of the four numbers in a proportion is called a *term* of the proportion. We number the terms of a proportion as follows:

First term ⟶ a, Second term ⟶ b, Third term ⟶ c, Fourth term ⟶ d

$$\frac{a}{b} = \frac{c}{d}$$

The first and fourth terms of a proportion are called the *extremes,* and the second and third terms of a proportion are called the *means.*

Means ⟶ b and c; Extremes ⟶ a and d

$$\frac{a}{b} = \frac{c}{d}$$

EXAMPLE 1 In the proportion $\frac{3}{4} = \frac{6}{8}$, name the four terms, the means, and the extremes.

SOLUTION The terms are numbered as follows:

First term = 3 Third term = 6
Second term = 4 Fourth term = 8

The means are 4 and 6; the extremes are 3 and 8.

PRACTICE PROBLEMS

1. In the proportion $\frac{2}{3} = \frac{6}{9}$, name the four terms, the means, and the extremes.

Answer

1. See solutions section.

The final thing we need to know about proportions is expressed in the following property.

B The Fundamental Property of Proportions

Fundamental Property of Proportions
In any proportion, the product of the extremes is equal to the product of the means. This property is also referred to as the means/extremes property, and in symbols, it looks like this:

$$\text{If } \frac{a}{b} = \frac{c}{d} \text{ then } ad = bc$$

2. Verify the fundamental property of proportions for the following proportions.

a. $\frac{5}{6} = \frac{15}{18}$

b. $\frac{13}{39} = \frac{1}{3}$

c. $\frac{\frac{2}{3}}{\frac{5}{3}} = \frac{2}{5}$

d. $\frac{0.12}{0.18} = \frac{2}{3}$

EXAMPLE 2 Verify the fundamental property of proportions for the following proportions.

a. $\frac{3}{4} = \frac{6}{8}$ **b.** $\frac{17}{34} = \frac{1}{2}$

SOLUTION We verify the fundamental property by finding the product of the means and the product of the extremes in each case.

Proportion	Product of the Means	Product of the Extremes
a. $\frac{3}{4} = \frac{6}{8}$	$4 \cdot 6 = 24$	$3 \cdot 8 = 24$
b. $\frac{17}{34} = \frac{1}{2}$	$34 \cdot 1 = 34$	$17 \cdot 2 = 34$

For each proportion the product of the means is equal to the product of the extremes.

We can use the fundamental property of proportions, along with a property we encountered in Section 6.3, to solve an equation that has the form of a proportion.

A Note on Multiplication Previously, we have used a multiplication dot to indicate multiplication, both with whole numbers and with variables. A more compact form for multiplication involving variables is simply to leave out the dot.

That is, $5 \cdot y = 5y$ and $10 \cdot x \cdot y = 10xy$.

3. Find the missing term:

a. $\frac{3}{4} = \frac{9}{x}$

b. $\frac{5}{8} = \frac{3}{x}$

In some of these problems you will be able to see what the solution is just by looking the problem over. In those cases it is still best to show all the work involved in solving the proportion. It is good practice for the more difficult problems.

EXAMPLE 3 Solve for x.

$$\frac{2}{3} = \frac{4}{x}$$

SOLUTION Applying the fundamental property of proportions, we have

If $\frac{2}{3} = \frac{4}{x}$

then $2 \cdot x = 3 \cdot 4$ The product of the extremes equals the product of the means

$2x = 12$ Multiply

The result is an equation. We know from Section 4.3 that we can divide both sides of an equation by the same nonzero number without changing the solution to the equation. In this case we divide both sides by 2 to solve for x:

Answer
2. See solutions section.

$$2x = 12$$

$$\frac{2x}{2} = \frac{12}{2} \quad \text{Divide both sides by 2}$$

$$x = 6 \quad \text{Simplify each side}$$

The solution is 6. We can check our work by using the fundamental property of proportions:

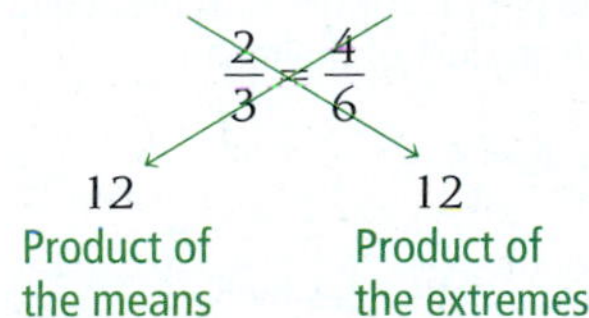

Because the product of the means and the product of the extremes are equal, our work is correct.

EXAMPLE 4 Solve for y: $\frac{5}{y} = \frac{10}{13}$

SOLUTION We apply the fundamental property and solve as we did in Example 3:

If	$\frac{5}{y} = \frac{10}{13}$	
then	$5 \cdot 13 = y \cdot 10$	The product of the extremes equals the product of the means
	$65 = 10y$	Multiply $5 \cdot 13$
	$\frac{65}{10} = \frac{10y}{10}$	Divide both sides by 10
	$6.5 = y$	$65 \div 10 = 6.5$

The solution is 6.5. We could check our result by substituting 6.5 for y in the original proportion and then finding the product of the means and the product of the extremes.

4. Solve for y: $\frac{2}{y} = \frac{8}{19}$

EXAMPLE 5 Find n if $\frac{n}{3} = \frac{0.4}{8}$.

SOLUTION We proceed as we did in the previous two examples:

If	$\frac{n}{3} = \frac{0.4}{8}$	
then	$n \cdot 8 = 3(0.4)$	The product of the extremes equals the product of the means
	$8n = 1.2$	$3(0.4) = 1.2$
	$\frac{8n}{8} = \frac{1.2}{8}$	Divide both sides by 8
	$n = 0.15$	$1.2 \div 8 = 0.15$

The missing term is 0.15.

5. Find n

a. $\frac{n}{6} = \frac{0.3}{15}$

b. $\frac{0.35}{n} = \frac{7}{100}$

Answers

3. a. 12 **b.** 4.8 **4.** 4.75

5. a. 0.12 **b.** 5

6. Solve for x:

a. $\dfrac{\frac{3}{4}}{7} = \dfrac{x}{8}$

b. $\dfrac{6}{\frac{3}{5}} = \dfrac{15}{x}$

7. Solve $\dfrac{b}{18} = 0.5$

Answers

6. a. $\frac{6}{7}$ b. $\frac{3}{2}$ 7. 9

EXAMPLE 6 Solve for x: $\dfrac{\frac{2}{3}}{5} = \dfrac{x}{6}$

SOLUTION We begin by multiplying the means and multiplying the extremes:

$$\text{If} \quad \frac{\frac{2}{3}}{5} = \frac{x}{6}$$

$$\text{then} \quad \frac{2}{3} \cdot 6 = 5 \cdot x \qquad \text{The product of the extremes equals the product of the means}$$

$$4 = 5 \cdot x \qquad \frac{2}{3} \cdot 6 = 4$$

$$\frac{4}{5} = \frac{\not{5} \cdot x}{\not{5}} \qquad \text{Divide both sides by 5}$$

$$\frac{4}{5} = x$$

The missing term is $\frac{4}{5}$, or 0.8.

EXAMPLE 7 Solve $\dfrac{b}{15} = 2$.

SOLUTION Since the number 2 can be written as the ratio of 2 to 1, we can write this equation as a proportion, and then solve as we have in the examples above.

$$\frac{b}{15} = 2$$

$$\frac{b}{15} = \frac{2}{1} \qquad \text{Write 2 as a ratio}$$

$$b \cdot 1 = 15 \cdot 2 \qquad \text{Product of the extremes equals Product of the means}$$

$$b = 30$$

The procedure for finding a missing term in a proportion is always the same. We first apply the fundamental property of proportions to find the product of the extremes and the product of the means. Then we solve the resulting equation.

Getting Ready for Class

After reading through the preceding section, respond in your own words and in complete sentences.

1. In your own words, give a definition of a *proportion.*
2. In the proportion $\frac{2}{5} = \frac{4}{x}$, name the means and the extremes.
3. State the Fundamental Property of Proportions in words and in symbols.
4. For the proportion $\frac{2}{5} = \frac{4}{x}$, find the product of the means and the product of the extremes.

Problem Set 6.3

A For each of the following proportions, name the means, name the extremes, and show that the product of the means is equal to the product of the extremes. [Examples 1, 2]

1. $\frac{1}{3} = \frac{5}{15}$

2. $\frac{6}{12} = \frac{1}{2}$

3. $\frac{10}{25} = \frac{2}{5}$

4. $\frac{5}{8} = \frac{10}{16}$

5. $\frac{\frac{1}{3}}{\frac{1}{2}} = \frac{4}{6}$

6. $\frac{2}{\frac{1}{4}} = \frac{4}{\frac{1}{2}}$

7. $\frac{0.5}{5} = \frac{1}{10}$

8. $\frac{0.3}{1.2} = \frac{1}{4}$

B Find the missing term in each of the following proportions. Set up each problem like the examples in this section. Write your answers as fractions in lowest terms. [Examples 3–7]

9. $\frac{2}{5} = \frac{4}{x}$

10. $\frac{3}{8} = \frac{9}{x}$

11. $\frac{1}{y} = \frac{5}{12}$

12. $\frac{2}{y} = \frac{6}{10}$

13. $\frac{x}{4} = \frac{3}{8}$

14. $\frac{x}{5} = \frac{7}{10}$

15. $\frac{5}{9} = \frac{x}{2}$

16. $\frac{3}{7} = \frac{x}{3}$

17. $\frac{3}{7} = \frac{3}{x}$

18. $\frac{2}{9} = \frac{2}{x}$

19. $\frac{x}{2} = 7$

20. $\frac{x}{3} = 10$

21. $\frac{\frac{1}{2}}{y} = \frac{\frac{1}{3}}{12}$

22. $\frac{\frac{2}{3}}{y} = \frac{\frac{1}{3}}{5}$

23. $\frac{n}{12} = \frac{\frac{1}{4}}{\frac{1}{2}}$

24. $\frac{n}{10} = \frac{\frac{3}{5}}{\frac{3}{8}}$

25. $\frac{10}{20} = \frac{20}{n}$

26. $\frac{8}{4} = \frac{4}{n}$

27. $\frac{x}{10} = \frac{10}{2}$ **28.** $\frac{x}{12} = \frac{12}{48}$ **29.** $\frac{y}{12} = 9$ **30.** $\frac{y}{16} = 0.75$ **31.** $\frac{0.4}{1.2} = \frac{1}{x}$ **32.** $\frac{5}{0.5} = \frac{20}{x}$

33. $\frac{0.3}{0.18} = \frac{n}{0.6}$ **34.** $\frac{0.01}{0.1} = \frac{n}{10}$ **35.** $\frac{0.5}{x} = \frac{1.4}{0.7}$ **36.** $\frac{0.3}{x} = \frac{2.4}{0.8}$

37. $\frac{168}{324} = \frac{56}{x}$ **38.** $\frac{280}{530} = \frac{112}{x}$ **39.** $\frac{429}{y} = \frac{858}{130}$ **40.** $\frac{573}{y} = \frac{2{,}292}{316}$

41. $\frac{n}{39} = \frac{533}{507}$ **42.** $\frac{n}{47} = \frac{1{,}003}{799}$ **43.** $\frac{756}{903} = \frac{x}{129}$ **44.** $\frac{321}{1{,}128} = \frac{x}{376}$

Getting Ready for the Next Section

Divide.

45. 360 ÷ 18 **46.** 2,700 ÷ 6

Multiply.

47. 3.5(85) **48.** 4.75(105)

Solve each equation.

49. $\frac{x}{10} = \frac{270}{6}$ **50.** $\frac{x}{45} = \frac{8}{18}$ **51.** $\frac{x}{25} = \frac{4}{20}$ **52.** $\frac{x}{3.5} = \frac{85}{1}$

Maintaining Your Skills

Give the place value of the 5 in each number.

53. 250.14 **54.** 2.5014

Add or subtract as indicated.

55. 2.3 + 0.18 + 24.036 **56.** 5 + 0.03 + 1.9 **57.** 3.18 − 2.79 **58.** 3.4 − 1.975

Applications of Proportions

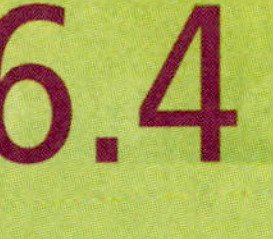

6.4

Objectives

A Use proportions to solve application problems.

Examples now playing at MathTV.com/books

Proportions can be used to solve a variety of word problems. The examples that follow show some of these word problems. In each case we will translate the word problem into a proportion and then solve the proportion using the method developed in this chapter.

 A Applications

EXAMPLE 1 A woman drives her car 270 miles in 6 hours. If she continues at the same rate, how far will she travel in 10 hours?

SOLUTION We let x represent the distance traveled in 10 hours. Using x, we translate the problem into the following proportion:

$$\begin{matrix} \text{Miles} \longrightarrow \\ \text{Hours} \longrightarrow \end{matrix} \frac{x}{10} = \frac{270}{6} \begin{matrix} \longleftarrow \text{Miles} \\ \longleftarrow \text{Hours} \end{matrix}$$

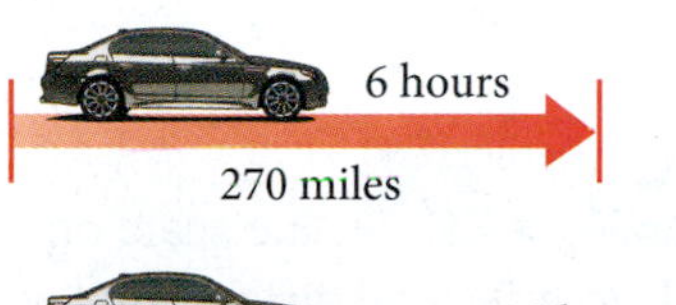

Notice that the two ratios in the proportion compare the same quantities. That is, both ratios compare miles to hours. In words this proportion says:

x miles is to 10 hours as 270 miles is to 6 hours

$$\frac{x}{10} = \frac{270}{6}$$

Next, we solve the proportion.

$$x \cdot 6 = 10 \cdot 270$$
$$x \cdot 6 = 2{,}700$$
$$\frac{x \cdot \cancel{6}}{\cancel{6}} = \frac{2{,}700}{6}$$
$$x = 450 \text{ miles}$$

If the woman continues at the same rate, she will travel 450 miles in 10 hours.

EXAMPLE 2 A baseball player gets 8 hits in the first 18 games of the season. If he continues at the same rate, how many hits will he get in 45 games?

SOLUTION We let x represent the number of hits he will get in 45 games. Then

x is to 45 as 8 is to 18

$$\begin{matrix} \text{Hits} \longrightarrow \\ \text{Games} \longrightarrow \end{matrix} \frac{x}{45} = \frac{8}{18} \begin{matrix} \longleftarrow \text{Hits} \\ \longleftarrow \text{Games} \end{matrix}$$

Notice again that the two ratios are comparing the same quantities, hits to games. We solve the proportion as follows:

$$18x = 360 \qquad 45 \cdot 8 = 360$$
$$\frac{\cancel{18}x}{\cancel{18}} = \frac{360}{18} \qquad \text{Divide both sides by 18}$$
$$x = 20 \qquad 360 \div 18 = 20$$

If he continues to hit at the rate of 8 hits in 18 games, he will get 20 hits in 45 games.

PRACTICE PROBLEMS

1. A man drives his car 288 miles in 6 hours. If he continues at the same rate, how far will he travel in:
   a. 10 hours
   b. 11 hours

2. A softball player gets 10 hits in the first 18 games of the season. If she continues at the same rate, how many hits will she get in:
   a. 54 games
   b. 27 games

Answers
1. **a.** 480 miles **b.** 528 miles
2. **a.** 30 hits **b.** 15 hits

3. A solution contains 8 milliliters of alcohol and 20 milliliters of water. If another solution is to have the same ratio of milliliters of alcohol to milliliters of water and must contain 35 milliliters of water, how much alcohol should it contain?

4. The scale on a map indicates that 1 inch on the map corresponds to an actual distance of 105 miles. Two cities are 4.75 inches apart on the map. What is the actual distance between the two cities?

Answers
3. 14 mL **4.** 498.75 mi

EXAMPLE 3 A solution contains 4 milliliters of alcohol and 20 milliliters of water. If another solution is to have the same ratio of milliliters of alcohol to milliliters of water and must contain 25 milliliters of water, how much alcohol should it contain?

SOLUTION We let x represent the number of milliliters of alcohol in the second solution. The problem translates to

x milliliters is to 25 milliliters as 4 milliliters is to 20 milliliters

$$\begin{array}{l} \text{Alcohol} \longrightarrow \\ \text{Water} \longrightarrow \end{array} \frac{x}{25} = \frac{4}{20} \begin{array}{l} \longleftarrow \text{Alcohol} \\ \longleftarrow \text{Water} \end{array}$$

$$20x = 100 \qquad 25 \cdot 4 = 100$$

$$\frac{20x}{20} = \frac{100}{20} \qquad \text{Divide both sides by 20}$$

$$x = 5 \text{ milliliters of alcohol} \qquad 100 \div 20 = 5$$

EXAMPLE 4 The scale on a map indicates that 1 inch on the map corresponds to an actual distance of 85 miles. Two cities are 3.5 inches apart on the map. What is the actual distance between the two cities?

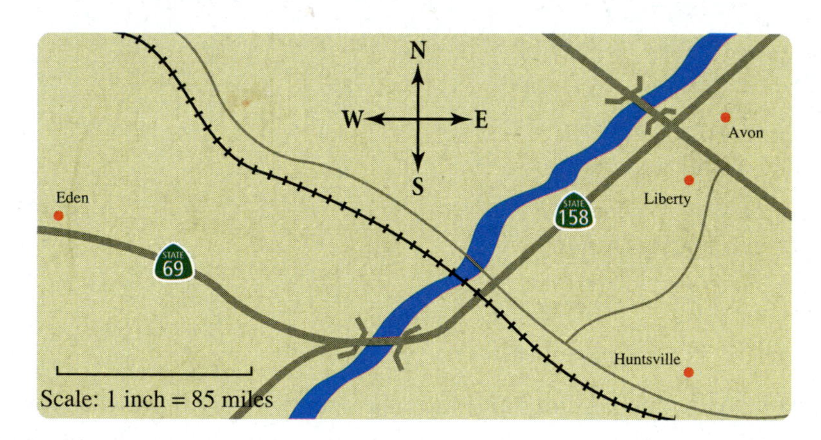

SOLUTION We let x represent the actual distance between the two cities. The proportion is

$$\begin{array}{l} \text{Miles} \longrightarrow \\ \text{Inches} \longrightarrow \end{array} \frac{x}{3.5} = \frac{85}{1} \begin{array}{l} \longleftarrow \text{Miles} \\ \longleftarrow \text{Inches} \end{array}$$

$$x \cdot 1 = 3.5(85)$$

$$x = 297.5 \text{ miles}$$

Getting Ready for Class

After reading through the preceding section, respond in your own words and in complete sentences.

1. Give an example, not found in the book, of a proportion problem you may encounter.
2. Write a word problem for the proportion $\frac{2}{5} = \frac{4}{x}$.
3. What does it mean to translate a word problem into a proportion?
4. Name some jobs that may frequently require solving proportion problems.

Problem Set 6.4

A Solve each of the following word problems by translating the statement into a proportion. Be sure to show the proportion used in each case. [Examples 1–4]

1. **Distance** A woman drives her car 235 miles in 5 hours. At this rate how far will she travel in 7 hours?

2. **Distance** An airplane flies 1,260 miles in 3 hours. How far will it fly in 5 hours?

3. **Basketball** A basketball player scores 162 points in 9 games. At this rate how many points will he score in 20 games?

4. **Football** In the first 4 games of the season, a football team scores a total of 68 points. At this rate how many points will the team score in 11 games?

5. **Mixture** A solution contains 8 pints of antifreeze and 5 pints of water. How many pints of water must be added to 24 pints of antifreeze to get a solution with the same concentration?

6. **Nutrition** If 10 ounces of a certain breakfast cereal contain 3 ounces of sugar, how many ounces of sugar do 25 ounces of the same cereal contain?

7. **Map Reading** The scale on a map indicates that 1 inch corresponds to an actual distance of 95 miles. Two cities are 4.5 inches apart on the map. What is the actual distance between the two cities?

8. **Map Reading** A map is drawn so that every 2.5 inches on the map corresponds to an actual distance of 100 miles. If the actual distance between two cities is 350 miles, how far apart are they on the map?

9. **Farming** A farmer knows that of every 50 eggs his chickens lay, only 45 will be marketable. If his chickens lay 1,000 eggs in a week, how many of them will be marketable?

10. **Manufacturing** Of every 17 parts manufactured by a certain machine, only 1 will be defective. How many parts were manufactured by the machine if 8 defective parts were found?

11. **Nursing** A patient is given a prescription of 10 pills. The total prescription contains 355 milligrams. How many milligrams is contained in each pill?

12. **Nursing** A child is given a prescription for 9 mg of a drug. If she has to take 3 chewable tablets, what is the strength of each tablet?

13. **Nursing** An oral medication has a dosage strength of 275 mg/5 mL. If a patient takes a dosage of 300 mg, how many milliliters does he take? Round to the nearest tenth

14. **Nursing** An atropine sulfate injection has a dosage strength of 0.1 mg/mL. If 4.5 mL was given to the patient, how many milligrams did she receive?

15. **Nursing** A tablet has a strength of 45 mg. If a patient is prescribed a dose of 112.5 mg, how many tablets does he take?

16. **Nursing** A tablet has a dosage strength of 35 mg. What was the prescribed dosage if the patient was told to take 1.5 tablets?

Model Trains The size of a model train relative to an actual train is referred to as its scale. Each scale is associated with a ratio as shown in the table. For example, an HO model train has a ratio of 1 to 87, meaning it is $\frac{1}{87}$ as large as an actual train.

17. **Length of a Boxcar** How long is an actual boxcar that has an HO scale model 5 inches long? Give your answer in inches, then divide by 12 to give the answer in feet.

18. **Length of a Flatcar** How long is an actual flatcar that has an LGB scale model 24 inches long? Give your answer in feet.

Scale	Ratio
LGB	1 to 22.5
#1	1 to 32
O	1 to 43.5
S	1 to 64
HO	1 to 87
TT	1 to 120

Spencer Grant/PhotoEdit

19. **Travel Expenses** A traveling salesman figures it costs 55¢ for every mile he drives his car. How much does it cost him a week to drive his car if he travels 570 miles a week?

20. **Travel Expenses** A family plans to drive their car during their annual vacation. The car can go 350 miles on a tank of gas, which is 18 gallons of gas. The vacation they have planned will cover 1,785 miles. How many gallons of gas will that take?

21. **Nutrition** A 9-ounce serving of pasta contains 159 grams of carbohydrates. How many grams of carbohydrates do 15 ounces of this pasta contain?

22. **Nutrition** If 100 grams of ice cream contains 13 grams of fat, how much fat is in 250 grams of ice cream?

23. **Travel Expenses** If a car travels 378.9 miles on 50 liters of gas, how many liters of gas will it take to go 692 miles if the car travels at the same rate? (Round to the nearest tenth.)

24. **Nutrition** If 125 grams of peas contain 26 grams of carbohydrates, how many grams of carbohydrates do 375 grams of peas contain?

25. **Elections** During a recent election, 47 of every 100 registered voters in a certain city voted. If there were 127,900 registered voters in that city, how many people voted?

26. **Map Reading** The scale on a map is drawn so that 4.5 inches corresponds to an actual distance of 250 miles. If two cities are 7.25 inches apart on the map, how many miles apart are they? (Round to the nearest tenth.)

27. **Students to Teachers** The chart shows the student to teacher ratio in the United States from 1975 to 2002. If a school had 1,400 students in 1985, how many teachers does the school have? Round to the nearest teacher.

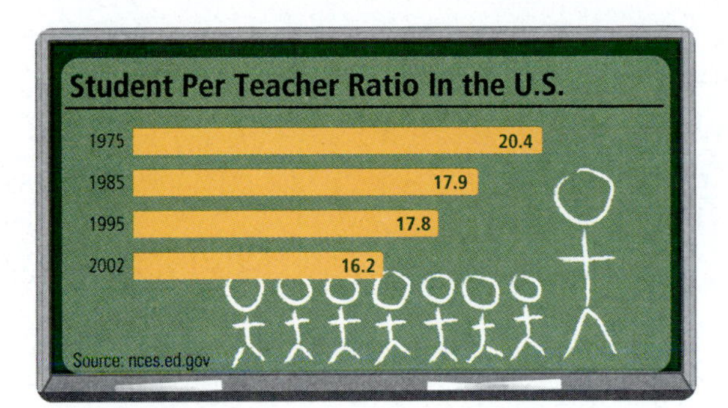

28. **Skyscrapers** The chart shows the heights of the three tallest buildings in the world. The ratio of feet to meters is given by 3.28/1. Using this information, convert the height of the Petronas Towers to meters. Round to the nearest hundredth.

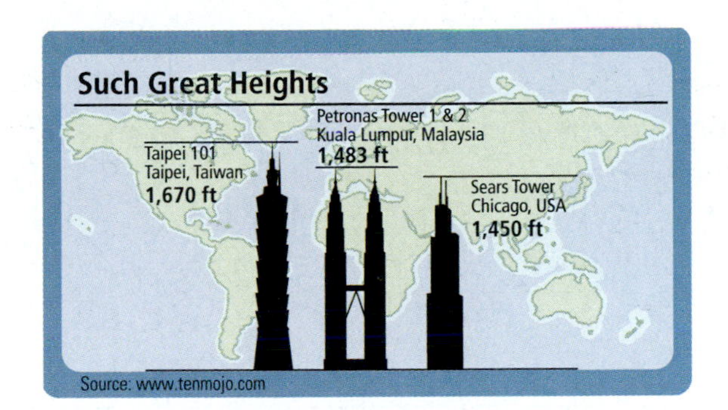

29. **Google Earth** The Google Earth image shows the western side of The Mall in Washington, D.C. If the scale indicates that one inch is 800 meters and the distance between the Lincoln Memorial and the World War II memorial is $\frac{17}{16}$ inches, what is the actual distance between the two landmarks?

30. **Google Earth** The Google Earth image shows Disney World in Florida. A scale indicates that one inch is 200 meters. If the distance between Splash Mountain and the Jungle Cruise is 190 meters, what is the distance on the map in inches?

Nursing Liquid medication is usually given in milligrams per milliliter. Use the information to find the amount a patient should take for a prescribe dosage.

31. A patient is prescribed a dosage of Ceclor® of 561 mg. The dosage strength is 187 mg per 5 mL. How many milliliters should he take?

32. A brand of amoxicillin has a dosage strength of 125 mg/5 mL. If a patient is prescribed a dosage of 25 mg, how many milliliters should she take?

Nursing For children, the amount of medicine prescribed is often determined by the child's mass. Usually it is calculated from the milligrams per kilogram per day listed on the medication's box.

33. How much should an 18 kg child be given a day if the dosage is 50 mg/kg/day?

34. How much should a 16.5 kg child be given a day if the dosage is 24 mg/kg/day?

Getting Ready for the Next Section

Simplify.

35. $\frac{320}{160}$

36. $21 \cdot 105$

37. $2{,}205 \div 15$

38. $\frac{48}{24}$

Solve each equation.

39. $\frac{x}{5} = \frac{28}{7}$

40. $\frac{x}{4} = \frac{6}{3}$

41. $\frac{x}{21} = \frac{105}{15}$

42. $\frac{b}{15} = 2$

Maintaining Your Skills

The problems below are a review of some of the concepts we covered previously.

Find the following products. (Multiply.)

43. 2.7×0.5

44. $(0.7)^2$

45. 3.18×1.2

46. $(0.3)^4$

Find the following quotients. (Divide.)

47. $2.8 \div 0.7$

48. $0.042 \div 0.21$

49. $24 \div 0.15$

50. $6.99 \div 2.33$

Divide and round answers to the nearest hundredth.

51. $5{,}679 \div 30.9$

52. $4{,}070 \div 64.2$

6.5 Similar Figures

This 8-foot-high bronze sculpture "Cellarman" in Napa, California, is an exact replica of the smaller, 12-inch sculpture. Both pieces are the product of artist Tim Lloyd of Arroyo Grande, California.

Courtesy of Timothy Lloyd Sculpture

Objectives

A Use proportions to find the lengths of sides of similar triangles.

B Use proportions to find the lengths of sides of other similar figures.

C Draw a figure similar to a given figure, given the length of one side.

D Use similar figures to solve application problems.

Examples now playing at MathTV.com/books

In mathematics, when two or more objects have the same shape, but are different sizes, we say they are similar. If two figures are similar, then their corresponding sides are proportional.

In order to give more details on what we mean by corresponding sides of similar figures, it will be helpful to introduce a simple way to label the parts of a triangle.

A Similar Triangles

Two triangles that have the same shape are similar when their corresponding sides are proportional, or have the same ratio. The triangles below are similar.

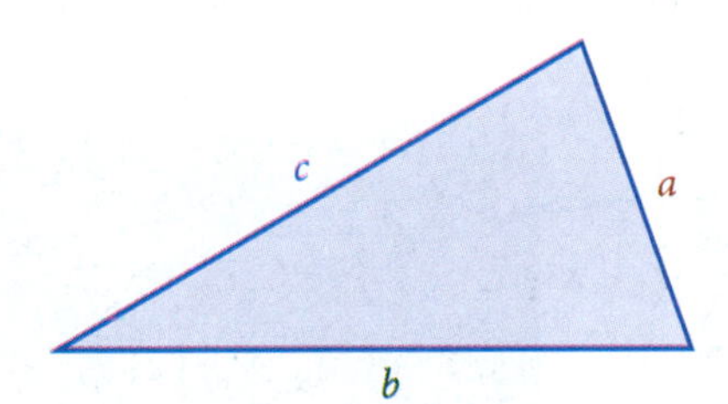

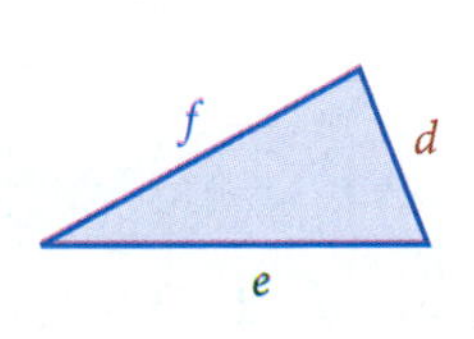

Corresponding Sides	**Ratio**
side a corresponds with side d	$\frac{a}{d}$
side b corresponds with side e	$\frac{b}{e}$
side c corresponds with side f	$\frac{c}{f}$

Because their corresponding sides are proportional, we write

$$\frac{a}{d} = \frac{b}{e} = \frac{c}{f}$$

LABELING TRIANGLES

One way to label the important parts of a triangle is to label the vertices with capital letters and the sides with lower-case letters.

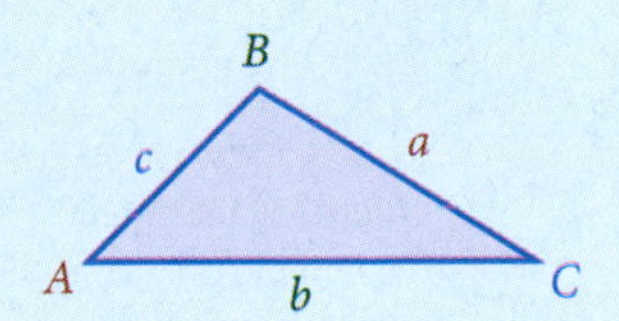

Notice that side a is opposite vertex A, side b is opposite vertex B, and side c is opposite vertex C. Also, because each vertex is the vertex of one of the angles of the triangle, we refer to the three interior angles as A, B, and C.

PRACTICE PROBLEMS

1. The two triangles below are similar. Find the missing side, x.

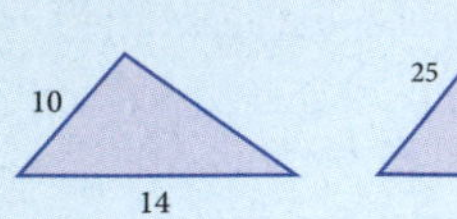

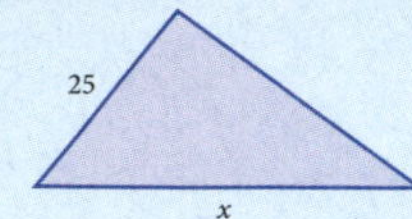

EXAMPLE 1 The two triangles below are similar. Find side x.

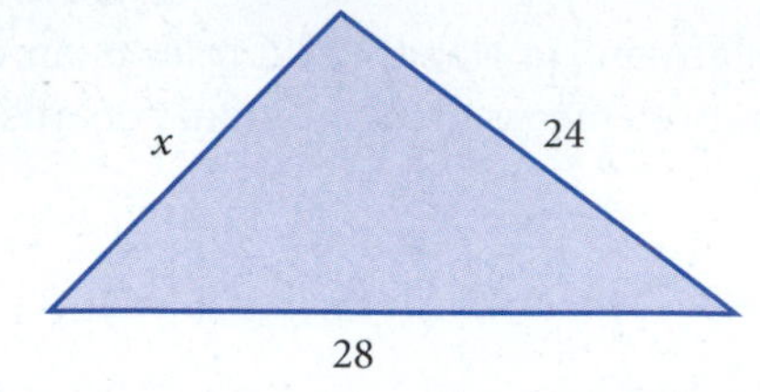

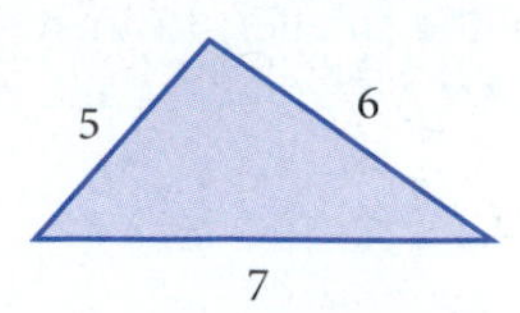

SOLUTION To find the length x, we set up a proportion of equal ratios. The ratio of x to 5 is equal to the ratio of 24 to 6 and to the ratio of 28 to 7. Algebraically we have

$$\frac{x}{5} = \frac{24}{6} \quad \text{and} \quad \frac{x}{5} = \frac{28}{7}$$

We can solve either proportion to get our answer. The first gives us

$$\frac{x}{5} = 4 \qquad \frac{24}{6} = 4$$

$$x = 4 \cdot 5 \qquad \text{Multiply both sides by 5}$$

$$x = 20 \qquad \text{Simplify}$$

B Other Similar Figures

When one shape or figure is either a reduced or enlarged copy of the same shape or figure, we consider them similar. For example, video viewed over the Internet was once confined to a small "postage stamp" size. Now it is common to see larger video over the Internet. Although the width and height have increased, the shape of the video has not changed.

2. Find the height, h, in pixels of a video clip proportional to those in Example 2 with a width of 360 pixels.

Note A pixel is the smallest dot made on a computer monitor. Many computer monitors have a width of 800 pixels and a height of 600 pixels.

EXAMPLE 2 The width and height of the two video clips are proportional. Find the height, h, in pixels of the larger video window.

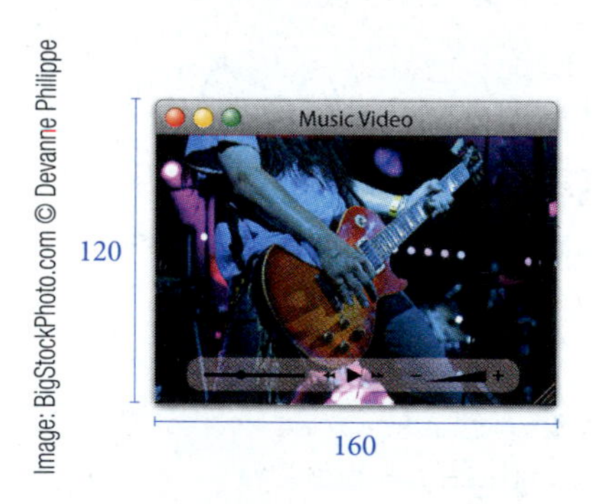

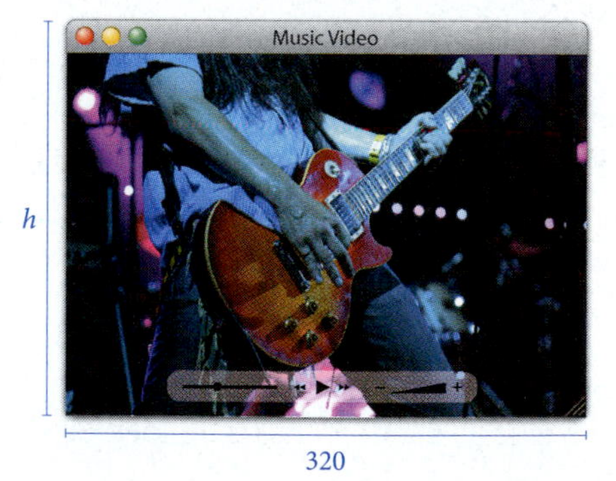

SOLUTION We write our proportion as the ratio of the height of the new video to the height of the old video is equal to the ratio of the width of the new video to the width of the old video:

$$\frac{h}{120} = \frac{320}{160}$$

$$\frac{h}{120} = 2$$

$$h = 2 \cdot 120$$

$$h = 240$$

The height of the larger video is 240 pixels.

Answers
1. 35 **2.** 270

C Drawing Similar Figures

EXAMPLE 3 Draw a triangle similar to triangle *ABC*, if *AC* is proportional to *DF*. Make *E* the third vertex of the new triangle.

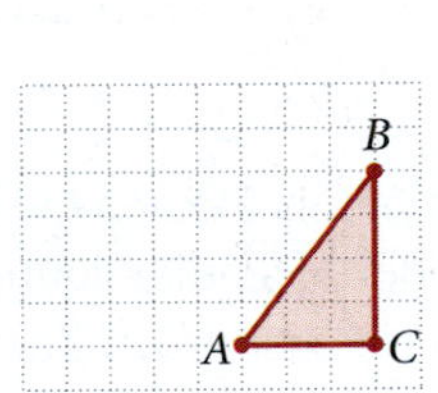

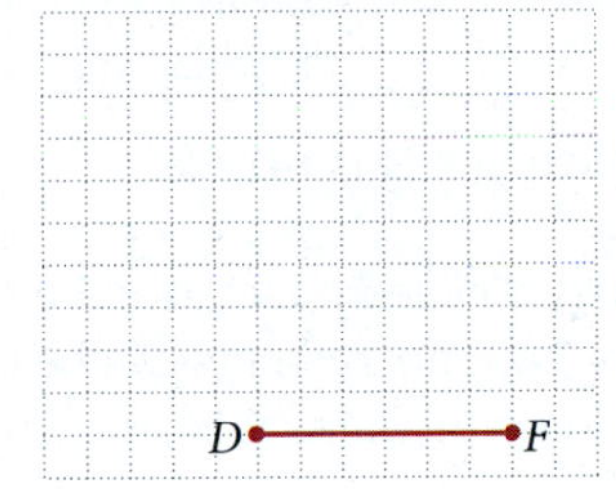

SOLUTION We see that *AC* is 3 units in length and *BC* has a length of 4 units. Since *AC* is proportional to *DF*, which has a length of 6 units, we set up a proportion to find the length *EF*.

$$\frac{EF}{BC} = \frac{DF}{AC}$$

$$\frac{EF}{4} = \frac{6}{3}$$

$$\frac{EF}{4} = 2$$

$$EF = 8$$

Now we can draw *EF* with a length of 8 units, then complete the triangle by drawing line *DE*.

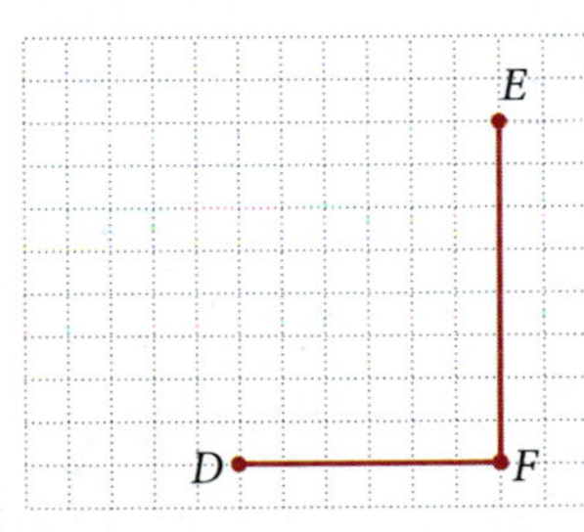

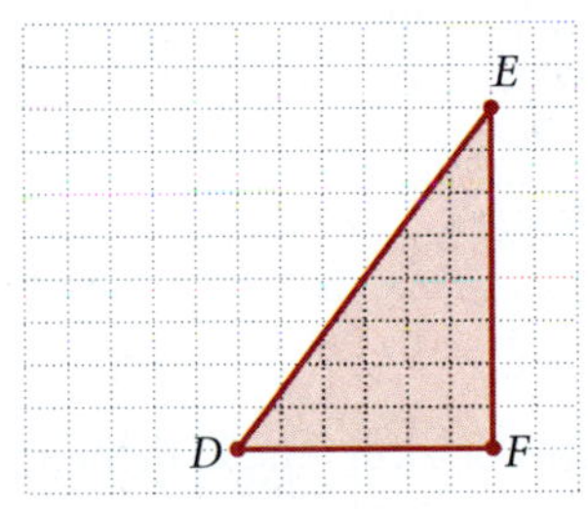

We have drawn triangle *DEF* similar to triangle *ABC*.

3. Draw a triangle similar to triangle *ABC*, if *AC* is proportional to *GI*.

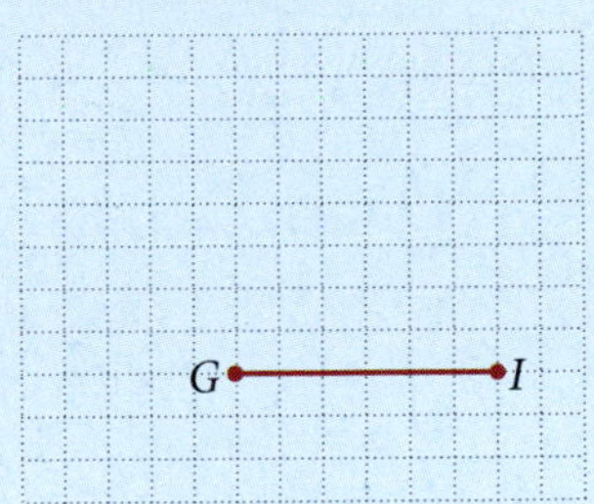

D Applications

EXAMPLE 4 A building casts a shadow of 105 feet while a 21-foot flagpole casts a shadow that is 15 feet. Find the height of the building.

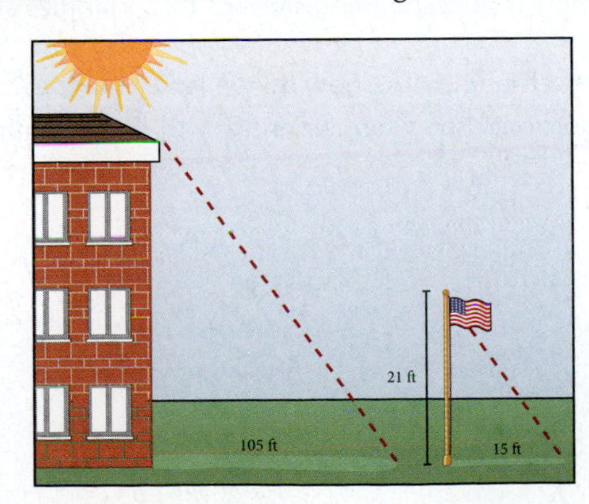

4. A building casts a shadow of 42 feet, while an 18-foot flagpole casts a shadow that is 12 feet. Find the height of the building.

Answer

3. See solutions section.

SOLUTION The figure shows both the building and the flagpole, along with their respective shadows. From the figure it is apparent that we have two similar triangles. Letting x = the height of the building, we have

$$\frac{x}{21} = \frac{105}{15}$$

$$15x = 2205 \quad \text{Extremes/means property}$$

$$x = 147 \quad \text{Divide both sides by 15}$$

The height of the building is 147 feet.

The Violin Family The instruments in the violin family include the bass, cello, viola, and violin. These instruments can be considered similar figures because the entire length of each instrument is proportional to its body length.

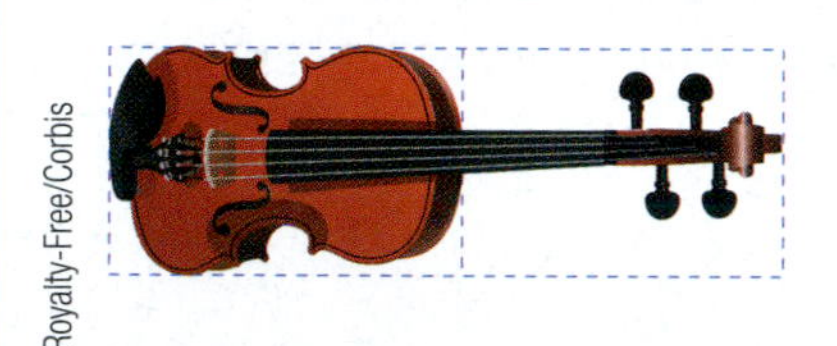

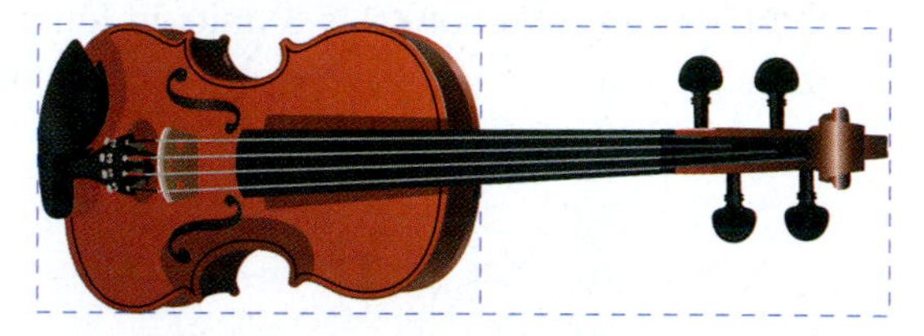

Royalty-Free/Corbis

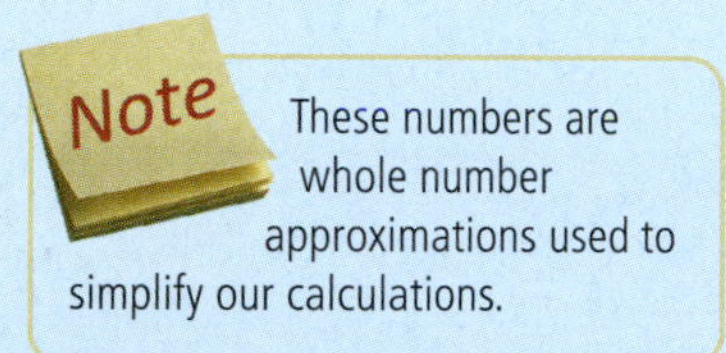

Note These numbers are whole number approximations used to simplify our calculations.

5. Find the body length of an instrument proportional to the violin family that has a total length of 32 inches.

EXAMPLE 5 The entire length of a violin is 24 inches, while the body length is 15 inches. Find the body length of a cello if the entire length is 48 inches.

SOLUTION Let b equal the body length of the cello, and set up the proportion.

$$\frac{b}{15} = \frac{48}{24}$$

$$\frac{b}{15} = 2$$

$$b = 2 \cdot 15$$

$$b = 30$$

The body length of a cello is 30 inches.

Getting Ready for Class

After reading through the preceding section, respond in your own words and in complete sentences.

1. What are similar figures?
2. How do we know if corresponding sides of two triangles are proportional?
3. When labeling a triangle ABC, how do we label the sides?
4. How are proportions used when working with similar figures?

Answers
4. 63 ft **5.** 20 in.

Problem Set 6.5

A In problems 1–4, for each pair of similar triangles, set up a proportion in order to find the unknown. [Example 1]

1.

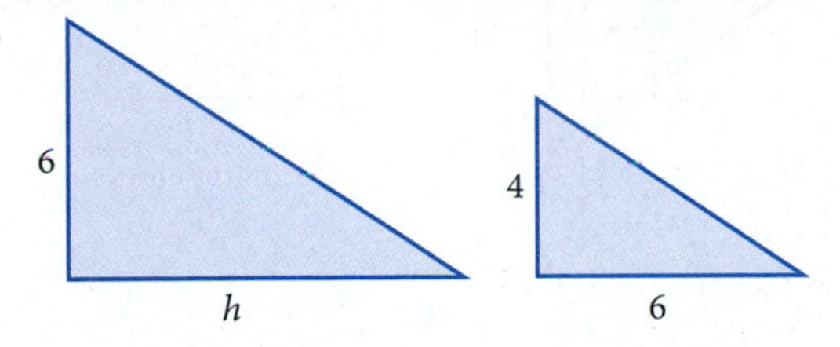

2.

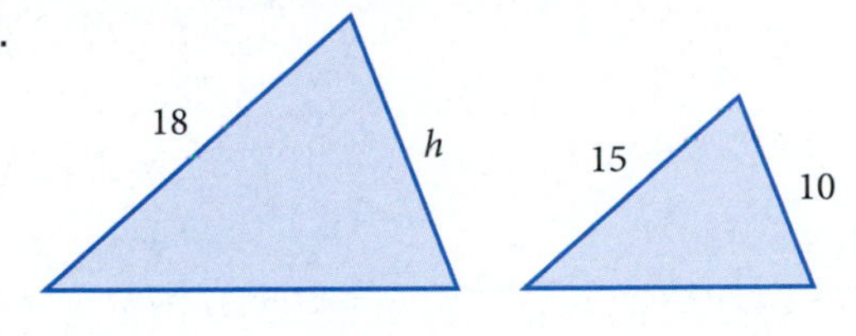

3.

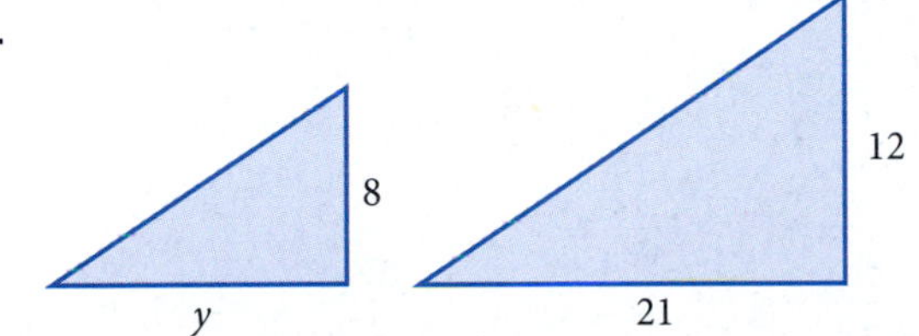

4.

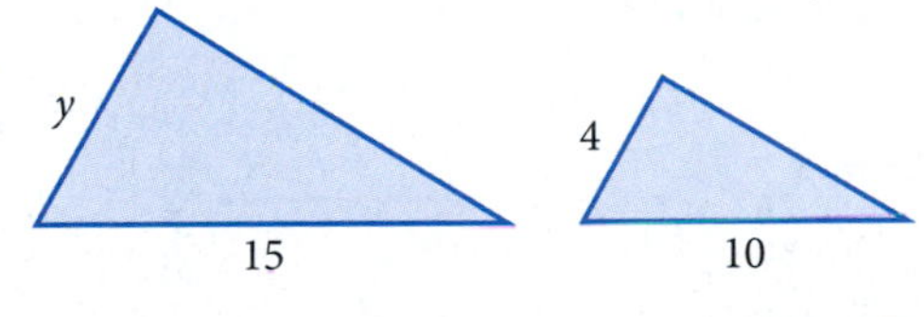

B In problems 5–10, for each pair of similar figures, set up a proportion in order to find the unknown. [Example 2]

5.

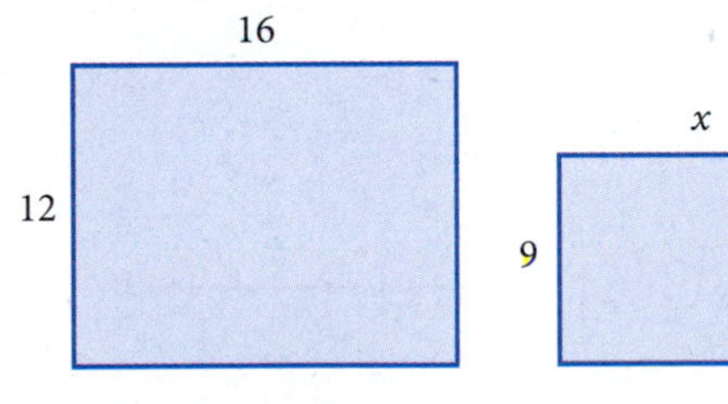

6.

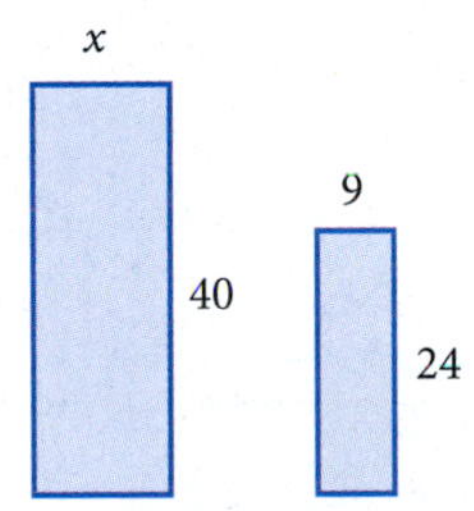

7.

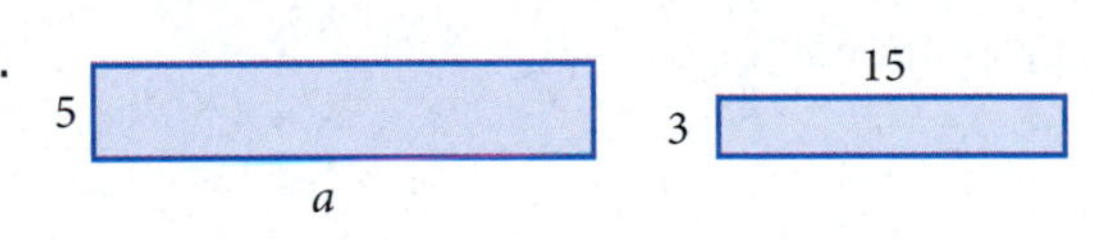

8.

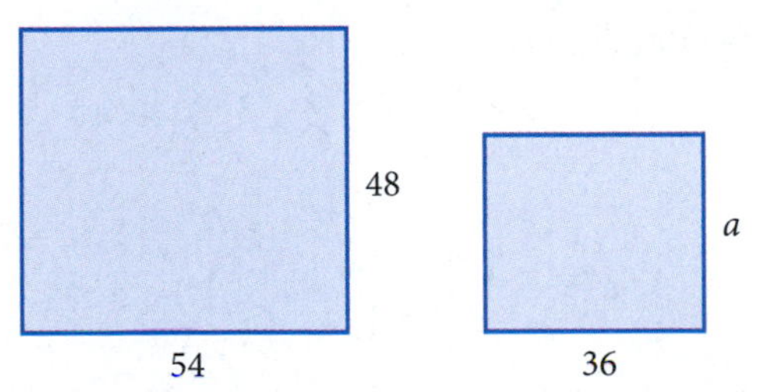

9.

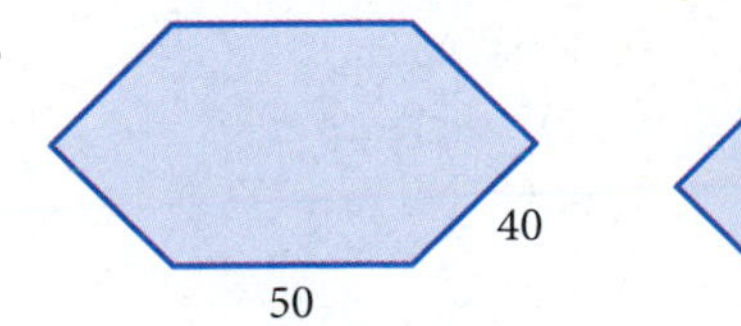

10.

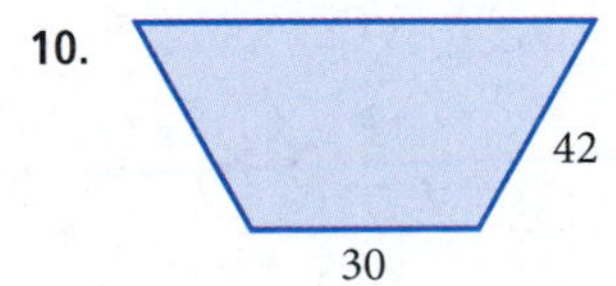

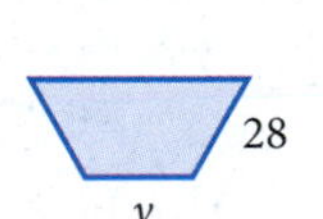

C For each problem, draw a figure on the grid on the right that is similar to the given figure. [Example 3]

11. *AC* is proportional to *DF*.

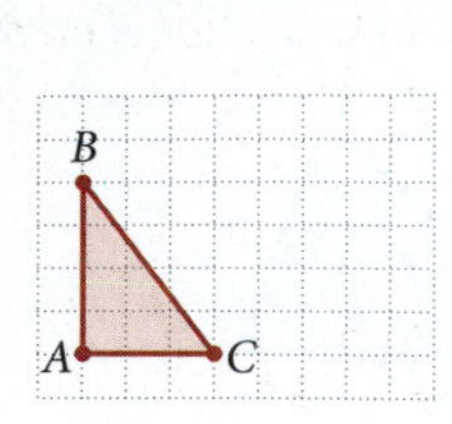

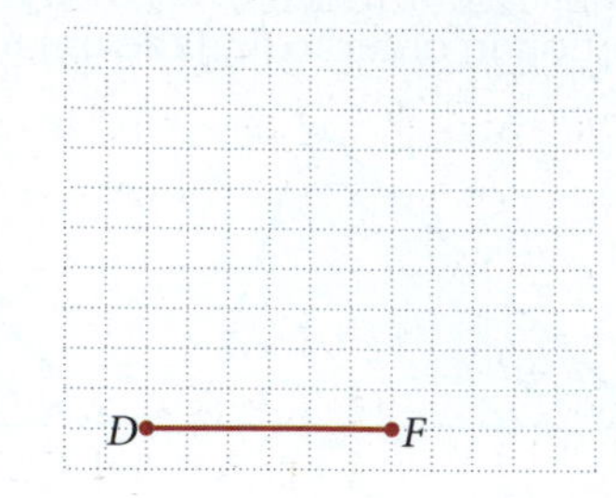

12. *AB* is proportional to *DE*.

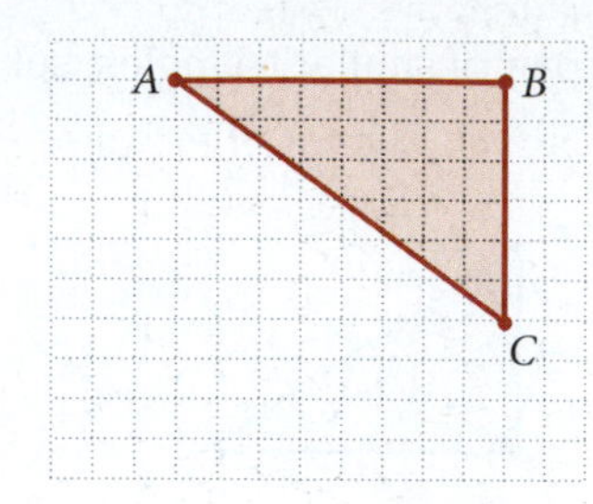

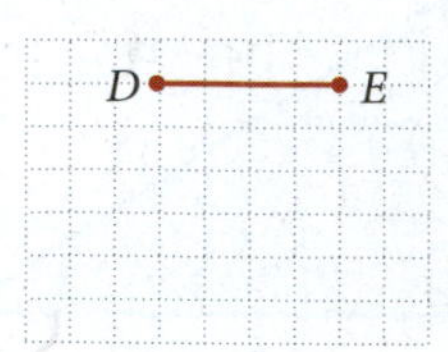

13. *BC* is proportional to *EF*.

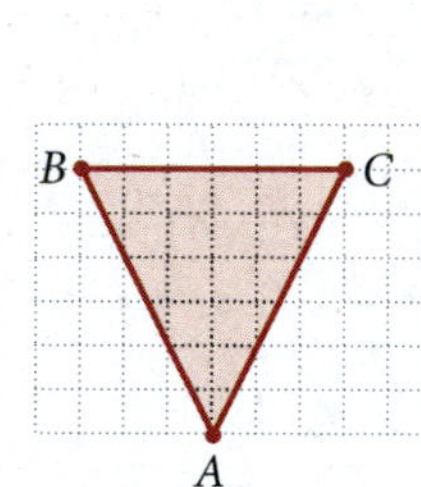

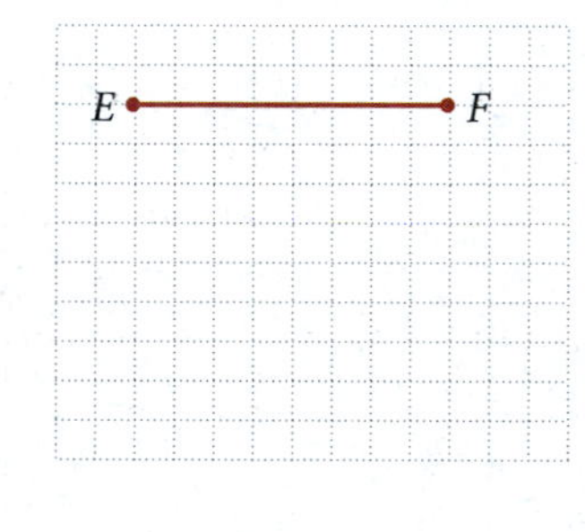

14. *AC* is proportional to *DF*.

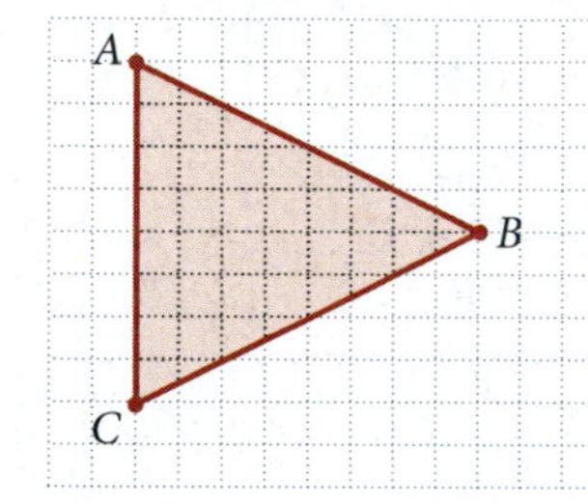

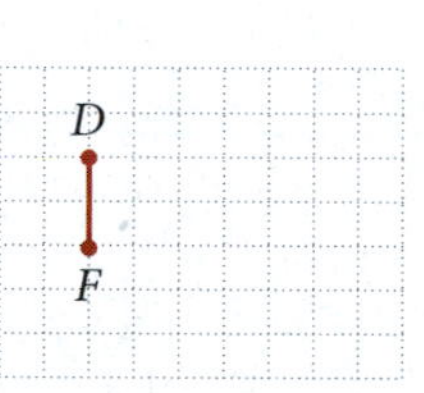

15. *DC* is proportional to *HG*.

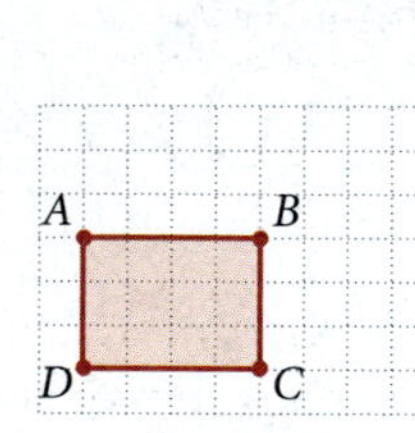

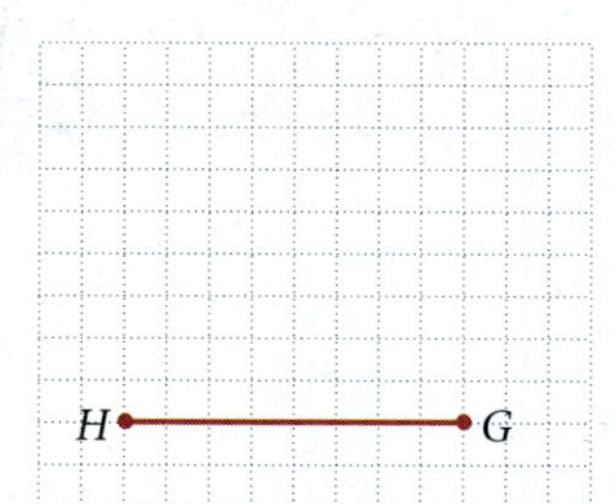

16. *AD* is proportional to *EH*.

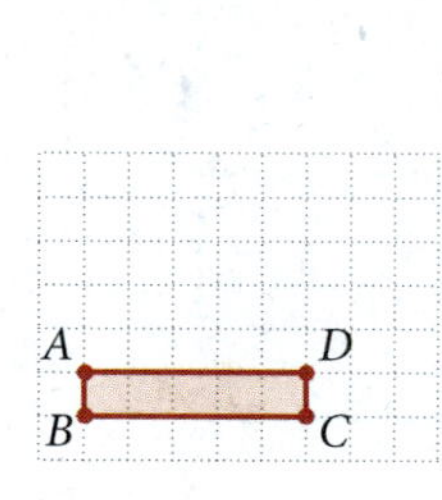

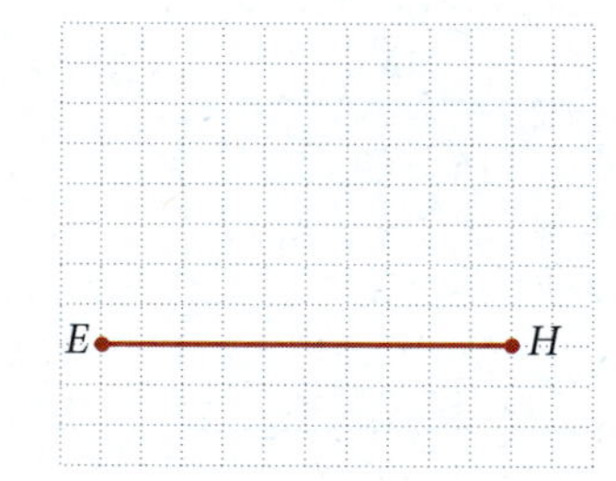

17. *AB* is proportional to *FG*.

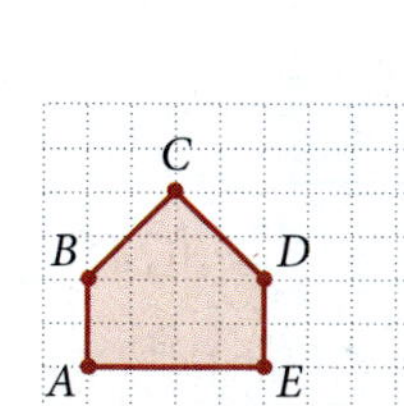

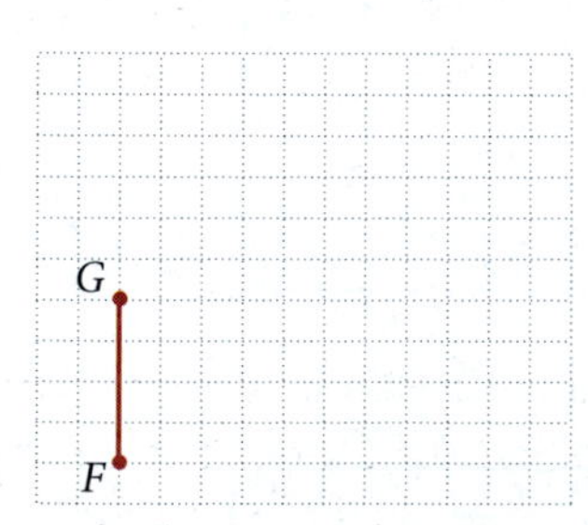

18. *BC* is proportional to *FG*.

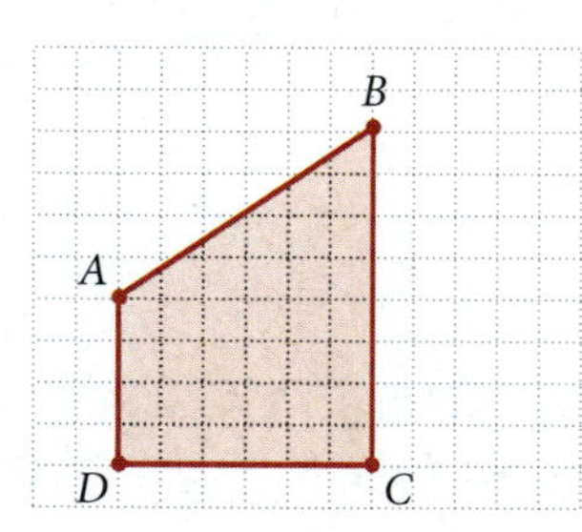

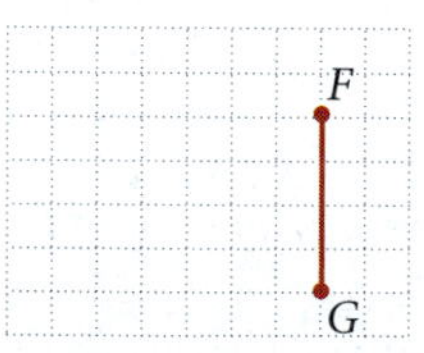

D Applying the Concepts [Examples 4, 5]

19. Length of a Bass The entire length of a violin is 24 inches, while its body length is 15 inches. The bass is an instrument proportional to the violin. If the total length of a bass is 72 inches, find its body length.

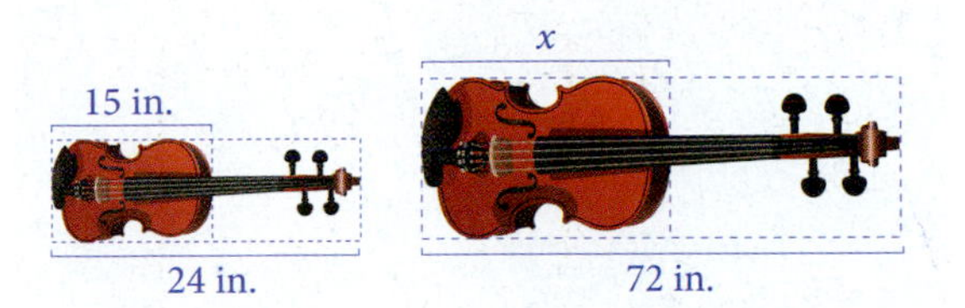

20. Length of an Instrument The entire length of a violin is 24 inches, while the body length is 15 inches. Another instrument proportional to the violin has a body length of 25 inches. What is the total length of this instrument?

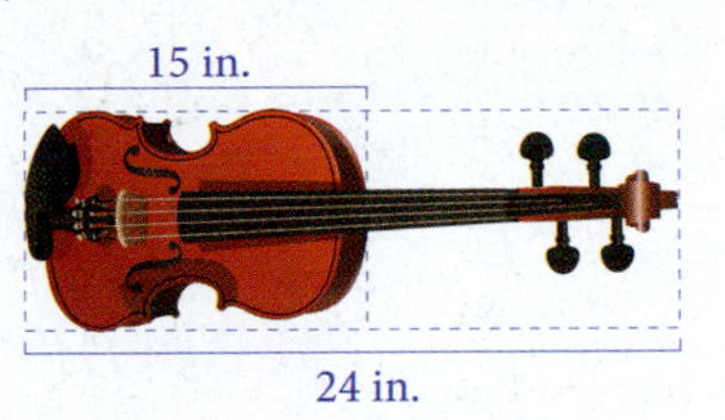

21. Video Resolution A new graphics card can increase the resolution of a computer's monitor. Suppose a monitor has a horizontal resolution of 800 pixels and a vertical resolution of 600 pixels. By adding a new graphics card, the resolutions remain in the same proportions, but the horizontal resolution increases to 1,280 pixels. What is the new vertical resolution?

22. Screen Resolution The display of a 20″ computer monitor is proportional to that of a 23″ monitor. A 20″ monitor has a horizontal resolution of 1,680 pixels and a vertical resolution of 1,050 pixels. If a 23″ monitor has a horizontal resolution of 1,920 pixels, what is its vertical resolution?

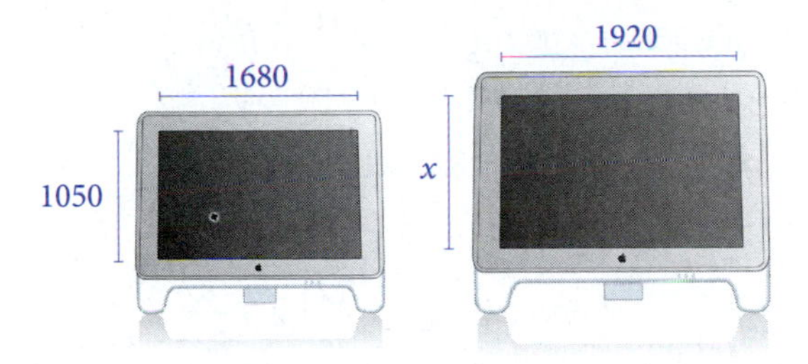

23. Screen Resolution The display of a 20″ computer monitor is proportional to that of a 17″ monitor. A 20″ monitor has a horizontal resolution of 1,680 pixels and a vertical resolution of 1,050 pixels. If a 17″ monitor has a vertical resolution of 900 pixels, what is its horizontal resolution?

24. Video Resolution A new graphics card can increase the resolution of a computer's monitor. Suppose a monitor has a horizontal resolution of 640 pixels and a vertical resolution of 480 pixels. By adding a new graphics card, the resolutions remain in the same proportions, but the vertical resolution increases to 786 pixels. What is the new horizontal resolution?

25. Height of a Tree A tree casts a shadow 38 feet long, while a 6-foot man casts a shadow 4 feet long. How tall is the tree?

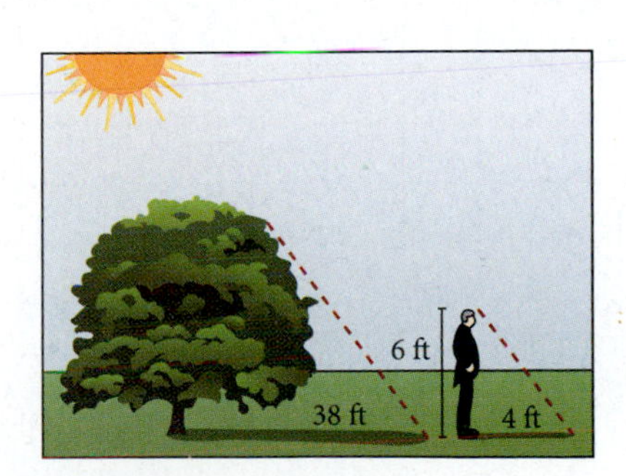

26. Height of a Building A building casts a shadow 128 feet long, while a 24-foot flagpole casts a shadow 32 feet long. How tall is the building?

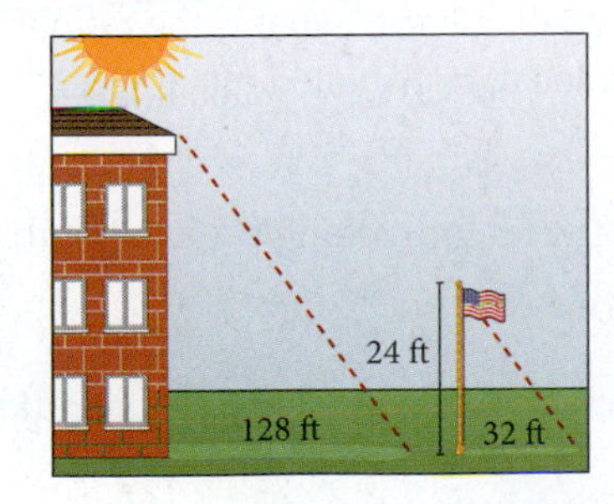

27. Eiffel Tower At the Paris Las Vegas Hotel is a replica of the Eiffel Tower in France. The heights of the tower in Las Vegas and the tower in France are 460 feet and 1,063 feet respectively. The base of the Eiffel Tower in France is 410 feet wide. What is the width of the base of the tower in Las Vegas? Round to the nearest foot.

28. Pyramids The Luxor Hotel in Las Vegas is almost an exact model of the pyramid of Khafre, the second largest Egyptian pyramid. The heights of the Luxor hotel and the pyramid of Khafre are 350 feet and 470 feet respectively. If the base of the pyramid in Khafre was 705 feet wide, what is the width of the base of the Luxor Hotel?

Maintaining Your Skills

The problems below are a review of the four basic operations with fractions and decimals.

Add.

29. $2.03 + 11.958 + 0.002$

30. $\frac{3}{4} + \frac{1}{6} + \frac{5}{8}$

Subtract.

31. $65.002 - 24.003$

32. $5\frac{1}{8} - 2\frac{5}{8}$

Multiply.

33. 42.18×0.0025

34. $7\frac{1}{7} \times 2\frac{1}{3}$

Divide.

35. $378.9 \div 21.05$

36. $12.25 \div \frac{3}{4}$

37. Find the sum of $2\frac{2}{3}$ and $1\frac{1}{2}$.

38. Find the difference of $2\frac{2}{3}$ and $1\frac{1}{2}$.

39. Find the product of $2\frac{2}{3}$ and $1\frac{1}{2}$.

40. Find the quotient of $2\frac{2}{3}$ and $1\frac{1}{2}$.

Extending the Concepts

41. The rectangles shown here are similar, with similar rectangles within.

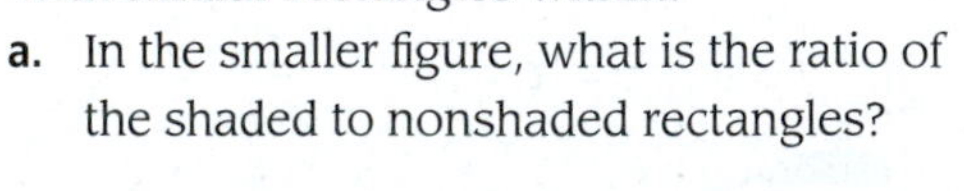

a. In the smaller figure, what is the ratio of the shaded to nonshaded rectangles?

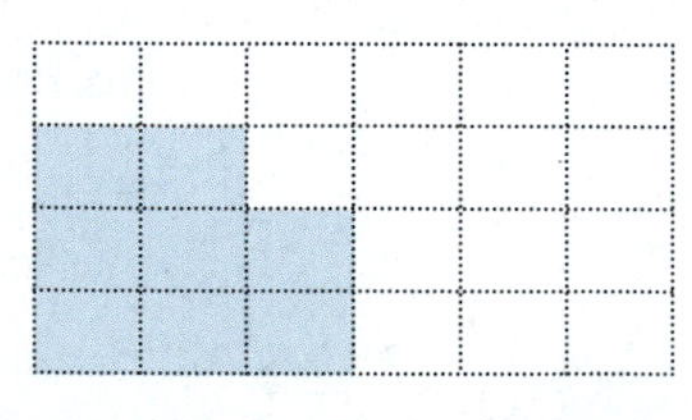

b. Shade the larger rectangle such that the ratio of shaded to nonshaded rectangles is $\frac{1}{2}$.

c. For each of the figures, what is the ratio of the shaded rectangles to total rectangles?

Chapter 6 Summary

Ratio [6.1]

The ratio of a to b is $\frac{a}{b}$. The ratio of two numbers is a way of comparing them using fraction notation.

EXAMPLES

1. The ratio of 6 to 8 is
$\frac{6}{8}$
which can be reduced to
$\frac{3}{4}$

Rates [6.2]

Whenever a ratio compares two quantities that have different units (and neither unit can be converted to the other), then the ratio is called a *rate*.

2. If a car travels 150 miles in 3 hours, then the ratio of miles to hours is considered a rate:

$$\frac{150 \text{ miles}}{3 \text{ hours}} = 50 \frac{\text{miles}}{\text{hour}} = 50 \text{ miles per hour}$$

Unit Pricing [6.2]

The unit price of an item is the ratio of price to quantity when the quantity is one unit.

3. If a 10-ounce package of frozen peas costs 69¢, then the price per ounce, or unit price, is

$$\frac{69 \text{ cents}}{10 \text{ ounces}} = 6.9 \frac{\text{cents}}{\text{ounce}} = 6.9 \text{ cents per ounce}$$

Proportion [6.3]

A proportion is an equation that indicates that two ratios are equal.

The numbers in a proportion are called *terms* and are numbered as follows:

First term → a, Second term → b, Third term → c, Fourth term → d

$$\frac{a}{b} = \frac{c}{d}$$

The first and fourth terms are called the *extremes*. The second and third terms are called the *means*.

$$\frac{a}{b} = \frac{c}{d}$$

Means → b, c; Extremes → a, d

4. The following is a proportion:
$\frac{6}{8} = \frac{3}{4}$

Fundamental Property of Proportions [6.3]

In any proportion the product of the extremes is equal to the product of the means. In symbols,

$$\text{If } \frac{a}{b} = \frac{c}{d} \quad \text{then} \quad ad = bc$$

Finding an Unknown Term in a Proportion [6.3]

5. Find x: $\frac{2}{5} = \frac{8}{x}$

$$2 \cdot x = 5 \cdot 8$$
$$2 \cdot x = 40$$
$$\frac{\not{2} \cdot x}{\not{2}} = \frac{40}{2}$$
$$x = 20$$

To find the unknown term in a proportion, we apply the fundamental property of proportions and solve the equation that results by dividing both sides by the number that is multiplied by the unknown. For instance, if we want to find the unknown in the proportion

$$\frac{2}{5} = \frac{8}{x}$$

we use the fundamental property of proportions to set the product of the extremes equal to the product of the means.

Using Proportions to Find Unknown Length with Similar Figures [6.5]

6. Find x.

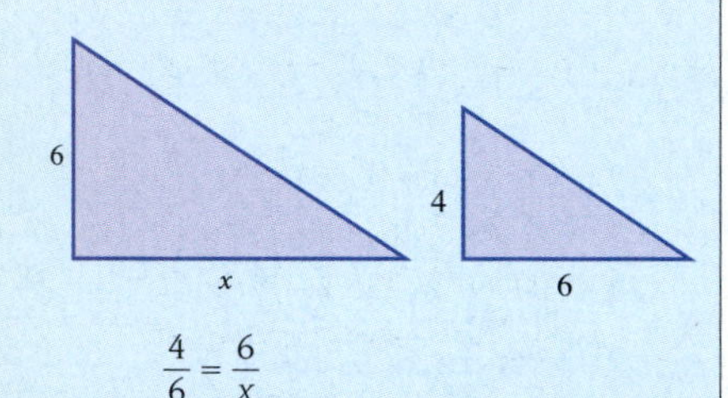

$$\frac{4}{6} = \frac{6}{x}$$
$$36 = 4x$$
$$9 = x$$

Two triangles that have the same shape are similar when their corresponding sides are proportional, or have the same ratio. The triangles below are similar.

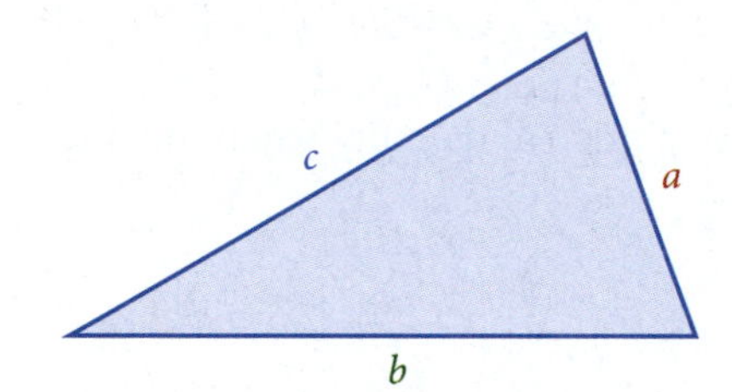

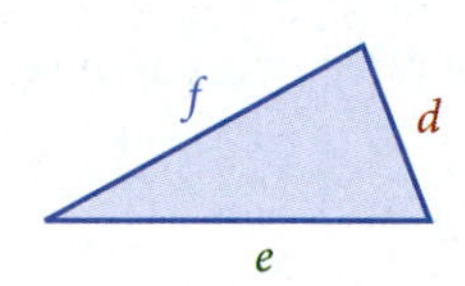

Corresponding Sides	Ratio
side a corresponds with side d	$\frac{a}{d}$
side b corresponds with side e	$\frac{b}{e}$
side c corresponds with side f	$\frac{c}{f}$

Because their corresponding sides are proportional, we write

$$\frac{a}{d} = \frac{b}{e} = \frac{c}{f}$$

COMMON MISTAKES

A common mistake when working with ratios is to write the numbers in the wrong order when writing the ratio as a fraction. For example, the ratio 3 to 5 is equivalent to the fraction $\frac{3}{5}$. It cannot be written as $\frac{5}{3}$.

Chapter 6 Review

The problems below form a comprehensive review of the material in this chapter. They can be used to study for exams. If you would like to take a practice test on this chapter, you can use the odd-numbered problems. Give yourself an hour and work as many of the odd-numbered problems as possible. When you are finished, or when an hour has passed, check your answers with the answers in the back of the book. You can use the even-numbered problems for a second practice test.

Write each of the following ratios as a fraction in lowest terms. [6.1]

1. 9 to 30

2. 30 to 9

3. $\frac{3}{7}$ to $\frac{4}{7}$

4. $\frac{8}{5}$ to $\frac{8}{9}$

5. $2\frac{1}{3}$ to $1\frac{2}{3}$

6. 3 to $2\frac{3}{4}$

7. 0.6 to 1.2

8. 0.03 to 0.24

9. $\frac{1}{5}$ to $\frac{3}{5}$

10. $\frac{2}{7}$ to $\frac{3}{7}$

The chart shows where each dollar spent on gasoline in the United States goes. Use the chart for problems 11–14. [6.1]

11. Ratio Find the ratio of money paid for taxes to money that goes to oil company profits.

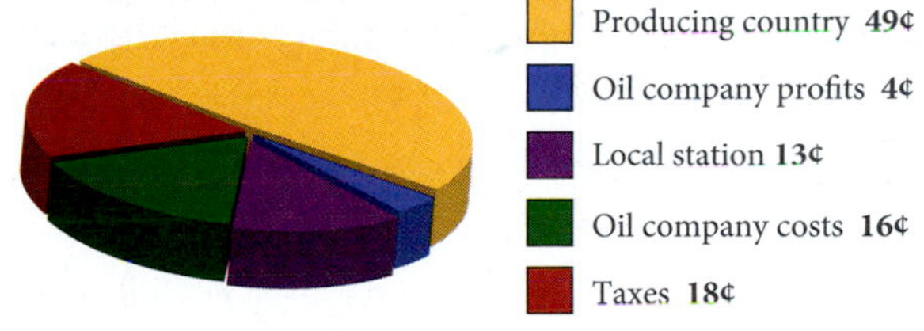

12. Ratio What is the ratio of the number of cents spent on oil company costs to the number of cents that goes to local stations?

13. Ratio Give the ratio of oil company profits to oil company costs.

14. Ratio Give the ratio of taxes to oil company costs and profits.

15. Gas Mileage A car travels 285 miles on 15 gallons of gas. What is the rate of gas mileage in miles per gallon? [6.2]

16. Speed of Sound If it takes 2.5 seconds for sound to travel 2,750 feet, what is the speed of sound in feet per second? [6.2]

17. Unit Price A certain brand of ice cream comes in two different-sized cartons with prices marked as shown. Give the unit price for each carton, and indicate which is the better buy.

64-ounce carton **$5.79**

32-ounce carton **$2.69**

18. Unit Price A 6-pack of store-brand soda is $1.25, while a 24-pack of name-brand soda is $5.99. Find the price per soda for each, and determine which is less expensive.

Find the missing term in each of the following proportions. [6.3]

19. $\frac{5}{7} = \frac{35}{x}$

20. $\frac{n}{18} = \frac{18}{54}$

21. $\frac{\frac{1}{2}}{10} = \frac{y}{2}$

22. $\frac{x}{1.8} = \frac{5}{1.5}$

23. Chemistry Suppose every 2,000 milliliters of a solution contains 24 milliliters of a certain drug. How many milliliters of solution are required to obtain 18 milliliters of the drug? [6.4]

24. Nutrition If $\frac{1}{2}$ cup of breakfast cereal contains 8 milligrams of calcium, how much calcium does $1\frac{1}{2}$ cups of the cereal contain? [6.4]

25. Weight Loss A man loses 8 pounds during the first 2 weeks of a diet. If he continues losing weight at the same rate, how long will it take him to lose 20 pounds? [6.4]

26. Men and Women If the ratio of men to women in a math class is 2 to 3, and there are 12 men in the class, how many women are in the class? [6.4]

27. Nursing A patient received a dosage of 7.5 mg of a certain medication. How many tablets must he take if the tablet strength is 2.5 mg?

28. Nursing A patient is told to take 300 mg of a certain medication daily. If he takes it in two sittings, how many milligrams is he taking each time he takes the medication?

29. Similar Triangles The triangles below are similar figures. Find x. [6.5]

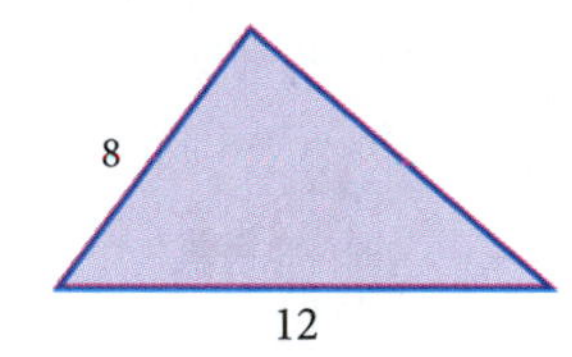

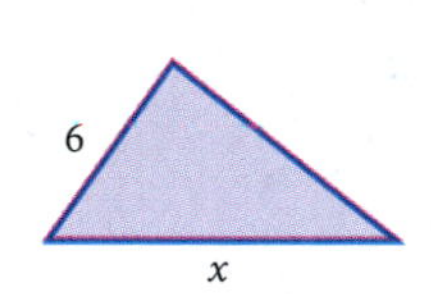

30. Find x if the two rectangles are similar.

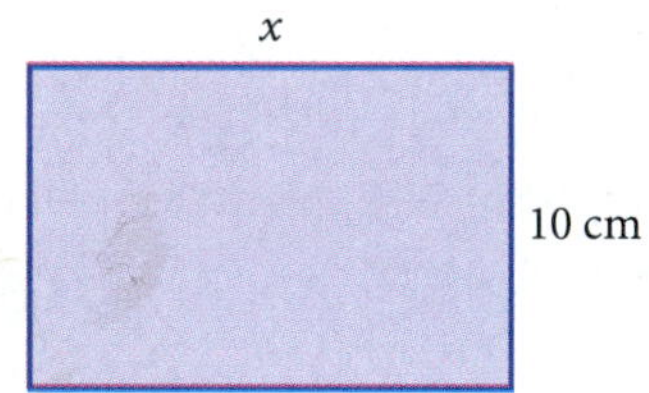

31. Video Size The width and height of the two video clips are proportional. Find the height, h, in pixels of the larger video window.

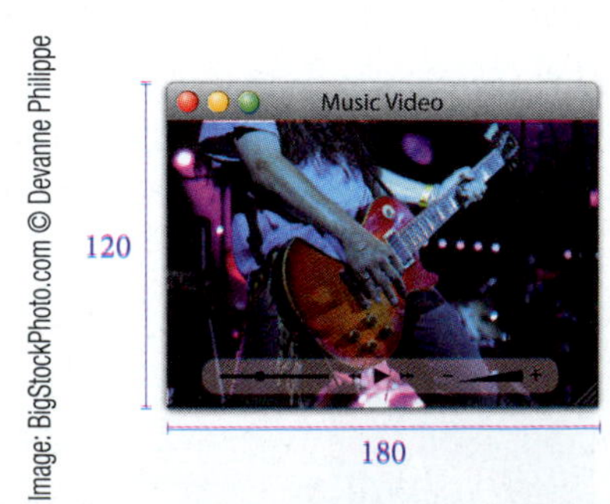

Chapter 6 Cumulative Review

Simplify.

1. $-4\left(\frac{x}{2} - 3\right)$

2. $x - \frac{5}{6}$

3. $613 - 297$

4. $\left(3\frac{1}{2} + \frac{1}{3}\right)\left(4\frac{1}{6} - \frac{2}{3}\right)$

5. $-\frac{2}{3}(3x + 9)$

6. $53(807)$

7. $\frac{5}{8} \div (-10)$

8. Round 37.6451 to the nearest hundredth.

9. Change $4\frac{7}{8}$ to an improper fraction.

10. Write the number 38,609 in words.

11. Identify the property or properties used in the following: $5(x + 9) = 5(x) + 5(9)$

Simplify:

12. $6(3)^3 - 9(2)^2$

13. $\sqrt{\frac{36}{49}}$

14. $4\sqrt{12} - 6\sqrt{27}$

15. $\frac{9(-4) + 8(3)}{-2^2}$

16. $\frac{9 - 5}{-9 + 5}$

17. $-(-6)$

Write each ratio as a fraction in lowest terms.

18. 24 seconds to 1 minute

19. $\frac{2}{3}$ to $\frac{3}{4}$

20. Find the value of the expression $2y - 15$ when $y = 6$.

21. Translate into symbols: Four times x subtracted from twice the sum of x and nine.

Solve.

22. $\frac{7}{8} = \frac{49}{x}$

23. $7x + 0.21 = x - 0.69$

24. $\frac{5}{6}x = 25$

25. $5(2x - 7) = 8x - 5 - 4x$

26. Find the perimeter and area of the figure below.

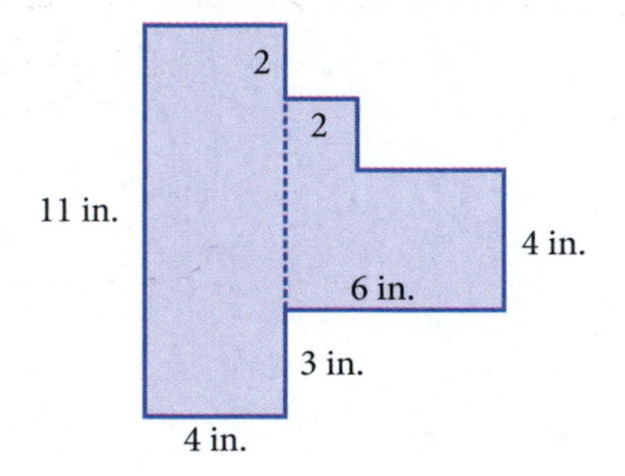

27. Find the perimeter of the figure below.

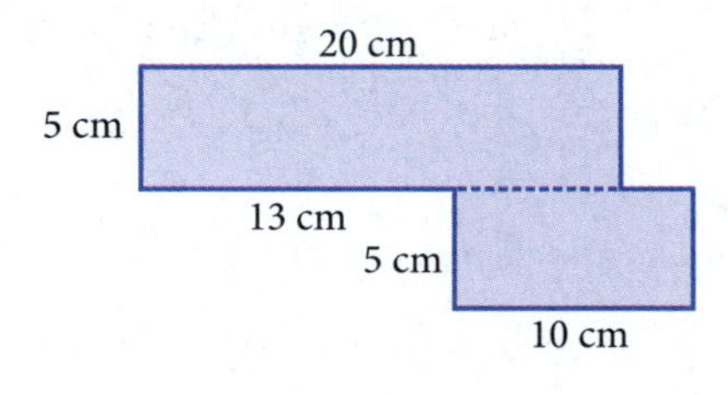

28. Write $\frac{14}{25}$ as a decimal.

29. Use the equation $2x - 7y = 10$ to find y when $x = -2$.

30. Reduce: $\frac{99xy}{36x}$

31. **Temperature** On Thursday, Arturo notices that the temperature reaches a high of 9° above 0 and a low of 8° below 0. What is the difference between the high and low temperatures for Thursday?

32. **Construction** A corrugated steel pipe has a radius of 3 feet and length of 20 feet.

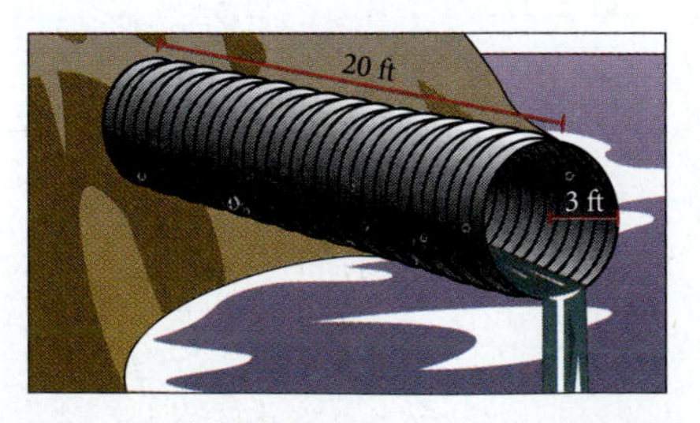

a. Find the circumference of the pipe. Use 3.14 for π.

b. Find the volume of the pipe. Use 3.14 for π.

33. Ratio If the ratio of men to women in a self-defense class is 3 to 4, and there are 15 men in the class, how many women are in the class?

34. Surfboard Length A surfing company decides that a surfboard would be more efficient if its length were reduced by $3\frac{5}{8}$ inches. If the original length was 7 feet $\frac{3}{16}$ inches, what will be the new length of the board (in inches)?

35. Average Distance A bicyclist on a cross-country trip travels 72 miles the first day, 113 miles the second day, 108 miles the third day, and 95 miles the fourth day. What is her average distance traveled during the four days?

36. Teaching A teacher lectures on five sections in two class periods. If she continues at the same rate, on how many sections can the teacher lecture in 60 class periods?

37. Unit Price A certain brand of ice cream comes in two different-sized cartons with prices marked as shown. Give the unit price for each carton, and indicate which is the better buy.

72-ounce carton
$6.10

48-ounce carton
$3.56

38. Model Plane This plane is from the Franklin Mint. It is a $\frac{1}{48}$ scale model of the F4U Corsair, the last propeller-driven fighter plane built by the United States. If the wingspan of the model is 10.25 inches, what is the wingspan of the actual plane? Give your answer in inches, then divide by 12 to give the answer in feet.

Franklin Mint

39. Find x if the two rectangles are similar.

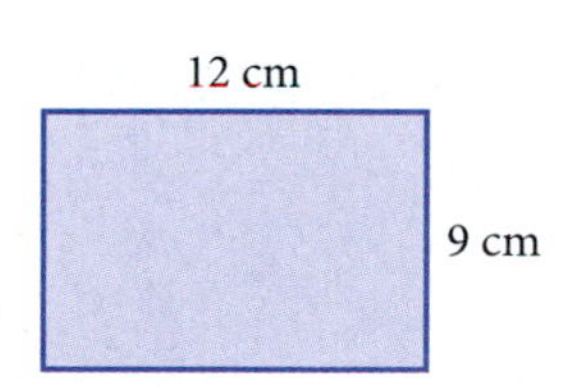

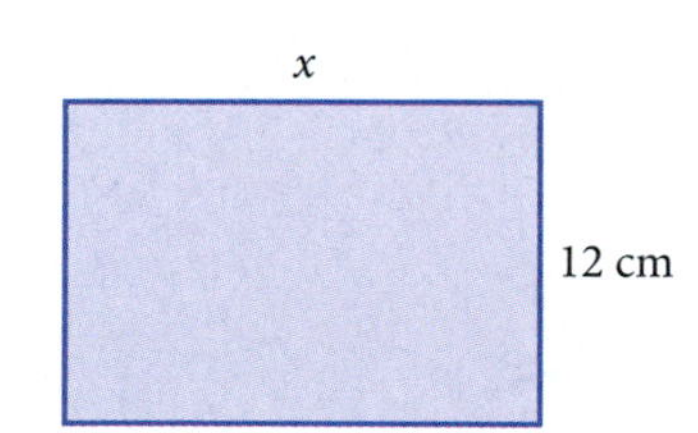

40. A 25-foot flagpole casts a shadow 15 feet long. At the same time, a building casts a shadow 21 feet long. Use similar triangles to find the height of the building to the nearest tenth.

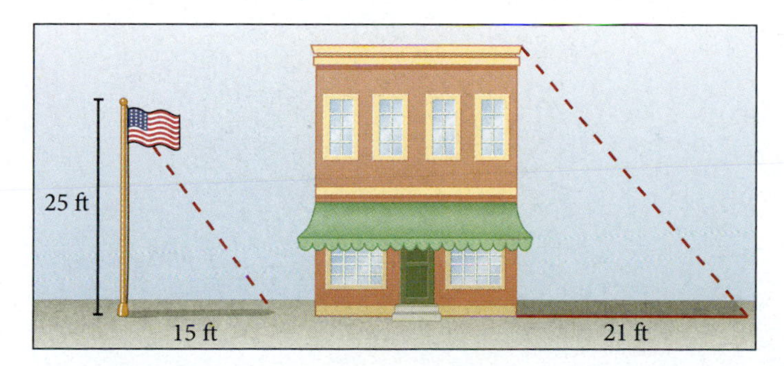

Chapter 6 Test

Write each ratio as a fraction in lowest terms.

1. 24 to 18

2. $\frac{3}{4}$ to $\frac{5}{6}$

3. 5 to $3\frac{1}{3}$

4. 0.18 to 0.6

5. $\frac{3}{11}$ to $\frac{5}{11}$

A family of three budgeted the following amounts for some of their monthly bills:

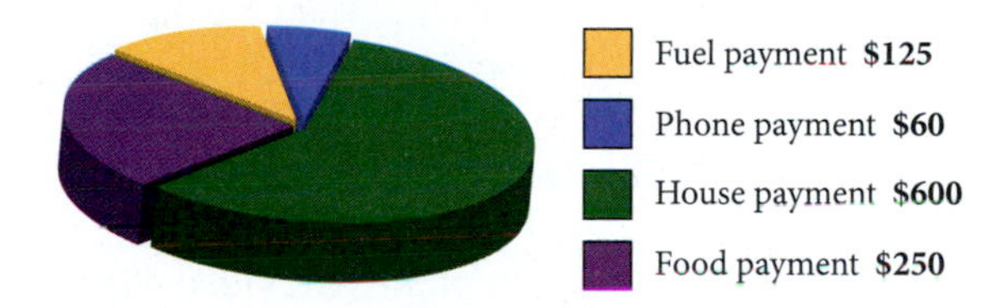

6. **Ratio** Find the ratio of house payment to fuel payment.

7. **Ratio** Find the ratio of phone payment to food payment.

8. **Gas Mileage** A car travels 414 miles on 18 gallons of gas. What is the rate of gas mileage in miles per gallon?

9. **Unit Price** A certain brand of frozen orange juice comes in two different-sized cans with prices marked as shown. Give the unit price for each can, and indicate which is the better buy.

16-ounce can
$2.59

12-ounce can
$1.89

Find the unknown term in each proportion.

10. $\frac{5}{6} = \frac{30}{x}$

11. $\frac{1.8}{6} = \frac{2.4}{x}$

12. Baseball A baseball player gets 9 hits in his first 21 games of the season. If he continues at the same rate, how many hits will he get in 56 games?

13. Map Reading The scale on a map indicates that 1 inch on the map corresponds to an actual distance of 60 miles. Two cities are $2\frac{1}{4}$ inches apart on the map. What is the actual distance between the two cities?

14. Model Trains Earlier we indicated that the size of a model train relative to an actual train is referred to as its scale. Each scale is associated with a ratio as shown in the table below. For example, an HO model train has a ratio of 1 to 87, meaning it is $\frac{1}{87}$ as large as an actual train.

Scale	Ratio
LGB	1 to 22.5
#1	1 to 32
O	1 to 43.5
S	1 to 64
HO	1 to 87
TT	1 to 120

Barry Rosenthal/Getty Images

a. If all six scales model the same boxcar, which one will have the largest model boxcar?

b. How many times larger is a boxcar that is O scale than a boxcar that is HO scale?

15. The triangles below are similar figures. Find h.

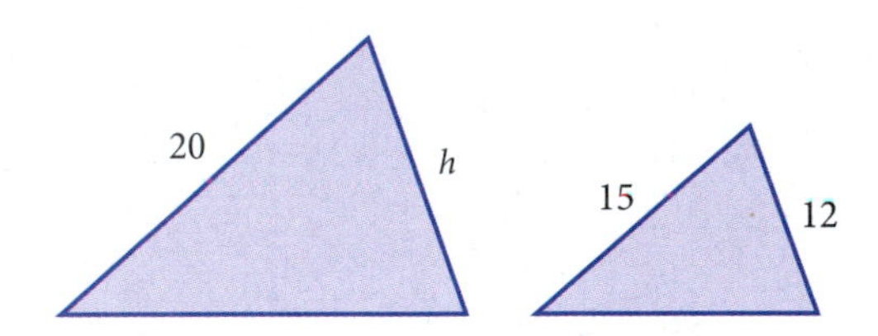

16. Video Size The width and height of the two video clips are proportional. Find the height, h, in pixels of the larger video window.

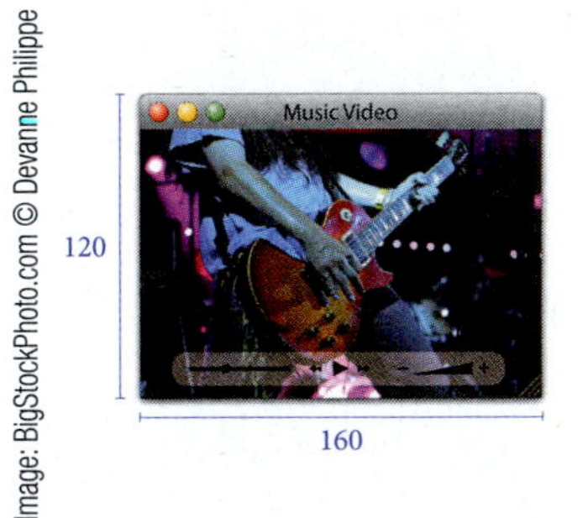

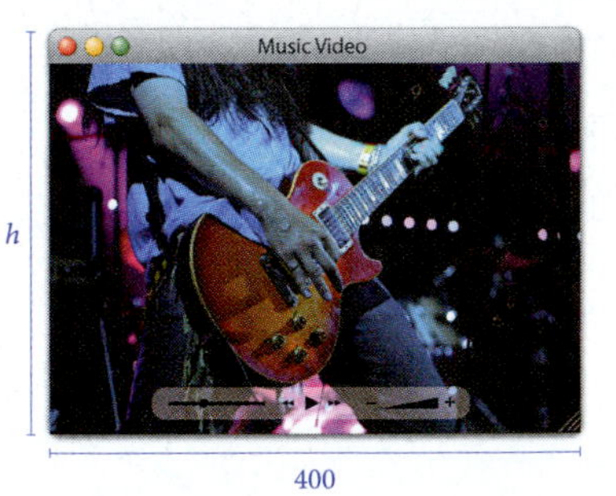

Image: BigStockPhoto.com © Devanne Philippe

Nursing Sometimes body surface area is used to calculate the necessary dosage for a patient.

17. The dosage for a drug is 15 mg/m^2. If an adult has a BSA of 1.8 m^2, what dosage should he take?

18. Find the dosage an adult should take if her BSA is 1.3 m^2 and the dosage strength is 25.5 mg/m^2.

Chapter 6 Projects

RATIO AND PROPORTION

GROUP PROJECT

Soil Texture

Number of People 2–3

Time Needed 8–12 minutes

Equipment Paper and pencil

Background Soil texture is defined as the relative proportions of sand, silt, and clay. The figure shows the relative sizes of each of these soil particles. People who study soil science, or work with soil, become very familiar with ratios.

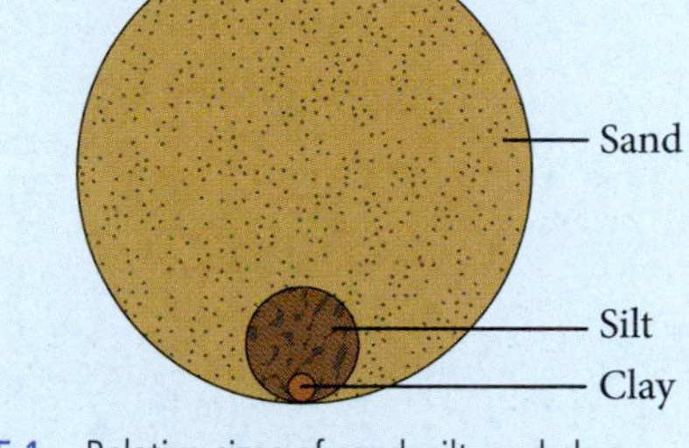

FIGURE 1 Relative sizes of sand, silt, and clay

Procedure A certain type of soil is one part silt, two parts clay, and three parts sand. Use your understanding of ratios and proportions to find the following ratios. Write these ratios as fractions.

1. Sand to total soil
2. Silt to total soil
3. Clay to total soil
4. What is the sum of the three fractions given in questions 1–3?
5. Let the 48 parts of the rectangle below each represent one cubic yard of the soil mixture above. Label each of the squares with either S (for sand), C (for clay), or T (for silt) based on the amount of each in 48 cubic yards of this soil.

RESEARCH PROJECT

The Golden Ratio

If you were going to design something with a rectangular shape—a television screen, a pool, or a house, for example—would one shape be more pleasing to the eye than another?

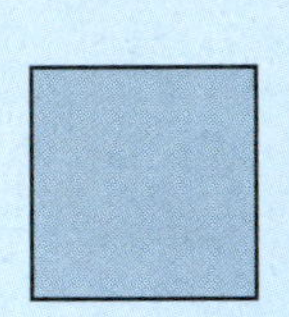
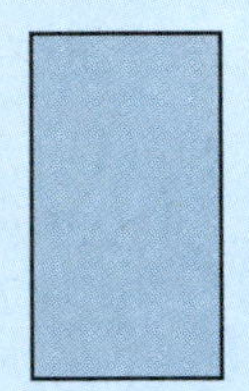
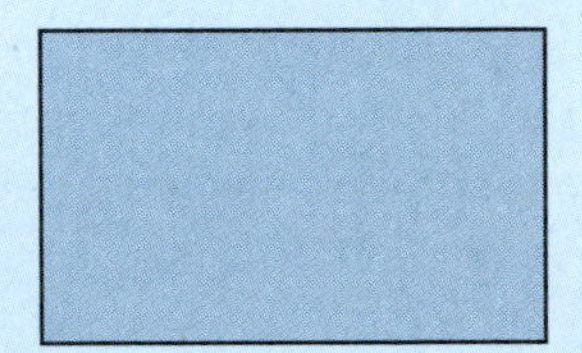

For many people, the most pleasing rectangles are rectangles in which the ratio of length to width is a number called the *golden ratio*, which we have written below.

$$\text{Golden Ratio} = \frac{\sqrt{5}+1}{2} \approx 1.6180339\ldots$$

Kevin Schafer/Corbis

Research the golden ratio in mathematics and give examples of where it is used in architecture and art. Then measure the length and width of some rectangles around you (TV/computer monitor screen, picture frame, math book, calculator, a dollar, notebook paper, etc.). Calculate the ratio of length to width and indicate which are close to the golden ratio.

Percent

7

Chapter Outline

Introduction

The eruption of Mount St. Helens in 1980 was the most catastrophic volcanic eruption in American history. The volcano is located in the state of Washington about 100 miles south of Seattle. It's eruption caused dozens of deaths, and destroyed almost 230 acres of forest, along with more than 200 homes. The effects on the carrying capacity of nearby rivers were devastating as well. As the rivers filled with debris and sediment, surrounding lands flooded, more vegetation was lost, and the fish population was greatly reduced.

Effects of Lava Flows on Rivers

	Carrying Capacity of the Cowlitz River (cubic feet per second)	Channel Depth of Columbia River
Prior to 1980	**76,000**	**40 feet**
After 1980 Eruption	**15,000**	**14 feet**
Percent Decrease	**80%**	**65%**

Source: US Forest Service

In this chapter we will work with fractions, decimals, and percents. We will see how percents are used in everyday applications, including volcanoes.

Chapter Pretest

The pretest below contains problems that are representative of the problems you will find in the chapter.

Change each percent to a decimal.

1. 68%

2. 2%

3. 21.5%

Change each decimal to a percent.

4. 0.39

5. 0.386

6. 3.98

Change each percent to a fraction or mixed number in lowest terms.

7. 33%

8. 45%

9. 8.5%

Change each fraction or mixed number to a percent.

10. $\frac{67}{100}$

11. $\frac{4}{5}$

12. $2\frac{1}{4}$

13. What number is 5% of 24?

14. What percent of 40 is 6?

15. 12 is 24% of what number?

Getting Ready for Chapter 7

The problems below review material covered previously that you need to know in order to be successful in Chapter 7. If you have any difficulty with the problems here, you need to go back and review before going on to Chapter 7.

Perform the indicated operations.

1. $136 + 5.44$

2. $300 - 75$

3. $1{,}793{,}000 - 315{,}568$

4. $\frac{65}{2} \times \frac{1}{100}$

5. 0.2×100

6. 4.89×100

7. $0.15 \cdot 63$

8. $\frac{35.2}{100}$

9. $3.62 \div 100$

10. $\frac{34}{0.29}$ (Round to the nearest tenth.)

11. $600 \times 0.04 \times \frac{60}{360}$

Reduce.

12. $\frac{36}{100}$

13. $\frac{45}{1000}$

14. Change $32\frac{1}{2}$ to an improper fraction.

Change each fraction or mixed number to a decimal.

15. $\frac{3}{8}$

16. $\frac{5}{12}$

17. $2\frac{1}{2}$

Solve.

18. $25 = 0.40 \cdot n$

19. $0.12n = 1{,}836$

20. $1.075x = 3{,}200$ (Round to the nearest hundredth.)

Percents, Decimals, and Fractions

7.1

Objectives

- **A** Change percents to fractions.
- **B** Change percents to decimals.
- **C** Change decimals to percents.
- **D** Change percents to fractions in lowest terms.
- **E** Change fractions to percents.

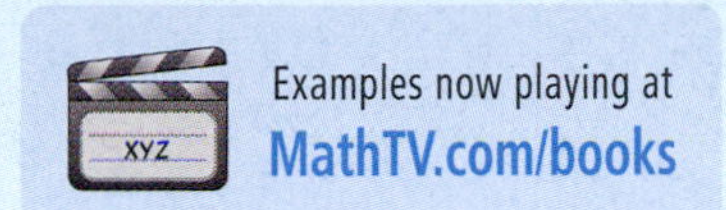

Introduction . . .

The sizes of categories in the pie chart below are given as percents. The whole pie chart is represented by 100%. In general, 100% of something is the whole thing.

In this section we will look at the meaning of percent. To begin, we learn to change decimals to percents and percents to decimals.

Factors Producing More Traffic Today

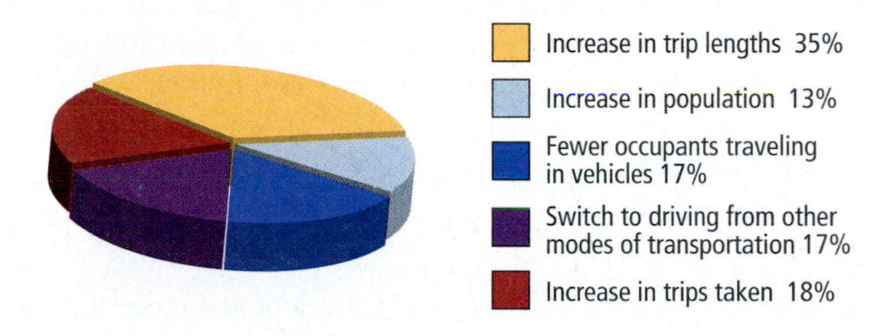

A The Meaning of Percent

Percent means "per hundred." Writing a number as a percent is a way of comparing the number with 100. For example, the number 42% (the % symbol is read "percent") is the same as 42 one-hundredths. That is:

$$42\% = \frac{42}{100}$$

Percents are really fractions (or ratios) with denominator 100.

Here are some examples that show the meaning of percent.

EXAMPLE 1 $50\% = \frac{50}{100}$

EXAMPLE 2 $75\% = \frac{75}{100}$

EXAMPLE 3 $25\% = \frac{25}{100}$

EXAMPLE 4 $33\% = \frac{33}{100}$

EXAMPLE 5 $6\% = \frac{6}{100}$

EXAMPLE 6 $160\% = \frac{160}{100}$

PRACTICE PROBLEMS

Write each number as an equivalent fraction without the % symbol.

1. 40%
2. 80%
3. 15%
4. 37%
5. 8%
6. 150%

Answers

1. $\frac{40}{100}$ 2. $\frac{80}{100}$ 3. $\frac{15}{100}$
4. $\frac{37}{100}$ 5. $\frac{8}{100}$ 6. $\frac{150}{100}$

B Changing Percents to Decimals

To change a percent to a decimal number, we simply use the meaning of percent.

7. Change to a decimal.
   a. 25.2%
   b. 2.52%

EXAMPLE 7 Change 35.2% to a decimal.

SOLUTION We drop the % symbol and write 35.2 over 100.

$$35.2\% = \frac{35.2}{100}$$ Use the meaning of % to convert to a fraction with denominator 100

$$= 0.352$$ Divide 35.2 by 100

We see from Example 7 that 35.2% is the same as the decimal 0.352. The result is that the % symbol has been dropped and the decimal point has been moved two places to the *left*. Because % always means "per hundred," we will always end up moving the decimal point two places to the left when we change percents to decimals. Because of this, we can write the following rule.

Rule
To change a percent to a decimal, drop the % symbol and move the decimal point two places to the *left*, inserting zeros as placeholders if needed.

Here are some examples to illustrate how to use this rule.

Change each percent to a decimal.
8. 40%
9. 80%
10. 15%
11. 5.6%
12. 4.86%
13. 0.6%
14. 0.58%

EXAMPLE 8 25% = 0.25

EXAMPLE 9 75% = 0.75 Notice that the results in Examples 8, 9, and 10 are consistent with the results in Examples 1, 2, and 3

EXAMPLE 10 50% = 0.50

EXAMPLE 11 6.8% = 0.068 Notice here that we put a 0 in front of the 6 so we can move the decimal point two places to the left

EXAMPLE 12 3.62% = 0.0362

EXAMPLE 13 0.4% = 0.004 This time we put two 0s in front of the 4 in order to be able to move the decimal point two places to the left

EXAMPLE 14 The cortisone cream shown here is 0.5% hydrocortisone. Writing this number as a decimal, we have

$$0.5\% = 0.005$$

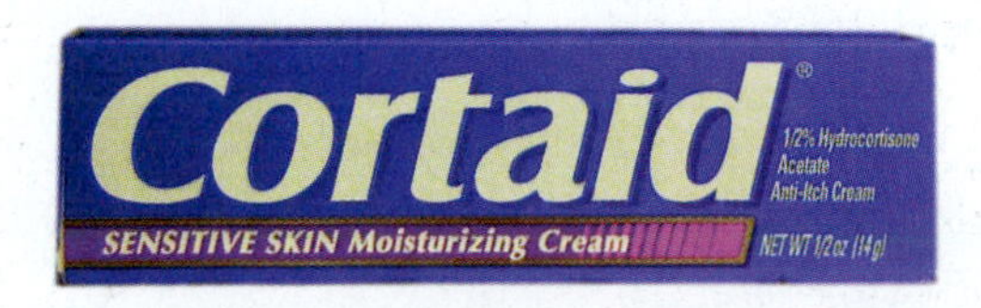

Answers
7. a. 0.252 b. 0.0252 8. 0.40
9. 0.80 10. 0.15 11. 0.056
12. 0.0486 13. 0.006 14. 0.0058

C Changing Decimals to Percents

Now we want to do the opposite of what we just did in Examples 7–14. We want to change decimals to percents. We know that 42% written as a decimal is 0.42, which means that in order to change 0.42 back to a percent, we must move the decimal point two places to the *right* and use the % symbol:

$0.42 = 42\%$ Notice that we don't show the new decimal point if it is at the end of the number

Rule

To change a decimal to a percent, we move the decimal point two places to the *right* and use the % symbol.

Examples 15–20 show how we use this rule.

EXAMPLE 15 $0.27 = 27\%$

EXAMPLE 16 $4.89 = 489\%$

EXAMPLE 17 $0.2 = 0.20 = 20\%$ Notice here that we put a 0 after the 2 so we can move the decimal point two places to the right

EXAMPLE 18 $0.09 = 09\% = 9\%$ Notice that we can drop the 0 at the left without changing the value of the number

EXAMPLE 19 $25 = 25.00 = 2{,}500\%$ Here, we put two 0s after the 5 so that we can move the decimal point two places to the right

EXAMPLE 20 A softball player has a batting average of 0.650. As a percent, this number is $0.650 = 65.0\%$.

Eyewire/Getty Images

As you can see from the examples above, percent is just a way of comparing numbers to 100. To multiply decimals by 100, we move the decimal point two places to the right. To divide by 100, we move the decimal point two places to the left. Because of this, it is a fairly simple procedure to change percents to decimals and decimals to percents.

Write each decimal as a percent.

15. 0.35

16. 5.77

17. 0.4

18. 0.03

19. 45

20. 0.69

Answers

15. 35% **16.** 577% **17.** 40%
18. 3% **19.** 4,500% **20.** 69%

D Changing Percents to Fractions

To change a percent to a fraction, drop the % symbol and write the original number over 100.

Who Pays Health Care Bills

Patient 19%

Private insurance 36%

Government 45%

EXAMPLE 21 The pie chart in the margin shows who pays health care bills. Change each percent to a fraction.

SOLUTION In each case, we drop the percent symbol and write the number over 100. Then we reduce to lowest terms if possible.

$$19\% = \frac{19}{100} \qquad 45\% = \frac{45}{100} = \frac{9}{20} \qquad 36\% = \frac{36}{100} = \frac{9}{25}$$

reduce　　reduce

21. Change 82% to a fraction in lowest terms.

22. Change 6.5% to a fraction in lowest terms.

EXAMPLE 22 Change 4.5% to a fraction in lowest terms.

SOLUTION We begin by writing 4.5 over 100:

$$4.5\% = \frac{4.5}{100}$$

We now multiply the numerator and the denominator by 10 so the numerator will be a whole number:

$$\frac{4.5}{100} = \frac{4.5 \times \mathbf{10}}{100 \times \mathbf{10}} \qquad \text{Multiply the numerator and the denominator by 10}$$

$$= \frac{45}{1{,}000}$$

$$= \frac{9}{200} \qquad \text{Reduce to lowest terms}$$

23. Change $42\frac{1}{2}\%$ to a fraction in lowest terms.

EXAMPLE 23 Change $32\frac{1}{2}\%$ to a fraction in lowest terms.

SOLUTION Writing $32\frac{1}{2}\%$ over 100 produces a complex fraction. We change $32\frac{1}{2}$ to an improper fraction and simplify:

$$32\frac{1}{2}\% = \frac{32\frac{1}{2}}{100}$$

$$= \frac{\frac{65}{2}}{100} \qquad \text{Change } 32\tfrac{1}{2} \text{ to the improper fraction } \tfrac{65}{2}$$

$$= \frac{65}{2} \times \frac{1}{100} \qquad \text{Dividing by 100 is the same as multiplying by } \tfrac{1}{100}$$

$$= \frac{\not{5} \cdot 13 \cdot 1}{2 \cdot \not{5} \cdot 20} \qquad \text{Multiplication}$$

$$= \frac{13}{40} \qquad \text{Reduce to lowest terms}$$

Note that we could have changed our original mixed number to a decimal first and then changed to a fraction:

$$32\frac{1}{2}\% = 32.5\% = \frac{32.5}{100} = \frac{32.5 \times 10}{100 \times 10} = \frac{325}{1000} = \frac{\not{5} \cdot \not{5} \cdot 13}{\not{5} \cdot \not{5} \cdot 40} = \frac{13}{40}$$

The result is the same in both cases.

Answers

21. $\frac{41}{50}$ **22.** $\frac{13}{200}$ **23.** $\frac{17}{40}$

E Changing Fractions to Percents

To change a fraction to a percent, we can change the fraction to a decimal and then change the decimal to a percent.

EXAMPLE 24 Suppose the price your bookstore pays for your textbook is $\frac{7}{10}$ of the price you pay for your textbook. Write $\frac{7}{10}$ as a percent.

SOLUTION We can change $\frac{7}{10}$ to a decimal by dividing 7 by 10:

$$10\overline{)7.0} = 0.7, \quad 7\,0, \quad 0$$

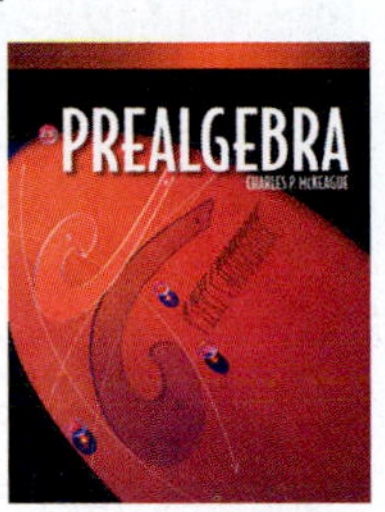

We then change the decimal 0.7 to a percent by moving the decimal point two places to the *right* and using the % symbol:

$$0.7 = 70\%$$

You may have noticed that we could have saved some time in Example 24 by simply writing $\frac{7}{10}$ as an equivalent fraction with denominator 100. That is:

$$\frac{7}{10} = \frac{7 \cdot \mathbf{10}}{10 \cdot \mathbf{10}} = \frac{70}{100} = 70\%$$

This is a good way to convert fractions like $\frac{7}{10}$ to percents. It works well for fractions with denominators of 2, 4, 5, 10, 20, 25, and 50, because they are easy to change to fractions with denominators of 100.

EXAMPLE 25 Change $\frac{3}{8}$ to a percent.

SOLUTION We write $\frac{3}{8}$ as a decimal by dividing 3 by 8. We then change the decimal to a percent by moving the decimal point two places to the right and using the % symbol.

$$\frac{3}{8} = 0.375 = 37.5\%$$

$$8\overline{)3.000} = .375; \quad 2\,4, \; 60, \; 56, \; 40, \; 40, \; 0$$

EXAMPLE 26 Change $\frac{5}{12}$ to a percent.

SOLUTION We begin by dividing 5 by 12:

$$12\overline{)5.0000} = .4166; \quad 4\,8, \; 20, \; 12, \; 80, \; 72, \; 80, \; 72$$

24. Change to a percent.
a. $\frac{9}{10}$
b. $\frac{9}{20}$

25. Change to a percent.
a. $\frac{5}{8}$
b. $\frac{9}{8}$

26. Change to a percent.
a. $\frac{7}{12}$
b. $\frac{13}{12}$

Answers
24. a. 90% **b.** 45%
25. a. 62.5% **b.** 112.5%

Note When rounding off, let's agree to round off to the nearest thousandth and then move the decimal point. Our answers in percent form will then be accurate to the nearest tenth of a percent, as in Example 26.

27. Change to a percent.

a. $3\frac{3}{4}$

b. $3\frac{7}{8}$

Because the 6s repeat indefinitely, we can use mixed number notation to write

$$\frac{5}{12} = 0.41\overline{6} = 41\frac{2}{3}\%$$

Or, rounding, we can write

$$\frac{5}{12} = 41.7\%$$ To the nearest tenth of a percent

EXAMPLE 27 Change $2\frac{1}{2}$ to a percent.

SOLUTION We first change to a decimal and then to a percent:

$$2\frac{1}{2} = 2.5$$

$$= 250\%$$

Table 1 lists some of the most commonly used fractions and decimals and their equivalent percents.

TABLE 1

Fraction	Decimal	Percent
$\frac{1}{2}$	0.5	50%
$\frac{1}{4}$	0.25	25%
$\frac{3}{4}$	0.75	75%
$\frac{1}{3}$	$0.\overline{3}$	$33\frac{1}{3}\%$
$\frac{2}{3}$	$0.\overline{6}$	$66\frac{2}{3}\%$
$\frac{1}{5}$	0.2	20%
$\frac{2}{5}$	0.4	40%
$\frac{3}{5}$	0.6	60%
$\frac{4}{5}$	0.8	80%

Getting Ready for Class

After reading through the preceding section, respond in your own words and in complete sentences.

1. What is the relationship between the word *percent* and the number *100*?
2. Explain in words how you would change 25% to a decimal.
3. Explain in words how you would change 25% to a fraction.
4. After reading this section you know that $\frac{1}{2}$, 0.5, and 50% are equivalent. Show mathematically why this is true.

Answers

26. a. $58\frac{1}{3}\% \approx 58.3\%$ b. $108\frac{1}{3}\% \approx 108.3\%$

27. a. 375% b. 387.5%

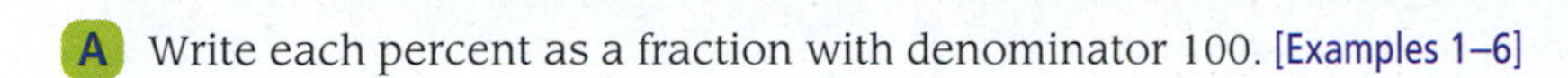

Problem Set 7.1

A Write each percent as a fraction with denominator 100. [Examples 1–6]

1. 20% **2.** 40% **3.** 60% **4.** 80% **5.** 24% **6.** 48%

7. 65% **8.** 35%

B Change each percent to a decimal. [Examples 7–14]

9. 23% **10.** 34% **11.** 92% **12.** 87% **13.** 9% **14.** 7%

15. 3.4% **16.** 5.8% **17.** 6.34% **18.** 7.25% **19.** 0.9% **20.** 0.6%

C Change each decimal to a percent. [Examples 15–20]

21. 0.23 **22.** 0.34 **23.** 0.92 **24.** 0.87 **25.** 0.45 **26.** 0.54

27. 0.03 **28.** 0.04 **29.** 0.6 **30.** 0.9 **31.** 0.8 **32.** 0.5

33. 0.27 **34.** 0.62 **35.** 1.23 **36.** 2.34

D Change each percent to a fraction in lowest terms. [Examples 21–23]

37. 60% **38.** 40% **39.** 75% **40.** 25% **41.** 4% **42.** 2%

43. 26.5% **44.** 34.2% **45.** 71.87% **46.** 63.6% **47.** 0.75% **48.** 0.45%

49. $6\frac{1}{4}\%$ **50.** $5\frac{1}{4}\%$ **51.** $33\frac{1}{3}\%$ **52.** $66\frac{2}{3}\%$

E Change each fraction or mixed number to a percent. [Examples 24–27]

53. $\frac{1}{2}$ **54.** $\frac{1}{4}$ **55.** $\frac{3}{4}$ **56.** $\frac{2}{3}$ **57.** $\frac{1}{3}$ **58.** $\frac{1}{5}$

59. $\frac{4}{5}$ **60.** $\frac{1}{6}$ **61.** $\frac{7}{8}$ **62.** $\frac{1}{8}$ **63.** $\frac{7}{50}$ **64.** $\frac{9}{25}$

65. $3\frac{1}{4}$ **66.** $2\frac{1}{8}$ **67.** $1\frac{1}{2}$ **68.** $1\frac{3}{4}$

69. $\frac{21}{43}$ to the nearest tenth of a percent

70. $\frac{36}{49}$ to the nearest tenth of a percent

Applying the Concepts

71. Mothers The chart shows the percentage of women who continue working after having a baby.

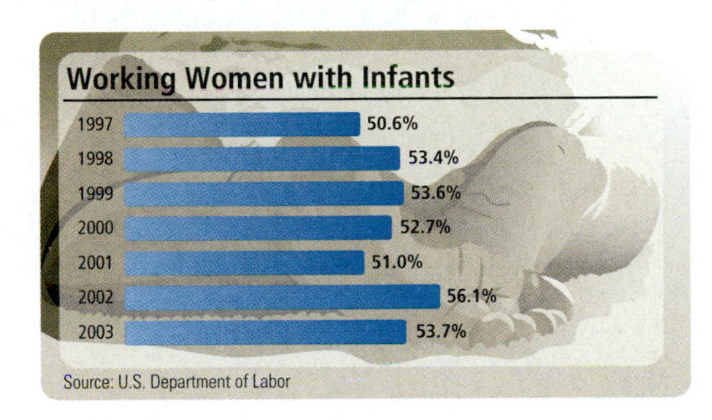

Using the chart, convert the percentage for the following years to a decimal.

a. 1997

b. 2000

c. 2003

72. U.S. Energy The pie chart shows where Americans get their energy.

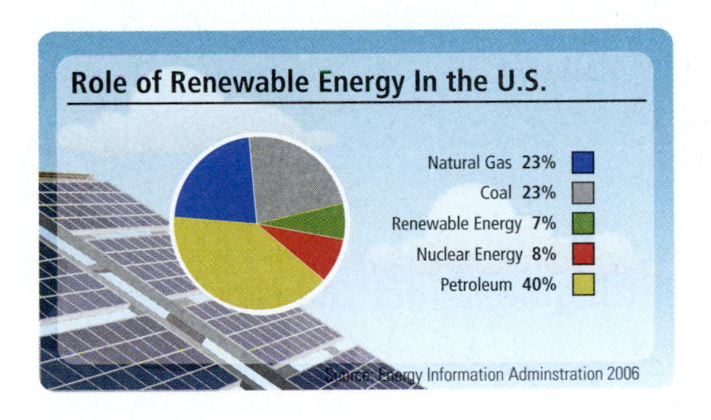

Using the chart, convert the percentage to a fraction for the following types of energy. Reduce to lowest terms.

a. Natural Gas

b. Nuclear Energy

c. Petroleum

73. Paying Bills According to Pew Research, a non-political organization that provides information on the issues, attitudes and trends shaping America, most people still pay their monthly bills by check.

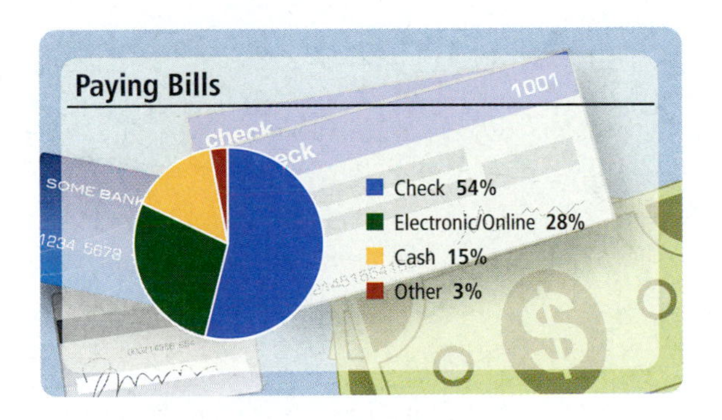

a. Convert each percent to a fraction.

b. Convert each percent to a decimal.

c. About how many times more likely are you to pay a bill with a check than by electronic or online methods?

74. Pizza Ingredients The pie chart below shows the decimal representation of each ingredient by weight that is used to make a sausage and mushroom pizza. We see that half of the pizza's weight comes from the crust. Change each decimal to a percent.

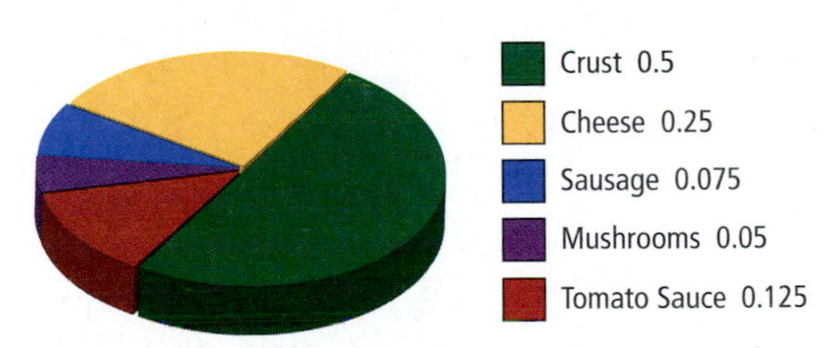

Calculator Problems

Use a calculator to write each fraction as a decimal, and then change the decimal to a percent. Round all answers to the nearest tenth of a percent.

75. $\frac{29}{37}$ **76.** $\frac{18}{83}$ **77.** $\frac{6}{51}$ **78.** $\frac{8}{95}$ **79.** $\frac{236}{327}$ **80.** $\frac{568}{732}$

Getting Ready for the Next Section

Multiply.

81. 0.25(74) **82.** 0.15(63) **83.** 0.435(25) **84.** 0.635(45)

Divide. Round the answers to the nearest thousandth, if necessary.

85. $\frac{21}{42}$ **86.** $\frac{21}{84}$ **87.** $\frac{25}{0.4}$ **88.** $\frac{31.9}{78}$

Solve for n.

89. $42n = 21$ **90.** $25 = 0.40n$

Maintaining Your Skills

Write as a decimal.

91. $\frac{1}{8}$ **92.** $\frac{3}{8}$ **93.** $\frac{5}{8}$ **94.** $\frac{7}{8}$

95. $\frac{1}{16}$ **96.** $\frac{3}{16}$ **97.** $\frac{5}{16}$ **98.** $\frac{7}{16}$

Divide.

99. $\frac{1}{8} \div \frac{1}{16}$ **100.** $\frac{3}{8} \div \frac{3}{16}$ **101.** $\frac{5}{8} \div \frac{5}{16}$ **102.** $\frac{7}{8} \div \frac{7}{16}$

103. $0.125 \div 0.0625$ **104.** $0.375 \div 0.1875$ **105.** $0.625 \div 0.3125$ **106.** $0.875 \div 0.4375$

7.2 Basic Percent Problems

Objectives

A Solve the three types of percent problems.

B Solve percent problems involving food labels.

C Solve percent problems using proportions.

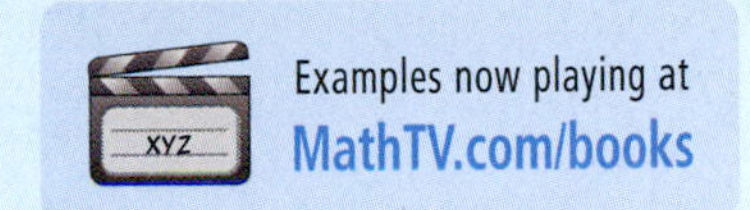

Introduction . . .

The American Dietetic Association (ADA) recommends eating foods in which the number of calories from fat is less than 30% of the total number of calories. Foods that satisfy this requirement are considered healthy foods. Is the nutrition label shown below from a food that the ADA would consider healthy? This is the type of question we will be able to answer after we have worked through the examples in this section.

Nutrition Facts

Serving Size 1/2 cup (65g)
Servings Per Container: 8

Amount Per Serving	
Calories 150	Calories from fat 90
	% Daily Value*
Total Fat 10g	**16%**
Saturated Fat 6g	**32%**
Cholesterol 35mg	**12%**
Sodium 30mg	**1%**
Total Carbohydrate 14g	**5%**
Dietary Fiber 0g	**0%**
Sugars 11g	
Protein 2g	
Vitamin A 6% •	Vitamin C 0%
Calcium 6% •	Iron 0%

*Percent Daily Values are based on a 2,000 calorie diet.

FIGURE 1 Nutrition label from vanilla ice cream

This section is concerned with three kinds of word problems that are associated with percents. Here is an example of each type:

Type A: What number is 15% of 63?
Type B: What percent of 42 is 21?
Type C: 25 is 40% of what number?

A Solving Percent Problems Using Equations

The first method we use to solve all three types of problems involves translating the sentences into equations and then solving the equations. The following translations are used to write the sentences as equations:

English	Mathematics
is	$=$
of	$\cdot$ (multiply)
a number	n
what number	n
what percent	n

The word *is* always translates to an $=$ sign. The word *of* almost always means multiply. The number we are looking for can be represented with a letter, such as n or x.

PRACTICE PROBLEMS

1. a. What number is 25% of 74?
   b. What number is 50% of 74?

2. a. What percent of 84 is 21?
   b. What percent of 84 is 42?

3. a. 35 is 40% of what number?
   b. 70 is 40% of what number?

Answers
1. a. 18.5 b. 37
2. a. 25% b. 50%
3. a. 87.5 b. 175

EXAMPLE 1

What number is 15% of 63?

SOLUTION We translate the sentence into an equation as follows:

What number is 15% *of* 63?
↓ ↓ ↓ ↓ ↓

$$n = 0.15 \cdot 63$$

To do arithmetic with percents, we have to change to decimals. That is why 15% is rewritten as 0.15. Solving the equation, we have

$$n = 0.15 \cdot 63$$
$$n = 9.45$$

15% of 63 is 9.45

EXAMPLE 2

What percent of 42 is 21?

SOLUTION We translate the sentence as follows:

What percent of 42 *is* 21?
↓ ↓ ↓ ↓ ↓

$$n \cdot 42 = 21$$

We solve for n by dividing both sides by 42.

$$\frac{n \cdot \cancel{42}}{\cancel{42}} = \frac{21}{42}$$
$$n = \frac{21}{42}$$
$$n = 0.50$$

Because the original problem asked for a percent, we change 0.50 to a percent:

$$n = 50\%$$

21 is 50% of 42

EXAMPLE 3

25 is 40% of what number?

SOLUTION Following the procedure from the first two examples, we have

25 *is* 40% *of what number*?
↓ ↓ ↓ ↓ ↓

$$25 = 0.40 \cdot n$$

Again, we changed 40% to 0.40 so we can do the arithmetic involved in the problem. Dividing both sides of the equation by 0.40, we have

$$\frac{25}{0.40} = \frac{\cancel{0.40} \cdot n}{\cancel{0.40}}$$
$$\frac{25}{0.40} = n$$
$$62.5 = n$$

25 is 40% of 62.5

As you can see, all three types of percent problems are solved in a similar manner. We write *is* as =, *of* as ·, and *what number* as n. The resulting equation is then solved to obtain the answer to the original question.

EXAMPLE 4 What number is 43.5% of 25?

$$n = 0.435 \cdot 25$$

$$n = 10.9 \qquad \text{Rounded to the nearest tenth}$$

10.9 is 43.5% of 25

EXAMPLE 5 What percent of 78 is 31.9?

$$n \cdot 78 = 31.9$$

$$\frac{n \cdot 78}{78} = \frac{31.9}{78}$$

$$n = \frac{31.9}{78}$$

$$n = 0.409 \qquad \text{Rounded to the nearest thousandth}$$

$$n = 40.9\%$$

40.9% of 78 is 31.9

EXAMPLE 6 34 is 29% of what number?

$$34 = 0.29 \cdot n$$

$$\frac{34}{0.29} = \frac{0.29 \cdot n}{0.29}$$

$$\frac{34}{0.29} = n$$

$$117.2 = n \qquad \text{Rounded to the nearest tenth}$$

34 is 29% of 117.2

B Food Labels

EXAMPLE 7 As we mentioned in the introduction to this section, the American Dietetic Association recommends eating foods in which the number of calories from fat is less than 30% of the total number of calories. According to the nutrition label below, what percent of the total number of calories is fat calories?

Nutrition Facts
Serving Size 1/2 cup (65g)
Servings Per Container: 8

Amount Per Serving	
Calories 150	Calories from fat 90
	% Daily Value*
Total Fat 10g	**16%**
Saturated Fat 6g	**32%**
Cholesterol 35mg	**12%**
Sodium 30mg	**1%**
Total Carbohydrate 14g	**5%**
Dietary Fiber 0g	**0%**
Sugars 11g	
Protein 2g	
Vitamin A 6% •	Vitamin C 0%
Calcium 6% •	Iron 0%

*Percent Daily Values are based on a 2,000 calorie diet.

FIGURE 2 Nutrition label from vanilla ice cream

---

4. What number is 63.5% of 45? (Round to the nearest tenth.)

5. What percent of 85 is 11.9?

6. 62 is 39% of what number? (Round to the nearest tenth.)

7. The nutrition label below is from a package of vanilla frozen yogurt. What percent of the total number of calories is fat calories? Round your answer to the nearest tenth of a percent.

Nutrition Facts
Serving Size 1/2 cup (98g)
Servings Per Container: 4

Amount Per Serving	
Calories 160	Calories from fat 25
	% Daily Value*
Total Fat 2.5g	**4%**
Saturated Fat 1.5g	**7%**
Cholesterol 45mg	**15%**
Sodium 55mg	**2%**
Total Carbohydrate 26g	**9%**
Dietary Fiber 0g	**0%**
Sugars 19g	
Protein 8g	
Vitamin A 0% •	Vitamin C 0%
Calcium 25% •	Iron 0%

*Percent Daily Values are based on a 2,000 calorie diet.

Answers
4. 28.6 **5.** 14% **6.** 159.0

SOLUTION To solve this problem, we must write the question in the form of one of the three basic percent problems shown in Examples 1–6. Because there are 90 calories from fat and a total of 150 calories, we can write the question this way: 90 is what percent of 150?

Now that we have written the question in the form of one of the basic percent problems, we simply translate it into an equation. Then we solve the equation.

$$90 \textit{ is what percent of } 150?$$

$$\downarrow \quad \downarrow \quad \downarrow$$

$$90 = n \cdot 150$$

$$\frac{90}{150} = n$$

$$n = 0.60 = 60\%$$

The number of calories from fat in this package of ice cream is 60% of the total number of calories. Thus the ADA would not consider this to be a healthy food.

C Solving Percent Problems Using Proportions

We can look at percent problems in terms of proportions also. For example, we know that 24% is the same as $\frac{24}{100}$, which reduces to $\frac{6}{25}$. That is

$$\underbrace{\frac{24}{100}}_{\text{24 is to 100}} \underset{\text{as}}{=} \underbrace{\frac{6}{25}}_{\text{6 is to 25}}$$

We can illustrate this visually with boxes of proportional lengths:

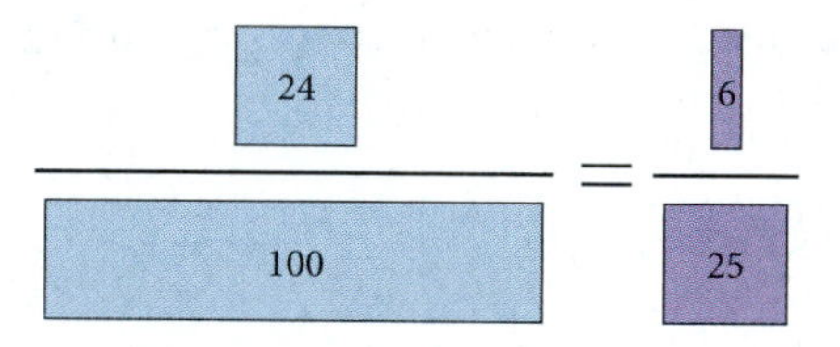

In general, we say

$$\underbrace{\frac{\text{Percent}}{100}}_{\text{Percent is to 100}} \underset{\text{as}}{=} \underbrace{\frac{\text{Amount}}{\text{Base}}}_{\text{Amount is to Base}}$$

Answer

7. 15.6% of the calories are from fat. (So far as fat content is concerned, the frozen yogurt is a healthier choice than the ice cream.)

EXAMPLE 8 What number is 15% of 63?

SOLUTION This is the same problem we worked in Example 1. We let n be the number in question. We reason that n will be smaller than 63 because it is only 15% of 63. The base is 63 and the amount is n. We compare n to 63 as we compare 15 to 100. Our proportion sets up as follows:

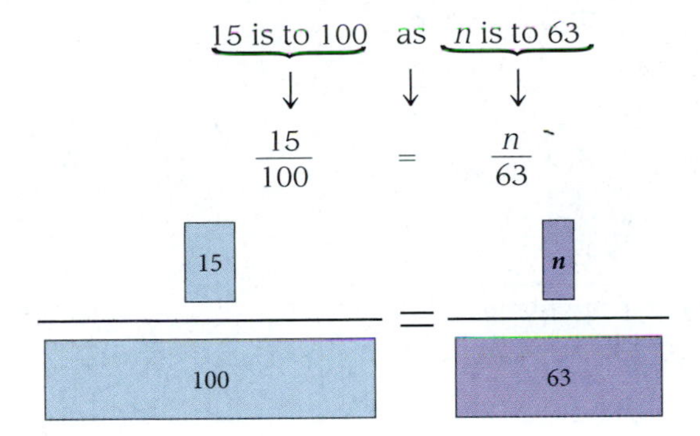

Solving the proportion, we have

$$15 \cdot 63 = 100n \quad \text{Extremes/means property}$$
$$945 = 100n \quad \text{Simplify the left side}$$
$$9.45 = n \quad \text{Divide each side by 100}$$

This gives us the same result we obtained in Example 1.

EXAMPLE 9 What percent of 42 is 21?

SOLUTION This is the same problem we worked in Example 2. We let n be the percent in question. The amount is 21 and the base is 42. We compare n to 100 as we compare 21 to 42. Here is our reasoning and proportion:

n is to 100 as 21 is to 42

$$\frac{n}{100} = \frac{21}{42}$$

Solving the proportion, we have

$$42n = 21 \cdot 100 \quad \text{Extremes/means property}$$
$$42n = 2{,}100 \quad \text{Simplify the right side}$$
$$n = 50 \quad \text{Divide each side by 42}$$

Since n is a percent, our answer is 50%, giving us the same result we obtained in Example 2.

8. Rework Practice Problem 1 using proportions.

9. Rework Practice Problem 2 using proportions.

Answers
8. a. 18.5 **b.** 37
9. a. 25% **b.** 50%

10. Rework Practice Problem 3 using proportions.

EXAMPLE 10 25 is 40% of what number?

SOLUTION This is the same problem we worked in Example 3. We let n be the number in question. The base is n and the amount is 25. We compare 25 to n as we compare 40 to 100. Our proportion sets up as follows:

$$\underbrace{40 \text{ is to } 100}\quad \text{as} \quad \underbrace{25 \text{ is to } n}$$

$$\frac{40}{100} = \frac{25}{n}$$

40 / 100 = 25 / n

Solving the proportion, we have

$40 \cdot n = 25 \cdot 100$ Extremes/means property

$40 \cdot n = 2{,}500$ Simplify the right side

$n = 62.5$ Divide each side by 40

So, 25 is 40% of 62.5, which is the same result we obtained in Example 3.

Note When you work the problems in the problem set, use whichever method you like, unless your instructor indicates that you are to use one method instead of the other.

Getting Ready for Class

After reading through the preceding section, respond in your own words and in complete sentences.

1. When we translate a sentence such as "What number is 15% of 63?" into symbols, what does each of the following translate to?

a. *is* b. *of* c. *what number*

2. Look at Example 1 in your text and answer the question below.

The number 9.45 is what percent of 63?

3. Show that the answer to the question below is the same as the answer to the question in Example 2 of your text.

The number 21 is what percent of 42?

4. If 21 is 50% of 42, then 21 is what percent of 84?

Answer

10. a. 87.5 **b.** 175

Problem Set 7.2

A C Solve each of the following problems. [Examples 1–6]

1. What number is 25% of 32?
2. What number is 10% of 80?
3. What number is 20% of 120?
4. What number is 15% of 75?
5. What number is 54% of 38?
6. What number is 72% of 200?
7. What number is 11% of 67?
8. What number is 2% of 49?
9. What percent of 24 is 12?
10. What percent of 80 is 20?
11. What percent of 50 is 5?
12. What percent of 20 is 4?
13. What percent of 36 is 9?
14. What percent of 70 is 14?
15. What percent of 8 is 6?
16. What percent of 16 is 9?
17. 32 is 50% of what number?
18. 16 is 20% of what number?
19. 10 is 20% of what number?
20. 11 is 25% of what number?
21. 37 is 4% of what number?
22. 90 is 80% of what number?
23. 8 is 2% of what number?
24. 6 is 3% of what number?

A C The following problems can be solved by the same method you used in Problems 1–24. [Examples 1–6]

25. What is 6.4% of 87?

26. What is 10% of 102?

27. 25% of what number is 30?

28. 10% of what number is 22?

29. 28% of 49 is what number?

30. 97% of 28 is what number?

31. 27 is 120% of what number?

32. 24 is 150% of what number?

33. 65 is what percent of 130?

34. 26 is what percent of 78?

35. What is 0.4% of 235,671?

36. What is 0.8% of 721,423?

37. 4.89% of 2,000 is what number?

38. 3.75% of 4,000 is what number?

39. Write a basic percent problem, the solution to which can be found by solving the equation $n = 0.25(350)$.

40. Write a basic percent problem, the solution to which can be found by solving the equation $n = 0.35(250)$.

41. Write a basic percent problem, the solution to which can be found by solving the equation $n \cdot 24 = 16$.

42. Write a basic percent problem, the solution to which can be found by solving the equation $n \cdot 16 = 24$.

43. Write a basic percent problem, the solution to which can be found by solving the equation $46 = 0.75 \cdot n$.

44. Write a basic percent problem, the solution to which can be found by solving the equation $75 = 0.46 \cdot n$.

B Applying the Concepts [Example 7]

Nutrition For each nutrition label in Problems 45–48, find what percent of the total number of calories comes from fat calories. Then indicate whether the label is from a food considered healthy by the American Dietetic Association. Round to the nearest tenth of a percent if necessary.

45. Spaghetti

Nutrition Facts	
Serving Size 2 oz. (56g per 1/8 of pkg) dry	
Servings Per Container: 8	
Amount Per Serving	
Calories 210	Calories from fat 10
	% Daily Value*
Total Fat 1g	**2%**
Saturated Fat 0g	**0%**
Polyunsaturated Fat 0.5g	
Monounsaturated Fat 0g	
Cholesterol 0mg	**0%**
Sodium 0mg	**0%**
Total Carbohydrate 42g	**14%**
Dietary Fiber 2g	**7%**
Sugars 3g	
Protein 7g	
Vitamin A 0% •	Vitamin C 0%
Calcium 0% •	Iron 10%
Thiamin 30% •	Riboflavin 10%
Niacin 15% •	
*Percent Daily Values are based on a 2,000 calorie diet	

46. Canned Italian tomatoes

Nutrition Facts	
Serving Size 1/2 cup (121g)	
Servings Per Container: about 3 1/2	
Amount Per Serving	
Calories 25	Calories from fat 0
	% Daily Value*
Total Fat 0g	**0%**
Saturated Fat 0g	**0%**
Cholesterol 0mg	**0%**
Sodium 300mg	**12%**
Potassium 145mg	**4%**
Total Carbohydrate 4g	**2%**
Dietary Fiber 1g	**4%**
Sugars 4g	
Protein 1g	
Vitamin A 20% •	Vitamin C 15%
Calcium 4% •	Iron 15%
*Percent Daily Values are based on a 2,000 calorie diet.	

47. Shredded Romano cheese

Nutrition Facts	
Serving Size 2 tsp (5g)	
Servings Per Container: 34	
Amount Per Serving	
Calories 20	Calories from fat 10
	% Daily Value*
Total Fat 1.5g	**2%**
Saturated Fat 1g	**5%**
Cholesterol 5mg	**2%**
Sodium 70mg	**3%**
Total Carbohydrate 0g	**0%**
Fiber 0g	**0%**
Sugars 0g	
Protein 2g	
Vitamin A 0% •	Vitamin C 0%
Calcium 4% •	Iron 0%
*Percent Daily Values are based on a 2,000 calorie diet.	

48. Tortilla chips

Nutrition Facts	
Serving Size 1 oz (28g/About 12 chips)	
Servings Per Container: about 2	
Amount Per Serving	
Calories 140	Calories from fat 60
	% Daily Value*
Total Fat 7g	**1%**
Saturated Fat 1g	**6%**
Cholesterol 0mg	**0%**
Sodium 170mg	**7%**
Total Carbohydrate 18g	**6%**
Dietary Fiber 1g	**4%**
Sugars less than 1g	
Protein 2g	
Vitamin A 0% •	Vitamin C 0%
Calcium 4% •	Iron 2%
*Percent Daily Values are based on a 2,000 calorie diet.	

Getting Ready for the Next Section

Solve each equation.

49. $96 = n \cdot 120$

50. $2{,}400 = 0.48 \cdot n$

51. $114 = 150n$

52. $3{,}360 = 0.42n$

53. What number is 80% of 60?

54. What number is 25% of 300?

Maintaining Your Skills

Multiply.

55. 2×0.125

56. 3×0.125

57. 4×0.125

58. 5×0.125

59. The sequence below is an arithmetic sequence in which each term is found by adding $\frac{1}{8}$ to the previous term. Find the next three numbers in the sequence.

$$\frac{1}{4}, \frac{3}{8}, \frac{1}{2}, \ldots$$

60. The sequence below is an arithmetic sequence in which each term is found by adding $\frac{1}{16}$ to the previous term. Find the next three numbers in the sequence.

$$\frac{1}{8}, \frac{3}{16}, \frac{1}{4}, \ldots$$

Simplify.

61. $\frac{1}{4} - \frac{1}{8} + \frac{1}{2} - \frac{3}{8}$

62. $\frac{7}{8} - \frac{3}{4} + \frac{5}{8} - \frac{1}{2}$

Write as a decimal.

63. $\frac{2}{8}$

64. $\frac{4}{8}$

65. $\frac{6}{8}$

66. $\frac{8}{8}$

67. $\frac{2}{16}$

68. $\frac{4}{16}$

69. $\frac{6}{16}$

70. $\frac{8}{16}$

Write in order from smallest to largest.

71. $\frac{3}{8}, \frac{1}{4}, \frac{5}{8}, \frac{1}{8}, \frac{1}{2}, \frac{3}{4}, \frac{7}{8}$

72. $\frac{3}{16}, \frac{1}{8}, \frac{1}{4}, \frac{3}{8}, \frac{7}{16}, \frac{1}{16}, \frac{1}{2}, \frac{5}{16}$

General Applications of Percent

7.3

Objectives

A Solve application problems involving percent.

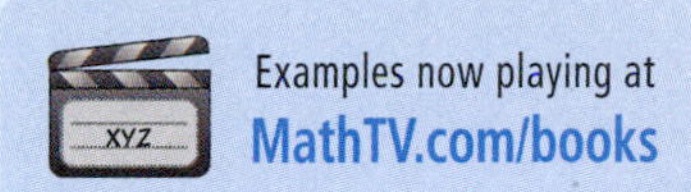

Introduction . . .

As you know from watching television and reading the newspaper, we encounter percents in many situations in everyday life. A recent newspaper article discussing the effects of a cholesterol-lowering drug stated that the drug in question "lowered levels of LDL cholesterol by an average of 35%." As we progress through this chapter, we will become more and more familiar with percent. As a result, we will be better equipped to understand statements like the one above concerning cholesterol.

In this section we continue our study of percent by doing more of the translations that were introduced in Section 7.2. The better you are at working the problems in Section 7.2, the easier it will be for you to get started on the problems in this section.

A Applications Involving Percent

EXAMPLE 1 On a 120-question test, a student answered 96 correctly. What percent of the problems did the student work correctly?

SOLUTION We have 96 correct answers out of a possible 120. The problem can be restated as

96 *is what percent of* 120?

$$96 = n \cdot 120$$

$$\frac{96}{120} = \frac{n \cdot \cancel{120}}{\cancel{120}} \quad \text{Divide both sides by 120}$$

$$n = \frac{96}{120} \quad \text{Switch the left and right sides of the equation}$$

$$n = 0.80 \quad \text{Divide 96 by 120}$$

$$= 80\% \quad \text{Rewrite as a percent}$$

When we write a test score as a percent, we are comparing the original score to an equivalent score on a 100-question test. That is, 96 correct out of 120 is the same as 80 correct out of 100.

EXAMPLE 2 How much HCl (hydrochloric acid) is in a 60-milliliter bottle that is marked 80% HCl?

SOLUTION If the bottle is marked 80% HCl, that means 80% of the solution is HCl and the rest is water. Because the bottle contains 60 milliliters, we can restate the question as:

What is 80% *of* 60?

$$n = 0.80 \cdot 60$$

$$n = 48$$

HCL 80%
60 ml

There are 48 milliliters of HCl in 60 milliliters of 80% HCl solution.

PRACTICE PROBLEMS

1. On a 150-question test, a student answered 114 correctly. What percent of the problems did the student work correctly?

2. How much HCl is in a 40-milliliter bottle that is marked 75% HCl?

Answers

1. 76% 2. 30 milliliters

3. If 42% of the students in a certain college are female and there are 3,360 female students, what is the total number of students in the college?

EXAMPLE 3 If 48% of the students in a certain college are female and there are 2,400 female students, what is the total number of students in the college?

SOLUTION We restate the problem as:

2,400 *is* 48% *of what number*?

$$2{,}400 = 0.48 \cdot n$$

$$\frac{2{,}400}{0.48} = \frac{0.48 \cdot n}{0.48} \quad \text{Divide both sides by 0.48}$$

$$n = \frac{2{,}400}{0.48} \quad \text{Switch the left and right sides of the equation}$$

$$n = 5{,}000$$

There are 5,000 students.

4. Suppose in Example 4 that 35% of the students receive a grade of A. How many of the 300 students is that?

EXAMPLE 4 If 25% of the students in elementary algebra courses receive a grade of A, and there are 300 students enrolled in elementary algebra this year, how many students will receive As?

SOLUTION After reading the question a few times, we find that it is the same as this question:

What number is 25% *of* 300?

$$n = 0.25 \cdot 300$$

$$n = 75$$

Thus, 75 students will receive A's in elementary algebra.

Almost all application problems involving percents can be restated as one of the three basic percent problems we listed in Section 7.2. It takes some practice before the restating of application problems becomes automatic. You may have to review Section 7.2 and Examples 1–4 above several times before you can translate word problems into mathematical expressions yourself.

Getting Ready for Class

After reading through the preceding section, respond in your own words and in complete sentences.

1. On the test mentioned in Example 1, how many questions would the student have answered correctly if she answered 40% of the questions correctly?
2. If the bottle in Example 2 contained 30 milliliters instead of 60, what would the answer be?
3. In Example 3, how many of the students were male?
4. How many of the students mentioned in Example 4 received a grade lower than A?

Answers

3. 8,000 students 4. 105 students

Problem Set 7.3

A Solve each of the following problems by first restating it as one of the three basic percent problems of Section 7.2. In each case, be sure to show the equation. [Examples 1–4]

1. **Test Scores** On a 120-question test a student answered 84 correctly. What percent of the problems did the student work correctly?

2. **Test Scores** An engineering student answered 81 questions correctly on a 90-question trigonometry test. What percent of the questions did she answer correctly? What percent were answered incorrectly?

3. **Basketball** A basketball player made 63 out of 75 free throws. What percent is this?

4. **Family Budget** A family spends $450 every month on food. If the family's income each month is $1,800, what percent of the family's income is spent on food?

5. **Chemistry** How much HCl (hydrochloric acid) is in a 60-milliliter bottle that is marked 75% HCl?

6. **Chemistry** How much acetic acid is in a 5-liter container of acetic acid and water that is marked 80% acetic acid? How much is water?

7. **Farming** A farmer owns 28 acres of land. Of the 28 acres, only 65% can be farmed. How many acres are available for farming? How many are not available for farming?

8. **Number of Students** Of the 420 students enrolled in a basic math class, only 30% are first-year students. How many are first-year students? How many are not?

9. **Determining a Tip** Servers and wait staff are often paid minimum wage and depend on tips for much of their income. It is common for tips to be 15% to 20% of the bill. After dinner at a local restaurant the total bill is $56.00. Since your service was above average you decide to give a 20% tip. Determine the amount of the tip you leave for your server.

10. **Determining a Tip** Suppose you decide to leave a 15% tip for services after your dinner out in the preceding problem. How much of a tip did you leave your server? How much smaller was the tip?

11. **Voting** In the 2004 Presidential election, George Bush received 53.25% of the total electoral votes and John Kerry received 46.75% of the total electoral votes. If there were 537 total votes cast by the Electoral College how many electoral votes did each candidate receive?

12. **Census Data** According to the U.S. Census Bureau, national population estimates grouped by age and gender for July, 2006, approximately 7.4% of the 147,512,152 males in our population are between the ages of 15 and 19 years old. How many males are in this age group?

13. **Bachelors** According to the U.S. Census Bureau data for the number of marriages in 2004 approximately 31.2% of the 109,830,000 males age 15 years or older have never been married. How many males age 15 years or older have never been married?

14. **Bachelorettes** According to the U.S. Census Bureau data for the number of marriages in 2004, approximately 25.8 % of the 117,677,000 females age 15 years or older have never been married. How many females age 15 years or older have never been married?

15. **Number of Students** If 48% of the students in a certain college are female and there are 1,440 female students, what is the total number of students in the college?

Tom Stewart/Corbis

16. **Mixture Problem** A solution of alcohol and water is 80% alcohol. The solution is found to contain 32 milliliters of alcohol. How many milliliters total (both alcohol and water) are in the solution?

17. **Number of Graduates** Suppose 60% of the graduating class in a certain high school goes on to college. If 240 students from this graduating class are going on to college, how many students are there in the graduating class?

18. **Defective Parts** In a shipment of airplane parts, 3% are known to be defective. If 15 parts are found to be defective, how many parts are in the shipment?

19. **Number of Students** There are 3,200 students at our school. If 52% of them are female, how many female students are there at our school?

20. **Number of Students** In a certain school, 75% of the students in first-year chemistry have had algebra. If there are 300 students in first-year chemistry, how many of them have had algebra?

21. **Population** In a city of 32,000 people, there are 10,000 people under 25 years of age. What percent of the population is under 25 years of age?

22. **Number of Students** If 45 people enrolled in a psychology course but only 35 completed it, what percent of the students completed the course? (Round to the nearest tenth of a percent.)

Calculator Problems

The following problems are similar to Problems 1–22. They should be set up the same way. Then the actual calculations should be done on a calculator.

23. Number of People Of 7,892 people attending an outdoor concert in Los Angeles, 3,972 are over 18 years of age. What percent is this? (Round to the nearest whole-number percent.)

24. Manufacturing A car manufacturer estimates that 25% of the new cars sold in one city have defective engine mounts. If 2,136 new cars are sold in that city, how many will have defective engine mounts?

25. Population The map shows the most populated cities in the United States. If the population of New York City is about 42% of the state's population, what is the approximate population of the state?

Where Is Everyone?

City	Population (millions)
Los Angeles, CA	3.80
San Diego, CA	1.26
Phoenix, AZ	1.37
Dallas, TX	1.21
Houston, TX	2.01
Chicago, IL	2.89
Philadelphia, PA	1.49
New York City, NY	8.08

Source: U.S. Census Bureau

26. Prom The graph shows how much girls plan to spend on the prom. If 5,086 girls were surveyed, how many are planning on spending less than $200 on the prom? Round to the nearest whole number.

Getting Ready for the Next Section

Multiply.

27. 0.06(550)

28. 0.06(625)

29. 0.03(289,500)

30. 0.03(115,900)

Divide. Write your answers as decimals.

31. $5.44 \div 0.04$

32. $4.35 \div 0.03$

33. $19.80 \div 396$

34. $11.82 \div 197$

35. $\frac{1,836}{0.12}$

36. $\frac{115}{0.1}$

37. $\frac{90}{600}$

38. $\frac{105}{750}$

Maintaining Your Skills

The problems below review multiplication with fractions and mixed numbers.

Multiply.

39. $\frac{1}{2} \cdot \frac{2}{5}$

40. $\frac{3}{4} \cdot \frac{1}{3}$

41. $\frac{3}{4} \cdot \frac{5}{9}$

42. $\frac{5}{6} \cdot \frac{12}{13}$

43. $2 \cdot \frac{3}{8}$

44. $3 \cdot \frac{5}{12}$

45. $1\frac{1}{4} \cdot \frac{8}{15}$

46. $2\frac{1}{3} \cdot \frac{9}{10}$

Extending the Concepts: Batting Averages

Batting averages in baseball are given as decimal numbers, rounded to the nearest thousandth. For example, at the end of June 2008, Milton Bradley had the highest batting average in the American League. At that time, he had 76 hits in 235 times at bat. His batting average was .323, which is found by dividing the number of hits by the number of times he was at bat and then rounding to the nearest thousandth.

$$\text{Batting average} = \frac{\text{number of hits}}{\text{number of times at bat}} = \frac{76}{235} = 0.323$$

Because we can write any decimal number as a percent, we can convert batting averages to percents and use our knowledge of percent to solve problems. Looking at Milton Bradley's batting average as a percent, we can say that he will get a hit 32.3% of the times he is at bat.

Each of the following problems can be solved by converting batting averages to percents and translating the problem into one of our three basic percent problems. (All numbers are from the end of June 2008.)

47. Chipper Jones had the highest batting average in the National League with 100 hits in 254 times at bat. What percent of the time Chipper Jones is at bat can we expect him to get a hit?

48. Sammy Sosa had 104 hits in 412 times at bat. What percent of the time can we expect Sosa to get a hit?

49. Barry Bonds was batting .276. If he had been at bat 340 times, how many hits did he have? (Remember his batting average has been rounded to the nearest thousandth.)

Peter DeSilva/Corbis Sygma

50. Joe Mauer was batting .321. If he had been at bat 265 times, how many hits did he have? (Remember, his batting average has been rounded to the nearest thousandth.)

51. How many hits must Milton Bradley have in his next 50 times at bat to maintain a batting average of at least .323?

52. How many hits must Chipper Jones have in his next 50 times at bat to maintain a batting average of at least .394?

Sales Tax and Commission

7.4

Objectives

A Solve application problems involving sales tax.

B Solve application problems involving commission.

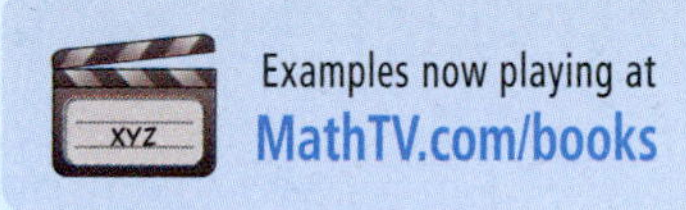

To solve the problems in this section, we will first restate them in terms of the problems we have already learned how to solve.

A Sales Tax

EXAMPLE 1 Suppose the sales tax rate in Mississippi is 6% of the purchase price. If the price of a refrigerator is \$550, how much sales tax must be paid?

SOLUTION Because the sales tax is 6% of the purchase price, and the purchase price is \$550, the problem can be restated as:

What is 6% of \$550?

We solve this problem, as we did in Section 7.2, by translating it into an equation:

What is 6% *of* \$550?

$$n = 0.06 \cdot 550$$

$$n = 33$$

The sales tax is \$33. The total price of the refrigerator would be

Purchase price		*Sales tax*		*Total price*
\$550	+	\$33	=	\$583

EXAMPLE 2 Suppose the sales tax rate is 4%. If the sales tax on a 10-speed bicycle is \$5.44, what is the purchase price, and what is the total price of the bicycle?

SOLUTION We know that 4% of the purchase price is \$5.44. We find the purchase price first by restating the problem as:

\$5.44 *is* 4% *of what number*?

$$5.44 = 0.04 \cdot n$$

We solve the equation by dividing both sides by 0.04:

$$\frac{5.44}{0.04} = \frac{\cancel{0.04} \cdot n}{\cancel{0.04}} \quad \text{Divide both sides by 0.04}$$

$$n = \frac{5.44}{0.04} \quad \text{Switch the left and right sides of the equation}$$

$$n = 136 \quad \text{Divide}$$

The purchase price is \$136. The total price is the sum of the purchase price and the sales tax.

Purchase price	=	\$136.00
Sales tax	=	+ 5.44
Total price	=	\$141.44

PRACTICE PROBLEMS

1. What is the sales tax on a new washing machine if the machine is purchased for \$625 and the sales tax rate is 6%?

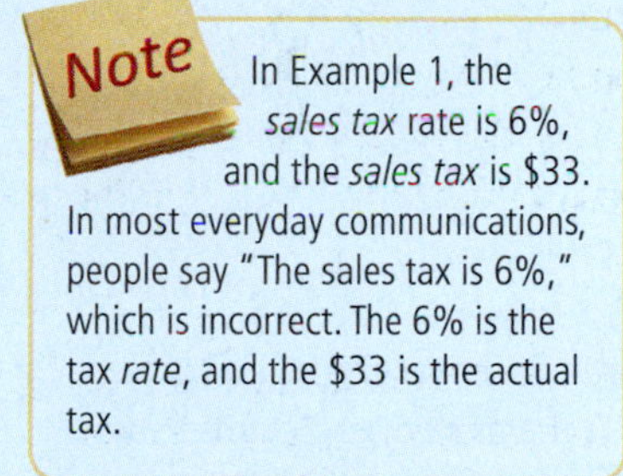

In Example 1, the *sales tax* rate is 6%, and the *sales tax* is \$33. In most everyday communications, people say "The sales tax is 6%," which is incorrect. The 6% is the tax *rate*, and the \$33 is the actual tax.

2. Suppose the sales tax rate is 3%. If the sales tax on a 10-speed bicycle is \$4.35, what is the purchase price, and what is the total price of the bicycle?

Answers
1. \$37.50 **2.** \$145; \$149.35

3. Suppose the purchase price of two speakers is $197 and the sales tax is $11.82. What is the sales tax rate?

EXAMPLE 3 Suppose the purchase price of a stereo system is $396 and the sales tax is $19.80. What is the sales tax rate?

SOLUTION We restate the problem as:

$19.80 *is what percent of* $396?

$$19.80 = n \cdot 396$$

To solve this equation, we divide both sides by 396:

$$\frac{19.80}{396} = \frac{n \cdot \cancel{396}}{\cancel{396}} \quad \text{Divide both sides by 396}$$

$$n = \frac{19.80}{396} \quad \text{Switch the left and right sides of the equation}$$

$$n = 0.05 \quad \text{Divide}$$

$$n = 5\% \quad 0.05 = 5\%$$

The sales tax rate is 5%.

B Commission

Many salespeople work on a *commission* basis. That is, their earnings are a percentage of the amount they sell. The *commission rate* is a percent, and the actual commission they receive is a dollar amount.

4. A real estate agent gets 3% of the price of each house she sells. If she sells a house for $115,000, how much money does she earn?

EXAMPLE 4 A real estate agent gets 3% of the price of each house she sells. If she sells a house for $289,500, how much money does she earn?

SOLUTION The commission is 3% of the price of the house, which is $289,500. We restate the problem as:

What is 3% *of* $289,500?

$$n = 0.03 \cdot 289{,}500$$

$$n = 8{,}685$$

The commission is $8,685.

5. An appliance salesperson's commission rate is 10%. If the commission on one of the ovens is $115, what is the purchase price of the oven?

EXAMPLE 5 Suppose a car salesperson's commission rate is 12%. If the commission on one of the cars is $1,836, what is the purchase price of the car?

SOLUTION 12% of the sales price is $1,836. The problem can be restated as:

12% *of what number is* $1,836?

$$0.12 \cdot n = 1{,}836$$

$$\frac{\cancel{0.12} \cdot n}{\cancel{0.12}} = \frac{1{,}836}{0.12} \quad \text{Divide both sides by 0.12}$$

$$n = 15{,}300$$

The car sells for $15,300.

Answers
3. 6% 4. $3,450 5. $1,150

EXAMPLE 6 If the commission on a \$600 dining room set is \$90, what is the commission rate?

SOLUTION The commission rate is a percentage of the selling price. What we want to know is:

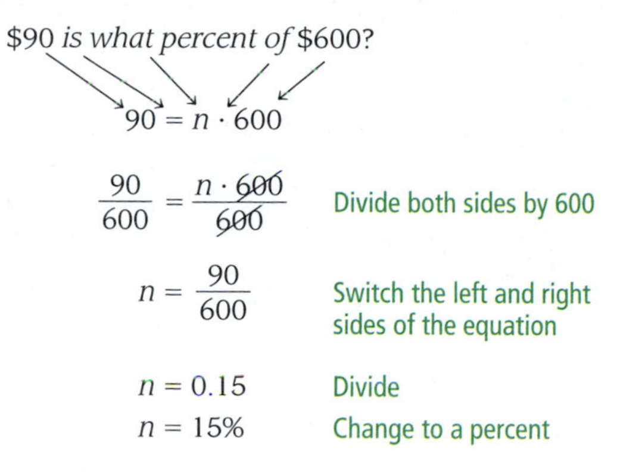

$$\frac{90}{600} = \frac{n \cdot \cancel{600}}{\cancel{600}} \quad \text{Divide both sides by 600}$$

$$n = \frac{90}{600} \quad \text{Switch the left and right sides of the equation}$$

$$n = 0.15 \quad \text{Divide}$$

$$n = 15\% \quad \text{Change to a percent}$$

The commission rate is 15%.

Getting Ready for Class

After reading through the preceding section, respond in your own words and in complete sentences.

1. Explain the difference between the sales tax and the sales tax rate.
2. Rework Example 1 using a sales tax rate of 7% instead of 6%.
3. Suppose the bicycle in Example 2 was purchased in California, where the sales tax rate in 2008 was 7.25%. How much more would the bicycle have cost?
4. Suppose the car salesperson in Example 5 receives a commission of \$3,672. Assuming the same commission rate of 12%, how much does this car sell for?

6. If the commission on a \$750 sofa is \$105, what is the commission rate?

Answer

6. 14%

Problem Set 7.4

A These problems should be solved by the method shown in this section. In each case show the equation needed to solve the problem. Write neatly, and show your work. [Examples 1–3]

1. **Sales Tax** Suppose the sales tax rate in Mississippi is 7% of the purchase price. If a new food processor sells for $750, how much is the sales tax?

2. **Sales Tax** If the sales tax rate is 5% of the purchase price, how much sales tax is paid on a television that sells for $980?

3. **Sales Tax and Purchase Price** Suppose the sales tax rate in Michigan is 6%. How much is the sales tax on a $45 concert ticket? What is the total price?

4. **Sales Tax and Purchase Price** Suppose the sales tax rate in Hawaii is 4%. How much tax is charged on a new car if the purchase price is $16,400? What is the total price?

5. **Total Price** The sales tax rate is 4%. If the sales tax on a 10-speed bicycle is $6, what is the purchase price? What is the total price?

6. **Total Price** The sales tax on a new microwave oven is $30. If the sales tax rate is 5%, what is the purchase price? What is the total price?

7. **Tax Rate** Suppose the purchase price of a dining room set is $450. If the sales tax is $22.50, what is the sales tax rate?

FURNITURE PLUS

RECEIPT

Dining Room Set	$450.00
Tax Rate	? %
Tax	$22.50
TOTAL	$472.50

Receipt #1007 07/18/08 4:15PM

8. **Tax Rate** If the purchase price of a bottle of California wine is $24 and the sales tax is $1.50, what is the sales tax rate?

OAKS WINERY

RECEIPT

California wine	$24.00
Tax Rate	? %
Tax	$1.50
TOTAL	$25.50

Receipt #128 07/30/08 2:32PM

9. **Energy** The chart shows the cost to install either solar panels or a wind turbine. A farmer is installing the equipment to generate energy from the wind. If he lives in a state that has a 6% sales tax rate, how much did the farmer pay in sales tax on the total equipment cost?

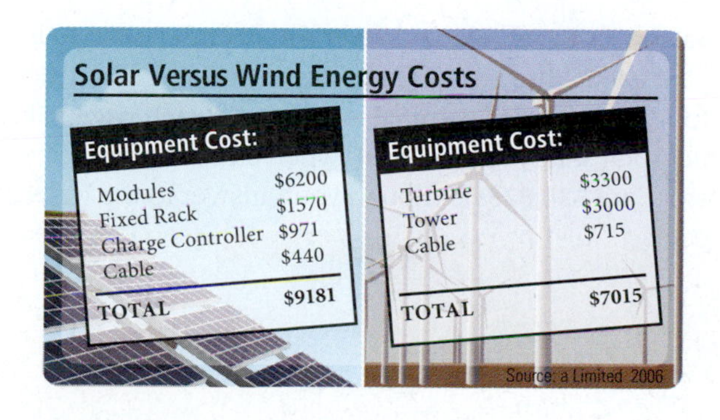

10. **Prom** The graph shows how much guys plan to spend on prom. The sum of the tax on all the expenses a guy had for prom was $15.75. If he lived in a state that has a sales tax rate of 7.5%, what spending bracket would he have been in?

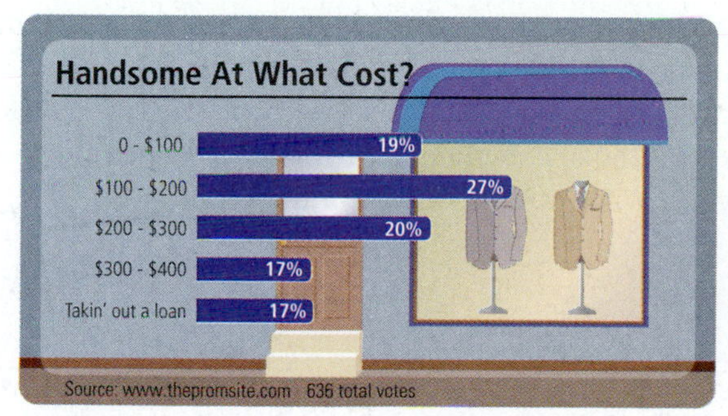

B [Examples 4–6]

11. **Commission** A real estate agent has a commission rate of 3%. If a piece of property sells for $94,000, what is her commission?

12. **Commission** A tire salesperson has a 12% commission rate. If he sells a set of radial tires for $400, what is his commission?

13. **Commission and Purchase Price** Suppose a salesperson gets a commission rate of 12% on the lawnmowers she sells. If the commission on one of the mowers is $24, what is the purchase price of the lawnmower?

14. **Commission and Purchase Price** If an appliance salesperson gets 9% commission on all the appliances she sells, what is the price of a refrigerator if her commission is $67.50?

15. **Commission Rate** If the commission on an $800 washer is $112, what is the commission rate?

16. **Commission Rate** A realtor makes a commission of $11,400 on a $190,000 house he sells. What is his commission rate?

17. **Phone Bill** You recently received your monthly phone bill for service in your local area. The total of the bill was $53.35. You pay $14.36 in surcharges and federal and local taxes. What percent of your phone bill is made up of surcharges and taxes? Round your answer to the nearest tenth of a percent.

18. **Wireless Phone Bill** You recently received your Verizon wireless phone bill for the month. The total monthly bill is $70.52. Included in that total is $13.27 in surcharges and taxes. What percent of your wireless bill goes towards surcharges and taxes? Round your answer to the nearest tenth of a percent.

19. **Gasoline Tax** New York state has one of the highest gasoline taxes in the country. If gas is currently selling at $4.27 for a gallon of regular gas and the tax rate is 14.7%, how much of the price of a gallon of gas goes towards taxes?

20. **Cigarette Tax** In an effort to encourage people to quit smoking, many states place a high tax on a pack of cigarettes. Nine states place a tax of $2.00 or more on a pack of cigarettes, with New Jersey being the highest at $2.575 per pack. If this is 39% of the cost of a pack of cigarettes in New Jersey, how much does a single pack cost?

21. **Salary Plus Commission** A computer salesperson earns a salary of $425 a week and a 6% commission on all sales over $4000 each week. Suppose she was able to sell $6,250 in computer parts and accessories one week. What was her salary for the week?

22. **Salary Plus Bonus** The manager for a computer store is paid a weekly salary of $650 plus a bonus amounting to 1.5% of the net earnings of the store each week. Find her total salary for the week when earnings for the store are $26,875.56. Round your answer to the nearest cent.

Calculator Problems

The following problems are similar to Problems 1–22. Set them up in the same way, but use a calculator for the calculations.

23. **Sales Tax** The sales tax rate on a certain item is 5.5%. If the purchase price is $216.95, how much is the sales tax? (Round to the nearest cent.)

24. **Purchase Price** If the sales tax rate is 4.75% and the sales tax is $18.95, what is the purchase price? What is the total price? (Both answers should be rounded to the nearest cent.)

25. **Tax Rate** The purchase price for a new suit is $229.50. If the sales tax is $10.33, what is the sales tax rate? (Round to the nearest tenth of a percent.)

26. **Commission** If the commission rate for a mobile home salesperson is 11%, what is the commission on the sale of a $15,794 mobile home?

27. **Selling Price** Suppose the commission rate on the sale of used cars is 13%. If the commission on one of the cars is $519.35, what did the car sell for?

28. **Commission Rate** If the commission on the sale of $79.40 worth of clothes is $14.29, what is the commission rate? (Round to the nearest percent.)

Getting Ready for the Next Section

Multiply.

29. 0.05(22,000)

30. 0.176(1,793,000)

31. 0.25(300)

32. 0.12(450)

Divide. Write your answers as decimals.

33. $4 \div 25$

34. $7 \div 35$

Subtract.

35. $25 - 21$

36. $1{,}793{,}000 - 315{,}568$

37. $450 - 54$

38. $300 - 75$

Add.

39. $396 + 19.8$

40. $22{,}000 + 1{,}100$

Maintaining Your Skills

The problems below review some basic concepts of division with fractions and mixed numbers.

Divide.

41. $\frac{1}{3} \div \frac{2}{3}$

42. $\frac{2}{3} \div \frac{1}{3}$

43. $2 \div \frac{3}{4}$

44. $3 \div \frac{1}{2}$

45. $\frac{3}{8} \div \frac{1}{4}$

46. $\frac{5}{9} \div \frac{2}{3}$

47. $2\frac{1}{4} \div \frac{1}{2}$

48. $1\frac{1}{4} \div 2\frac{1}{2}$

Extending the Concepts: Luxury Taxes

In 1990, Congress passed a law, which took effect on January 1, 1991, requiring an additional tax of 10% on a portion of the purchase price of certain luxury items. (In 1996 the law was amended so that this tax on luxury automobiles expired in 2003.) For expensive cars, it was paid on the part of the purchase price that exceeded $30,000. For example, if you purchased a Jaguar XJ-S for $53,000, you would pay a luxury tax of 10% of $23,000, because the purchase price, $53,000, is $23,000 above $30,000.

49. If you purchased a Jaguar XJ-S for $53,000 on February 1, 1991, in California, where the sales tax rate was 6%, how much would you pay in luxury tax and how much would you pay in sales tax?

50. If you purchased a Mercedes 300E for $43,500 on January 20, 1991, in California, where the sales tax rate was 6%, how much more would you pay in sales tax than luxury tax?

51. How much would you have saved if you had purchased the Jaguar mentioned in Problem 49 on December 31, 1990?

52. How much would you have saved if you bought a car with a purchase price of $45,000 on December 31, 1990, instead of January 1, 1991?

53. Suppose you bought a car in 1991. How much did you save on a car with a sticker price of $31,500, if you persuaded the car dealer to reduce the price to $29,900?

54. Suppose you bought a car in 1991. One of the cars you were interested in had a sticker price of $35,500, while another had a sticker price of $28,500. If you expected to pay full price for either car, how much did you save if you bought the less expensive car?

Percent Increase or Decrease and Discount

7.5

Objectives

A Find the percent increase.
B Find the percent decrease.
C Solve application problems involving the rate of discount.

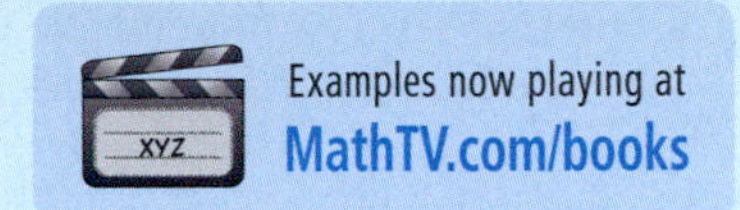

The table and bar chart below show some statistics compiled by insurance companies regarding stopping distances for automobiles traveling at 20 miles per hour on ice.

	Stopping Distance	Percent Decrease
Regular tires	150 ft	0
Snow tires	151 ft	−1%
Studded snow tires	120 ft	20%
Reinforced tire chains	75 ft	50%

Source: Copyrighted table courtesy of *The Casualty Adjuster's Guide*

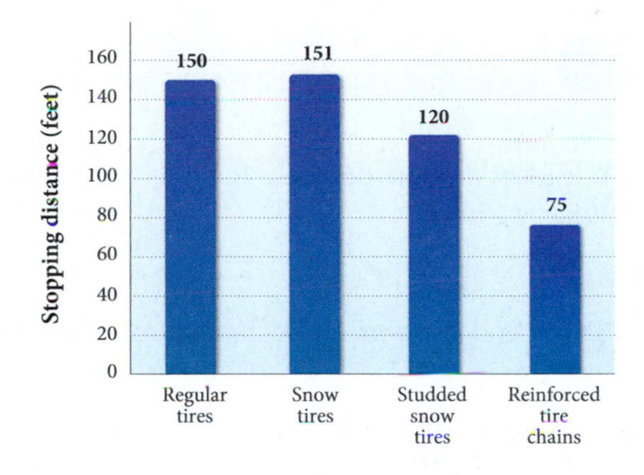

Many times it is more effective to state increases or decreases as percents, rather than the actual number, because with percent we are comparing everything to 100.

A Percent Increase

EXAMPLE 1 If a person earns \$22,000 a year and gets a 5% increase in salary, what is the new salary?

SOLUTION We can find the dollar amount of the salary increase by finding 5% of \$22,000:

$$0.05 \times 22{,}000 = 1{,}100$$

The increase in salary is \$1,100. The new salary is the old salary plus the raise:

	\$22,000	Old salary
+	1,100	Raise (5% of \$22,000)
	\$23,100	New salary

PRACTICE PROBLEMS

1. A person earning \$18,000 a year gets a 7% increase in salary. What is the new salary?

Answer
1. \$19,260

2. In 1986, there were approximately 271,000 drunk drivers under correctional supervision (prison, jail, or probation). By 1997, that number had increased 89%. How many drunk drivers were under correctional supervision in 1997? Round to the nearest thousand.

3. Shoes that usually sell for $35 are on sale for $28. What is the percent decrease in price?

4. During a sale, a microwave oven that usually sells for $550 is marked "15% off." What is the discount? What is the sale price?

Answers
2. 512,000 **3.** 20%

B Percent Decrease

EXAMPLE 2 In 1986, there were approximately 1,793,000 arrests for driving under the influence of alcohol or drugs (DUI) in the United States. By 1997, the number of arrests for DUI had decreased 17.6% from the 1986 number. How many people were arrested for DUI in 1997? Round the answer to the nearest thousand.

SOLUTION The decrease in the number of arrests is 17.6% of 1,793,000, or

$$0.176 \times 1{,}793{,}000 = 315{,}568$$

Subtracting this number from 1,793,000, we have the number of DUI arrests in 1997.

1,793,000	Number of arrests in 1986
− 315,568	Decrease of 17.6%
1,477,432	Number of arrests in 1997

To the nearest thousand, there were approximately 1,477,000 arrests for DUI in 1997.

EXAMPLE 3 Shoes that usually sell for $25 are on sale for $21. What is the percent decrease in price?

SOLUTION We must first find the decrease in price. Subtracting the sale price from the original price, we have

$$\$25 - \$21 = \$4$$

The decrease is $4. To find the percent decrease (from the original price), we have

$4 *is what percent of* $25?

$$4 = n \cdot 25$$

$$\frac{4}{25} = \frac{n \cdot \cancel{25}}{\cancel{25}} \quad \text{Divide both sides by 25}$$

$$n = \frac{4}{25} \quad \text{Switch the left and right sides of the equation}$$

$$n = 0.16 \quad \text{Divide}$$

$$n = 16\% \quad \text{Change to a percent}$$

The shoes that sold for $25 have been reduced by 16% to $21. In a problem like this, $25 is the *original* (or *marked*) price, $21 is the *sale price*, $4 is the *discount*, and 16% is the *rate of discount.*

C Discount Rate

EXAMPLE 4 During a clearance sale, a suit that usually sells for $300 is marked "25% off." What is the discount? What is the sale price?

SOLUTION To find the discount, we restate the problem as:

What is 25% *of* 300?

$$n = 0.25 \cdot 300$$

$$n = 75$$

The discount is \$75. The sale price is the original price less the discount:

$$\begin{array}{rl} \$300 & \text{Original price} \\ -\quad 75 & \text{Less the discount (25\% of \$300)} \\ \hline \$225 & \text{Sale price} \end{array}$$

EXAMPLE 5 A man buys a washing machine on sale. The machine usually sells for \$450, but it is on sale at 12% off. If the sales tax rate is 5%, how much is the total bill for the washer?

SOLUTION First we have to find the sale price of the washing machine, and we begin by finding the discount:

What is 12% *of* \$450?

$$\begin{aligned} n &= 0.12 \cdot 450 \\ n &= 54 \end{aligned}$$

The washing machine is marked down \$54. The sale price is

$$\begin{array}{rl} \$450 & \text{Original price} \\ -\quad 54 & \text{Discount (12\% of \$450)} \\ \hline \$396 & \text{Sale price} \end{array}$$

Because the sales tax rate is 5%, we find the sales tax as follows:

What is 5% *of* 396?

$$\begin{aligned} n &= 0.05 \cdot 396 \\ n &= 19.80 \end{aligned}$$

The sales tax is \$19.80. The total price the man pays for the washing machine is

$$\begin{array}{rl} \$396.00 & \text{Sale price} \\ +\quad 19.80 & \text{Sales tax} \\ \hline \$415.80 & \text{Total price} \end{array}$$

5. A woman buys a new coat on sale. The coat usually sells for \$45, but it is on sale at 15% off. If the sales tax rate is 5%, how much is the total bill for the coat?

Getting Ready for Class

After reading through the preceding section, respond in your own words and in complete sentences.

1. Suppose the person mentioned in Example 1 was earning \$32,000 per year and received the same percent increase in salary. How much more would the raise have been?
2. Suppose the shoes mentioned in Example 3 were on sale for \$20, instead of \$21. Calculate the new percent decrease in price.
3. Suppose a store owner pays \$225 for a suit, and then marks it up \$75, to \$300. Find the percent increase in price.
4. Compare your answer to Problem 3 above with the problem given in Example 4 of your text. Do you think it is generally true that a 25% discount is equivalent to a $33\frac{1}{3}\%$ markup?

Answer
4. \$82.50; \$467.50 **5.** \$40.16

Problem Set 7.5

A B Solve each of these problems using the method developed in this section. [Examples 1–3]

1. **Salary Increase** If a person earns $23,000 a year and gets a 7% increase in salary, what is the new salary?

2. **Salary Increase** A computer programmer's yearly income of $57,000 is increased by 8%. What is the dollar amount of the increase, and what is her new salary?

3. **Tuition Increase** The yearly tuition at a college is presently $6,000. Next year it is expected to increase by 17%. What will the tuition at this school be next year?

4. **Price Increase** A market increased the price of cheese selling for $4.98 per pound by 3%. What is the new price for a pound of cheese? (Round to the nearest cent.)

5. **Car Value** In one year a new car decreased in value by 20%. If it sold for $16,500 when it was new, what was it worth after 1 year?

6. **Calorie Content** A certain light beer has 20% fewer calories than the regular beer. If the regular beer has 120 calories per bottle, how many calories are in the same-sized bottle of the light beer?

7. **Salary Increase** A person earning $3,500 a month gets a raise of $350 per month. What is the percent increase in salary?

8. **Rate Increase** A student reader is making $6.50 per hour and gets a $0.70 raise. What is the percent increase? (Round to the nearest tenth of a percent.)

9. **Shoe Sale** Shoes that usually sell for $25 are on sale for $20. What is the percent decrease in price?

10. **Enrollment Decrease** The enrollment in a certain elementary school was 410 in 2007. In 2008, the enrollment in the same school was 328. Find the percent decrease in enrollment from 2007 to 2008.

11. **Students to Teachers** The chart shows the student to teacher ratio in the United States from 1975 to 2002. What is the percent decrease in student to teacher ratio from 1975 to 2002? Round to the nearest percent.

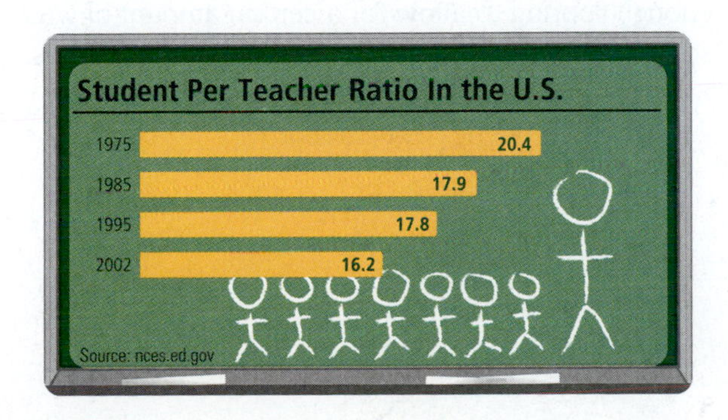

12. **Health Care** The graph shows the rising cost of health care. What is the percent increase in health care costs from 2002 to 2014?

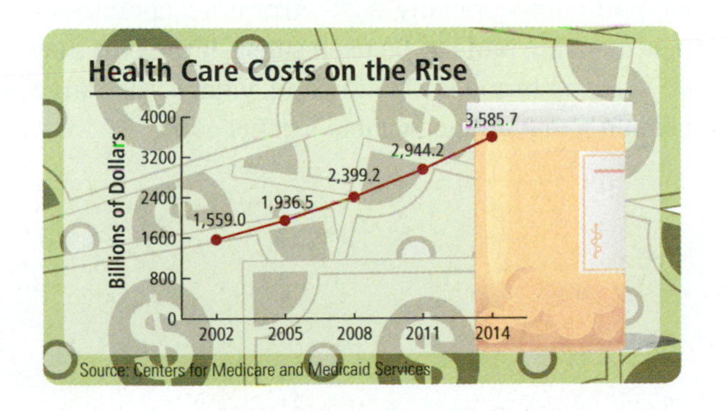

C [Examples 4, 5]

13. **Discount** During a clearance sale, a three-piece suit that usually sells for $300 is marked "15% off." What is the discount? What is the sale price?

14. **Sale Price** On opening day, a new music store offers a 12% discount on all electric guitars. If the regular price on a guitar is $550, what is the sale price?

15. **Total Price** A man buys a washing machine that is on sale. The washing machine usually sells for $450 but is on sale at 20% off. If the sales tax rate in his state is 6%, how much is the total bill for the washer?

16. **Total Price** A bedroom set that normally sells for $1,450 is on sale for 10% off. If the sales tax rate is 5%, what is the total price of the bedroom set if it is bought while on sale?

17. **Real Estate Market** In 2006 the average price of a home began to fall in most real estate markets across the country. The median price of a single family home in the U.S. was $227,000 in 2006. The median price is now $195,500. By what percent did the median price of a single family home drop? Round your answer to the nearest tenth of a percent.

18. **Deep Discount** When buying some of today's newest electronic gadgets, good things come to those who wait. When Apple released its new iPhone in the summer of 2007, an 8GB model sold for $499. In July 2008, Apple released its new iPhone 3G. The 8GB model sells for $199. What is the percent decrease in price for this new model? Round your answer to the nearest tenth of a percent.

19. **Losing Weight** According to the Centers for Disease Control and Prevention (CDC), more than 60% of U.S. adults are overweight, and about 15% of children and adolescents ages 6 to 19 are overweight. Your friend decides to go on a diet and goes from 155 pounds to 130 pounds over a 4 month period. What was her percentage weight loss? Round your answer to the nearest percent.

20. **Ordering Online** You are in the market for a new laptop. The model that you wish to purchase is $1,500 in a local store. However, you decide to buy the computer over the Internet for $1200. You will need to pay shipping charges of $59 plus the 6% local sales tax. Taking into account taxes and shipping charges, what percentage do you save by ordering it online? Round your answer to the nearest tenth of a percent.

21. **Product Error** When manufacturing a product, a certain amount of variation (or error) can occur in the process and still create a part or product that is useable. For one particular company, a 3% error is acceptable for their machine parts to be used safely. If the part they are manufacturing is 22.5 in. long, what is the range of measures that are acceptable for this part?

22. **Home Remodeling** You have decided to update your house by laying a new wood floor in your living room. Your floor has an area of 440 sq ft. You decide to buy enough flooring to allow for a certain amount of waste so you purchase 470 sq ft of wood flooring materials. Express your waste allowance as a percent. Round your result to the nearest percent.

Calculator Problems

Set up the following problems the same way you set up Problems 1–22. Then use a calculator to do the calculations.

23. Salary Increase A teacher making $43,752 per year gets a 6.5% raise. What is the new salary?

24. Utility Increase A homeowner had a $95.90 electric bill in December. In January the bill was $107.40. Find the percent increase in the electric bill from December to January. (Round to the nearest whole number.)

25. Soccer The rules for soccer state that the playing field must be from 100 to 120 yards long and 55 to 75 yards wide. The 1999 Women's World Cup was played at the Rose Bowl on a playing field 116 yards long and 72 yards wide. The diagram below shows the smallest possible soccer field, the largest possible soccer field, and the soccer field at the Rose Bowl.

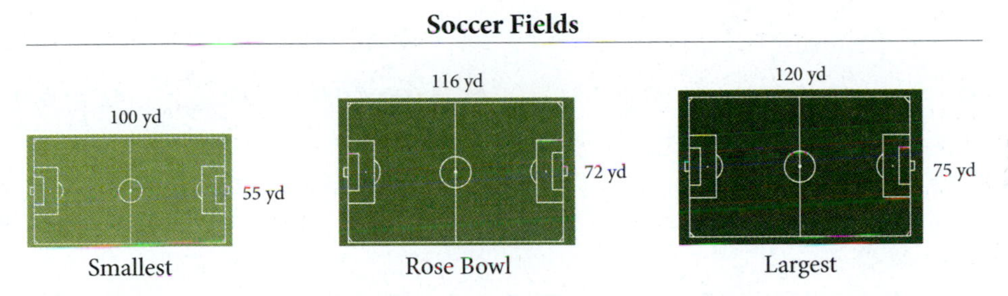

a. **Percent Increase** A team plays on the smallest field, then plays in the Rose Bowl. What is the percent increase in the area of the playing field from the smallest field to the Rose Bowl? Round to the nearest tenth of a percent.

b. **Percent Increase** A team plays a soccer game in the Rose Bowl. The next game is on a field with the largest dimensions. What is the percent increase in the area of the playing field from the Rose Bowl to the largest field? Round to the nearest tenth of a percent.

26. Football The diagrams below show the dimensions of playing fields for the National Football League (NFL), the Canadian Football League (CFL), and Arena Football.

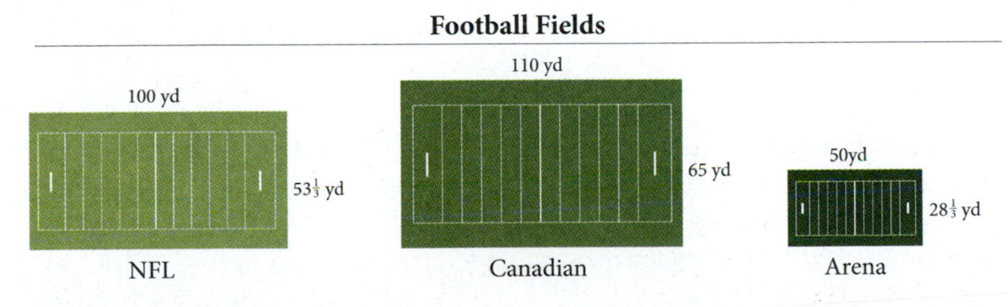

a. **Percent Increase** In 1999 Kurt Warner made a successful transition from Arena Football to the NFL, winning the Most Valuable Player award. What was the percent increase in the area of the fields he played on in moving from Arena Football to the NFL? Round to the nearest percent.

b. **Percent Decrease** Doug Flutie played in the Canadian Football League before moving to the NFL. What was the percent decrease in the area of the fields he played on in moving from the CFL to the NFL? Round to the nearest tenth of a percent.

Getting Ready for the Next Section

Multiply. Round to nearest hundredth if necessary.

27. 0.07(2,000)

28. 0.12(8,000)

29. $600(0.04)\left(\frac{1}{6}\right)$

30. $900(0.06)\left(\frac{1}{4}\right)$

31. $10{,}150(0.06)\left(\frac{1}{4}\right)$

32. $10{,}302.25(0.06)\left(\frac{1}{4}\right)$

Add.

33. 3,210 + 224.7

34. 900 + 13.50

35. 10,000 +150

36. 10,150 + 152.25

37. 10,302.25 + 154.53

38. 10,456.78 + 156.85

Simplify.

39. 2,000 + 0.07(2,000)

40. 8,000 + 0.12(8,000)

41. 3,000 + 0.07(3,000)

42. 9,000 + 0.12(9,000)

Maintaining Your Skills

The problems below review some basic concepts of addition of fractions and mixed numbers. Add each of the following and reduce all answers to lowest terms.

43. $\frac{1}{3} + \frac{2}{3}$

44. $\frac{3}{8} + \frac{1}{8}$

45. $\frac{1}{2} + \frac{1}{4}$

46. $\frac{1}{5} + \frac{3}{10}$

47. $\frac{3}{4} + \frac{2}{3}$

48. $\frac{3}{8} + \frac{1}{6}$

49. $2\frac{1}{2} + 3\frac{1}{2}$

50. $3\frac{1}{4} + 2\frac{1}{8}$

Interest

7.6

Objectives

A Solve simple interest problems.

B Solve compound interest problems.

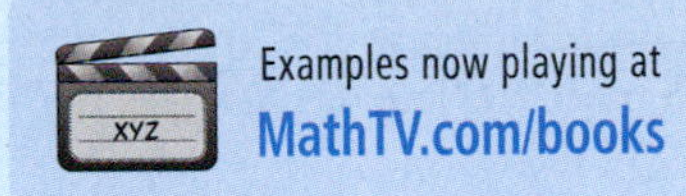

Anyone who has borrowed money from a bank or other lending institution, or who has invested money in a savings account, is aware of *interest.* Interest is the amount of money paid for the use of money. If we put \$500 in a savings account that pays 6% annually, the interest will be 6% of \$500, or 0.06(500) = \$30. The amount we invest (\$500) is called the *principal,* the percent (6%) is the *interest rate,* and the money earned (\$30) is the *interest.*

EXAMPLE 1 A man invests \$2,000 in a savings plan that pays 7% per year. How much money will be in the account at the end of 1 year?

SOLUTION We first find the interest by taking 7% of the principal, \$2,000:

$$\begin{aligned} \text{Interest} &= 0.07(\$2{,}000) \\ &= \$140 \end{aligned}$$

The interest earned in 1 year is \$140. The total amount of money in the account at the end of a year is the original amount plus the \$140 interest:

\$2,000	Original investment (principal)
+ 140	Interest (7% of \$2,000)
\$2,140	Amount after 1 year

The amount in the account after 1 year is \$2,140.

EXAMPLE 2 A farmer borrows \$8,000 from his local bank at 12%. How much does he pay back to the bank at the end of the year to pay off the loan?

SOLUTION The interest he pays on the \$8,000 is

$$\begin{aligned} \text{Interest} &= 0.12(\$8{,}000) \\ &= \$960 \end{aligned}$$

At the end of the year, he must pay back the original amount he borrowed (\$8,000) plus the interest at 12%:

\$8,000	Amount borrowed (principal)
+ 960	Interest at 12%
\$8,960	Total amount to pay back

The total amount that the farmer pays back is \$8,960.

A Simple Interest

There are many situations in which interest on a loan is figured on other than a yearly basis. Many short-term loans are for only 30 or 60 days. In these cases we can use a formula to calculate the interest that has accumulated. This type of interest is called *simple interest.* The formula is

$$I = P \cdot R \cdot T$$

where

I = Interest
P = Principal
R = Interest rate (this is the percent)
T = Time (in years, 1 year = 360 days)

PRACTICE PROBLEMS

1. A man invests \$3,000 in a savings plan that pays 8% per year. How much money will be in the account at the end of 1 year?

2. If a woman borrows \$7,500 from her local bank at 12% interest, how much does she pay back to the bank if she pays off the loan in 1 year?

Answers
1. \$3,240 2. \$8,400

We could have used this formula to find the interest in Examples 1 and 2. In those two cases, T is 1. When the length of time is in days rather than years, it is common practice to use 360 days for 1 year, and we write T as a fraction. Examples 3 and 4 illustrate this procedure.

EXAMPLE 3 A student takes out an emergency loan for tuition, books, and supplies. The loan is for \$600 at an interest rate of 4%. How much interest does the student pay if the loan is paid back in 60 days?

SOLUTION The principal P is \$600, the rate R is 4% = 0.04, and the time T is $\frac{60}{360}$. Notice that T must be given in years, and 60 days = $\frac{60}{360}$ year. Applying the formula, we have

$$I = P \cdot R \cdot T$$
$$I = 600 \times 0.04 \times \frac{60}{360}$$
$$I = 600 \times 0.04 \times \frac{1}{6} \qquad \frac{60}{360} = \frac{1}{6}$$
$$I = 4 \qquad \text{Multiplication}$$

The interest is \$4.

EXAMPLE 4 A woman deposits \$900 in an account that pays 6% annually. If she withdraws all the money in the account after 90 days, how much does she withdraw?

SOLUTION We have $P = \$900$, $R = 0.06$, and $T = 90$ days $= \frac{90}{360}$ year. Using these numbers in the formula, we have

$$I = P \cdot R \cdot T$$
$$I = 900 \times 0.06 \times \frac{90}{360}$$
$$I = 900 \times 0.06 \times \frac{1}{4} \qquad \frac{90}{360} = \frac{1}{4}$$
$$I = 13.5 \qquad \text{Multiplication}$$

The interest earned in 90 days is \$13.50. If the woman withdraws all the money in her account, she will withdraw

\$900.00	Original amount (principal)
+ 13.50	Interest for 90 days
\$913.50	Total amount withdrawn

The woman will withdraw \$913.50.

B Compound Interest

A second common kind of interest is *compound interest.* Compound interest includes interest paid on interest. We can use what we know about simple interest to help us solve problems involving compound interest.

EXAMPLE 5 A homemaker puts \$3,000 into a savings account that pays 7% compounded annually. How much money is in the account at the end of 2 years?

SOLUTION Because the account pays 7% annually, the simple interest at the end of 1 year is 7% of \$3,000:

3. Another student takes out a loan like the one in Example 3. This loan is for \$700 at 4%. How much interest does this student pay if the loan is paid back in 90 days?

4. Suppose \$1,200 is deposited in an account that pays 9.5% interest per year. If all the money is withdrawn after 120 days, how much money is withdrawn?

5. If \$5,000 is put into an account that pays 6% compounded annually, how much money is in the account at the end of 2 years?

Answers
3. \$7 **4.** \$1,238

$$\text{Interest after 1 year} = 0.07(\$3{,}000) = \$210$$

Because the interest is paid annually, at the end of 1 year the total amount of money in the account is

\$3,000	Original amount
+ 210	Interest for 1 year
\$3,210	Total in account after 1 year

The interest paid for the second year is 7% of this new total, or

$$\text{Interest paid the second year} = 0.07(\$3{,}210) = \$224.70$$

At the end of 2 years, the total in the account is

\$3,210.00	Amount at the beginning of year 2
+ 224.70	Interest paid for year 2
\$3,434.70	Account after 2 years

At the end of 2 years, the account totals \$3,434.70. The total interest earned during this 2-year period is \$210 (first year) + \$224.70 (second year) = \$434.70.

Note If the interest earned in Example 5 were calculated using the formula for simple interest, $I = P \cdot R \cdot T$, the amount of money in the account at the end of two years would be \$3,420.00.

You may have heard of savings and loan companies that offer interest rates that are compounded quarterly. If the interest rate is, say, 6% and it is compounded quarterly, then after every 90 days ($\frac{1}{4}$ of a year) the interest is added to the account. If it is compounded semiannually, then the interest is added to the account every 6 months. Most accounts have interest rates that are compounded daily, which means the simple interest is computed daily and added to the account.

EXAMPLE 6 If \$10,000 is invested in a savings account that pays 6% compounded quarterly, how much is in the account at the end of a year?

SOLUTION The interest for the first quarter ($\frac{1}{4}$ of a year) is calculated using the formula for simple interest:

$$I = P \cdot R \cdot T$$

$$I = \$10{,}000 \times 0.06 \times \frac{1}{4} \qquad \text{First quarter}$$

$$I = \$150$$

At the end of the first quarter, this interest is added to the original principal. The new principal is \$10,000 + \$150 = \$10,150. Again we apply the formula to calculate the interest for the second quarter:

$$I = \$10{,}150 \times 0.06 \times \frac{1}{4} \qquad \text{Second quarter}$$

$$I = \$152.25$$

The principal at the end of the second quarter is \$10,150 + \$152.25 = \$10,302.25. The interest earned during the third quarter is

$$I = \$10{,}302.25 \times 0.06 \times \frac{1}{4} \qquad \text{Third quarter}$$

$$I = \$154.53 \qquad \text{To the nearest cent}$$

6. If \$20,000 is invested in an account that pays 8% compounded quarterly, how much is in the account at the end of a year?

Answer
5. \$5,618

The new principal is \$10,302.25 + \$154.53 = \$10,456.78. Interest for the fourth quarter is

$$I = \$10{,}456.78 \times 0.06 \times \frac{1}{4} \quad \text{Fourth quarter}$$

$$I = \$156.85 \quad \text{To the nearest cent}$$

The total amount of money in this account at the end of 1 year is

$$\$10{,}456.78 + \$156.85 = \$10{,}613.63$$

USING TECHNOLOGY

Compound Interest from a Formula

We can summarize the work above with a formula that allows us to calculate compound interest for any interest rate and any number of compounding periods. If we invest P dollars at an annual interest rate r, compounded n times a year, then the amount of money in the account after t years is given by the formula

$$A = P\left(1 + \frac{r}{n}\right)^{nt}$$

Using numbers from Example 6 to illustrate, we have

P = Principal = \$10,000
r = annual interest rate = 0.06
n = number of compounding periods = 4 (interest is compounded quarterly)
t = number of years = 1

Substituting these numbers into the formula above, we have

$$A = 10{,}000\left(1 + \frac{0.06}{4}\right)^{4\cdot 1}$$
$$= 10{,}000(1 + 0.015)^4$$
$$= 10{,}000(1.015)^4$$

To simplify this last expression on a calculator, we have

Scientific calculator: 10,000 [×] 1.015 [y^x] 4 [=]

Graphing calculator: 10,000 [×] 1.015 [^] 4 [ENTER]

In either case, the answer is \$10,613.63551, which rounds to \$10,613.64.

Note The reason that this answer is different from the result we obtained in Example 6 is that, in Example 6, we rounded each calculation as we did it. The calculator will keep all the digits in all of the intermediate calculations.

Getting Ready for Class

After reading through the preceding section, respond in your own words and in complete sentences.

1. Suppose the man in Example 1 invested \$3,000, instead of \$2,000, in the savings plan. How much more interest would he have earned?
2. How much does the student in Example 3 pay back if the loan is paid off after a year, instead of after 60 days?
3. Suppose the homemaker mentioned in Example 5 invests \$3,000 in an account that pays only $3\frac{1}{2}$% compounded annually. How much is in the account at the end of 2 years?
4. In Example 6, how much money would the account contain at the end of 1 year if it were compounded annually, instead of quarterly?

Answer
6. \$21,648.64

Problem Set 7.6

A These problems are similar to the examples found in this section. They should be set up and solved in the same way. (Problems 1–12 involve simple interest.) [Examples 1–4]

1. **Savings Account** A man invests $2,000 in a savings plan that pays 8% per year. How much money will be in the account at the end of 1 year?

2. **Savings Account** How much simple interest is earned on $5,000 if it is invested for 1 year at 5%?

3. **Savings Account** A savings account pays 7% per year. How much interest will $9,500 invested in such an account earn in a year?

4. **Savings Account** A local bank pays 5.5% annual interest on all savings accounts. If $600 is invested in this type of account, how much will be in the account at the end of a year?

5. **Bank Loan** A farmer borrows $8,000 from his local bank at 7%. How much does he pay back to the bank at the end of the year when he pays off the loan?

6. **Bank Loan** If $400 is borrowed at a rate of 12% for 1 year, how much is the interest?

7. **Bank Loan** A bank lends one of its customers $2,000 at 8% for 1 year. If the customer pays the loan back at the end of the year, how much does he pay the bank?

8. **Bank Loan** If a loan of $2,000 at 20% for 1 year is to be paid back in one payment at the end of the year, how much does the borrower pay the bank?

9. **Student Loan** A student takes out an emergency loan for tuition, books, and supplies. The loan is for $600 with an annual interest rate of 5%. How much interest does the student pay if the loan is paid back in 60 days?

10. **Short-Term Loan** If a loan of $1,200 at 9% is paid off in 90 days, what is the interest?

11. **Savings Account** A woman deposits $800 in a savings account that pays 5%. If she withdraws all the money in the account after 120 days, how much does she withdraw?

12. **Savings Account** $1,800 is deposited in a savings account that pays 6%. If the money is withdrawn at the end of 30 days, how much interest is earned?

B The problems that follow involve compound interest. [Examples 5, 6]

Compound Interest The chart shows the interest rates for various CD accounts.

13. Last year Samuel invested $400 in a 6-month CD. If the interest is compounded quarterly, how much was in the account at the end of 6 months? Round to the nearest cent.

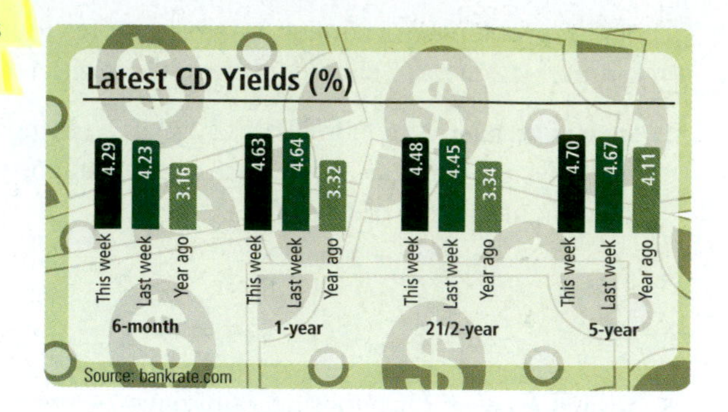

14. If Alice deposited $200 in a $2\frac{1}{2}$ year CD account earlier this week, what will the account make at the end of its term if interest is compounded quarterly. Use the compound interest formula and round to the nearest cent.

15. Compound Interest A woman puts $5,000 into a savings account that pays 6% compounded annually. How much money is in the account at the end of 2 years?

16. Compound Interest A savings account pays 5% compounded annually. If $10,000 is deposited in the account, how much is in the account after 2 years?

17. Compound Interest If $8,000 is invested in a savings account that pays 5% compounded quarterly, how much is in the account at the end of a year?

18. Compound Interest Suppose $1,200 is invested in a savings account that pays 6% compounded semiannually. How much is in the account at the end of $1\frac{1}{2}$ years?

Calculator Problems

The following problems should be set up in the same way in which Problems 1–18 have been set up. Then the calculations should be done on a calculator.

19. Savings Account A woman invests $917.26 in a savings account that pays 6.25% annually. How much is in the account at the end of a year?

20. Business Loan The owner of a clothing store borrows $6,210 for 1 year at 11.5% interest. If he pays the loan back at the end of the year, how much does he pay back?

21. Compound Interest Suppose $10,000 is invested in each account below. In each case find the amount of money in the account at the end of 5 years.

a. Annual interest rate = 6%, compounded quarterly

b. Annual interest rate = 6%, compounded monthly

c. Annual interest rate = 5%, compounded quarterly

d. Annual interest rate = 5%, compounded monthly

22. Compound Interest Suppose $5,000 is invested in each account below. In each case find the amount of money in the account at the end of 10 years.

a. Annual interest rate = 5%, compounded quarterly

b. Annual interest rate = 6%, compounded quarterly

c. Annual interest rate = 7%, compounded quarterly

d. Annual interest rate = 8%, compounded quarterly

Getting Ready for the Next Section

Change to percent.

23. $\frac{75}{250}$ **24.** $\frac{150}{250}$ **25.** $\frac{400}{2,400}$ **26.** $\frac{200}{2,400}$

Multiply.

27. 0.3(360) **28.** 0.4(360) **29.** 0.45(360) **30.** 0.15(360)

Divide.

31. $40 \div 5$ **32.** $45 \div 5$ **33.** $15 \div 5$ **34.** $5 \div 5$

Maintaining Your Skills

The problems below will allow you to review subtraction of fractions and mixed numbers.

35. $\frac{3}{4} - \frac{1}{4}$ **36.** $\frac{9}{10} - \frac{7}{10}$ **37.** $\frac{5}{8} - \frac{1}{4}$ **38.** $\frac{7}{10} - \frac{1}{5}$

39. $2 - \frac{4}{3}$ **40.** $2 + \frac{4}{3}$ **41.** $1 + \frac{1}{2}$ **42.** $1 - \frac{1}{2}$

43. $\frac{1}{3} - \frac{1}{4}$ **44.** $\frac{9}{12} - \frac{1}{5}$ **45.** $3\frac{1}{4} - 2$ **46.** $5\frac{1}{6} - 3\frac{1}{4}$

47. Find the sum of $\frac{8}{15}$ and $\frac{8}{35}$.

48. Find the difference of $\frac{8}{15}$ and $\frac{8}{35}$.

49. Find the product of $\frac{8}{15}$ and $\frac{8}{35}$.

50. Find the quotient of $\frac{8}{15}$ and $\frac{8}{35}$.

Extending the Concepts

The following problems are percent problems. Use any of the methods developed in this chapter to solve them.

51. Credit Card Debt Student credit-card debt is at an all-time high. Consolidated Credit Counseling Services Inc. reports that 20% of all college freshman got their first credit card in high school and nearly 40% sign up for one in their first year at college. Suppose your credit card company charges 1.3% in finance charges per month on the average daily balance in your credit card account. If your average daily balance for this month is $2,367.90 determine the finance charge for the month.

52. Finding Your Interest Rate In early January, your bank sent out a form called a 1099-INT, which summarizes the amount of interest you have received on a savings account for the previous year. If you received $72 interest for the year on an account in which you started with $1,200, determine the annual interest rate paid by your bank.

53. Movie Making The bar chart below shows the production costs for each of the first four *Star Wars* movies. Find the percent increase in production costs from each *Star Wars* movie to the next. Round your results to the nearest tenth.

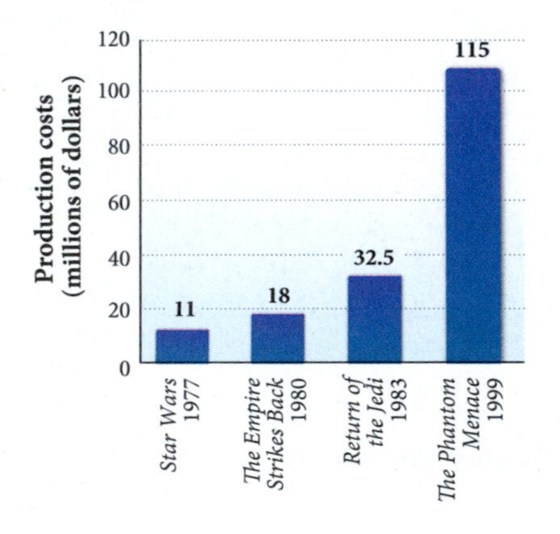

Douglas Kirkland/Corbis

54. Movie Making The table below shows how much money each of the first four *Star Wars* movies brought in during the first weekend they were shown. Find the percent increase in opening weekend income from each *Star Wars* movie to the next. Round to the nearest percent.

Opening Weekend Income	
Star Wars (1977)	$1,554,000
The Empire Strikes Back (1980)	$6,415,000
Return of the Jedi (1983)	$30,490,000
The Phantom Menace (1999)	$64,810,000

Pie Charts

Objectives

A Read a pie chart.
B Construct a pie chart.

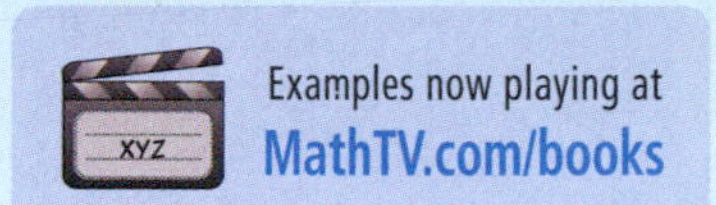

Pie charts are another way in which to visualize numerical information. They lend themselves well to information that adds up to 100% and are very common in the world around us. In fact, it is hard to pick up a newspaper or magazine without seeing a pie chart. As the diagram below shows, even a computer will represent the amount of free space and used space on one of its disks by using a pie chart.

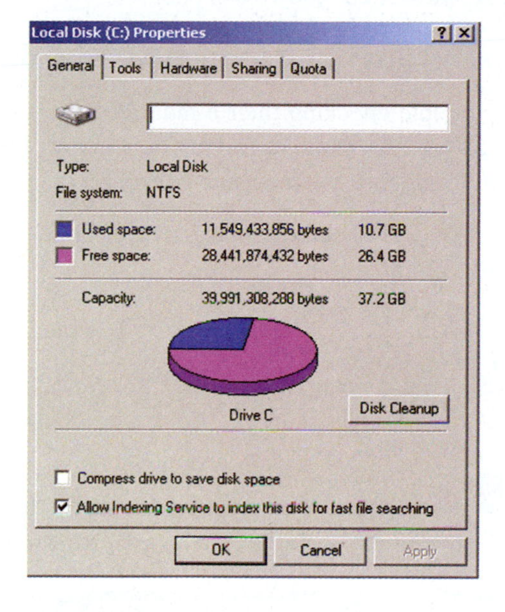

A Reading a Pie Chart

Some of this introductory material will be review. We want to begin our study of pie charts by reading information from pie charts.

EXAMPLE 1 The pie chart shows the class rank of the members of a drama club. Use the pie chart to answer the following questions.

a. Find the total membership of the club.
b. Find the ratio of freshmen to total number of members.
c. Find the ratio of seniors to juniors.

SOLUTION **a.** To find the total membership in the club, we add the numbers in all sections of the pie chart.

$$9 + 11 + 15 + 10 = 45 \text{ members}$$

b. The ratio of freshmen to total members is

$$\frac{\text{number of freshmen}}{\text{total number of members}} = \frac{11}{45}$$

c. The ratio of seniors to juniors is

$$\frac{\text{number of seniors}}{\text{number of juniors}} = \frac{9}{15} = \frac{3}{5}$$

PRACTICE PROBLEMS

1. Work Example 1 again if one more junior joins the club.

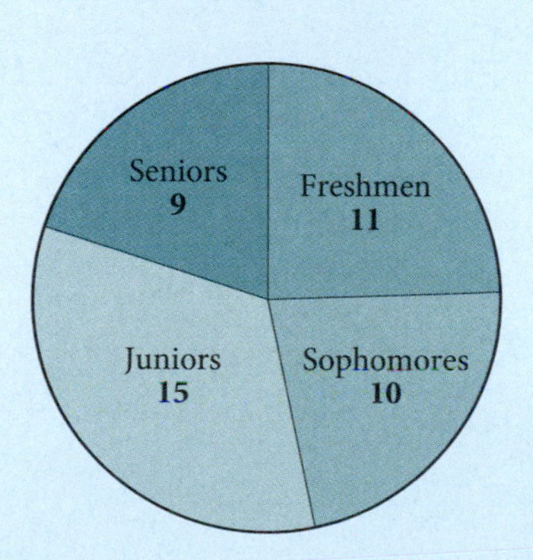

Answer

1. a. 46 **b.** $\frac{11}{46}$ **c.** $\frac{9}{16}$

2. Work Example 2 again if 600 people responded to the survey.

Source: UCLA Center for Communication Policy

EXAMPLE 2 The pie chart shows the results of a survey on how often people check their e-mail. Use the pie chart to answer the following questions. Suppose 500 people participated in the survey.

a. How many people in the survey check their e-mail daily?

b. How many people check their e-mail once a week or less often?

SOLUTION **a.** To find out how many people in the survey check their e-mail daily, we need to find 76% of 500.

$0.76(500) = 380$ of the people surveyed check their e-mail daily

b. The people checking their e-mail weekly or less often account for $23\% + 1\% = 24\%$. To find out how many of the 500 people are in this category, we must find 24% of 500.

$0.24(500) = 120$ of the people surveyed check their e-mail weekly or less often

B Constructing Pie Charts

3. Construct a pie chart that shows the used space and free space on a 256-MB flash memory stick that contains 102 MB of data.

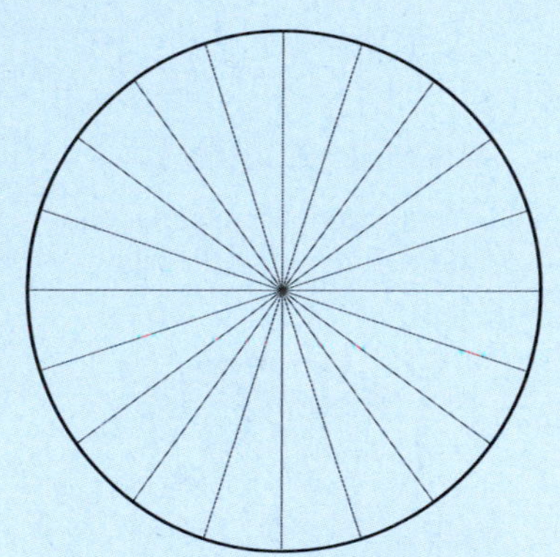

PIE CHART TEMPLATE Each slice is 5% of the area of the circle.

EXAMPLE 3 Construct a pie chart that shows the free space and used space for a 256-MB flash memory stick that contains 77 MB of data.

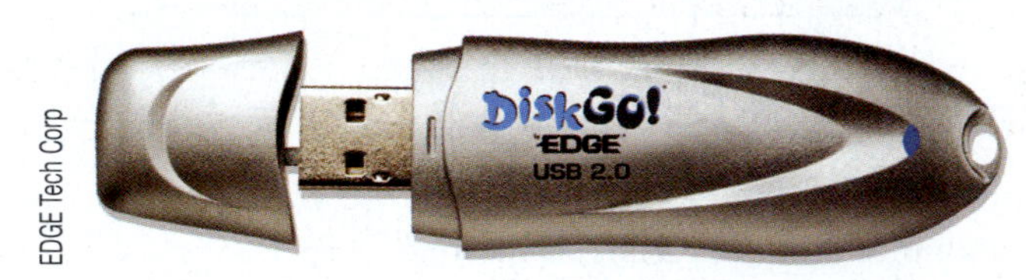

SOLUTION 1 **Using a Template** As mentioned previously, pie charts are constructed with percents. Therefore we must first convert data to percents. To find the percent of used space, we divide the amount of used space by the amount of total space. We have

$$\frac{77}{256} = 0.30078 \text{ which is 30\% to the nearest percent}$$

The area of each section of the template on the left is 5% of the area of the whole circle. If we shade 6 sections of the template, we will have shaded 30% of the area of the whole circle.

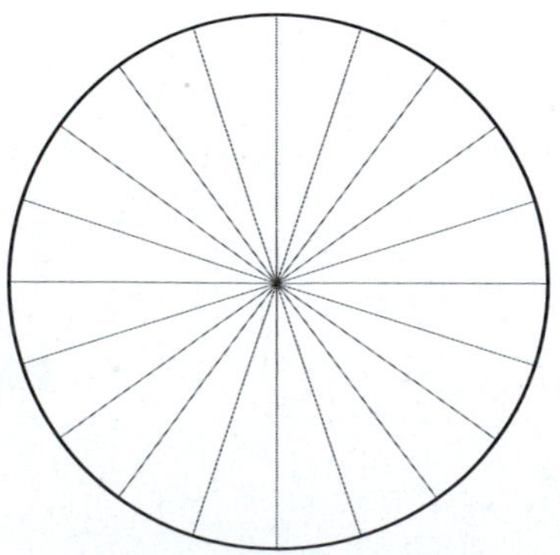

PIE CHART TEMPLATE Each slice is 5% of the area of the circle.

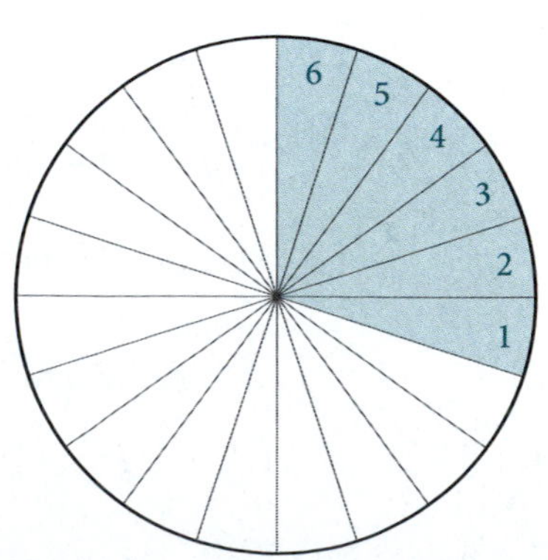

CREATING A PIE CHART To shade 30% of the circle, we shade 6 sections of the template.

Answer

2. **a.** 456 people **b.** 144 people

The shaded area represents 30%, which is the amount of used disk space. The rest of the circle must represent the 70% free space on the disk. Shading each area with a different color and labeling each, we have our pie chart.

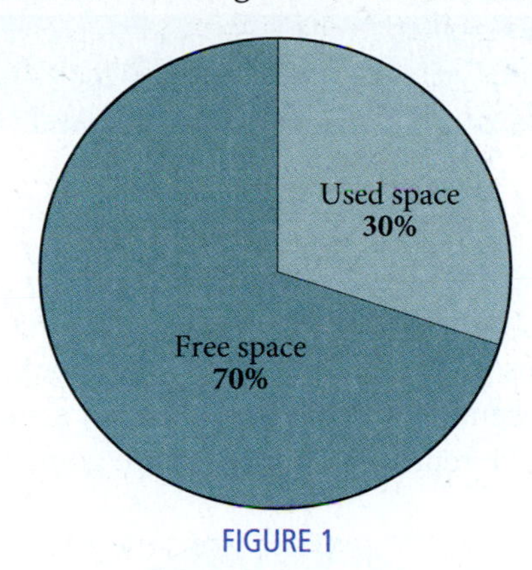

FIGURE 1

SOLUTION 2 **Using a Protractor** Since a pie chart is a circle, and a circle contains 360°, we must now convert our data to degrees. We do this by multiplying our percents in decimal form by 360. We have

$$(0.30)360° = 108°$$

Now we place a protractor on top of a circle. First we draw a line from the center of the circle to 0° as shown in Figure 2. Now we measure and mark 108° from our starting point, as shown in Figure 3.

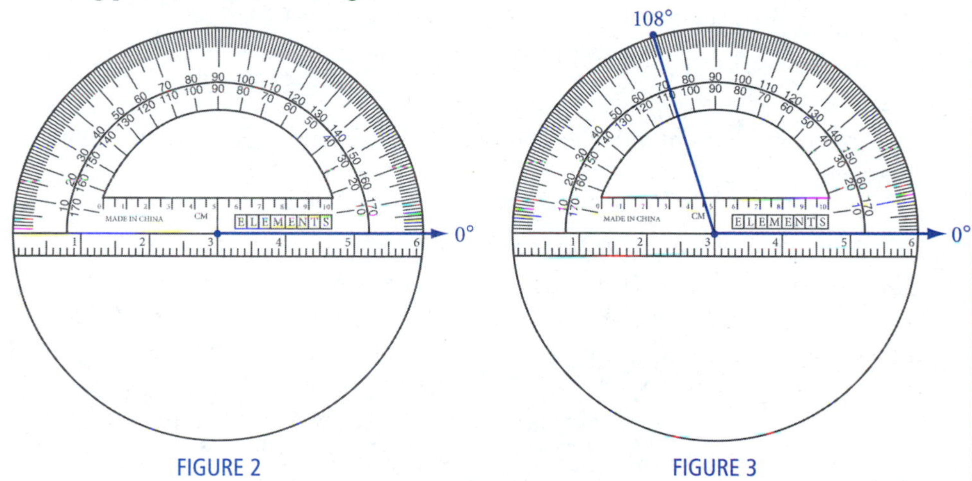

FIGURE 2 FIGURE 3

Finally we draw a line from the center of the circle to this mark, as shown in Figure 4. Then we shade and label the two regions as shown in Figure 5.

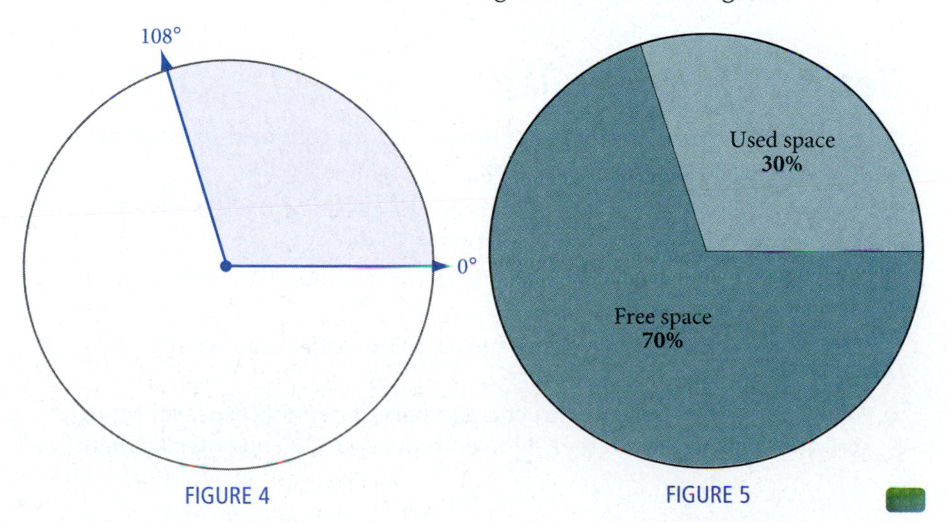

FIGURE 4 FIGURE 5

Answer
3. See solutions section.

4. The table below shows how the expenses for a paperback novel are divided. Use the information in the table to construct a pie chart.

Expense	Percent of Price
Bookstore	45%
Publisher	50%
Author	5%

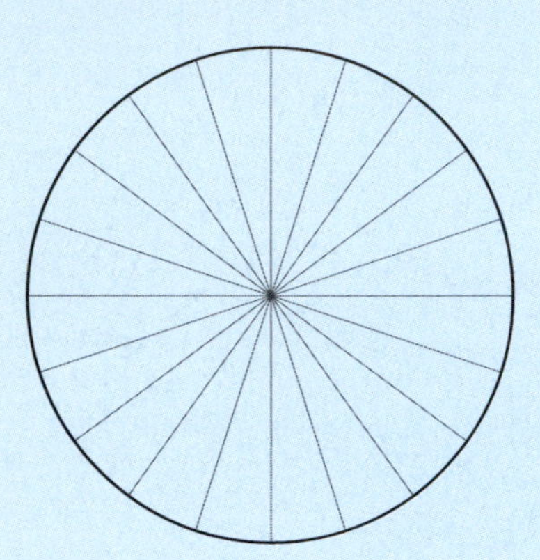

PIE CHART TEMPLATE Each slice is 5% of the area of the circle.

Answer

4. See solutions section.

EXAMPLE 4

Construct a pie chart from the information in the following table.

WHERE DOES YOUR TEXTBOOK MONEY GO?	
Expense	Percent of Price
Bookstore	40%
Publisher	45%
Author	15%

SOLUTION Since our template uses sections that each represent 5% of the circle, we shade 8 sections, representing 40%, for the bookstore's share. Then we shade 9 sections, representing 45% for the publisher's share. We should have 3 sections remaining, which represent the 15% share going to the author.

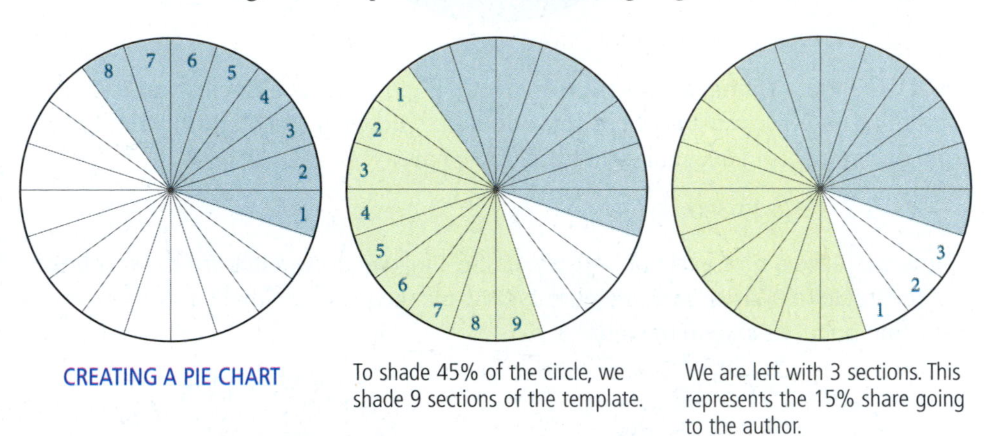

CREATING A PIE CHART To shade 45% of the circle, we shade 9 sections of the template. We are left with 3 sections. This represents the 15% share going to the author.

We label each section with the appropriate information, and our pie chart is complete.

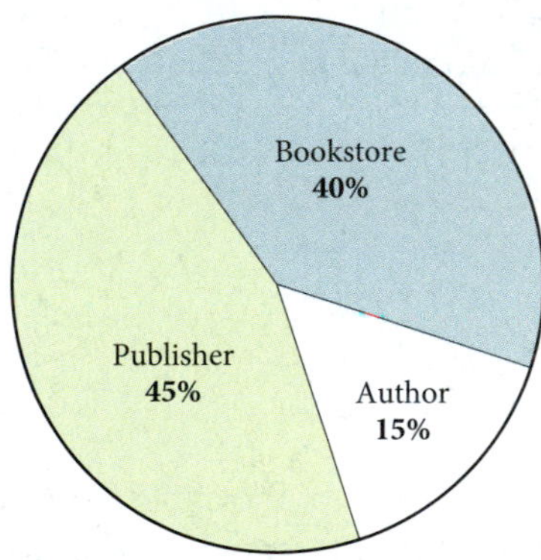

Getting Ready for Class

After reading through the preceding section, respond in your own words and in complete sentences.

1. If a circle is divided into 20 equal slices, then each of the slices is what percent of the total area enclosed by the circle?
2. If a 250 MB computer drive contains 75 MB of data, then how much of the drive is free space?
3. If a 250 MB computer drive contains 75 MB of data, then what percent of the drive contains data?
4. Explain how you would construct a pie chart of monthly expenses for a person who spends $700 on rent, $200 on food, and $100 on entertainment.

Problem Set 7.7

A [Examples 1, 2]

1. **High School Seniors with Jobs** The pie chart shows the results of surveying 200 high school seniors to find out how many hours they worked per week at a job.

   a. Find the ratio of students who work more than 15 hours a week to total students.

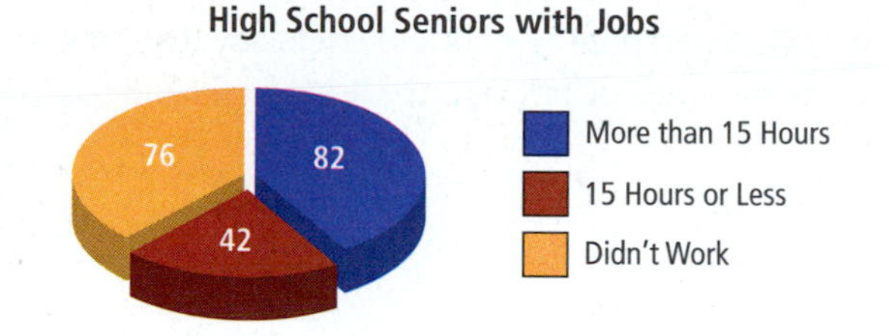

   b. Find the ratio of students who don't have a job to students who work more than 15 hours a week.

   c. Find the ratio of students with jobs to total students.

   d. Find the ratio of students with jobs to students without jobs.

2. **Favorite Dip Flavor** The pie chart shows the results of a survey on favorite dip flavor.

   a. What is the most preferred dip flavor?

Favorite Dip Flavor

7% · 37% · 56%

Ranch · Dill · Onion

   b. Which dip flavor is preferred second most?

   c. Which dip flavor is least preferred?

   d. What percentage of people preferred ranch?

   e. What percentage of people preferred onion or dill?

   f. If 50 people responded to the survey, how many people preferred ranch?

   g. If 50 people responded to the survey, how many people preferred dill? (Round your answer to the nearest whole number.)

3. **Food Dropped on the Floor** The pie chart shows the results of a survey about eating food that has been dropped on the floor. Participants were asked whether they eat food that has been on the floor for 3, 5, or 10 seconds.

   a. What percentage of people say it is not safe to eat food dropped on the floor?

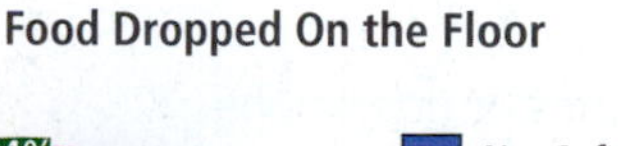

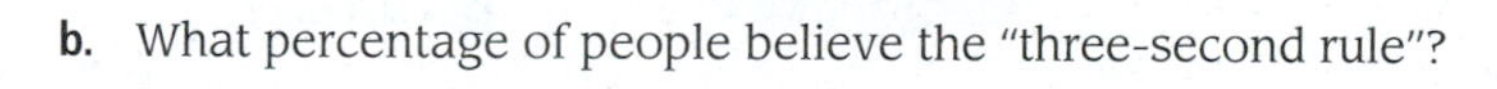

   b. What percentage of people believe the "three-second rule"?

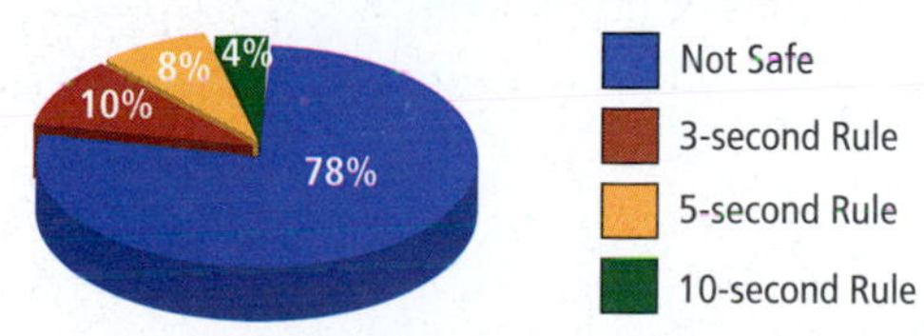

   c. What percentage of people will eat food that stays on the floor for five seconds or less?

   d. What percentage of people will eat foot that stays on the floor for ten seconds or less?

4. Talking to Our Dogs A survey showed that most dog owners talk to their dogs.

a. What percentage of dog owners say they never talk to their dogs?

b. What percentage of dog owners say they talk to their dogs all the time?

c. What percentage of dog owners say they talk to their dogs sometimes or not often?

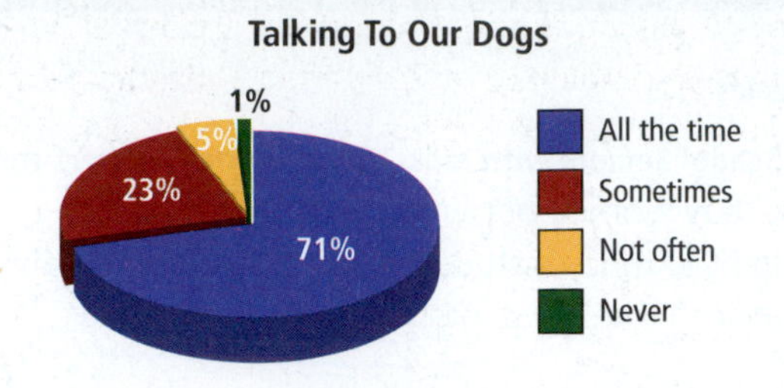

5. Monthly Car Payments Suppose 3,000 people responded to a survey on car loan payments, the results of which are shown in the pie chart. Find the number of people whose monthly payments would be the following:

a. $700 or more

b. Less than $300

c. $500 or more

d. $300 to $699

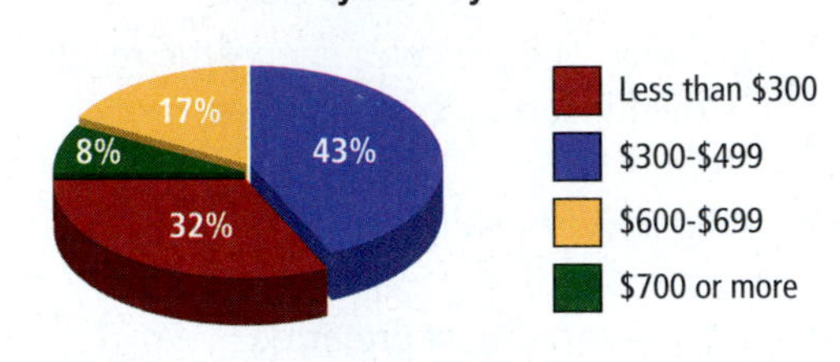

6. Where Workers Say Germs Lurk A survey asked workers where they thought the most germ-contaminated spot in the workplace was. Suppose the survey took place at a large company with 4,200 employees. Use the pie chart to determine the number of employees who would vote for each of the following as the most germ-contaminated areas.

a. Keyboards

b. Doorknobs

c. Restrooms or other

d. Telephones or doorknobs

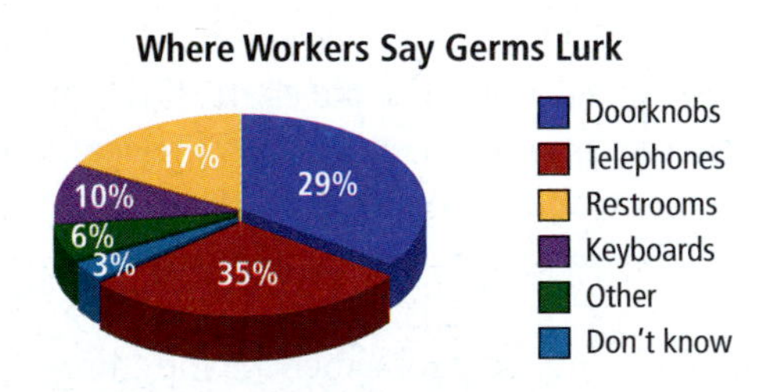

B [Examples 3, 4]

7. **Grade Distribution** Student scores, for a class of 20, on a recent math test are shown in the table below. Construct a pie chart that shows the number of As, Bs, and Cs earned on the test. Use the template provided here or use a protractor.

GRADE DISTRIBUTION	
Grade	Number
A	5
B	8
C	7
Total	20

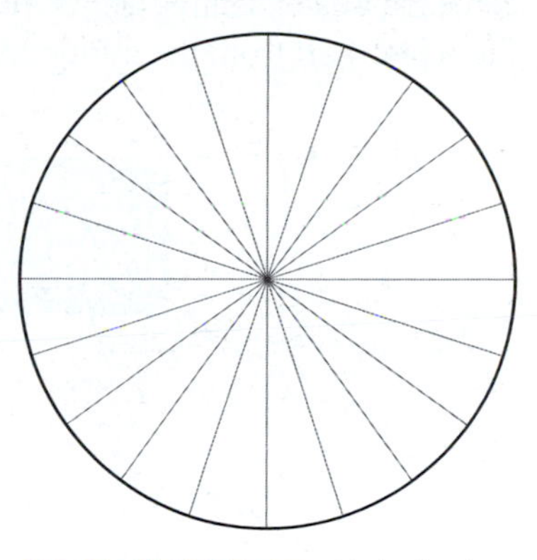

PIE CHART TEMPLATE Each slice is 5% of the area of the circle.

8. **Building Sizes** The Lean and Mean Gym Company recently ran a promotion for their four locations in the county. The table shows the locations along with the amount of square feet at each location. Use the information in the table to construct a pie chart, using the template provided here or using a protractor.

GYM LOCATION AND SIZE	
Location	Square Feet
Downtown	35,000
Uptown	85,000
Lakeside	25,000
Mall	75,000
Total	220,000

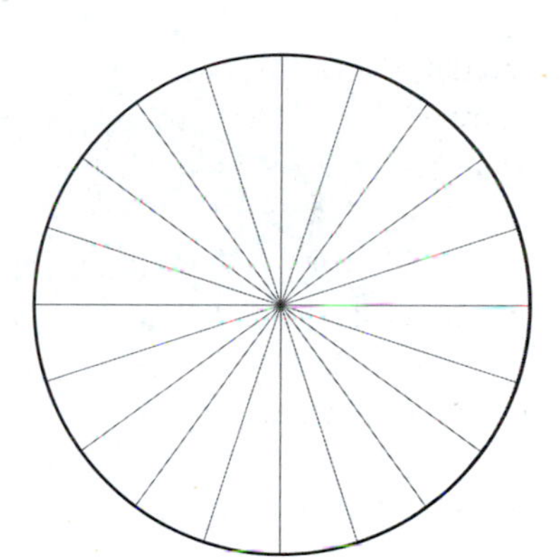

PIE CHART TEMPLATE Each slice is 5% of the area of the circle.

9. **Room Sizes** Scott and Amy are building their dream house. The size of the house will be 2,400 square feet. The table below shows the size of each room. Use the information in the table to construct a pie chart, using the template provided here or using a protractor.

ROOM SIZES	
Room	Square Feet
Kitchen	400
Dining room	310
Bedrooms	890
Living room	600
Bathrooms	200
Total	2,400

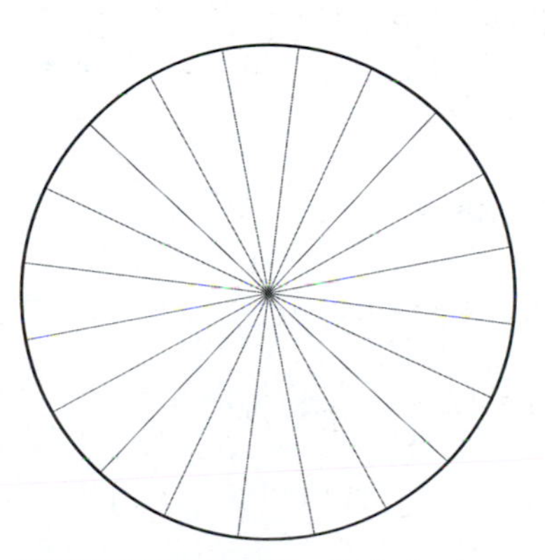

PIE CHART TEMPLATE Each slice is 5% of the area of the circle.

10. Airline Seating The table below gives the number of seats in each of the three classes of seating on an American Airlines Boeing 777 airliner. Create a pie chart from the information in the table.

AIRLINE SEATING	
Seating Class	Number Of Seats
First	18
Business	42
Coach	163

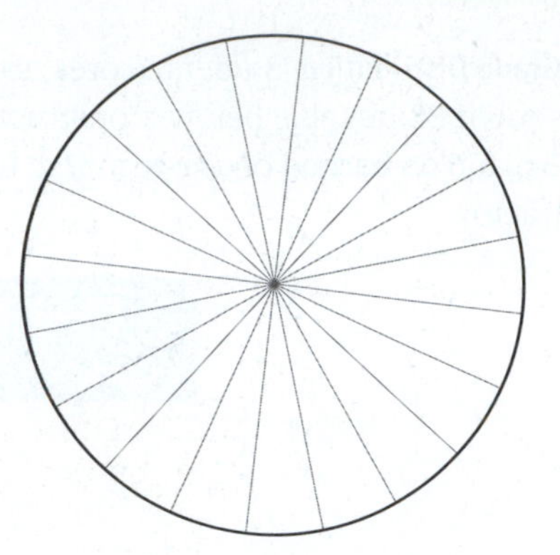

PIE CHART TEMPLATE Each slice is 5% of the area of the circle.

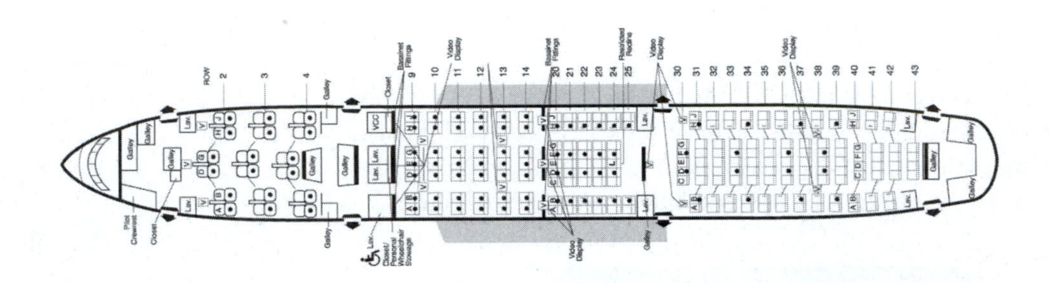

Maintaining Your Skills

Multiply.

11. $8 \cdot \frac{1}{3}$

12. $9 \cdot \frac{1}{3}$

13. $25 \cdot \frac{1}{1{,}000}$

14. $25 \cdot \frac{1}{100}$

15. $36.5 \cdot \frac{1}{100} \cdot 10$

16. $36.5 \cdot \frac{1}{1{,}000} \cdot 100$

17. $248 \cdot \frac{1}{10} \cdot \frac{1}{10}$

18. $969 \cdot \frac{1}{10} \cdot \frac{1}{10}$

19. $48 \cdot \frac{1}{12} \cdot \frac{1}{3}$

20. $56 \cdot \frac{1}{12} \cdot \frac{1}{2}$

Chapter 7 Summary

The Meaning of Percent [7.1]

Percent means "per hundred." It is a way of comparing numbers to the number 100.

EXAMPLES

1. 42% means 42 per hundred or $\frac{42}{100}$.

Changing Percents to Decimals [7.1]

To change a percent to a decimal, drop the percent symbol (%), and move the decimal point two places to the *left.*

2. $75\% = 0.75$

Changing Decimals to Percents [7.1]

To change a decimal to a percent, move the decimal point two places to the *right,* and use the % symbol.

3. $0.25 = 25\%$

Changing Percents to Fractions [7.1]

To change a percent to a fraction, drop the % symbol, and use a denominator of 100. Reduce the resulting fraction to lowest terms if necessary.

4. $6\% = \frac{6}{100} = \frac{3}{50}$

Changing Fractions to Percents [7.1]

To change a fraction to a percent, either write the fraction as a decimal and then change the decimal to a percent, or write the fraction as an equivalent fraction with denominator 100, drop the 100, and use the % symbol.

5. $\frac{3}{4} = 0.75 = 75\%$

or

$\frac{9}{10} = \frac{90}{100} = 90\%$

Basic Word Problems Involving Percents [7.2]

There are three basic types of word problems:

Type A: What number is 14% of 68?

Type B: What percent of 75 is 25?

Type C: 25 is 40% of what number?

To solve them, we write *is* as =, *of* as · (multiply), and *what number* or *what percent* as n. We then solve the resulting equation to find the answer to the original question.

6. Translating to equations, we have:
Type A: $n = 0.14(68)$
Type B: $75n = 25$
Type C: $25 = 0.40n$

Applications of Percent [7.3, 7.4, 7.5, 7.6]

There are many different kinds of application problems involving percent. They include problems on income tax, sales tax, commission, discount, percent increase and decrease, and interest. Generally, to solve these problems, we restate them as an equivalent problem of Type A, B, or C from the previous page. Problems involving simple interest can be solved using the formula

$$I = P \cdot R \cdot T$$

where I = interest, P = principal, R = interest rate, and T = time (in years). It is standard procedure with simple interest problems to use 360 days = 1 year.

Pie Charts [7.7]

A pie chart is another way to give a visual representation of the information in a table.

Seating Class	Number Of Seats
First	18
Business	42
Coach	163

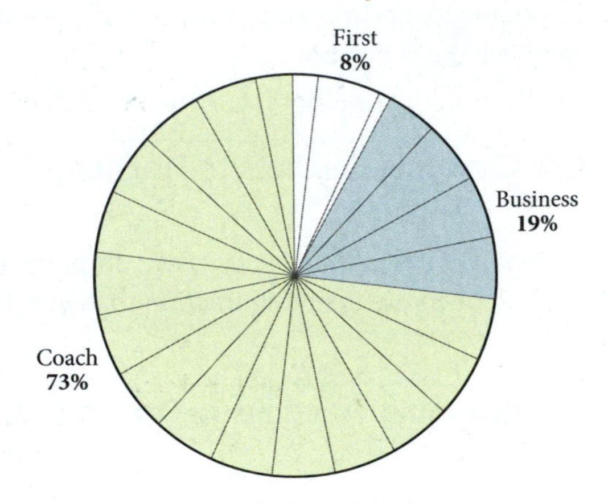

COMMON MISTAKES

1. A common mistake is forgetting to change a percent to a decimal when working problems that involve percents in the calculations. We always change percents to decimals before doing any calculations.
2. Moving the decimal point in the wrong direction when converting percents to decimals or decimals to percents is another common mistake. Remember, *percent* means "per hundred." Rewriting a number expressed as a percent as a decimal will make the numerical part smaller.

   25% = 0.25

Chapter 7 Review

Write each percent as a decimal. [7.1]

1. 35%
2. 17.8%
3. 5%
4. 0.2%

Write each decimal as a percent. [7.1]

5. 0.95
6. 0.8
7. 0.495
8. 1.65

Write each percent as a fraction or mixed number in lowest terms. [7.1]

9. 75%
10. 4%
11. 145%
12. 2.5%

Write each fraction or mixed number as a percent. [7.1]

13. $\frac{3}{10}$
14. $\frac{5}{8}$
15. $\frac{2}{3}$
16. $4\frac{3}{4}$

Solve the following problems. [7.2]

17. What number is 60% of 28?
18. What number is 122% of 55?
19. What percent of 38 is 19?
20. What percent of 19 is 38?
21. 24 is 30% of what number?
22. 16 is 8% of what number?
23. **Survey** Suppose 45 out of 60 people surveyed believe a college education will increase a person's earning potential. What percent believe this? [7.3]
24. **Discount** A lawnmower that usually sells for $175 is marked down to $140. What is the discount? What is the discount rate? [7.5]

25. Total Price A sewing machine that normally sells for \$600 is on sale for 25% off. If the sales tax rate is 6%, what is the total price of the sewing machine if it is purchased during the sale? [7.4, 7.5]

26. Home Mortgage If the interest rate on a home mortgage is 9%, then each month you pay 0.75% of the unpaid balance in interest. If the unpaid balance on one such loan is \$60,000 at the beginning of a month, how much interest must be paid that month? [7.6]

27. Percent Increase At the beginning of the summer, the price for a gallon of regular gasoline is \$4.25. By the end of summer, the price has increased 16%. What is the new price of a gallon of regular gasoline? Round to the nearest cent. [7.5]

28. Percent Decrease A gallon of regular gasoline is selling for \$1.45 in September. If the price decreases 14% in October, what is the new price for a gallon of regular gasoline? Round to the nearest cent. [7.5]

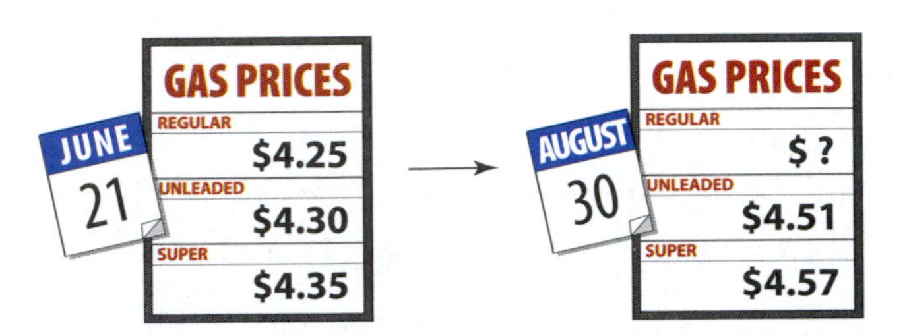

29. Medical Costs The table shows the average yearly cost of visits to the doctor, as reported in *USA Today*. What is the percent increase in cost from 1990 to 2000? Round to the nearest tenth of a percent. [7.5]

30. Commission A real estate agent gets a commission of 6% on all houses he sells. If his total sales for December are \$420,000, how much money does he make? [7.4]

MEDICAL COSTS

Year	Average Annual Cost
1990	\$583
1995	\$739
2000	\$906
2005	\$1,172

31. Discount A washing machine that usually sells for \$300 is marked down to \$240. What is the discount? What is the discount rate? [7.5]

32. Total Price A tennis racket that normally sells for \$240 is on sale for 25% off. If the sales tax rate is 5%, what is the total price of the tennis racket if it is purchased during the sale? [7.4]

33. Simple Interest If $1,800 is invested at 7% simple interest for 120 days, how much interest is earned? [7.6]

34. Compound Interest How much interest will be earned on a savings account that pays 8% compounded semiannually if $1,000 is invested for 2 years? [7.6]

36. Airline Seating The table below gives the number of seats in each of the three classes of seating on a United Airlines Boeing 777 airliner. Create a pie chart from the information in the table.

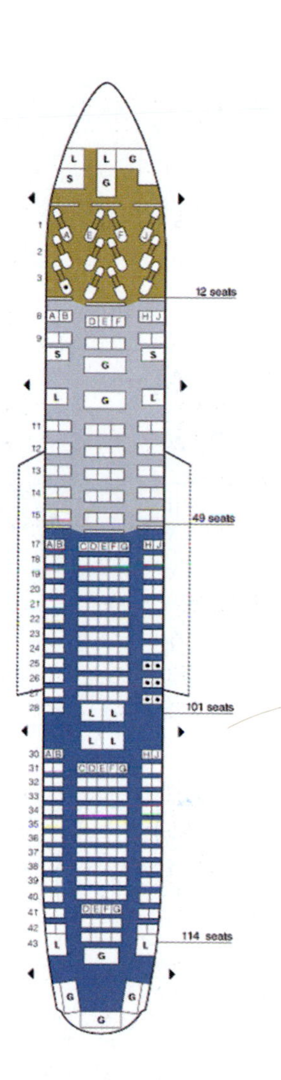

Seating Class	Number Of Seats
First	12
Business	49
Coach	215

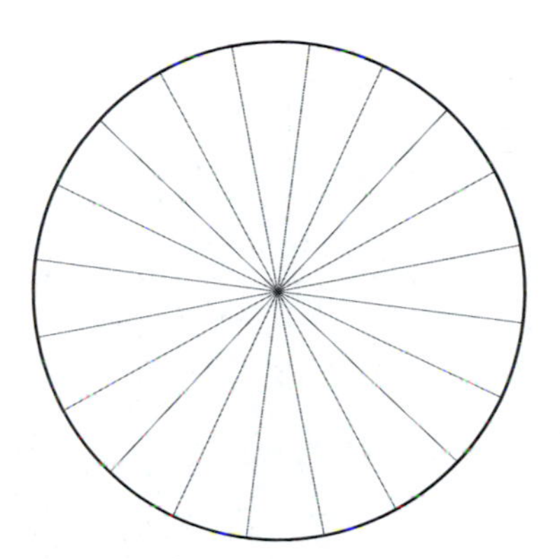

PIE CHART TEMPLATE Each slice is 5% of the area of the circle.

36. Digital Storage Space A storage device can hold 250 MB of data. The amount of space available on the disk is 100 MB. Use this information to construct a pie chart. [7.7]

Total Space	250 MB
Used space	150 MB
Free space	100 MB

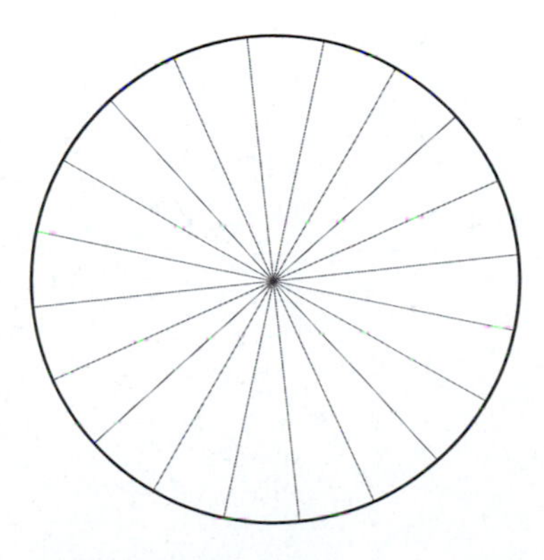

PIE CHART TEMPLATE Each slice is 5% of the area of the circle.

Chapter 7 Cumulative Review

Simplify.

1. $\begin{array}{r} 3{,}261 \\ 1{,}018 \\ +\ 297 \\ \hline \end{array}$

2. $3{,}201 - 1{,}596$

3. $\begin{array}{r} 240 \\ \times\ 21 \\ \hline \end{array}$

4. $16\overline{)1{,}448}$

5. $265.02 \div 6.31$

6. $\frac{5}{9} + \frac{1}{9}$

7. $-\frac{2}{7} \cdot \frac{3}{8}$

8. $\left(-\frac{2}{3}\right)^3$

9. $5 \cdot 3\frac{1}{2}$

10. $6 + (-2.15)$

11. $-4.8(2.01)$

12. $3\frac{1}{4} - 1\frac{1}{2}$

13. $\frac{9}{28} \div \frac{3}{16}$

14. $\frac{1}{4} - \frac{1}{5}$

15. $6.237 - (-1.38)$

16. $\frac{2}{5}(1.3) + \frac{1}{5}(2.1)$

17. $3\sqrt{100} - \sqrt{81}$

18. $11 - \frac{5}{6} \div \frac{2}{3}$

19. $3(7x + 4) - 12$

20. $19x - 3 + 4x + 10$

Solve.

21. $\frac{x}{20} = \frac{5}{4}$

22. $3x - 8 = 7x + 20$

23. Find the perimeter and area of the figure below.

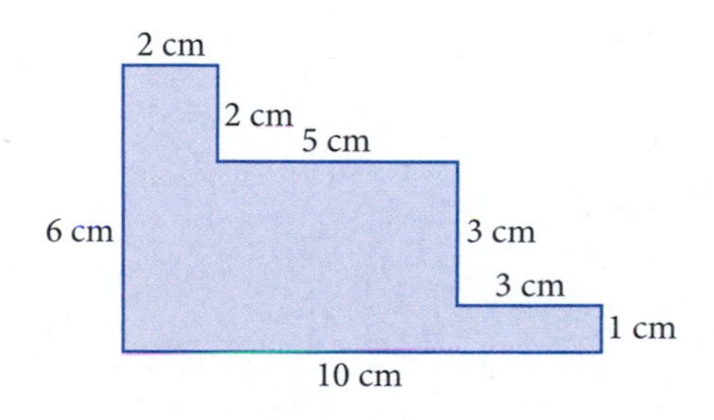

24. **Circumference of a Stamp** The stamp shown here was issued in Germany in 2000 to commemorate the 100th anniversary of soccer. Find the circumference of the circle if the radius is 14.5 millimeters.

25. Factor 264 into a product of primes.

26. Find $\frac{1}{8}$ of the quotient of 160 and 10.

27. Write the ratio 25 to 20 as a fraction in lowest terms.

28. Convert 60% to a fraction.

29. Write 3.5 as a percent.

30. Write 125% as a decimal.

31. What number is 4% of 15?

32. 3.2 is 20% of what number?

33. Discount A $20 sweater is on sale for 15% off. Find the sale price of the sweater.

34. Sales Tax The sales tax on a $25 purchase is $1.75. What is the sales tax rate?

35. Lemonade Stand Some kids sell 32 cups of lemonade for 75¢ each. How much money do they take in?

36. Geometry Find the perimeter and area of a 6.2 inch by 5 inch rectangle.

37. Model Car The model below is a $\frac{1}{18}$ size model of a 1955 Buick taxi cab. The model is 26.5 centimeters long. How long is the actual 1955 Buick?

Courtesy of Intermarket USA

38. Unit Price A 6-pack of store-brand soda is $1.25, while a 24-pack of name-brand soda is $5.99. Find the price per soda for each, and determine which is less expensive. Round prices to the nearest tenth of a cent.

39. Model Trains The water tower shown here is part of a G-gauge model railroad. The height of the model is $15\frac{3}{4}$ inches. How tall is the actual water tower if the model is $\frac{1}{29}$ scale?

Courtesy of Aristo-Craft Trains

Chapter 7 Test

Write each percent as a decimal.

1. 18%

2. 4%

3. 0.5%

Write each decimal as a percent.

4. 0.45

5. 0.7

6. 1.35

Write each percent as a fraction or a mixed number in lowest terms.

7. 65%

8. 146%

9. 3.5%

Write each number as a percent.

10. $\frac{7}{20}$

11. $\frac{3}{8}$

12. $1\frac{3}{4}$

13. What number is 75% of 60?

14. What percent of 40 is 18?

15. 16 is 20% of what number?

16. Driver's Test On a 25-question driver's test, a student answered 23 questions correctly. What percent of the questions did the student answer correctly?

17. Commission A salesperson gets an 8% commission rate on all computers she sells. If she sells $12,000 in computers in 1 day, what is her commission?

18. Discount A washing machine that usually sells for $250 is marked down to $210. What is the discount? What is the discount rate?

19. Total Price A tennis racket that normally sells for $280 is on sale for 25% off. If the sales tax rate is 5%, what is the total price of the tennis racket if it is purchased during the sale?

20. Simple Interest If $5,000 is invested at 8% simple interest for 3 months, how much interest is earned?

21. Compound Interest How much interest will be earned on a savings account that pays 10% compounded annually, if $12,000 is invested for 2 years?

Chapter 7 Projects

PERCENTS

GROUP PROJECT

Group Project

Number of People 2

Time Needed 5 minutes

Equipment Pencil, paper, and calculator.

Background All of us spend time buying clothes and eating meals at restaurants. In all of these situations, it is good practice to check receipts. This project is intended to give you practice creating receipts of your own.

Procedure Fill in the missing parts of each receipt.

SALES RECEIPT	
Jeans	29.99
Sales Tax (7.75%)	
Total	

SALES RECEIPT	
2 Buffet Dinners @ 9.99	19.98
Discount (10%)	
Total	

SALES RECEIPT	
Computer	400.00
Discount: 30% off	
Discounted Price	
Sales Tax (6%)	
Total	

SALES RECEIPT	
Couch	
Sales Tax (7%)	
Total	588.50

RESEARCH PROJECT

Credit-Card Debt

Credit-card companies are now offering credit-cards to college students who would not be able to get a card under normal credit-card criteria (due to lack of credit history and low income). The credit-card industry sees young people as a valuable market because research shows that they remain loyal to their first cards as they grow older. Nellie Mae, the student loan agency, found that 78% of college students had credit cards in 2000. For many of these students, lack of financial experience or education leads to serious debt. According to Nellie Mae, undergraduates with credit-cards carried an average balance of $2,748 in 2000. Half of credit-card-carrying college students don't pay their balances in full every month. Choose a credit-card and find out the minimum monthly payment and the APR (annual percentage rate). Compute the minimum monthly payment and interest charges for a balance of $2,748.

Stockbyte/SuperStock

Measurement

8

Chapter Outline

Introduction

The Google Earth image here shows the Nile River in Africa. The Nile is the longest river in the world, measuring 4,160 miles and stretching across ten different countries. Rivers across the world serve as important means of transportation, particularly in less developed countries, like those in Africa.

Nile River

	English Units	Metric Units
Length	4,160 mi	6,695 km
Nile Delta Area	1,004 mi^2	36,000 km^2
Flow Rate (monsoon season)	285,829 ft^3/s	8,100 m^3/s
Average Summer Temperature	86°F	30°C

Source: http://www.worldwildlife.org

In this chapter we look at the process we use to convert from one set of units, such as miles per hour, to another set of units, such as kilometers per hour. You will be interested to know that regardless of the units in question, the method we use is the same in all cases. The method is called *unit analysis* and it is the foundation of this chapter.

Chapter Pretest

The pretest below contains problems that are representative of the problems you will find in the chapter.

Make the following conversions.

1. 8 ft to inches

2. 90 in. to yards

3. 32 m to centimeters

4. 61 mm to centimeters

5. 30 yd^2 to square feet

6. 432 in^2 to square feet

7. 3,840 acres to square miles

8. 1.4 m^2 to square centimeters

9. 3 gallons to quarts

10. 72 pints to gallons

11. 251 mL to liters

12. 4 lb to ounces

13. 2,142 mg to grams

14. 9 m to yards

15. 3 gal to liters

16. 104°F to degrees Celsius

17. The speed limit on a certain road is 45 miles/hour. Convert this to feet/second.

18. If meat costs $3.05 per pound, how much will 2 lb 4 oz cost?

Getting Ready for Chapter 8

The problems below review material covered previously that you need to know in order to be successful in Chapter 8. If you have any difficulty with the problems here, you need to go back and review before going on to Chapter 8.

Write each of the following ratios as a fraction in lowest terms.

1. 12 to 30

2. 5,280 to 1,320

Simplify.

3. 12×16

4. 50×250

5. $75 \times 43{,}560$

6. $100 \times 3 \times 12$

7. 2.49×3.75

8. $5 \times 28 \times 1.36$

9. $8 \times \frac{1}{3}$

10. $25 \times \frac{1}{1000}$

11. $\frac{1800}{4}$

12. $256 \div 640$

13. $\frac{80.5}{1.61}$

14. $\frac{36.5 \times 10}{100}$

15. $\frac{1100 \times 60 \times 60}{5280}$

16. $10 \cdot \frac{12}{5}$

17. $\frac{2 \times 1000}{16.39}$ (Round to the nearest whole number.)

18. $\frac{5(102 - 32)}{9}$ (Round to the nearest tenth.)

19. Convert $\frac{12}{16}$ to a decimal.

20. Find the perimeter and area of a 24 in. × 36 in. poster.

Unit Analysis I: Length

8.1

Objectives

A Convert between lengths in the U.S. system.

B Convert between lengths in the metric system.

C Solve application problems involving unit analysis.

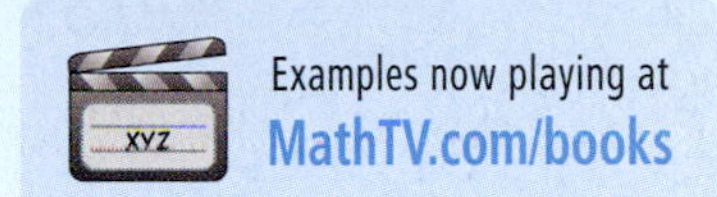

Introduction . . .

In this section we will become more familiar with the units used to measure length. We will look at the U.S. system of measurement and the metric system of measurement.

A U.S. Units of Length

Measuring the length of an object is done by assigning a number to its length. To let other people know what that number represents, we include with it a unit of measure. The most common units used to represent length in the U.S. system are inches, feet, yards, and miles. The basic unit of length is the foot. The other units are defined in terms of feet, as Table 1 shows.

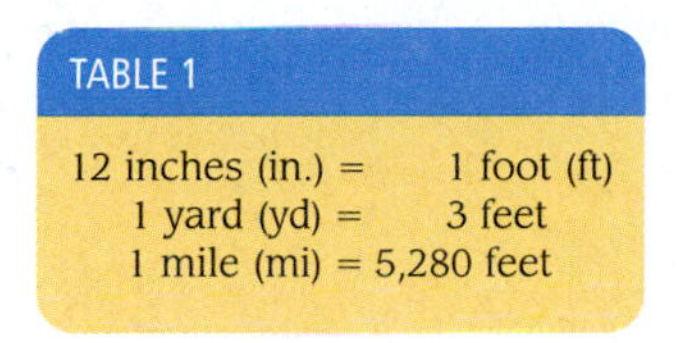

TABLE 1

12 inches (in.) =	1 foot (ft)
1 yard (yd) =	3 feet
1 mile (mi) =	5,280 feet

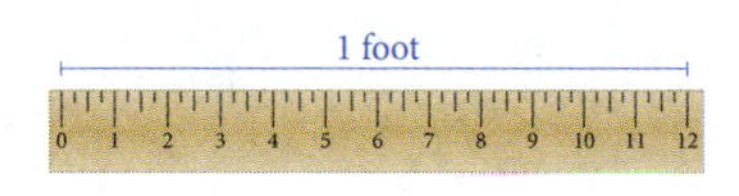

As you can see from the table, the abbreviations for inches, feet, yards, and miles are in., ft, yd, and mi, respectively. What we haven't indicated, even though you may not have realized it, is what 1 foot represents. We have defined all our units associated with length in terms of feet, but we haven't said what a foot is.

There is a long history of the evolution of what is now called a foot. At different times in the past, a foot has represented different arbitrary lengths. Currently, a foot is defined to be exactly 0.3048 meter (the basic measure of length in the metric system), where a meter is 1,650,763.73 wavelengths of the orange-red line in the spectrum of krypton-86 in a vacuum (this doesn't mean much to me either). The reason a foot and a meter are defined this way is that we always want them to measure the same length. Because the wavelength of the orange-red line in the spectrum of krypton-86 will always remain the same, so will the length that a foot represents.

Now that we have said what we mean by 1 foot (even though we may not understand the technical definition), we can go on and look at some examples that involve converting from one kind of unit to another.

EXAMPLE 1 Convert 5 feet to inches.

SOLUTION Because 1 foot = 12 inches, we can multiply 5 by 12 inches to get

$$5 \text{ feet} = 5 \times 12 \text{ inches}$$
$$= 60 \text{ inches}$$

This method of converting from feet to inches probably seems fairly simple. But as we go further in this chapter, the conversions from one kind of unit to another will become more complicated. For these more complicated problems, we need another way to show conversions so that we can be certain to end them with the correct unit of measure. For example, since 1 ft = 12 in., we can say that there are 12 in. per 1 ft or 1 ft per 12 in. That is:

$$\frac{12 \text{ in.}}{1 \text{ ft}} \longleftarrow \text{Per} \qquad \text{or} \qquad \frac{1 \text{ ft}}{12 \text{ in.}} \longleftarrow \text{Per}$$

PRACTICE PROBLEMS

1. Convert 8 feet to inches.

Answer

1. 96 in.

We call the expressions $\frac{12 \text{ in.}}{1 \text{ ft}}$ and $\frac{1 \text{ ft}}{12 \text{ in.}}$ *conversion factors*. The fraction bar is read as "per." Both these conversion factors are really just the number 1. That is:

$$\frac{12 \text{ in.}}{1 \text{ ft}} = \frac{12 \text{ in.}}{12 \text{ in.}} = 1$$

We already know that multiplying a number by 1 leaves the number unchanged. So, to convert from one unit to the other, we can multiply by one of the conversion factors without changing value. Both the conversion factors above say the same thing about the units feet and inches. They both indicate that there are 12 inches in every foot. The one we choose to multiply by depends on what units we are starting with and what units we want to end up with. If we start with feet and we want to end up with inches, we multiply by the conversion factor

$$\frac{12 \text{ in.}}{1 \text{ ft}}$$

The units of feet will divide out and leave us with inches.

$$\begin{aligned} 5 \text{ feet} &= 5 \cancel{\text{ft}} \times \frac{12 \text{ in.}}{1 \cancel{\text{ft}}} \\ &= 5 \times 12 \text{ in.} \\ &= 60 \text{ in.} \end{aligned}$$

The key to this method of conversion lies in setting the problem up so that the correct units divide out to simplify the expression. We are treating units such as feet in the same way we treated factors when reducing fractions. If a factor is common to the numerator and the denominator, we can divide it out and simplify the fraction. The same idea holds for units such as feet.

We can rewrite Table 1 so that it shows the conversion factors associated with units of length, as shown in Table 2.

Note We will use this method of converting from one kind of unit to another throughout the rest of this chapter. You should practice using it until you are comfortable with it and can use it correctly. However, it is not the only method of converting units. You may see shortcuts that will allow you to get results more quickly. Use shortcuts if you wish, so long as you can consistently get correct answers and are not using your shortcuts because you don't understand our method of conversion. Use the method of conversion as given here until you are good at it; then use shortcuts if you want to.

TABLE 2

UNITS OF LENGTH IN THE U.S. SYSTEM

The Relationship Between	Is	To Convert From One To The Other, Multiply By
feet and inches	12 in. = 1 ft	$\frac{12 \text{ in.}}{1 \text{ ft}}$ or $\frac{1 \text{ ft}}{12 \text{ in.}}$
feet and yards	1 yd = 3 ft	$\frac{3 \text{ ft}}{1 \text{ yd}}$ or $\frac{1 \text{ yd}}{3 \text{ ft}}$
feet and miles	1 mi = 5,280 ft	$\frac{5{,}280 \text{ ft}}{1 \text{ mi}}$ or $\frac{1 \text{ mi}}{5{,}280 \text{ ft}}$

EXAMPLE 2 The most common ceiling height in houses is 8 feet. How many yards is this?

8 ft

2. The roof of a two-story house is 26 feet above the ground. How many yards is this?

SOLUTION To convert 8 feet to yards, we multiply by the conversion factor $\frac{1 \text{ yd}}{3 \text{ ft}}$ so that feet will divide out and we will be left with yards.

$$8 \text{ ft} = 8 \text{ ft} \times \frac{1 \text{ yd}}{3 \text{ ft}} \qquad \text{Multiply by correct conversion factor}$$

$$= \frac{8}{3} \text{ yd} \qquad 8 \times \frac{1}{3} = \frac{8}{3}$$

$$= 2\tfrac{2}{3} \text{ yd} \qquad \text{Or 2.67 yd to the nearest hundredth}$$

EXAMPLE 3 A football field is 100 yards long. How many inches long is a football field?

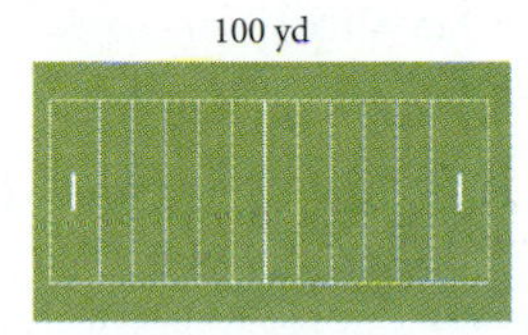

SOLUTION In this example we must convert yards to feet and then feet to inches. (To make this example more interesting, we are pretending we don't know that there are 36 inches in a yard.) We choose the conversion factors that will allow all the units except inches to divide out.

$$100 \text{ yd} = 100 \text{ yd} \times \frac{3 \text{ ft}}{1 \text{ yd}} \times \frac{12 \text{ in.}}{1 \text{ ft}}$$

$$= 100 \times 3 \times 12 \text{ in.}$$

$$= 3{,}600 \text{ in.}$$

B Metric Units of Length

In the metric system the standard unit of length is a meter. A meter is a little longer than a yard (about 3.4 inches longer). The other units of length in the metric system are written in terms of a meter. The metric system uses prefixes to indicate what part of the basic unit of measure is being used. For example, in *milli*meter the prefix *milli* means "one thousandth" of a meter. Table 3 gives the meanings of the most common metric prefixes.

TABLE 3
THE MEANING OF METRIC PREFIXES

Prefix	Meaning
milli	0.001
centi	0.01
deci	0.1
deka	10
hecto	100
kilo	1,000

We can use these prefixes to write the other units of length and conversion factors for the metric system, as given in Table 4.

3. How many inches are in 220 yards?

Answers

2. $8\frac{2}{3}$ yd, or 8.67 yd
3. 7,920 in.

TABLE 4

METRIC UNITS OF LENGTH

The Relationship Between	Is	To Convert From One To The Other, Multiply By
millimeters (mm) and meters (m)	1,000 mm = 1 m	$\frac{1{,}000 \text{ mm}}{1 \text{ m}}$ or $\frac{1 \text{ m}}{1{,}000 \text{ mm}}$
centimeters (cm) and meters	100 cm = 1 m	$\frac{100 \text{ cm}}{1 \text{ m}}$ or $\frac{1 \text{ m}}{100 \text{ cm}}$
decimeters (dm) and meters	10 dm = 1 m	$\frac{10 \text{ dm}}{1 \text{ m}}$ or $\frac{1 \text{ m}}{10 \text{ dm}}$
dekameters (dam) and meters	1 dam = 10 m	$\frac{10 \text{ m}}{1 \text{ dam}}$ or $\frac{1 \text{ dam}}{10 \text{ m}}$
hectometers (hm) and meters	1 hm = 100 m	$\frac{100 \text{ m}}{1 \text{ hm}}$ or $\frac{1 \text{ hm}}{100 \text{ m}}$
kilometers (km) and meters	1 km = 1,000 m	$\frac{1{,}000 \text{ m}}{1 \text{ km}}$ or $\frac{1 \text{ km}}{1{,}000 \text{ m}}$

We use the same method to convert between units in the metric system as we did with the U.S. system. We choose the conversion factor that will allow the units we start with to divide out, leaving the units we want to end up with.

4. Convert 67 centimeters to meters.

EXAMPLE 4 Convert 25 millimeters to meters.

SOLUTION To convert from millimeters to meters, we multiply by the conversion factor $\frac{1 \text{ m}}{1{,}000 \text{ mm}}$:

$$\begin{aligned} 25 \text{ mm} &= 25 \cancel{\text{mm}} \times \frac{1 \text{ m}}{1{,}000 \cancel{\text{mm}}} \\ &= \frac{25 \text{ m}}{1{,}000} \\ &= 0.025 \text{ m} \end{aligned}$$

5. Convert 78.4 mm to decimeters.

EXAMPLE 5 Convert 36.5 centimeters to decimeters.

SOLUTION We convert centimeters to meters and then meters to decimeters:

$$\begin{aligned} 36.5 \text{ cm} &= 36.5 \cancel{\text{cm}} \times \frac{1 \cancel{\text{m}}}{100 \cancel{\text{cm}}} \times \frac{10 \text{ dm}}{1 \cancel{\text{m}}} \\ &= \frac{36.5 \times 10}{100} \text{dm} \\ &= 3.65 \text{ dm} \end{aligned}$$

The most common units of length in the metric system are millimeters, centimeters, meters, and kilometers. The other units of length we have listed in our table of metric lengths are not as widely used. The method we have used to convert from one unit of length to another in Examples 2–5 is called *unit analysis*. If you take a chemistry class, you will see it used many times. The same is true of many other science classes as well.

Answers
4. 0.67 m **5.** 0.784 dm

We can summarize the procedure used in unit analysis with the following steps:

Strategy Unit Analysis

Step 1: Identify the units you are starting with.

Step 2: Identify the units you want to end with.

Step 3: Find conversion factors that will bridge the starting units and the ending units.

Step 4: Set up the multiplication problem so that all units except the units you want to end with will divide out.

C Applications

EXAMPLE 6 A sheep rancher is making new lambing pens for the upcoming lambing season. Each pen is a rectangle 6 feet wide and 8 feet long. The fencing material he wants to use sells for $1.36 per foot. If he is planning to build five separate lambing pens (they are separate because he wants a walkway between them), how much will he have to spend for fencing material?

SOLUTION To find the amount of fencing material he needs for one pen, we find the perimeter of a pen.

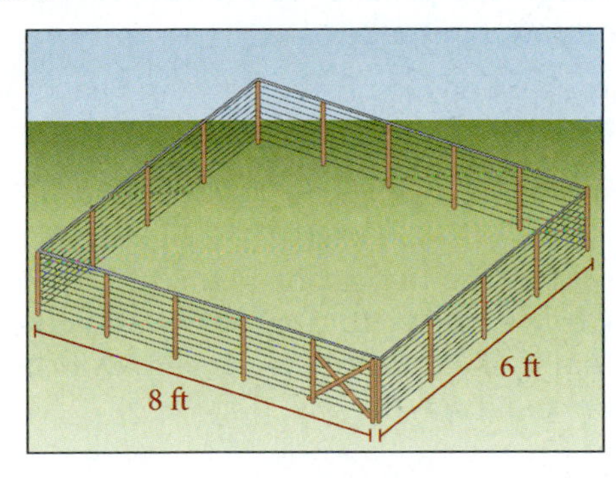

$$\text{Perimeter} = 6 + 6 + 8 + 8 = 28 \text{ feet}$$

We set up the solution to the problem using unit analysis. Our starting unit is *pens* and our ending unit is *dollars*. Here are the conversion factors that will form a bridge between pens and dollars:

$$1 \text{ pen} = 28 \text{ feet of fencing}$$
$$1 \text{ foot of fencing} = 1.36 \text{ dollars}$$

Next we write the multiplication problem, using the conversion factors, that will allow all the units except dollars to divide out:

$$5 \text{ pens} = 5 \cancel{\text{pens}} \times \frac{28 \cancel{\text{feet of fencing}}}{1 \cancel{\text{pen}}} \times \frac{1.36 \text{ dollars}}{1 \cancel{\text{foot of fencing}}}$$

$$= 5 \times 28 \times 1.36 \text{ dollars}$$
$$= \$190.40$$

6. The rancher in Example 6 decides to build six pens instead of five and upgrades his fencing material so that it costs $1.72 per foot. How much does it cost him to build the six pens?

Answer
6. $288.96

7. Assume that the mistake in the advertisement is that feet per second should read feet per minute. Is 1,100 feet per minute a reasonable speed for a chair lift?

EXAMPLE 7 A number of years ago, a ski resort in Vermont advertised their new high-speed chair lift as "the world's fastest chair lift, with a speed of 1,100 feet per second." Show why the speed cannot be correct.

SOLUTION To solve this problem, we can convert feet per second into miles per hour, a unit of measure we are more familiar with on an intuitive level. Here are the conversion factors we will use:

$$1 \text{ mile} = 5{,}280 \text{ feet}$$

$$1 \text{ hour} = 60 \text{ minutes}$$

$$1 \text{ minute} = 60 \text{ seconds}$$

$$1{,}100 \text{ ft/second} = \frac{1{,}100 \cancel{\text{feet}}}{1 \cancel{\text{second}}} \times \frac{1 \text{ mile}}{5{,}280 \cancel{\text{feet}}} \times \frac{60 \cancel{\text{seconds}}}{1 \cancel{\text{minute}}} \times \frac{60 \cancel{\text{minutes}}}{1 \text{ hour}}$$

$$= \frac{1{,}100 \times 60 \times 60 \text{ miles}}{5{,}280 \text{ hours}}$$

$$= 750 \text{ miles/hour}$$

Getting Ready for Class

After reading through the preceding section, respond in your own words and in complete sentences.

1. Write the relationship between feet and miles. That is, write an equality that shows how many feet are in every mile.
2. Give the metric prefix that means "one hundredth."
3. Give the metric prefix that is equivalent to 1,000.
4. As you know from reading the section in the text, conversion factors are ratios. Write the conversion factor that will allow you to convert from inches to feet. That is, if we wanted to convert 27 inches to feet, what conversion factor would we use?

Answer

7. 12.5 mi/hr is a reasonable speed for a chair lift.

Problem Set 8.1

A Make the following conversions in the U.S. system by multiplying by the appropriate conversion factor. Write your answers as whole numbers or mixed numbers. [Examples 1–3]

1. 5 ft to inches

2. 9 ft to inches

3. 10 ft to inches

4. 20 ft to inches

5. 2 yd to feet

6. 8 yd to feet

7. 4.5 yd to inches

8. 9.5 yd to inches

9. 27 in. to feet

10. 36 in. to feet

11. 2.5 mi to feet

12. 6.75 mi to feet

13. 48 in. to yards

14. 56 in. to yards

B Make the following conversions in the metric system by multiplying by the appropriate conversion factor. Write your answers as whole numbers or decimals. [Examples 4, 5]

15. 18 m to centimeters

16. 18 m to millimeters

17. 4.8 km to meters

18. 8.9 km to meters

19. 5 dm to centimeters

20. 12 dm to millimeters

21. 248 m to kilometers

22. 969 m to kilometers

23. 67 cm to millimeters

24. 67 mm to centimeters

25. 3,498 cm to meters

26. 4,388 dm to meters

27. 63.4 cm to decimeters

28. 89.5 cm to decimeters

C Applying the Concepts [Examples 6, 7]

29. Mountains The map shows the heights of the tallest mountains in the world. According to the map, K2 is 28,238 ft. Convert this to miles. Round to the nearest tenth of a mile.

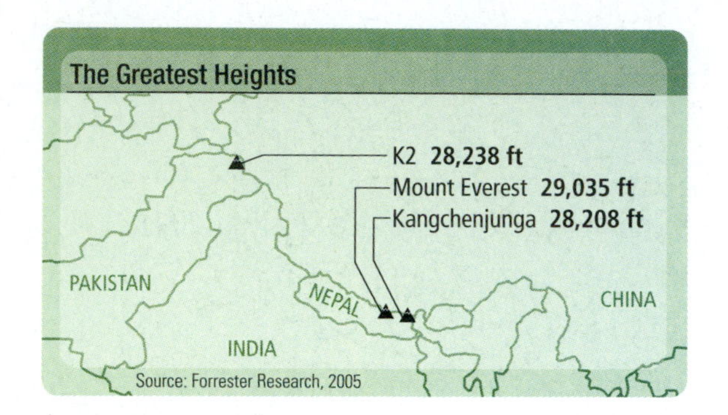

30. Classroom Energy The chart shows how much energy is wasted in the classroom by leaving appliances on.

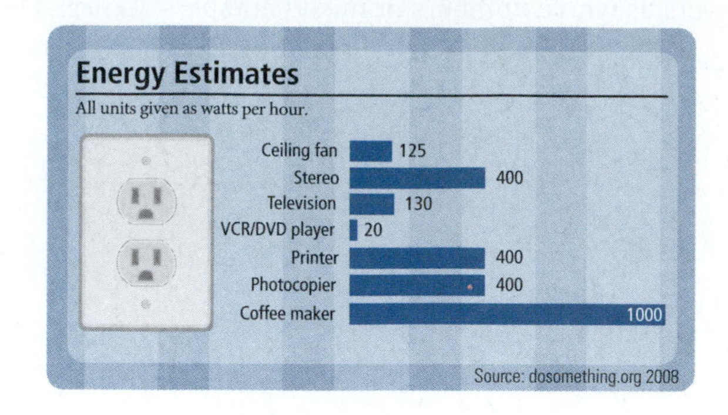

Convert the the wattage of the following appliances to kilowatts.

a. Ceiling fan
b. VCR/DVD player
c. Coffee maker

31. Softball If the distance between first and second base in softball is 60 feet, how many yards is it from first to second base?

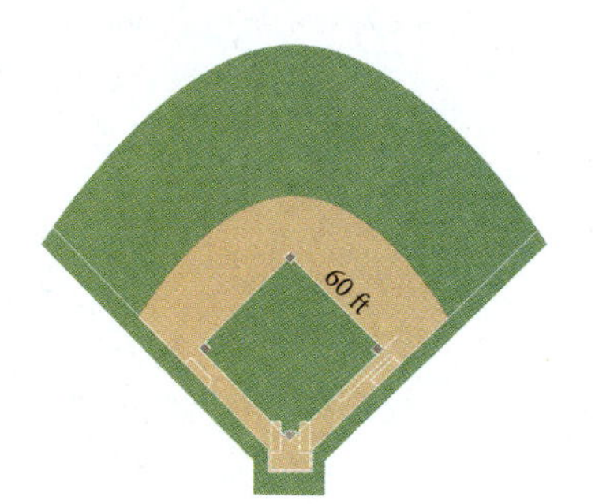

32. Notebook Width Standard-sized notebook paper is 21.6 centimeters wide. Express this width in millimeters.

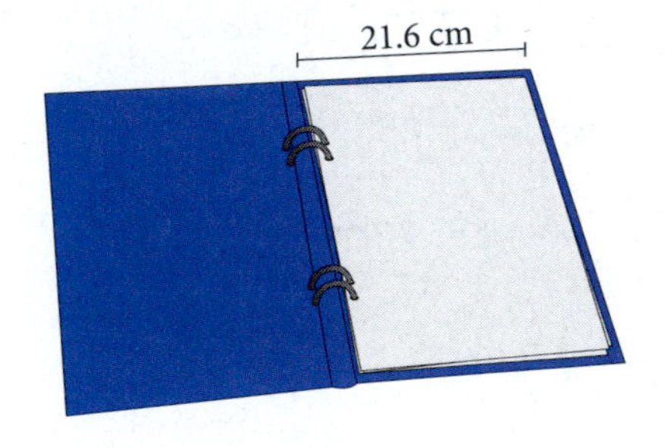

33. High Jump If a person high jumps 6 feet 8 inches, how many inches is the jump?

34. Desk Width A desk is 48 inches wide. What is the width in yards?

35. Ceiling Height Suppose the ceiling of a home is 2.44 meters above the floor. Express the height of the ceiling in centimeters.

36. Tower Height A transmitting tower is 100 feet tall. How many inches is that?

37. Surveying A unit of measure sometimes used in surveying is the *chain*. There are 80 chains in 1 mile. How many chains are in 37 miles?

38. Surveying Another unit of measure used in surveying is a *link;* 1 link is about 8 inches. About how many links are there in 5 feet?

39. Metric System A very small unit of measure in the metric system is the *micron* (abbreviated μm). There are 1,000 μm in 1 millimeter. How many microns are in 12 centimeters?

40. Metric System Another very small unit of measure in the metric system is the *angstrom* (abbreviated Å). There are 10,000,000 Å in 1 millimeter. How many angstroms are in 15 decimeters?

41. **Horse Racing** In horse racing, 1 *furlong* is 220 yards. How many feet are in 12 furlongs?

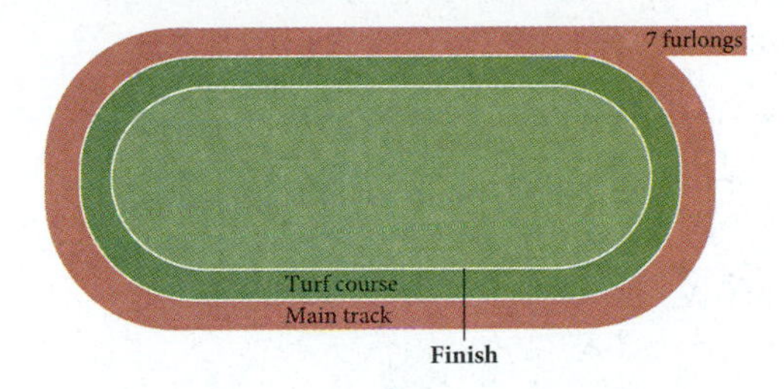

42. **Speed of a Bullet** A bullet from a machine gun on a B-17 Flying Fortress in World War II had a muzzle speed of 1,750 feet/second. Convert 1,750 feet/second to miles/hour. (Round to the nearest whole number.)

Courtesy of the U.S. Air Force Museum

43. **Speed Limit** The maximum speed limit on part of Highway 101 in California is 55 miles/hour. Convert 55 miles/hour to feet/second. (Round to the nearest tenth.)

44. **Speed Limit** The maximum speed limit on part of Highway 5 in California is 65 miles/hour. Convert 65 miles/hour to feet/second. (Round to the nearest tenth.)

45. **Track and Field** A person who runs the 100-yard dash in 10.5 seconds has an average speed of 9.52 yards/second. Convert 9.52 yards/second to miles/hour. (Round to the nearest tenth.)

46. **Track and Field** A person who runs a mile in 8 minutes has an average speed of 0.125 miles/minute. Convert 0.125 miles/minute to miles/hour.

47. **Speed of a Bullet** The bullet from a rifle leaves the barrel traveling 1,500 feet/second. Convert 1,500 feet/second to miles/hour. (Round to the nearest whole number.)

48. **Sailing** A *fathom* is 6 feet. How many yards are in 19 fathoms?

Calculator Problems

Set up the following conversions as you have been doing. Then perform the calculations on a calculator.

49. Change 751 miles to feet.

50. Change 639.87 centimeters to meters.

51. Change 4,982 yards to inches.

52. Change 379 millimeters to kilometers.

53. Mount Whitney is the highest point in California. It is 14,494 feet above sea level. Give its height in miles to the nearest tenth.

54. The tallest mountain in the United States is Mount McKinley in Alaska. It is 20,320 feet tall. Give its height in miles to the nearest tenth.

55. California has 3,427 miles of shoreline. How many feet is this?

56. The tip of the TV tower at the top of the Empire State Building in New York City is 1,472 feet above the ground. Express this height in miles to the nearest hundredth.

Getting Ready for the Next Section

Perform the indicated operations.

57. 12×12

58. 36×24

59. $1 \times 4 \times 2$

60. $5 \times 4 \times 2$

61. $10 \times 10 \times 10$

62. $100 \times 100 \times 100$

63. $75 \times 43{,}560$

64. $55 \times 43{,}560$

65. $864 \div 144$

66. $1{,}728 \div 144$

67. $256 \div 640$

68. $960 \div 240$

69. $45 \times \frac{9}{1}$

70. $36 \times \frac{9}{1}$

71. $1{,}800 \times \frac{1}{4}$

72. $2{,}000 \times \frac{1}{4} \times \frac{1}{10}$

73. 1.5×30

74. 1.5×45

75. $2.2 \times 1{,}000$

76. $3.5 \times 1{,}000$

77. 67.5×9

78. 43.5×9

Maintaining Your Skills

Write your answers as whole numbers, proper fractions, or mixed numbers.

Find each product. (Multiply.)

79. $\frac{2}{3} \cdot \frac{1}{2}$

80. $\frac{7}{9} \cdot \frac{3}{14}$

81. $8 \cdot \frac{3}{4}$

82. $12 \cdot \frac{1}{3}$

83. $1\frac{1}{2} \cdot 2\frac{1}{3}$

84. $\frac{1}{6} \cdot 4\frac{2}{3}$

Find each quotient. (Divide.)

85. $\frac{3}{4} \div \frac{1}{8}$

86. $\frac{3}{5} \div \frac{6}{25}$

87. $4 \div \frac{2}{3}$

88. $1 \div \frac{1}{3}$

89. $1\frac{3}{4} \div 2\frac{1}{2}$

90. $\frac{9}{8} \div 1\frac{7}{8}$

Extending the Concepts

91. Fitness Walking The guidelines for fitness now indicate that a person who walks 10,000 steps daily is physically fit. According to *The Walking Site* on the Internet, "The average person's stride length is approximately 2.5 feet long. That means it takes just over 2,000 steps to walk one mile, and 10,000 steps is close to 5 miles." Use your knowledge of unit analysis to determine if these facts are correct.

Unit Analysis II: Area and Volume

8.2

Objectives

A Convert between areas using the U.S. system.

B Convert between areas using the metric system.

C Convert between volumes using the U.S. system.

D Convert between volumes using the metric system.

Figure 1 below gives a summary of the geometric objects we have worked with in previous chapters, along with the formulas for finding the area of each object.

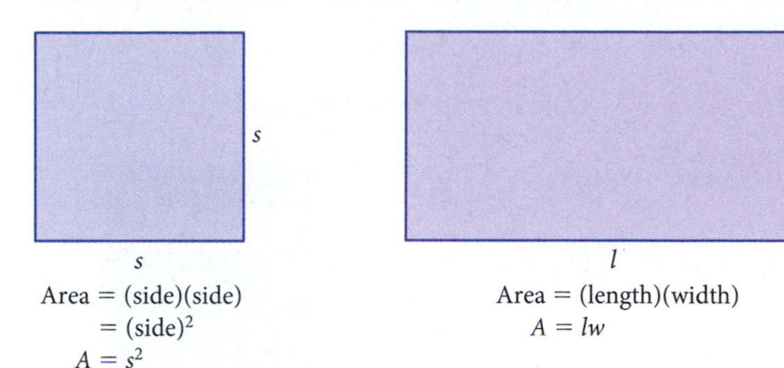

Area = (side)(side)
= $(\text{side})^2$
$A = s^2$

Area = (length)(width)
$A = lw$

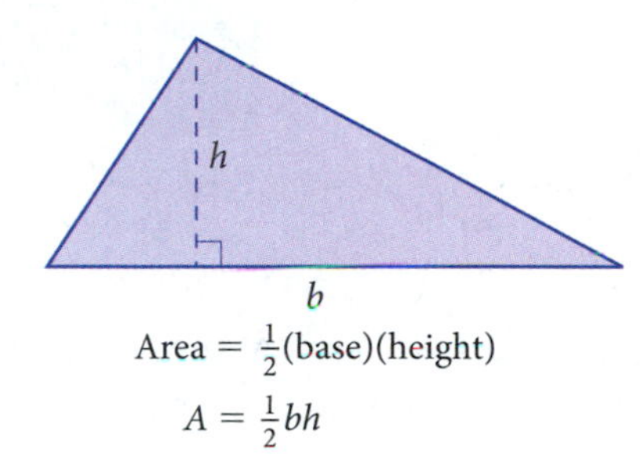

Area = $\frac{1}{2}$(base)(height)
$A = \frac{1}{2}bh$

FIGURE 1 Areas of common geometric shapes

A Conversion Factors in the U.S. System

EXAMPLE 1 Find the number of square inches in 1 square foot.

SOLUTION We can think of 1 square foot as 1 ft² = 1 ft × ft. To convert from feet to inches, we use the conversion factor 1 foot = 12 inches. Because the unit foot appears twice in 1 ft², we multiply by our conversion factor twice.

$$1 \text{ ft}^2 = 1 \cancel{\text{ft}} \times \cancel{\text{ft}} \times \frac{12 \text{ in.}}{1 \cancel{\text{ft}}} \times \frac{12 \text{ in.}}{1 \cancel{\text{ft}}} = 12 \times 12 \text{ in.} \times \text{in.} = 144 \text{ in}^2$$

Now that we know that 1 ft² is the same as 144 in², we can use this fact as a conversion factor to convert between square feet and square inches. Depending on which units we are converting from, we would use either

$$\frac{144 \text{ in}^2}{1 \text{ ft}^2} \quad \text{or} \quad \frac{1 \text{ ft}^2}{144 \text{ in}^2}$$

PRACTICE PROBLEMS

1. Find the number of square feet in 1 square yard.

Answer

1. 1 yd² = 9 ft²

2. If the poster in Example 2 is surrounded by a frame 6 inches wide, find the number of square feet of wall space covered by the framed poster.

EXAMPLE 2 A rectangular poster measures 36 inches by 24 inches. How many square feet of wall space will the poster cover?

SOLUTION One way to work this problem is to find the number of square inches the poster covers, and then convert square inches to square feet.

$$\text{Area of poster} = \text{length} \times \text{width} = 36 \text{ in.} \times 24 \text{ in.} = 864 \text{ in}^2$$

To finish the problem, we convert square inches to square feet:

$$\begin{aligned} 864 \text{ in}^2 &= 864 \cancel{\text{in}^2} \times \frac{1 \text{ ft}^2}{144 \cancel{\text{in}^2}} \\ &= \frac{864}{144} \text{ ft}^2 \\ &= 6 \text{ ft}^2 \end{aligned}$$

Table 1 gives the most common units of area in the U.S. system of measurement, along with the corresponding conversion factors.

TABLE 1

U.S. UNITS OF AREA

The Relationship Between	Is	To Convert From One To The Other, Multiply By
square inches and square feet	$144 \text{ in}^2 = 1 \text{ ft}^2$	$\frac{144 \text{ in}^2}{1 \text{ ft}^2}$ or $\frac{1 \text{ ft}^2}{144 \text{ in}^2}$
square yards and square feet	$9 \text{ ft}^2 = 1 \text{ yd}^2$	$\frac{9 \text{ ft}^2}{1 \text{ yd}^2}$ or $\frac{1 \text{ yd}^2}{9 \text{ ft}^2}$
acres and square feet	$1 \text{ acre} = 43{,}560 \text{ ft}^2$	$\frac{43{,}560 \text{ ft}^2}{1 \text{ acre}}$ or $\frac{1 \text{ acre}}{43{,}560 \text{ ft}^2}$
acres and square miles	$640 \text{ acres} = 1 \text{ mi}^2$	$\frac{640 \text{ acres}}{1 \text{ mi}^2}$ or $\frac{1 \text{ mi}^2}{640 \text{ acres}}$

3. The same dressmaker orders a bolt of material that is 1.5 yards wide and 45 yards long. How many square feet of material were ordered?

EXAMPLE 3 A dressmaker orders a bolt of material that is 1.5 yards wide and 30 yards long. How many square feet of material were ordered?

SOLUTION The area of the material in square yards is

$$\begin{aligned} A &= 1.5 \times 30 \\ &= 45 \text{ yd}^2 \end{aligned}$$

Converting this to square feet, we have

$$\begin{aligned} 45 \text{ yd}^2 &= 45 \cancel{\text{yd}^2} \times \frac{9 \text{ ft}^2}{1 \cancel{\text{yd}^2}} \\ &= 405 \text{ ft}^2 \end{aligned}$$

Answers

2. 12 ft^2 3. 607.5 ft^2

EXAMPLE 4 A farmer has 75 acres of land. How many square feet of land does the farmer have?

SOLUTION Changing acres to square feet, we have

$$75 \text{ acres} = 75 \text{ acres} \times \frac{43{,}560 \text{ ft}^2}{1 \text{ acre}}$$
$$= 75 \times 43{,}560 \text{ ft}^2$$
$$= 3{,}267{,}000 \text{ ft}^2$$

4. A farmer has 55 acres of land. How many square feet of land does the farmer have?

EXAMPLE 5 A new shopping center is to be constructed on 256 acres of land. How many square miles is this?

SOLUTION Multiplying by the conversion factor that will allow acres to divide out, we have

$$256 \text{ acres} = 256 \text{ acres} \times \frac{1 \text{ mi}^2}{640 \text{ acres}}$$
$$= \frac{256}{640} \text{ mi}^2$$
$$= 0.4 \text{ mi}^2$$

5. A school is to be constructed on 960 acres of land. How many square miles is this?

B Area: The Metric System

Units of area in the metric system are considerably simpler than those in the U.S. system because metric units are given in terms of powers of 10. Table 2 lists the conversion factors that are most commonly used.

TABLE 2

METRIC UNITS OF AREA

The Relationship Between	Is	To Convert From One To The Other, Multiply By
square millimeters and square centimeters	$1 \text{ cm}^2 = 100 \text{ mm}^2$	$\frac{100 \text{ mm}^2}{1 \text{ cm}^2}$ or $\frac{1 \text{ cm}^2}{100 \text{ mm}^2}$
square centimeters and square decimeters	$1 \text{ dm}^2 = 100 \text{ cm}^2$	$\frac{100 \text{ cm}^2}{1 \text{ dm}^2}$ or $\frac{1 \text{ dm}^2}{100 \text{ cm}^2}$
square decimeters and square meters	$1 \text{ m}^2 = 100 \text{ dm}^2$	$\frac{100 \text{ dm}^2}{1 \text{ m}^2}$ or $\frac{1 \text{ m}^2}{100 \text{ dm}^2}$
square meters and ares (a)	$1 \text{ a} = 100 \text{ m}^2$	$\frac{100 \text{ m}^2}{1 \text{ a}}$ or $\frac{1 \text{ a}}{100 \text{ m}^2}$
ares and hectares (ha)	$1 \text{ ha} = 100 \text{ a}$	$\frac{100 \text{ a}}{1 \text{ ha}}$ or $\frac{1 \text{ ha}}{100 \text{ a}}$

Answers
4. $2{,}395{,}800 \text{ ft}^2$
5. 1.5 mi^2

6. How many square centimeters are in 1 square meter?

EXAMPLE 6 How many square millimeters are in 1 square meter?

SOLUTION We start with 1 m² and end up with square millimeters:

$$1 \text{ m}^2 = 1 \cancel{\text{m}^2} \times \frac{100 \cancel{\text{dm}^2}}{1 \cancel{\text{m}^2}} \times \frac{100 \cancel{\text{cm}^2}}{1 \cancel{\text{dm}^2}} \times \frac{100 \text{ mm}^2}{1 \cancel{\text{cm}^2}}$$

$$= 100 \times 100 \times 100 \text{ mm}^2$$
$$= 1{,}000{,}000 \text{ mm}^2$$

C Volume: The U.S. System

Table 3 lists the units of volume in the U.S. system and their conversion factors.

TABLE 3

UNITS OF VOLUME IN THE U.S. SYSTEM

The Relationship Between	Is	To Convert From One To The Other, Multiply By
cubic inches (in³) and cubic feet (ft³)	1 ft³ = 1,728 in³	$\frac{1{,}728 \text{ in}^3}{1 \text{ ft}^3}$ or $\frac{1 \text{ ft}^3}{1{,}728 \text{ in}^3}$
cubic feet and cubic yards (yd³)	1 yd³ = 27 ft³	$\frac{27 \text{ ft}^3}{1 \text{ yd}^3}$ or $\frac{1 \text{ yd}^3}{27 \text{ ft}^3}$
fluid ounces (fl oz) and pints (pt)	1 pt = 16 fl oz	$\frac{16 \text{ fl oz}}{1 \text{ pt}}$ or $\frac{1 \text{ pt}}{16 \text{ fl oz}}$
pints and quarts (qt)	1 qt = 2 pt	$\frac{2 \text{ pt}}{1 \text{ qt}}$ or $\frac{1 \text{ qt}}{2 \text{ pt}}$
quarts and gallons (gal)	1 gal = 4 qt	$\frac{4 \text{ qt}}{1 \text{ gal}}$ or $\frac{1 \text{ gal}}{4 \text{ qt}}$

7. How many pints are in a 5-gallon pail?

EXAMPLE 7 What is the capacity (volume) in pints of a 1-gallon container of milk?

SOLUTION We change from gallons to quarts and then quarts to pints by multiplying by the appropriate conversion factors as given in Table 3.

$$1 \text{ gal} = 1 \cancel{\text{gal}} \times \frac{4 \cancel{\text{qt}}}{1 \cancel{\text{gal}}} \times \frac{2 \text{ pt}}{1 \cancel{\text{qt}}}$$

$$= 1 \times 4 \times 2 \text{ pt}$$

$$= 8 \text{ pt}$$

A 1-gallon container has the same capacity as 8 one-pint containers.

Answers
6. 10,000 cm² 7. 40 pt

EXAMPLE 8 A dairy herd produces 1,800 quarts of milk each day. How many gallons is this equivalent to?

SOLUTION Converting 1,800 quarts to gallons, we have

$$\begin{aligned} 1{,}800 \text{ qt} &= 1{,}800 \cancel{\text{qt}} \times \frac{1 \text{ gal}}{4 \cancel{\text{qt}}} \\ &= \frac{1{,}800}{4} \text{ gal} \\ &= 450 \text{ gal} \end{aligned}$$

We see that 1,800 quarts is equivalent to 450 gallons.

8. A dairy herd produces 2,000 quarts of milk each day. How many 10-gallon containers will this milk fill?

D Volume: The Metric System

In the metric system the basic unit of measure for volume is the liter. A liter is the volume enclosed by a cube that is 10 cm on each edge, as shown in Figure 2. We can see that a liter is equivalent to 1,000 cm³.

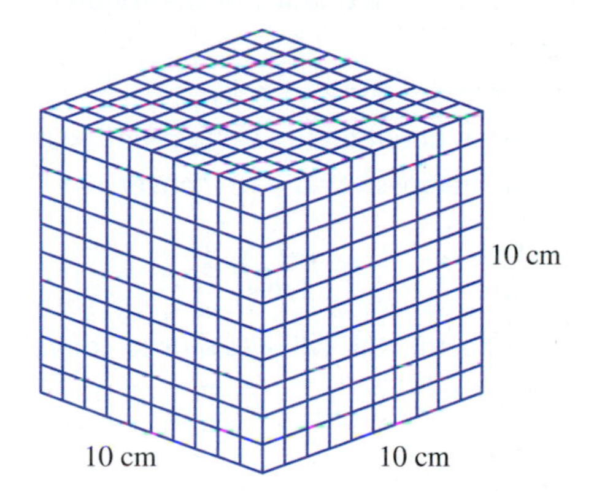

$$\begin{aligned} 1 \text{ liter} &= 10 \text{ cm} \times 10 \text{ cm} \times 10 \text{ cm} \\ &= 1{,}000 \text{ cm}^3 \end{aligned}$$

FIGURE 2

The other units of volume in the metric system use the same prefixes we encountered previously. The units with prefixes centi, deci, and deka are not as common as the others, so in Table 4 we include only liters, milliliters, hectoliters, and kiloliters.

TABLE 4

METRIC UNITS OF VOLUME

The Relationship Between	Is	To Convert From One To The Other, Multiply By
milliliters (mL) and liters	1 liter (L) = 1,000 mL	$\frac{1{,}000 \text{ mL}}{1 \text{ liter}}$ or $\frac{1 \text{ liter}}{1{,}000 \text{ mL}}$
hectoliters (hL) and liters	100 liters = 1 hL	$\frac{100 \text{ liters}}{1 \text{ hL}}$ or $\frac{1 \text{ hL}}{100 \text{ liters}}$
kiloliters (kL) and liters	1,000 liters (L) = 1 kL	$\frac{1{,}000 \text{ liters}}{1 \text{ kL}}$ or $\frac{1 \text{ kL}}{1{,}000 \text{ liters}}$

Note As you can see from the table and the discussion above, a cubic centimeter (cm³) and a milliliter (mL) are equal. Both are one thousandth of a liter. It is also common in some fields (like medicine) to abbreviate the term cubic centimeter as cc. Although we will use the notation mL when discussing volume in the metric system, you should be aware that

$1 \text{ mL} = 1 \text{ cm}^3 = 1 \text{ cc}$.

Answer

8. 50 containers

9. A 3.5-liter engine will have a volume of how many milliliters?

Here is an example of conversion from one unit of volume to another in the metric system.

EXAMPLE 9 A sports car has a 2.2-liter engine. What is the displacement (volume) of the engine in milliliters?

SOLUTION Using the appropriate conversion factor from Table 4, we have

$$2.2 \text{ liters} = 2.2 \text{ liters} \times \frac{1{,}000 \text{ mL}}{1 \text{ liter}}$$

$$= 2.2 \times 1{,}000 \text{ mL}$$

$$= 2{,}200 \text{ mL}$$

Getting Ready for Class

After reading through the preceding section, respond in your own words and in complete sentences.

1. Write the formula for the area of each of the following:
   a. a square of side *s*.
   b. a rectangle with length *l* and width *w*.
2. What is the relationship between square feet and square inches?
3. Fill in the numerators below so that each conversion factor is equal to 1.

   a. $\frac{\text{qt}}{1 \text{ gal}}$ b. $\frac{\text{mL}}{1 \text{ liter}}$ c. $\frac{\text{acres}}{1 \text{ mi}^2}$

4. Write the conversion factor that will allow us to convert from square yards to square feet.

Answer

9. 3,500 mL

Problem Set 8.2

A Use the tables given in this section to make the following conversions. Be sure to show the conversion factor used in each case. [Examples 1–5]

1. 3 ft^2 to square inches

2. 5 ft^2 to square inches

3. 288 in^2 to square feet

4. 720 in^2 to square feet

5. 30 acres to square feet

6. 92 acres to square feet

7. 2 mi^2 to acres

8. 7 mi^2 to acres

9. 1,920 acres to square miles

10. 3,200 acres to square miles

11. 12 yd^2 to square feet

12. 20 yd^2 to square feet

B [Example 6]

13. 17 cm^2 to square millimeters

14. 150 mm^2 to square centimeters

15. 2.8 m^2 to square centimeters

16. 10 dm^2 to square millimeters

17. 1,200 mm^2 to square meters

18. 19.79 cm^2 to square meters

19. 5 a to square meters

20. 12 a to square centimeters

21. 7 ha to ares

22. 3.6 ha to ares

23. 342 a to hectares

24. 986 a to hectares

C D Make the following conversions using the conversion factors given in Tables 3 and 4. [Examples 7–9]

25. 5 yd^3 to cubic feet

26. 3.8 yd^3 to cubic feet

27. 3 pt to fluid ounces

28. 8 pt to fluid ounces

29. 2 gal to quarts

30. 12 gal to quarts

31. 2.5 gal to pints

32. 7 gal to pints

33. 15 qt to fluid ounces

34. 5.9 qt to fluid ounces

35. 64 pt to gallons

36. 256 pt to gallons

37. 12 pt to quarts

38. 18 pt to quarts

39. 243 ft^3 to cubic yards

40. 864 ft^3 to cubic yards

41. 5 L to milliliters

42. 9.6 L to milliliters

43. 127 mL to liters

44. 93.8 mL to liters

45. 4 kL to milliliters

46. 3 kL to milliliters

47. 14.92 kL to liters

48. 4.71 kL to liters

Applying the Concepts

49. Google Earth The Google Earth map shows Yellowstone National Park. If the area of the park is roughly 3,402 square miles, how many acres does the park cover?

50. Google Earth The Google Earth image shows an aerial view of a crop circle found near Wroughton, England. If the crop circle has a radius of about 59 meters, how many ares does it cover? Round to the nearest are.

51. Swimming Pool A public swimming pool measures 100 meters by 30 meters and is rectangular. What is the area of the pool in ares?

52. Construction A family decides to put tiles in the entryway of their home. The entryway has an area of 6 square meters. If each tile is 5 centimeters by 5 centimeters, how many tiles will it take to cover the entryway?

53. Landscaping A landscaper is putting in a brick patio. The area of the patio is 110 square meters. If the bricks measure 10 centimeters by 20 centimeters, how many bricks will it take to make the patio? Assume no space between bricks.

54. Sewing A dressmaker is using a pattern that requires 2 square yards of material. If the material is on a bolt that is 54 inches wide, how long a piece of material must be cut from the bolt to be sure there is enough material for the pattern?

55. Filling Coffee Cups If a regular-size coffee cup holds about $\frac{1}{2}$ pint, about how many cups can be filled from a 1-gallon coffee maker?

56. Filling Glasses If a regular-size drinking glass holds about 0.25 liter of liquid, how many glasses can be filled from a 750-milliliter container?

57. Capacity of a Refrigerator A refrigerator has a capacity of 20 cubic feet. What is the capacity of the refrigerator in cubic inches?

58. Volume of a Tank The gasoline tank on a car holds 18 gallons of gas. What is the volume of the tank in quarts?

59. Filling Glasses How many 8-fluid-ounce glasses of water will it take to fill a 3-gallon aquarium?

60. Filling a Container How many 5-milliliter test tubes filled with water will it take to fill a 1-liter container?

Calculator Problems

Set up the following problems as you have been doing. Then use a calculator to perform the actual calculations. Round answers to two decimal places where appropriate.

61. Geography Lake Superior is the largest of the Great Lakes. It covers 31,700 square miles of area. What is the area of Lake Superior in acres?

62. Geography The state of California consists of 156,360 square miles of land and 2,330 square miles of water. Write the total area (both land and water) in acres.

63. Geography Death Valley National Monument contains 2,067,795 acres of land. How many square miles is this?

64. Geography The Badlands National Monument in South Dakota was established in 1929. It covers 243,302 acres of land. What is the area in square miles?

65. Convert 93.4 qt to gallons.

66. Convert 7,362 fl oz to gallons.

67. How many cubic feet are contained in 796 cubic yards?

68. The engine of a car has a displacement of 440 cubic inches. What is the displacement in cubic feet?

Getting Ready for the Next Section

Perform the indicated operations.

69. 12×16

70. 15×16

71. $3 \times 2{,}000$

72. $5 \times 2{,}000$

73. $3 \times 1{,}000 \times 100$

74. $5 \times 1{,}000 \times 100$

75. $12{,}500 \times \frac{1}{1{,}000}$

76. $15{,}000 \times \frac{1}{1{,}000}$

Maintaining Your Skills

The following problems review addition and subtraction with fractions and mixed numbers.

77. $\frac{3}{8} + \frac{1}{4}$

78. $\frac{1}{2} + \frac{1}{4}$

79. $3\frac{1}{2} + 5\frac{1}{2}$

80. $6\frac{7}{8} + 1\frac{5}{8}$

81. $\frac{7}{15} - \frac{2}{15}$

82. $\frac{5}{8} - \frac{1}{4}$

83. $\frac{5}{36} - \frac{1}{48}$

84. $\frac{7}{39} - \frac{2}{65}$

Unit Analysis III: Weight

8.3

Objectives

A Convert between weights using the U.S. system.

B Convert between weights using the metric system.

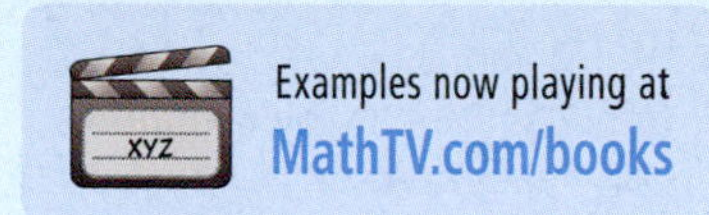

A Weights: The U.S. System

The most common units of weight in the U.S. system are ounces, pounds, and tons. The relationships among these units are given in Table 1.

TABLE 1

UNITS OF WEIGHT IN THE U.S. SYSTEM

The Relationship Between	Is	To Convert From One To The Other, Multiply By
ounces (oz) and pounds (lb)	1 lb = 16 oz	$\frac{16\text{ oz}}{1\text{ lb}}$ or $\frac{1\text{ lb}}{16\text{ oz}}$
pounds and tons (T)	1 T = 2,000 lb	$\frac{2{,}000\text{ lb}}{1\text{ T}}$ or $\frac{1\text{ T}}{2{,}000\text{ lb}}$

EXAMPLE 1 Convert 12 pounds to ounces.

SOLUTION Using the conversion factor from the table, and applying the method we have been using, we have

$$\begin{aligned} 12\text{ lb} &= 12\text{ }\cancel{\text{lb}} \times \frac{16\text{ oz}}{1\text{ }\cancel{\text{lb}}} \\ &= 12 \times 16\text{ oz} \\ &= 192\text{ oz} \end{aligned}$$

12 pounds is equivalent to 192 ounces.

EXAMPLE 2 Convert 3 tons to pounds.

SOLUTION We use the conversion factor from the table. We have

$$\begin{aligned} 3\text{ T} &= 3\text{ }\cancel{\text{T}} \times \frac{2{,}000\text{ lb}}{1\text{ }\cancel{\text{T}}} \\ &= 6{,}000\text{ lb} \end{aligned}$$

6,000 pounds is the equivalent of 3 tons.

PRACTICE PROBLEMS

1. Convert 15 pounds to ounces.

2. Convert 5 tons to pounds.

B Weights: The Metric System

In the metric system the basic unit of weight is a gram. We use the same prefixes we have already used to write the other units of weight in terms of grams. Table 2 lists the most common metric units of weight and their conversion factors.

TABLE 2

METRIC UNITS OF WEIGHT

The Relationship Between	Is	To Convert From One To The Other, Multiply By
milligrams (mg) and grams (g)	1 g = 1,000 mg	$\frac{1{,}000\text{ mg}}{1\text{ g}}$ or $\frac{1\text{ g}}{1{,}000\text{ mg}}$
centigrams (cg) and grams	1 g = 100 cg	$\frac{100\text{ cg}}{1\text{ g}}$ or $\frac{1\text{ g}}{100\text{ cg}}$
kilograms (kg) and grams	1,000 g = 1 kg	$\frac{1{,}000\text{ g}}{1\text{ kg}}$ or $\frac{1\text{ kg}}{1{,}000\text{ g}}$
metric tons (t) and kilograms	1,000 kg = 1 t	$\frac{1{,}000\text{ kg}}{1\text{ t}}$ or $\frac{1\text{ t}}{1{,}000\text{ kg}}$

Answers
1. 240 oz **2.** 10,000 lb

3. Convert 5 kilograms to milligrams.

EXAMPLE 3 Convert 3 kilograms to centigrams.

SOLUTION We convert kilograms to grams and then grams to centigrams:

$$3 \text{ kg} = 3 \cancel{\text{kg}} \times \frac{1{,}000 \cancel{\text{g}}}{1 \cancel{\text{kg}}} \times \frac{100 \text{ cg}}{1 \cancel{\text{g}}}$$

$$= 3 \times 1{,}000 \times 100 \text{ cg}$$

$$= 300{,}000 \text{ cg}$$

4. A bottle of vitamin C contains 75 tablets. If each tablet contains 200 milligrams of vitamin C, what is the total number of grams of vitamin C in the bottle?

EXAMPLE 4 A bottle of vitamin C contains 50 tablets. Each tablet contains 250 milligrams of vitamin C. What is the total number of grams of vitamin C in the bottle?

SOLUTION We begin by finding the total number of milligrams of vitamin C in the bottle. Since there are 50 tablets, and each contains 250 mg of vitamin C, we can multiply 50 by 250 to get the total number of milligrams of vitamin C:

$$\text{Milligrams of vitamin C} = 50 \times 250 \text{ mg}$$

$$= 12{,}500 \text{ mg}$$

Next we convert 12,500 mg to grams:

$$12{,}500 \text{ mg} = 12{,}500 \cancel{\text{mg}} \times \frac{1 \text{ g}}{1{,}000 \cancel{\text{mg}}}$$

$$= \frac{12{,}500}{1{,}000} \text{g}$$

$$= 12.5 \text{ g}$$

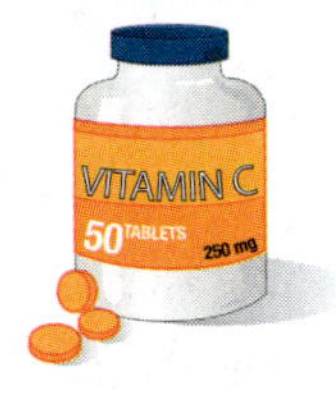

The bottle contains 12.5 g of vitamin C.

Getting Ready for Class

After reading through the preceding section, respond in your own words and in complete sentences.

1. What is the relationship between pounds and ounces?
2. Write the conversion factor used to convert from pounds to ounces.
3. Write the conversion factor used to convert from milligrams to grams.
4. What is the relationship between grams and kilograms?

Answers
3. 5,000,000 mg 4. 15 g

Problem Set 8.3

A Use the conversion factors in Tables 1 and 2 to make the following conversions. [Examples 1, 2]

1. 8 lb to ounces
2. 5 lb to ounces
3. 2 T to pounds
4. 5 T to pounds
5. 192 oz to pounds
6. 176 oz to pounds
7. 1,800 lb to tons
8. 10,200 lb to tons
9. 1 T to ounces
10. 3 T to ounces
11. $3\frac{1}{2}$ lb to ounces
12. $5\frac{1}{4}$ lb to ounces
13. $6\frac{1}{2}$ T to pounds
14. $4\frac{1}{5}$ T to pounds
15. 2 kg to grams

B [Examples 3, 4]

16. 5 kg to grams
17. 4 cg to milligrams
18. 3 cg to milligrams
19. 2 kg to centigrams
20. 5 kg to centigrams
21. 5.08 g to centigrams
22. 7.14 g to centigrams
23. 450 cg to grams
24. 979 cg to grams
25. 478.95 mg to centigrams
26. 659.43 mg to centigrams
27. 1,578 mg to grams
28. 1,979 mg to grams
29. 42,000 cg to kilograms
30. 97,000 cg to kilograms

Applying the Concepts

31. Fish Oil A bottle of fish oil contains 60 soft gels, each containing 800 mg of the omega-3 fatty acid. How many total grams of the omega-3 fatty acid are in this bottle?

32. Fish Oil A bottle of fish oil contains 50 soft gels, each containing 300 mg of the omega-6 fatty acid. How many total grams of the omega-6 fatty acid are in this bottle?

33. B-Complex A certain B-complex vitamin supplement contains 50 mg of riboflavin, or vitamin B_2. A bottle contains 80 vitamins. How many total grams of riboflavin are in this bottle?

34. B-Complex A certain B-complex vitamin supplement contains 30 mg of thiamine, or vitamin B_1. A bottle contains 80 vitamins. How many total grams of thiamine are in this bottle?

35. Aspirin A bottle of low-strength aspirin contains 120 tablets. Each tablet contains 81 mg of aspirin. How many total grams of aspirin are in this bottle?

36. Aspirin A bottle of maximum-strength aspirin contains 90 tablets. Each tablet contains 500 mg of aspirin. How many total grams of aspirin are in this bottle?

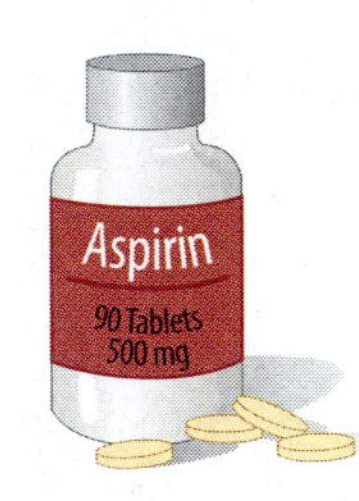

37. Vitamin C A certain brand of vitamin C contains 500 mg per tablet. A bottle contains 240 tablets. How many total grams of vitamin C are in this bottle?

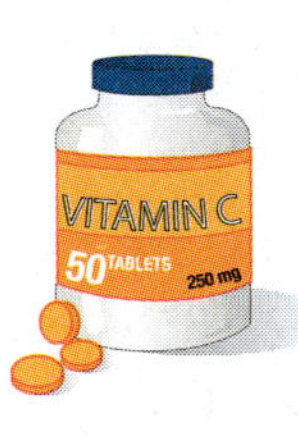

38. Vitamin C A certain brand of vitamin C contains 600 mg per tablet. A bottle contains 150 vitamins. How many total grams of vitamin C are in this bottle?

Coca-Cola Bottles The soft drink Coke is sold throughout the world. Although the size of the bottle varies between different countries, a "six-pack" is sold everywhere. For each of the problems below, find the number of liters in a "6-pack" from the given bottle size.

Country	Bottle size	Liters in a 6-pack
39. Estonia	500 mL	
40. Israel	350 mL	
41. Jordan	250 mL	
42. Kenya	300 mL	

Paul A. Souders/Corbis

43. Nursing A patient is prescribed a dosage of Ceclor® of 561 mg. How many grams is the dosage?

44. Nursing A patient is prescribed a dosage of 425 mg. How many grams is the dosage?

45. Nursing Dilatrate®-SR comes in 40 milligram capsules. Use this information to determine how many capsules should be given for the prescribed dosages.

a. 120 mg

b. 40 mg

c. 80 mg

46. Nursing A brand of methyldopa comes in 250 milligram tablets. Use this information to determine how many capsules should be given for the prescribed dosages.

a. 0.125 gram

b. 750 milligrams

c. 0.5 gram

Getting Ready for the Next Section

Perform the indicated operations.

47. 8×2.54

48. 9×3.28

49. $3 \times 1.06 \times 2$

50. $3 \times 5 \times 3.79$

51. $80.5 \div 1.61$

52. $96.6 \div 1.61$

53. $125 \div 2.50$

54. $165 \div 2.20$

55. $2{,}000 \div 16.39$
(Round your answer to the nearest whole number.)

56. $2{,}200 \div 16.39$
(Round your answer to the nearest whole number.)

57. $\frac{9}{5}(120) + 32$

58. $\frac{9}{5}(40) + 32$

59. $\frac{5(102 - 30)}{9}$

60. $\frac{5(105 - 42)}{9}$

Maintaining Your Skills

Write each decimal as an equivalent proper fraction or mixed number.

61. 0.18 **62.** 0.04 **63.** 0.09 **64.** 0.045

65. 0.8 **66.** 0.08 **67.** 1.75 **68.** 3.125

Write each fraction or mixed number as a decimal.

69. $\frac{3}{4}$ **70.** $\frac{9}{10}$ **71.** $\frac{17}{20}$ **72.** $\frac{1}{8}$

73. $\frac{3}{5}$ **74.** $\frac{7}{8}$ **75.** $3\frac{5}{8}$ **76.** $1\frac{1}{16}$

Use the definition of exponents to simplify each expression.

77. $\left(\frac{1}{2}\right)^3$ **78.** $\left(\frac{5}{9}\right)^2$ **79.** $\left(2\frac{1}{2}\right)^2$ **80.** $\left(\frac{1}{3}\right)^4$

81. $(0.5)^3$ **82.** $(0.05)^3$ **83.** $(2.5)^2$ **84.** $(0.5)^4$

Converting Between the Two Systems and Temperature

8.4

Objectives

A Convert between the two systems.

B Convert temperatures between the Fahrenheit and Celsius scales.

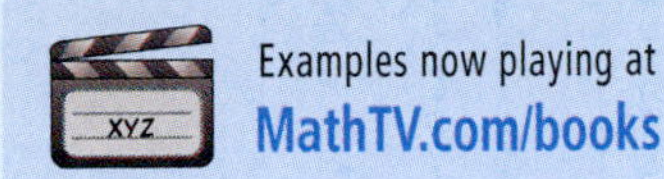

A Converting Between the U.S. and Metric Systems

Because most of us have always used the U.S. system of measurement in our everyday lives, we are much more familiar with it on an intuitive level than we are with the metric system. We have an intuitive idea of how long feet and inches are, how much a pound weighs, and what a square yard of material looks like. The metric system is actually much easier to use than the U.S. system. The reason some of us have such a hard time with the metric system is that we don't have the feel for it that we do for the U.S. system. We have trouble visualizing how long a meter is or how much a gram weighs. The following list is intended to give you something to associate with each basic unit of measurement in the metric system:

1. A meter is just a little longer than a yard.
2. The length of the edge of a sugar cube is about 1 centimeter.
3. A liter is just a little larger than a quart.
4. A sugar cube has a volume of approximately 1 milliliter.
5. A paper clip weighs about 1 gram.
6. A 2-pound can of coffee weighs about 1 kilogram.

TABLE 1

ACTUAL CONVERSION FACTORS BETWEEN THE METRIC AND U.S. SYSTEMS OF MEASUREMENT

The Relationship Between	Is	To Convert From One To The Other, Multiply By
Length		
inches and centimeters	2.54 cm = 1 in.	$\frac{2.54\text{ cm}}{1\text{ in.}}$ or $\frac{1\text{ in.}}{2.54\text{ cm}}$
feet and meters	1 m = 3.28 ft	$\frac{3.28\text{ ft}}{1\text{ m}}$ or $\frac{1\text{ m}}{3.28\text{ ft}}$
miles and kilometers	1.61 km = 1 mi	$\frac{1.61\text{ km}}{1\text{ mi}}$ or $\frac{1\text{ mi}}{1.61\text{ km}}$
Area		
square inches and square centimeters	$6.45\text{ cm}^2 = 1\text{ in}^2$	$\frac{6.45\text{ cm}^2}{1\text{ in}^2}$ or $\frac{1\text{ in}^2}{6.45\text{ cm}^2}$
square meters and square yards	$1.196\text{ yd}^2 = 1\text{ m}^2$	$\frac{1.196\text{ yd}^2}{1\text{ m}^2}$ or $\frac{1\text{ m}^2}{1.196\text{ yd}^2}$
acres and hectares	1 ha = 2.47 acres	$\frac{2.47\text{ acres}}{1\text{ ha}}$ or $\frac{1\text{ ha}}{2.47\text{ acres}}$
Volume		
cubic inches and milliliters	$16.39\text{ mL} = 1\text{ in}^3$	$\frac{16.39\text{ mL}}{1\text{ in}^3}$ or $\frac{1\text{ in}^3}{16.39\text{ mL}}$
liters and quarts	1.06 qt = 1 liter	$\frac{1.06\text{ qt}}{1\text{ liter}}$ or $\frac{1\text{ liter}}{1.06\text{ qt}}$
gallons and liters	3.79 liters = 1 gal	$\frac{3.79\text{ liters}}{1\text{ gal}}$ or $\frac{1\text{ gal}}{3.79\text{ liters}}$
Weight		
ounces and grams	28.3 g = 1 oz	$\frac{28.3\text{ g}}{1\text{ oz}}$ or $\frac{1\text{ oz}}{28.3\text{ g}}$
kilograms and pounds	2.20 lb = 1 kg	$\frac{2.20\text{ lb}}{1\text{ kg}}$ or $\frac{1\text{ kg}}{2.20\text{ lb}}$

There are many other conversion factors that we could have included in Table 1. We have listed only the most common ones. Almost all of them are approximations. That is, most of the conversion factors are decimals that have been rounded to the nearest hundredth. If we want more accuracy, we obtain a table that has more digits in the conversion factors.

PRACTICE PROBLEMS

1. Convert 10 inches to centimeters.

EXAMPLE 1 Convert 8 inches to centimeters.

SOLUTION Choosing the appropriate conversion factor from Table 1, we have

$$8 \text{ in.} = 8 \cancel{\text{in.}} \times \frac{2.54 \text{ cm}}{1 \cancel{\text{in.}}}$$
$$= 8 \times 2.54 \text{ cm}$$
$$= 20.32 \text{ cm}$$

2. Convert 16.4 feet to meters.

EXAMPLE 2 Convert 80.5 kilometers to miles.

SOLUTION Using the conversion factor that takes us from kilometers to miles, we have

$$80.5 \text{ km} = 80.5 \cancel{\text{km}} \times \frac{1 \text{ mi}}{1.61 \cancel{\text{km}}}$$
$$= \frac{80.5}{1.61} \text{mi}$$
$$= 50 \text{ mi}$$

So 50 miles is equivalent to 80.5 kilometers. If we travel at 50 miles per hour in a car, we are moving at the rate of 80.5 kilometers per hour.

3. Convert 10 liters to gallons. Round to the nearest hundredth.

EXAMPLE 3 Convert 3 liters to pints.

SOLUTION Because Table 1 doesn't list a conversion factor that will take us directly from liters to pints, we first convert liters to quarts, and then convert quarts to pints.

$$3 \text{ liters} = 3 \cancel{\text{liters}} \times \frac{1.06 \cancel{\text{qt}}}{1 \cancel{\text{liter}}} \times \frac{2 \text{ pt}}{1 \cancel{\text{qt}}}$$
$$= 3 \times 1.06 \times 2 \text{ pt}$$
$$= 6.36 \text{ pt}$$

4. The engine in a car has a 2.2-liter displacement. What is the displacement in cubic inches (to the nearest cubic inch)?

EXAMPLE 4 The engine in a car has a 2-liter displacement. What is the displacement in cubic inches?

SOLUTION We convert liters to milliliters and then milliliters to cubic inches:

$$2 \text{ liters} = 2 \cancel{\text{liters}} \times \frac{1{,}000 \cancel{\text{mL}}}{1 \cancel{\text{liter}}} \times \frac{1 \text{ in}^3}{16.39 \cancel{\text{mL}}}$$
$$= \frac{2 \times 1{,}000}{16.39} \text{ in}^3 \quad \text{This calculation should be done on a calculator}$$
$$= 122 \text{ in}^3 \quad \text{To the nearest cubic inch}$$

Answers
1. 25.4 cm 2. 5 m
3. 2.64 gal 4. 134 in^3

EXAMPLE 5 If a person weighs 125 pounds, what is her weight in kilograms?

SOLUTION Converting from pounds to kilograms, we have

$$125 \text{ lb} = 125 \text{ lb} \times \frac{1 \text{ kg}}{2.20 \text{ lb}}$$
$$= \frac{125}{2.20} \text{kg}$$
$$= 56.8 \text{ kg} \quad \text{To the nearest tenth}$$

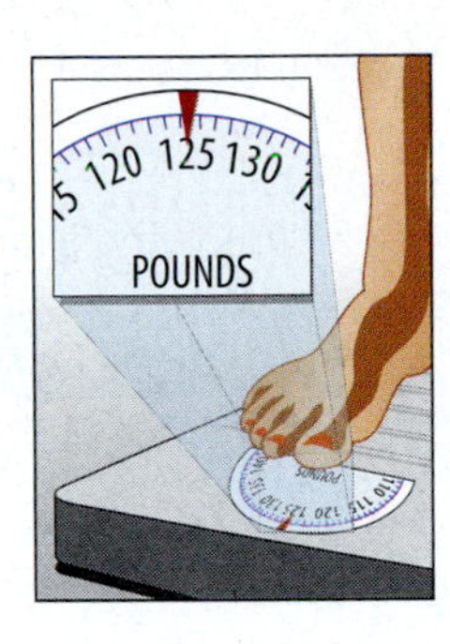

5. A person who weighs 165 pounds weighs how many kilograms?

B Temperature

We end this section with a discussion of temperature in both systems of measurement.

In the U.S. system we measure temperature on the Fahrenheit scale. On this scale, water boils at 212 degrees and freezes at 32 degrees. When we write 32 degrees measured on the Fahrenheit scale, we use the notation

32°F (read, "32 degrees Fahrenheit")

In the metric system the scale we use to measure temperature is the Celsius scale (formerly called the centigrade scale). On this scale, water boils at 100 degrees and freezes at 0 degrees. When we write 100 degrees measured on the Celsius scale, we use the notation

100°C (read, "100 degrees Celsius")

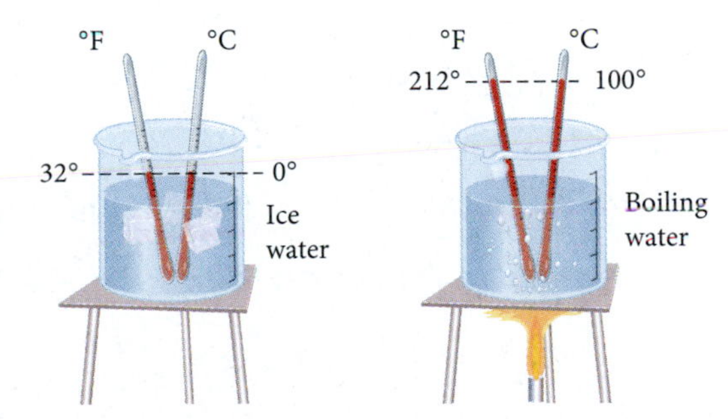

Table 2 is intended to give you a feel for the relationship between the two temperature scales. Table 3 gives the formulas, in both symbols and words, that are used to convert between the two scales.

TABLE 2

Situation	Temperature Fahrenheit	Temperature Celsius
Water freezes	32°F	0°C
Room temperature	68°F	20°C
Normal body temperature	98.6°F	37°C
Water boils	212°F	100°C
Bake cookies	350°F	176.7°C
Broil meat	554°F	290°C

Answer
5. 75 kg

TABLE 3

To Convert From	Formula In Symbols	Formula In Words
Fahrenheit to Celsius	$C = \frac{5(F - 32)}{9}$	Subtract 32, multiply by 5, and then divide by 9.
Celsius to Fahrenheit	$F = \frac{9}{5}C + 32$	Multiply by $\frac{9}{5}$, and then add 32.

The following examples show how we use the formulas given in Table 3.

6. Convert 40°C to degrees Fahrenheit.

EXAMPLE 6 Convert 120°C to degrees Fahrenheit.

SOLUTION We use the formula

$$F = \frac{9}{5}C + 32$$

and replace C with 120:

When $C = 120$

the formula $F = \frac{9}{5}C + 32$

becomes $F = \frac{9}{5}(120) + 32$

$F = 216 + 32$

$F = 248$

We see that 120°C is equivalent to 248°F; they both mean the same temperature.

7. A child is running a temperature of 101.6°F. What is her temperature, to the nearest tenth of a degree, on the Celsius scale?

EXAMPLE 7 A man with the flu has a temperature of 102°F. What is his temperature on the Celsius scale?

SOLUTION When $F = 102$

the formula $C = \frac{5(F - 32)}{9}$

becomes $C = \frac{5(102 - 32)}{9}$

$C = \frac{5(70)}{9}$

$C = 38.9$ Rounded to the nearest tenth

The man's temperature, rounded to the nearest tenth, is 38.9°C on the Celsius scale.

Getting Ready for Class

After reading through the preceding section, respond in your own words and in complete sentences.

1. Write the equality that gives the relationship between centimeters and inches.
2. Write the equality that gives the relationship between grams and ounces.
3. Fill in the numerators below so that each conversion factor is equal to 1.
   a. $\frac{\text{ft}}{1 \text{ meter}}$ b. $\frac{\text{qt}}{1 \text{ liter}}$ c. $\frac{\text{lb}}{1 \text{ kg}}$
4. Is it a hot day if the temperature outside is 37°C?

Answers
6. 104°F 7. 38.7°C

Problem Set 8.4

A B Use Tables 1 and 3 to make the following conversions. [Examples 1–7]

1. 6 in. to centimeters
2. 1 ft to centimeters
3. 4 m to feet
4. 2 km to feet
5. 6 m to yards
6. 15 mi to kilometers
7. 20 mi to meters (round to the nearest hundred meters)
8. 600 m to yards
9. 5 m^2 to square yards (round to the nearest hundredth)
10. 2 in^2 to square centimeters (round to the nearest tenth)
11. 10 ha to acres
12. 50 a to acres
13. 500 in^3 to milliliters
14. 400 in^3 to liters
15. 2 L to quarts
16. 15 L to quarts
17. 20 gal to liters
18. 15 gal to liters
19. 12 oz to grams
20. 1 lb to grams (round to the nearest 10 grams)
21. 15 kg to pounds
22. 10 kg to ounces
23. 185°C to degrees Fahrenheit
24. 20°C to degrees Fahrenheit
25. 86°F to degrees Celsius
26. 122°F to degrees Celsius

Applying the Concepts

27. Temperature The chart shows the temperatures for some of the world's hottest places. Convert the temperature in Al'Aziziyah to Celsius.

Heating Up

136.4°F Al'Aziziyah, Libya
134.0°F Greenland Ranch, Death Valley, United States
131.0°F Ghudamis, Libya
131.0°F Kebili, Tunisia
130.1°F Tombouctou, Mali

160 140 120 100 80 60 40

Source: Aneki.com

28. Google Earth The Google Earth image is of Lake Clark National Park in Alaska. Lake Clark has an average temperature of 40 degrees Fahrenheit. What is its average temperature in Celsius to the nearest degree?

Nursing Liquid medication is usually given in milligrams per milliliter. Use the information to find the amount a patient should take for a prescribed dosage.

29. Vantin© has a dosage strength of 100 mg/5 mL. If a patient is prescribed a dosage of 150 mg, how many milliliters should she take?

30. A brand of amoxicillin has a dosage strength of 125 mg/5 mL. If a patient is prescribed a dosage of 25 mg, how many milliliters should she take?

Calculator Problems

Set up the following problems as we have set up the examples in this section. Then use a calculator for the calculations and round your answers to the nearest hundredth.

31. 10 cm to inches

32. 100 mi to kilometers

33. 25 ft to meters

34. 400 mL to cubic inches

35. 49 qt to liters

36. 65 L to gallons

37. 500 g to ounces

38. 100 lb to kilograms

39. **Weight** Give your weight in kilograms.

40. **Height** Give your height in meters and centimeters.

41. **Sports** The 100-yard dash is a popular race in track. How far is 100 yards in meters?

42. **Engine Displacement** A 351-cubic-inch engine has a displacement of how many liters?

43. **Sewing** 25 square yards of material is how many square meters?

44. **Weight** How many grams does a 5 lb 4 oz roast weigh?

45. **Speed** 55 miles per hour is equivalent to how many kilometers per hour?

46. **Capacity** A 1-quart container holds how many liters?

47. **Sports** A high jumper jumps 6 ft 8 in. How many meters is this?

48. **Farming** A farmer owns 57 acres of land. How many hectares is that?

49. **Body Temperature** A person has a temperature of 101°F. What is the person's temperature, to the nearest tenth, on the Celsius scale?

50. **Air Temperature** If the temperature outside is 30°C, is it a better day for water skiing or for snow skiing?

Getting Ready for the Next Section

Perform the indicated operations.

51. $15 + 60$

52. $25 + 60$

53. $\begin{array}{r} 37 \\ +\ 45 \\ \hline \end{array}$

54. $\begin{array}{r} 27 \\ +\ 46 \\ \hline \end{array}$

55. $3 + 0.25$

56. $2 + 0.75$

57. $82 - 60$

58. $73 - 60$

59. $\begin{array}{r} 75 \\ -\ 34 \\ \hline \end{array}$

60. $\begin{array}{r} 85 \\ -\ 42 \\ \hline \end{array}$

61. 12×4

62. 8×4

63. $3 \times 60 + 15$

64. $2 \times 65 + 45$

65. $3 + 17 \times \frac{1}{65}$

66. $2 + 45 \times \frac{1}{60}$

67. If fish costs \$6.00 per pound, find the cost of 15 pounds.

68. If fish costs \$5.00 per pound, find the cost of 14 pounds.

Maintaining Your Skills

Find the mean and the range for each set of numbers.

69. 5, 7, 9, 11

70. 6, 8, 10, 12

71. 1, 4, 5, 10, 10

72. 2, 4, 4, 6, 9

Find the median and the range for each set of numbers.

73. 15, 18, 21, 24, 29

74. 20, 30, 35, 45, 50

75. 32, 38, 42, 48

76. 53, 61, 67, 75

Find the mode and the range for each set of numbers.

77. 20, 15, 14, 13, 14, 18

78. 17, 31, 31, 26, 31, 29

79. A student has quiz scores of 65, 72, 70, 88, 70, and 73. Find each of the following:

a. mean score

b. median score

c. mode of the scores

d. range of scores

80. A person has bowling scores of 207, 224, 195, 207, 185, and 182. Find each of the following:

a. mean score

b. median score

c. mode of the scores

d. range of scores

Extending the Concepts

Nursing For children, the amount of medicine prescribed is often determined by the child's weight. Usually, it is calculated from the milligrams per kilogram per day listed on the medication's box.

81. Ceclor® has a dosage strength of 250 mg/mL. How much should a 42 lb child be given a day if the dosage is 20 mg/kg/day? How many milliliters is that?

8.5 Operations with Time and Mixed Units

Objectives

A Convert mixed units to a single unit.

B Add and subtract mixed units.

C Use multiplication with mixed units.

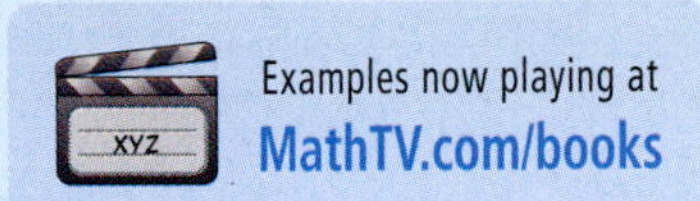

Many occupations require the use of a time card. A time card records the number of hours and minutes at work. At the end of a work week the hours and minutes are totaled separately, and then the minutes are converted to hours.

In this section we will perform operations with mixed units of measure. Mixed units are used when we use 2 hours 30 minutes, rather than 2 and a half hours, or 5 feet 9 inches, rather than five and three-quarter feet. As you will see, many of these types of problems arise in everyday life.

A Converting Time to Single Units

The Relationship Between	is	To Convert from One to the Other, Multiply by
minutes and seconds	1 min = 60 sec	$\frac{1 \text{ min}}{60 \text{ sec}}$ or $\frac{60 \text{ sec}}{1 \text{ min}}$
hours and minutes	1 hr = 60 min	$\frac{1 \text{ hr}}{60 \text{ min}}$ or $\frac{60 \text{ min}}{1 \text{ hr}}$

EXAMPLE 1 Convert 3 hours 15 minutes to

a. Minutes **b.** Hours

SOLUTION **a.** To convert to minutes, we multiply the hours by the conversion factor and then add minutes:

$$\begin{aligned} 3 \text{ hr } 15 \text{ min} &= 3 \text{ hr} \times \frac{60 \text{ min}}{1 \text{ hr}} + 15 \text{ min} \\ &= 180 \text{ min} + 15 \text{ min} \\ &= 195 \text{ min} \end{aligned}$$

b. To convert to hours, we multiply the minutes by the conversion factor and then add hours:

$$\begin{aligned} 3 \text{ hr } 15 \text{ min} &= 3 \text{ hr} + 15 \text{ min} \times \frac{1 \text{ hr}}{60 \text{ min}} \\ &= 3 \text{ hr} + 0.25 \text{ hr} \\ &= 3.25 \text{ hr} \end{aligned}$$

B Addition and Subtraction with Mixed Units

EXAMPLE 2 Add 5 minutes 37 seconds and 7 minutes 45 seconds.

SOLUTION First, we align the units properly

$$\begin{array}{rr} 5 \text{ min} & 37 \text{ sec} \\ + \ 7 \text{ min} & 45 \text{ sec} \\ \hline 12 \text{ min} & 82 \text{ sec} \end{array}$$

Since there are 60 seconds in every minute, we write 82 seconds as 1 minute 22 seconds. We have

$$\begin{aligned} 12 \text{ min } 82 \text{ sec} &= 12 \text{ min} + 1 \text{ min } 22 \text{ sec} \\ &= 13 \text{ min } 22 \text{ sec} \end{aligned}$$

PRACTICE PROBLEMS

1. Convert 2 hours 45 minutes to
   a. Minutes **b.** Hours

2. Add 4 min. 27 sec. and 8 min. 46 sec.

Answers

1. **a** 165 minutes **b.** 2.75 hours
2. 13 min 13 sec

The idea of adding the units separately is similar to adding mixed fractions. That is, we align the whole numbers with the whole numbers and the fractions with the fractions.

Similarly, when we subtract units of time, we "borrow" 60 seconds from the minutes column, or 60 minutes from the hours column.

3. Subtract 42 min from 6 hr 25 min.

EXAMPLE 3 Subtract 34 minutes from 8 hours 15 minutes.

SOLUTION Again, we first line up the numbers in the hours column, and then the numbers in the minutes column:

$$\begin{array}{rr} 8\text{ hr} & 15\text{ min} \\ - & 34\text{ min} \\ \hline \end{array} \qquad \begin{array}{rr} 7\text{ hr} & 75\text{ min} \\ - & 34\text{ min} \\ \hline 7\text{ hr} & 41\text{ min} \end{array}$$

C Multiplication with Mixed Units

Next we see how to multiply and divide using units of measure.

4. Rob is purchasing 4 halibut. The fish cost $5.00 per pound, and each weighs 3 lb 8 oz. What is the cost of the fish?

EXAMPLE 4 Jake purchases 4 halibut. The fish cost $6.00 per pound, and each weighs 3 lb 12 oz. What is the cost of the fish?

SOLUTION First, we multiply each unit by 4:

$$\begin{array}{rr} 3\text{ lb} & 12\text{ oz} \\ \times & 4 \\ \hline 12\text{ lb} & 48\text{ oz} \end{array}$$

To convert the 48 ounces to pounds, we multiply the ounces by the conversion factor.

$$\begin{aligned} 12\text{ lb }48\text{ oz} &= 12\text{ lb} + 48\text{ oz} \times \frac{1\text{ lb}}{16\text{ oz}} \\ &= 12\text{ lb} + 3\text{ lb} \\ &= 15\text{ lb} \end{aligned}$$

Finally, we multiply the 15 lb and $6.00/lb for a total price of $90.00

Getting Ready for Class

After reading through the preceding section, respond in your own words and in complete sentences.

1. Explain the difference between saying *2 and a half hours* and saying *2 hours and 50 minutes.*
2. How are operations with mixed units of measure similar to operations with mixed numbers?
3. Why do we borrow a 60 from the minutes column for the seconds column when subtracting in Example 3?
4. Give an example of when you may have to use multiplication with mixed units of measure.

Answers
3. 5 hr 43 min 4. $70

Problem Set 8.5

A Use the tables of conversion factors given in this section and other sections in this chapter to make the following conversions. (Round your answers to the nearest hundredth.) [Example 1]

1. 4 hours 30 minutes to
 a. Minutes
 b. Hours

2. 2 hours 45 minutes to
 a. Minutes
 b. Hours

3. 5 hours 20 minutes to
 a. Minutes
 b. Hours

4. 4 hours 40 minutes to
 a. Minutes
 b. Hours

5. 6 minutes 30 seconds to
 a. Seconds
 b. Minutes

6. 8 minutes 45 seconds to
 a. Seconds
 b. Minutes

7. 5 minutes 20 seconds to
 a. Seconds
 b. Minutes

8. 4 minutes 40 seconds to
 a. Seconds
 b. Minutes

9. 2 pounds 8 ounces to
 a. Ounces
 b. Pounds

10. 3 pounds 4 ounces to
 a. Ounces
 b. Pounds

11. 4 pounds 12 ounces to
 a. Ounces
 b. Pounds

12. 5 pounds 16 ounces to
 a. Ounces
 b. Pounds

13. 4 feet 6 inches to
 a. Inches
 b. Feet

14. 3 feet 3 inches to
 a. Inches
 b. Feet

15. 5 feet 9 inches to
 a. Inches
 b. Feet

16. 3 feet 4 inches to
 a. Inches
 b. Feet

17. 2 gallons 1 quart
 a. Quarts
 b. Gallons

18. 3 gallons 2 quarts
 a. Quarts
 b. Gallons

B Perform the indicated operation. Again, remember to use the appropriate conversion factor. [Examples 2, 3]

19. Add 4 hours 47 minutes and 6 hours 13 minutes.

20. Add 5 hours 39 minutes and 2 hours 21 minutes.

21. Add 8 feet 10 inches and 13 feet 6 inches

22. Add 16 feet 7 inches and 7 feet 9 inches.

23. Add 4 pounds 12 ounces and 6 pounds 4 ounces.

24. Add 11 pounds 9 ounces and 3 pounds 7 ounces.

25. Subtract 2 hours 35 minutes from 8 hours 15 minutes.

26. Subtract 3 hours 47 minutes from 5 hours 33 minutes.

27. Subtract 3 hours 43 minutes from 7 hours 30 minutes.

28. Subtract 1 hour 44 minutes from 6 hours 22 minutes.

29. Subtract 4 hours 17 minutes from 5 hours 9 minutes.

30. Subtract 2 hours 54 minutes from 3 hours 7 minutes.

Applying the Concepts

31. Fifth Avenue Mile The chart shows the times of the five fastest runners for 2005's Continental Airlines Fifth Avenue Mile. How much faster was Craig Mottram than Rui Silva?

Fastest on Fifth

Continental Airlines Fifth Avenue Mile

Craig Mottram, AUS	3:49.90
Alan Webb, USA	3:51.40
Elkanah Angwenyi, KEN	3:54.30
Anthony Famiglietti, USA	3:57.10
Rui Silva, POR	3:57.40

Source: www.coolrunning.com, 2005

32. Cars The chart shows the fastest cars in America. Convert the speed of the Ford GT to feet per second. Round to the nearest tenth.

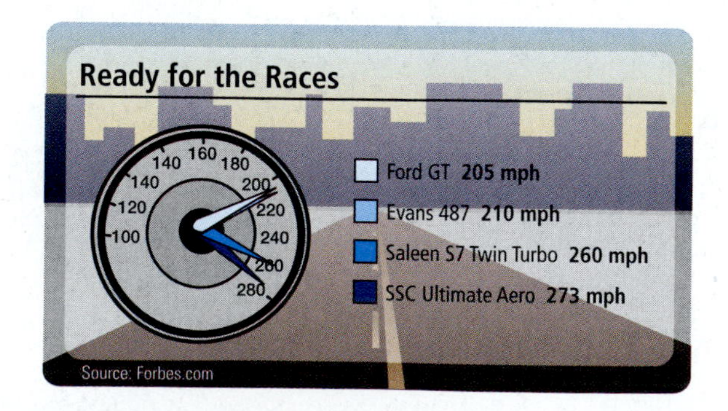

Triathlon The Ironman Triathlon World Championship, held each October in Kona on the island of Hawaii, consists of three parts: a 2.4-mile ocean swim, a 112-mile bike race, and a 26.2-mile marathon. The table shows the results from the 2003 event.

Triathlete	Swim Time (Hr:Min:Sec)	Bike Time (Hr:Min:Sec)	Run Time (Hr:Min:Sec)	Total Time (Hr:Min:Sec)
Peter Reid	0:50:36	4:40:04	2:47:38	
Lori Bowden	0:56:51	5:09:00	3:02:10	

Sanford/Agliolo/Corbis

33. Fill in the total time column.

34. How much faster was Peter's total time than Lori's?

35. How much faster was Peter than Lori in the swim?

36. How much faster was Peter than Lori in the run?

37. Cost of Fish Fredrick is purchasing four whole salmon. The fish cost $4.00 per pound, and each weighs 6 lb 8 oz. What is the cost of the fish?

38. Cost of Steak Mike is purchasing eight top sirloin steaks. The meat costs $4.00 per pound, and each steak weighs 1 lb 4 oz. What is the total cost of the steaks?

39. Stationary Bike Maggie rides a stationary bike for 1 hour and 15 minutes, 4 days a week. After 2 weeks, how many hours has she spent riding the stationary bike?

40. Gardening Scott works in his garden for 1 hour and 5 minutes, 3 days a week. After 4 weeks, how many hours has Scott spent gardening?

41. Cost of Fabric Allison is making a quilt. She buys 3 yards and 1 foot each of six different fabrics. The fabrics cost $7.50 a yard. How much will Allison spend?

42. Cost of Lumber Trish is building a fence. She buys six fence posts at the lumberyard, each measuring 5 ft 4 in. The lumber costs $3 per foot. How much will Trish spend?

43. Cost of Avocados Jacqueline is buying six avocados. Each avocado weighs 8 oz. How much will they cost her if avocados cost $2.00 a pound?

44. Cost of Apples Mary is purchasing 12 apples. Each apple weighs 4 oz. If the cost of the apples is $1.50 a pound, how much will Mary pay?

Maintaining Your Skills

45. Caffeine Content The following bar chart shows the amount of caffeine in five different soft drinks. Use the information in the bar chart to fill in the table.

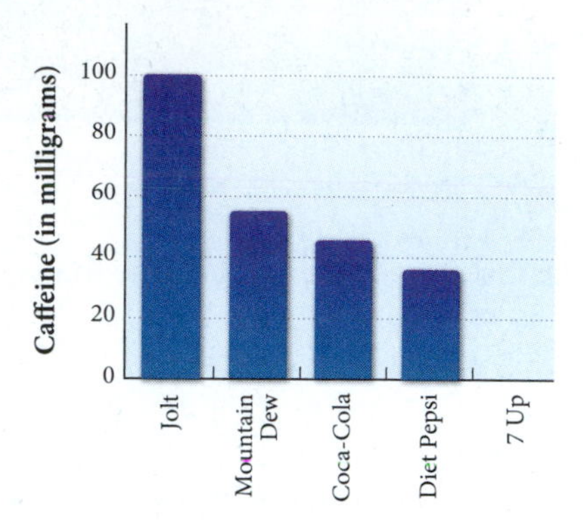

CAFFEINE CONTENT IN SOFT DRINKS	
Drink	Caffeine (In Milligrams)
Jolt	
Mountain Dew	
Coca-Cola	
Diet Pepsi	
7 Up	

46. Exercise The following bar chart shows the number of calories burned in 1 hour of exercise by a person who weighs 150 pounds. Use the information in the bar chart to fill in the table.

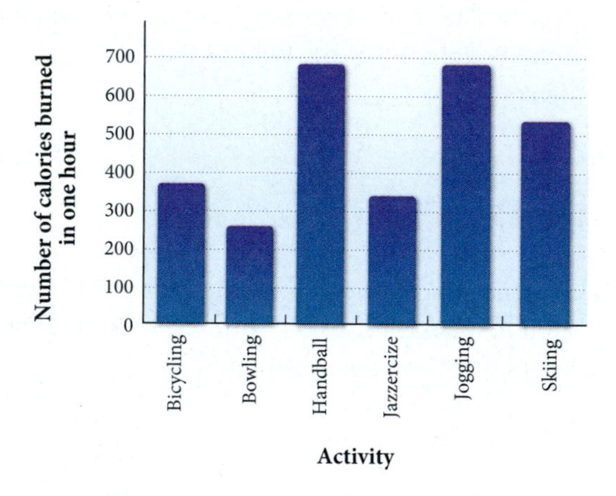

CALORIES BURNED BY A 150-POUND PERSON IN ONE HOUR	
Activity	Calories
Bicycling	
Bowling	
Handball	
Jazzercise	
Jogging	
Skiing	

Extending the Concepts

47. In 2003, the horse Funny Cide won the Kentucky Derby with a time of 2:01.19, or two minutes and 1.19 seconds. The record time for the Kentucky Derby is still held by Secretariat, who won the race with a time of 1:59.40 in 1973. How much faster did Secretariat run than Funny Cide 30 years later?

48. In 2003, the horse Empire Maker won the Belmont Stakes with a time of 2:28.20, or two minutes and 28.2 seconds. The record time for the Belmont Stakes is still held by Secretariat, who won the race with a time of 2:24.00 in 1973. How much faster did Secretariat run in 1973 than Empire Maker 30 years later?

Chapter 8 Summary

Conversion Factors [8.1, 8.2, 8.3, 8.4, 8.5]

To convert from one kind of unit to another, we choose an appropriate conversion factor from one of the tables given in this chapter. For example, if we want to convert 5 feet to inches, we look for conversion factors that give the relationship between feet and inches. There are two conversion factors for feet and inches:

$$\frac{12 \text{ in.}}{1 \text{ ft}} \quad \text{and} \quad \frac{1 \text{ ft}}{12 \text{ in.}}$$

EXAMPLES

1. Convert 5 feet to inches.

$$5 \text{ ft} = 5 \cancel{\text{ft}} \times \frac{12 \text{ in.}}{1 \cancel{\text{ft}}}$$
$$= 5 \times 12 \text{ in.}$$
$$= 60 \text{ in.}$$

Length [8.1]

U.S. SYSTEM		
The Relationship Between	**Is**	**To Convert From One To The Other, Multiply By**
feet and inches	12 in. = 1 ft	$\frac{12 \text{ in.}}{1 \text{ ft}}$ or $\frac{1 \text{ ft}}{12 \text{ in.}}$
feet and yards	1 yd = 3 ft	$\frac{3 \text{ ft}}{1 \text{ yd}}$ or $\frac{1 \text{ yd}}{3 \text{ ft}}$
feet and miles	1 mi = 5,280 ft	$\frac{5{,}280 \text{ ft}}{1 \text{ mi}}$ or $\frac{1 \text{ mi}}{5{,}280 \text{ ft}}$

2. Convert 8 feet to yards.

$$8 \text{ ft} = 8 \cancel{\text{ft}} \times \frac{1 \text{ yd}}{3 \cancel{\text{ft}}}$$
$$= \frac{8}{3} \text{ yd}$$
$$= 2\tfrac{2}{3} \text{ yd}$$

METRIC SYSTEM		
The Relationship Between	**Is**	**To Convert From One To The Other, Multiply By**
millimeters (mm) and meters (m)	1,000 mm = 1 m	$\frac{1{,}000 \text{ mm}}{1 \text{ m}}$ or $\frac{1 \text{ m}}{1{,}000 \text{ mm}}$
centimeters (cm) and meters	100 cm = 1 m	$\frac{100 \text{ cm}}{1 \text{ m}}$ or $\frac{1 \text{ m}}{100 \text{ cm}}$
decimeters (dm) and meters	10 dm = 1 m	$\frac{10 \text{ dm}}{1 \text{ m}}$ or $\frac{1 \text{ m}}{10 \text{ dm}}$
dekameters (dam) and meters	1 dam = 10 m	$\frac{10 \text{ m}}{1 \text{ dam}}$ or $\frac{1 \text{ dam}}{10 \text{ m}}$
hectometers (hm) and meters	1 hm = 100 m	$\frac{100 \text{ m}}{1 \text{ hm}}$ or $\frac{1 \text{ hm}}{100 \text{ m}}$
kilometers (km) and meters	1 km = 1,000 m	$\frac{1{,}000 \text{ m}}{1 \text{ km}}$ or $\frac{1 \text{ km}}{1{,}000 \text{ m}}$

3. Convert 25 millimeters to meters.

$$25 \text{ mm} = 25 \cancel{\text{mm}} \times \frac{1 \text{ m}}{1{,}000 \cancel{\text{mm}}}$$
$$= \frac{25 \text{ m}}{1{,}000}$$
$$= 0.025 \text{ m}$$

Area [8.2]

4. Convert 256 acres to square miles.

$$256 \text{ acres} = 256 \text{ acres} \times \frac{1 \text{ mi}^2}{640 \text{ acres}}$$
$$= \frac{256}{640} \text{ mi}^2$$
$$= 0.4 \text{ mi}^2$$

U.S. SYSTEM		
The Relationship Between	**Is**	**To Convert From One To The Other, Multiply By**
square inches and square feet	$144 \text{ in}^2 = 1 \text{ ft}^2$	$\frac{144 \text{ in}^2}{1 \text{ ft}^2}$ or $\frac{1 \text{ ft}^2}{144 \text{ in}^2}$
square yards and square feet	$9 \text{ ft}^2 = 1 \text{ yd}^2$	$\frac{9 \text{ ft}^2}{1 \text{ yd}^2}$ or $\frac{1 \text{ yd}^2}{9 \text{ ft}^2}$
acres and square feet	$1 \text{ acre} = 43{,}560 \text{ ft}^2$	$\frac{43{,}560 \text{ ft}^2}{1 \text{ acre}}$ or $\frac{1 \text{ acre}}{43{,}560 \text{ ft}^2}$
acres and square miles	$640 \text{ acres} = 1 \text{ mi}^2$	$\frac{640 \text{ acres}}{1 \text{ mi}^2}$ or $\frac{1 \text{ mi}^2}{640 \text{ acres}}$

METRIC SYSTEM		
The Relationship Between	**Is**	**To Convert From One To The Other, Multiply By**
square millimeters and square centimeters	$1 \text{ cm}^2 = 100 \text{ mm}^2$	$\frac{100 \text{ mm}^2}{1 \text{ cm}^2}$ or $\frac{1 \text{ cm}^2}{100 \text{ mm}^2}$
square centimeters and square decimeters	$1 \text{ dm}^2 = 100 \text{ cm}^2$	$\frac{100 \text{ cm}^2}{1 \text{ dm}^2}$ or $\frac{1 \text{ dm}^2}{100 \text{ cm}^2}$
square decimeters and square meters	$1 \text{ m}^2 = 100 \text{ dm}^2$	$\frac{100 \text{ dm}^2}{1 \text{ m}^2}$ or $\frac{1 \text{ m}^2}{100 \text{ dm}^2}$
square meters and ares (a)	$1 \text{ a} = 100 \text{ m}^2$	$\frac{100 \text{ m}^2}{1 \text{ a}}$ or $\frac{1 \text{ a}}{100 \text{ m}^2}$
ares and hectares (ha)	$1 \text{ ha} = 100 \text{ a}$	$\frac{100 \text{ a}}{1 \text{ ha}}$ or $\frac{1 \text{ ha}}{100 \text{ a}}$

Volume [8.2]

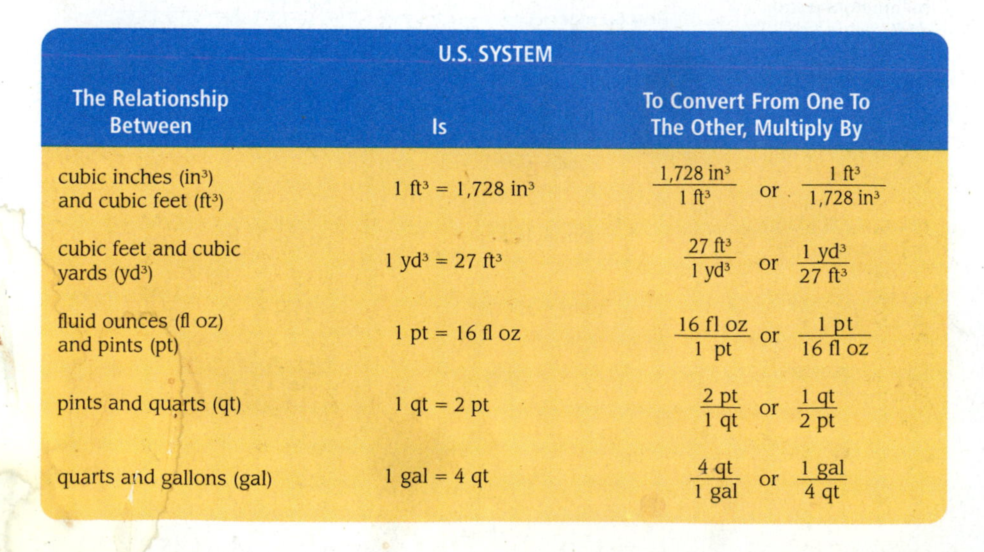

U.S. SYSTEM		
The Relationship Between	**Is**	**To Convert From One To The Other, Multiply By**
cubic inches (in³) and cubic feet (ft³)	$1 \text{ ft}^3 = 1{,}728 \text{ in}^3$	$\frac{1{,}728 \text{ in}^3}{1 \text{ ft}^3}$ or $\frac{1 \text{ ft}^3}{1{,}728 \text{ in}^3}$
cubic feet and cubic yards (yd³)	$1 \text{ yd}^3 = 27 \text{ ft}^3$	$\frac{27 \text{ ft}^3}{1 \text{ yd}^3}$ or $\frac{1 \text{ yd}^3}{27 \text{ ft}^3}$
fluid ounces (fl oz) and pints (pt)	$1 \text{ pt} = 16 \text{ fl oz}$	$\frac{16 \text{ fl oz}}{1 \text{ pt}}$ or $\frac{1 \text{ pt}}{16 \text{ fl oz}}$
pints and quarts (qt)	$1 \text{ qt} = 2 \text{ pt}$	$\frac{2 \text{ pt}}{1 \text{ qt}}$ or $\frac{1 \text{ qt}}{2 \text{ pt}}$
quarts and gallons (gal)	$1 \text{ gal} = 4 \text{ qt}$	$\frac{4 \text{ qt}}{1 \text{ gal}}$ or $\frac{1 \text{ gal}}{4 \text{ qt}}$

METRIC SYSTEM		
The Relationship Between	Is	To Convert From One To The Other, Multiply By
milliliters (mL) and liters	1 liter (L) = 1,000 mL	$\frac{1{,}000 \text{ mL}}{1 \text{ liter}}$ or $\frac{1 \text{ liter}}{1{,}000 \text{ mL}}$
hectoliters (hL) and liters	100 liters = 1 hL	$\frac{100 \text{ liters}}{1 \text{ hL}}$ or $\frac{1 \text{ hL}}{100 \text{ liters}}$
kiloliters (kL) and liters	1,000 liters (L) = 1 kL	$\frac{1{,}000 \text{ liters}}{1 \text{ kL}}$ or $\frac{1 \text{ kL}}{1{,}000 \text{ liters}}$

5. Convert 2.2 liters to milliliters.

$$2.2 \text{ liters} = 2.2 \text{ liters} \times \frac{1{,}000 \text{ mL}}{1 \text{ liter}} = 2.2 \times 1{,}000 \text{ mL} = 2{,}200 \text{ mL}$$

Weight [8.3]

U.S. SYSTEM		
The Relationship Between	Is	To Convert From One To The Other, Multiply By
ounces (oz) and pounds (lb)	1 lb = 16 oz	$\frac{16 \text{ oz}}{1 \text{ lb}}$ or $\frac{1 \text{ lb}}{16 \text{ oz}}$
pounds and tons (T)	1 T = 2,000 lb	$\frac{2{,}000 \text{ lb}}{1 \text{ T}}$ or $\frac{1 \text{ T}}{2{,}000 \text{ lb}}$

6. Convert 12 pounds to ounces.

$$12 \text{ lb} = 12 \text{ lb} \times \frac{16 \text{ oz}}{1 \text{ lb}} = 12 \times 16 \text{ oz} = 192 \text{ oz}$$

METRIC SYSTEM		
The Relationship Between	Is	To Convert From One To The Other, Multiply By
milligrams (mg) and grams (g)	1 g = 1,000 mg	$\frac{1{,}000 \text{ mg}}{1 \text{ g}}$ or $\frac{1 \text{ g}}{1{,}000 \text{ mg}}$
centigrams (cg) and grams	1 g = 100 cg	$\frac{100 \text{ cg}}{1 \text{ g}}$ or $\frac{1 \text{ g}}{100 \text{ cg}}$
kilograms (kg) and grams	1,000 g = 1 kg	$\frac{1{,}000 \text{ g}}{1 \text{ kg}}$ or $\frac{1 \text{ kg}}{1{,}000 \text{ g}}$
metric tons (t) and kilograms	1,000 kg = 1 t	$\frac{1{,}000 \text{ kg}}{1 \text{ t}}$ or $\frac{1 \text{ t}}{1{,}000 \text{ kg}}$

7. Convert 3 kilograms to centigrams.

$$3 \text{ kg} = 3 \text{ kg} \times \frac{1{,}000 \text{ g}}{1 \text{ kg}} \times \frac{100 \text{ cg}}{1 \text{ g}} = 3 \times 1{,}000 \times 100 \text{ cg} = 300{,}000 \text{ cg}$$

Converting Between the Systems [8.4]

8. Convert 8 inches to centimeters.

$$8\text{ in.} = 8\not{\text{in.}} \times \frac{2.54\text{ cm}}{1\not{\text{in.}}}$$
$$= 8 \times 2.54\text{ cm}$$
$$= 20.32\text{ cm}$$

CONVERSION FACTORS

The Relationship Between	Is	To Convert From One To The Other, Multiply By
Length		
inches and centimeters	2.54 cm = 1 in.	$\frac{2.54\text{ cm}}{1\text{ in.}}$ or $\frac{1\text{ in.}}{2.54\text{ cm}}$
feet and meters	1 m = 3.28 ft	$\frac{3.28\text{ ft}}{1\text{ m}}$ or $\frac{1\text{ m}}{3.28\text{ ft}}$
miles and kilometers	1.61 km = 1 mi	$\frac{1.61\text{ km}}{1\text{ mi}}$ or $\frac{1\text{ mi}}{1.61\text{ km}}$
Area		
square inches and square centimeters	6.45 cm^2 = 1 in^2	$\frac{6.45\text{ cm}^2}{1\text{ in}^2}$ or $\frac{1\text{ in}^2}{6.45\text{ cm}^2}$
square meters and square yards	1.196 yd^2 = 1 m^2	$\frac{1.196\text{ yd}^2}{1\text{ m}^2}$ or $\frac{1\text{ m}^2}{1.196\text{ yd}^2}$
acres and hectares	1 ha = 2.47 acres	$\frac{2.47\text{ acres}}{1\text{ ha}}$ or $\frac{1\text{ ha}}{2.47\text{ acres}}$
Volume		
cubic inches and milliliters	16.39 mL = 1 in^3	$\frac{16.39\text{ mL}}{1\text{ in}^3}$ or $\frac{1\text{ in}^3}{16.39\text{ mL}}$
liters and quarts	1.06 qt = 1 liter	$\frac{1.06\text{ qt}}{1\text{ liter}}$ or $\frac{1\text{ liter}}{1.06\text{ qt}}$
gallons and liters	3.79 liters = 1 gal	$\frac{3.79\text{ liters}}{1\text{ gal}}$ or $\frac{1\text{ gal}}{3.79\text{ liters}}$
Weight		
ounces and grams	28.3 g = 1 oz	$\frac{28.3\text{ g}}{1\text{ oz}}$ or $\frac{1\text{ oz}}{28.3\text{ g}}$
kilograms and pounds	2.20 lb = 1 kg	$\frac{2.20\text{ lb}}{1\text{ kg}}$ or $\frac{1\text{ kg}}{2.20\text{ lb}}$

Temperature [8.4]

9. Convert 120°C to degrees Fahrenheit.

$$F = \frac{9}{5}C + 32$$
$$F = \frac{9}{5}(120) + 32$$
$$F = 216 + 32$$
$$F = 248$$

To Convert From	Formula In Symbols	Formula In Words
Fahrenheit to Celsius	$C = \frac{5(F - 32)}{9}$	Subtract 32, multiply by 5, and then divide by 9.
Celsius to Fahrenheit	$F = \frac{9}{5}C + 32$	Multiply by $\frac{9}{5}$, and then add 32.

Time [8.5]

10. Convert 3 hours 45 minutes to minutes.

$$= 3\not{\text{hr}} \times \frac{60\text{ min}}{1\not{\text{hr}}} + 45\text{ min}$$
$$= 180\text{ min} + 45\text{ min}$$
$$= 225\text{ min}$$

The Relationship Between	Is	To Convert From One To The Other, Multiply By
minutes and seconds	1 min = 60 sec	$\frac{1\text{ min}}{60\text{ sec}}$ or $\frac{60\text{ sec}}{1\text{ min}}$
hours and minutes	1 hr = 60 min	$\frac{1\text{ hr}}{60\text{ min}}$ or $\frac{60\text{ min}}{1\text{ hr}}$

Chapter 8 Review

Use the tables given in this chapter to make the following conversions. [8.1-8.4]

1. 12 ft to inches
2. 18 ft to yards
3. 49 cm to meters
4. 2 km to decimeters
5. 10 acres to square feet
6. 7,800 m^2 to ares
7. 4 ft^2 to square inches
8. 7 qt to pints
9. 24 qt to gallons
10. 5 L to milliliters
11. 8 lb to ounces
12. 2 lb 4 oz to ounces
13. 5 kg to grams
14. 5 t to kilograms
15. 4 in. to centimeters
16. 7 mi to kilometers
17. 7 L to quarts
18. 5 gal to liters
19. 5 oz to grams
20. 9 kg to pounds
21. 120°C to degrees Fahrenheit
22. 122°F to degrees Celsius

Work the following problems. Round answers to the nearest hundredth where necessary.

23. A case of soft drinks holds 24 cans. If each can holds 355 ml, how many liters are there in the whole case? [8.2]

24. Change 862 mi to feet. [8.1]

25. Glacier Bay National Monument covers 2,805,269 acres. What is the area in square miles? [8.2]

26. How many ounces does a 134-lb person weigh? [8.3]

27. Change 250 mi to kilometers. [8.1]

28. How many grams is 7 lb 8 oz? [8.4]

29. Construction A 12-square-meter patio is to be built using bricks that measure 10 centimeters by 20 centimeters. How many bricks will be needed to cover the patio? [8.2]

30. Capacity If a regular drinking glass holds 0.25 liter of liquid, how many glasses can be filled from a 6.5-liter container? [8.2]

31. Filling an Aquarium How many 8-fluid-ounce glasses of water will it take to fill a 5-gallon aquarium? [8.2]

32. Comparing Area On April 3, 2000, *USA Today* changed the size of its paper. Previous to this date, each page of the paper was $13\frac{1}{2}$ inches wide and $22\frac{1}{4}$ inches long, giving each page an area of $300\frac{3}{8}$ in^2. Convert this area to square feet. [8.2]

33. Speed The instrument display below shows a speed of 188 kilometers per hour. What is the speed in miles per hour? Round to the nearest whole number. [8.4]

34. Volcanoes Pyroclastic flows are high speed avalanches of volcanic gases and ash that accompany some volcano eruptions. Pyroclastic flows have been known to travel at more than 80 kilometers per hour.

USGS

a. Convert 80 km/hr to miles per hour. Round to the nearest whole number.

b. Could you outrun a pyroclastic flow on foot, on a bicycle, or in a car?

35. Speed A race car is traveling at 200 miles per hour. What is the speed in kilometers per hour? [8.4]

36. 4 hours 45 minutes to [8.5]

a. Minutes

b. Hours

37. Add 4 pounds 4 ounces and 8 pounds 12 ounces. [8.5]

38. Cost of Fish. Mark is purchasing two whole salmon. The fish cost $5.00 per pound, and each weighs 12 lb 8 oz. What is the cost of the fish? [8.5]

Chapter 8 Cumulative Review

Simplify:

1. $\begin{array}{r} 230 \\ 4{,}976 \\ +\ 349 \\ \hline \end{array}$

2. 55(101)

3. $1845 \div 25$

4. $0.25\overline{)1.845}$

5. $\left(\frac{4}{5}\right)^3$

6. $3.846 + 5.09$

7. 3.4(0.34)

8. $-\frac{8}{15} \cdot \frac{5}{16}$

9. $\frac{7}{12} \div -\frac{21}{4}$

10. $-\frac{5}{9} + \frac{2}{27}$

11. $3\frac{1}{5} - 1\frac{1}{10}$

12. $24 \cdot 1\frac{1}{2}$

Solve each equation.

13. $4y - 16 = -2y + 26$

14. $-\frac{2}{3}(3x + 9) = 12$

15. $\frac{2}{x} + \frac{1}{3} = \frac{2}{3}$

16. Add $2\frac{1}{4}$ to 3.75.

17. Find the product of $\frac{3}{4}$ and $1\frac{1}{2}$.

18. Translate and simplify: The sum of twice 9 and 2.

19. Write the ratio of 12 to 36 as a fraction in lowest terms.

20. Convert 20 feet per second to miles per hour. Round to the nearest tenth.

21. Convert 2.5 square miles to acres.

22. Write $\frac{5}{8}$ as a percent.

23. Convert 55% to a fraction.

24. Solve the equation $\frac{3}{x} = \frac{6}{7}$.

25. Write 30,405 in expanded form.

26. What number is 30% of 700?

27. 26 is what percent of 39?

28. $\frac{3}{4}$ is 75% of what number?

29. If 1 kilogram is 2.2 pounds, convert 10 kilograms to pounds.

30. If 1 mile is 1.61 kilometers, convert 50 miles per hour to kilometers per hour.

31. Use the formula $F = \frac{9}{5}C + 32$ to convert 50°C to degrees Fahrenheit.

32. Textbook Prices A used copy of a textbook sells for \$56.25, which is 75% of the price of a new book. What is the price of a new book?

33. Cost of Avocados An avocado grower is going to sell avocados at a local farmer's market. If each avocado weighs 12 ounces and they sell for \$1.50 per pound, how much money will the grower earn for selling 400 avocados?

34. Pythagorean Theorem Use the Pythagorean theorem and the diagram of the square softball diamond to find the distance from third base to first base. Round to the nearest tenth.

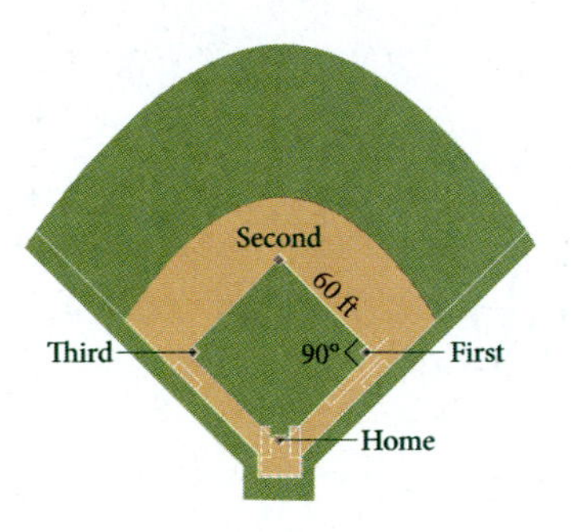

35. Ferris Wheel The first Ferris wheel was built by George Ferris in 1893. It had 36 carriages, each of which held 60 people. The diameter of the wheel was 250 feet, and a ride once around the wheel took 20 minutes.

a. If all the carriages are full, how many people can ride on the wheel at one time?

b. What is the circumference of the wheel? Use 3.14 for π.

c. Give the speed of a rider on the wheel in feet per minute.

d. Give the speed of a rider on the wheel in miles per hour if there are 5,280 feet in 1 mile. Round to the nearest hundredth.

Bettmann/Corbis

Chapter 8 Test

Use the tables in the chapter to make the following conversions.

1. 7 yd to feet

2. 750 m to kilometers

3. 3 acres to square feet

4. 432 in^2 to square feet

5. 10 L to milliliters

6. 5 mi to kilometers

7. 10 L to quarts

8. 80°F to degrees Celsius (round to the nearest tenth)

Work the following problems. Round answers to the nearest hundredth.

9. How many gallons are there in a 1-liter bottle of cola?

10. Change 579 yd to inches.

11. A car engine has a displacement of 409 in^3. What is the displacement in cubic feet?

12. Change 75 qt to liters.

13. Change 245 ft to meters.

14. How many liters are contained in an 8-quart container?

15. Construction A 40-square-foot pantry floor is to be tiled using tiles that measure 8 inches by 8 inches. How many tiles will be needed to cover the pantry floor?

16. Filling an Aquarium How many 12-fluid-ounce glasses of water will it take to fill a 6-gallon aquarium?

17. 5 hours 30 minutes to

a. Minutes

b. Hours

18. Add 3 pounds 4 ounces and 7 pounds 12 ounces.

Chapter 8 Projects

MEASUREMENT

GROUP PROJECT

Body Mass Index

Number of People 2

Time Needed 25 minutes

Equipment Pencil, paper, and calculator

Background Body mass index (BMI) is computed by using a mathematical formula in which one's weight in kilograms is divided by the square of one's height in meters. According to the Centers for Disease Control and Prevention, a healthy BMI for adults is between 18.5 and 24.9. Children aged 2–20 have a healthy BMI if they are in the 5th to 84th percentile for their age and sex. A high BMI is predictive of cardiovascular disease.

Height / Weight	4′10″	5′2″	5′9″	6′1″
100				
120				
140				
200				

Procedure Complete the given BMI chart using the following conversion factors.

1 inch = 2.54 cm, 1 meter = 100 cm, 1 kg = 2.2 lb

Example 5′4″, 120 lbs

1. Convert height to inches.

$$5 \text{ feet} \times \frac{12 \text{ in.}}{1 \text{ ft}} = 60 \text{ in.}$$

$$5'4'' = 64 \text{ in.}$$

Then, convert height to meters.

$$64 \text{ in.} \times \frac{2.54 \text{ cm}}{1 \text{ in.}} = 162.56 \text{ cm}$$

$$162.56 \text{ cm} \times \frac{1 \text{ m}}{100 \text{ cm}} = 1.6256 \text{ m}$$

2. Convert weight to kilograms.

$$120 \text{ lbs} \times \frac{1 \text{ kg}}{2.2 \text{ lbs}} \approx 54.5 \text{ kg}$$

3. Compute $\frac{\text{weight in kg}}{(\text{height in m})^2}$.

$$\frac{54.5}{(1.6256)^2} \approx 21$$

RESEARCH PROJECT

Richard Alfred Tapia

Richard A. Tapia is a mathematician and professor at Rice University in Houston, Texas, where he is Noah Harding Professor of Computational and Applied Mathematics. His parents immigrated from Mexico, separately, as teenagers to provide better educational opportunities for themselves and future generations. Born in Los Angeles, Tapia was the first in his family to attend college. In addition to being internationally known for his research, Tapia has helped his department at Rice become a national leader in awarding Ph.D. degrees to women and minority recipients. Research the life and work of Dr. Tapia. Summarize your results in an essay.

Courtesy of Rice University

Exponents and Polynomials

Chapter Outline

Introduction

Yosemite National Park, located in east central California, covers approximately 761,000 acres. The park is known for its spectacular granite rock formations, the most famous of which is Half Dome, shown in the Google Earth image above. Each year, Yosemite National Park is visited by over 3.5 million people who enjoy camping, hiking, sightseeing, and other outdoor activities.

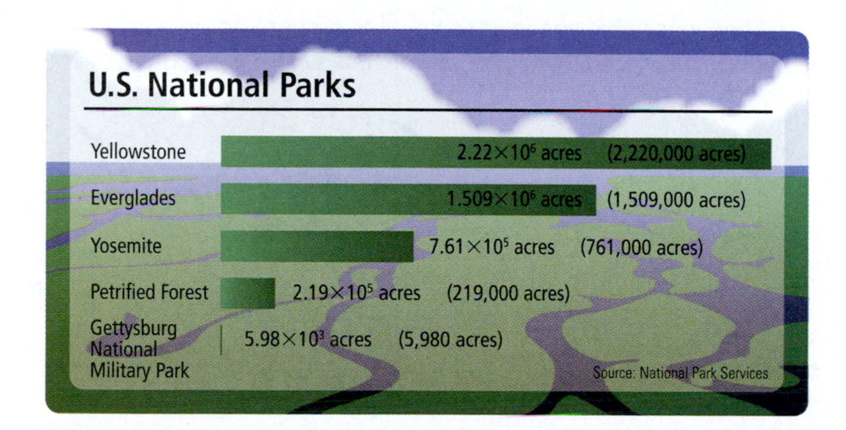

As you can see, the sizes of the National Parks in the diagram are given in both scientific notation and standard form. In this chapter we will use exponents to write very large or very small numbers in scientific notation.

Chapter Pretest

The pretest below contains problems that are representative of the problems you will find in the chapter.

Multiply.

1. $a^3 \cdot a^5$

2. $(5x^4)^2$

3. $(3ab^5)^3(2a^7b^2)$

Add.

4. $(2y^2 + 6y - 3) + (y^2 - 5y + 1)$

5. $(a^3 - 3a^2 + a) + (2a^2 + 2a - 1)$

Subtract.

6. $(7t^2 + 4t + 5) - (2t^2 + t + 1)$

7. $(4x^3 + 6x^2 + 3) - (x^3 - 4x - 9)$

8. Find the value of $2y^2 - 3y + 1$ when $y = -2$.

Multiply.

9. $x^3(x^2 + x^4)$

10. $5xy(4xy - 3)$

11. $(x + 4)(3x - 2)$

12. $(2x + 5)^2$

Write each expression with positive exponents, then simplify.

13. $(-2)^{-3}$

14. $x^{-4} \cdot x^{-5}$

15. $\dfrac{10^{-3}}{10^{-4}}$

16. $(9a^{-6})(-2a^{-1})$

Getting Ready for Chapter 9

The problems below review material covered previously that you need to know in order to be successful in Chapter 9. If you have any difficulty with the problems here, you need to go back and review before going on to Chapter 9.

Simplify.

1. $3 + 4 + 2$

2. $-4 + 1$

3. $9 + (-12)$

4. $-3 + (-4)$

5. $5 - 8$

6. $-3 - 5$

7. $-8 - (-6)$

8. $16 \cdot 125$

9. 2^6

10. $(-3)^2$

11. $8.64 \times 10{,}000$

12. $376{,}000 \div 100{,}000$

13. $1.3(5)$

14. $\dfrac{(6.8)(3.9)}{7.8}$

15. $-2x + 3x$

16. $-4y - 7y$

17. $7(2x - 3)$

18. $(x^2)(x^4)$

19. $(3xy)^3$

20. Evaluate $-7x + 2$ when $x = -2$.

Multiplication Properties of Exponents

9.1

In this section we will extend our work with exponents to develop some general properties of exponents that we can use to simplify expressions. Each of the properties we will develop has its foundation in the definition for exponents.

A Property 1 for Exponents

EXAMPLE 1 Multiply: $x^2 \cdot x^4$

SOLUTION We write x^2 as $x \cdot x$ and x^4 as $x \cdot x \cdot x \cdot x$ and then use the definition of exponents to write the answer with just one exponent.

$$\begin{aligned} x^2 \cdot x^4 &= (x \cdot x)(x \cdot x \cdot x \cdot x) && \text{Expand exponents} \\ &= x \cdot x \cdot x \cdot x \cdot x \cdot x \\ &= x^6 && \text{Write with a single exponent} \end{aligned}$$

Notice that the two exponents in the original problem given in Example 1 add up to the exponent in the answer: $2 + 4 = 6$. We use this result as justification for writing our first property of exponents.

Property 1 for Exponents

If r and s are any two whole numbers and a is an integer, then

$$a^r \cdot a^s = a^{r+s}$$

In words: To multiply two expressions with the same base, add exponents and use the common base.

EXAMPLE 2 Multiply: $5x^3 \cdot 3x^2$

SOLUTION We apply the commutative and associative properties so that the numbers 5 and 3 are together, and the x's are also. Then we use Property 1 for exponents to add exponents.

$$\begin{aligned} 5x^3 \cdot 3x^2 &= (5 \cdot 3)(x^3 \cdot x^2) && \text{Commutative and associative properties} \\ &= (5 \cdot 3)(x^{3+2}) && \text{Property 1 for exponents} \\ &= 15x^5 && \text{Multiply 3 and 5; add 3 and 2} \end{aligned}$$

EXAMPLE 3 Multiply: $4x^3 \cdot 3x^4 \cdot 5x^2$

SOLUTION Here we have the product of three expressions with exponents. Using the same steps shown in Example 2, we have

$$\begin{aligned} 4x^3 \cdot 3x^4 \cdot 5x^2 &= (4 \cdot 3 \cdot 5)(x^3 \cdot x^4 \cdot x^2) \\ &= (4 \cdot 3 \cdot 5)(x^{3+4+2}) \\ &= 60x^9 \end{aligned}$$

Objectives

A Understand and apply Property 1 for exponents.

B Understand and apply Property 2 for exponents.

C Understand and apply Property 3 for exponents.

D Simplify expressions using two or more properties of exponents.

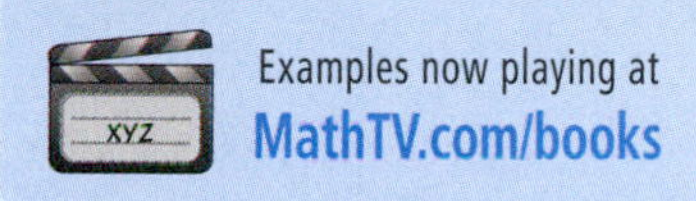

PRACTICE PROBLEMS

1. Multiply: $x^3 \cdot x^5$

2. Multiply: $4x^2 \cdot 6x^5$

3. Multiply: $2x^5 \cdot 3x^4 \cdot 4x^3$

Answers

1. x^8 2. $24x^7$ 3. $24x^{12}$

4. Multiply: $(10xy^2)(9x^3y^5)$

EXAMPLE 4 Multiply: $(2x^2y^3)(7x^4y)$

SOLUTION This time we have two different variables to work with. We use the commutative and associative properties to group the numbers together, the x's together, and the y's together.

$$\begin{aligned}(2x^2y^3)(7x^4y) &= (2 \cdot 7)(x^2 \cdot x^4)(y^3 \cdot y)\\ &= (2 \cdot 7)(x^{2+4})(y^{3+1})\\ &= 14x^6y^4\end{aligned}$$

B Property 2 for Exponents

Another common expression in algebra is that of a power raised to another power—for example, $(x^2)^3$, $(10^4)^5$, and $(y^6)^7$. To see how these expressions can be written with a single exponent, we return to the definition of exponents.

$$\begin{aligned}(x^2)^3 &= x^2 \cdot x^2 \cdot x^2 && \text{Definition of exponents}\\ &= x^{2+2+2} && \text{Property 1 for exponents}\\ &= x^6 && \text{Add exponents}\end{aligned}$$

As you can see, to obtain the exponent in the answer, we multiply the exponents in the original problem. The result leads us to our second property of exponents.

Property 2 for Exponents

If r and s are any two whole numbers and a is an integer, then

$$(a^r)^s = a^{r \cdot s}$$

In words: A power raised to another power is the base raised to the product of the powers.

5. Simplify the expression: $(2^2)^3$

EXAMPLE 5 Simplify the expression $(2^3)^2$.

SOLUTION We apply Property 2 by multiplying the exponents; then we simplify if we can.

$$\begin{aligned}(2^3)^2 &= 2^{3 \cdot 2} && \text{Property 2 for exponents}\\ &= 2^6 && \text{Multiply exponents}\\ &= 64\end{aligned}$$

6. Simplify: $(x^3)^4 \cdot (x^5)^2$

EXAMPLE 6 Simplify: $(x^5)^6 \cdot (x^4)^7$

SOLUTION In this case we apply Property 2 for exponents first to write each expression with a single exponent. Then we apply Property 1 to simplify further.

$$\begin{aligned}(x^5)^6 \cdot (x^4)^7 &= x^{5 \cdot 6} \cdot x^{4 \cdot 7} && \text{Property 2 for exponents}\\ &= x^{30} \cdot x^{28} && \text{Multiply exponents}\\ &= x^{30+28} && \text{Property 1 for exponents}\\ &= x^{58} && \text{Add exponents}\end{aligned}$$

C Property 3 for Exponents

The third, and final, property of exponents for this section covers the situation that occurs when we have the product of more than one number or variable, all raised to a power. Expressions such as $(4x)^2$, $(2ab)^3$, and $(5x^2y^3)^5$ all fall into this category.

Answers
4. $90x^4y^7$ 5. 64 6. x^{22}

$(4x)^2 = (4x)(4x)$ Definition of exponents
$= (4 \cdot 4)(x \cdot x)$ Commutative and associative properties
$= 4^2 \cdot x^2$ Definition of exponents

As you can see, both the numbers, 4 and x, contained within the parentheses end up raised to the second power.

Property 3 for Exponents
If r is a whole number and a and b are integers, then

$$(a \cdot b)^r = a^r \cdot b^r$$

EXAMPLE 7 Simplify: $(3xy)^4$

SOLUTION Applying our new property, we have

$(3xy)^4 = 3^4 \cdot x^4 \cdot y^4$ Property 3 for exponents
$= 81x^4y^4$ Simplify

D Combining Two or More Properties of Exponents

EXAMPLE 8 Simplify: $(2x^3y^2)^5$

SOLUTION In this case we will have to apply Property 3 first and then Property 2, in order to simplify the expression.

$(2x^3y^2)^5 = 2^5(x^3)^5(y^2)^5$ Property 3 for exponents
$= 32x^{15}y^{10}$ Property 2 for exponents

EXAMPLE 9 Simplify: $(4x^3y)^2(5x^2y^4)^3$

SOLUTION This simplification will require all three properties of exponents, as well as the commutative and associative properties of multiplication. Here are all the steps:

$(4x^3y)^2(5x^2y^4)^3 = 4^2(x^3)^2y^2 \cdot 5^3(x^2)^3(y^4)^3$ Property 3 for exponents
$= 16x^6y^2 \cdot 125x^6y^{12}$ Property 2 for exponents
$= (16 \cdot 125)(x^6x^6)(y^2y^{12})$ Commutative and associative properties
$= 2{,}000x^{12}y^{14}$ Property 2 for exponents

Getting Ready for Class

After reading through the preceding section, respond in your own words and in complete sentences.

1. When do we add the exponents on two expressions?
2. Under what conditions do we multiply two exponents?
3. The problem below is wrong. Correct the right-hand side.
   $(2x)^3 = 2x^3$
4. The problem below is wrong. Correct the right-hand side.
   $(x^4)^3 = x^7$

7. Simplify: $(5xy)^3$

8. Simplify: $(4x^5y^2)^3$

9. Simplify: $(2x^3y^4)^3(3xy^5)^2$

Answers
7. $125x^3y^3$ **8.** $64x^{15}y^6$
9. $72x^{11}y^{22}$

Problem Set 9.1

A Multiply the following expressions. [Examples 1–4]

1. $x^3 \cdot x^5$
2. $x^4 \cdot x^7$
3. $a^3 \cdot a^5$
4. $a^7 \cdot a^5$
5. $y^{10} \cdot y^5$
6. $y^{12} \cdot y^8$
7. $x^4 \cdot x$
8. $x^6 \cdot x$
9. $x^2 \cdot x^3 \cdot x^4$
10. $x^3 \cdot x^5 \cdot x^7$
11. $a^6 \cdot a^4 \cdot a^2$
12. $a^5 \cdot a^4 \cdot a^3$
13. $y^3 \cdot y^2 \cdot y$
14. $y^5 \cdot y^3 \cdot y$
15. $2x^4 \cdot 7x^5$
16. $6x^2 \cdot 9x^4$
17. $3y^5 \cdot 8y^3$
18. $2y^4 \cdot 7y^4$
19. $7x \cdot 3x$
20. $6x \cdot 2x$
21. $4x^3 \cdot 6x^5 \cdot 8x^7$
22. $3x^4 \cdot 5x^6 \cdot 7x^8$
23. $2a \cdot 3a \cdot 4a$
24. $10a \cdot 9a \cdot 8a$
25. $5y^4 \cdot 5y^4 \cdot 5y^4$
26. $4y^2 \cdot 4y^2 \cdot 4y^2$
27. $(3x^2y^7)(2x^3y)$
28. $(4x^5y^2)(3x^2y^2)$
29. $(7a^3b^2)(8ab^3)$
30. $(9a^4b^4)(10ab^3)$
31. $(2ab^3)(3a^2b^2)(4a^3b)$
32. $(8a^4b^4)(5a^3b^5)(a^2b^6)$
33. $(7a^3b^5)(4a^4b^4)(a^5b^3)$
34. $(6a^6b^3)(3a^3b^6)(ab)$
35. $(6x^2y^3)(-5x^3y^5)$
36. $(-2x^5y^2)(-4x^3y^4)$

B **C** Simplify the following expressions. [Examples 5–7]

37. $(a^3)^7$
38. $(a^7)^3$
39. $(5y^4)^3$
40. $(4y^2)^3$
41. $(9x^4)^2$
42. $(10x^5)^2$
43. $(2ab)^3$
44. $(3ab)^2$
45. $(5x^2y^3)^4$
46. $(3x^3y^2)^4$
47. $(2x^2y^5)^5$
48. $(3x^3y^4)^4$

D Simplify the following expressions. [Examples 8, 9]

49. $(a^3)^7 \cdot (a^4)^2$
50. $(a^4)^5 \cdot (a^3)^6$
51. $(3x^3y^5)^2(2x^2y^4)^3$
52. $(5x^4y^3)^2(2x^3y^2)^3$
53. $(5a^2b^4)^2(7a^8b)$
54. $(9a^5b^6)^2(8a^5b)$
55. $(a^2b^4)^3(ab^5)^4(a^4b)^2$
56. $(a^5b^3)^2(ab^3)^4(a^3b)^5$

57. Complete the following table.

Number x	Square x^2
1	
2	
3	
4	
5	
6	
7	

58. Complete the following table.

Number x	Square x^2
$\frac{1}{2}$	
$\frac{1}{3}$	
$\frac{1}{4}$	
$\frac{1}{5}$	
$\frac{1}{6}$	
$\frac{1}{7}$	

59. Complete the following table.

Number x	Square x^2
−2.5	
−1.5	
−0.5	
0	
0.5	
1.5	
2.5	

60. Complete the following table.

Number x	Square x^2
$-\frac{5}{2}$	
$-\frac{3}{2}$	
$-\frac{1}{2}$	
0	
$\frac{1}{2}$	
$\frac{3}{2}$	
$\frac{5}{2}$	

Maintaining Your Skills

Find the perimeter of each figure.

61.

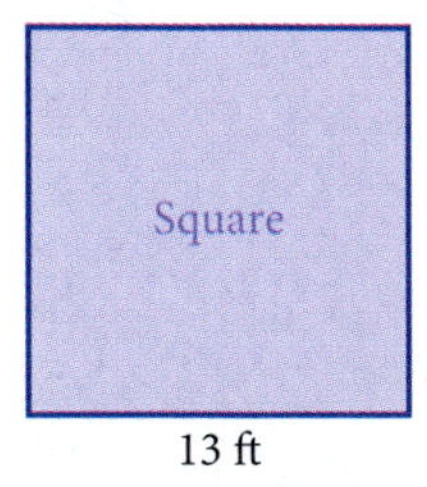

62.

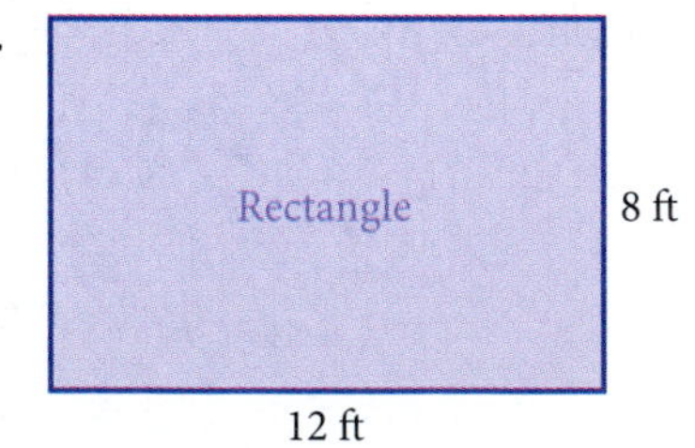

63.

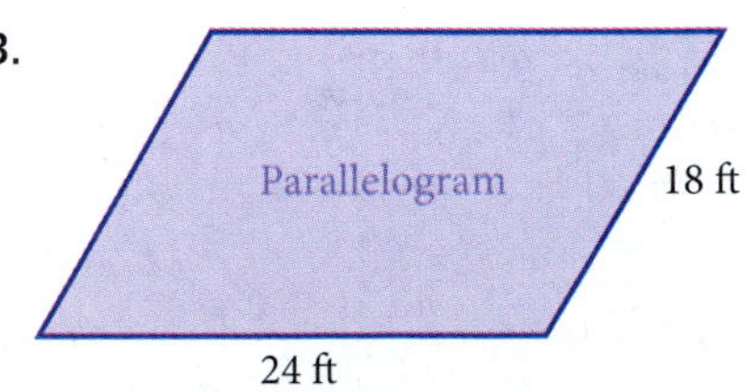

64.

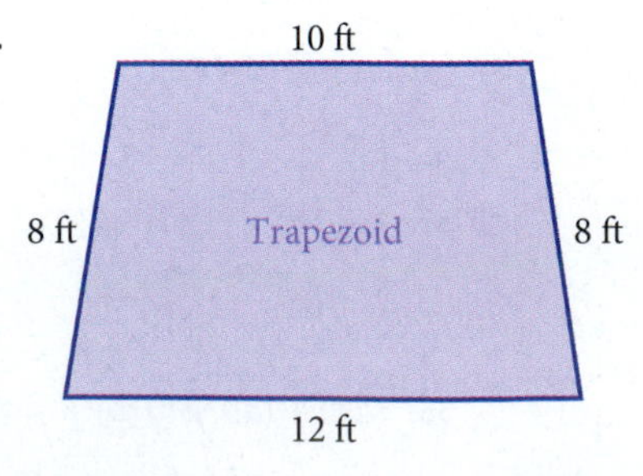

65.

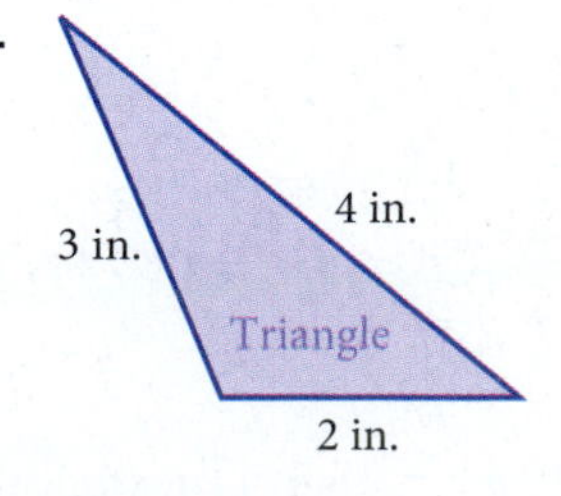

66.

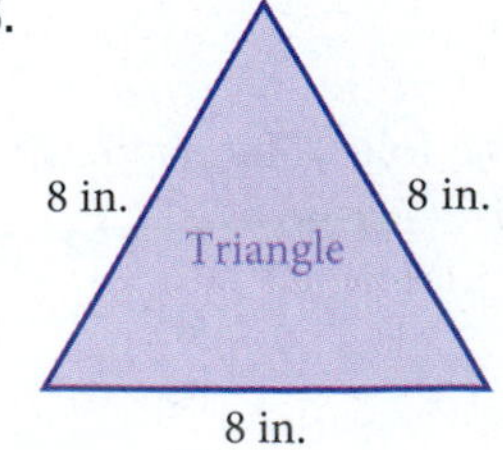

Adding and Subtracting Polynomials

9.2

Objectives

A Add polynomials.

B Subtract polynomials.

C Find the value of a polynomial.

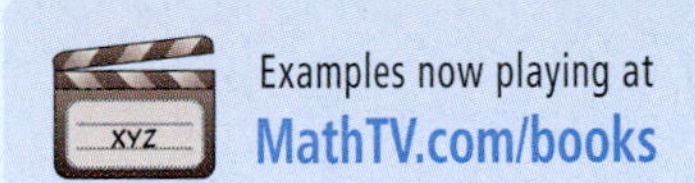

Previously we wrote numbers in expanded form to show the place value of each of the digits. For example, the number 456 in expanded form looks like this:

$$456 = 4 \cdot 100 + 5 \cdot 10 + 6 \cdot 1$$

If we replace 100 with 10^2, we have

$$456 = 4 \cdot 10^2 + 5 \cdot 10 + 6 \cdot 1$$

If we replace the 10's with x's, we get what is called a *polynomial.* It looks like this:

$$4x^2 + 5x + 6$$

Polynomials are to algebra what whole numbers written in expanded form are to arithmetic. As in other expressions in algebra, we can use any variable we choose. Here are some other examples of polynomials:

$$5a + 3 \qquad y^2 - 2y + 4 \qquad x^3 - 2x^2 + 5x - 1$$

A Adding Polynomials

When we add two whole numbers, we add in columns. That is, if we add 234 and 345, we write one number under the other and add the numbers in the ones column, then the numbers in the tens column, and finally the numbers in the hundreds column. Here is how it looks:

$$\begin{array}{rr} 234 & 2 \cdot 10^2 + 3 \cdot 10 + 4 \\ \underline{345} & \underline{3 \cdot 10^2 + 4 \cdot 10 + 5} \\ 579 & 5 \cdot 10^2 + 7 \cdot 10 + 9 \end{array}$$

We add polynomials in the same manner. If we want to add $2x^2 + 3x + 4$ and $3x^2 + 4x + 5$, we write one polynomial under the other, and then add in columns:

$$\begin{array}{r} 2x^2 + 3x + 4 \\ \underline{3x^2 + 4x + 5} \\ 5x^2 + 7x + 9 \end{array}$$

The sum of the two polynomials is the polynomial $5x^2 + 7x + 9$. We add only the digits. Notice that the variable parts (the x's) stay the same, just as the powers of 10 did when we added 234 and 345. The reason we add the numbers, while the variable parts of each term stay the same, can be explained with the commutative, associative, and distributive properties. Here is the same problem again, but this time we show the properties in use:

$$\begin{aligned} &(2x^2 + 3x + 4) + (3x^2 + 4x + 5) \\ &\quad = (2x^2 + 3x^2) + (3x + 4x) + (4 + 5) && \text{Commutative and associative properties} \\ &\quad = (2 + 3)x^2 + (3 + 4)x + (4 + 5) && \text{Distributive property} \\ &\quad = 5x^2 + 7x + 9 && \text{Addition} \end{aligned}$$

PRACTICE PROBLEMS

1. Add $5x^2 - 3x + 2$ and $2x^2 + 10x - 9$.

EXAMPLE 1 Add $3x^2 - 2x + 1$ and $5x^2 + 3x - 4$.

SOLUTION We will work the problem two ways. First, we write one polynomial under the other, and add in columns:

$$\begin{array}{r} 3x^2 - 2x + 1 \\ \underline{5x^2 + 3x - 4} \\ 8x^2 + 1x - 3 \end{array}$$

The sum of the two polynomials is $8x^2 + x - 3$.

Next we add horizontally, showing the commutative and associative properties in the first step, then the distributive property in the second step:

$$\begin{aligned} &(3x^2 - 2x + 1) + (5x^2 + 3x - 4) \\ &\quad = (3x^2 + 5x^2) + (-2x + 3x) + (1 - 4) \\ &\quad = (3 + 5)x^2 + (-2 + 3)x + (1 - 4) \\ &\quad = 8x^2 + 1x + (-3) \\ &\quad = 8x^2 + x - 3 \end{aligned}$$

2. Add $3y^2 + 9y - 5$ and $6y^2 - 4$.

EXAMPLE 2 Add $6y^2 + 4y - 3$ and $2y^2 - 4$.

SOLUTION We write one polynomial under the other, so that the terms with y^2 line up, and the terms without any y's line up:

$$\begin{array}{r} 6y^2 + 4y - 3 \\ \underline{2y^2 \qquad - 4} \\ 8y^2 + 4y - 7 \end{array}$$

The same problem, written horizontally, looks like this:

$$\begin{aligned} (6y^2 + 4y - 3) + (2y^2 - 4) &= (6y^2 + 2y^2) + 4y + (-3 - 4) \\ &= (6 + 2)y^2 + 4y + (-3 - 4) \\ &= 8y^2 + 4y + (-7) \\ &= 8y^2 + 4y - 7 \end{aligned}$$

3. Add $8x^3 + 4x^2 + 3x + 2$ and $4x^2 + 5x + 6$.

EXAMPLE 3 Add $2x^3 + 5x^2 + 3x + 4$ and $3x^2 + 2x + 1$.

SOLUTION Showing only the vertical method, we line up terms with the same variable part and add:

$$\begin{array}{r} 2x^3 + 5x^2 + 3x + 4 \\ \underline{3x^2 + 2x + 1} \\ 2x^3 + 8x^2 + 5x + 5 \end{array}$$

B Subtracting Polynomials

If there is a negative sign directly preceding the parentheses surrounding a polynomial, we may remove the parentheses and preceding negative sign by applying the distributive property. For example, to remove the parentheses from the expression

$$-(3x + 4)$$

we think of the negative sign as representing -1. Doing so allows us to apply the distributive property:

$$\begin{aligned} -(3x + 4) &= -1(3x + 4) \\ &= -1(3x) + (-1)(4) && \text{Distributive property} \\ &= -3x + (-4) && \text{Multiply} \\ &= -3x - 4 && \text{Simplify} \end{aligned}$$

Answers

1. $7x^2 + 7x - 7$
2. $9y^2 + 9y - 9$
3. $8x^3 + 8x^2 + 8x + 8$

As a result, the sign of each term inside the parentheses changes. Without showing all the steps involved in removing the parentheses, here are some more examples:

$$-(2x - 8) = -2x + 8$$
$$-(x^2 + 2x + 3) = -x^2 - 2x - 3$$
$$-(-4x^2 + 5x - 7) = 4x^2 - 5x + 7$$
$$-(3y^3 - 6y^2 + 7y - 3) = -3y^3 + 6y^2 - 7y + 3$$

In each case we remove the parentheses and preceding negative sign by changing the sign of each term that is found within the parentheses.

EXAMPLE 4 Subtract: $(6x^2 - 3x + 5) - (3x^2 + 2x - 3)$

SOLUTION Because subtraction is addition of the opposite, we simply change the sign of each term of the second polynomial and then add. With subtraction, there is less chance of making mistakes if we add horizontally, rather than in columns.

$$\begin{aligned} &(6x^2 - 3x + 5) - (3x^2 + 2x - 3) && \text{Subtraction} \\ &= 6x^2 - 3x + 5 - 3x^2 - 2x + 3 && \text{Addition of the opposite} \\ &= (6x^2 - 3x^2) + (-3x - 2x) + (5 + 3) \\ &= (6 - 3)x^2 + (-3 - 2)x + (5 + 3) && \text{Distributive property} \\ &= 3x^2 - 5x + 8 \end{aligned}$$

EXAMPLE 5 Subtract: $(2y^3 - 3y^2 - 4y - 2) - (3y^3 - 6y^2 + 7y - 3)$

SOLUTION Again, to subtract, we add the opposite of the polynomial that follows the subtraction sign; that is, we change the sign of each term in the second polynomial. Then we combine similar terms.

$$\begin{aligned} &(2y^3 - 3y^2 - 4y - 2) - (3y^3 - 6y^2 + 7y - 3) && \text{Subtraction} \\ &= 2y^3 - 3y^2 - 4y - 2 - 3y^3 + 6y^2 - 7y + 3 \\ &= -y^3 + 3y^2 - 11y + 1 \end{aligned}$$

EXAMPLE 6 Subtract $4x^2 - 3x + 1$ from $-3x^2 + 5x - 2$.

SOLUTION We must supply our own subtraction sign and write the two polynomials in the correct order.

$$\begin{aligned} &(-3x^2 + 5x - 2) - (4x^2 - 3x + 1) \\ &= -3x^2 + 5x - 2 - 4x^2 + 3x - 1 \\ &= -7x^2 + 8x - 3 \end{aligned}$$

C Finding the Value of a Polynomial

The last topic we want to consider in this section is finding the value of a polynomial for a given value of the variable.

EXAMPLE 7 Find the value of $4x^2 - 7x + 2$ when $x = -2$.

SOLUTION When $x = -2$, the polynomial $4x^2 - 7x + 2$ becomes

$$\begin{aligned} 4(-2)^2 - 7(-2) + 2 &= 4(4) + 14 + 2 \\ &= 16 + 14 + 2 \\ &= 32 \end{aligned}$$

4. Subtract:
$(5x^2 - 2x + 7) - (4x^2 + 8x - 4)$

5. Subtract: $(3y^3 - 2y^2 + 7y - 6) - (8y^3 - 6y^2 + 4y - 8)$

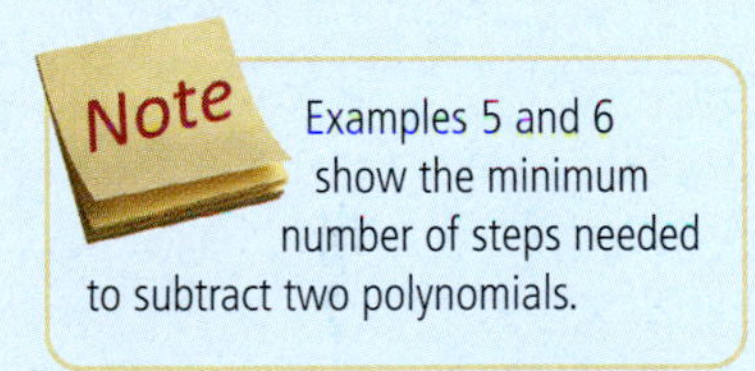

Examples 5 and 6 show the minimum number of steps needed to subtract two polynomials.

6. Subtract $6x^2 - 2x + 5$ from $-2x^2 + 5x - 1$.

7. Find the value of $5x^2 - 3x + 8$ when $x = -3$.

Answers
4. $x^2 - 10x + 11$
5. $-5y^3 + 4y^2 + 3y + 2$
6. $-8x^2 + 7x - 6$ **7.** 62

8. Fill in the tables

a.

n	1	2	3	4
$3n$				

b.

n	1	2	3	4
n^3				

More about Sequences

As the next example indicates, when we substitute the counting numbers, in order, into algebraic expressions, we form some of the sequences of numbers that we studied previously. To review, recall that the sequence of counting numbers (also called the sequence of positive integers) is

$$\text{Counting numbers} = 1, 2, 3, \ldots$$

EXAMPLE 8 Fill in the tables below to find the sequences formed by substituting the first four counting numbers into the expressions $2n$ and n^2.

a.

n	1	2	3	4
$2n$				

b.

n	1	2	3	4
n^2				

SOLUTION Proceeding as we did in the previous example, we substitute the numbers 1, 2, 3, and 4 into the given expressions.

a. When $n = 1$, $2n = 2 \cdot 1 = 2$.
When $n = 2$, $2n = 2 \cdot 2 = 4$.
When $n = 3$, $2n = 2 \cdot 3 = 6$.
When $n = 4$, $2n = 2 \cdot 4 = 8$.

As you can see, the expression $2n$ produces the sequence of even numbers when n is replaced by the counting numbers. Placing these results into our first table gives us

n	1	2	3	4
$2n$	2	4	6	8

b. The expression n^2 produces the sequence of squares when n is replaced by 1, 2, 3, and 4. In table form we have

n	1	2	3	4
n^2	1	4	9	16

Getting Ready for Class

After reading through the preceding section, respond in your own words and in complete sentences.

1. Describe how to add polynomials.
2. Describe how to subtract polynomials.
3. Make up an addition problem involving two polynomials so that the answer is the polynomial $8x^2 + 5x - 3$.
4. Find the value of the polynomial $8x^2 + 5x - 3$ when x is 2.

Answer

8.

a.

n	1	2	3	4
$3n$	3	6	9	12

b.

n	1	2	3	4
n^3	1	8	27	64

Problem Set 9.2

A Add the following polynomials. [Examples 1–3]

1. $(3x^2 + 2x + 5) + (3x^2 + 4x + 3)$

2. $(x^2 + 3x + 7) + (x^2 + 4x + 5)$

3. $(3a^2 - 4a + 2) + (2a^2 - 5a + 6)$

4. $(5a^2 - 2a + 1) + (3a^2 - 4a + 2)$

5. $(6x^3 - 5x^2 + 3x) + (9x^2 - 6x + 2)$

6. $(5x^3 + 2x^2 + 4x) + (2x^2 + 5x + 1)$

7. $(7t^2 - 4t - 5) + (6t^2 + 4t + 3)$

8. $(8t^2 - 3t - 2) + (7t^2 + 3t + 9)$

9. $(20x^3 - 12x^2 + x - 5) + (13x^3 + 2x^2 - 5x + 1)$

10. $(30x^3 + 3x^2 - 10x + 3) + (12x^3 - 3x^2 + 10x - 3)$

B Subtract the following polynomials. [Examples 4–6]

11. $(4x^2 + 3x + 2) - (2x^2 + 5x + 1)$

12. $(8x^2 + 4x + 1) - (6x^2 + 2x + 3)$

13. $(9y^3 - 8y^2 - 4) - (5y^3 - 3y + 8)$

14. $(12y^2 - 2y + 9) - (10y^3 + 3y^2 - 9)$

15. $(4x - 5) - (3x + 2) - (5x - 3)$

16. $(8x + 4) - (5x - 1) - (7x - 5)$

17. Subtract $10x^2 + 20x - 50$ from $11x^2 - 10x + 14$.

18. Subtract $2x^2 - 3x + 5$ from $4x^2 - 6x + 12$.

19. Subtract $3y^2 + 7y - 15$ from $10y^2 + 10y + 10$.

20. Subtract $15y^2 - 7y + 3$ from $15y^2 - 7y - 3$.

C Evaluate the following expressions for the given value. [Example 7]

21. Find the value of $x^2 - 10x + 25$ when x is 2.

22. Find the value of $(x - 5)^2$ when x is 2.

23. Find the value of $a^2 + 6a + 9$ when a is -3.

24. Find the value of $(a + 3)^2$ when a is -3.

25. Find the value of $4y^2 + 12y + 9$ when $y = -1$.

26. Find the value of $(2y + 3)^2$ when $y = -1$.

Substitute 1, 2, 3, and 4 for n in each of the following expressions.

27. $3n - 1$

28. $3n + 1$

29. $(n + 1)^2$

30. $n^2 + 1$

31. $n^2 + 9$

32. $(n + 3)^2$

Fill in each table.

33.

n	1	2	3	4
$4n$				

34.

n	1	2	3	4
n^4				

35.

n	1	2	3	4
$n - 3$				

36.

n	1	2	3	4
$2n - 5$				

Maintaining Your Skills

Find the area of each figure.

37.

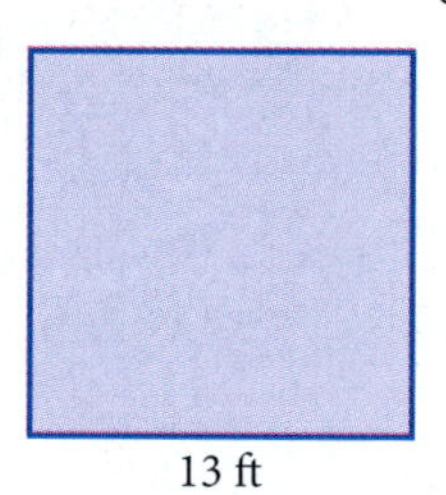

38.

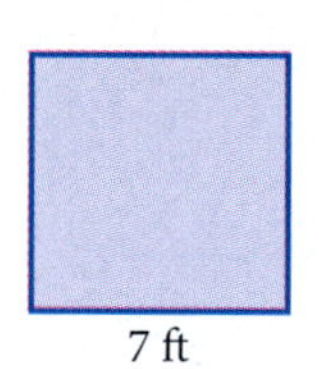

39.

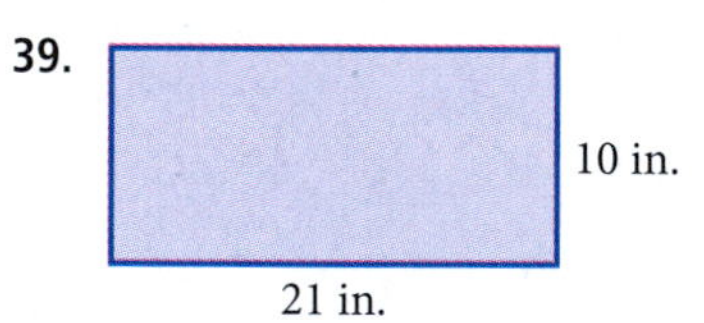

40.

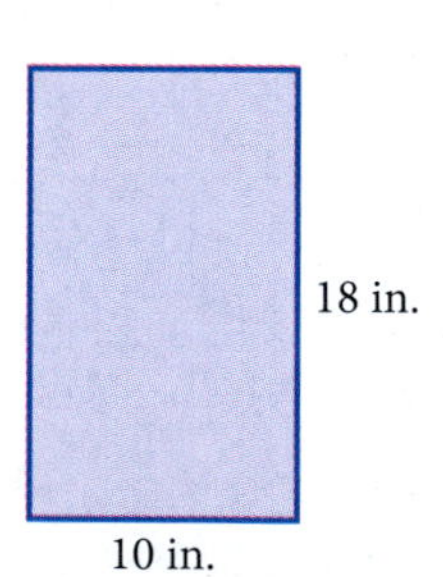

41.

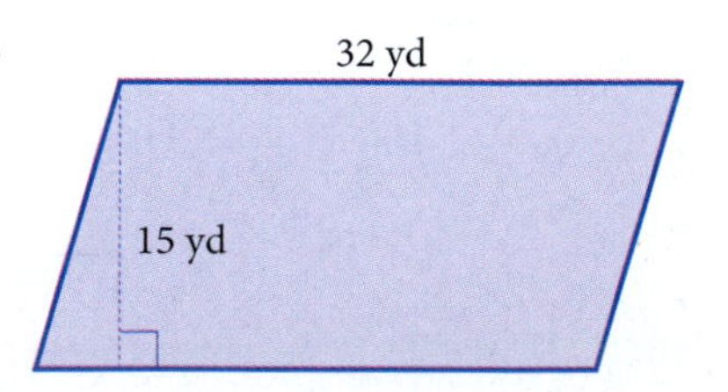

42.

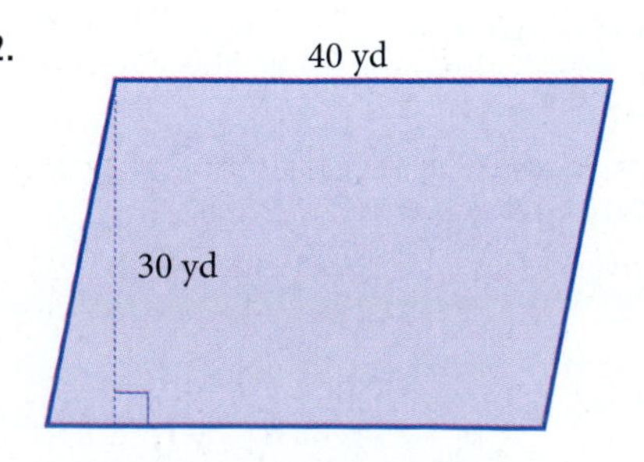

Multiplying Polynomials: An Introduction

9.3

Objectives

A Multiply polynomials algebraically.
B Multiply binomials.
C Multiply polynomials geometrically.

Examples now playing at **MathTV.com/books**

A Multiplying Polynomials Algebraically

Recall that the distributive property allows us to multiply across parentheses when a sum or difference is enclosed within the parentheses. That is:

$$a(b + c) = a \cdot b + a \cdot c$$

We can use the distributive property to multiply polynomials.

EXAMPLE 1 Multiply: $x^2(x^3 + x^4)$

SOLUTION Applying the distributive property, we have

$$\begin{aligned} \mathbf{x^2}(x^3 + x^4) &= \mathbf{x^2} \cdot x^3 + \mathbf{x^2} \cdot x^4 && \text{Distributive property} \\ &= x^5 + x^6 \end{aligned}$$

The distributive property works for multiplication from the right as well as the left. That is, we can also write the distributive property this way:

$$(b + c)a = b \cdot a + c \cdot a$$

EXAMPLE 2 Multiply: $(x^3 + x^4)x^2$

SOLUTION Because multiplication is a commutative operation, we should expect to obtain the same answer as in Example 1 above.

$$\begin{aligned} (x^3 + x^4)\mathbf{x^2} &= x^3 \cdot \mathbf{x^2} + x^4 \cdot \mathbf{x^2} \\ &= x^5 + x^6 \end{aligned}$$

EXAMPLE 3 Multiply: $4x^3(6x^2 - 8)$

SOLUTION The distributive property allows us to multiply $4x^3$ by both $6x^2$ and 8:

$$\begin{aligned} \mathbf{4x^3}(6x^2 - 8) &= \mathbf{4x^3} \cdot 6x^2 - \mathbf{4x^3} \cdot 8 \\ &= (4 \cdot 6)(x^3 \cdot x^2) - (4 \cdot 8)x^3 \\ &= 24x^5 - 32x^3 \end{aligned}$$

EXAMPLE 4 Multiply: $2a^4b^2(3a^3 + 4b^5)$

SOLUTION Applying the distributive property as we did in the previous two examples, we have

$$\begin{aligned} \mathbf{2a^4b^2}(3a^3 + 4b^5) &= \mathbf{2a^4b^2} \cdot 3a^3 + \mathbf{2a^4b^2} \cdot 4b^5 \\ &= (2 \cdot 3)(a^4 \cdot a^3)b^2 + (2 \cdot 4)(a^4)(b^2 \cdot b^5) \\ &= 6a^7b^2 + 8a^4b^7 \end{aligned}$$

B Multiplying Binomials

Polynomials with exactly two terms are called *binomials*. We multiply binomials by applying the distributive property.

PRACTICE PROBLEMS

1. Multiply: $x^3(x^5 + x^7)$
2. Multiply: $(x^5 + x^7)x^3$
3. Multiply: $5x^2(6x^3 - 4)$
4. Multiply: $5a^3b^5(2a^2 + 7b^2)$

Answers
1. $x^8 + x^{10}$ 2. $x^8 + x^{10}$
3. $30x^5 - 20x^2$
4. $10a^5b^5 + 35a^3b^7$

5. Multiply: $(x + 2)(x + 6)$

6. Multiply: $(x - 2)(x + 6)$

7. Multiply: $(3x - 2)(5x + 4)$

8. Expand and multiply: $(x + 3)^2$

Answers
5. $x^2 + 8x + 12$
6. $x^2 + 4x - 12$ **7.** $15x^2 + 2x - 8$
8. $x^2 + 6x + 9$

EXAMPLE 5

Multiply: $(x + 3)(x + 5)$

SOLUTION We can think of the first binomial, $x + 3$, as a single number. (Remember, for any value of x, $x + 3$ will be just a number.) We apply the distributive property by multiplying $x + 3$ times both x and 5.

$$\mathbf{(x + 3)}(x + 5) = \mathbf{(x + 3)} \cdot x + \mathbf{(x + 3)} \cdot 5$$

Next, we apply the distributive property again to multiply x times both x and 3, and 5 times both x and 3.

$$= x \cdot x + 3 \cdot x + x \cdot 5 + 3 \cdot 5$$
$$= x^2 + 3x + 5x + 15$$

The last thing to do is to combine the similar terms $3x$ and $5x$ to get $8x$. (Remember, this is also an application of the distributive property.)

$$= x^2 + 8x + 15$$

EXAMPLE 6

Multiply: $(x - 3)(x + 5)$

SOLUTION The only difference between the binomials in this example and those in Example 5 is the subtraction sign in $x - 3$. The steps in multiplying are exactly the same.

$(x - 3)(x + 5) = (x - 3) \cdot x + (x - 3) \cdot 5$	Multiply $x - 3$ times both x and 5
$= x \cdot x - 3 \cdot x + x \cdot 5 - 3 \cdot 5$	Distributive property two more times
$= x^2 - 3x + 5x - 15$	Simplify each term
$= x^2 + 2x - 15$	$-3x + 5x = 2x$

EXAMPLE 7

Multiply: $(2x - 3)(4x + 7)$

SOLUTION Using the same steps shown in Examples 5 and 6, we have

$$(2x - 3)(4x + 7) = (2x - 3) \cdot 4x + (2x - 3) \cdot 7$$
$$= 2x \cdot 4x - 3 \cdot 4x + 2x \cdot 7 - 3 \cdot 7$$
$$= 8x^2 - 12x + 14x - 21$$
$$= 8x^2 + 2x - 21$$

Our next two examples show how we raise binomials to the second power.

EXAMPLE 8

Expand and multiply: $(x + 5)^2$

SOLUTION We use the definition of exponents to write $(x + 5)^2$ as $(x + 5)(x + 5)$. Then we multiply as we did in the previous examples.

$(x + 5)^2 = (x + 5)(x + 5)$	Definition of exponents
$= (x + 5) \cdot x + (x + 5) \cdot 5$	Distributive property
$= x \cdot x + 5 \cdot x + x \cdot 5 + 5 \cdot 5$	Distributive property
$= x^2 + 5x + 5x + 25$	Simplify each term
$= x^2 + 10x + 25$	$5x + 5x = 10x$

EXAMPLE 9 Expand and multiply: $(3x - 7)^2$

SOLUTION We know that $(3x - 7)^2 = (3x - 7)(3x - 7)$. It will be easier to apply the distributive property to this last expression if we think of the second $3x - 7$ as $3x + (-7)$. In doing so we will also be less likely to make a mistake in our signs. (Try the problem without changing subtraction to addition of the opposite, and see how your answer compares to the answer in this example.)

$$\begin{aligned}(3x - 7)(3x - 7) &= (3x - 7)[3x + (-7)] \\ &= (3x - 7) \cdot 3x + (3x - 7)(-7) \\ &= 3x \cdot 3x - 7 \cdot 3x + 3x(-7) - 7(-7) \\ &= 9x^2 - 21x - 21x + 49 \\ &= 9x^2 - 42x + 49\end{aligned}$$

C Multiplying Polynomials Geometrically

Suppose we have a rectangle with length $x + 3$ and width $x + 2$. Remember, the letter x is used to represent a number, so $x + 3$ and $x + 2$ are just numbers. Here is a diagram:

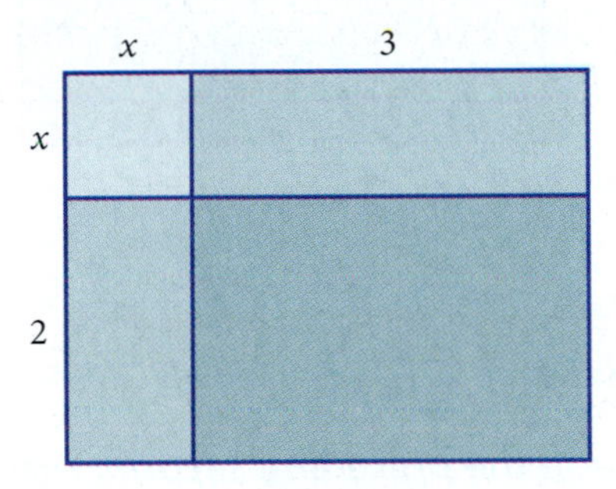

The area of the whole rectangle is the length times the width, or

$$\text{Total area} = (x + 3)(x + 2)$$

But we can also find the total area by first finding the area of each smaller rectangle and then adding these smaller areas together. The area of each rectangle is its length times its width, as shown in the following diagram:

x 3
x x^2 3x
2 2x 6

Because the total area $(x + 3)(x + 2)$ must be the same as the sum of the smaller areas, we have:

$$\begin{aligned}(x + 3)(x + 2) &= x^2 + 2x + 3x + 6 \\ &= x^2 + 5x + 6 \qquad \text{Add } 2x \text{ and } 3x \text{ to get } 5x\end{aligned}$$

The polynomial $x^2 + 5x + 6$ is the product of the two polynomials $x + 3$ and $x + 2$. Here are some more examples.

9. Expand and multiply: $(3x - 5)^2$

Answer

9. $9x^2 - 30x + 25$

10. Find the product of $3x + 7$ and $2x + 5$ by using the following diagram:

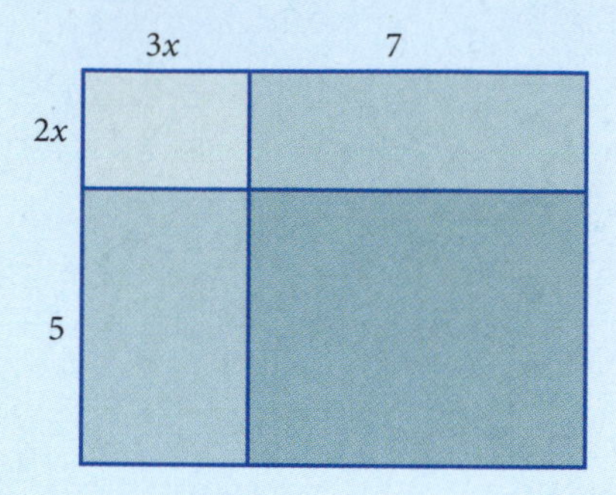

EXAMPLE 10 Find the product of $2x + 5$ and $3x + 2$ by using the following diagram:

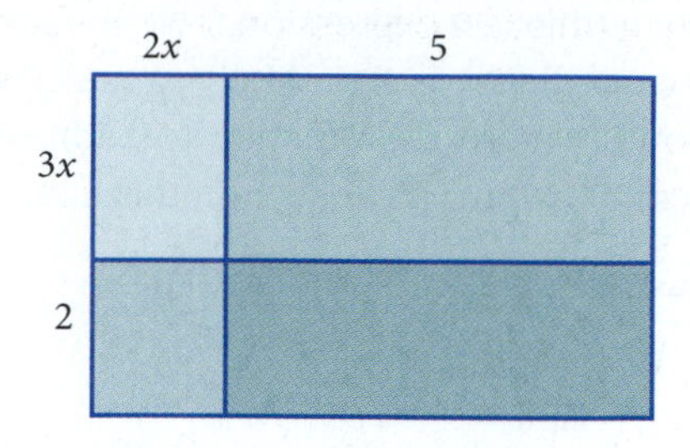

SOLUTION We fill in each of the smaller rectangles by multiplying length times width in each case:

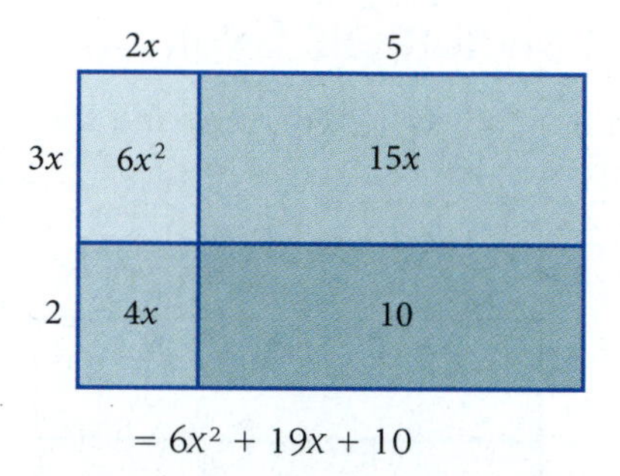

$$= 6x^2 + 19x + 10$$

Getting Ready for Class

After reading through the preceding section, respond in your own words and in complete sentences.

1. Describe how to multiply two polynomials algebraically.
2. Describe how to multiply two polynomials geometrically.
3. There is a mistake in the problem below. What is it?

$$x^2(x^3 + x^4) = x^5 + x^6 = x^{11}$$

4. The problem below is wrong. Correct the right-hand side.

$$(x + 5)^2 = x^2 + 25$$

Answer

10. $6x^2 + 29x + 35$

Problem Set 9.3

A B Multiply the following expressions. [Examples 1–7]

1. $x^2(x^4 + x^3)$
2. $x^3(x^2 + x^4)$
3. $(x^4 + x^3)x^2$
4. $(x^4 + x^2)x^3$
5. $2x^4(3x^2 - 7)$
6. $5x^3(2x^2 - 8)$
7. $4x^2y^3(3x^4 + 2y^3)$
8. $8x^2y^2(5x^3 + 4y^3)$
9. $(3x^2 - 7)2x^4$
10. $(2x^2 - 8)5x^3$
11. $3xy(4xy - 9)$
12. $2xy(3xy - 10)$
13. $(4xy - 9)3xy$
14. $(3xy - 10)2xy$
15. $2x^4y^5(3xy^2 + 2x^2y + 5)$
16. $5x^2y^2(2x^3y + 4xy^3 + 7)$
17. $(x + 2)(x + 3)$
18. $(x + 3)(x + 4)$
19. $(x - 2)(x + 3)$
20. $(x - 3)(x + 4)$
21. $(2x + 5)(3x + 2)$
22. $(3x + 5)(2x + 3)$
23. $(3x - 6)(x + 4)$
24. $(5x - 2)(x + 6)$
25. $(4x - 3)(4x + 3)$
26. $(7x - 3)(7x + 3)$
27. $(3x - 1)(4x + 1)$
28. $(6x - 1)(3x + 1)$
29. $(2x - 5)(3x - 4)$
30. $(7x - 1)(4x - 5)$

C Use the diagram in each problem to help multiply the polynomials. [Example 10]

31. $(x + 4)(x + 2)$

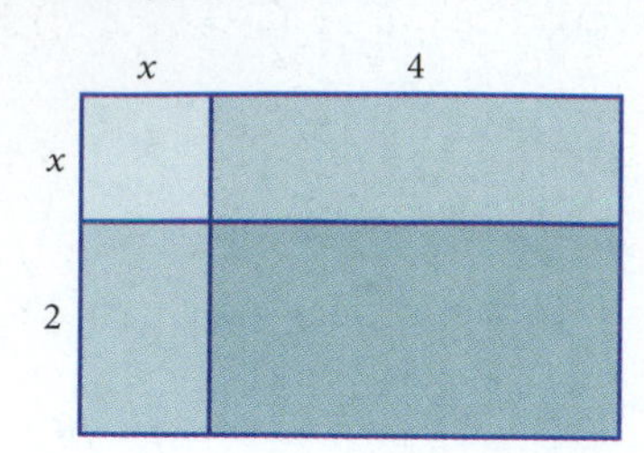

32. $(x + 1)(x + 3)$

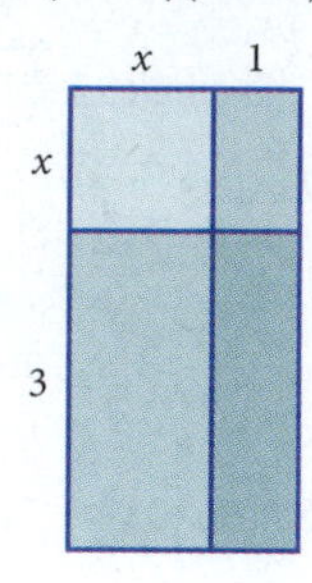

33. $(2x + 3)(3x + 2)$

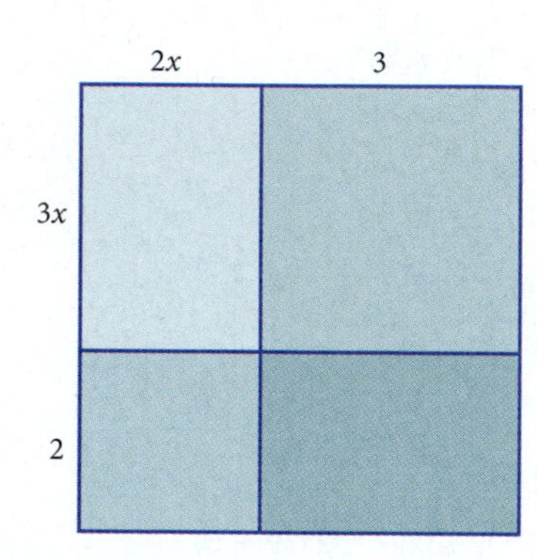

34. $(5x + 4)(6x + 1)$

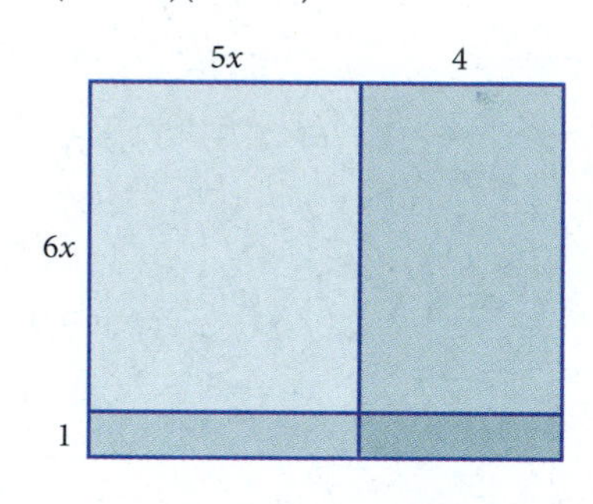

35. $(7x + 2)(3x + 4)$

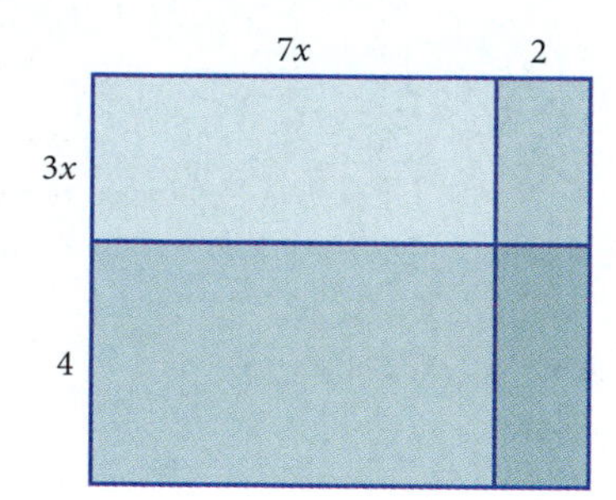

36. $(3x + 5)(2x + 5)$

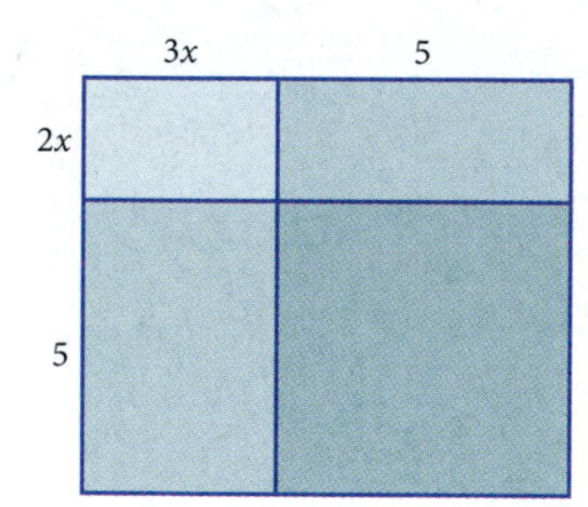

B Expand and multiply. [Examples 8, 9]

37. $(x + 3)^2$ **38.** $(x + 4)^2$ **39.** $(x + 6)^2$ **40.** $(x + 7)^2$

41. $(3x + 2)^2$ **42.** $(4x + 5)^2$ **43.** $(2x + 3)^2$ **44.** $(5x + 4)^2$

45. $(7x + 6)^2$ **46.** $(6x + 7)^2$ **47.** $(4x + 1)^2$ **48.** $(3x + 1)^2$

49. $(x - 5)^2$ **50.** $(x - 4)^2$ **51.** $(2x - 4)^2$ **52.** $(5x - 2)^2$

Fill in the following tables.

53.

x	$(x+3)^2$	x^2+9	x^2+6x+9
1			
2			
3			
4			

54.

x	$(x-5)^2$	x^2+25	$x^2-10x+25$
1			
2			
3			
4			

C Simplify each expression by filling in the diagrams and then using the results to find areas. [Example 10]

55. $(x+3)^2$

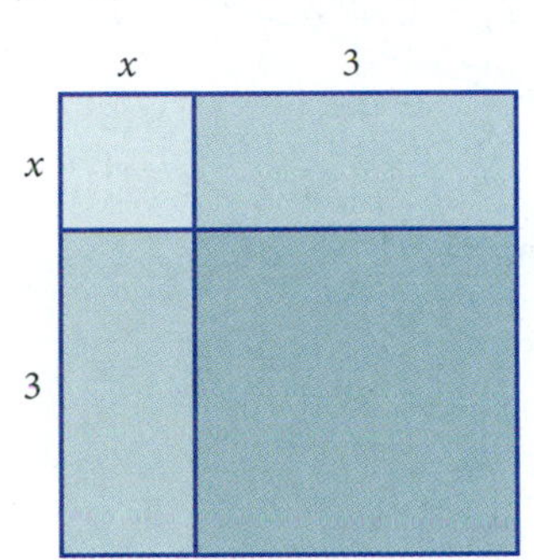

56. $(x+4)^2$

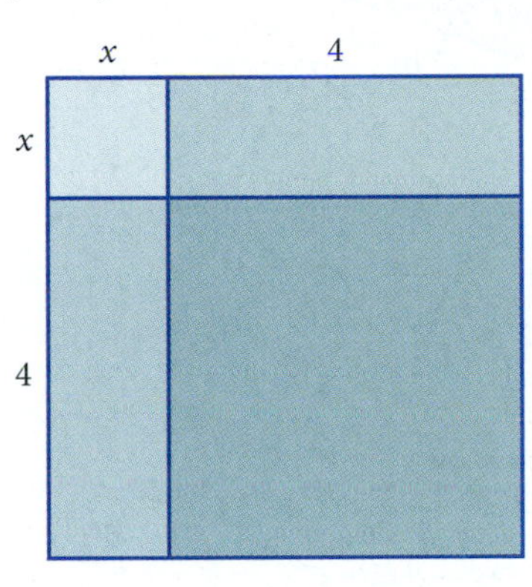

57. $(x+5)^2$

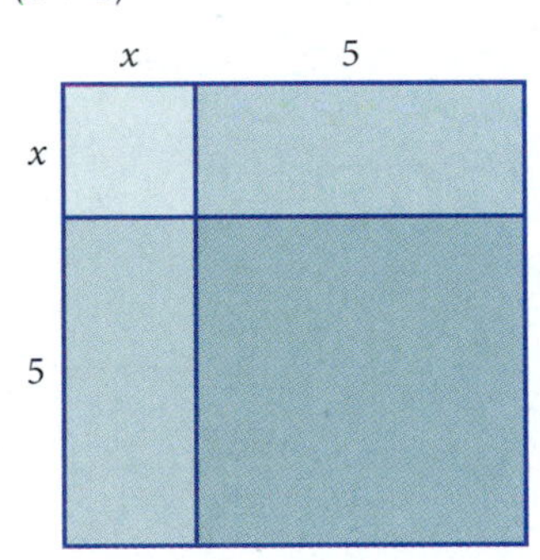

58. $(x+1)^2$

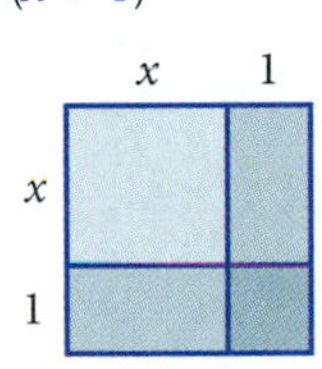

59. $(2x+3)^2$

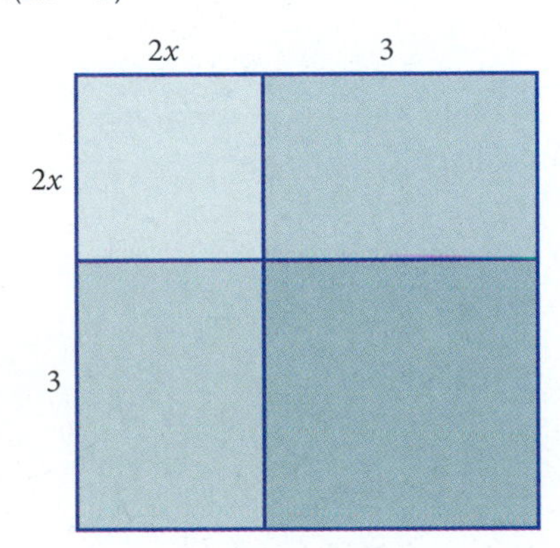

60. $(3x+2)^2$

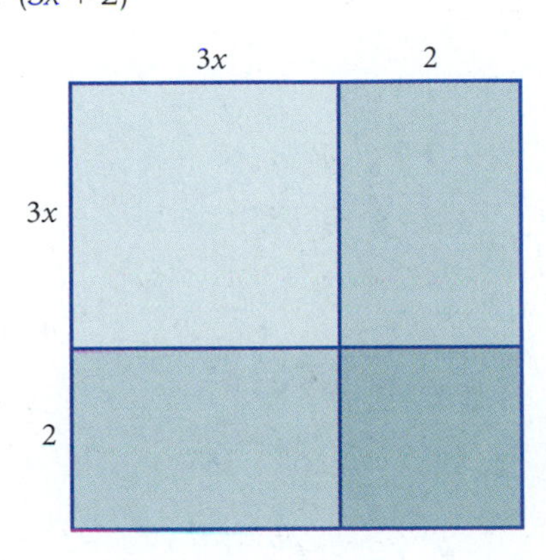

Maintaining Your Skills

Find the volume and the surface area of each figure.

61.

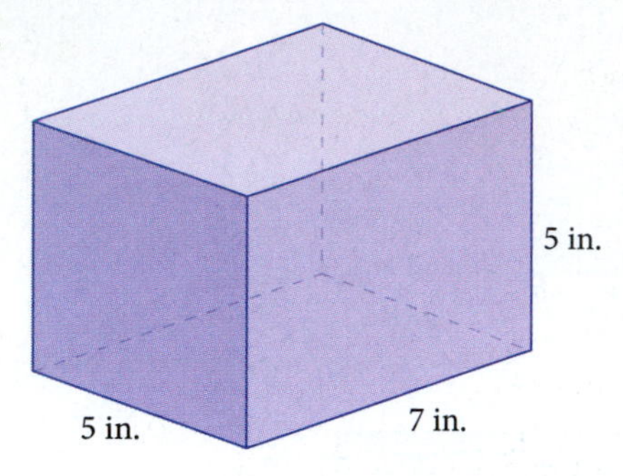

62.

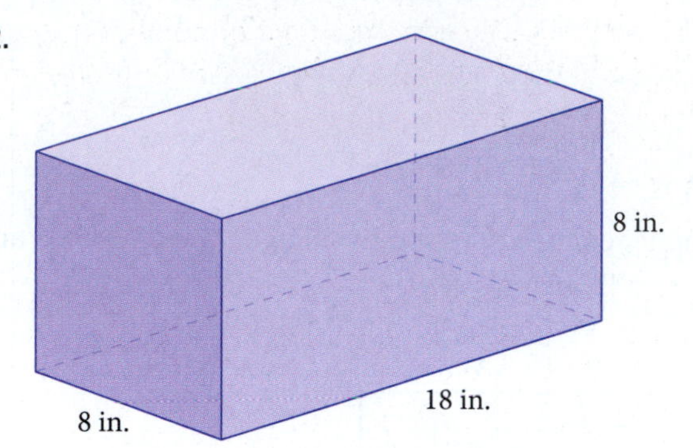

63.

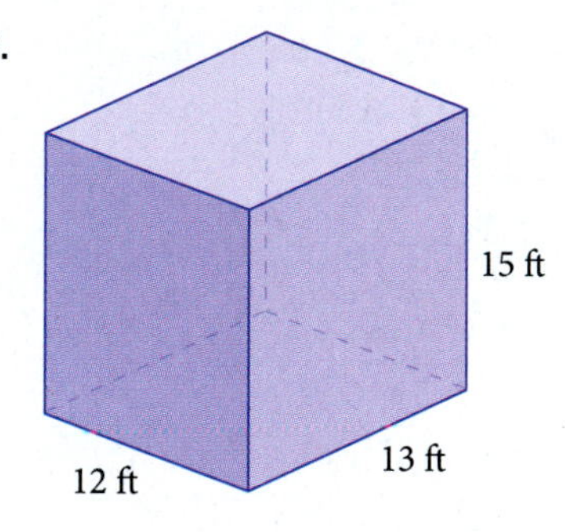

64.

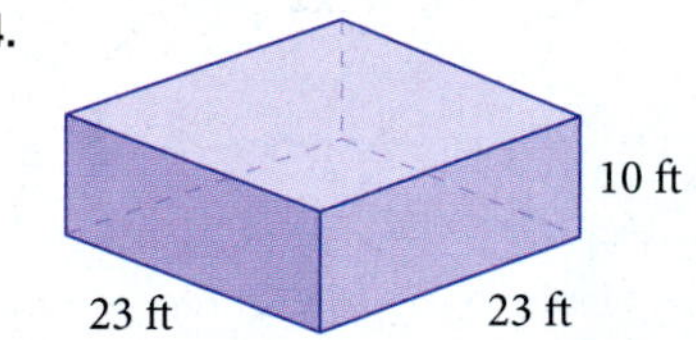

Negative Exponents

9.4

Objectives

A Simplify expressions containing negative exponents.

B Understand and apply the division property for exponents.

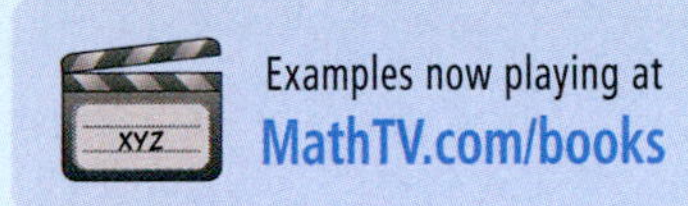

Up to this point, all the exponents we have worked with have been positive numbers. We want to extend the type of numbers we can use for exponents to include negative numbers as well. The definition that follows allows us to work with exponents that are negative numbers.

 Negative Exponents

Definition

Negative Exponents If r is a positive integer and a is any number other than zero, then

$$a^{-r} = \frac{1}{a^r}$$

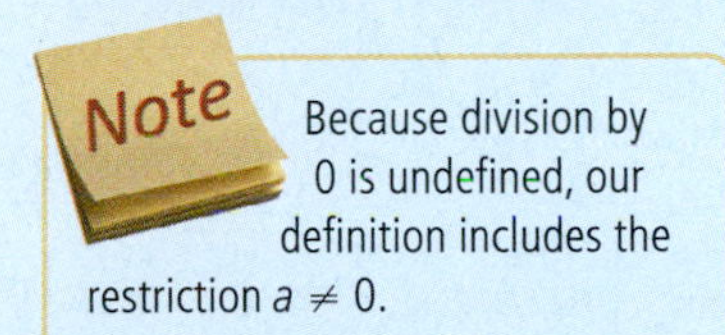

Because division by 0 is undefined, our definition includes the restriction $a \neq 0$.

The definition indicates that negative exponents give us reciprocals, as the following examples illustrate.

EXAMPLE 1 Simplify: 2^{-3}

SOLUTION The first step is to rewrite the expression with a positive exponent, by using our definition. After that, we simplify.

$$2^{-3} = \frac{1}{2^3} \quad \text{Definition of negative exponents}$$

$$= \frac{1}{8} \quad \text{The cube of 2 is 8}$$

EXAMPLE 2 Simplify: 5^{-2}

SOLUTION First we write the expression with a positive exponent, then we simplify.

$$5^{-2} = \frac{1}{5^2} \quad \text{Definition of negative exponents}$$

$$= \frac{1}{25} \quad \text{The square of 5 is 25}$$

As you can see from our first two examples, when we apply the definition for negative exponents to expressions containing negative exponents, we end up with reciprocals.

EXAMPLE 3 Simplify: $(-3)^{-2}$

SOLUTION The fact that the base in this problem is negative does not change the procedure we use to simplify:

$$(-3)^{-2} = \frac{1}{(-3)^2} \quad \text{Definition of negative exponents}$$

$$= \frac{1}{9} \quad \text{The square of } -3 \text{ is 9}$$

To check your understanding of negative exponents, look over the two lines below.

$$2^1 = 2 \qquad 2^2 = 4 \qquad 2^3 = 8 \qquad 2^4 = 16$$

$$2^{-1} = \frac{1}{2} \qquad 2^{-2} = \frac{1}{4} \qquad 2^{-3} = \frac{1}{8} \qquad 2^{-4} = \frac{1}{16}$$

PRACTICE PROBLEMS

1. Simplify: 2^{-4}

2. Simplify: 4^{-2}

3. Simplify: $(-2)^{-3}$

Answers

1. $\frac{1}{16}$ **2.** $\frac{1}{16}$ **3.** $-\frac{1}{8}$

4. Simplify: $3^4 \cdot 3^{-7}$

5. Simplify: $x^6 \cdot x^{-10}$

Answers

4. $\frac{1}{27}$ 5. $\frac{1}{x^4}$

The properties of exponents we have developed so far hold for negative exponents as well as positive exponents. For example, Property 1, our multiplication property for exponents, is still written as

$$a^r \cdot a^s = a^{r+s}$$

but now r and s can be negative numbers also.

EXAMPLE 4 Simplify: $2^5 \cdot 2^{-7}$

SOLUTION This is multiplication with the same base, so we add exponents.

$$\begin{aligned} 2^5 \cdot 2^{-7} &= 2^{5+(-7)} && \text{Property 1 for exponents} \\ &= 2^{-2} && \text{Addition} \\ &= \frac{1}{2^2} && \text{Definition of negative exponents} \\ &= \frac{1}{4} && \text{The square of 2 is 4} \end{aligned}$$

When we simplify expressions containing negative exponents, let's agree that the final expression contains only positive exponents.

EXAMPLE 5 Simplify: $x^9 \cdot x^{-12}$

SOLUTION Again, because we have the product of two expressions with the same base, we use Property 1 for exponents to add exponents.

$$\begin{aligned} x^9 \cdot x^{-12} &= x^{9+(-12)} && \text{Property 1 for exponents} \\ &= x^{-3} && \text{Add exponents} \\ &= \frac{1}{x^3} && \text{Definition of negative exponents} \end{aligned}$$

B Division with Exponents

To develop our next property of exponents, we use the definition for positive exponents. Consider the expression $\frac{x^6}{x^4}$. We can simplify by expanding the numerator and denominator and then reducing to lowest terms by dividing out common factors.

$$\begin{aligned} \frac{x^6}{x^4} &= \frac{x \cdot x \cdot x \cdot x \cdot x \cdot x}{x \cdot x \cdot x \cdot x} && \text{Expand numerator and denominator} \\ &= \frac{\cancel{x} \cdot \cancel{x} \cdot \cancel{x} \cdot \cancel{x} \cdot x \cdot x}{\cancel{x} \cdot \cancel{x} \cdot \cancel{x} \cdot \cancel{x}} && \text{Divide out common factors} \\ &= x \cdot x \\ &= x^2 && \text{Write answer with exponent 2} \end{aligned}$$

Note that the exponent in the answer is the difference of the exponents in the original problem. More specifically, if we subtract the exponent in the denominator from the exponent in the numerator, we obtain the exponent in the answer. This discussion leads us to another property of exponents.

Division Property for Expressions Containing Exponents

If a is any number other than zero, and r and s are integers, then

$$\frac{a^r}{a^s} = a^{r-s}$$

In words: To divide two numbers with the same base, subtract the exponent in the denominator from the exponent in the numerator, and use the common base as the base in the answer.

EXAMPLE 6 Simplify: $\frac{2^5}{2^8}$

SOLUTION Using our new property, we subtract the exponent in the denominator from the exponent in the numerator. The result is an expression containing a negative exponent.

$$\begin{aligned} \frac{2^5}{2^8} &= 2^{5-8} && \text{Division property for exponents} \\ &= 2^{-3} && \text{Subtraction} \\ &= \frac{1}{2^3} && \text{Definition of negative exponents} \\ &= \frac{1}{8} && \text{The cube of 2 is 8} \end{aligned}$$

To further justify our new property of exponents, we can rework the problem shown in Example 6, without using the division property for exponents, to see that we obtain the same answer.

$$\begin{aligned} \frac{2^5}{2^8} &= \frac{2 \cdot 2 \cdot 2 \cdot 2 \cdot 2}{2 \cdot 2 \cdot 2 \cdot 2 \cdot 2 \cdot 2 \cdot 2 \cdot 2} && \text{Expand numerator and denominator} \\ &= \frac{\not{2} \cdot \not{2} \cdot \not{2} \cdot \not{2} \cdot \not{2}}{\not{2} \cdot \not{2} \cdot \not{2} \cdot \not{2} \cdot \not{2} \cdot 2 \cdot 2 \cdot 2} && \text{Divide out common factors} \\ &= \frac{1}{2 \cdot 2 \cdot 2} \\ &= \frac{1}{8} && \text{Multiply} \end{aligned}$$

As you can see, the answer matches the answer obtained by using the division property for exponents.

EXAMPLE 7 Divide: $10^{-3} \div 10^5$

SOLUTION We begin by writing the division problem in fractional form. Then we apply our division property.

$$\begin{aligned} 10^{-3} \div 10^5 &= \frac{10^{-3}}{10^5} && \text{Write problem in fractional form} \\ &= 10^{-3-5} && \text{Division property for exponents} \\ &= 10^{-8} && \text{Subtraction} \\ &= \frac{1}{10^8} && \text{Definition of negative exponents} \end{aligned}$$

We can leave the answer in exponential form, as it is, or we can expand the denominator to obtain

$$= \frac{1}{100{,}000{,}000}$$

6. Simplify: $\frac{2^6}{2^8}$

7. Divide: $10^{-5} \div 10^3$

Answers

6. $\frac{1}{4}$ 7. $\frac{1}{10^8}$

8. Simplify: $\frac{10^{-6}}{10^{-8}}$

EXAMPLE 8 Simplify: $\frac{10^{-8}}{10^{-6}}$

SOLUTION Again, we are dividing expressions that have the same base. To find the exponent on the answer, we subtract the exponent in the denominator from the exponent in the numerator.

$$\frac{10^{-8}}{10^{-6}} = 10^{-8-(-6)} \quad \text{Division property for exponents}$$

$$= 10^{-2} \quad \text{Subtraction}$$

$$= \frac{1}{10^2} \quad \text{Definition of negative exponents}$$

$$= \frac{1}{100} \quad \text{Answer in expanded form}$$

9. Simplify: $4x^{-3} \cdot 7x$

EXAMPLE 9 Simplify: $3x^{-4} \cdot 5x$

SOLUTION To begin to simplify this expression, we regroup using the commutative and associative properties. That way, the numbers 3 and 5 are grouped together, as are the powers of x. Note also how we write x as x^1, so we can see the exponent.

$$3x^{-4} \cdot 5x = 3x^{-4} \cdot 5x^1 \quad \text{Write } x \text{ as } x^1$$

$$= (3 \cdot 5)(x^{-4} \cdot x^1) \quad \text{Commutative and associative properties}$$

$$= 15x^{-4+1} \quad \text{Multiply 3 and 5, then add exponents}$$

$$= 15x^{-3} \quad \text{The sum of } -4 \text{ and 1 is } -3$$

$$= \frac{15}{x^3} \quad \text{Definition of negative exponents}$$

10. Simplify: $\frac{x^{-5} \cdot x^2}{x^{-8}}$

EXAMPLE 10 Simplify: $\frac{x^{-4} \cdot x^7}{x^{-2}}$

SOLUTION We simplify the numerator first by adding exponents.

$$\frac{x^{-4} \cdot x^7}{x^{-2}} = \frac{x^{-4+7}}{x^{-2}} \quad \text{Multiplication property for exponents}$$

$$= \frac{x^3}{x^{-2}} \quad \text{The sum of } -4 \text{ and 7 is 3}$$

$$= x^{3-(-2)} \quad \text{Division property for exponents}$$

$$= x^5 \quad \text{Subtracting } -2 \text{ is equivalent to adding } +2$$

Getting Ready for Class

After reading through the preceding section, respond in your own words and in complete sentences.

1. Negative exponents are ________.
2. Do negative exponents result in negative numbers?
3. How do you divide exponents with the same base?
4. The problem below is wrong. Correct the right-hand side.

$$\frac{x^7}{x^8} = x$$

Answers

8. 100 **9.** $\frac{28}{x^2}$ **10.** x^5

Problem Set 9.4

A Write each expression with positive exponents, then simplify. [Examples 1–5]

1. 2^{-5}

2. 3^{-4}

3. 10^{-2}

4. 10^{-3}

5. x^{-3}

6. x^{-4}

7. $(-4)^{-2}$

8. $(-2)^{-4}$

9. $(-5)^{-3}$

10. $(-4)^{-3}$

11. $2^{6} \cdot 2^{-8}$

12. $3^{5} \cdot 3^{-8}$

13. $10^{-2} \cdot 10^{5}$

14. $10^{-3} \cdot 10^{6}$

15. $x^{-4} \cdot x^{-3}$

16. $x^{-3} \cdot x^{-7}$

17. $3^{8} \cdot 3^{-6}$

18. $2^{9} \cdot 2^{-6}$

19. $2 \cdot 2^{-4}$

20. $3^{-5} \cdot 3$

21. $10^{-3} \cdot 10^{8} \cdot 10^{-2}$

22. $10^{-3} \cdot 10^{9} \cdot 10^{-4}$

23. $x^{-5} \cdot x^{-4} \cdot x^{-3}$

24. $x^{-7} \cdot x^{-9} \cdot x^{-6}$

25. $2^{-3} \cdot 2 \cdot 2^{3}$

26. $3 \cdot 3^{5} \cdot 3^{-5}$

B Divide as indicated. Write your answer using only positive exponents. [Examples 6–10]

27. $\frac{2^9}{2^7}$

28. $\frac{3^8}{3^4}$

29. $\frac{2^7}{2^9}$

30. $\frac{3^5}{3^7}$

31. $\frac{x^4}{x^3}$

32. $\frac{x^7}{x^5}$

33. $\frac{x^3}{x^4}$

34. $\frac{x^5}{x^7}$

35. $10^{-2} \div 10^5$ **36.** $10^{-4} \div 10^2$ **37.** $10^{-5} \div 10^2$ **38.** $10^{-3} \div 10$

39. $\frac{10^{-2}}{10^{-5}}$ **40.** $\frac{10^{-4}}{10^{-7}}$ **41.** $x^{12} \div x^{-12}$ **42.** $x^6 \div x^{-6}$

43. $2^{12} \div 2^{10}$ **44.** $3^{11} \div 3^9$ **45.** $\frac{10}{10^{-3}}$ **46.** $\frac{10^2}{10^{-5}}$

A B Simplify each expression. Write your answer using only positive exponents.

47. $3x^{-2} \cdot 5x^7$ **48.** $5x^{-3} \cdot 8x^6$ **49.** $2x^{-2} \cdot 3x^{-3}$ **50.** $4x^{-6} \cdot 5x^{-4}$

51. $(4x^{-4})(-3x^{-3})$ **52.** $(3x^2)(-4x^{-1})$ **53.** $7x \cdot 3x^{-4} \cdot 2x^5$ **54.** $6x \cdot 2x^{-3} \cdot 3x^4$

55. $\frac{x^{-3} \cdot x^7}{x^{-1}}$ **56.** $\frac{x^{-4} \cdot x^6}{x^{-2}}$ **57.** $\frac{x^{-2} \cdot x^{-8}}{x^{-12}}$ **58.** $\frac{x^{-5} \cdot x^{-7}}{x^{-15}}$

Maintaining Your Skills

Make the following conversions.

59. 6 minutes 45 seconds to
  a. Seconds
  b. Minutes

60. 4 minutes 15 seconds to
  a. Seconds
  b. Minutes

61. 3 pounds 8 ounces to
  a. Ounces
  b. Pounds

62. 5 pounds 4 ounces to
  a. Ounces
  b. Pounds

63. 5 feet 9 inches to
  a. Inches
  b. Feet

64. 6 feet 3 inches to
  a. Inches
  b. Feet

Scientific Notation

9.5

Objectives

A Write numbers in scientific notation.

B Convert numbers written in scientific notation to standard form.

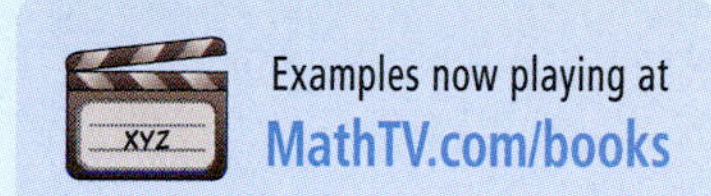

There are many disciplines that deal with very large numbers and others that deal with very small numbers. For example, in astronomy, distances commonly are given in light-years. A light-year is the distance that light will travel in one year. It is approximately

$$5,880,000,000,000 \text{ miles}$$

It can be difficult to perform calculations with numbers in this form because of the number of zeros present. Scientific notation provides a way of writing very large, or very small, numbers in a more manageable form.

A Scientific Notation

Definition

A number is in scientific notation when it is written as the product of a number between 1 and 10 and an integer power of 10. A number written in scientific notation has the form

$$n \times 10^r$$

where $1 \le n < 10$ and $r =$ an integer.

EXAMPLE 1 The speed of light is 186,000 miles per second. Write 186,000 in scientific notation.

SOLUTION To write this number in scientific notation, we rewrite it as the product of a number between 1 and 10 and a power of 10. To do so, we move the decimal point 5 places to the left so that it appears between the 1 and the 8, giving us 1.86. Then we multiply this number by 10^5. The number that results has the same value as our original number but is written in scientific notation. Here is our result:

$$186,000 = 1.86 \times 10^5$$

Both numbers have exactly the same value. The number on the left is written in *standard form,* while the number on the right is written in scientific notation.

PRACTICE PROBLEMS

1. Write 27,500 in scientific notation.

B Converting to Standard Form

EXAMPLE 2 If your pulse rate is 60 beats per minute, then your heart will beat 8.64×10^4 times each day. Write 8.64×10^4 in standard form.

SOLUTION Because 10^4 is 10,000, we can think of this as simply a multiplication problem. That is,

$$8.64 \times 10^4 = 8.64 \times 10,000 = 86,400$$

Looking over our result, we can think of the exponent 4 as indicating the number of places we need to move the decimal point to write our number in standard form. Because our exponent is positive 4, we move the decimal point from its original position, between the 8 and the 6, four places to the right. If we need to add any zeros on the right we do so. The result is the standard form of our number, 86,400.

2. Write 7.89×10^5 in standard form.

Answers

1. 2.75×10^4 2. 789,000

Next, we turn our attention to writing small numbers in scientific notation. To do so, we use the negative exponents developed in the previous section. For example, the number 0.00075, when written in scientific notation, is equivalent to 7.5×10^{-4}. Here's why:

$$7.5 \times 10^{-4} = 7.5 \times \frac{1}{10^4} = 7.5 \times \frac{1}{10{,}000} = \frac{7.5}{10{,}000} = 0.00075$$

The table below lists some other numbers both in scientific notation and in standard form.

3. Fill in the missing numbers in the table below:

	Standard Form		Scientific Notation
a.	24,500	=	
b.		=	5.6×10^5
c.	0.000789	=	
d.		=	4.8×10^{-3}

EXAMPLE 3 Each pair of numbers in the table below is equal.

Standard Form		Scientific Notation
376,000	=	3.76×10^5
49,500	=	4.95×10^4
3,200	=	3.2×10^3
591	=	5.91×10^2
46	=	4.6×10^1
8	=	8×10^0
0.47	=	4.7×10^{-1}
0.093	=	9.3×10^{-2}
0.00688	=	6.88×10^{-3}
0.0002	=	2×10^{-4}
0.000098	=	9.8×10^{-5}

As we read across the table, for each pair of numbers, notice how the decimal point in the number on the right is placed so that the number containing the decimal point is always a number between 1 and 10. Correspondingly, the exponent on 10 keeps track of how many places the decimal point was moved in converting from standard form to scientific notation. In general, when the exponent is positive, we are working with a large number. On the other hand, when the exponent is negative, we are working with a small number. (By small number, we mean a number that is less than 1, but larger than 0.)

We end this section with a diagram that shows two numbers, one large and one small, that are converted to scientific notation.

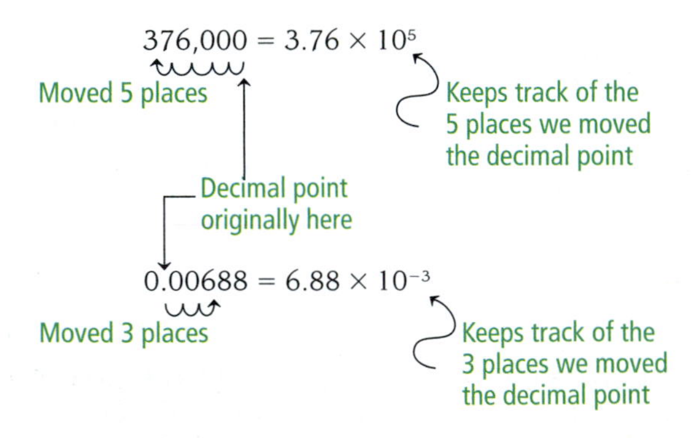

Getting Ready for Class

After reading through the preceding section, respond in your own words and in complete sentences.

1. What is scientific notation?
2. What types of numbers are frequently written with scientific notation?
3. In writing the distance, in miles, to the sun, would we use a positive power of ten or a negative power of ten?
4. In writing the weight, in kilograms, of a paper clip, would we use a positive power of ten or a negative power of ten?

Answer

3. a. 2.45×10^4
   b. 560,000
   c. 7.89×10^{-4}
   d. 0.0048

Problem Set 9.5

A Write each number in scientific notation. [Examples 1, 3]

1. 425,000 **2.** 635,000 **3.** 6,780,000 **4.** 5,490,000

5. 11,000 **6.** 29,000 **7.** 89,000,000 **8.** 37,000,000

B Write each number in standard form. [Examples 2, 3]

9. 3.84×10^4 **10.** 3.84×10^7 **11.** 5.71×10^7 **12.** 5.71×10^5

13. 3.3×10^3 **14.** 3.3×10^2 **15.** 8.913×10^7 **16.** 8.913×10^5

A Write each number in scientific notation. [Examples 1, 3]

17. 0.00035 **18.** 0.0000035 **19.** 0.0007 **20.** 0.007

21. 0.06035 **22.** 0.0006035 **23.** 0.1276 **24.** 0.001276

B Write each number in standard form. [Examples 2, 3]

25. 8.3×10^{-4} **26.** 8.3×10^{-7} **27.** 6.25×10^{-2} **28.** 6.25×10^{-4}

29. 3.125×10^{-1} **30.** 3.125×10^{-2} **31.** 5×10^{-3} **32.** 5×10^{-5}

Applying the Concepts

Super Bowl Advertising and Viewers The cost of a 30-second television ad along with the approximate number of viewers for four different Super Bowls is shown below. Complete the table by writing the ad cost in scientific notation, and the number of viewers in standard form.

Year	Super Bowl	Ad Cost	Ad Cost in Scientific Notation	Viewers in Scientific Notation	Number of Viewers
33. 1967	I	\$42,000		4.0×10^6	
34. 1977	XI	\$162,000		6.2×10^7	
35. 1987	XXI	\$575,000		8.7×10^7	
36. 1997	XXXI	\$1,200,000		8.8×10^7	

Galilean Moons The planet Jupiter has about 60 known moons. In the year 1610 Galileo first discovered the four largest moons of Jupiter, Io, Europa, Ganymede, and Callisto. These moons are known as the Galilean moons. Each moon has a unique period, or the time it takes to make a trip around Jupiter. Fill in the tables below.

NASA

37.

Jupiter's Moon	Period (seconds)	
Io	153,000	
Europa		3.07×10^5
Ganymede	618,000	
Callisto		1.44×10^6

38.

Jupiter's Moon	Distance from Jupiter (kilometers)	
Io	422,000	
Europa		6.17×10^5
Ganymede	1,070,000	
Callisto		1.88×10^6

Computer Science The smallest amount of data that a computer can hold is measured in bits. A byte is the next largest unit and is equal to 8, or 2^3, bits. Fill in the table below.

	Number of Bytes	
Unit	Exponential Form	Scientific Notation
39. Kilobyte	$2^{10} = 1{,}024$	
40. Megabyte	$2^{20} \approx 1{,}048{,}000$	
41. Gigabyte	$2^{30} \approx 1{,}074{,}000{,}000$	
42. Terabyte	$2^{40} \approx 1{,}099{,}500{,}000{,}000$	

Maintaining Your Skills

Simplify the following expressions.

43. $7x - 3x$

44. $11y + 6y$

45. $5x + 3x - 11x$

46. $3x - 7x + 2x$

47. $7x + 3x - x$

48. $12x - 8x + 5x$

49. $3a - 14a + 5a$

50. $2a - 7a - 10a$

More About Scientific Notation

9.6

Objectives

A Multiply and divide numbers written in scientific notation.

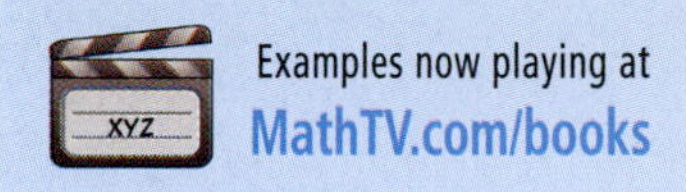

A Multiplication and Division with Numbers Written in Scientific Notation

In this section, we extend our work with scientific notation to include multiplication and division with numbers written in scientific notation. To work the problems in this section, we use the material presented in the previous two sections, along with the commutative and associative properties of multiplication and the rule for multiplication with fractions. Here is our first example.

EXAMPLE 1 Multiply: $(3.5 \times 10^8)(2.2 \times 10^{-5})$

SOLUTION First we apply the commutative and associative properties to rearrange the numbers, so that the decimal numbers are grouped together and the powers of 10 are also.

$$(3.5 \times 10^8)(2.2 \times 10^{-5}) = (3.5)(2.2) \times (10^8)(10^{-5})$$

Next, we multiply the decimal numbers together and then the powers of ten. To multiply the powers of ten, we add exponents.

$$= 7.7 \times 10^{8+(-5)}$$
$$= 7.7 \times 10^3$$

EXAMPLE 2 Find the product of 130,000,000 and 0.000005. Write your answer in scientific notation.

SOLUTION We begin by writing both numbers in scientific notation. Then we proceed as we did in Example 1: We group the numbers between 1 and 10 separately from the powers of 10.

$$\begin{aligned}(130{,}000{,}000)(0.000005) &= (1.3 \times 10^8)(5 \times 10^{-6})\\ &= (1.3)(5) \times (10^8)(10^{-6})\\ &= 6.5 \times 10^2\end{aligned}$$

Our next examples involve division with numbers in scientific notation.

EXAMPLE 3 Divide: $\dfrac{8 \times 10^3}{4 \times 10^{-6}}$

SOLUTION To separate the numbers between 1 and 10 from the powers of 10, we "undo" the multiplication and write the problem as the product of two fractions. Doing so looks like this:

$$\frac{8 \times 10^3}{4 \times 10^{-6}} = \frac{8}{4} \times \frac{10^3}{10^{-6}} \quad \text{Write as two separate fractions}$$

Next, we divide 8 by 4 to obtain 2. Then we divide 10^3 by 10^{-6} by subtracting exponents.

$$= 2 \times 10^{3-(-6)} \quad \text{Divide}$$
$$= 2 \times 10^9$$

PRACTICE PROBLEMS

1. Multiply: $(2.5 \times 10^6)(1.4 \times 10^2)$
2. Find the product of 2,200,000 and 0.00015.
3. Divide: $\dfrac{6 \times 10^5}{2 \times 10^{-4}}$

Answers
1. 3.5×10^8 **2.** 3.3×10^2
3. 3×10^9

4. Divide: $\frac{0.0038}{19{,}000{,}000}$

5. Simplify: $\frac{(6.8 \times 10^{-4})(3.9 \times 10^{2})}{7.8 \times 10^{-6}}$

6. Simplify: $\frac{(0.000035)(45{,}000)}{0.000075}$

Answers
4. 2×10^{-10} **5.** 3.4×10^{4}
6. 2.1×10^{4}

EXAMPLE 4 Divide: $\frac{0.00045}{1{,}500{,}000}$

SOLUTION To begin, write each number in scientific notation.

$$\frac{0.00045}{1{,}500{,}000} = \frac{4.5 \times 10^{-4}}{1.5 \times 10^{6}} \quad \text{Write numbers in scientific notation}$$

Next, as in the previous example, we write the problem as two separate fractions in order to group the numbers between 1 and 10 together, as well as the powers of 10.

$$= \frac{4.5}{1.5} \times \frac{10^{-4}}{10^{6}} \quad \text{Write as two separate fractions}$$

$$= 3 \times 10^{-4-6} \quad \text{Divide}$$

$$= 3 \times 10^{-10} \quad -4 - 6 = -4 + (-6) = -10$$

EXAMPLE 5 Simplify: $\frac{(6.8 \times 10^{5})(3.9 \times 10^{-7})}{7.8 \times 10^{-4}}$

SOLUTION We group the numbers between 1 and 10 separately from the powers of 10:

$$\frac{(6.8 \times 10^{5})(3.9 \times 10^{-7})}{7.8 \times 10^{-4}} = \frac{(6.8)(3.9)}{7.8} \times \frac{(10^{5})(10^{-7})}{10^{-4}}$$

$$= 3.4 \times 10^{5+(-7)-(-4)}$$

$$= 3.4 \times 10^{2}$$

EXAMPLE 6 Simplify: $\frac{(35{,}000)(0.0045)}{7{,}500{,}000}$

SOLUTION We write each number in scientific notation, and then we proceed as we have in the examples above.

$$\frac{(35{,}000)(0.0045)}{7{,}500{,}000} = \frac{(3.5 \times 10^{4})(4.5 \times 10^{-3})}{7.5 \times 10^{6}}$$

$$= \frac{(3.5)(4.5)}{7.5} \times \frac{(10^{4})(10^{-3})}{10^{6}}$$

$$= 2.1 \times 10^{4+(-3)-6}$$

$$= 2.1 \times 10^{-5}$$

Getting Ready for Class

After reading through the preceding section, respond in your own words and in complete sentences.

1. How is the commutative property used when multiplying or dividing with scientific notation?
2. How is the associative property used when multiplying or dividing with scientific notation?
3. The following problem is not complete. Why not?

$$(6.0 \times 10^{3})(5.2 \times 10^{4}) = 31.2 \times 10^{7}$$

4. Explain why it is easier to multiply very large or very small numbers when they are written in scientific notation, rather than standard form.

Problem Set 9.6

A Find each product. Write all answers in scientific notation. [Examples 1, 2]

1. $(2 \times 10^4)(3 \times 10^6)$

2. $(3 \times 10^3)(1 \times 10^5)$

3. $(2.5 \times 10^7)(6 \times 10^3)$

4. $(3.8 \times 10^6)(5 \times 10^3)$

5. $(7.2 \times 10^3)(9.5 \times 10^{-6})$

6. $(8.5 \times 10^5)(4.2 \times 10^{-9})$

7. $(36{,}000)(450{,}000)$

8. $(25{,}000)(620{,}000)$

9. $(4{,}200)(0.00009)$

10. $(0.0000065)(86{,}000)$

Find each quotient. Write all answers in scientific notation. [Examples 3, 4]

11. $\dfrac{3.6 \times 10^5}{1.8 \times 10^2}$

12. $\dfrac{9.3 \times 10^{15}}{3.0 \times 10^5}$

13. $\dfrac{8.4 \times 10^{-6}}{2.1 \times 10^3}$

14. $\dfrac{6.0 \times 10^{-10}}{1.5 \times 10^3}$

15. $\dfrac{3.5 \times 10^5}{7.0 \times 10^{-10}}$

16. $\dfrac{1.6 \times 10^7}{8.0 \times 10^{-14}}$

17. $\dfrac{540{,}000}{9{,}000}$

18. $\dfrac{750{,}000{,}000}{250{,}000}$

19. $\dfrac{0.00092}{46{,}000}$

20. $\dfrac{0.00000047}{235{,}000}$

Simplify each expression, and write all answers in scientific notation. [Examples 5, 6]

21. $\dfrac{(3 \times 10^7)(8 \times 10^4)}{6 \times 10^5}$

22. $\dfrac{(4 \times 10^9)(6 \times 10^5)}{8 \times 10^3}$

23. $\dfrac{(2 \times 10^{-3})(6 \times 10^{-5})}{3 \times 10^{-4}}$

24. $\dfrac{(4 \times 10^{-5})(9 \times 10^{-10})}{6 \times 10^{-6}}$

25. $\dfrac{(3.5 \times 10^{-4})(4.2 \times 10^5)}{7 \times 10^3}$

26. $\dfrac{(2.4 \times 10^{-6})(3.6 \times 10^3)}{9 \times 10^5}$

27. $\dfrac{(0.00087)(40{,}000)}{1{,}160{,}000}$

28. $\dfrac{(0.0045)(24{,}000)}{270{,}000}$

29. $\dfrac{(525)(0.0000032)}{0.0025}$

30. $\dfrac{(465)(0.000004)}{0.0093}$

Applying the Concepts

Super Bowl Advertising and Viewers The cost of a 30-second television ad along with the approximate number of viewers for four different Super Bowls is shown below. Complete the table by finding the cost per viewer by dividing the ad cost by the number of viewers.

Year	Super Bowl	Ad Cost	Number of Viewers	Cost per Viewer
31. 1967	I	$\$4.2 \times 10^4$	4.0×10^6	
32. 1977	XI	$\$1.62 \times 10^5$	6.2×10^7	
33. 1987	XXI	$\$5.75 \times 10^5$	8.7×10^7	
34. 1997	XXXI	$\$1.2 \times 10^6$	8.8×10^7	

35. Pyramids and Scientific Notation The Great Pyramid at Giza is one of the largest and oldest man-made structures in the world. It weighs over 10^{10} kilograms. If each stone making up the pyramid weighs approximately 4,000 kilograms, how many stones make up the structure? Write your answer in scientific notation.

36. Technology A CD-ROM holds about 700 megabytes, or 7.0×10^8 bytes, of data. A four-and-a-half minute song downloaded from the Internet uses 5 megabytes, or 5.0×10^6 bytes, of storage space. How many four-and-a-half minute songs can fit on a CD?

Image: BigStockPhoto.com © Devanne Philippe

Maintaining Your Skills

Find the value of each of the following expressions when $x = 4$.

37. $3x - 5$

38. $-2 - 2x$

39. $-5x + 6$

40. $4x + 7$

Find the value of each of the following expressions when $x = -3$.

41. $3x - 5$

42. $-2 - 2x$

43. $-5x + 6$

44. $4x + 7$

Chapter 9 Summary

Property 1 for Exponents [9.1]

If r and s are any two whole numbers and a is an integer, then

$$a^r \cdot a^s = a^{r+s}$$

In words: To multiply two expressions with the same base, add exponents and use the common base.

EXAMPLES

1. $x^3 \cdot x^4 \cdot x^2 = x^{3+4+2}$
$= x^9$

Property 2 for Exponents [9.1]

If r and s are any two whole numbers and a is an integer, then

$$(a^r)^s = a^{r \cdot s}$$

In words: A power raised to another power is the base raised to the product of the powers.

2. $(2^3)^2 = 2^{3 \cdot 2} = 2^6$

Property 3 for Exponents [9.1]

If r is a whole number and a and b are integers, then

$$(a \cdot b)^r = a^r \cdot b^r$$

3. $(3xy)^4 = 3^4 \cdot x^4 \cdot y^4$
$= 81x^4y^4$

Adding Polynomials [9.2]

We add polynomials by writing one polynomial under the other and then adding in columns.

4. $$\begin{array}{r} 2x^2 + 3x + 4 \\ \underline{3x^2 + 4x + 5} \\ 5x^2 + 7x + 9 \end{array}$$

Subtracting Polynomials [9.2]

Because subtraction is addition of the opposite, we simply change the sign of each term of the second polynomial and then add. With subtraction, there is less chance of making mistakes if we add horizontally, rather than in columns.

5. $(6x^2 - 3x + 5) - (3x^2 + 2x - 3)$
$= 6x^2 - 3x + 5 - 3x^2 - 2x + 3$
$= (6x^2 - 3x^2) + (-3x - 2x) + (5 + 3)$
$= (6 - 3)x^2 + (-3 - 2)x + (5 + 3)$
$= 3x^2 - 5x + 8$

Multiplying Polynomials [9.3]

We use the distributive property to multiply polynomials.

6. $\mathbf{x^2}(x^3 + x^4) = \mathbf{x^2} \cdot x^3 + \mathbf{x^2} \cdot x^4$
$= x^5 + x^6$

7. $\mathbf{(2x - 3)}(4x + 7)$
$= \mathbf{(2x - 3)} \cdot 4x + \mathbf{(2x - 3)} \cdot 7$
$= 2x \cdot 4x - 3 \cdot 4x + 2x \cdot 7 - 3 \cdot 7$
$= 8x^2 - 12x + 14x - 21$
$= 8x^2 + 2x - 21$

Negative Exponents [9.4]

8. $2^{-3} = \frac{1}{2^3} = \frac{1}{8}$

If r is a positive number and a is any number other than zero, then $a^{-r} = \frac{1}{a^r}$.

Division Property for Expressions Containing Exponents [9.4]

9. $\frac{2^4}{2^7} = 2^{4-7}$
$= 2^{-3}$
$= \frac{1}{2^3}$
$= \frac{1}{8}$

If a is any number other than zero, and r and s are integers, then

$$\frac{a^r}{a^s} = a^{r-s}$$

In words: To divide two numbers with the same base, subtract the exponent in the denominator from the exponent in the numerator, and use the common base as the base in the answer.

Scientific Notation [9.5, 9.6]

10. $768{,}000 = 7.68 \times 10^5$
$0.00039 = 3.9 \times 10^{-4}$

A number is in scientific notation when it is written as the product of a number between 1 and 10 and an integer power of 10.

Chapter 9 Review

Multiply: [9.1]

1. $3y^6 \cdot 5y^8$
2. $7y^3 \cdot 2y^6 \cdot 4y^9$
3. $(9x^5y^2)(6x^3y^7)$
4. $(4a^2b^4)(3a^6b^3)(7a^3b^6)$
5. $(x^4)^5$
6. $(y^2)^3(y^4)^2$
7. $(5x^3y^2)^4$
8. $(3a^4b^3)^2(2a^2b)^3$

Add or subtract the following polynomials as indicated. [9.2]

9. $(3t^2 - 2t - 5) + (6t^2 + 5t - 2)$
10. $(4y^2 - 5y + 10) + (8y^2 - 9y - 7)$
11. $(8x^2 - 3x + 6) - (5x^2 + 7x + 4)$
12. $(7y^2 - 9y - 4) - (2y^2 - 6y - 5)$

Find the value of the following polynomials when $x = 2$. [9.2]

13. $x^2 + 4x + 4$
14. $2x^2 - 6x - 5$

Multiply. [9.3]

15. $2x^3(3x^4 - 8)$
16. $3a^3b^4(5a^2b - 2a^4b^5)$
17. $(3x - 5)(x + 3)$
18. $(6x + 4)(3x + 2)$

Expand and multiply. [9.3]

19. $(x - 3)^2$
20. $(4y + 7)^2$

Find the product of the given binomials by using the diagrams. [9.3]

21. $(x + 3)$ and $(x + 5)$

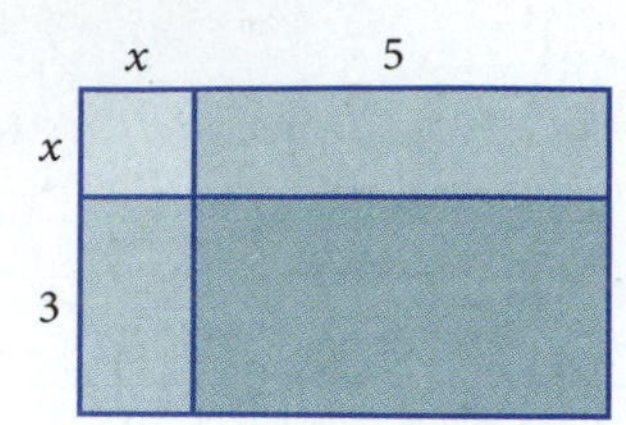

22. $(3x + 7)$ and $(5x + 2)$

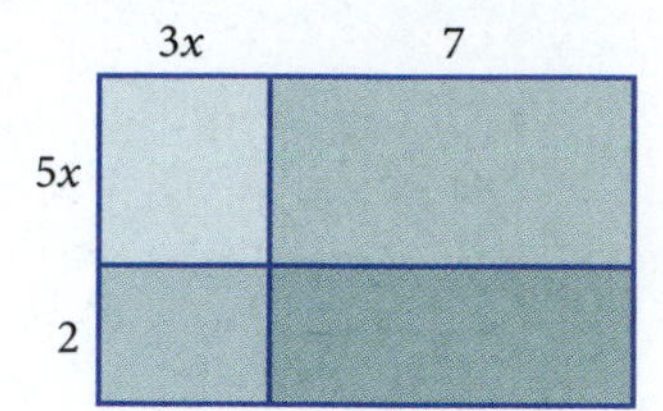

Simplify each expression completely. Your answer should have only positive exponents. [9.4]

23. 5^{-3}

24. $(-4)^{-2}$

25. $3^5 \cdot 3^{-7}$

26. $4^{-8} \cdot 4^5$

27. $y^2 \cdot y^{-6}$

28. $x^{-3} \cdot x^8 \cdot x^{-6}$

29. $\dfrac{x^7}{x^9}$

30. $\dfrac{x^7}{x^{-9}}$

31. $\dfrac{2^5}{2^8}$

32. $\dfrac{3^{-6}}{3^{-10}}$

33. $(4x^{-3})(-5x^{-5})$

34. $\dfrac{x^{-6} \cdot x^{-2}}{x^{-10}}$

Write each number in scientific notation. [9.5]

35. 73,800,000

36. 0.002935

Write each number in standard form. [9.5]

37. 4.4×10^9

38. 1.6×10^{-5}

Simplify each expression and write all answers in scientific notation. [9.6]

39. $(4.8 \times 10^7)(3.6 \times 10^{-5})$

40. $(460{,}000)(0.00056)$

41. $\dfrac{6.8 \times 10^6}{3.4 \times 10^{-8}}$

42. $\dfrac{0.000025}{0.00005}$

43. $\dfrac{(4.5 \times 10^{-6})(2.4 \times 10^5)}{1.2 \times 10^6}$

44. $\dfrac{(392)(500{,}000)}{0.0007}$

Chapter 9 Cumulative Review

Simplify.

1. $5{,}309 + 687$

2. $8 + \frac{5}{x}$

3. $11.09 - 6.531$

4. $4\frac{1}{8} - 1\frac{3}{4}$

Multiply.

5. $(3x - 5)(6x + 7)$

6. $5a^3(2a^2 - 7)$

Divide.

7. $314\overline{)13{,}188}$

8. $\frac{6}{32} \div \frac{9}{48}$

9. Round the number 435,906 to the nearest ten thousand.

10. Write 0.48 as a fraction in lowest terms.

11. Change $\frac{76}{12}$ to a mixed number in lowest terms.

12. Use the equation $3x + 4y = 24$ to find x when $y = -3$.

13. Write the decimal 0.8 as a percent.

14. Convert 7 kilograms to pounds (1 kg = 2.21 lb).

15. Write 124% as a fraction or mixed number in lowest terms.

16. What percent of 60 is 21?

Simplify.

17. $\sqrt{32xy^2}$

18. $8x + 9 - 9x - 14$

19. $-|-7|$

20. $\frac{-3(-8) + 4(-2)}{11 - 9}$

21. $19 - 5(7 - 4)$

22. $5\sqrt{49} + 3\sqrt{81}$

Solve.

23. $\frac{3}{8}y = 21$

24. $-3(2x - 1) = 3(x + 5)$

25. $\frac{3.6}{4} = \frac{4.5}{x}$

26. Write the following ratio as a fraction in lowest terms: 0.04 to 0.32

27. Subtract $2x - 9$ from $6x - 8$.

28. Identify the property or properties used in the following: $(6 + 8) + 2 = 6 + (8 + 2)$.

29. Surface Area Find the surface area of a rectangular solid with length 7 inches, width 3 inches, and height 2 inches.

30. Age Ben is 8 years older than Ryan. In 6 years the sum of their ages will be 38. How old are they now?

31. Gas Mileage A truck travels 432 miles on 27 gallons of gas. What is the rate of gas mileage in miles per gallon?

32. Discount A surfboard that usually sells for $400 is marked down to $320. What is the discount? What is the discount rate?

33. Geometry Find the length of the hypotenuse of a right triangle with sides of 5 and 12 meters.

34. Cost of Coffee If coffee costs $6.40 per pound, how much will 2 lb 4 oz cost?

35. Interest If $1,400 is invested at 6% simple interest for 90 days, how much interest is earned?

36. Wildflower Seeds C.J. works in a nursery, and one of his tasks is filling packets of wildflower seeds. If each packet is to contain $\frac{1}{4}$ pound of seeds, how many packets can be filled from 16 pounds of seeds?

37. Checking Account Balance Rosa has a balance of $469 in her checking account when she writes a check for $376 for her car payment. Then she writes another check for $138 for textbooks. Write a subtraction problem that gives the new balance in her checking account. What is the new balance?

38. Commission A car stereo salesperson receives a commission of 8% of all units he sells. If his total sales for March are $9,800, how much money in commission will he make?

39. Volume How many 8-fluid-ounce glasses of water will it take to fill a 15-gallon aquarium?

40. Internet Access Speed The table below gives the speed of the most common modems used for Internet access. The abbreviation bps stands for bits per second. Use the template to construct a bar chart of the information in the table.

MODEM SPEEDS

Modem Type	Speed (bps)
28K	28,000
56K	56,000
ISDN	128,000
Cable	512,000
DSL	786,000

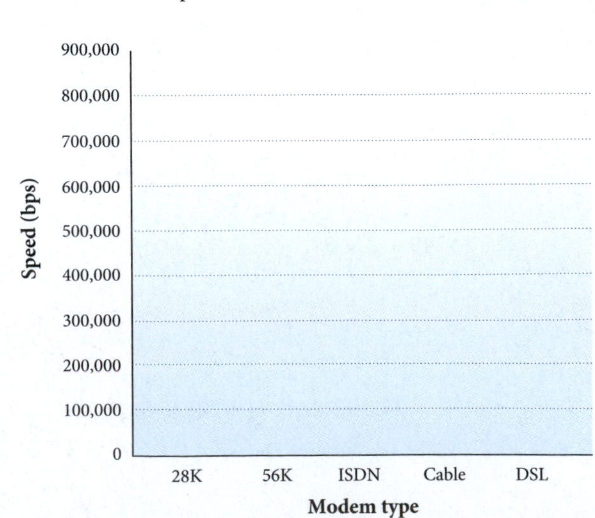

Chapter 9 Test

Simplify each of the following expressions.

1. $3x^5 \cdot 4x^4$
2. $(4x^3y^2)(6x^4y^5)$
3. $(a^4)^3(a^7)^2$
4. $(6x^3y^2)^3$
5. $(4a^4b^3)^3(2a^2b)^2$

Add or subtract the following polynomials as indicated.

6. $(4t^2 + 7t + 3) + (9t^2 - 6t + 4)$
7. $(6a^2 - 7a - 4) - (2a^2 - 8a + 5)$

Find the value of the following polynomials when $x = 4$.

8. $x^2 + 4x + 4$
9. $2x^2 - 6x - 5$

Multiply.

10. $3x^6(4x^4 - 6)$
11. $(3x - 5)(2x + 6)$
12. $(4y - 3)^2$

Simplify each expression completely. Your answer should have only positive exponents.

13. $4^5 \cdot 4^{-7}$
14. $x^{-1} \cdot x^5 \cdot x^{-8}$
15. $\frac{5^{-6}}{5^{-8}}$
16. $\frac{y^{-8} \cdot y^3}{y^{-7}}$
17. Write 0.0000385 in scientific notation.
18. Write 6.75×10^6 in standard form.

Simplify each expression completely. Write your answer using scientific notation.

19. $(2.7 \times 10^6)(6.3 \times 10^{-4})$

20. $\dfrac{36{,}500{,}000}{73{,}000}$

21. $\dfrac{6.8 \times 10^{-8}}{1.7 \times 10^{-4}}$

22. $\dfrac{(640)(0.000000032)}{0.008}$

23. Technology A broad band cable connection can download information from the Internet at 3 megabits per second, or 3.0×10^6 bits per second. A music video streams over the Internet at 250 kilobits per second, or 2.5×10^5 bits per second. How many music videos can be played at the same time over this connection?

Image: BigStockPhoto.com © Devanne Philippe

24. Technology A portable music player holds 20 gigabytes, or 2.0×10^{10} bytes, of data. A four-and-a-half minute song downloaded from the Internet uses 5 megabytes, or 5.0×10^6 bytes, of storage space. How many four-and-a-half minute songs can fit on this portable music player?

Chapter 9 Projects

EXPONENTS AND POLYNOMIALS

GROUP PROJECT

Discovering Pascal's Triangle

Number of People 3

Time Needed 20 minutes

Equipment Paper and pencils

Background The triangular array of numbers shown here is known as Pascal's triangle, after the French philosopher Blaise Pascal (1623–1662).

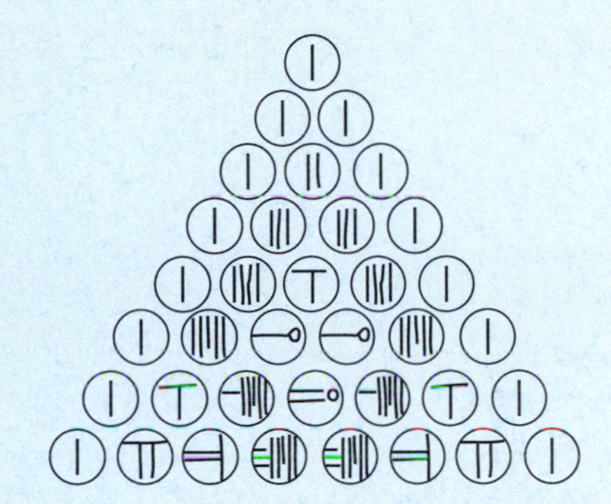

Pascal's triangle in Japanese (1781)

$$
\begin{array}{ccccccccccc}
 & & & & & 1 & & & & & \\
 & & & & 1 & & 1 & & & & \\
 & & & 1 & & 2 & & 1 & & & \\
 & & 1 & & 3 & & 3 & & 1 & & \\
 & 1 & & 4 & & 6 & & 4 & & 1 & \\
1 & & 5 & & 10 & & 10 & & 5 & & 1
\end{array}
$$

Procedure Look at Pascal's triangle and discover how the numbers in each row of the triangle are obtained from the numbers in the row above it.

1. Once you have discovered how to extend the triangle, write the next two rows.
2. Pascal's triangle can be linked to the Fibonacci sequence by rewriting Pascal's triangle so that the 1's on the left side of the triangle line up under one another and the other columns are equally spaced to the right of the first column. Rewrite Pascal's triangle as indicated and then look along the diagonals of the new array until you discover how the Fibonacci sequence can be obtained from it.
3. The diagram above shows Pascal's triangle as written in Japanese in 1781. Use your knowledge of Pascal's triangle to translate the numbers written in Japanese into our number system. Then write down the Japanese numbers from 1 to 20.

RESEARCH PROJECT

Binomial Expansions

The title on the following diagram is *Binomial Expansions* because each line gives the expansion of the binomial $x + y$ raised to a whole-number power.

Binomial Expansions

$$(x + y)^0 = 1$$

$$(x + y)^1 = x + y$$

$$(x + y)^2 = x^2 + 2xy + y^2$$

$$(x + y)^3 = x^3 + 3x^2y + 3xy^2 + y^3$$

$$(x + y)^4 =$$

$$(x + y)^5 =$$

The fourth row in the diagram was completed by expanding $(x + y)^3$ using the methods developed in this chapter. Next, complete the diagram by expanding the binomials $(x + y)^4$ and $(x + y)^5$ using the multiplication procedures you have learned in this chapter. Finally, study the completed diagram until you see patterns that will allow you to continue the diagram one more row without using multiplication. (One pattern that you will see is Pascal's triangle, which we mentioned in the preceding group project.) When you are finished, write an essay in which you describe what you have done and the results you have obtained.

Appendix A

RESOURCES

By gathering resources early in the term, before you need help, the information about these resources will be available to you when they are needed.

Instructor

Knowing the contact information for your instructor is very important. You may already have this information from the course syllabus. It is a good idea to write it down again.

Name ______________________ Office Location ______________________
Available Hours: M __________ T __________ W __________ TH __________ F __________
Phone Number ______________ ext. __________ E-mail Address ______________________ @ __________

Tutoring Center

Many schools offer tutoring, free of charge to their students. If this is the case at your school, find out when and where tutoring is offered.

Tutoring Location ______________________ Phone Number ______________ ext. __________
Available Hours: M __________ T __________ W __________ TH __________ F __________

Computer Lab

Many schools offer a computer lab where students can use the online resources and software available with their textbook. Other students using the same software and websites as you can be very helpful. Find out where the computer lab at your school is located.

Computer Lab Location ______________________ Phone Number ______________ ext. __________
Available Hours: M __________ T __________ W __________ TH __________ F __________

Videos

A complete set of videos is available to your school. These videos feature the author of your textbook presenting full-length, 15- to 20-minute lessons from every section of your textbook. If you miss class, or find yourself behind, these tapes will prove very useful.

Videotape Location ______________________ Phone Number ______________ ext. __________
Available Hours: M __________ T __________ W __________ TH __________ F __________

Classmates

Form a study group and meet on a regular basis. When you meet try to speak to each other using proper mathematical language. That is, use the words that you see in the definition and property boxes in your textbook.

Name ______________________ Phone ______________ E-mail ______________________
Name ______________________ Phone ______________ E-mail ______________________
Name ______________________ Phone ______________ E-mail ______________________

Appendix B

ONE HUNDRED ADDITION FACTS

The following 100 problems should be done mentally. You should be able to find these sums quickly and accurately. Do all 100 problems, and then check your answers. Make a list of each problem you missed, and then go over the list as many times as it takes to memorize the correct answers. Once this has been done, go back and work all 100 problems again. Repeat this process until you get all 100 problems correct.

Add.

1. $\begin{array}{r} 0 \\ +4 \\ \hline \end{array}$	**2.** $\begin{array}{r} 1 \\ +9 \\ \hline \end{array}$	**3.** $\begin{array}{r} 2 \\ +7 \\ \hline \end{array}$	**4.** $\begin{array}{r} 0 \\ +3 \\ \hline \end{array}$	**5.** $\begin{array}{r} 4 \\ +1 \\ \hline \end{array}$	**6.** $\begin{array}{r} 2 \\ +5 \\ \hline \end{array}$	**7.** $\begin{array}{r} 5 \\ +6 \\ \hline \end{array}$	**8.** $\begin{array}{r} 2 \\ +6 \\ \hline \end{array}$
9. $\begin{array}{r} 9 \\ +8 \\ \hline \end{array}$	**10.** $\begin{array}{r} 4 \\ +9 \\ \hline \end{array}$	**11.** $\begin{array}{r} 3 \\ +7 \\ \hline \end{array}$	**12.** $\begin{array}{r} 9 \\ +4 \\ \hline \end{array}$	**13.** $\begin{array}{r} 4 \\ +5 \\ \hline \end{array}$	**14.** $\begin{array}{r} 1 \\ +4 \\ \hline \end{array}$	**15.** $\begin{array}{r} 0 \\ +1 \\ \hline \end{array}$	**16.** $\begin{array}{r} 6 \\ +4 \\ \hline \end{array}$
17. $\begin{array}{r} 5 \\ +3 \\ \hline \end{array}$	**18.** $\begin{array}{r} 9 \\ +5 \\ \hline \end{array}$	**19.** $\begin{array}{r} 5 \\ +8 \\ \hline \end{array}$	**20.** $\begin{array}{r} 0 \\ +7 \\ \hline \end{array}$	**21.** $\begin{array}{r} 5 \\ +1 \\ \hline \end{array}$	**22.** $\begin{array}{r} 1 \\ +0 \\ \hline \end{array}$	**23.** $\begin{array}{r} 6 \\ +5 \\ \hline \end{array}$	**24.** $\begin{array}{r} 5 \\ +2 \\ \hline \end{array}$
25. $\begin{array}{r} 5 \\ +7 \\ \hline \end{array}$	**26.** $\begin{array}{r} 3 \\ +6 \\ \hline \end{array}$	**27.** $\begin{array}{r} 7 \\ +0 \\ \hline \end{array}$	**28.** $\begin{array}{r} 4 \\ +6 \\ \hline \end{array}$	**29.** $\begin{array}{r} 1 \\ +5 \\ \hline \end{array}$	**30.** $\begin{array}{r} 3 \\ +1 \\ \hline \end{array}$	**31.** $\begin{array}{r} 1 \\ +7 \\ \hline \end{array}$	**32.** $\begin{array}{r} 7 \\ +6 \\ \hline \end{array}$
33. $\begin{array}{r} 3 \\ +3 \\ \hline \end{array}$	**34.** $\begin{array}{r} 8 \\ +0 \\ \hline \end{array}$	**35.** $\begin{array}{r} 1 \\ +8 \\ \hline \end{array}$	**36.** $\begin{array}{r} 5 \\ +4 \\ \hline \end{array}$	**37.** $\begin{array}{r} 2 \\ +8 \\ \hline \end{array}$	**38.** $\begin{array}{r} 0 \\ +5 \\ \hline \end{array}$	**39.** $\begin{array}{r} 5 \\ +9 \\ \hline \end{array}$	**40.** $\begin{array}{r} 9 \\ +9 \\ \hline \end{array}$
41. $\begin{array}{r} 0 \\ +0 \\ \hline \end{array}$	**42.** $\begin{array}{r} 6 \\ +9 \\ \hline \end{array}$	**43.** $\begin{array}{r} 7 \\ +4 \\ \hline \end{array}$	**44.** $\begin{array}{r} 8 \\ +2 \\ \hline \end{array}$	**45.** $\begin{array}{r} 7 \\ +3 \\ \hline \end{array}$	**46.** $\begin{array}{r} 8 \\ +5 \\ \hline \end{array}$	**47.** $\begin{array}{r} 8 \\ +4 \\ \hline \end{array}$	**48.** $\begin{array}{r} 6 \\ +7 \\ \hline \end{array}$
49. $\begin{array}{r} 8 \\ +1 \\ \hline \end{array}$	**50.** $\begin{array}{r} 1 \\ +3 \\ \hline \end{array}$	**51.** $\begin{array}{r} 7 \\ +5 \\ \hline \end{array}$	**52.** $\begin{array}{r} 0 \\ +1 \\ \hline \end{array}$	**53.** $\begin{array}{r} 8 \\ +3 \\ \hline \end{array}$	**54.** $\begin{array}{r} 6 \\ +6 \\ \hline \end{array}$	**55.** $\begin{array}{r} 9 \\ +6 \\ \hline \end{array}$	**56.** $\begin{array}{r} 8 \\ +9 \\ \hline \end{array}$
57. $\begin{array}{r} 3 \\ +0 \\ \hline \end{array}$	**58.** $\begin{array}{r} 6 \\ +2 \\ \hline \end{array}$	**59.** $\begin{array}{r} 2 \\ +1 \\ \hline \end{array}$	**60.** $\begin{array}{r} 6 \\ +0 \\ \hline \end{array}$	**61.** $\begin{array}{r} 4 \\ +8 \\ \hline \end{array}$	**62.** $\begin{array}{r} 6 \\ +1 \\ \hline \end{array}$	**63.** $\begin{array}{r} 2 \\ +0 \\ \hline \end{array}$	**64.** $\begin{array}{r} 7 \\ +9 \\ \hline \end{array}$
65. $\begin{array}{r} 2 \\ +9 \\ \hline \end{array}$	**66.** $\begin{array}{r} 9 \\ +3 \\ \hline \end{array}$	**67.** $\begin{array}{r} 3 \\ +8 \\ \hline \end{array}$	**68.** $\begin{array}{r} 7 \\ +2 \\ \hline \end{array}$	**69.** $\begin{array}{r} 8 \\ +8 \\ \hline \end{array}$	**70.** $\begin{array}{r} 4 \\ +4 \\ \hline \end{array}$	**71.** $\begin{array}{r} 8 \\ +7 \\ \hline \end{array}$	**72.** $\begin{array}{r} 2 \\ +4 \\ \hline \end{array}$
73. $\begin{array}{r} 3 \\ +2 \\ \hline \end{array}$	**74.** $\begin{array}{r} 4 \\ +7 \\ \hline \end{array}$	**75.** $\begin{array}{r} 9 \\ +7 \\ \hline \end{array}$	**76.** $\begin{array}{r} 1 \\ +2 \\ \hline \end{array}$	**77.** $\begin{array}{r} 6 \\ +3 \\ \hline \end{array}$	**78.** $\begin{array}{r} 2 \\ +2 \\ \hline \end{array}$	**79.** $\begin{array}{r} 9 \\ +2 \\ \hline \end{array}$	**80.** $\begin{array}{r} 9 \\ +1 \\ \hline \end{array}$
81. $\begin{array}{r} 3 \\ +5 \\ \hline \end{array}$	**82.** $\begin{array}{r} 1 \\ +1 \\ \hline \end{array}$	**83.** $\begin{array}{r} 5 \\ +5 \\ \hline \end{array}$	**84.** $\begin{array}{r} 7 \\ +7 \\ \hline \end{array}$	**85.** $\begin{array}{r} 0 \\ +8 \\ \hline \end{array}$	**86.** $\begin{array}{r} 7 \\ +8 \\ \hline \end{array}$	**87.** $\begin{array}{r} 2 \\ +3 \\ \hline \end{array}$	**88.** $\begin{array}{r} 3 \\ +9 \\ \hline \end{array}$
89. $\begin{array}{r} 6 \\ +8 \\ \hline \end{array}$	**90.** $\begin{array}{r} 4 \\ +0 \\ \hline \end{array}$	**91.** $\begin{array}{r} 0 \\ +6 \\ \hline \end{array}$	**92.** $\begin{array}{r} 4 \\ +3 \\ \hline \end{array}$	**93.** $\begin{array}{r} 7 \\ +1 \\ \hline \end{array}$	**94.** $\begin{array}{r} 0 \\ +9 \\ \hline \end{array}$	**95.** $\begin{array}{r} 9 \\ +0 \\ \hline \end{array}$	**96.** $\begin{array}{r} 8 \\ +6 \\ \hline \end{array}$
97. $\begin{array}{r} 3 \\ +4 \\ \hline \end{array}$	**98.** $\begin{array}{r} 5 \\ +0 \\ \hline \end{array}$	**99.** $\begin{array}{r} 1 \\ +6 \\ \hline \end{array}$	**100.** $\begin{array}{r} 4 \\ +2 \\ \hline \end{array}$				

Appendix C

ONE HUNDRED MULTIPLICATION FACTS

The following 100 problems should be done mentally. You should be able to find these products quickly and accurately. Do all 100 problems, and then check your answers. Make a list of each problem you missed, and then go over the list as many times as it takes to memorize the correct answers. Once this has been done, go back and work all 100 problems again. Repeat this process until you get all 100 problems correct.

Multiply.

1. 3×4	**2.** 5×4	**3.** 0×5	**4.** 0×6	**5.** 2×5	**6.** 1×7	**7.** 4×4	**8.** 7×2
9. 7×1	**10.** 5×9	**11.** 2×6	**12.** 5×3	**13.** 3×5	**14.** 4×5	**15.** 8×1	**16.** 3×7
17. 5×8	**18.** 0×3	**19.** 6×0	**20.** 0×4	**21.** 4×6	**22.** 8×2	**23.** 5×7	**24.** 0×9
25. 2×7	**26.** 6×1	**27.** 1×8	**28.** 4×7	**29.** 3×8	**30.** 6×8	**31.** 5×4	**32.** 7×3
33. 0×8	**34.** 4×8	**35.** 2×9	**36.** 5×6	**37.** 1×9	**38.** 2×8	**39.** 0×7	**40.** 6×2
41. 4×9	**42.** 8×0	**43.** 3×0	**44.** 9×6	**45.** 3×4	**46.** 8×3	**47.** 7×9	**48.** 1×9
49. 1×3	**50.** 9×9	**51.** 2×4	**52.** 8×7	**53.** 5×2	**54.** 7×8	**55.** 1×4	**56.** 0×2
57. 9×8	**58.** 2×3	**59.** 9×5	**60.** 7×7	**61.** 6×5	**62.** 0×1	**63.** 9×3	**64.** 8×9
65. 9×7	**66.** 6×4	**67.** 3×1	**68.** 1×5	**69.** 9×6	**70.** 1×2	**71.** 4×1	**72.** 2×2
73. 6×3	**74.** 5×1	**75.** 9×2	**76.** 6×9	**77.** 7×0	**78.** 4×0	**79.** 7×6	**80.** 5×0
81. 0×0	**82.** 3×9	**83.** 1×1	**84.** 8×6	**85.** 9×0	**86.** 7×4	**87.** 8×4	**88.** 2×0
89. 9×1	**90.** 8×5	**91.** 4×2	**92.** 8×3	**93.** 7×5	**94.** 2×1	**95.** 6×6	**96.** 3×2
97. 4×3	**98.** 1×6	**99.** 6×7	**100.** 1×0				

Solutions to Selected Practice Problems

Solutions to all practice problems that require more than one step are shown here. Before you look back here to see where you have made a mistake, you should try the problem you are working on twice. If you do not get the correct answer the second time you work the problem, then the solution here should show you where you went wrong.

Chapter 1

Section 1.2

1. $$\begin{array}{r} 63 \\ +25 \\ \hline 88 \end{array}$$

2. $$\begin{array}{r} 342 \\ +605 \\ \hline 947 \end{array}$$

3. a. $$\begin{array}{r} {\scriptstyle 1} \\ 375 \\ 121 \\ +473 \\ \hline 969 \end{array}$$

b. $$\begin{array}{r} {\scriptstyle 22} \\ 495 \\ 699 \\ +978 \\ \hline 2{,}172 \end{array}$$

4. a. $$\begin{array}{r} {\scriptstyle 11\ 11} \\ 57{,}904 \\ 7{,}193 \\ 655 \\ \hline 65{,}752 \end{array}$$

b. $$\begin{array}{r} {\scriptstyle 1\ 1\ 2\ 3} \\ 68{,}495 \\ 7{,}236 \\ 878 \\ 29 \\ 5 \\ \hline 76{,}643 \end{array}$$

7. a. $6 + 2 + 4 + 8 + 3 = (6 + 4) + (2 + 8) + 3 = 10 + 10 + 3 = 23$
b. $24 + 17 + 36 + 13 = (24 + 36) + (17 + 13) = 60 + 30 = 90$

8. a. $n = 8$, since $8 + 9 = 17$
b. $n = 8$, since $8 + 2 = 10$
c. $n = 1$, since $8 + 1 = 9$
d. $n = 6$, since $16 = 6 + 10$

9. a. $7 + 7 + 7 + 7 = 28$ ft **b.** $88 + 88 + 33 + 33 = 242$ in. **c.** $44 + 66 + 77 = 187$ yd

Section 1.3

5. a.

Food	\$ 5,296
Car	4,847
Total	\$10,143 = \$10,140 to the nearest ten dollars

b.

Savings	\$2,149
Taxes	6,137
Total	\$8,286 = \$8,300 to the nearest hundred dollars

c.

House	\$10,200
Taxes	6,137
Misc.	6,142
Car	4,847
Savings	2,149
Total	\$29,475 = \$29,000 to the nearest thousand dollars

6. a. We round each of the four numbers in the sum to the nearest thousand, and then we add the rounded numbers.

$$\begin{array}{rlr} 5{,}287 & \text{rounds to} & 5{,}000 \\ 2{,}561 & \text{rounds to} & 3{,}000 \\ 888 & \text{rounds to} & 1{,}000 \\ +4{,}898 & \text{rounds to} & +\,5{,}000 \\ \hline & & 14{,}000 \end{array}$$

We estimate the answer to this problem to be approximately 14,000. The actual answer, found by adding the original, unrounded numbers, is 13,634.

b. We round each of the four numbers in the sum to the nearest thousand, and then we add the rounded numbers.

$$\begin{array}{rlr} 702 & \text{rounds to} & 1{,}000 \\ 2{,}944 & \text{rounds to} & 4{,}000 \\ 1{,}001 & \text{rounds to} & 1{,}000 \\ +3{,}500 & \text{rounds to} & +\,4{,}000 \\ \hline & & 10{,}000 \end{array}$$

We estimate the answer to this problem to be approximately 10,000. The actual answer, found by adding the original, unrounded numbers, is 9,147.

Section 1.4

1. a. $$\begin{array}{r} 684 \\ -431 \\ \hline 253 \end{array}$$

b. $$\begin{array}{r} 7{,}406 \\ -3{,}405 \\ \hline 4{,}001 \end{array}$$

2. a. $$\begin{array}{r} 6{,}857 \\ -405 \\ \hline 6{,}452 \end{array}$$

b. $$\begin{array}{r} 345 \\ -234 \\ \hline 111 \end{array}$$

3. a.

$$\begin{array}{rl} 63 = & 6 \text{ tens} + 3 \text{ ones} = 5 \text{ tens} + 13 \text{ ones} \\ -47 = & 4 \text{ tens} \quad 7 \text{ ones} = 4 \text{ tens} \quad 7 \text{ ones} \\ \hline & \qquad\qquad\qquad\quad 1 \text{ ten} + 6 \text{ ones} \end{array}$$

Answer: 16

b.

$$\begin{array}{rl} 532 = & 5 \text{ hundreds} + 3 \text{ tens} + 2 \text{ ones} = 5 \text{ hundreds} + 2 \text{ tens} + 12 \text{ ones} \\ -403 = & 4 \text{ hundreds} \quad 0 \text{ tens} \quad 3 \text{ tens} = 4 \text{ hundreds} \quad 0 \text{ tens} \quad 3 \text{ ones} \\ \hline & \qquad\qquad\qquad\qquad\qquad\qquad 1 \text{ hundred} + 2 \text{ tens} + 9 \text{ ones} \end{array}$$

Answer: 129

4. a. $\begin{array}{r} \overset{5\,15}{\not6\not56} \\ -283 \\ \hline 373 \end{array}$ **b.** $\begin{array}{r} \overset{2\,16}{\not3,\not729} \\ -1,749 \\ \hline 1,980 \end{array}$

Section 1.5

1. a. $4 \cdot 70 = 70 + 70 + 70 + 70$
$= 280$

b. $4 \cdot 700 = 700 + 700 + 700 + 700$
$= 2,800$

c. $4 \cdot 7,000 = 7,000 + 7,000 + 7,000 + 7,000$
$= 28,000$

4. a. $\begin{array}{r} \overset{5}{57} \\ \times\ 8 \\ \hline 456 \end{array}$ **b.** $\begin{array}{r} \overset{5}{570} \\ \times\ 8 \\ \hline 4,560 \end{array}$

5. a. $\begin{array}{rl} 45 & \\ \times 62 & \\ \hline 90 & \leftarrow 2(45) = 90 \\ +2,700 & \leftarrow 60(45) = 2,700 \\ \hline 2,790 & \end{array}$

b. $\begin{array}{rl} \overset{1}{620} & \\ \times 45 & \\ \hline 3,100 & \leftarrow 5(620) = 3,100 \\ +24,800 & \leftarrow 40(620) = 24,800 \\ \hline 27,900 & \end{array}$

6. a. $\begin{array}{rl} 356 & \\ \times 641 & \\ \hline 356 & \leftarrow 1(356) = 356 \\ 14,240 & \leftarrow 40(356) = 14,240 \\ 213,600 & \leftarrow 600(356) = 213,600 \\ \hline 228,196 & \end{array}$

b. $\begin{array}{rl} 3,560 & \\ \times 641 & \\ \hline 3,560 & \leftarrow 1(3,560) = 3,560 \\ 142,200 & \leftarrow 40(3,560) = 142,400 \\ 2,136,000 & \leftarrow 600(3,560) = 2,136,000 \\ \hline 2,281,960 & \end{array}$

7. $\begin{array}{r} 365 \\ \times 550 \\ \hline 18,250 \\ 182,500 \\ \hline 200,750 \text{ mg} \end{array}$

8. 36($12) = $432 Total weekly earnings
$432 − $109 = $323 Take-home pay

9. Fat: 3(10) = 30 grams of fat; sodium: 3(160) = 480 milligrams of sodium

10. Bowling for 3 hours burns 3(265) = 795 calories. Eating two bags of chips means you are consuming 2(3)(160) = 960 calories. No; bowling won't burn all the calories.

Section 1.6

1. a. $\begin{array}{r} 74 \\ 4\overline{)296} \\ \underline{28}\downarrow \\ 16 \\ \underline{16} \\ 0 \end{array}$ **b.** $\begin{array}{r} 740 \\ 4\overline{)2,960} \\ \underline{28}\downarrow \\ 16 \\ \underline{16}\downarrow \\ 00 \end{array}$

2. a. $\begin{array}{r} 283 \\ 24\overline{)6,792} \\ \underline{48}\downarrow \\ 199 \\ \underline{192}\downarrow \\ 72 \\ \underline{72} \\ 0 \end{array}$ **b.** $\begin{array}{r} 2,830 \\ 24\overline{)67,920} \\ \underline{48}\downarrow \\ 199 \\ \underline{192}\downarrow \\ 72 \\ \underline{72}\downarrow \\ 00 \end{array}$

3. $\begin{array}{r} 208 \\ 9\overline{)1,872} \\ \underline{18}\downarrow \\ 07 \\ \underline{0}\downarrow \\ 72 \\ \underline{72} \\ 0 \end{array}$

4. a. $\begin{array}{r} 69 \text{ R } 20, \text{ or } 69\frac{20}{27} \\ 27\overline{)1,883} \\ \underline{162}\downarrow \\ 263 \\ \underline{243} \\ 20 \end{array}$ **b.** $\begin{array}{r} 104 \text{ R } 11, \text{ or } 104\frac{11}{18} \\ 18\overline{)1,883} \\ \underline{18}\downarrow \\ 08 \\ \underline{0}\downarrow \\ 83 \\ \underline{72} \\ 11 \end{array}$

5. $\begin{array}{r} 156 \\ 12\overline{)1,872} \\ \underline{12}\downarrow \\ 67 \\ \underline{60}\downarrow \\ 72 \\ \underline{72} \\ 0 \end{array}$ The family spent $156 per day

Section 1.7

1. Base 5, exponent 2; 5 to the second power, or 5 squared **2.** Base 2, exponent 3; 2 to the third power, or 2 cubed
3. Base 1, exponent 4; 1 to the fourth power **4.** $5^2 = 5 \cdot 5 = 25$ **5.** $9^2 = 9 \cdot 9 = 81$ **6.** $2^3 = 2 \cdot 2 \cdot 2 = 8$
7. $1^4 = 1 \cdot 1 \cdot 1 \cdot 1 = 1$ **8.** $2^5 = 2 \cdot 2 \cdot 2 \cdot 2 \cdot 2 = 32$ **9.** $7^1 = 7$ **10.** $4^1 = 4$ **11.** $9^0 = 1$ **12.** $1^0 = 1$

13. a. $5 \cdot 7 - 3 \cdot 6 = 35 - 18$
$= 17$

b. $5 \cdot 70 - 3 \cdot 60 = 350 - 180$
$= 170$

14. $7 + 3(6 + 4) = 7 + 3(10)$
$= 7 + 30$
$= 37$

15. a. $28 \div 7 - 3 = 4 - 3$
$= 1$

b. $6 \cdot 3^2 + 64 \div 2^4 - 2 = 6 \cdot 9 + 64 \div 16 - 2$
$= 54 + 4 - 2$
$= 58 - 2$
$= 56$

16. **a.** $5 + 3[24 - 5(6 - 2)] = 5 + 3[24 - 5(4)]$
$= 5 + 3[24 - 20]$
$= 5 + 3[4]$
$= 5 + 12$
$= 17$

b. $50 + 30[240 - 50(6 - 2)] = 50 + 30[240 - 50(4)]$
$= 50 + 30[240 - 200]$
$= 50 + 30(40)$
$= 50 + 1,200$
$= 1,250$

17.
187
273
150
173
227
1010

$\frac{1010}{5} = 202$ miles

18. First we place them in order from smallest to largest
150 173 187 227 273
Because there are 5 numbers, the one in the middle, 187, is the median.

19. The numbers are already in order from smallest to largest. Because there is an even number of numbers, we find the mean of the middle two:

$$\frac{42,635 + 44,475}{2} = \frac{87,110}{2} = 43,555$$

The median is $43,555.

20. The most frequently occurring score is 74. It occurs three times. The mode is 74.

Section 1.8

1. $A = bh = 3 \cdot 2 = 6 \text{ cm}^2$
2. $A = lw = 70 \cdot 35 = 2,450 \text{ mm}^2$
3. $A = 38 \cdot 13 + 24 \cdot 27$
$= 494 + 648$
$= 1142 \text{ ft}^2$

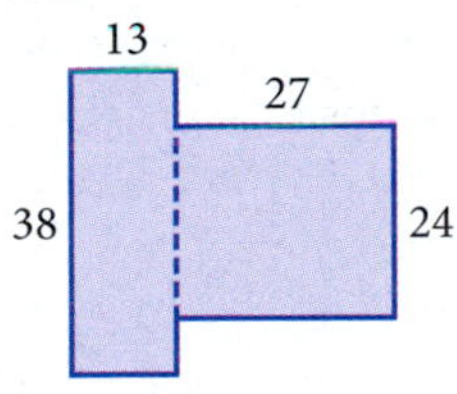

4. $V = 15 \cdot 12 \cdot 8 = 1,440 \text{ ft}^3$
5. **a.** Surface area $= 2(15 \cdot 8) + 2(8 \cdot 12) + (15 \cdot 12) = 612 \text{ ft}^2$
   b. Two gallons will cover it, with some paint left over.

Chapter 2

Section 2.1

9.

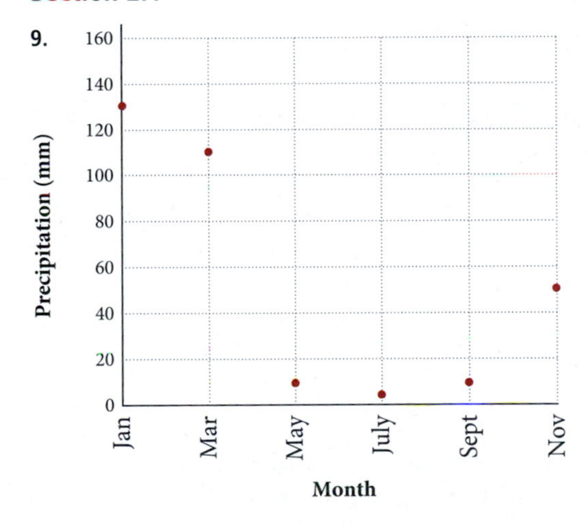

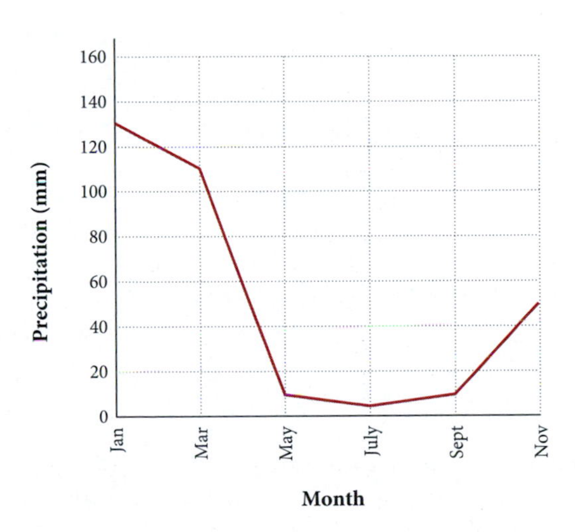

Section 2.2

1. $2 + (-5) = -3$
2. $-2 + 5 = 3$
3. $-2 + (-5) = -7$
4. $2 + 6 = 8$
5. $2 + (-6) = -4$
6. $-2 + 6 = 4$
7. $-2 + (-6) = -8$
8. $15 + 12 = 27$
$15 + (-12) = 3$
$-15 + 12 = -3$
$-15 + (-12) = -27$
9. $12 + (-3) + (-7) + 5 = 9 + (-7) + 5$
$= 2 + 5$
$= 7$
10. $[-2 + (-12)] + [7 + (-5)] = [-14] + [2]$
$= -12$

Section 2.3

1. $7 - 3 = 7 + (-3) = 4$
2. $-7 - 3 = -7 + (-3) = -10$
3. $-8 - 6 = -8 + (-6) = -14$
4. $10 - (-6) = 10 + 6 = 16$
5. $-10 - (-15) = -10 + 15 = 5$
6. a. $8 - 5 = 8 + (-5) = 3$ b. $-8 - 5 = -8 + (-5) = -13$ c. $8 - (-5) = 8 + 5 = 13$ d. $-8 - (-5) = -8 + 5 = -3$ e. $12 - 10 = 12 + (-10) = 2$ f. $-12 - 10 = -12 + (-10) = -22$ g. $12 - (-10) = 12 + 10 = 22$ h. $-12 - (-10) = -12 + 10 = -2$
7. $-4 + 6 - 7 = -4 + 6 + (-7) = 2 + (-7) = -5$
8. $15 - (-5) - 8 = 15 + 5 + (-8) = 20 + (-8) = 12$
9. $-8 - 2 = -8 + (-2) = -10$
10. $7 - (-5) = 7 + 5 = 12$
11. $-8 - (-6) = -8 + 6 = -2$
12. $42 - (-42) = 42 + 42 = 84°F$

Section 2.4

1. $2(-6) = (-6) + (-6) = -12$
2. $-2(6) = 6(-2) = (-2) + (-2) + (-2) + (-2) + (-2) + (-2) = -12$
3. $-2(-6) = 12$

10. $-5(2)(-4) = -10(-4) = 40$
11. a. $(-8)^2 = (-8)(-8) = 64$ b. $-8^2 = -8 \cdot 8 = -64$ c. $(-3)^3 = (-3)(-3)(-3) = -27$ d. $-3^3 = -3 \cdot 3 \cdot 3 = -27$
12. $-2[5 + (-8)] = -2[-3] = 6$
13. $-3 + 4(-7 + 3) = -3 + 4(-4) = -3 + (-16) = -19$
14. $-3(5) + 4(-4) = -15 + (-16) = -31$
15. $-2(3 - 5) - 7(-2 - 4) = -2(-2) - 7(-6) = 4 - (-42) = 4 + 42 = 46$
16. $(-6 - 1)(4 - 9) = (-7)(-5) = 35$

Section 2.5

6. $\frac{8(-5)}{-4} = \frac{-40}{-4} = 10$
7. $\frac{-20 + 6(-2)}{7 - 11} = \frac{-20 + (-12)}{-4} = \frac{-32}{-4} = 8$
8. $-3(4^2) + 10 \div (-5) = -3(16) + 10 \div (-5) = -48 + (-2) = -50$
9. $-80 \div 2 \div 10 = -40 \div 10 = -4$

Section 2.6

1. $5(7a) = (5 \cdot 7)a = 35a$
2. $-3(9x) = (-3 \cdot 9)x = -27x$
3. $5(-8y) = [5(-8)]y = -40y$
4. $6 + (9 + x) = (6 + 9) + x = 15 + x$
5. $(3x + 7) + 4 = 3x + (7 + 4) = 3x + 11$
6. $6(x + 4) = 6(x) + 6(4) = 6x + 24$
7. $7(a - 5) = 7(a) - 7(5) = 7a - 35$
8. $6(4x + 5) = 6(4x) + 6(5) = (6 \cdot 4)x + 6(5) = 24x + 30$
9. $3(8a - 4) = 3(8a) - 3(4) = 24a - 12$
10. $8(3x + 4y) = 8(3x) + 8(4y) = 24x + 32y$
11. $A = s^2 = 12^2 = 144 \text{ ft}^2$
    $P = 4s = 4(12) = 48 \text{ ft}$
12. $A = lw = 100(53) = 5{,}300 \text{ yd}^2$
    $P = 2l + 2w = 2(100) + 2(53) = 200 + 106 = 306 \text{ yd}$

Chapter 3

Section 3.1

6. $\frac{2}{3} = \frac{2 \cdot 4}{3 \cdot 4} = \frac{8}{12}$ **7.** $\frac{2}{3} = \frac{2 \cdot 4x}{3 \cdot 4x} = \frac{8x}{12x}$ **8.** $\frac{15}{20} = \frac{15 \div 5}{20 \div 5} = \frac{3}{4}$ **10.** $\frac{1}{3} \cdot \frac{4}{4} = \frac{4}{12}; \frac{1}{6} \cdot \frac{2}{2} = \frac{2}{12}; \frac{1}{4} \cdot \frac{3}{3} = \frac{3}{12}$

Section 3.2

1. 37 and 59 are prime numbers; 39 is divisible by 3 and 13; 51 is divisible by 3 and 17.

2. a. $90 = 9 \cdot 10$
$= 3 \cdot 3 \cdot 2 \cdot 5$
$= 2 \cdot 3^2 \cdot 5$

b. $900 = 9 \cdot 100$
$= 3 \cdot 3 \cdot 25 \cdot 4$
$= 3 \cdot 3 \cdot 5 \cdot 5 \cdot 2 \cdot 2$
$= 2^2 \cdot 3^2 \cdot 5^2$

4. $\frac{12}{18} = \frac{12 \div 6}{18 \div 6} = \frac{2}{3}$ **5.** $\frac{15}{20} = \frac{3 \cdot \cancel{5}}{2 \cdot 2 \cdot \cancel{5}} = \frac{3}{4}$

6. a. $\frac{30}{35} = \frac{2 \cdot 3 \cdot \cancel{5}}{\cancel{5} \cdot 7} = \frac{6}{7}$ **b.** $\frac{300}{350} = \frac{\cancel{2} \cdot 2 \cdot 3 \cdot \cancel{5} \cdot \cancel{5}}{\cancel{2} \cdot \cancel{5} \cdot \cancel{5} \cdot 7} = \frac{6}{7}$ **7. a.** $\frac{8}{72} = \frac{\cancel{2} \cdot \cancel{2} \cdot \cancel{2} \cdot 1}{\cancel{2} \cdot \cancel{2} \cdot \cancel{2} \cdot 3 \cdot 3} = \frac{1}{9}$ **b.** $\frac{16}{144} = \frac{\cancel{2} \cdot \cancel{2} \cdot \cancel{2} \cdot \cancel{2}}{\cancel{2} \cdot \cancel{2} \cdot \cancel{2} \cdot \cancel{2} \cdot 3 \cdot 3} = \frac{1}{9}$

8. $\frac{5}{50} = \frac{\cancel{5} \cdot 1}{\cancel{5} \cdot 2 \cdot 5} = \frac{1}{10}$ **9.** $\frac{120}{25} = \frac{2 \cdot 2 \cdot 2 \cdot 3 \cdot \cancel{5}}{5 \cdot \cancel{5}} = \frac{24}{5}$ **10.** $\frac{54x}{90xy} = \frac{\cancel{2} \cdot \cancel{3} \cdot \cancel{3} \cdot 3 \cdot \cancel{x}}{\cancel{2} \cdot \cancel{3} \cdot \cancel{3} \cdot 5 \cdot \cancel{x} \cdot y} = \frac{3}{5y}$ **11.** $\frac{306a^2}{228a} = \frac{\cancel{2} \cdot \cancel{3} \cdot 3 \cdot 17 \cdot \cancel{a} \cdot a}{\cancel{2} \cdot 2 \cdot \cancel{3} \cdot 19 \cdot \cancel{a}} = \frac{51a}{38}$

Section 3.3

1. $\frac{2}{3} \cdot \frac{5}{9} = \frac{10}{27}$

2. $-\frac{2}{5} \cdot 7 = -\frac{2}{5} \cdot \frac{7}{1}$
$= -\frac{14}{5}$

3. $\frac{1}{3}\left(\frac{4}{5} \cdot \frac{1}{3}\right) = \frac{1}{3}\left(\frac{4}{15}\right)$
$= \frac{4}{45}$

4. $\frac{1}{4}(4y) = \left(\frac{1}{4} \cdot 4\right)y$
$= 1 \cdot y$
$= y$

5. $\frac{1}{2}(2x - 4) = \frac{1}{2}(2x) - \frac{1}{2}(4)$
$= x - 2$

6. a. $\frac{12}{25} \cdot \frac{5}{6} = \frac{12 \cdot 5}{25 \cdot 6}$
$= \frac{(2 \cdot \cancel{2} \cdot \cancel{3}) \cdot \cancel{5}}{(\cancel{5} \cdot 5) \cdot (\cancel{2} \cdot \cancel{3})}$
$= \frac{2}{5}$

b. $\frac{12}{25} \cdot \frac{50}{60} = \frac{12 \cdot 50}{25 \cdot 60}$
$= \frac{(\cancel{2} \cdot \cancel{2} \cdot \cancel{3}) \cdot (2 \cdot \cancel{5} \cdot \cancel{5})}{(\cancel{5} \cdot \cancel{5}) \cdot (\cancel{2} \cdot \cancel{2} \cdot \cancel{3} \cdot 5)}$
$= \frac{2}{5}$

7. a. $\frac{8}{3} \cdot \frac{9}{24} = \frac{8 \cdot 9}{3 \cdot 24}$
$= \frac{(\cancel{2} \cdot \cancel{2} \cdot \cancel{2}) \cdot (\cancel{3} \cdot \cancel{3})}{\cancel{3} \cdot (\cancel{2} \cdot \cancel{2} \cdot \cancel{2} \cdot \cancel{3})}$
$= \frac{1}{1}$
$= 1$

b. $\frac{8}{30} \cdot \frac{90}{24} = \frac{8 \cdot 90}{30 \cdot 24}$
$= \frac{(\cancel{2} \cdot \cancel{2} \cdot \cancel{2}) \cdot (\cancel{2} \cdot \cancel{3} \cdot \cancel{3} \cdot \cancel{5})}{(\cancel{2} \cdot \cancel{3} \cdot \cancel{5}) \cdot (\cancel{2} \cdot \cancel{2} \cdot \cancel{2} \cdot \cancel{3})}$
$= \frac{1}{1}$
$= 1$

8. $\frac{yz^2}{x} \cdot \frac{x^3}{yz} = \frac{\cancel{y} \cdot \cancel{z} \cdot z \cdot \cancel{x} \cdot x \cdot x}{\cancel{x} \cdot \cancel{y} \cdot \cancel{z}}$
$= \frac{z \cdot x \cdot x}{1}$
$= x^2z$

9. $\frac{3}{4} \cdot \frac{8}{3} \cdot \frac{1}{6} = \frac{3 \cdot 8 \cdot 1}{4 \cdot 3 \cdot 6}$
$= \frac{\cancel{3} \cdot (\cancel{2} \cdot \cancel{2} \cdot \cancel{2}) \cdot 1}{(\cancel{2} \cdot \cancel{2}) \cdot \cancel{3} \cdot (\cancel{2} \cdot 3)}$
$= \frac{1}{3}$

10. $\left(\frac{2}{3}\right)^2 = \frac{2}{3} \cdot \frac{2}{3}$
$= \frac{4}{9}$

11. a. $\left(\frac{3}{4}\right)^2 \cdot \frac{1}{2} = \frac{3}{4} \cdot \frac{3}{4} \cdot \frac{1}{2}$
$= \frac{9}{32}$

b. $\left(\frac{2}{3}\right)^3 \cdot \frac{9}{8} = \frac{2}{3} \cdot \frac{2}{3} \cdot \frac{2}{3} \cdot \frac{9}{8}$
$= \frac{2 \cdot 2 \cdot 2 \cdot 9}{3 \cdot 3 \cdot 3 \cdot 8}$
$= \frac{2 \cdot 2 \cdot 2 \cdot (3 \cdot 3)}{3 \cdot 3 \cdot 3 \cdot (2 \cdot 2 \cdot 2)}$
$= \frac{\cancel{2} \cdot \cancel{2} \cdot \cancel{2} \cdot (\cancel{3} \cdot \cancel{3})}{3 \cdot \cancel{3} \cdot \cancel{3} \cdot (\cancel{2} \cdot \cancel{2} \cdot \cancel{2})}$
$= \frac{1}{3}$

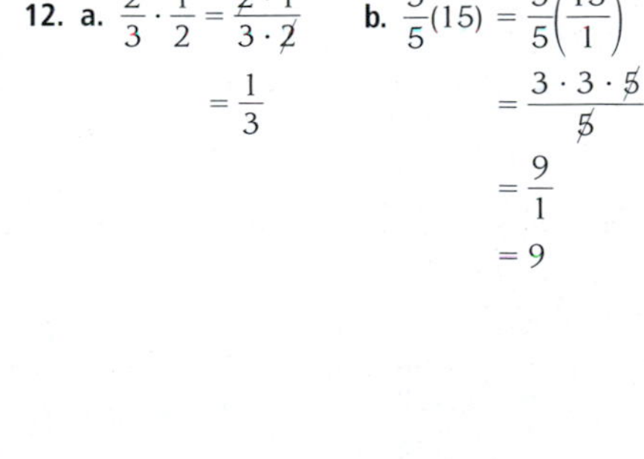

12. a. $\frac{2}{3} \cdot \frac{1}{2} = \frac{\cancel{2} \cdot 1}{3 \cdot \cancel{2}}$
$= \frac{1}{3}$

b. $\frac{3}{5}(15) = \frac{3}{5}\left(\frac{15}{1}\right)$
$= \frac{3 \cdot 3 \cdot \cancel{5}}{\cancel{5}}$
$= \frac{9}{1}$
$= 9$

13. a. $\frac{2}{3}(12) = \frac{2}{3}\left(\frac{12}{1}\right)$
$= \frac{2 \cdot 2 \cdot 2 \cdot \cancel{3}}{\cancel{3} \cdot 1}$
$= \frac{8}{1}$
$= 8$

b. $\frac{2}{3}(120) = \frac{2}{3}\left(\frac{120}{1}\right)$
$= \frac{2 \cdot 2 \cdot 2 \cdot 2 \cdot \cancel{3} \cdot 5}{\cancel{3}}$
$= \frac{80}{1}$
$= 80$

14. $A = \frac{1}{2}(7)(10)$
$= 35 \text{ in}^2$

15.

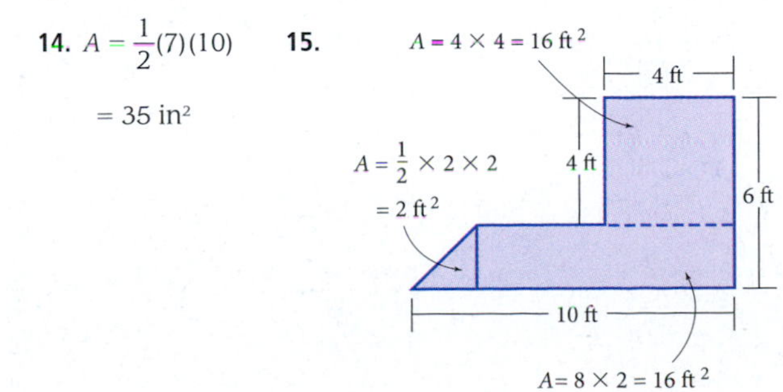

Total area $= 2 + 16 + 16 = 34 \text{ ft}^2$

Section 3.4

1. a. $$\begin{aligned}\frac{1}{3} \div \frac{1}{6} &= \frac{1}{3} \cdot \frac{6}{1}\\ &= \frac{6}{3}\\ &= 2\end{aligned}$$

b. $$\begin{aligned}\frac{1}{30} \div \frac{1}{60} &= \frac{1}{30} \cdot \frac{60}{1}\\ &= \frac{1 \cdot 2 \cdot 2 \cdot 3 \cdot 5}{2 \cdot 3 \cdot 5 \cdot 1}\\ &= \frac{2}{1}\\ &= 2\end{aligned}$$

2. $$\begin{aligned}\frac{5}{9} \div \frac{10}{3} &= \frac{5}{9} \cdot \frac{3}{10}\\ &= \frac{\cancel{5} \cdot \cancel{3}}{\cancel{3} \cdot 3 \cdot 2 \cdot \cancel{5}}\\ &= \frac{1}{6}\end{aligned}$$

3. a. $$\begin{aligned}\frac{3}{4} \div 3 &= \frac{3}{4} \cdot \frac{1}{3}\\ &= \frac{1}{4}\end{aligned}$$

b. $$\begin{aligned}\frac{3}{5} \div 3 &= \frac{3}{5} \cdot \frac{1}{3}\\ &= \frac{1}{5}\end{aligned}$$

c. $$\begin{aligned}\frac{3}{7} \div 3 &= \frac{3}{7} \cdot \frac{1}{3}\\ &= \frac{1}{7}\end{aligned}$$

4. $$\begin{aligned}4 \div \left(\frac{1}{5}\right) &= 4(5)\\ &= 20\end{aligned}$$

5. a. $$\begin{aligned}\frac{5}{32} \div \frac{10}{42} &= \frac{5}{32} \cdot \frac{42}{10}\\ &= \frac{\cancel{5} \cdot (\cancel{2} \cdot 3 \cdot 7)}{\cancel{2} \cdot 2 \cdot 2 \cdot 2 \cdot 2 \cdot 2 \cdot \cancel{5}}\\ &= \frac{21}{32}\end{aligned}$$

b. $$\begin{aligned}\frac{15}{32} \div \frac{30}{42} &= \frac{15}{32} \cdot \frac{42}{30}\\ &= \frac{(\cancel{3} \cdot \cancel{5}) \cdot (\cancel{2} \cdot 3 \cdot 7)}{(\cancel{2} \cdot 2 \cdot 2 \cdot 2 \cdot 2) \cdot (2 \cdot \cancel{3} \cdot \cancel{5})}\\ &= \frac{21}{32}\end{aligned}$$

6. $$\begin{aligned}\frac{12}{25} \div 6 &= \frac{12}{25} \cdot \frac{1}{6}\\ &= \frac{\cancel{2} \cdot 2 \cdot \cancel{3} \cdot 1}{5 \cdot 5 \cdot \cancel{2} \cdot \cancel{3}}\\ &= \frac{2}{25}\end{aligned}$$

b. $$\begin{aligned}\frac{24}{25} \div 6 &= \frac{24}{25} \cdot \frac{1}{6}\\ &= \frac{\cancel{2} \cdot 2 \cdot 2 \cdot \cancel{3} \cdot 1}{(5 \cdot 5) \cdot (\cancel{2} \cdot \cancel{3})}\\ &= \frac{4}{25}\end{aligned}$$

7. a. $$\begin{aligned}12 \div \left(\frac{4}{3}\right) &= 12\left(\frac{3}{4}\right)\\ &= 9\end{aligned}$$

b. $$\begin{aligned}12 \div \frac{4}{5} &= 12\left(\frac{5}{4}\right)\\ &= 15\end{aligned}$$

c. $$\begin{aligned}12 \div \frac{4}{7} &= 12\left(\frac{7}{4}\right)\\ &= 21\end{aligned}$$

8. $$\begin{aligned}\frac{x^3}{y} \div \frac{x^2}{y^2} &= \frac{x^3}{y} \cdot \frac{y^2}{x^2}\\ &= \frac{\cancel{x} \cdot \cancel{x} \cdot x \cdot y \cdot \cancel{y}}{\cancel{y} \cdot \cancel{x} \cdot \cancel{x}}\\ &= \frac{x \cdot y}{1}\\ &= xy\end{aligned}$$

9. $$\begin{aligned}\frac{5}{4} \div \frac{1}{8} + 8 &= \frac{5}{4} \cdot \frac{8}{1} + 8\\ &= 10 + 8\\ &= 18\end{aligned}$$

10. $$\begin{aligned}18 \div \left(\frac{3}{5}\right)^2 + 48 \div \left(\frac{2}{5}\right)^2 &= 18 \div \frac{9}{25} + 48 \div \frac{4}{25}\\ &= 18 \cdot \frac{25}{9} + 48 \cdot \frac{25}{4}\\ &= 50 + 300\\ &= 350\end{aligned}$$

11. $$\begin{aligned}12 \div \frac{3}{4} &= 12 \cdot \frac{4}{3}\\ &= 4 \cdot 4\\ &= 16 \text{ blankets}\end{aligned}$$

Section 3.5

1. $$\begin{aligned}\frac{3}{10} + \frac{1}{10} &= \frac{3+1}{10}\\ &= \frac{4}{10}\\ &= \frac{2}{5}\end{aligned}$$

2. $$\begin{aligned}\frac{a+5}{12} + \frac{3}{12} &= \frac{a+5+3}{12}\\ &= \frac{a+8}{12}\end{aligned}$$

3. $$\begin{aligned}\frac{8}{7} - \frac{5}{7} &= \frac{8-5}{7}\\ &= \frac{3}{7}\end{aligned}$$

4. $$\begin{aligned}\frac{5}{9} + \frac{8}{9} + \frac{5}{9} &= \frac{5+8+5}{9}\\ &= \frac{18}{9}\\ &= 2\end{aligned}$$

5. a. $$\left.\begin{aligned}18 &= 2 \cdot 3 \cdot 3\\ 14 &= 2 \cdot 7\end{aligned}\right\} \quad \begin{aligned}\text{LCD} &= 2 \cdot 3 \cdot 3 \cdot 7\\ &= 126\end{aligned}$$

b. $$\left.\begin{aligned}36 &= 2 \cdot 2 \cdot 3 \cdot 3\\ 28 &= 2 \cdot 2 \cdot 7\end{aligned}\right\} \quad \begin{aligned}\text{LCD} &= 2 \cdot 2 \cdot 3 \cdot 3 \cdot 7\\ &= 252\end{aligned}$$

8. a. $$\begin{aligned}\frac{2}{9} + \frac{4}{15} &= \frac{2 \cdot 5}{9 \cdot 5} + \frac{4 \cdot 3}{15 \cdot 3}\\ &= \frac{10}{45} + \frac{12}{45}\\ &= \frac{22}{45}\end{aligned}$$

b. $$\begin{aligned}\frac{2}{27} + \frac{4}{45} &= \frac{2 \cdot 5}{27 \cdot 5} + \frac{4 \cdot 3}{45 \cdot 3}\\ &= \frac{10}{135} + \frac{12}{135}\\ &= \frac{22}{135}\end{aligned}$$

9. LCD = 100; $$\begin{aligned}\frac{8}{25} - \frac{3}{20} &= \frac{8 \cdot 4}{25 \cdot 4} - \frac{3 \cdot 5}{20 \cdot 5}\\ &= \frac{32}{100} - \frac{15}{100}\\ &= \frac{17}{100}\end{aligned}$$

10. LCD = 20; $$\begin{aligned}\frac{3}{4} - \frac{x}{5} &= \frac{3 \cdot 5}{4 \cdot 5} - \frac{x \cdot 4}{5 \cdot 4}\\ &= \frac{15}{20} - \frac{4x}{20}\\ &= \frac{15 - 4x}{20} = 6\end{aligned}$$

11. a. LCD = 36; $$\begin{aligned}\frac{1}{9} + \frac{1}{4} + \frac{1}{6} &= \frac{1 \cdot 4}{9 \cdot 4} + \frac{1 \cdot 9}{4 \cdot 9} + \frac{1 \cdot 6}{6 \cdot 6}\\ &= \frac{4}{36} + \frac{9}{36} + \frac{6}{36}\\ &= \frac{19}{36}\end{aligned}$$

b. $$\begin{aligned}\frac{1}{90} + \frac{1}{40} + \frac{1}{60} &= \frac{1 \cdot 4}{90 \cdot 4} + \frac{1 \cdot 9}{40 \cdot 9} + \frac{1 \cdot 6}{60 \cdot 6}\\ &= \frac{4}{360} + \frac{9}{360} + \frac{6}{360}\\ &= \frac{19}{360}\end{aligned}$$

12. $$\begin{aligned}2 - \frac{3}{4} &= \frac{2}{1} - \frac{3}{4}\\ &= \frac{2 \cdot 4}{1 \cdot 4} - \frac{3}{4}\\ &= \frac{8}{4} - \frac{3}{4}\\ &= \frac{5}{4}\end{aligned}$$

13. $$\begin{aligned}\frac{5}{x} + \frac{2}{3} &= \frac{5 \cdot 3}{x \cdot 3} + \frac{2 \cdot x}{3 \cdot x}\\ &= \frac{15}{3x} + \frac{2x}{3x}\\ &= \frac{15 + 2x}{3x}\end{aligned}$$

14. $$\begin{aligned}\frac{x}{4} + \frac{x}{2} &= \frac{x}{4} + \frac{x \cdot 2}{2 \cdot 2}\\ &= \frac{x}{4} + \frac{2x}{4}\\ &= \frac{3x}{4}\end{aligned}$$

Section 3.6

1. $5\frac{2}{3} = 5 + \frac{2}{3} = \frac{5}{1} + \frac{2}{3} = \frac{5 \cdot 3}{1 \cdot 3} + \frac{2}{3} = \frac{15}{3} + \frac{2}{3} = \frac{17}{3}$

2. $3\frac{1}{6} = 3 + \frac{1}{6} = \frac{3}{1} + \frac{1}{6} = \frac{3 \cdot 6}{1 \cdot 6} + \frac{1}{6} = \frac{18}{6} + \frac{1}{6} = \frac{19}{6}$

3. $5\frac{2}{3} = \frac{(3 \cdot 5) + 2}{3} = \frac{17}{3}$

4. $6\frac{4}{9} = \frac{(9 \cdot 6) + 4}{9} = \frac{58}{9}$

5. $3\overline{)11}$ quotient 3; -9, remainder 2, so $\frac{11}{3} = 3\frac{2}{3}$

6. $5\overline{)14}$ quotient 2; -10, remainder 4, so $\frac{14}{5} = 2\frac{4}{5}$

7. $26\overline{)207}$ quotient 7; -182, remainder 25, so $\frac{207}{26} = 7\frac{25}{26}$

8. $3 + \frac{2}{x} = \frac{3}{1} + \frac{2}{x} = \frac{3 \cdot x}{1 \cdot x} + \frac{2}{x} = \frac{3x}{x} + \frac{2}{x} = \frac{3x + 2}{x}$

9. $x - \frac{2}{3} = \frac{x}{1} - \frac{2}{3} = \frac{x \cdot 3}{1 \cdot 3} - \frac{2}{3} = \frac{3x}{3} - \frac{2}{3} = \frac{3x - 2}{3}$

Section 3.7

1. $2\frac{3}{4} \cdot 4\frac{1}{3} = \frac{11}{4} \cdot \frac{13}{3} = \frac{143}{12} = 11\frac{11}{12}$

2. $2 \cdot 3\frac{5}{8} = \frac{2}{1} \cdot \frac{29}{8} = \frac{58}{8} = 7\frac{2}{8} = 7\frac{1}{4}$

3. $1\frac{3}{5} \div 3\frac{2}{5} = \frac{8}{5} \div \frac{17}{5} = \frac{8}{5} \cdot \frac{5}{17} = \frac{8}{17}$

4. $4\frac{5}{8} \div 2 = \frac{37}{8} \div \frac{2}{1} = \frac{37}{8} \cdot \frac{1}{2} = \frac{37}{16} = 2\frac{5}{16}$

Section 3.8

1. $3\frac{2}{3} + 2\frac{1}{4} = 3 + \frac{2}{3} + 2 + \frac{1}{4} = (3 + 2) + \left(\frac{2}{3} + \frac{1}{4}\right) = 5 + \left(\frac{2 \cdot 4}{3 \cdot 4} + \frac{1 \cdot 3}{4 \cdot 3}\right) = 5 + \left(\frac{8}{12} + \frac{3}{12}\right) = 5 + \frac{11}{12} = 5\frac{11}{12}$

2.
$5\frac{3}{4} = 5\frac{3 \cdot 5}{4 \cdot 5} = 5\frac{15}{20}$
$+\,6\frac{4}{5} = 6\frac{4 \cdot 4}{5 \cdot 4} = 6\frac{16}{20}$
$11\frac{31}{20} = 11 + 1\frac{11}{20} = 12\frac{11}{20}$

3.
$6\frac{3}{4} = 6\frac{3 \cdot 2}{4 \cdot 2} = 6\frac{6}{8}$
$+\,2\frac{7}{8} = 2\frac{7}{8} = 2\frac{7}{8}$
$8\frac{13}{8} = 9\frac{5}{8}$

4.
$2\frac{1}{3} = 2\frac{1 \cdot 4}{3 \cdot 4} = 2\frac{4}{12}$
$1\frac{1}{4} = 1\frac{1 \cdot 3}{4 \cdot 3} = 1\frac{3}{12}$
$+\,3\frac{11}{12} = 3\frac{11}{12} = 3\frac{11}{12}$
$6\frac{18}{12} = 7\frac{6}{12} = 7\frac{1}{2}$

5.
$4\frac{7}{8}$
$-1\frac{5}{8}$
$3\frac{2}{8} = 3\frac{1}{4}$

6.
$12\frac{7}{10} = 12\frac{7}{10} = 12\frac{7}{10}$
$-\,7\frac{2}{5} = -7\frac{2 \cdot 2}{5 \cdot 2} = -7\frac{4}{10}$
$5\frac{3}{10}$

7.
$10 = 9\frac{7}{7}$
$-\,5\frac{4}{7} = -5\frac{4}{7}$
$4\frac{3}{7}$

8.
$6\frac{1}{3} = \left(5 + \frac{3}{3}\right) + \frac{1}{3} = 5\frac{4}{3}$
$-2\frac{2}{3} = -2\frac{2}{3} = -2\frac{2}{3}$
$3\frac{2}{3}$

9.
$6\frac{3}{4} = 6\frac{3 \cdot 3}{4 \cdot 3} = 6\frac{9}{12} = 5\frac{21}{12}$
$-2\frac{5}{6} = -2\frac{5 \cdot 2}{6 \cdot 2} = -2\frac{10}{12} = -2\frac{10}{12}$
$3\frac{11}{12}$

Section 3.9

1. $4 + \left(1\frac{1}{2}\right)\left(2\frac{3}{4}\right) = 4 + \left(\frac{3}{2}\right)\left(\frac{11}{4}\right) = 4 + \frac{33}{8} = \frac{32}{8} + \frac{33}{8} = \frac{65}{8} = 8\frac{1}{8}$

2. $\left(\frac{2}{3} + \frac{1}{6}\right)\left(2\frac{5}{6} + 1\frac{1}{3}\right) = \left(\frac{5}{6}\right)\left(4\frac{1}{6}\right) = \frac{5}{6}\left(\frac{25}{6}\right) = \frac{125}{36} = 3\frac{17}{36}$

3. $\frac{3}{7} + \frac{1}{3}\left(1\frac{1}{2} + 4\frac{1}{2}\right)^2 = \frac{3}{7} + \frac{1}{3}(6)^2 = \frac{3}{7} + \frac{1}{3}(36) = \frac{3}{7} + 12 = 12\frac{3}{7}$

4. $\dfrac{\frac{2}{3}}{\frac{5}{9}} = \frac{2}{3} \div \frac{5}{9} = \frac{2}{3} \cdot \frac{9}{5} = \frac{18}{15} = \frac{6}{5} = 1\frac{1}{5}$

5. $\dfrac{\frac{1}{2} + \frac{3}{4}}{\frac{2}{3} - \frac{1}{4}} = \dfrac{12\left(\frac{1}{2} + \frac{3}{4}\right)}{12\left(\frac{2}{3} - \frac{1}{4}\right)} = \dfrac{12 \cdot \frac{1}{2} + 12 \cdot \frac{3}{4}}{12 \cdot \frac{2}{3} - 12 \cdot \frac{1}{4}} = \dfrac{6 + 9}{8 - 3} = \frac{15}{5} = 3$

6. $\dfrac{4 + \frac{2}{3}}{3 - \frac{1}{4}} = \dfrac{12\left(4 + \frac{2}{3}\right)}{12\left(3 - \frac{1}{4}\right)} = \dfrac{12 \cdot 4 + 12 \cdot \frac{2}{3}}{12 \cdot 3 - 12 \cdot \frac{1}{4}} = \dfrac{48 + 8}{36 - 3} = \frac{56}{33} = 1\frac{23}{33}$

7. $\dfrac{12\frac{1}{3}}{6\frac{2}{3}} = 12\frac{1}{3} \div 6\frac{2}{3} = \frac{37}{3} \div \frac{20}{3} = \frac{37}{3} \cdot \frac{3}{20} = \frac{37}{20} = 1\frac{17}{20}$

Chapter 4

Section 4.1

1. $6(x + 4) = 6(x) + 6(4) = 6x + 24$

2. $-3(2x + 4) = -3(2x) + (-3)(4) = -6x + (-12) = -6x - 12$

3. $\frac{1}{2}(2x - 4) = \frac{1}{2}(2x) - \frac{1}{2}(4) = x - 2$

4. $6x - 2 + 3x + 8 = 6x + 3x + (-2) + 8 = 9x + 6$

5. $2(4x + 3) + 7 = 2(4x) + 2(3) + 7 = 8x + 6 + 7 = 8x + 13$

6. $3(2x + 1) + 5(4x - 3) = 3(2x) + 3(1) + 5(4x) - 5(3) = 6x + 3 + 20x - 15 = 26x - 12$

10. $A = lw = 25(8 + 2x) = 25(8) + 25(2x) = 200 + 50x$

11. **a.** $x = 90° - 45° = 45°$
b. $x = 180° - 60° = 120°$

Section 4.2

1. When $x = 3$
the equation $5x - 4 = 11$
becomes $5(3) - 4 = 11$
or $15 - 4 = 11$
$11 = 11$

2. When $a = -3$
the equation $6a - 3 = 2a + 4$
becomes $6(-3) - 3 = 2(-3) + 4$
$-18 - 3 = -6 + 4$
$-21 = -2$
This is a false statement, so $a = -3$ is not a solution.

3. $x + 5 = -2$
$x + 5 + (-5) = -2 + (-5)$
$x + 0 = -7$
$x = -7$

4. $a - 2 = 7$
$a - 2 + 2 = 7 + 2$
$a + 0 = 9$
$a = 9$

5. $y + 6 - 2 = 8 - 9$
$y + 4 = -1$
$y + 4 + (-4) = -1 + (-4)$
$y + 0 = -5$
$y = -5$

6. $5x - 3 - 4x = 4 - 7$
$x - 3 = -3$
$x - 3 + 3 = -3 + 3$
$x + 0 = 0$
$x = 0$

7. $-5 - 7 = x + 2$
$-12 = x + 2$
$-12 + (-2) = x + 2 + (-2)$
$-14 = x + 0$
$-14 = x$

8. $a - \frac{2}{3} = \frac{5}{6}$
$a - \frac{2}{3} + \frac{2}{3} = \frac{5}{6} + \frac{2}{3}$
$a = \frac{9}{6} = \frac{3}{2}$

9. $5(3a - 4) - 14a = 25$
$15a - 20 - 14a = 25$
$a - 20 = 25$
$a - 20 + 20 = 25 + 20$
$a = 45$

Section 4.3

1. $\frac{1}{3}x = 5$

$3 \cdot \frac{1}{3}x = 3 \cdot 5$

$x = 15$

2. $\frac{1}{5}a + 3 = 7$

$\frac{1}{5}a + 3 + (-3) = 7 + (-3)$

$\frac{1}{5}a = 4$

$5 \cdot \frac{1}{5}a = 5 \cdot 4$

$a = 20$

3. $\frac{3}{5}y = 6$

$\frac{5}{3} \cdot \frac{3}{5}y = \frac{5}{3} \cdot 6$

$y = 10$

4. $-\frac{3}{4}x = \frac{6}{5}$

$-\frac{4}{3}(-\frac{3}{4}x) = -\frac{4}{3} \cdot \frac{6}{5}$

$x = -\frac{8}{5}$

5. $6x = -42$

$\frac{6x}{6} = -\frac{42}{6}$

$x = -7$

6. $-5x + 6 = -14$

$-5x + 6 + (-6) = -14 + (-6)$

$-5x = -20$

$\frac{-5x}{-5} = \frac{-20}{-5}$

$x = 4$

7. $3x - 7x + 5 = 3 - 18$

$-4x + 5 = -15$

$-4x + 5 + (-5) = -15 + (-5)$

$-4x = -20$

$\frac{-4x}{-4} = \frac{-20}{-4}$

$x = 5$

8. $-5 + 4 = 2x - 11 + 3x$

$-1 = 5x - 11$

$-1 + 11 = 5x - 11 + 11$

$10 = 5x$

$\frac{10}{5} = \frac{5x}{5}$

$2 = x$

Section 4.4

1. $4(x + 3) = -8$

$4x + 12 = -8$

$4x + 12 + (-12) = -8 + (-12)$

$4x = -20$

$\frac{4x}{4} = \frac{-20}{4}$

$x = -5$

2. $6a + 7 = 4a - 3$

$6a + (-4a) + 7 = 4a + (-4a) - 3$

$2a + 7 = -3$

$2a + 7 + (-7) = -3 + (-7)$

$2a = -10$

$\frac{2a}{2} = \frac{-10}{2}$

$a = -5$

3. $5(x - 2) + 3 = -12$

$5x - 10 + 3 = -12$

$5x - 7 = -12$

$5x - 7 + 7 = -12 + 7$

$5x = -5$

$\frac{5x}{5} = \frac{-5}{5}$

$x = -1$

4. $3(4x - 5) + 6 = 3x + 9$

$12x - 15 + 6 = 3x + 9$

$12x - 9 = 3x + 9$

$12x + (-3x) - 9 = 3x + (-3x) + 9$

$9x - 9 = 9$

$9x - 9 + 9 = 9 + 9$

$9x = 18$

$\frac{9x}{9} = \frac{18}{9}$

$x = 2$

5. $\frac{x}{3} + \frac{x}{6} = 9$

$6\left(\frac{x}{3} + \frac{x}{6}\right) = 6(9)$

$6\left(\frac{x}{3}\right) + 6\left(\frac{x}{6}\right) = 6(9)$

$2x + x = 54$

$3x = 54$

$x = 18$

6. $3x + \frac{1}{4} = \frac{5}{8}$

$8\left(3x + \frac{1}{4}\right) = 8\left(\frac{5}{8}\right)$

$8(3x) + 8\left(\frac{1}{4}\right) = 8\left(\frac{5}{8}\right)$

$24x + 2 = 5$

$24x = 3$

$x = \frac{1}{8}$

7. $\frac{4}{x} + 3 = \frac{11}{5}$

$5x\left(\frac{4}{x} + 3\right) = 5x\left(\frac{11}{5}\right)$

$5x\left(\frac{4}{x}\right) + 5x(3) = 5x\left(\frac{11}{5}\right)$

$20 + 15x = 11x$

$20 = -4x$

$-5 = x$

Section 4.5

1. Step 1 ***Read and list.***

Known items: The numbers 3 and 10

Unknown item: The number in question

Step 2 ***Assign a variable and translate the information.***

Let x = the number asked for in the problem.

Then "The sum of a number and 3" translates to $x + 3$.

Step 3 ***Reread and write an equation.***

The sum of x and 3 is 10.

$x + 3 \quad = 10$

Step 4 ***Solve the equation.***

$x + 3 = 10$

$x = 7$

Step 5 ***Write your answer.***

The number is 7.

Step 6 ***Reread and check.***

The sum of **7** and 3 is 10.

2. Step 1 ***Read and list.***

Known items: The numbers 4 and 34, twice a number, and three times a number

Unknown item: The number in question

Step 2 ***Assign a variable and translate the information.***

Let x = the number asked for in the problem.

Then "The sum of twice a number and three times the number" translates to $2x + 3x$.

Step 3 ***Reread and write an equation.***

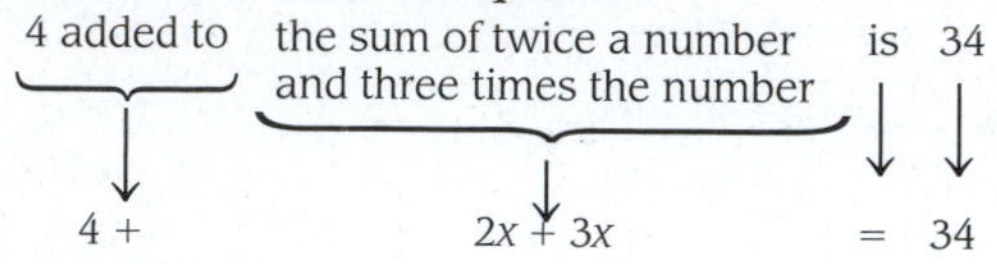

Step 4 ***Solve the equation.***

$$4 + 2x + 3x = 34$$
$$5x + 4 = 34$$
$$5x = 30$$
$$x = 6$$

Step 5 ***Write your answer.***

The number is 6.

Step 6 ***Reread and check.***

Twice **6** is 12 and three times **6** is 18. Their sum is 12 + 18 = 30. Four added to this is 34. Therefore, 4 added to the sum of twice **6** and three times **6** is 34.

3. Step 1 ***Read and list.***

Known items: Length is twice width; perimeter 42 cm

Unknown items: The length and the width

Step 2 ***Assign a variable and translate the information.***

Let x = the width. Since the length is twice the width, the length must be $2x$. Here is a picture.

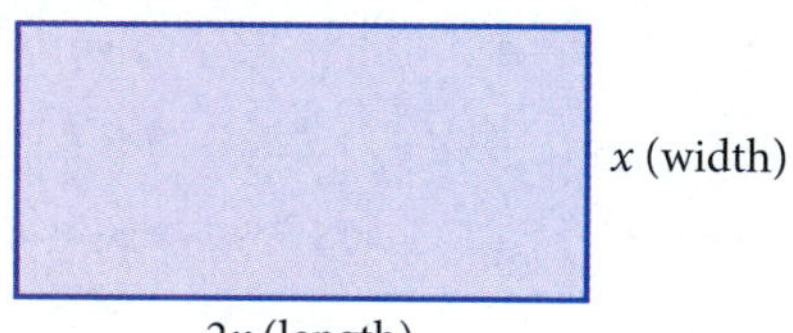

Step 3 ***Reread and write an equation.***

The perimeter is the sum of the sides, and is given as 42; therefore,

$$x + x + 2x + 2x = 42$$

Step 4 ***Solve the equation.***

$$x + x + 2x + 2x = 42$$
$$6x = 42$$
$$x = 7$$

Step 5 ***Write your answer.***

The width is 7 centimeters and the length is

2(7) = 14 centimeters

Step 6 ***Reread and check.***

The length, 14, is twice the width, 7. The perimeter is 7 + 7 + 14 + 14 = 42 centimeters.

4. Step 1 ***Read and list.***

Known items: Three angles are in a triangle. One is 3 times the smallest. The largest is 5 times the smallest.

Unknown item: The three angles

Step 2 ***Assign a variable and translate the information.***

Let x = the smallest angle. The other two angles are $3x$ and $5x$.

Step 3 ***Reread and write an equation.***

The three angles must add up to 180°, so

$$x + 3x + 5x = 180°$$

Step 4 ***Solve the equation.***

$x + 3x + 5x = 180°$

$9x = 180°$ Add similar terms on left side

$x = 20°$ Divide each side by 9

Step 5 ***Write the answer.***

The three angles are, $20°$, $3(20°) = 60°$, and $5(20°) = 100°$.

Step 6 ***Reread and check.***

The sum of the three angles is $20° + 60° + 100° = 180°$. One angle is 3 times the smallest, while the largest is 5 times the smallest.

5. Step 1 ***Read and list.***

Known items: Joyce is 21 years older than Travis. Six years from now their ages will add to 49.

Unknown items: Their ages now

Step 2 ***Assign a variable and translate the information.***

Let x = Travis's age now; since Joyce is 21 years older than that, she is presently $x + 21$ years old.

Step 3 ***Reread and write an equation.***

	Now	in 6 years
Joyce	$x + 21$	$x + 27$
Travis	x	$x + 6$

$x + 27 + x + 6 = 49$

Step 4 ***Solve the equation.***

$$x + 27 + x + 6 = 49$$
$$2x + 33 = 49$$
$$2x = 16$$
$$x = 8$$

Step 5 ***Write your answer.***

Travis is now 8 years old, and Joyce is $8 + 21 = 29$ years old.

Step 6 ***Reread and check.***

Joyce is 21 years older than Travis. In six years, Joyce will be 35 years old and Travis will be 14 years old. At that time, the sum of their ages will be $35 + 14 = 49$.

Section 4.6

1. When $P = 80$ and $w = 6$

the formula $P = 2l + 2w$

becomes $80 = 2l + 2(6)$

$$80 = 2l + 12$$
$$68 = 2l$$
$$34 = l$$

The length is 34 feet.

2. When $F = 77$

the formula $C = \frac{5}{9}(F - 32)$

becomes $C = \frac{5}{9}(77 - 32)$

$$= \frac{5}{9}(45)$$
$$= \frac{5}{9} \cdot \frac{45}{1}$$
$$= \frac{225}{9}$$

$= 25$ degrees Celsius

3. When $x = 0$

the formula $y = 2x + 6$

becomes $y = 2 \cdot 0 + 6$

$$= 0 + 6$$
$$= 6$$

4. When $x = -3$
the formula $2x + 3y = 4$
becomes $2(-3) + 3y = 4$
$-6 + 3y = 4$
$3y = 10$
$y = \frac{10}{3}$

5. a. $11 - 9 = 2$ hr
b. $d = 60$ mi/hr $\cdot$ 2 hr
$d = 60 \cdot 2$
$d = 120$ miles

6. With $x = 35°$ we use the formulas for finding the complement and the supplement of an angle:
The complement of 35° is 90° − 35° = 55°
The supplement is 35° is 180° − 35° = 145°

Section 4.7

1. When $x = 0$
the equation $3x + 5y = 15$
becomes $3 \cdot 0 + 5y = 15$
$5y = 15$
$y = 3$

which gives (0, 3) as one solution

When $y = 0$
the equation $3x + 5y = 15$
becomes $3x + 5 \cdot 0 = 15$
$3x = 15$
$x = 5$

which means (5, 0) is a second solution

When $x = -5$
the equation $3x + 5y = 15$
becomes $3(-5) + 5y = 15$
$-15 + 5y = 15$
$5y = 30$
$y = 6$

which gives (−5, 6) as a third solution

2. When $x = 2$, we have
$5 \cdot 2 + 2y = 20$
$10 + 2y = 20$
$2y = 10$
$y = 5$

When $x = 0$, we have
$5 \cdot 0 + 2y = 20$
$2y = 20$
$y = 10$

When $y = 5$, we have
$5x + 2 \cdot 5 = 20$
$5x + 10 = 20$
$5x = 10$
$x = 2$

When $y = 0$, we have
$5x + 2 \cdot 0 = 20$
$5x = 20$
$x = 4$

3. When $x = 0$, we have
$y = \frac{1}{2} \cdot 0 + 1$
$y = 1$

When $x = 4$, we have
$y = \frac{1}{2} \cdot 4 + 1$
$y = 2 + 1$
$y = 3$

When $y = 7$, we have
$7 = \frac{1}{2}x + 1$
$6 = \frac{1}{2}x$
$12 = x$

When $y = -3$, we have
$-3 = \frac{1}{2}x + 1$
$-4 = \frac{1}{2}x$
$-8 = x$

4. (1, 5) is not a solution. When we substitute 1 for x and 5 for y into $y = 5x - 6$, we get a false statement.
$5 = 5 \cdot 1 - 6$
$5 = 5 - 6$
$5 = -1$ a false statement

(2, 4) is a solution. Substituting 2 for x and 4 for y in the equation $y = 5x - 6$ yields a true statement.
$4 = 5 \cdot 2 - 6$
$4 = 10 - 6$
$4 = 4$ a true statement

Section 4.8

1.

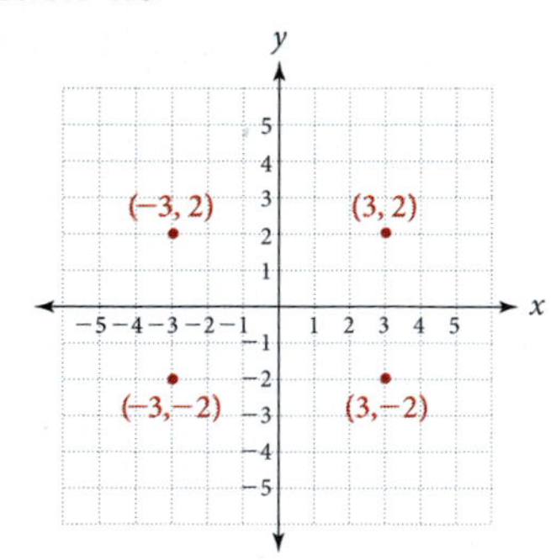

2.

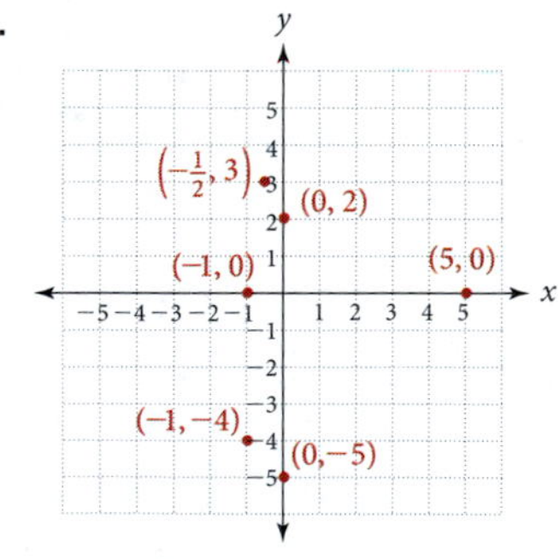

Section 4.9

1.

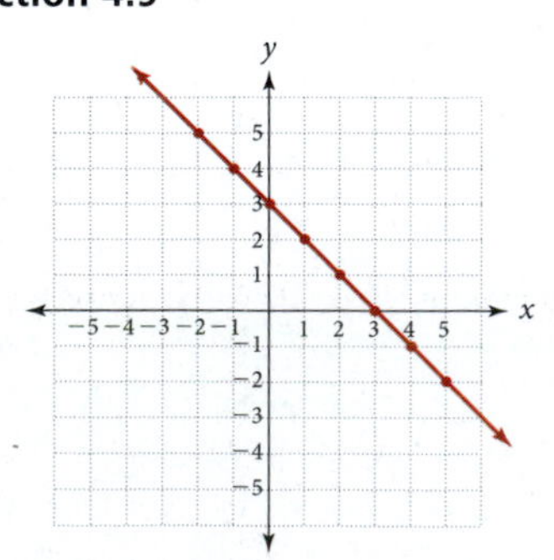

All the points shown on the graph have coordinates that add to 3.

2. When $x = 1, y = -2(1) + 1$
$y = -2 + 1$
$y = -1$
$(1, -1)$ is one solution
When $x = 0, y = -2(0) + 1$
$y = 0 + 1$
$y = 1$
$(0, 1)$ is a second solution
When $x = -1, y = -2(-1) + 1$
$y = 2 + 1$
$y = 3$
$(-1, 3)$ is our third solution

3. Let $x = -1$: $y = 2(-1) - 3$
$y = -2 - 3$
$y = -5$
$(-1, -5)$ is one solution
Let $x = 0$: $y = 2(0) - 3$
$y = 0 - 3$
$y = -3$
$(0, -3)$ is another solution
Let $x = 2$: $y = 2(2) - 3$
$y = 4 - 3$
$y = 1$
$(2, 1)$ is a third solution

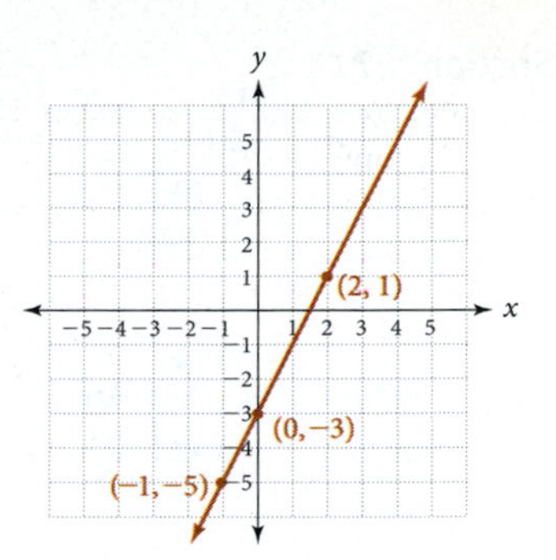

4. Let $x = 0$: $3 \cdot 0 - 2y = 6$
$0 - 2y = 6$
$-2y = 6$
$y = -3$
$(0, -3)$ is one solution
Let $y = 0$: $3x - 2 \cdot 0 = 6$
$3x - 0 = 6$
$3x = 6$
$x = 2$
$(2, 0)$ is a second solution
The third point is up to you to find. Substituting -2, 4, or 6 for x will make your work easier. Do you know why?

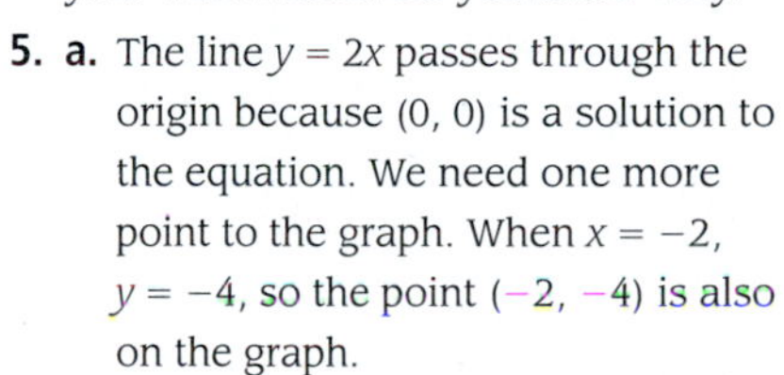
5. a. The line $y = 2x$ passes through the origin because $(0, 0)$ is a solution to the equation. We need one more point to the graph. When $x = -2$, $y = -4$, so the point $(-2, -4)$ is also on the graph.

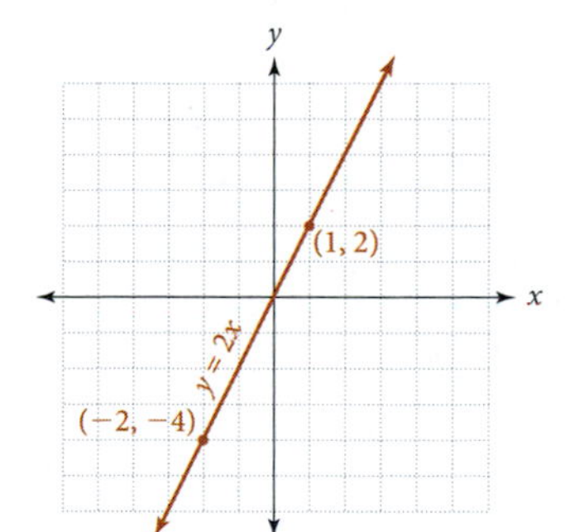

b. The line $x = -3$ is the set of points whose x-coordinate is -3. Since the variable y does not appear in the equation, it can have any value. The graph of $x = -3$ is the vertical line shown here.

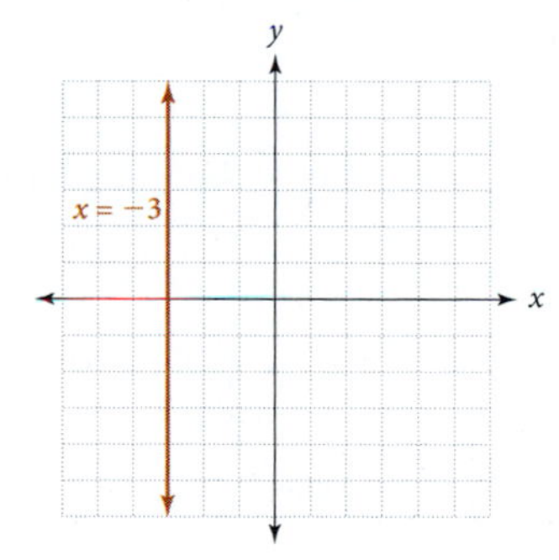

c. The line $y = 2$ is the set of points whose y-coordinate is 2. Since the variable x does not appear in the equation, it can have any value. The graph of $y = 2$ is the horizontal line shown here.

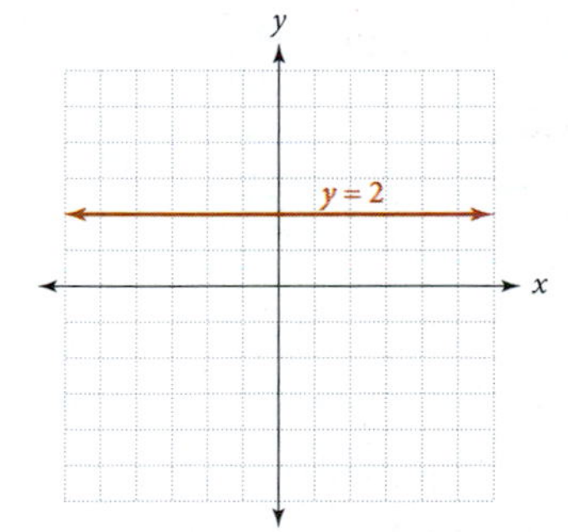

Chapter 5

Section 5.1

1. $700 + 80 + 5 + \frac{4}{10} + \frac{6}{100} + \frac{2}{1{,}000}$ **2. a.** Six hundredths **b.** Seven tenths **c.** Eight thousandths

3. a. Five and six hundredths **b.** Four and seven tenths **c.** Three and eight thousandths

4. a. Five and ninety-eight hundredths **b.** Five and ninety-eight thousandths **5.** Three hundred five and four hundred six thousandths

Section 5.2

1. $$\begin{array}{rl} 38.45 & = 38\frac{45}{100} = 38\frac{450}{1,000} \\ +456.073 & = 456\frac{73}{1,000} = 456\frac{73}{1,000} \\ \hline & 494\frac{523}{1,000} = 494.523 \end{array}$$

b. $$\begin{array}{rl} 38.045 & = 38\frac{45}{1,000} = 38\frac{45}{1,000} \\ +456.73 & = 456\frac{73}{1,000} = 456\frac{730}{1,000} \\ \hline & 494\frac{775}{1,000} = 494.775 \end{array}$$

2. $$\begin{array}{r} 78.674 \\ -23.431 \\ \hline 55.243 \end{array}$$

3. $$\begin{array}{r} 16.000 \\ 0.033 \\ 4.600 \\ +\ 0.080 \\ \hline 20.713 \end{array}$$

4. a. $$\begin{array}{r} 6.70 \\ -2.05 \\ \hline 4.65 \end{array}$$ **b.** $$\begin{array}{r} 6.7000 \\ -2.0563 \\ \hline 4.6437 \end{array}$$

5. $$\begin{array}{r} 7.000 \\ +3.567 \\ \hline 10.567 \end{array} \text{ and } \begin{array}{r} 10.567 \\ -\ 5.890 \\ \hline 4.677 \end{array}$$

8. $$\begin{array}{r} \$10.00 \\ -\ \ 9.56 \\ \hline \$\ \ .44 \end{array}$$ 1 quarter + 2 dimes + 4 pennies = 0.25 + 0.20 + 0.04 = 0.49, which is too much change. One of the dimes should be a nickel. Tell the clerk that you have been given too much change.

9. a. $P = 1.38 + 1.38 + 1.38 + 1.38 = 5.52$ in.

b. $P = 6.6 + 4.7 + 4.7 = 16.0$ cm

Section 5.3

1. a. $0.4 \times 0.6 = \frac{4}{10} \times \frac{6}{10} = \frac{24}{100} = 0.24$

b. $0.04 \times 0.06 = \frac{4}{100} \times \frac{6}{100} = \frac{24}{10,000} = 0.0024$

2. a. $0.5 \times 0.007 = \frac{5}{10} \times \frac{7}{1,000} = \frac{35}{10,000} = 0.0035$

b. $0.05 \times 0.07 = \frac{5}{100} \times \frac{7}{100} = \frac{35}{10,000} = 0.0035$

3. a. $3.5 \times 0.04 = 3\frac{5}{10} \times \frac{4}{100} = \frac{35}{10} \times \frac{4}{100} = \frac{140}{1,000} = \frac{14}{100} = 0.14$

b. $0.35 \times 0.4 = \frac{35}{100} \times \frac{4}{10} = \frac{140}{1,000} = \frac{14}{100} = 0.14$

4. a. $3 + 2 = 5$ digits to the right **b.** $2 + 4 = 6$ digits to the right

5. a. $$\begin{array}{r} 4.03 \\ \times 5.22 \\ \hline 806 \\ 8060 \\ 201500 \\ \hline 21.0366 \end{array}$$ **b.** $$\begin{array}{r} 40.3 \\ \times 0.522 \\ \hline 806 \\ 8060 \\ 201500 \\ \hline 21.0366 \end{array}$$

8. a. $80 \times 6 = 480$

b. $40 \times 180 = 7,200$

c. $8^2 = 64$

9. a. $0.03(5.5 + 0.02) = 0.03(5.52) = 0.1656$

b. $0.03(0.55 + 0.002) = 0.03(0.552) = 0.01656$

10. a. $5.7 + 14(2.4)^2 = 5.7 + 14(5.76) = 5.7 + 80.64 = 86.34$

b. $0.57 + 1.4(2.4)^2 = 0.57 + 1.4(5.76) = 0.57 + 8.064 = 8.634$

11. a. $A = s^2 = 1.38^2 = 1.90 \text{ in}^2$

b. $A = lw = (39.6)(25.1) = 993.96 \text{ mm}^2$

12. $6.82(36) + 10.23(14) = 245.52 + 143.22 = \$388.74$

13. $C = 3.14(3) = 9.42$ cm

14. a. $C = \pi d \approx (3.14)(0.92) \approx 2.89$ in.

b. $C = 2\pi r \approx 2(3.14)(13.20) \approx 82.90$ mm

15. Radius $= \frac{1}{2}(20) = 10$ ft

$A = \pi r^2 = (3.14)(10)^2 = 314 \text{ ft}^2$

16. Radius $= 2(0.125) = 0.250$

$V = \pi r^2 h = (3.14)(0.250)^2(6) = 1.178 \text{ in}^3$

Section 5.4

1. $$\begin{array}{r} 154.2 \\ 30\overline{)4,626.0} \\ \underline{30} \\ 162 \\ \underline{150} \\ 126 \\ \underline{120} \\ 60 \\ \underline{60} \\ 0 \end{array}$$

2. a. $$\begin{array}{r} 6.7 \\ 5\overline{)33.5} \\ \underline{30} \\ 35 \\ \underline{35} \\ 0 \end{array}$$ **b.** $$\begin{array}{r} 6.9 \\ 5\overline{)34.5} \\ \underline{30} \\ 45 \\ \underline{45} \\ 0 \end{array}$$ **c.** $$\begin{array}{r} 7.1 \\ 5\overline{)35.5} \\ \underline{35} \\ 05 \\ \underline{5} \\ 0 \end{array}$$

3. a. $$\begin{array}{r} 2.636 \\ 18\overline{)47.448} \\ \underline{36} \\ 114 \\ \underline{108} \\ 64 \\ \underline{54} \\ 108 \\ \underline{108} \\ 0 \end{array}$$ **b.** $$\begin{array}{r} 26.36 \\ 18\overline{)474.48} \\ \underline{36} \\ 114 \\ \underline{108} \\ 64 \\ \underline{54} \\ 108 \\ \underline{108} \\ 0 \end{array}$$

4. a.

$$\begin{array}{r} 45.54 \\ 25\overline{)1{,}138.50} \\ \underline{1\,00} \\ 138 \\ \underline{125} \\ 13\,5 \\ \underline{12\,5} \\ 1\,00 \\ \underline{1\,00} \\ 0 \end{array}$$

b.

$$\begin{array}{r} 4.554 \\ 25\overline{)113.850} \\ \underline{100} \\ 13\,8 \\ \underline{12\,5} \\ 1\,35 \\ \underline{1\,25} \\ 100 \\ \underline{100} \\ 0 \end{array}$$

5. a.

$$\begin{array}{r} 3.15 \\ 4.2.\overline{)13.2.30} \\ \underline{126} \\ 6\,3 \\ \underline{4\,2} \\ 2\,10 \\ \underline{2\,10} \\ 0 \end{array}$$

b.

$$\begin{array}{r} 31.5 \\ 0.42.\overline{)13.23.0} \\ \underline{12\,6} \\ 63 \\ \underline{42} \\ 21\,0 \\ \underline{21\,0} \\ 0 \end{array}$$

6.

$$\begin{array}{r} 1.422 \\ 0.32.\overline{)0.45.530} \\ \underline{32} \\ 13\,5 \\ \underline{12\,8} \\ 73 \\ \underline{64} \\ 90 \\ \underline{64} \\ 26 \end{array}$$

Answer to nearest hundredth is 1.42

7. a.

$$\begin{array}{r} 316.66 \\ 0.06.\overline{)19.00.00} \\ \underline{18} \\ 1\,0 \\ \underline{6} \\ 40 \\ \underline{36} \\ 4\,0 \\ \underline{3\,6} \\ 40 \\ \underline{36} \\ 4 \end{array}$$

Answer to nearest tenth is 316.7

b.

$$\begin{array}{r} 31.66 \\ 0.06.\overline{)1.90.00} \\ \underline{1\,8} \\ 10 \\ \underline{6} \\ 4\,0 \\ \underline{3\,6} \\ 40 \\ \underline{36} \\ 4 \end{array}$$

Answer to nearest tenth is 31.7

8.

$$\begin{array}{r} 28.5 \text{ hours} \\ 6.54.\overline{)186.39.0} \\ \underline{130\,8} \\ 55\,59 \\ \underline{52\,32} \\ 3\,27\,0 \\ \underline{3\,27\,0} \\ 0 \end{array}$$

9. $\dfrac{4.39 - 0.43}{0.33} = \dfrac{3.96}{0.33}$

$= 12$ additional minutes

The call was 13 minutes long.

10.

Class	units	grade	value	grade points
Algebra	5	B	3	$5 \times 3 = 15$
Chemistry	4	B	3	$4 \times 3 = 12$
English	3	A	4	$3 \times 4 = 12$
History	3	B	3	$3 \times 3 = 9$
Total Units:	15		Total Grade Points:	48

$\text{GPA} = \dfrac{48}{15} = 3.20$

Section 5.5

1. a.

$$\begin{array}{r} .4 \\ 5\overline{)2.0} \\ \underline{2\,0} \\ 0 \end{array}$$

so $\dfrac{2}{5} = 0.4$

b.

$$\begin{array}{r} 0.6 \\ 5\overline{)3.0} \\ \underline{3\,0} \\ 0 \end{array}$$

so $\dfrac{3}{5} = 0.6$

c.

$$\begin{array}{r} .8 \\ 5\overline{)4.0} \\ \underline{4\,0} \\ 0 \end{array}$$

so $\dfrac{4}{5} = 0.8$

2. a.

$$\begin{array}{r} 0.9166 \\ 12\overline{)11.0000} \\ \underline{10\,8} \\ 20 \\ \underline{12} \\ 80 \\ \underline{72} \\ 80 \\ \underline{72} \\ 8 \end{array}$$

so $\dfrac{11}{12} = 0.917$ to the nearest thousandth

b.

$$\begin{array}{r} .9230 \\ 13\overline{)12.0000} \\ \underline{11\,7} \\ 30 \\ \underline{26} \\ 40 \\ \underline{39} \\ 10 \\ \underline{0} \\ 10 \end{array}$$

so $\dfrac{12}{13} = 0.923$ to the nearest thousandth

3. $11\overline{)5.0000}$ gives 0.4545

44
60
55
50
44
60
55
5

so $\dfrac{5}{11} = 0.\overline{45}$

4. a. $0.48 = \dfrac{48}{100} = \dfrac{12}{25}$ **b.** $0.048 = \dfrac{48}{1{,}000} = \dfrac{6}{125}$

5. $0.025 = \dfrac{25}{1{,}000} = \dfrac{1}{40}$

6. $12.8 = 12\dfrac{8}{10} = 12\dfrac{4}{5}$

7.
$$\begin{aligned}\frac{14}{25}(2.43 + 0.27) &= 0.56(2.43 + 0.27)\\ &= 0.56(2.70)\\ &= 1.512\end{aligned}$$

8.
$$\begin{aligned}\frac{1}{4} + 0.25\left(\frac{3}{5}\right) &= \frac{1}{4} + \frac{1}{4}\left(\frac{3}{5}\right)\\ &= \frac{1}{4} + \frac{3}{20}\\ &= \frac{5}{20} + \frac{3}{20}\\ &= \frac{8}{20}\\ &= \frac{2}{5} \text{ or } 0.4\end{aligned}$$

9.
$$\begin{aligned}\left(\frac{1}{3}\right)^3(5.4) + \left(\frac{1}{5}\right)^2(2.5) &= \frac{1}{27}(5.4) + \frac{1}{25}(2.5)\\ &= 0.2 + 0.1\\ &= 0.3\end{aligned}$$

10.
$$\begin{aligned}35.50 - \frac{1}{4}(35.50) &= \frac{3}{4}(35.50) = 26.625\\ &= \$26.63 \text{ to the nearest cent}\end{aligned}$$

11. $A = \dfrac{1}{2}bh = \dfrac{1}{2}(6.6)(3.3) = 10.89 \text{ cm}^2$

12.
$$\begin{aligned}V &= \pi r^2 h + \frac{1}{2}\cdot\frac{4}{3}\pi r^3\\ &= (3.14)(10)^2(10) + \frac{1}{2}\cdot\frac{4}{3}(3.14)(10)^3\\ &= 3{,}140 + \frac{2}{3}(3{,}140)\\ &= 3{,}140 + 2{,}093.3\\ &= 5{,}233.3 \text{ in}^3\end{aligned}$$

Section 5.6

1.
$$\begin{aligned}x - 3.4 &= 6.7\\ x - 3.4 + \mathbf{3.4} &= 6.7 + \mathbf{3.4}\\ x + 0 &= 10.1\\ x &= 10.1\end{aligned}$$

2.
$$\begin{aligned}4y &= 3.48\\ \frac{4y}{\mathbf{4}} &= \frac{3.48}{\mathbf{4}}\\ y &= 0.87\end{aligned}$$

3.
$$\begin{aligned}\frac{1}{5}x - 2.4 &= 8.3\\ \frac{1}{5}x - 2.4 + \mathbf{2.4} &= 8.3 + \mathbf{2.4}\\ \frac{1}{5}x &= 10.7\\ \mathbf{5}\left(\frac{1}{5}x\right) &= \mathbf{5}(10.7)\\ x &= 53.5\end{aligned}$$

4.
$$\begin{aligned}7a - 0.18 &= 2a + 0.77\\ 7a + (\mathbf{-2a}) - 0.18 &= 2a + (\mathbf{-2a}) + 0.77\\ 5a - 0.18 &= 0.77\\ 5a - 0.18 + \mathbf{0.18} &= 0.77 + \mathbf{0.18}\\ 5a &= 0.95\\ \frac{5a}{\mathbf{5}} &= \frac{0.95}{\mathbf{5}}\\ a &= 0.19\end{aligned}$$

5. Let x = the number of miles driven
$$\begin{aligned}2(11) + 0.16x &= 40.88\\ 22 + 0.16x &= 40.88\\ 22 + (\mathbf{-22}) + 0.16x &= 40.88 + (\mathbf{-22})\\ 0.16x &= 18.88\\ \frac{0.16x}{\mathbf{0.16}} &= \frac{18.88}{\mathbf{0.16}}\\ x &= 118 \text{ miles}\end{aligned}$$

6. Let x = the number of quarters. Then the number of dimes is $x + 7$.

	quarters	dimes
Number of	x	$x + 7$
Value of	$0.25x$	$0.10(x + 7)$

$$\begin{aligned}0.25x + 0.10(x + 7) &= 1.75\\ 0.25x + 0.10x + 0.70 &= 1.75\\ 0.35x + 0.70 &= 1.75\\ 0.35x &= 1.05\\ x &= 3\end{aligned}$$

She has 3 quarters and 10 dimes.

Section 5.7

1. $4\sqrt{25} = 4 \cdot 5 = 20$

2. $\sqrt{36} + \sqrt{4} = 6 + 2 = 8$

3. $\sqrt{\frac{36}{100}} = \frac{6}{10} = \frac{3}{5}$

4. $14\sqrt{36} = 14 \cdot 6 = 84$

5. $\sqrt{81} - \sqrt{25} = 9 - 5 = 4$

6. $\sqrt{\frac{64}{121}} = \frac{8}{11}$

7. $5\sqrt{14} \approx 5(3.7416574)$
$= 18.708287$
$= 18.7083$ to the nearest ten thousandth

8. $\sqrt{405} + \sqrt{147} \approx 20.124612 + 12.124356$
$= 32.248968$
$= 32.25$ to the nearest hundredth

9. $\sqrt{\frac{7}{12}} \approx \sqrt{0.5833333}$
≈ 0.7637626
$= 0.764$ to the nearest thousandth

10. a. $c = \sqrt{5^2 + 5^2}$
$= \sqrt{25 + 25}$
$= \sqrt{50}$
$c = 7.07$ ft (to the nearest hundredth)

b. $c = \sqrt{16^2 + 12^2}$
$= \sqrt{256 + 144}$
$= \sqrt{400}$
$c = 20$ cm

11. $c = \sqrt{12^2 + 5^2} = \sqrt{144 + 25}$
$= \sqrt{169}$
$= 13$ ft

Section 5.8

1. $\sqrt{63} = \sqrt{3 \cdot 3 \cdot 7}$
$= \sqrt{3 \cdot 3} \cdot \sqrt{7}$
$= 3\sqrt{7}$

2. $\sqrt{45x^2} = \sqrt{3 \cdot 3 \cdot 5 \cdot x \cdot x}$
$= 3 \cdot x \cdot \sqrt{5}$
$= 3x\sqrt{5}$

3. $\sqrt{300} = \sqrt{10 \cdot 10 \cdot 3}$
$= 10\sqrt{3}$

4. $\sqrt{50x^3} = \sqrt{5 \cdot 5 \cdot 2 \cdot x \cdot x \cdot x}$
$= 5 \cdot x \cdot \sqrt{2x}$
$= 5x\sqrt{2x}$

Section 5.9

1. $5\sqrt{3} - 2\sqrt{3} = (5 - 2)\sqrt{3} = 3\sqrt{3}$

2. $7\sqrt{y} + 3\sqrt{y} = (7 + 3)\sqrt{y} = 10\sqrt{y}$

3. $8\sqrt{5} - 2\sqrt{5} + 9\sqrt{5} = (8 - 2 + 9)\sqrt{5} = 15\sqrt{5}$

4. $\sqrt{27} + \sqrt{75} - \sqrt{12} = \sqrt{3 \cdot 3 \cdot 3} + \sqrt{5 \cdot 5 \cdot 3} - \sqrt{2 \cdot 2 \cdot 3}$
$= 3\sqrt{3} + 5\sqrt{3} - 2\sqrt{3}$
$= (3 + 5 - 2)\sqrt{3}$
$= 6\sqrt{3}$

5. $5\sqrt{45} - 3\sqrt{20} = 5\sqrt{3 \cdot 3 \cdot 5} - 3\sqrt{2 \cdot 2 \cdot 5}$
$= 5 \cdot 3\sqrt{5} - 3 \cdot 2\sqrt{5}$
$= 15\sqrt{5} - 6\sqrt{5}$
$= (15 - 6)\sqrt{5}$
$= 9\sqrt{5}$

6. $7\sqrt{18x^3} - 3\sqrt{50x^3} = 7\sqrt{3 \cdot 3 \cdot 2 \cdot x \cdot x \cdot x} - 3\sqrt{5 \cdot 5 \cdot 2 \cdot x \cdot x \cdot x}$
$= 7 \cdot 3 \cdot x\sqrt{2x} - 3 \cdot 5 \cdot x\sqrt{2x}$
$= 21x\sqrt{2x} - 15x\sqrt{2x}$
$= 6x\sqrt{2x}$

Chapter 6

Section 6.1

1. a. $\frac{32}{48} = \frac{2}{3}$ b. $\frac{3.2}{4.8} = \frac{2}{3}$ c. $\frac{0.32}{0.48} = \frac{2}{3}$

2. a. $\frac{\frac{3}{5}}{\frac{9}{10}} = \frac{3}{5} \cdot \frac{10}{9} = \frac{2}{3}$ b. $\frac{0.6}{0.9} = \frac{2}{3}$

3. a. $\frac{0.06}{0.12} = \frac{0.06 \times 100}{0.12 \times 100} = \frac{6}{12} = \frac{1}{2}$ b. $\frac{600}{1{,}200} = \frac{1}{2}$

4. $\frac{12}{16} = \frac{3}{4}$

5. Alcohol to water: $\frac{4}{12} = \frac{1}{3}$; water to alcohol: $\frac{12}{4} = \frac{3}{1}$; water to total solution: $\frac{12}{16} = \frac{3}{4}$

Section 6.2

1. $\frac{107 \text{ miles}}{2 \text{ hours}} = 53.5$ miles/hour

2. $\frac{192 \text{ miles}}{6 \text{ gallons}} = 32$ miles/gallon

3. $\frac{48¢}{5.5 \text{ ounces}} = 8.7¢$/ounce; $\frac{75¢}{11.5 \text{ ounces}} = 6.5¢$/ounce; $\frac{219¢}{46 \text{ ounces}} = 4.8¢$/ounce
(Answers are rounded to the nearest tenth.)

Section 6.3

1. First term = 2, second term = 3, third term = 6, fourth term = 9, means: 3 and 6; extremes: 2 and 9

2. **a.** $5 \cdot 18 = 90$ $6 \cdot 15 = 90$ **b.** $13 \cdot 3 = 39$ $39 \cdot 1 = 39$ **c.** $\frac{2}{3} \cdot 5 = \frac{10}{3}$ $\frac{5}{3} \cdot 2 = \frac{10}{3}$ **d.** $0.12(3) = 0.36$ $0.18(2) = 0.36$

3. **a.** $3 \cdot x = 4 \cdot 9$

$$3 \cdot x = 36$$
$$\frac{\cancel{3} \cdot x}{\cancel{3}} = \frac{36}{3}$$
$$x = 12$$

b. $5 \cdot x = 8 \cdot 3$

$$5 \cdot x = 24$$
$$\frac{\cancel{5} \cdot x}{\cancel{5}} = \frac{24}{5}$$
$$x = 4.8$$

4. $2 \cdot 19 = 8 \cdot y$

$$38 = 8 \cdot y$$
$$\frac{38}{8} = \frac{\cancel{8} \cdot y}{\cancel{8}}$$
$$4.75 = y$$

5. **a.** $15 \cdot n = 6(0.3)$

$$15 \cdot n = 1.8$$
$$\frac{\cancel{15} \cdot n}{\cancel{15}} = \frac{1.8}{15}$$
$$n = 0.12$$

b. $0.35(100) = n \cdot 7$

$$35 = n \cdot 7$$
$$\frac{35}{7} = \frac{n \cdot \cancel{7}}{\cancel{7}}$$
$$5 = n$$

6. **a.** $\frac{3}{4} \cdot 8 = 7 \cdot x$

$$6 = 7 \cdot x$$
$$\frac{6}{7} = \frac{\cancel{7} \cdot x}{\cancel{7}}$$
$$\frac{6}{7} = x$$

b. $6 \cdot x = \frac{3}{5} \cdot 15$

$$6 \cdot x = 9$$
$$\frac{\cancel{6} \cdot x}{\cancel{6}} = \frac{9}{6}$$
$$x = \frac{3}{2}$$

7. $\frac{b}{18} = \frac{0.5}{1}$

$$b \cdot 1 = 18(0.5)$$
$$b = 9$$

Section 6.4

1. **a.** $\frac{x}{10} = \frac{288}{6}$

$$x \cdot 6 = 10 \cdot 288$$
$$x \cdot 6 = 2{,}880$$
$$\frac{x \cdot \cancel{6}}{\cancel{6}} = \frac{2{,}880}{6}$$
$$x = 480 \text{ miles}$$

b. $\frac{x}{11} = \frac{288}{6}$

$$x \cdot 6 = 11 \cdot 288$$
$$x \cdot 6 = 3{,}168$$
$$\frac{x \cdot \cancel{6}}{\cancel{6}} = \frac{3{,}168}{6}$$
$$x = 528 \text{ miles}$$

2. **a.** $\frac{x}{54} = \frac{10}{18}$

$$x \cdot 18 = 54 \cdot 10$$
$$x \cdot 18 = 540$$
$$\frac{x \cdot \cancel{18}}{\cancel{18}} = \frac{540}{18}$$
$$x = 30 \text{ hits}$$

b. $\frac{x}{27} = \frac{10}{18}$

$$x \cdot 18 = 27 \cdot 10$$
$$x \cdot 18 = 270$$
$$\frac{x \cdot \cancel{18}}{\cancel{18}} = \frac{270}{18}$$
$$x = 15 \text{ hits}$$

3. $\frac{x}{35} = \frac{8}{20}$

$$x \cdot 20 = 35 \cdot 8$$
$$x \cdot 20 = 280$$
$$\frac{x \cdot \cancel{20}}{\cancel{20}} = \frac{280}{20}$$
$$x = 14 \text{ milliliters of alcohol}$$

4. $\frac{x}{4.75} = \frac{105}{1}$

$$x \cdot 1 = 4.75(105)$$
$$x = 498.75 \text{ miles}$$

Section 6.5

1. $\frac{x}{14} = \frac{25}{10}$

$$10x = 350$$
$$x = 35$$

2. $\frac{h}{120} = \frac{360}{160}$

$$\frac{h}{120} = \frac{9}{4} \text{ reduce to lowest terms}$$
$$4h = 1080$$
$$h = 270 \text{ pixels}$$

3. We see AC has a length of 3 units and BC has a length of 4 units. Since AC is proportional to GI, which has a length of 9 units, we set up a proportion to find the length of a new side, HI.

$$\frac{HI}{BC} = \frac{GI}{AC}$$
$$\frac{HI}{4} = \frac{9}{3}$$
$$\frac{HI}{4} = 3$$
$$HI = 12$$

Now we can draw HI with length 12 units, and complete the triangle by drawing line GH

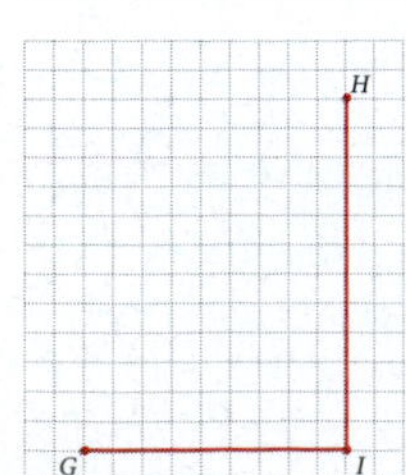

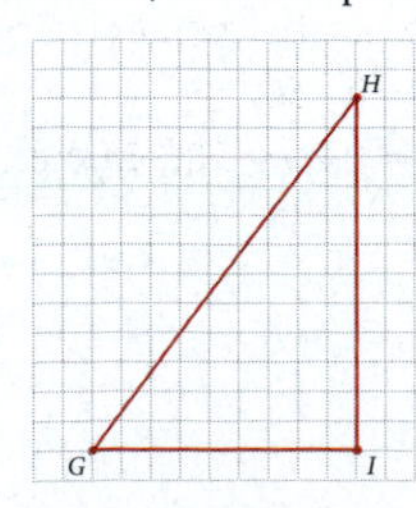

4. $\frac{x}{18} = \frac{42}{12}$

$$12x = 756$$
$$x = 63 \text{ ft}$$

5. $\frac{b}{15} = \frac{32}{24}$

$$24b = 480$$
$$b = 20 \text{ in.}$$

Chapter 7

Section 7.1

21. $\frac{82}{100} = \frac{41}{50}$ **22.** $\frac{6.5}{100} = \frac{65}{1000} = \frac{13}{200}$ **23.** $\frac{42\frac{1}{2}}{100} = \frac{\frac{85}{2}}{100} = \frac{85}{2} \cdot \frac{1}{100} = \frac{85}{200} = \frac{17}{40}$ **24. a.** $\frac{9}{10} = 0.9 = 90\%$ **b.** $\frac{9}{20} = 0.45 = 45\%$

25. a. $\frac{5}{8} = 0.625 = 62.5\%$ **b.** $\frac{9}{8} = 1.125 = 112.5\%$ **26. a.** $\frac{7}{12} = 0.58\overline{3} = 58\frac{1}{3}\%$ or $\approx 58.3\%$ **b.** $\frac{13}{12} = 1.08\overline{3} = 108\frac{1}{3}\%$ or $\approx 108.3\%$

27. a. $3\frac{3}{4} = 3.75 = 375\%$ **b.** $3\frac{7}{8} = 3.875 = 387.5\%$

Section 7.2

1. a. $n = 0.25(74) = 18.5$ **b.** $n = 0.50(74) = 37$ **2. a.** $n \cdot 84 = 21$; $\frac{n \cdot \cancel{84}}{\cancel{84}} = \frac{21}{84}$; $n = 0.25$; $n = 25\%$ **b.** $n \cdot 84 = 42$; $\frac{n \cdot \cancel{84}}{\cancel{84}} = \frac{42}{84}$; $n = 0.50$; $n = 50\%$ **3. a.** $35 = 0.40 \cdot n$; $\frac{35}{0.40} = \frac{\cancel{0.40} \cdot n}{\cancel{0.40}}$; $87.5 = n$ **b.** $70 = 0.40 \cdot n$; $\frac{70}{0.40} = \frac{\cancel{0.40} \cdot n}{\cancel{0.40}}$; $175 = n$

4. $n = 0.635(45)$; $n \approx 28.6$ **5.** $n \cdot 85 = 11.9$; $\frac{n \cdot \cancel{85}}{\cancel{85}} = \frac{11.9}{85}$; $n = 0.14$; $n = 14\%$ **6.** $62 = 0.39 \cdot n$; $\frac{62}{0.39} = \frac{\cancel{0.39} \cdot n}{\cancel{0.39}}$; $159.0 \approx n$ **7.** 25 is what percent of 160? $25 = n \cdot 160$; $\frac{25}{160} = n$; $n = 0.156 = 15.6\%$ to the nearest tenth of a percent

8. a. $\frac{25}{100} = \frac{n}{74}$; $25 \cdot 74 = 100 \cdot n$; $1850 = 100 \cdot n$; $18.5 = n$ **b.** $\frac{50}{100} = \frac{n}{74}$; $50 \cdot 74 = 100 \cdot n$; $3{,}700 = 100 \cdot n$; $37 = n$ **9. a.** $\frac{n}{100} = \frac{21}{84}$; $84 \cdot n = 21 \cdot 100$; $84 \cdot n = 2100$; $n = 25$; 25% **b.** $\frac{n}{100} = \frac{42}{84}$; $84 \cdot n = 42 \cdot 100$; $84 \cdot n = 4{,}200$; $n = 50$; 50% **10. a.** $\frac{40}{100} = \frac{35}{n}$; $40 \cdot n = 35 \cdot 100$; $40 \cdot n = 3500$; $n = 87.5$ **b.** $\frac{40}{100} = \frac{70}{n}$; $40 \cdot n = 70 \cdot 100$; $40 \cdot n = 7{,}000$; $n = 175$

Section 7.3

1. 114 is what percent of 150? $114 = n \cdot 150$; $\frac{114}{150} = \frac{n \cdot \cancel{150}}{\cancel{150}}$; $n = 0.76$; $n = 76\%$ **2.** What is 75% of 40? $n = 0.75(40)$; $n = 30$ milliliters HCl **3.** 3,360 is 42% of what number? $3{,}360 = 0.42 \cdot n$; $\frac{3{,}360}{0.42} = \frac{\cancel{0.42} \cdot n}{\cancel{0.42}}$; $n = 8{,}000$ students **4.** What is 35% of 300? $n = 0.35(300) = 105$ students

Section 7.4

1. What is 6% of \$625? $n = 0.06(625)$; $n = \$37.50$ **2.** \$4.35 is 3% of what number? $4.35 = 0.03 \cdot n$; $n = \$145$ Purchase price; Total price = \$145 + \$4.35 = \$149.35 **3.** \$11.82 is what percent of \$197? $11.82 = n \cdot 197$; $n = 0.06 = 6\%$ **4.** What is 3% of \$115,000? $n = 0.03(115{,}000)$; $n = \$3{,}450$

5. 10% of what number is \$115? $0.10 \cdot n = 115$; $n = \$1{,}150$ **6.** \$105 is what percent of \$750? $105 = n \cdot 750$; $n = 0.14 = 14\%$

Section 7.5

1. $0.07(18{,}000) = 1{,}260$
\$18,000 Old salary
+ 1,260 Raise
\$19,260 New salary

2. $0.89(271{,}000) = 241{,}190$
271,000 Drunk drivers in 1986
+241,190 Increase
512,190 = 512,000 to the nearest thousand

3. \$35 − \$28 = \$7 Decrease
\$7 is what percent of \$35?
$7 = n \cdot 35$
$n = 0.20 = 20\%$ Decrease

4. What is 15% of \$550?
$n = 0.15(550)$
$n =$ \$82.50 Discount

\$550.00	Original price
− 82.50	Less discount
\$467.50	Sale price

5. What is 15% of \$45?
$n = 0.15(45)$
$n =$ \$6.75 Discount

\$45.00	Original price
− 6.75	Less discount
\$38.25	Sale price

What is 5% of \$38.25?
$n = 0.05(38.25)$
$n =$ \$1.91 to the nearest cent

\$38.25	Sale price
+ 1.91	Sales tax
\$40.16	Total price

Section 7.6

1. Interest = 0.08(\$3,000)
= \$240

\$3,000	Principal
+ 240	Interest
\$3,240	Amount after 1 year

2. Interest = 0.12(\$7,500)
= \$900

\$7,500	Principal
+ 900	Interest
\$8,400	Total amount to pay back

3. $I = P \cdot R \cdot T$

$$I = 700 \times 0.04 \times \frac{90}{360}$$

$$I = 700 \times 0.04 \times \frac{1}{4}$$

$I =$ \$7 Interest

4. $I = P \cdot R \cdot T$

$$I = 1{,}200 \times 0.095 \times \frac{120}{360}$$

$$I = 1{,}200 \times 0.095 \times \frac{1}{3}$$

$I =$ \$38 Interest

\$1,200	Principal
+ 38	Interest
\$1,238	Total amount withdrawn

5. Interest after 1 year is
0.06(\$5,000) = \$300
Total in account after 1 year is

\$5,000	Principal
+ 300	Interest
\$5,300	

Interest paid the second year is
0.06 (\$5,300) = \$318
Total in account after 2 years is

\$5,300	Principal
+ 318	Interest
\$5,618	

6. Interest at the end of first quarter

$$I = \$20{,}000 \times 0.08 \times \frac{1}{4} = \$400$$

Total in account at end of first quarter
\$20,000 + \$400 = \$20,400
Interest for the second quarter

$$I = \$20{,}400 \times 0.08 \times \frac{1}{4} = \$408$$

Total in account at end of second quarter
\$20,400 + \$408 = \$20,808

Interest for the third quarter

$$I = \$20{,}808 \times 0.08 \times \frac{1}{4} = \$416.16$$

Total in account at the end of third quarter
\$20,808 + \$416.16 = \$21,224.16
Interest for the fourth quarter

$$I = \$21{,}224.16 \times 0.08 \times \frac{1}{4} = \$424.48 \text{ to the nearest cent}$$

Total in account at end of 1 year

\$21,224.16
+ 424.48
\$21,648.64

Section 7.7

1. a. $9 + 11 + 16 + 10 = 46$

2. a. 0.76(600) = 456 people
b. 0.24(600) = 144 people

3.

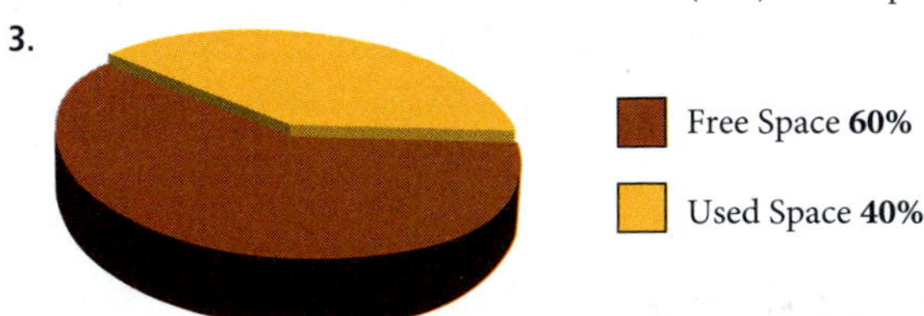

4.

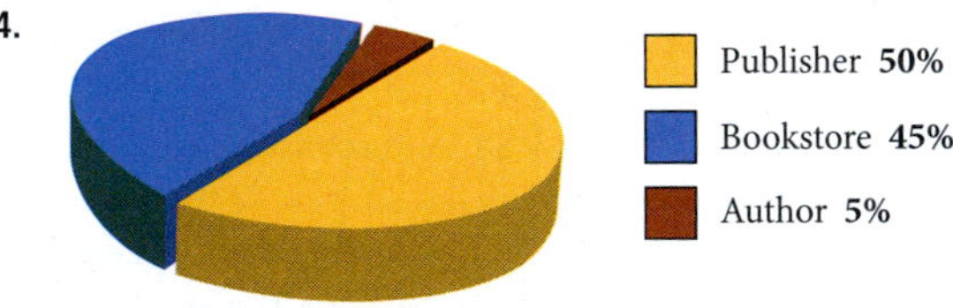

Chapter 8

Section 8.1

1. $8 \text{ ft} = 8 \times 12 \text{ in.}$
$$= 96 \text{ in.}$$

2. $26 \text{ ft} = 26 \cancel{\text{ft}} \times \frac{1 \text{ yd}}{3 \cancel{\text{ft}}}$
$$= \frac{26}{3} \text{ yd}$$
$$= 8\frac{2}{3} \text{ yd, or } 8.67 \text{ yd}$$

3. $220 \text{ yd} = 220 \cancel{\text{yd}} \times \frac{3 \cancel{\text{ft}}}{1 \cancel{\text{yd}}} \times \frac{12 \text{ in.}}{1 \cancel{\text{ft}}}$
$$= 220 \times 3 \times 12 \text{ in.}$$
$$= 7{,}920 \text{ in.}$$

4. $67 \text{ cm} = 67 \cancel{\text{cm}} \times \frac{1 \text{ m}}{100 \cancel{\text{cm}}}$
$$= \frac{67 \text{ m}}{100}$$
$$= 0.67 \text{ m}$$

5. $78.4 \text{ mm} = 78.4 \cancel{\text{mm}} \times \frac{1 \cancel{\text{m}}}{1{,}000 \cancel{\text{mm}}} \times \frac{10 \text{ dm}}{1 \cancel{\text{m}}}$
$$= \frac{78.4 \times 10}{1{,}000} \text{ dm}$$
$$= 0.784 \text{ dm}$$

6. $6 \text{ pens} = 6 \cancel{\text{pens}} \times \frac{28 \text{ feet of fencing}}{1 \cancel{\text{pen}}} \times \frac{1.72 \text{ dollars}}{1 \text{ foot of fencing}}$
$$= 6 \times 28 \times 1.72 \text{ dollars}$$
$$= \$288.96$$

7. $1{,}100 \text{ feet per minute} = \frac{1{,}100 \cancel{\text{feet}}}{1 \text{ minute}} \cdot \frac{1 \text{ mile}}{5{,}280 \cancel{\text{feet}}} \cdot \frac{60 \text{ minutes}}{1 \text{ hour}}$
$$= \frac{1{,}100 \cdot 60 \text{ miles}}{5{,}280 \text{ hours}}$$
$= 12.5$ miles per hour, which is a reasonable speed for a chair lift.

Section 8.2

1. $1 \text{ yd}^2 = 1 \cancel{\text{yd}} \times \cancel{\text{yd}} \times \frac{3 \text{ ft}}{1 \cancel{\text{yd}}} \times \frac{3 \text{ ft}}{1 \cancel{\text{yd}}} = 3 \times 3 \text{ ft} \times \text{ ft} = 9 \text{ ft}^2$

2. Length = 36 in. + 12 in. = 48 in.; Width = 24 in. + 12 in. = 36 in.;
Area = 48 in. × 36 in. = $1{,}728 \text{ in}^2$
Area in square feet $= 1{,}728 \cancel{\text{in}^2} \times \frac{1 \text{ ft}^2}{144 \cancel{\text{in}^2}} = \frac{1{,}728}{144} \text{ ft}^2 = 12 \text{ ft}^2$

3. $A = 1.5 \times 45$
$$= 67.5 \text{ yd}^2$$
$$67.5 \text{ yd}^2 = 67.5 \cancel{\text{yd}^2} \times \frac{9 \text{ ft}^2}{1 \cancel{\text{yd}^2}}$$
$$= 607.5 \text{ ft}^2$$

4. $55 \text{ acres} = 55 \cancel{\text{acres}} \times \frac{43{,}560 \text{ ft}^2}{1 \cancel{\text{acre}}}$
$$= 55 \times 43{,}560 \text{ ft}^2$$
$$= 2{,}395{,}800 \text{ ft}^2$$

5. $960 \text{ acres} = 960 \cancel{\text{acres}} \times \frac{1 \text{ mi}^2}{640 \cancel{\text{acres}}}$
$$= \frac{960}{640} \text{ mi}^2$$
$$= 1.5 \text{ mi}^2$$

6. $1 \text{ m}^2 = 1 \cancel{\text{m}^2} \times \frac{100 \cancel{\text{dm}^2}}{1 \cancel{\text{m}^2}} \times \frac{100 \text{ cm}^2}{1 \cancel{\text{dm}^2}}$
$$= 10{,}000 \text{ cm}^2$$

7. $5 \text{ gal} = 5 \cancel{\text{gal}} \times \frac{4 \cancel{\text{qt}}}{1 \cancel{\text{gal}}} \times \frac{2 \text{ pt}}{1 \cancel{\text{qt}}}$
$$= 5 \times 4 \times 2 \text{ qt}$$
$$= 40 \text{ pt}$$

8. $2{,}000 \text{ qt} = 2{,}000 \cancel{\text{qt}} \times \frac{1 \text{ gal}}{4 \cancel{\text{qt}}}$
$$= \frac{2{,}000}{4} \text{ gal}$$
$$= 500 \text{ gal}$$
The number of 10-gal containers in 500 gal is $\frac{500}{10} = 50$ containers.

9. $3.5 \text{ liters} = 3.5 \cancel{\text{liters}} \times \frac{1{,}000 \text{ mL}}{1 \cancel{\text{liter}}}$
$$= 3.5 \times 1{,}000 \text{ mL}$$
$$= 3{,}500 \text{ mL}$$

Section 8.3

1. $15 \text{ lb} = 15 \cancel{\text{lb}} \times \frac{16 \text{ oz}}{1 \cancel{\text{lb}}}$
$$= 15 \times 16 \text{ oz}$$
$$= 240 \text{ oz}$$

2. $5 \text{ T} = 5 \text{ T} \times \frac{2{,}000 \text{ lb}}{1 \text{ T}}$
$$= 10{,}000 \text{ lb}$$
10,000 lb is the equivalent of 5 tons

3. $5 \text{ kg} = 5 \cancel{\text{kg}} \times \frac{1{,}000 \cancel{\text{g}}}{1 \cancel{\text{kg}}} \times \frac{1{,}000 \text{ mg}}{1 \cancel{\text{g}}}$
$$= 5 \times 1{,}000 \times 1{,}000 \text{ mg}$$
$$= 5{,}000{,}000 \text{ mg}$$

4. Total number of milligrams in bottle = 75 × 200 = 15,000 mg.
$$15{,}000 \text{ mg} = 15{,}000 \cancel{\text{mg}} \times \frac{1 \text{ g}}{1{,}000 \cancel{\text{mg}}}$$
$$= \frac{15{,}000}{1{,}000} \text{ g}$$
$$= 15 \text{ g}$$

Section 8.4

1. $10 \text{ in.} = 10 \cancel{\text{in.}} \times \frac{2.54 \text{ cm}}{1 \cancel{\text{in.}}}$
$$= 10 \times 2.54 \text{ cm}$$
$$= 25.4 \text{ cm}$$

2. $16.4 \text{ ft} = 16.4 \cancel{\text{ft}} \times \frac{1 \text{ m}}{3.28 \cancel{\text{ft}}}$
$$= \frac{16.4}{3.28} \text{ m}$$
$$= 5 \text{ m}$$

3. $10 \text{ liters} = 10 \cancel{\text{liters}} \times \frac{1 \text{ gal}}{3.79 \cancel{\text{liters}}}$
$$= \frac{10}{3.79} \text{ gal}$$
$= 2.64$ gal (rounded to the nearest hundredth)

4. $2.2 \text{ liters} = 2.2 \text{ liters} \times \frac{1{,}000 \text{ mL}}{1 \text{ liter}} \times \frac{1 \text{ in}^3}{16.39 \text{ mL}}$

$= \frac{2.2 \times 1{,}000}{16.39} \text{ in}^3$

$= 134 \text{ in}^3$ (rounded to the nearest cubic inch)

5. $165 \text{ lb} = 165 \text{ lb} \times \frac{1 \text{ kg}}{2.20 \text{ lb}}$

$= \frac{165}{2.20} \text{ kg}$

$= 75 \text{ kg}$

6. $F = \frac{9}{5}(40) + 32$

$= 72 + 32$

$= 104°F$

7. $C = \frac{5(101.6 - 32)}{9}$

$= 38.7°C$ (rounded to the nearest tenth)

Section 8.5

1. a. $2 \text{ hr } 45 \text{ min} = 2 \text{ hr} \times \frac{60 \text{ min}}{1 \text{ hr}} + 45 \text{ min}$

$= 120 \text{ min} + 45 \text{ min}$

$= 165 \text{ min}$

b. $2 \text{ hr } 45 \text{ min} = 2 \text{ hr} + 45 \text{ min} \times \frac{1 \text{ hr}}{60 \text{ min}}$

$= 2 \text{ hr} + 0.75 \text{ hr}$

$= 2.75 \text{ hr}$

2.

	min	sec
	4 min	27 sec
+	8 min	45 sec
	12 min	73 sec

Since there are 60 seconds in every minute, we write 73 seconds as 1 minute 13 seconds. We have

12 min 73 sec = 12 min + 1 min 13 sec

= 13 min 13 sec

3.

	hr	min		hr	min
	6 hr	25 min	=	5 hr	85 min
−		42 min			− 42 min
				5 hr	43 min

4. First, we multiply each unit by 4:

	lb	oz
	3 lb	8 oz
×		4
	12 lb	32 oz

To convert the 32 ounces to pounds, we multiply the ounces by the conversion factor

$12 \text{ lb } 32 \text{ oz} = 12 \text{ lb} + 32 \text{ oz} \times \frac{1 \text{ lb}}{16 \text{ oz}}$

$= 12 \text{ lb} + 2 \text{ lbs}$

$= 14 \text{ lb}$

Finally, we multiply the 14 lb and $5.00 for a total price of $70.00.

Chapter 9

Section 9.1

1. $x^3 \cdot x^5 = (x \cdot x \cdot x)(x \cdot x \cdot x \cdot x \cdot x)$

$= x \cdot x \cdot x \cdot x \cdot x \cdot x \cdot x \cdot x$

$= x^8$

2. $4x^2 \cdot 6x^5 = (4 \cdot 6)(x^2 \cdot x^5)$

$= (4 \cdot 6)(x^{2+5})$

$= 24x^7$

3. $2x^5 \cdot 3x^4 \cdot 4x^3 = (2 \cdot 3 \cdot 4)(x^5 \cdot x^4 \cdot x^3)$

$= (2 \cdot 3 \cdot 4)(x^{5+4+3})$

$= 24x^{12}$

4. $(10xy^2)(9x^3y^5) = (10 \cdot 9)(x \cdot x^3)(y^2 \cdot y^5)$

$= (10 \cdot 9)(x^{1+3})(y^{2+5})$

$= 90x^4y^7$

5. $(2^2)^3 = 2^{2 \cdot 3}$

$= 2^6$

$= 64$

6. $(x^3)^4 \cdot (x^5)^2 = x^{3 \cdot 4} \cdot x^{5 \cdot 2}$

$= x^{12} \cdot x^{10}$

$= x^{12+10}$

$= x^{22}$

7. $(5xy)^3 = 5^3 \cdot x^3 \cdot y^3$

$= 125x^3y^3$

8. $(4x^5y^2)^3 = 4^3(x^5)^3(y^2)^3$

$= 64x^{15}y^6$

9. $(2x^3y^4)^3(3xy^5)^2 = 2^3(x^3)^3(y^4)^3 \cdot 3^2x^2(y^5)^2$

$= 8x^9y^{12} \cdot 9x^2y^{10}$

$= (8 \cdot 9)(x^9x^2)(y^{12}y^{10})$

$= 72x^{11}y^{22}$

Section 9.2

1. $(5x^2 - 3x + 2) + (2x^2 + 10x - 9) = (5x^2 + 2x^2) + (-3x + 10x) + (2 - 9)$
$= (5 + 2)x^2 + (-3 + 10)x + (2 - 9)$
$= 7x^2 + 7x - 7$

$$\begin{array}{r} 5x^2 - 3x + 2 \\ \underline{2x^2 + 10x - 9} \\ 7x^2 + 7x - 7 \end{array}$$

2. $(3y^2 + 9y - 5) + (6y^2 - 4) = (3y^2 + 6y^2) + 9y + (-5 - 4)$
$= (3 + 6)y^2 + 9y + (-5 - 4)$
$= 9y^2 + 9y - 9$

$$\begin{array}{r} 3y^2 + 9y - 5 \\ \underline{6y^2 \qquad - 4} \\ 9y^2 + 9y - 9 \end{array}$$

3. $(8x^3 + 4x^2 + 3x + 2) + (4x^2 + 5x + 6) = 8x^3 + (4x^2 + 4x^2) + (3x + 5x) + (2 + 6)$
$= 8x^3 + (4 + 4)x^2 + (3 + 5)x + (2 + 6)$
$= 8x^3 + 8x^2 + 8x + 8$

$$\begin{array}{r} 8x^3 + 4x^2 + 3x + 2 \\ \underline{4x^2 + 5x + 6} \\ 8x^3 + 8x^2 + 8x + 8 \end{array}$$

4. $(5x^2 - 2x + 7) - (4x^2 + 8x - 4) = 5x^2 - 2x + 7 - 4x^2 - 8x + 4$
$= (5x^2 - 4x^2) + (-2x - 8x) + (7 + 4)$
$= (5 - 4)x^2 + (-2 - 8)x + (7 + 4)$
$= 1x^2 - 10x + 11$
$= x^2 - 10x + 11$

5. $(3y^3 - 2y^2 + 7y - 6) - (8y^3 - 6y^2 + 4y - 8) = 3y^3 - 2y^2 + 7y - 6 - 8y^3 + 6y^2 - 4y + 8$
$= (3y^3 - 8y^3) + (-2y^2 + 6y^2) + (7y - 4y) + (-6 + 8)$
$= (3 - 8)y^3 + (-2 + 6)y^2 + (7 - 4)y + (-6 + 8)$
$= -5y^3 + 4y^2 + 3y + 2$

6. $(-2x^2 + 5x - 1) - (6x^2 - 2x + 5) = -2x^2 + 5x - 1 - 6x^2 + 2x - 5$
$= (-2x^2 - 6x^2) + (5x + 2x) + (-1 - 5)$
$= -8x^2 + 7x - 6$

7. When $x = -3$
the polynomial $5x^2 - 3x + 8$
becomes $5(-3)^2 - 3(-3) + 8 = 5(9) + 9 + 8$
$= 45 + 9 + 8$
$= 62$

8. a. When $n = 1$, $3n = 3 \cdot 1 = 3$
When $n = 2$, $3n = 3 \cdot 2 = 6$
When $n = 3$, $3n = 3 \cdot 3 = 9$
When $n = 4$, $3n = 3 \cdot 4 = 12$

b. When $n = 1$, $n^3 = 1^3 = 1$
When $n = 2$, $n^3 = 2^3 = 8$
When $n = 3$, $n^3 = 3^3 = 27$
When $n = 4$, $n^3 = 4^3 = 64$

Section 9.3

1. $x^3(x^5 + x^7) = x^3 \cdot x^5 + x^3 \cdot x^7$
$= x^8 + x^{10}$

2. $(x^5 + x^7)x^3 = x^5 \cdot x^3 + x^7 \cdot x^3$
$= x^8 + x^{10}$

3. $5x^2(6x^3 - 4) = 5x^2 \cdot 6x^3 - 5x^2 \cdot 4$
$= (5 \cdot 6)(x^2 \cdot x^3) - (5 \cdot 4)x^2$
$= 30x^5 - 20x^2$

4. $5a^3b^5(2a^2 + 7b^2) = 5a^3b^5 \cdot 2a^2 + 5a^3b^5 \cdot 7b^2$
$= (5 \cdot 2)(a^3 \cdot a^2)b^5 + (5 \cdot 7)(a^3)(b^5 \cdot b^2)$
$= 10a^5b^5 + 35a^3b^7$

5. $(x + 2)(x + 6) = (x + 2)x + (x + 2)6$
$= x \cdot x + 2 \cdot x + x \cdot 6 + 2 \cdot 6$
$= x^2 + 2x + 6x + 12$
$= x^2 + 8x + 12$

6. $(x - 2)(x + 6) = (x - 2)x + (x - 2)6$
$= x \cdot x - 2 \cdot x + x \cdot 6 - 2 \cdot 6$
$= x^2 - 2x + 6x - 12$
$= x^2 + 4x - 12$

7. $(3x - 2)(5x + 4) = (3x - 2) \cdot 5x + (3x - 2) \cdot 4$
$= 3x \cdot 5x - 2 \cdot 5x + 3x \cdot 4 - 2 \cdot 4$
$= 15x^2 - 10x + 12x - 8$
$= 15x^2 + 2x - 8$

8. $(x + 3)^2 = (x + 3)(x + 3)$
$= (x + 3) \cdot x + (x + 3) \cdot 3$
$= x \cdot x + 3 \cdot x + x \cdot 3 + 3 \cdot 3$
$= x^2 + 3x + 3x + 9$
$= x^2 + 6x + 9$

9. $(3x - 5)^2 = (3x - 5)(3x - 5)$
$= (3x - 5)[3x + (-5)]$
$= (3x - 5) \cdot 3x + (3x - 5)(-5)$
$= 3x \cdot 3x - 5 \cdot 3x + 3x(-5) - 5(-5)$
$= 9x^2 - 15x - 15x + 25$
$= 9x^2 - 30x + 25$

10.

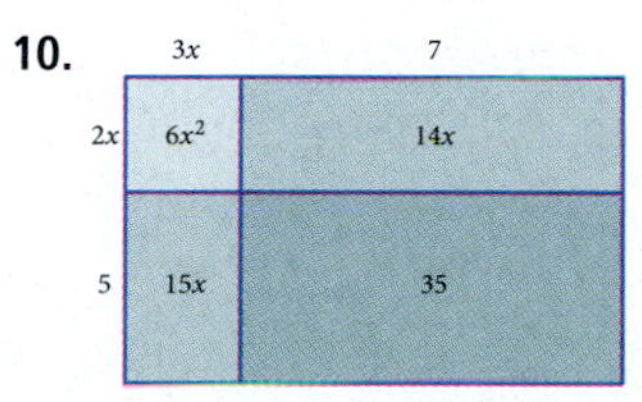

$6x^2 + 14x + 15x + 35 = 6x^2 + 29x + 35$

Section 9.4

1. $2^{-4} = \dfrac{1}{2^4}$
$= \dfrac{1}{16}$

2. $4^{-2} = \dfrac{1}{4^2}$
$= \dfrac{1}{16}$

3. $(-2)^{-3} = \dfrac{1}{(-2)^3}$
$= \dfrac{1}{-8}$
$= -\dfrac{1}{8}$

4. $3^4 \cdot 3^{-7} = 3^{4+(-7)}$
$= 3^{-3}$
$= \dfrac{1}{3^3}$
$= \dfrac{1}{27}$

5. $x^6 \cdot x^{-10} = x^{6+(-10)}$
$= x^{-4}$
$= \dfrac{1}{x^4}$

6. $\frac{2^6}{2^8} = 2^{6-8}$
$= 2^{-2}$
$= \frac{1}{2^2}$
$= \frac{1}{4}$

7. $\frac{10^{-5}}{10^3} = 10^{-5-3}$
$= 10^{-8}$
$\frac{1}{10^8} = \frac{1}{100{,}000{,}000}$

8. $\frac{10^{-6}}{10^{-8}} = 10^{-6-(-8)}$
$= 10^{-6+8}$
$= 10^2 = 100$

9. $4x^{-3} \cdot 7x = 4x^{-3} \cdot 7x^1$
$= (4 \cdot 7)(x^{-3} \cdot x^1)$
$= 28x^{-3+1}$
$= 28x^{-2} = \frac{28}{x^2}$

10. $\frac{x^{-5} \cdot x^2}{x^{-8}} = \frac{x^{-5+2}}{x^{-8}}$
$= \frac{x^{-3}}{x^{-8}}$
$= x^{-3-(-8)}$
$= x^{-3+8}$
$= x^5$

Section 9.6

1. $(2.5 \times 10^6)(1.4 \times 10^2) = (2.5)(1.4) \times (10^6)(10^2)$
$= 3.5 \times 10^{6+2}$
$= 3.5 \times 10^8$

2. $(2{,}200{,}000)(0.00015) = (2.2 \times 10^6)(1.5 \times 10^{-4})$
$= (2.2)(1.5) \times (10^6)(10^{-4})$
$= 3.3 \times 10^{6+(-4)}$
$= 3.3 \times 10^2$

3. $\frac{6 \times 10^5}{2 \times 10^{-4}} = \frac{6}{2} \times \frac{10^5}{10^{-4}}$
$= 3.0 \times 10^{5-(-4)}$
$= 3.0 \times 10^{5+4}$
$= 3.0 \times 10^9$

4. $\frac{0.0038}{19{,}000{,}000} = \frac{3.8 \times 10^{-3}}{1.9 \times 10^7}$
$= \frac{3.8}{1.9} \times \frac{10^{-3}}{10^7}$
$= 2.0 \times 10^{-3-7}$
$= 2.0 \times 10^{-10}$

5. $\frac{(6.8 \times 10^{-4})(3.9 \times 10^2)}{7.8 \times 10^{-6}} = \frac{(6.8)(3.9)}{(7.8)} \times \frac{(10^{-4})(10^2)}{10^{-6}}$
$= 3.4 \times 10^{-4+2-(-6)}$
$= 3.4 \times 10^4$

6. $\frac{(0.000035)(45{,}000)}{0.000075} = \frac{(3.5 \times 10^{-5})(4.5 \times 10^4)}{7.5 \times 10^{-5}}$
$= \frac{(3.5)(4.5)}{(7.5)} \times \frac{(10^{-5})(10^4)}{10^{-5}}$
$= 2.1 \times 10^{-5+4-(-5)}$
$= 2.1 \times 10^4$

Answers to Odd-Numbered Problems

Chapter 1

Problem Set 1.1

1. 8 ones, 7 tens **3.** 5 ones, 4 tens **5.** 8 ones, 4 tens, 3 hundreds **7.** 8 ones, 0 tens, 6 hundreds **9.** 8 ones, 7 tens, 3 hundreds, 2 thousands **11.** 9 ones, 6 tens, 5 hundreds, 3 thousands, 7 ten thousands, 2 hundred thousands **13.** Ten thousands **15.** Hundred millions **17.** Ones **19.** Hundred thousands **21.** 600 + 50 + 8 **23.** 60 + 8 **25.** 4,000 + 500 + 80 + 7 **27.** 30,000 + 2,000 + 600 + 70 + 4 **29.** 3,000,000 + 400,000 + 60,000 + 2,000 + 500 + 70 + 7 **31.** 400 + 7 **33.** 30,000 + 60 + 8 **35.** 3,000,000 + 4,000 + 8 **37.** Twenty-nine **39.** Forty **41.** Five hundred seventy-three **43.** Seven hundred seven **45.** Seven hundred seventy **47.** Twenty-three thousand, five hundred forty **49.** Three thousand, four **51.** Three thousand, forty **53.** One hundred four million, sixty-five thousand, seven hundred eighty **55.** Five billion, three million, forty thousand, eight **57.** Two million, five hundred forty-six thousand, seven hundred thirty-one **59.** 325 **61.** 5,432 **63.** 86,762 **65.** 2,000,200 **67.** 2,002,200 **69. a.** Twenty-eight thousand, six hundred thirty-one **b.** Ninety-three thousand, three hundred thirty-three **71.** Hundred thousands **73.** Three million, one hundred seventy-three thousand, four hundred three **75.** Twenty-one thousand, four hundred eighty **77.** Seven hundred fifty dollars and no cents **79.** 304,000,000 **81.** One hundred twenty-seven million **83.** 36,000,000 **85.** Ten million, nine hundred thousand

Problem Set 1.2

1. 15 **3.** 14 **5.** 24 **7.** 15 **9.** 20 **11.** 68 **13.** 98 **15.** 7,297 **17.** 6,487 **19.** 96 **21.** 7,449 **23.** 65 **25.** 102 **27.** 875 **29.** 829 **31.** 10,391 **33.** 16,204 **35.** 155,554 **37.** 111,110 **39.** 17,391 **41.** 14,892 **43.** 180 **45.** 2,220 **47.** 18,285 **49.**

First Number a	Second Number b	Their Sum $a + b$
61	38	99
63	36	99
65	34	99
67	32	99

51.

First Number a	Second Number b	Their Sum $a + b$
9	16	25
36	64	100
81	144	225
144	256	400

53. $9 + 5$ **55.** $8 + 3$ **57.** $4 + 6$ **59.** $1 + (2 + 3)$ **61.** $2 + (1 + 6)$ **63.** $(1 + 9) + 1$ **65.** $4 + (n + 1)$ **67.** $n = 4$ **69.** $n = 5$ **71.** $n = 8$ **73.** $n = 8$ **75.** The sum of 4 and 9 **77.** The sum of 8 and 1 **79.** The sum of 2 and 3 is 5. **81. a.** $5 + 2$ **b.** $8 + 3$ **83. a.** $m + 1$ **b.** $m + n$ **85.** 12 in. **87.** 16 ft **89.** 26 yd **91.** 18 in. **93. a.** 150 **b.** 1,125 **95.** $349 **97. a.** $62,377.00 **b.** $55,177.00 **c.** $7,200.00

Problem Set 1.3

1. 40 **3.** 50 **5.** 50 **7.** 80 **9.** 460 **11.** 470 **13.** 56,780 **15.** 4,500 **17.** 500 **19.** 800 **21.** 900 **23.** 1,100 **25.** 5,000 **27.** 39,600 **29.** 5,000 **31.** 10,000 **33.** 1,000 **35.** 658,000 **37.** 510,000 **39.** 3,789,000

Original Number	Rounded to the Nearest Ten	Hundred	Thousand
41. 7,821	7,820	7,800	8,000
43. 5,999	6,000	6,000	6,000
45. 10,985	10,990	11,000	11,000
47. 99,999	100,000	100,000	100,000

49. 1,200 **51.** 1,900 **53.** 58,000 **55.** 33,400 **57.** 190,000 **59.** 81,400 **61. a.** 4,265,997 babies **b.** No **c.** 2,300,000 babies **d.** 112,000 babies **63.** 160 miles per hour **65.** Answers will vary, but 70 miles per hour is a good estimate.

67.

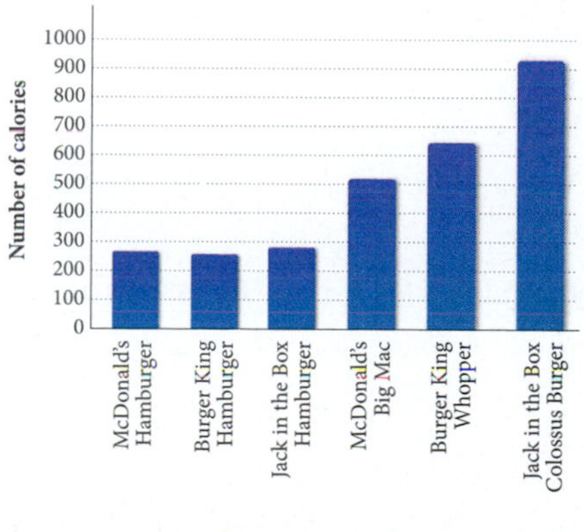

Problem Set 1.4

1. 32 **3.** 22 **5.** 10 **7.** 111 **9.** 312 **11.** 403 **13.** 1,111 **15.** 4,544 **17.** 15 **19.** 33 **21.** 5 **23.** 33 **25.** 95 **27.** 152 **29.** 274 **31.** 488 **33.** 538 **35.** 163 **37.** 1,610 **39.** 46,083

41.

First Number a	Second Number b	The Difference of a and b $a - b$
25	15	10
24	16	8
23	17	6
22	18	4

43.

First Number a	Second Number b	The Difference of a and b $a - b$
400	256	144
400	144	256
225	144	81
225	81	144

45. The difference of 10 and 2 **47.** The difference of a and 6 **49.** The difference of 8 and 2 is 6. **51.** 3 **53.** 8 **55.** 23 **57.** $8 - 3$ **59.** $y - 9$ **61.** $3 - 2 = 1$ **63.** $37 - 9x = 10$ **65.** $2y - 15x = 24$ **67.** $(x + 2) - (x + 1) = 1$ **69.** \$255 **71.** \$172,500 **73.** 91 mph **75.** 173 GB

77. a.

State	Energy (MegaWatts)
Texas	2,768
California	2,361
Iowa	936
Washington	818

b. 407 MW

Problem Set 1.5

1. 300 **3.** 600 **5.** 3,000 **7.** 5,000 **9.** 21,000 **11.** 81,000 **13.** 100 **15.** 228 **17.** 36 **19.** 1,440 **21.** 950 **23.** 1,725 **25.** 121 **27.** 1,552 **29.** 4,200 **31.** 66,248 **33.** 279,200 **35.** 12,321 **37.** 106,400 **39.** 198,592 **41.** 612,928 **43.** 333,180 **45.** 18,053,805 **47.** 263,646,976

49.

First Number a	Second Number b	Their Product ab
11	11	121
11	22	242
22	22	484
22	44	968

51.

First Number a	Second Number b	Their Product ab
25	10	250
25	100	2,500
25	1,000	25,000
25	10,000	250,000

53.

First Number a	Second Number b	Their Product ab
12	20	240
36	20	720
12	40	480
36	40	1,440

55. The product of 6 and 7 **57.** The product of 2 and n

59. The product of 9 and 7 is 63. **61.** $7 \cdot n$ **63.** $6 \cdot 7 = 42$ **65.** $0 \cdot 6 = 0$ **67.** Products: $9 \cdot 7$ and 63 **69.** Products: 4(4) and 16 **71.** Factors: 2, 3, and 4 **73.** Factors: 2, 2, and 3 **75.** 9(5) **77.** $7 \cdot 6$ **79.** $(2 \cdot 7) \cdot 6$ **81.** $(3 \times 9) \times 1$ **83.** $7(2) + 7(3) = 35$ **85.** $9(4) + 9(7) = 99$ **87.** $3x + 3$ **89.** $2x + 10$ **91.** $n = 3$ **93.** $n = 9$ **95.** $n = 0$ **97.** 2,860 mi **99.** \$7.18 **101.** 148,800 jets **103.** 2,081 calories **105.** 280 calories **107.** Yes **109.** 8,000 **111.** 1,500,000 **113.** 1,400,000 **115.** 40 **117.** 54

Problem Set 1.6

1. $6 \div 3$ **3.** $45 \div 9$ **5.** $r \div s$ **7.** $20 \div 4 = 5$ **9.** $2 \cdot 3 = 6$ **11.** $9 \cdot 4 = 36$ **13.** $6 \cdot 8 = 48$ **15.** $7 \cdot 4 = 28$ **17.** 5 **19.** 8 **21.** Undefined **23.** 45 **25.** 23 **27.** 1,530 **29.** 1,350 **31.** 18,000 **33.** 16,680 **35.** a **37.** b **39.** 1 **41.** 2 **43.** 4 **45.** 6 **47.** 45 **49.** 49 **51.** 432 **53.** 1,438 **55.** 705 **57.** 3,020

59.

First Number a	Second Number b	The Quotient of a and b $\frac{a}{b}$
100	25	4
100	26	3 R 22
100	27	3 R 19
100	28	3 R 16

61. 61 R 4 **63.** 90 R 1 **65.** 13 R 7 **67.** 234 R 6

69. 402 R 4 **71.** 35 R 35 **73.** \$3,525 **75.** 79¢ **77.** 3 bottles **79.** 6 glasses, with 2 ounces left over **81.** \$3,900,000 **83.** 665 mg **85.** 5 mi

Problem Set 1.7

1. Base 4; exponent 5 **3.** Base 3; exponent 6 **5.** Base 8; exponent 2 **7.** Base 9; exponent 1 **9.** Base 4; exponent 0 **11.** 36 **13.** 8 **15.** 1 **17.** 1 **19.** 81 **21.** 10 **23.** 12 **25.** 1 **27.** 12 **29.** 100 **31.** 4 **33.** 43 **35.** 16 **37.** 84 **39.** 14 **41.** 74 **43.** 12,768 **45.** 104 **47.** 416 **49.** 66 **51.** 21 **53.** 7 **55.** 16 **57.** 84 **59.** 40 **61.** 41 **63.** 18 **65.** 405 **67.** 124 **69.** 11 **71.** 91 **73.** 7 **75.** $8(4 + 2) = 48$ **77.** $2(10 + 3) = 26$ **79.** $3(3 + 4) + 4 = 25$ **81.** $(20 \div 2) - 9 = 1$ **83.** $(8 \cdot 5) + (5 \cdot 4) = 60$ **85.** Mean = 3; range = 4 **87.** Mean = 6; range = 10 **89.** Median = 11; range = 10 **91.** Median = 50; range = 90 **93.** Mode = 18; range = 59 **95.** 255 calories **97.** 465 calories **99.** 30 calories **101.** Big Mac has twice the calories. **103.** **a.** 78 **b.** 76 **c.** 76 **d.** 47 **105.** Mean = 6,881 students; range = 819 students **107.** **a.** 126 **b.** 126.5 **c.** 130 **d.** 28 **109.** **a.** \$3.73 **b.** \$3.60 **c.** \$0.30 **111.** 4 **113.** 16

Problem Set 1.8

1. 25 cm^2 **3.** 336 m^2 **5.** 60 ft^2 **7.** 45m^2 **9.** 16 cm^2 **11.** 2,200 ft^2 **13.** 945 cm^2 **15.** 100 in^2 **17.** Volume = 64 cm^3; surface area = 96 cm^2 **19.** Volume = 84 ft^3; surface area = 108 ft^2 **21.** 420 ft^3 **23.** 124 tiles **25.** The area increases from 25 ft^2 to 49 ft^2, which is an increase of 24 ft^2. **27.** 8,509 mm^2 **29.** 720 mm^2 **31.** 1,352 ft^3 **33.** **a.**

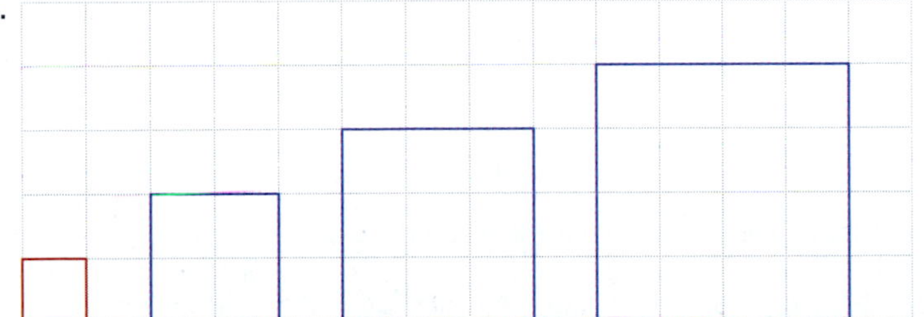

b.

PERIMETERS OF SQUARES	
Length of each Side (in Centimeters)	Perimeter (in Centimeters)
1	4
2	8
3	12
4	16

AREAS OF SQUARES	
Length of each Side (in Centimeters)	Area (in Square Centimeters)
1	1
2	4
3	9
4	16

35. 7 ft **37.** 9 ft

Chapter 1 Review

1. One thousand, three hundred seventy-six **3.** 5,245,652 **5.** 1,000,000 + 20,000 + 5,000 + 600 + 30 + 9 **7.** d **9.** c **11.** b **13.** g **15.** 749 **17.** 8,272 **19.** 314 **21.** 3,149 **23.** 584 **25.** 3,717 **27.** 173 **29.** 428 **31.** 3,781,090 **33.** 3,800,000 **35.** 79 **37.** 222 **39.** 8 **41.** 32 **43.** Mean = 79; median = 79 **45.** $3(4 + 6) = 30$ **47.** $2(17 - 5) = 24$ **49.** \$488 **51.** Smallest: 310 yd; Rose Bowl: 376 yd; largest: 390 yd **53.** \$2,032 **55.** \$1,938 **57.** \$532 **59.** 1,470 calories **61.** 250 more calories **63.** No **65.** Answers will vary. **67.** Volume = 160 cm^3; surface area = 184 cm^2

Chapter 1 Test

1. Twenty thousand, three hundred forty-seven [1.1C] **2.** 2,045,006 [1.1D] **3.** 100,000 + 20,000 + 3,000 + 400 + 7 [1.1B] **4.** f [1.2B] **5.** c [1.5B] **6.** a [1.2B] **7.** e [1.5B] **8.** 876 [1.2A] **9.** 16,383 [1.1A] **10.** 524 [1.4A] **11.** 3,085 [1.4B] **12.** 1,674 [1.5A] **13.** 22,258 [1.5A] **14.** 85 [1.6B] **15.** 21 [1.6B] **16.** 520,000 [1.3A] **17.** 11 [1.7B] **18.** 4 [1.7C] **19.** 107 [1.5B] **20.** $3x - 6$ [1.5B] **21.** \$264,300; \$142,000; \$125,000 [1.7D] **22.** $2(11 + 7) = 36$ [1.6A] **23.** $(20 \div 5) + 9 = 13$ [1.6A] **24.**

Urban Area	Average Hours in Gridlock Per Year
Los Angeles	52
Washington	35
Seattle-Everett	32
Atlanta	34
Boston	29

[1.3B]

25. Perimeter = 14 ft; area = 12 ft^2 [1.8A] **26.** 70 cm^3; 118 cm^2 [1.8A]

Chapter 2

Getting Ready for Chapter 2

1. $(5 \cdot 3)2$ **2.** $4(7 - 2)$ **3.** 2 **4.** 9 **5.** 5 **6.** 0 **7.** 125 **8.** 18 **9.** 6 **10.** 2 **11.** 3 **12.** 1 **13.** 3 **14.** 7 **15.** 3 **16.** 4 **17.** 24 in. **18.** 7,500 ft^2

Problem Set 2.1

1. 4 is less than 7. **3.** 5 is greater than −2. **5.** −10 is less than −3. **7.** 0 is greater than −4. **9.** $30 > -30$ **11.** $-10 < 0$
13. $-3 > -15$ **15.** $3 < 7$ **17.** $7 > -5$ **19.** $-6 < 0$ **21.** $-12 < -2$ **23.** $-\frac{1}{2} > -\frac{3}{4}$ **25.** $-0.75 < 0.25$ **27.** $-0.1 < -0.01$
29. $-3 < |6|$ **31.** $15 > |-4|$ **33.** $|-2| < |-7|$ **35.** 2 **37.** 100 **39.** 8 **41.** 231 **43.** $\frac{3}{4}$ **45.** 200 **47.** 8 **49.** 231
51. −3 **53.** 2 **55.** −75 **57.** 0 **59.** 121 **61.** −555 **63.** 2 **65.** 8 **67.** −2 **69.** −8 **71.** 0 **73.** Positive
75. −100 **77.** −20 **79.** −360 **81.** −450 feet **83.** −3,060 feet **85.** −\$5,000; −\$2,750 **87.** −61°F, −51°F **89.** −5°F, −15°F
91. New Orleans −6:00 GMT **93.** −7°F **95.** 10°F and 25-mph wind
97.

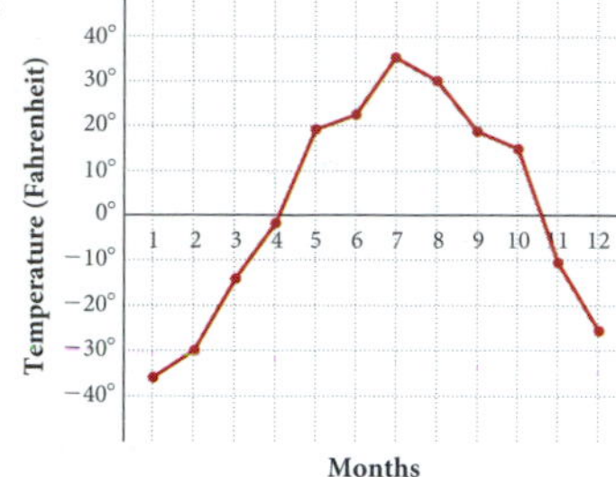

99. 25 **101.** 5 **103.** 6 **105.** 19 **107.** 4,313 **109.** 56 **111.** $5 + 3$
113. $(7 + 2) + 6$ **115.** $x + 4$ **117.** $y + 5$ **119.** −3
121. The opposite of a number is the number that is the same distance from 0, but on the opposite side of 0.
123. The opposite of the absolute value of −3. It simplifies to −3.

Problem Set 2.2

1. 5 **3.** 1 **5.** −2 **7.** −6 **9.** 4 **11.** 4 **13.** −9 **15.** 15 **17.** −3 **19.** −11 **21.** −7 **23.** −3 **25.** −16 **27.** −8
29. −127 **31.** 49 **33.** 34 **35.**

First Number a	Second Number b	Their Sum $a+b$
5	−3	2
5	−4	1
5	−5	0
5	−6	−1
5	−7	−2

37.

First Number x	Second Number y	Their Sum $x+y$
−5	−3	−8
−5	−4	−9
−5	−5	−10
−5	−6	−11
−5	−7	−12

39. 10 **41.** −445 **43.** 107 **45.** −20 **47.** −17 **49.** −50 **51.** −7 **53.** 3 **55.** 50 **57.** −73 **59.** −11 **61.** 17
63. −21 **65.** −5 **67.** −4 **69.** 7 **71.** 10 **73.** a **75.** b **77.** d **79.** c **81.** 380 feet above the trailhead **83.** \$10
85. −2 **87.** 4 **89.** $-\frac{2}{5}$ **91.** 30 **93.** −60.3 **95.** 2 **97.** 3 **99.** 604 **101.** 0 **103.** $10 - x$ **105.** $y - 17$

Problem Set 2.3

1. 2 **3.** 2 **5.** −8 **7.** −5 **9.** 7 **11.** 12 **13.** 3 **15.** −7 **17.** −3 **19.** −13 **21.** −50 **23.** −100 **25.** 399
27. −21 **29.**

First number x	second number y	the difference of x and y $x - y$
8	6	2
8	7	1
8	8	0
8	9	−1
8	10	−2

31.

First number x	second number y	the difference of x and y $x - y$
8	−6	14
8	−7	15
8	−8	16
8	−9	17
8	−10	18

33. −7 **35.** −9 **37.** −14 **39.** −65 **41.** −11 **43.** 202 **45.** −400 **47.** 11 **49.** −4 **51.** 8 **53.** 6 **55.** b
57. a **59.** −100 **61.** −16 degrees **63.** 7,603 feet **65.** 100 items **67.** \$1,760 **69.** \$3,009 **71.** $-11 - (-22) = 11°$ F
73. $3 - (-24) = 27°$ F **75.** $60 - (-26) = 86°$ F **77.** $-14 - (-26) = 12°$ F **79.** 30 **81.** 36 **83.** 64 **85.** 48 **87.** 41
89. 40 **91.** 17 **93.** 32 **95.** 25 **97.** 72 **99.** $3 \cdot 5$ **101.** $7x$ **103.** $5(3)$ **105.** $(5 \cdot 7) \cdot 8$ **107.** $2(3) + 2(4) = 6 + 8 = 14$
109. $3 - 5 \neq 5 - 3$ **113.** −5, −10 **115.** 2, 6

Problem Set 2.4

1. −56 **3.** −60 **5.** 56 **7.** 81 **9.** −24 **11.** 30 **13.** 0 **15.** −24 **17.** 24 **19.** −6 **21. a.** 16 **b.** −16
23. a. −125 **b.** −125 **25. a.** 16 **b.** −16

27.

Number x	Square x^2
−3	9
−2	4
−1	1
0	0
1	1
2	4
3	9

29.

First Number x	Second Number y	Their Product xy
6	2	12
6	1	6
6	0	0
6	−1	−6
6	−2	−12

31. −4 **33.** 50 **35.** 1 **37.** −35 **39.** −22 **41.** −30 **43.** −25 **45.** 9 **47.** −13 **49.** −11 **51.** 19 **53.** 6 **55.** −6 **57.** −4 **59.** −17 **61.** a **63.** d **65.** a **67.** b **69.** −6°C **71.** −16 degrees **73.** \$400 remains **75.** 7 **77.** 5 **79.** −5 **81.** 9 **83.** 4 **85.** 17 **87.** 405 **89.** $12 \div 6$ or $\frac{12}{6}$ **91.** $6 \div 3 = 2$ **93.** $5 \cdot 2 = 10$ **95.** 89 **97.** −54, 162 **99.** 54, −162 **101.** 44 **103.** 19 **105.** −17

Problem Set 2.5

1. −3 **3.** −5 **5.** 3 **7.** 2 **9.** −4 **11.** −2 **13.** 0 **15.** −5

17.

First Number a	Second Number b	The Quotient of a and b $\frac{a}{b}$
100	−5	−20
100	−10	−10
100	−25	−4
100	−50	−2

19.

First Number a	Second Number b	The Quotient of a and b $\frac{a}{b}$
−100	−5	20
−100	5	−20
100	−5	−20
100	5	20

21. −5 **23.** 35 **25.** 6 **27.** 1 **29.** −6 **31.** −2 **33.** −1 **35.** −1 **37.** 2 **39.** −3 **41.** −7 **43.** 30 **45.** 4 **47.** −5 **49.** −20 **51.** −5 **53.** −1 **55.** c **57.** a **59.** d **61.** −\$9,810.50

63.

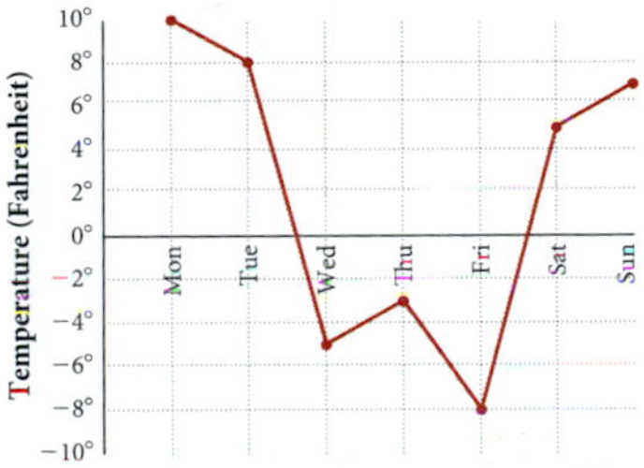

Temperature (Fahrenheit)
10° 8° 6° 4° 2° 0° −2° −4° −6° −8° −10°
Mon Tue Wed Thu Fri Sat Sun

65. $x + 3$ **67.** $(5 + 7) + a$ **69.** $(3 \cdot 4)y$ **71.** $5(3) + 5(7)$ **73.** 36 **75.** 64 **77.** 350

79. 7,500 **81.** 4 **83.** −12 **85.** 12 **87.** −32 **89.** 32 **91.** −2 **93.** −4 **95.** 4 **97.** −1 **99.** −1

Problem Set 2.6

1. $20a$ **3.** $48a$ **5.** $-18x$ **7.** $-27x$ **9.** $-10y$ **11.** $-60y$ **13.** $5 + x$ **15.** $13 + x$ **17.** $10 + y$ **19.** $8 + y$ **21.** $5x + 6$ **23.** $6y + 7$ **25.** $12a + 21$ **27.** $7x + 28$ **29.** $7x + 35$ **31.** $6a - 42$ **33.** $2x - 2y$ **35.** $20 + 4x$ **37.** $6x + 15$ **39.** $18a + 6$ **41.** $12x - 6y$ **43.** $35 - 20y$ **45.** $8x$ **47.** $4a$ **49.** $4x$ **51.** $5y$ **53.** $-10a$ **55.** $-5x$ **57.** $x(\$3{,}300 - \$300) - \$250 = \$3{,}000x - \$250$ **59.** $A = 36$ ft²; $P = 24$ ft **61.** $A = 81$ in²; $P = 36$ in. **63.** $A = 200$ in²; $P = 60$ in. **65.** $A = 300$ ft²; $P = 74$ ft **67.** 20° C **69.** 5° C **71.** −10° C

Chapter 2 Review

1. −17 **3.** 4.6 **5.** −6 **7.** 2 **9.** 4 **11.** 6 **13.** −2 **15.** −971 **17.** −12 **19.** 7 **21.** −20 **23.** −1,736 **25.** −3 **27.** 2 **29.** 36 **31.** −8 **33.** −5 **35.** −8 **37.** −1 **39.** −4 **41.** −42 **43.** −11 **45.** −27 **47.** 2 **49.** False **51.** True **53.** False **55.** \$58 + (−\$86) = −\$28 **57.** 24° **59.** $3x + 12$ **61.** $-21a$ **63.** $4x + 12$ **65.** $21y - 56$ **67.** $3x$ **69.** $4y$

Chapter 2 Cumulative Review

1. 7,714 [1.2A] **2.** 217 [2.2B] **3.** 2,269 [1.4B] **4.** −217 [2.3B] **5.** 45,084 [1.5A] **6.** 42 [2.4A] **7.** $10x - 40$ [1.5B], [2.6B] **8.** $-24x - 27$[1.5B], [2.6B] **9.** 68 R 3 or 68 3/15 [1.6D] **10.** 33 [1.6B] **11.** 4 [2.5A] **12.** −5 [2.5A] **13.** 0 [2.4B] **14.** 72 [2.5B] **15.** −11 [2.1B] **16.** 9 [2.1C] **17.** −27 [2.4B] **18.** 49 [2.4B] **19.** −15 [2.4B] **20.** 2 [2.5B] **21.** 12 [2.5B] **22.** $-4 > -6$ [2.1A] **23.** $|3| < |-5|$ [2.1A] **24.** Commutative and associative properties of addition [1.2B] **25.** $2(19 - 7) = 24$ [1.5C]

26. [1.3A]

OPENING DAY ATTENDANCE RECORDS			
Game	Date	Attendance	To the Nearest Hundred
Montreal at Colorado	4/9/93	80,227	80,200
San Francisco at Los Angeles	4/7/58	78,672	78,700
Detroit at Cleveland	4/7/73	74,420	74,400
St. Louis at Cleveland	4/20/48	73,163	73,200

27. [1.3B]

Speed (mi/hr)	Distance (ft)
20	22
30	49
40	88
50	137
60	198
70	269
80	352

28. 1 point gain [2.3B] **29.** \$7/hr [2.5C] **30.** 79 points [2.6D] **31.** 23 degrees [2.3B] **32.** $P = 32$ in.; $A = 64$ in^2 [1.1D], [1.8A] **33.** \$1,750 [1.5C] **34.** 34 [2.6A]

Chapter 2 Test

1. -14 [2.1C] **2.** 5 [2.1C] **3.** $-1 > -4$ [2.1A] **4.** $|-4| > |2|$ [2.1A] **5.** 7 [2.1C] **6.** -2 [2.1B] **7.** -9 [2.2B] **8.** -6 [2.3A] **9.** -21 [2.2B] **10.** -36 [2.3A] **11.** 42 [2.4A] **12.** 54 [2.4A] **13.** -5 [2.5A] **14.** 5 [2.5A] **15.** 9 [2.5B] **16.** -8 [2.5B] **17.** -11 [2.5B] **18.** -15 [2.5B] **19.** 7 [2.5B] **20.** -4 [2.5B] **21.** -61 [2.2B] **22.** -7 [2.3A] **23.** 24 [2.4A] **24.** -5 [2.5A]

25. [2.1D]

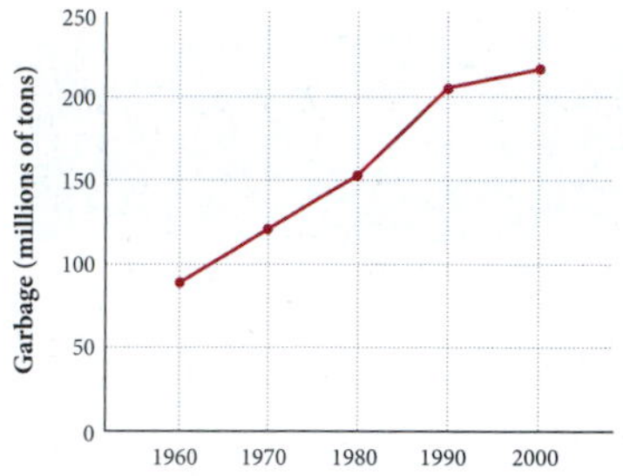

Garbage (millions of tons)
0
50
100
150
200
250
1960
1970
1980
1990
2000

26. $-\$35$ [2.2C] **27.** 25° [2.3B] **28.** $7x - 35$ [2.6B] **29.** $20x - 4$ [2.6B] **30.** $18x - 16y$ [2.6B] **31.** $32x$ [2.6C] **32.** $8a$ [2.6C]

Chapter 3

Getting Ready for Chapter 3

1. a. $>$ **b.** $>$ **c.** $<$ **d.** $>$ **2.** $10x$ **3.** $a + 2$ **4.** 16 **5.** 23 **6.** 16 **7.** 0 **8.** 24 **9.** 101 **10.** 5 **11.** 2 R 3 **12.** 8 R 16 **13.** 201 R 5 **14.** 507 R 10 **15.** $(2 + 3) \cdot 7$ **16.** $2^2 \cdot 3^3$

Problem Set 3.1

1. 1 **3.** 2 **5.** x **7.** a **9.** 5 **11.** 1 **13.** 12 **15.**

Numerator a	Denominator b	Fraction $\frac{a}{b}$
3	5	$\frac{3}{5}$
1	7	$\frac{1}{7}$
x	y	$\frac{x}{y}$
$x + 1$	x	$\frac{x+1}{x}$

17. $\frac{3}{4}, \frac{1}{2}, \frac{9}{10}$

19. True **21.** False **23.** $\frac{3}{4}$ **25.** $\frac{43}{47}$ **27.** $\frac{4}{3}$ **29.** $\frac{13}{17}$ **31.** $\frac{4}{6}$ **33.** $\frac{5}{6}$ **35.** $\frac{8}{12}$ **37.** $\frac{8}{12}$ **39.** $\frac{2x}{12x}$ **41.** $\frac{48a}{24a}$ **43.** $\frac{120a}{24a}$

45.

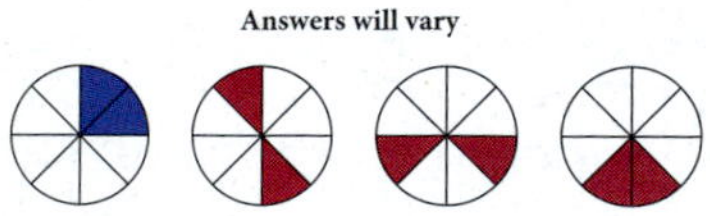

Answers will vary

47. 3 **49.** 2 **51.** 37 **53. a.** $\frac{1}{2}$ **b.** $\frac{1}{2}$ **c.** $\frac{1}{4}$ **d.** $\frac{1}{4}$

55.–63.

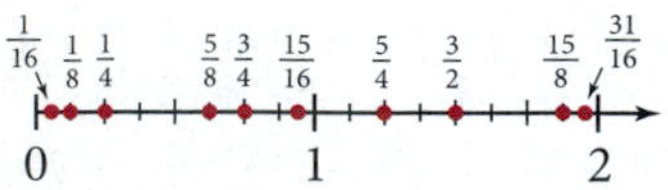

1/16
1/8
1/4
5/8
3/4
15/16
5/4
3/2
15/8
31/16
0
1
2

65. $\frac{1}{20} < \frac{4}{25} < \frac{3}{10} < \frac{2}{5}$ **67.** $\frac{3}{12}$

69.

How Often Workers Send Non-Work-Related E-Mail From the Office	Fraction of Respondents Saying Yes
never	$\frac{4}{25}$
1 to 5 times a day	$\frac{47}{100}$
5 to 10 times a day	$\frac{8}{25}$
more than 10 times a day	$\frac{1}{20}$

71. $\frac{4}{5}$ **73.** $\frac{311}{500}$ **75.** $\frac{19}{33}$ **77. a.** $\frac{90}{360}$ **b.** $\frac{45}{360}$ **c.** $\frac{180}{360}$ **d.** $\frac{270}{360}$

79. d **81.** a **83.** 108 **85.** 60 **87.** 4 **89.** 5 **91.** 7 **93.** 51 **95.** 23 **97.** 32 **99.** 16 **101.** 18

Problem Set 3.2

1. Prime **3.** Composite; 3, 5, and 7 are factors **5.** Composite; 3 is a factor **7.** Prime **9.** $2^2 \cdot 3$ **11.** 3^4 **13.** $5 \cdot 43$ **15.** $3 \cdot 5$ **17.** $\frac{1}{2}$ **19.** $\frac{2}{3}$ **21.** $\frac{4}{5}$ **23.** $\frac{9}{5}$ **25.** $\frac{7}{11}$ **27.** $\frac{3x}{5}$ **29.** $\frac{1}{7}$ **31.** $\frac{7}{9}$ **33.** $\frac{7x}{5}$ **35.** $\frac{a}{5}$ **37.** $\frac{11}{7}$ **39.** $\frac{5z}{3}$ **41.** $\frac{8x}{9y}$ **43.** $\frac{42}{55}$ **45.** $\frac{17ac}{19b}$ **47.** $\frac{14}{33}$ **49. a.** $\frac{2}{17}$ **b.** $\frac{3}{26}$ **c.** $\frac{1}{9}$ **d.** $\frac{3}{28}$ **e.** $\frac{2}{19}$ **51. a.** $\frac{1}{45}$ **b.** $\frac{1}{30}$ **c.** $\frac{1}{18}$ **d.** $\frac{1}{15}$ **e.** $\frac{1}{10}$ **53. a.** $\frac{1}{3}$ **b.** $\frac{5}{6}$ **c.** $\frac{1}{5}$ **55.** $\frac{9}{16}$ **57.–59.**

$\frac{1}{2} = \frac{2}{4} = \frac{4}{8} = \frac{8}{16}$ $\quad$ $\frac{3}{2} = \frac{6}{4} = \frac{12}{8} = \frac{24}{16}$

0 1 2

$\frac{1}{4} = \frac{2}{8} = \frac{4}{16}$ $\quad$ $\frac{5}{4} = \frac{10}{8} = \frac{20}{16}$

61. $\frac{106}{115}$ **63.** $\frac{1}{3}$ **65.** $\frac{8}{25}$ **67.** $\frac{1}{3}$ **69.** $\frac{37}{70}$

71. $\frac{1}{3}$ **73.** $\frac{1}{8}$ **75.** 3 **77.** 45 **79.** 25 **81.** $2^2 \cdot 5 \cdot 3$ **83.** $2^2 \cdot 5 \cdot 3$ **85.** 9 **87.** 25 **89.** 12 **91.** 18 **93.** 42 **95.** 53

Problem Set 3.3

1. $\frac{8}{15}$ **3.** $\frac{7}{8}$ **5.** -1 **7.** $\frac{27}{4}$ **9.** $\frac{3}{x}$ **11.** 1 **13.** $\frac{1}{24}$ **15.** $\frac{24}{125}$ **17.** 1

19.

First Number x	Second Number y	Their Product xy
$\frac{1}{2}$	$\frac{2}{3}$	$\frac{1}{3}$
$\frac{2}{3}$	$\frac{3}{4}$	$\frac{1}{2}$
$\frac{3}{4}$	$\frac{4}{5}$	$\frac{3}{5}$
$\frac{5}{a}$	$\frac{a}{6}$	$\frac{5}{6}$

21.

First Number x	Second Number y	Their Product xy
$\frac{1}{2}$	30	15
$\frac{1}{5}$	30	6
$\frac{1}{6}$	30	5
$\frac{1}{15}$	30	2

23. $\frac{3}{5}$ **25.** 9 **27.** 1 **29.** 8 **31.** $\frac{1}{15}$ **33.** $\frac{ac^2}{b}$ **35.** x **37.** y **39.** a **41.** $\frac{4}{9}$ **43.** $\frac{9}{16}$ **45.** $\frac{1}{4}$ **47.** $-\frac{8}{27}$ **49.** $\frac{1}{2}$ **51.** $\frac{9}{100}$ **53.** 3 **55.** $6x + 3$ **57.** $-2x - 4$ **59.** $\frac{5}{2}a + \frac{4}{3}$ **61.** $-2y - \frac{5}{2}$ **63.** $\frac{2}{3}x - 8$ **65.** 24 **67.** 4 **69.** 9 **71.** $\frac{3}{10}$; numerator should be 3, not 4. **73. a.**

Number x	Square x^2
1	1
2	4
3	9
4	16
5	25
6	36
7	49
8	64

b. Either ***larger*** or ***greater*** will work. **75.** 133 in^2

77. $\frac{4}{9}$ ft² **79.** 3 yd² **81.** 138 in² **83. a.** $\frac{5}{4}$ in. **b.** $\frac{7}{10}$ in. **c.** 3 in. **85.** 126,500 ft³ **87.** About 8 million **89.** 846 **91.** Canada: > 1,668,000 Venezuela: 1,251,000 Iraq: < 834,000 **93.** $\frac{1}{27}$ **95.** $\frac{8}{27}$ **97.** 2 **99.** 3 **101.** 2 **103.** 5 **105.** 3 **107.** $\frac{4}{3}$ **109.** 3 **111.** $\frac{1}{7}$ **113.** 100 **115.** 9 **117.** 18 **119.** 8

Problem Set 3.4

1. $\frac{15}{4}$ **3.** $-\frac{4}{3}$ **5.** -9 **7.** 200 **9.** $-\frac{3}{8}$ **11.** 1 **13.** $\frac{49}{64}$ **15.** $-\frac{3}{4}$ **17.** $\frac{15}{16}$ **19.** $\frac{1}{6}$ **21.** 6 **23.** $\frac{x}{y}$ **25.** ab **27.** $4a$ **29.** 9 **31.** $\frac{4}{5}$ **33.** $\frac{15}{22}$ **35.** 40 **37.** $\frac{7}{10}$ **39.** 13 **41.** 12 **43.** 186 **45.** 646 **47.** $\frac{3}{5}$ **49.** 40 **51.** $3 \div \frac{1}{5} = 3 \cdot \frac{5}{1} = 3 \cdot 5$ **53.** 490 feet **55.** 14 blankets **57.** 6 **59.** 28 cartons **61.** 20 lots **63.** $\frac{3}{6}$ **65.** $\frac{9}{6}$ **67.** $\frac{4}{12}$ **69.** $\frac{8}{12}$ **71.** $\frac{14}{30}$ **73.** $\frac{18}{30}$ **75.** $\frac{12}{24}$ **77.** $\frac{4}{24}$ **79.** $\frac{15}{36}$ **81.** $\frac{9}{36}$

83.

Number	Rounded to the Nearest		
	Ten	Hundred	Thousand
74	70	100	0
747	750	700	1,000
474	470	500	0

85. b

Problem Set 3.5

1. $\frac{2}{3}$ **3.** $-\frac{1}{4}$ **5.** $\frac{1}{2}$ **7.** $\frac{x-1}{3}$ **9.** $\frac{3}{2}$ **11.** $\frac{x+6}{2}$ **13.** $-\frac{3}{5}$ **15.** $\frac{10}{a}$

17.

First Number a	Second Number b	The Sum of a and b $a+b$
$\frac{1}{2}$	$\frac{1}{3}$	$\frac{5}{6}$
$\frac{1}{3}$	$\frac{1}{4}$	$\frac{7}{12}$
$\frac{1}{4}$	$\frac{1}{5}$	$\frac{9}{20}$
$\frac{1}{5}$	$\frac{1}{6}$	$\frac{11}{30}$

19.

First Number a	Second Number b	The Sum of a and b $a+b$
$\frac{1}{12}$	$\frac{1}{2}$	$\frac{7}{12}$
$\frac{1}{12}$	$\frac{1}{3}$	$\frac{5}{12}$
$\frac{1}{12}$	$\frac{1}{4}$	$\frac{4}{12} = \frac{1}{3}$
$\frac{1}{12}$	$\frac{1}{6}$	$\frac{3}{12} = \frac{1}{4}$

21. $\frac{7}{9}$ **23.** $\frac{7}{3}$ **25.** $\frac{7}{4}$ **27.** $\frac{7}{6}$ **29.** $\frac{5x+4}{20}$ **31.** $\frac{10+3x}{5x}$ **33.** $\frac{19}{24}$ **35.** $\frac{13}{60}$ **37.** $\frac{10a+1}{100}$ **39.** $\frac{3x+28}{7x}$ **41.** $\frac{29}{35}$ **43.** $\frac{949}{1,260}$ **45.** $\frac{13}{420}$ **47.** $\frac{41}{24}$ **49.** $\frac{7y+24}{6y}$ **51.** $\frac{5}{4}$ **53.** $\frac{88}{9}$ **55.** $\frac{3}{4}$ **57.** $\frac{1}{4}$ **59.** 19 **61.** 3 **63.** $\frac{160}{63}$ **65.** $\frac{5}{8}$ **67.** $\frac{2}{3}x$ **69.** $-\frac{1}{4}x$ **71.** $\frac{14}{15}x$ **73.** $\frac{11}{12}x$ **75.** $\frac{41}{40}x$ **77.** $\frac{7}{20}$ inch **79.** $\frac{9}{2}$ pints **81.** $\frac{61}{400}$ **83.** $\frac{5}{3}$ hours **85.** $\frac{1}{3}$ **87.** 10 lots **89.** $\frac{3}{2}$ in. **91.** $\frac{9}{5}$ ft **93.** $\frac{7}{3}$ **95.** 3 **97.** 59 **99.** $\frac{16}{8}$ **101.** $\frac{8}{8}$ **103.** $\frac{11}{4}$ **105.** $\frac{17}{8}$ **107.** $\frac{9}{8}$ **109.** 2 R 3 **111.** 8 R 16 **113.** $\frac{9}{10}$ **115.** 8 **117.** 3 **119.** 2 **121.** $\frac{2}{7}$ **123.** $\frac{15}{22}$

Problem Set 3.6

1. $\frac{14}{3}$ **3.** $\frac{21}{4}$ **5.** $\frac{13}{8}$ **7.** $\frac{47}{3}$ **9.** $\frac{104}{21}$ **11.** $\frac{427}{33}$ **13.** $1\frac{1}{8}$ **15.** $4\frac{3}{4}$ **17.** $4\frac{5}{6}$ **19.** $3\frac{1}{4}$ **21.** $4\frac{1}{27}$ **23.** $28\frac{8}{15}$ **25.** $\frac{8x+3}{x}$ **27.** $\frac{2y-5}{y}$ **29.** $\frac{6x+5}{6}$ **31.** $\frac{3a-1}{3}$ **33.** $\frac{10x+3}{2x}$ **35.** $\frac{9x-2}{3x}$ **37.** $\frac{11}{4}$ **39.** $\frac{37}{8}$ **41.** $\frac{14}{5}$ **43.** $\frac{9}{40}$ **45.** $\frac{3}{8}$ **47.** $\frac{32}{35}$ **49.** $\frac{4}{7}$ **51.** $12x - 8$ **53.** $28x - 49$ **55.** $\frac{4}{5}$ **57.** 9 **59.** 98

Problem Set 3.7

1. $5\frac{1}{10}$ **3.** $13\frac{2}{3}$ **5.** $6\frac{93}{100}$ **7.** $5\frac{5}{6}$ **9.** $9\frac{3}{4}$ **11.** $3\frac{1}{5}$ **13.** $12\frac{1}{2}$ **15.** $9\frac{9}{20}$ **17.** $\frac{32}{45}$ **19.** $1\frac{2}{3}$ **21.** 4 **23.** $4\frac{3}{10}$ **25.** $\frac{1}{10}$ **27.** $3\frac{1}{5}$ **29.** $2\frac{1}{8}$ **31.** $7\frac{1}{2}$ **33.** $\frac{11}{13}$ **35.** $5\frac{1}{2}$ cups **37.** $1\frac{1}{3}$ **39.** $2,687\frac{1}{5}$ cents **41.** $163\frac{3}{4}$ mi **43.** $6\frac{758}{1,207}$ shares **45.** $\$2,516\frac{2}{3}$ **47.** $182\frac{9}{16}$ ft² **49.** $3\frac{7}{16}$ mi²

51. Can 1 contains $157\frac{1}{2}$ calories, whereas Can 2 contains $87\frac{1}{2}$ calories. Can 1 contains 70 more calories than Can 2.
53. Can 1 contains 1,960 milligrams of sodium, whereas Can 2 contains 1,050 milligrams of sodium. Can 1 contains 910 more milligrams of sodium than Can 2.
55. a. $\frac{10}{15}$ **b.** $\frac{3}{15}$ **c.** $\frac{9}{15}$ **d.** $\frac{5}{15}$ **57. a.** $\frac{5}{20}$ **b.** $\frac{12}{20}$ **c.** $\frac{18}{20}$ **d.** $\frac{2}{20}$ **59.** $\frac{13}{15}$ **61.** $\frac{14}{9} = 1\frac{5}{9}$ **63.** $\frac{3}{5}$ **65.** $\frac{3}{14}$ **67.** $2\frac{1}{4}$
69. $3\frac{1}{16}$

Problem Set 3.8

1. $5\frac{4}{5}$ **3.** $12\frac{2}{5}$ **5.** $3\frac{4}{9}$ **7.** 12 **9.** $1\frac{3}{8}$ **11.** $14\frac{1}{6}$ **13.** $4\frac{1}{12}$ **15.** $2\frac{1}{12}$ **17.** $26\frac{7}{12}$ **19.** 12 **21.** $2\frac{1}{2}$ **23.** $8\frac{6}{7}$ **25.** $3\frac{3}{8}$
27. $10\frac{4}{15}$ **29.** $2\frac{1}{15}$ **31.** $21\frac{17}{20}$ **33.** 9 **35.** $18\frac{1}{10}$ **37.** 14 **39.** 17 **41.** $24\frac{1}{24}$ **43.** $27\frac{6}{7}$ **45.** $6\frac{1}{4}$ **47.** $9\frac{7}{10}$ **49.** $5\frac{1}{2}$
51. $\frac{2}{3}$ **53.** $1\frac{11}{12}$ **55.** $3\frac{11}{12}$ **57.** $5\frac{19}{20}$ **59.** $5\frac{1}{2}$ **61.** $\frac{13}{24}$ **63.** $3\frac{1}{2}$ **65.** $\$2\frac{1}{2}$ **67.** \$250 **69.** \$300
71. a. NFL: $P = 306\frac{2}{3}$ yd; Canadian: $P = 350$ yd; Arena: $P = 156\frac{2}{3}$ yd
b. NFL: $A = 5{,}333\frac{1}{3}$ sq yd; Canadian: $A = 7{,}150$ sq yd; Arena: $A = 1{,}416\frac{2}{3}$ sq yd
73. $31\frac{1}{6}$ in. **75.** $4\frac{63}{64}$ **77.** 2 **79.** $\frac{11}{8} = 1\frac{3}{8}$ **81.** $3\frac{5}{8}$ **83.** 17 **85.** 14 **87.** 104 **89.** 96 **91.** 40 **93.** $3\frac{1}{2}$
95. $5\frac{29}{40}$ **97.** $34\frac{4}{5}$

Problem Set 3.9

1. 7 **3.** 7 **5.** 2 **7.** 35 **9.** $\frac{7}{8}$ **11.** $8\frac{1}{3}$ **13.** $\frac{11}{36}$ **15.** $3\frac{2}{3}$ **17.** $6\frac{3}{8}$ **19.** $4\frac{5}{12}$ **21.** $\frac{8}{9}$ **23.** $\frac{1}{2}$ **25.** $1\frac{1}{10}$ **27.** 5
29. $\frac{3}{5}$ **31.** $\frac{7}{11}$ **33.** 5 **35.** $\frac{17}{28}$ **37.** $1\frac{7}{16}$ **39.** $\frac{13}{22}$ **41.** $\frac{5}{22}$ **43.** $\frac{15}{16}$ **45.** $1\frac{5}{17}$ **47.** $\frac{3}{29}$ **49.** $1\frac{34}{67}$ **51.** $\frac{346}{441}$
53. $5\frac{2}{5}$ **55.** 8 **57.** 5 miles^2 **59.** $115\frac{2}{3}$ yd **61.** $\frac{2}{3}$ **63.** $\frac{1}{6}$ **65.** $\frac{1}{7}$ **67.** $9\frac{7}{9}$

Chapter 3 Review

1. $\frac{3}{4}$ **3.** $\frac{11a^2}{7}$ **5.** x **7.** $\frac{8}{21}$ **9.** $\frac{2}{3}$ **11.** a^2b **13.** $\frac{1}{2}$ **15.** $-\frac{7}{2}$ **17.** $\frac{1}{36}$ **19.** $\frac{29}{8}$ **21.** $\frac{4x-3}{4}$ **23.** $\frac{8}{13}$ **25.** 12
27. $17\frac{11}{12}$ **29.** $11\frac{2}{3}$ **31.** 5 **33.** $\frac{1}{2}$ **35.** 20 items **37.** 9 **39.** $1\frac{7}{8}$ cups **41.** $10\frac{1}{2}$ tablespoons
43. Area = $25\frac{1}{5}$ ft^2; perimeter = 28 ft

Chapter 3 Cumulative Review

1. 17 [1.2B] **2.** 373 [1.2A] **3.** 1844 [1.4A] **4.** 8 [1.4A] **5.** $-6x$ [1.4A] **6.** $8\frac{19}{48}$ [3.8A] **7.** $2x$ [3.3A] **8.** $22n$ [1.5A]
9. 137,280 [1.5A] **10.** 3,375,650x [1.5A] **11.** 7 [1.7C] **12.** $\frac{2}{15}$ [3.3A] **13.** 4 [3.3A] **14.** $\frac{2}{5}$ [3.3A] **15.** 3 [3.3A] **16.** 6 [1.7C]
17. 16 [1.7C] **18.** $-\frac{9}{128}$ [1.7C, 3.3A] **19.** $\frac{1}{7}$ [3.2C] **20.** $3\frac{5}{33}$ [3.6B] **21.** 1 [1.6B] **22.** $\frac{105}{136}$ [3.9B] **23.** $\frac{81}{125}$ [3.5B]
24. $6\frac{6}{11}$ [2.9A, 3.4C] **25.** $14\frac{7}{12}$ [3.7A, 3.7B] **26.** 1090 [1.3A] **27.** $\frac{55}{36}$ [3.5B] **28.** $\frac{9x}{39x}$ [3.2C] **29.** $2\frac{19}{36}$ [3.7A] **30.** 100 [2.6A]
31. 192 in^2 [1.8A] **32.** [1.3A]

MEDICAL COSTS

Year	Average Annual Cost	Cost to the Nearest Hundred
1990	\$583	\$600
1995	\$739	\$700
2000	\$906	\$900
2005	\$1,172	\$1,200

33. $\frac{2}{7xy^3}$ [3.2C] **34.** 22 cm^2 [3.3B]
35. $-\frac{26}{45}$ [3.8A] **36.** Thirty thousand, seven hundred sixty; 30,000 + 700 + 60 [1.1B, 1.1C] **37.** 70 pictures [3.7C]
38. [1.3B]

Stopping distance (ft)
Speed (mi/hr)

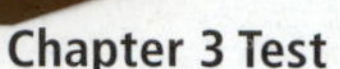

Chapter 3 Test

1. [3.1C]

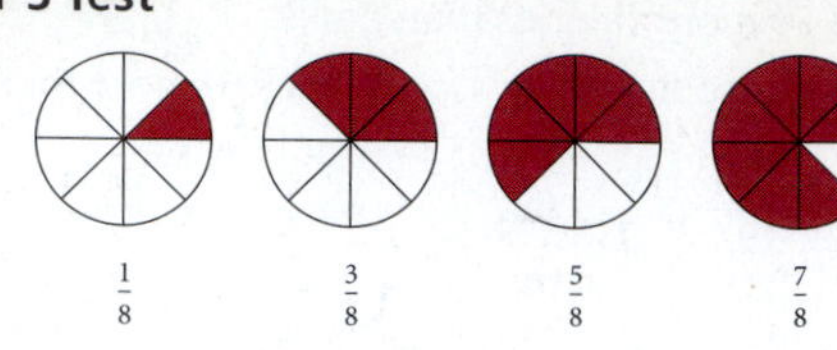

2. a. $\frac{2}{3}$ **b.** $\frac{13y}{5}$ [3.2C] **3.** $18x$ [3.7A] **4.** $\frac{8}{35}$ [3.3A] **5.** $\frac{8}{27}$ [3.4A]

6. $-\frac{1}{10}$ [3.4A] **7.** $\frac{2}{5}$ [3.5A] **8.** $\frac{3}{x}$ [3.5A] **9.** $-\frac{23}{5}$ [3.5B] **10.** $\frac{15 + 2x}{5x}$ [3.5B] **11.** $\frac{47}{36}$ [3.5B] **12.** $\frac{37}{7}$ [3.6A] **13.** $8\frac{3}{5}$ [3.6B]
14. $\frac{5x + 4}{x}$ [3.6D] **15.** $\frac{9}{2}$ [3.7B] **16.** $9\frac{17}{24}$ [3.8A] **17.** $3\frac{2}{3}$ [3.8B] **18.** $16\frac{3}{4}$ [3.9A] **19.** $9\frac{11}{12}$ [3.9A] **20.** $\frac{1}{2}$ [3.9B]
21. 40 grapefruit [3.2D] **22.** $27\frac{1}{2}$ in. [3.8C] **23.** $9\frac{1}{3}$ cups [3.7C] **24.** $3\frac{2}{15}$ ft [3.7C] **25.** Area = $23\frac{1}{3}$ ft²; perimeter = $23\frac{1}{3}$ ft

Chapter 4

Getting Ready for Chapter 4

1. 5 **2.** 135 **3.** −6 **4.** $\frac{11}{8}$ **5.** −20 **6.** −2 **7.** 18 **8.** 1 **9.** $-\frac{2}{3}$ **10.** 13 **11.** 7 **12.** 35 **13.** $10x$ **14.** $-12x$
15. $4x - 20$ **16.** x **17.** $x + 2$ **18.** $P = 8x$

Problem Set 4.1

1. $10x$ **3.** y **5.** $3a$ **7.** $8x + 16$ **9.** $6a + 14$ **11.** $x + 2$ **13.** $6x + 11$ **15.** $2x + 2$ **17.** $-a + 12$ **19.** $4y - 4$
21. $-2x + 4$ **23.** $8x - 6$ **25.** $5a + 9$ **27.** $-x + 3$ **29.** $17y + 3$ **31.** $a - 3$ **33.** $6x + 16$ **35.** $10x - 11$ **37.** $19y + 32$
39. $30y - 18$ **41.** $6x + 14$ **43.** $27a + 5$ **45.** 14 **47.** 27 **49.** −19 **51.** 7 **53.** 1 **55.** 18 **57.** 12 **59.** −10 **61.** 28
63. 40 **65.** 26 **67.** 4 **69.** 3 **71.** 0 **73.** 15 **75.** 6 **77.** $6(x + 4) = 6x + 24$ **79.** $4x + 4$ **81.** $10x - 4$ **83.** 5°
85. 55°; complementary angles **87.** 20°; complementary angles
89. a. Yes **b.** No, he should earn $108 for working 9 hours **c.** No, he should earn $84 for working 7 hours **d.** Yes
91. a. 32°F **b.** 22°F **c.** −18°F **93. a.** $27 **b.** $47 **95.** 0 **97.** −6 **99.** −3 **101.** $\frac{11}{8}$ **103.** 0 **105.** x **107.** $y - 2$
109. −9 **111.** 6 **113.** a **115.** c **117.** c **119.** $-\frac{2}{3}$ **121.** 18 **123.** $-\frac{1}{12}$

Problem Set 4.2

1. Yes **3.** Yes **5.** No **7.** Yes **9.** No **11.** 6 **13.** 11 **15.** −15 **17.** 1 **19.** −3 **21.** −1 **23.** $\frac{7}{5}$ **25.** −100 **27.** −4
29. −3 **31.** 2 **33.** −6 **35.** −6 **37.** −1 **39.** −2 **41.** −16 **43.** −3 **45.** 10 **47.** $x = 4$ **49.** $x = 12$ **51.** 58° celsius
53. 67° **55. a.** 225 **b.** $11,125 **57.** $\frac{1}{4}$ **59.** 2 **61.** $\frac{3}{2}$ **63.** 1 **65.** 1 **67.** 1 **69.** 1 **71.** x **73.** $7x - 11$ **75.** 2
77. $\frac{5}{14}$ **79.** $-\frac{1}{15}$ **81.** $-\frac{7}{6}$ **83.** Equation: $x + 12 = 30$; $x = 18$ **85.** Equation: $8 - 5 = x + 7$; $x = -4$

Problem Set 4.3

1. 8 **3.** −6 **5.** −6 **7.** 6 **9.** 16 **11.** 16 **13.** $-\frac{3}{2}$ **15.** −7 **17.** −8 **19.** 6 **21.** 2 **23.** 3 **25.** 4 **27.** −24
29. 12 **31.** −8 **33.** 15 **35.** 3 **37.** −1 **39.** 1 **41.** 3 **43.** 1 **45.** −1 **47.** $-\frac{1}{3}$ **49.** −14 **51.** 9 **53.** 8
55. 308 ft/s **57.** $35.00 **59.** 97,500,000 or 97.5 million viewers **61.** $2x + 5 = 19$; $x = 7$ **63.** $5x - 6 = -9$; $x = -\frac{3}{5}$
65. $6a - 16$ **67.** $-15x + 3$ **69.** $3y - 9$ **71.** $16x - 6$

Problem Set 4.4

1. 3 **3.** 2 **5.** −3 **7.** −4 **9.** 1 **11.** 0 **13.** 2 **15.** −2 **17.** 3 **19.** −1 **21.** 7 **23.** −3 **25.** 1 **27.** −2 **29.** 4
31. 10 **33.** −5 **35.** $-\frac{1}{12}$ **37.** −3 **39.** 20 **41.** −1 **43.** 5 **45.** 4 **47.** 10 **49.** 5 **51.** 1,250 feet **53.** 13
55. 30 hours **57.** $x + 2$ **59.** $2x$ **61.** $2(x + 6)$ **63.** $x - 4$ **65.** $2x + 5$
67. a. $4(x + 100) + 6x = 2{,}400$ **b.** $x = 200$ **c.** 500 people

Problem Set 4.5

1. $x + 3$ **3.** $2x + 1$ **5.** $5x - 6$ **7.** $3(x + 1)$ **9.** $5(3x + 4)$ **11.** The number is 2. **13.** The number is −2. **15.** The number is 3.
17. The number is 5. **19.** The number is −2. **21.** The length is 10 m and the width is 5 m. **23.** The length of one side is 8 cm.
25. The measures of the angles are 45°, 45°, and 90°. **27.** The angles are 30°, 60°, and 90°.
29. Patrick is 33 years old, and Pat is 53 years old. **31.** Sue is 35 years old, and Dale is 39 years old. **33.** $x = 8, y = 6, z = 9$
35. 39 hours **37.** 60 miles per hour **39.** 35 **41.** 65 **53.** 2 **55.** 14 **57.** $-\frac{2}{3}$

Problem Set 4.6

1. 704 ft² **3.** $\frac{9}{8}$ in² **5.** $240 **7.** $285 **9.** 12 ft **11.** 8 ft **13.** $140 **15.** 58 in. **17.** $3\frac{1}{4} = \frac{13}{4}$ ft **19.** $C = 100$°C; yes
21. $C = 20$°C; yes **23.** 0°C **25.** 5°C

27.

AGE (YEARS)	MAXIMUM HEART RATE (BEATS PER MINUTE)
18	202
19	201
20	200
21	199
22	198
23	197

29.

RESTING HEART RATE	TRAINING HEART RATE
60	144
65	146
70	148
75	150
80	152
85	154

31. a. 4 hrs **b.** 220 miles

33. a. 4 hours **b.** 65 mph **35.** 360 in³ **37.** 1 yd³ **39.** $y = 7$ **41.** $y = -3$ **43.** $y = -2$ **45.** $x = 3$ **47.** $x = 5$ **49.** $x = 8$ **51.** $y = 1$ **53.** $y = 3$ **55.** $y = \frac{5}{2}$ **57.** $y = 0$ **59.** $y = \frac{13}{3}$ **61.** $y = 4$ **63.** $x = 0$ **65.** $x = \frac{13}{4}$ **67.** $x = 3$ **69.** Complement: 45°; supplement: 135° **71.** Complement: 59°; supplement: 149° **73. a.** 13,330 kilobytes **b.** 2,962 kilobytes **75. a.** About 58° Celcius **b.** 328 K **77.** $\frac{3}{4}$ **79.** 3 **81.** 2 **83.** 6

Problem Set 4.7

1. (0, 4), (3, 1), (−2, 6) **3.** (0, 3), (2, 2), (18, −6) **5.** (0, 4), (3, 0), (−3, 8) **7.** (1, 1), (0, −3), (5, 17) **9.** (0, 3), (2, 7), (−2, −1) **11.** (2, 14), $\left(\frac{6}{7}, 6\right)$, (0, 0) **13.** (0, 0), (−2, 4), (2, −4) **15.** (0, 0), (2, 1), (4, 2) **17.** (−2, 3), (0, 2), (2, 1)

19.

x	y
2	3
3	2
1	4
0	5

21.

x	y
0	0
−1	4
−2	8
1	−4

23.

x	y
−2	−6
2	0
4	3
0	−3

25.

x	y
−1	−7
1	5
−2	−13
0	−1

27. b

29. (0, −2) **31.** (0, 3) **33.** (0, 0), (5, −5), (−3, 3) **35.** (1, 5) **37.** 96 watts **39.**

α	θ
0°	90°
30°	60°
45°	45°
60°	30°
75°	15°
90°	0°

41. $A = 6$ ft²; $P = 12$ ft **43.** $A = 12$ in²; $P = 18$ in.

Problem Set 4.8

1.–17.

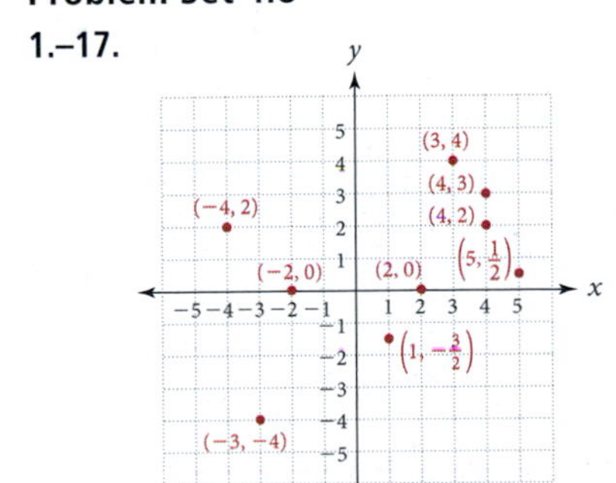

19. (2, 2) **21.** (−3, 2) **23.** (3, −3) **25.** (−4, 0) **27.** yes **29.** No **31.** Yes

33. No **35.** (1985, 20), (1990, 34), (1995, 45) **37. a.** (5, 40), (10, 80), (20, 160) **b.** \$320 **c.** 30 hours **d.** No, if she works 35 hours, she should be paid \$280. **39.** xy^2 **41.** $\frac{1}{xy}$ **43.** $\frac{4x + 15}{20}$ **45.** $\frac{3x - 1}{x}$ **47.** $A = (1, 2)$, $B = (6, 7)$ **49.** $A = (2, 2)$, $B = (2, 5)$, $C = (7, 5)$

Problem Set 4.9

1.

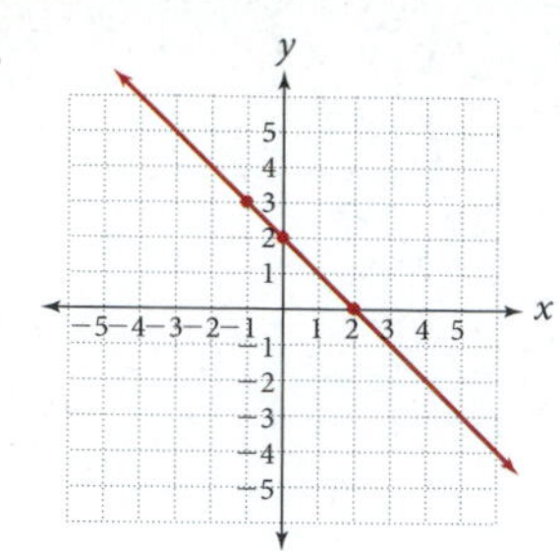

3.

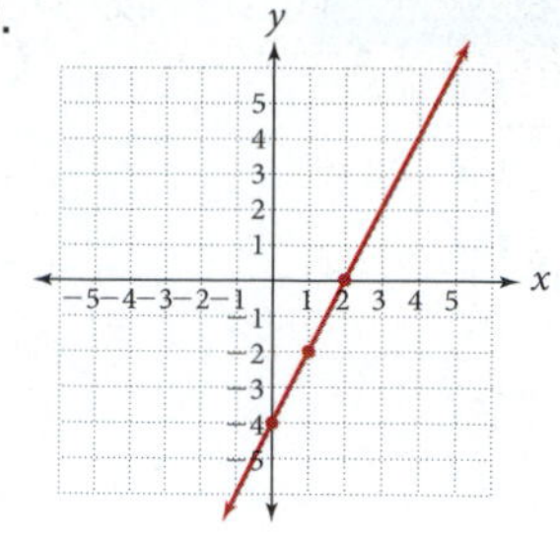

5.

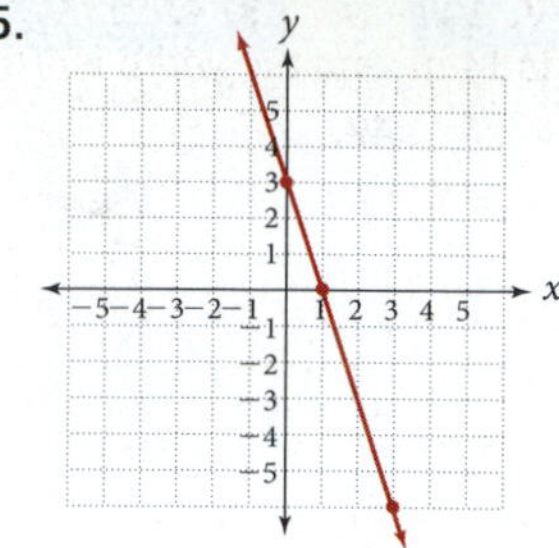

7.

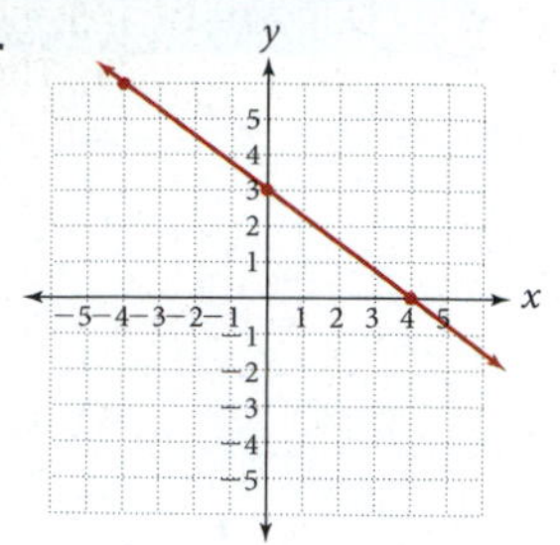

9.

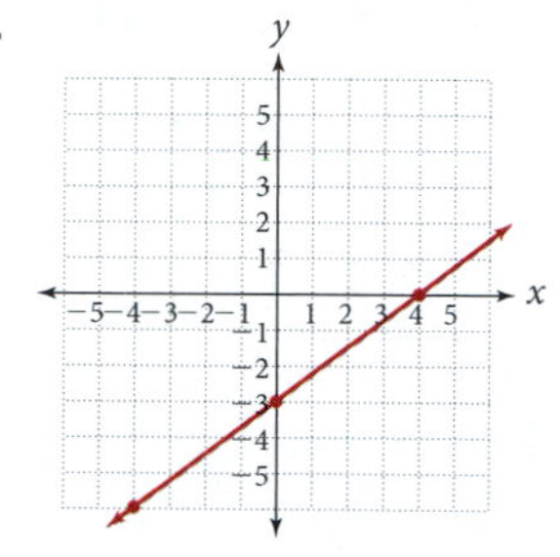

11.

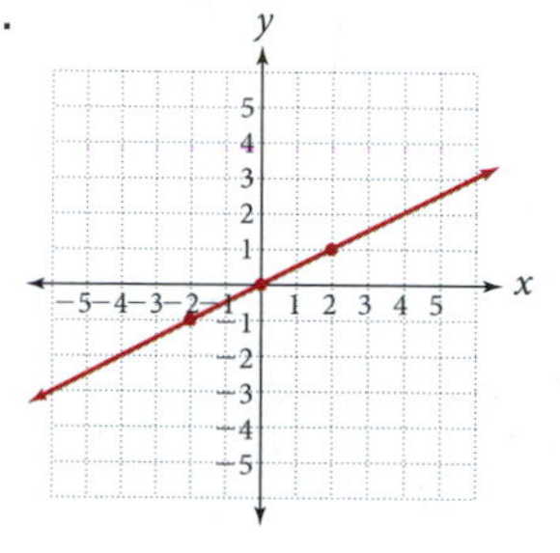

13.

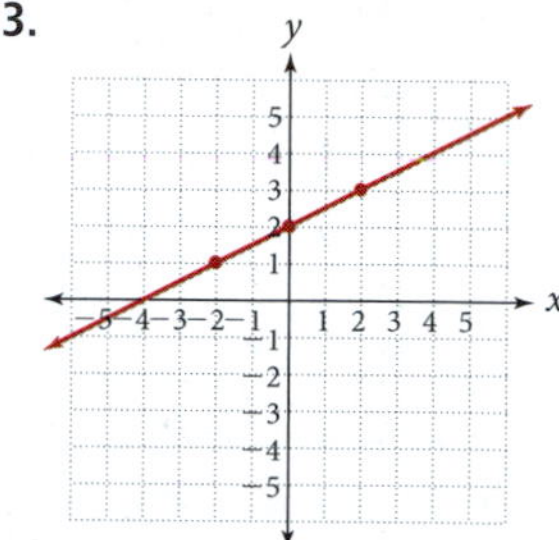

15.

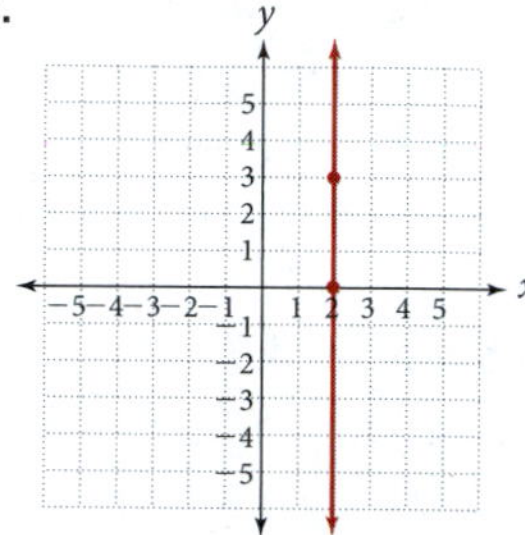

17.

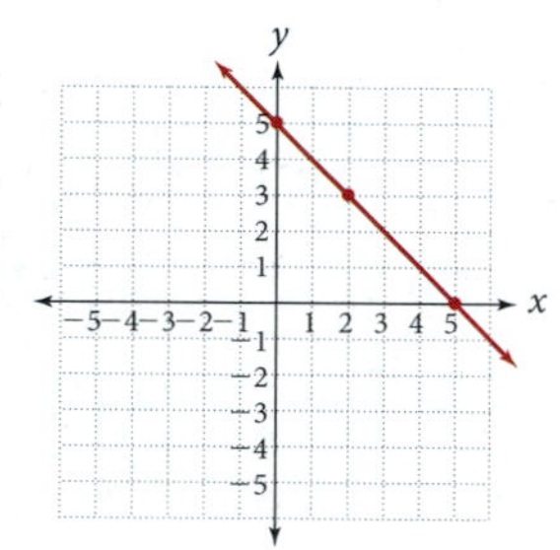

19.

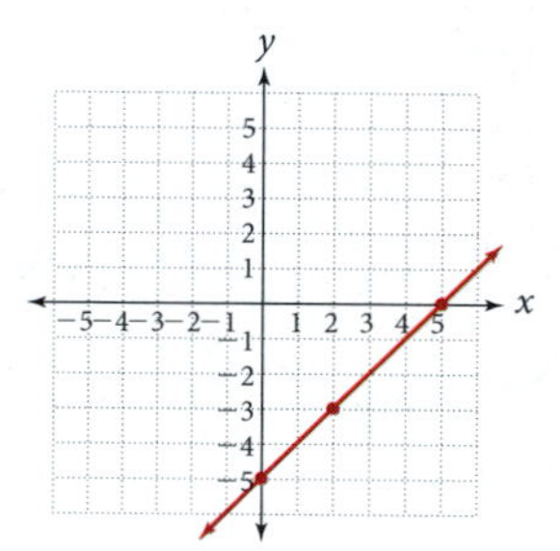

21.

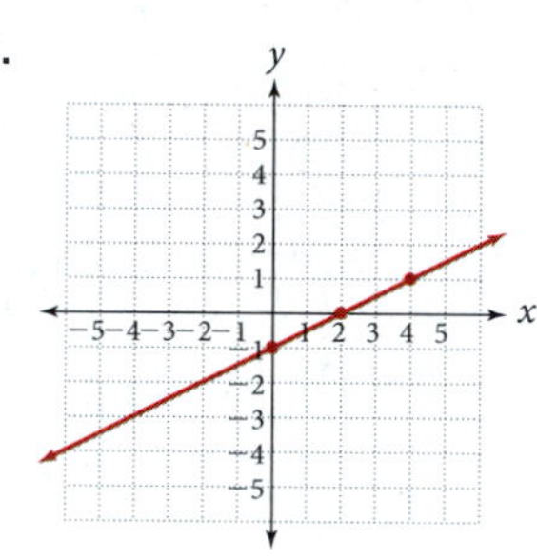

23.

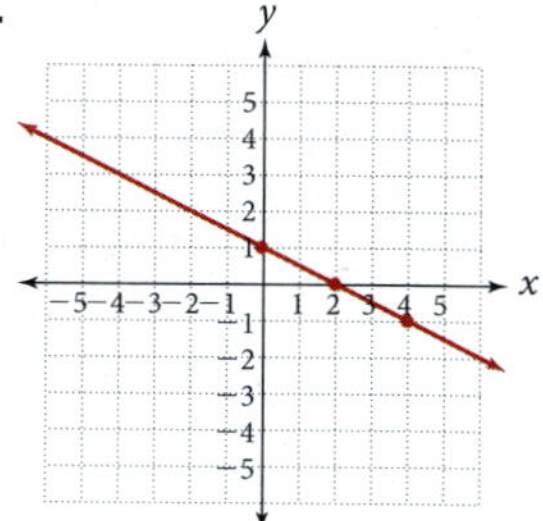

25.

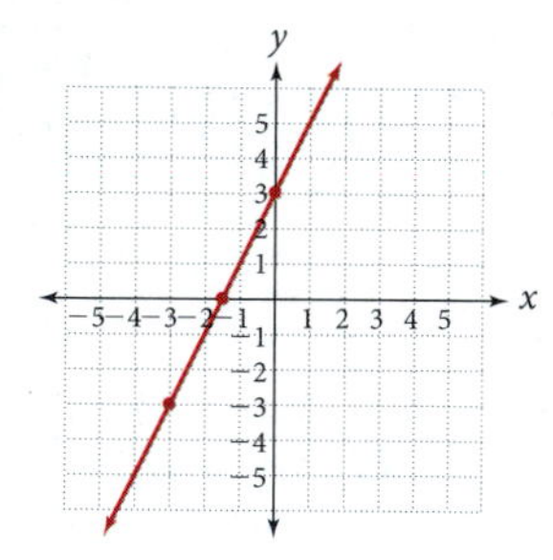

27.

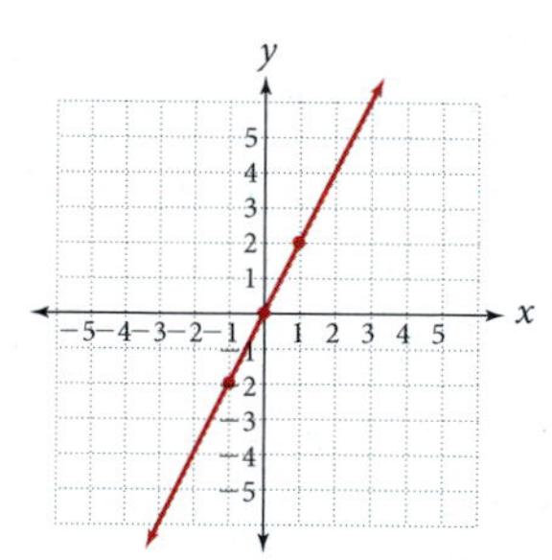

29.

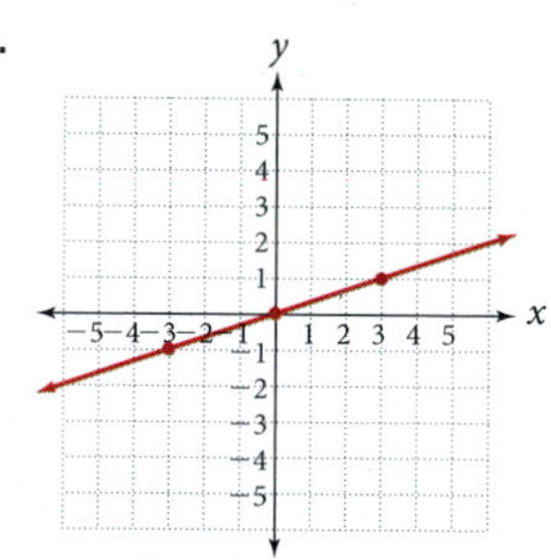

31.

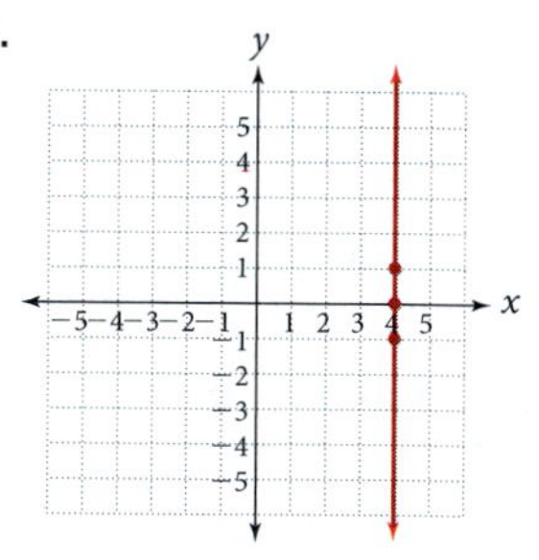

33.

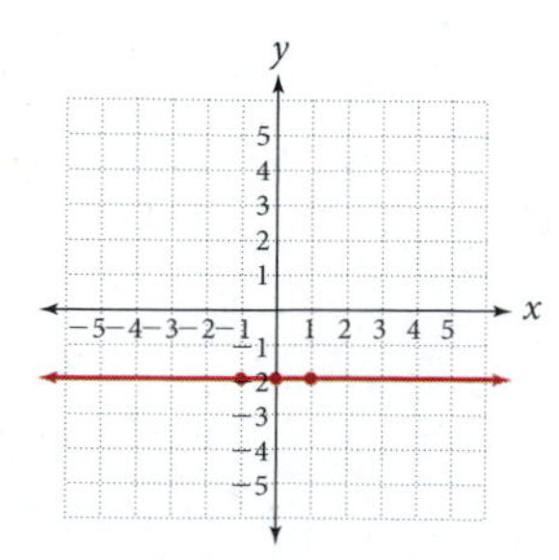

35. b

37.

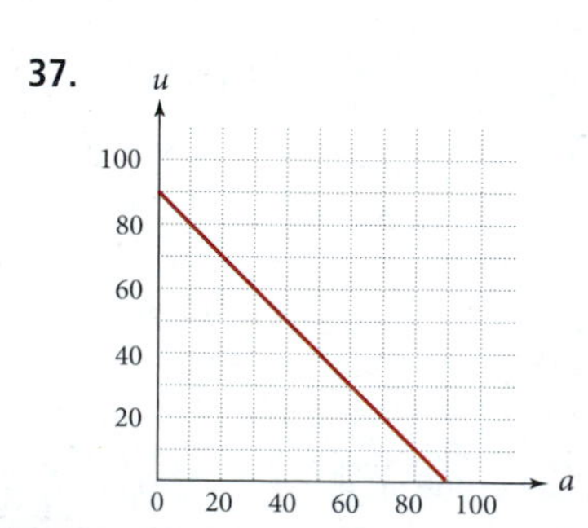

39.

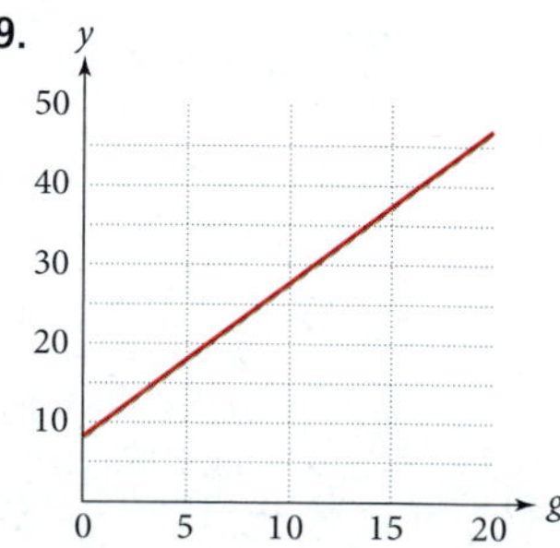

41. 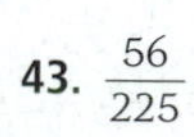 $\frac{33}{50}$ **43.** $\frac{56}{225}$ **45.** $\frac{1}{100}$

Chapter 4 Review

1. $17x$ **3.** $11a - 3$ **5.** $5x + 4$ **7.** $10a - 4$ **9.** 42 **11.** 1 **13.** Yes **15.** 9 **17.** 3 **19.** 9 **21.** 5 **23.** -1 **25.** $-\frac{1}{8}$ **27.** -2 **29.** -4 **31.** The length is 14 m, and the width is 7 m. **33.** $y = 6$ **35.** $y = 3$ **37.** $x = 0$ **39.** $x = 2$

41.–47.

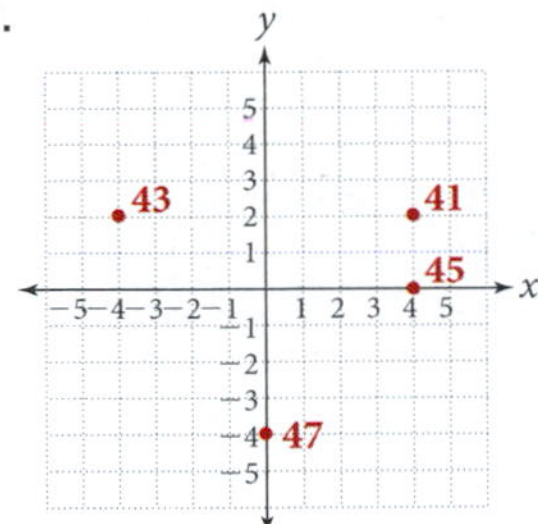

49.

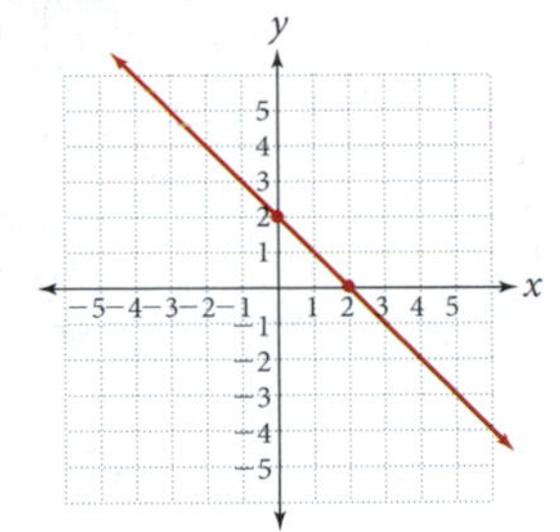

51.

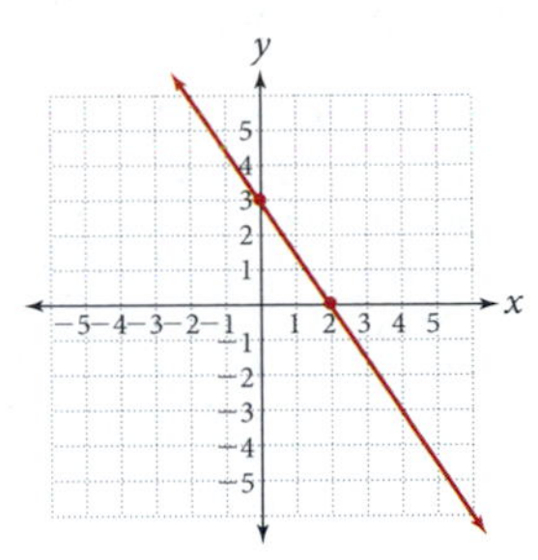

23.

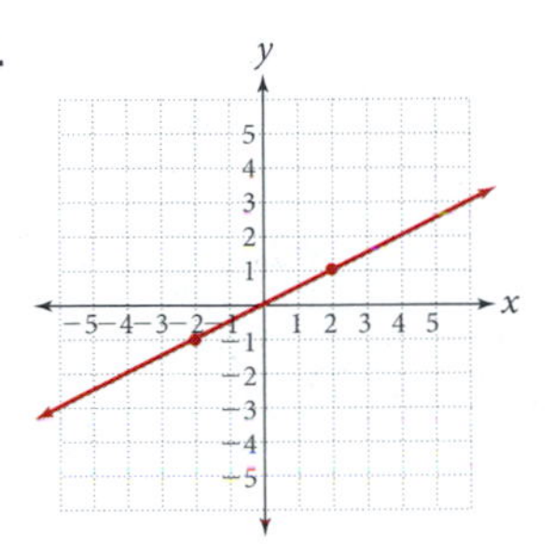

55.

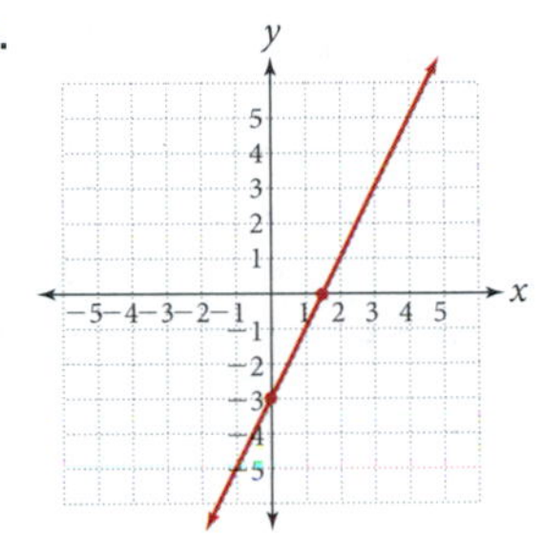

57.

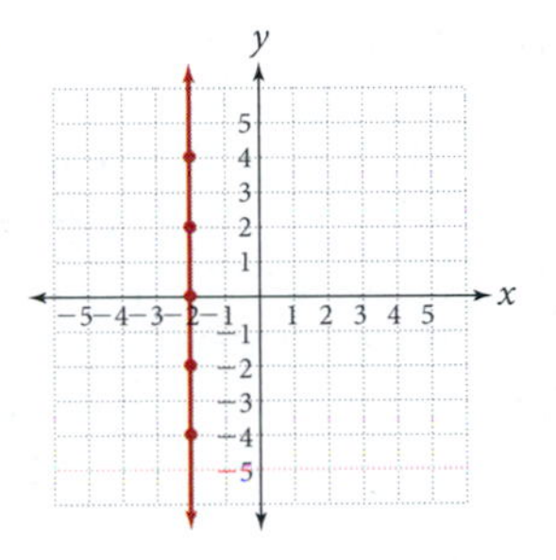

Chapter 4 Cumulative Review

1. 16,759 [1.2A] **2.** 2001 [1.4B] **3.** 12 [1.6B] **4.** 126 [1.5B] **5.** $490\frac{1}{32}$ [1.5D] **6.** 256 [1.7B] **7.** 21 [1.7C] **8.** 7 [3.2C] **9.** -38 [2.2B] **10.** 106 [2.2B] **11.** 13 [2.5A] **12.** -2 [1.7C] **13.** 3 [1.7C] **14.** 2 [1.7C] **15.** -8 [1.7C] **16.** $\frac{1}{256}$ [1.7C] **17.** 153 [1.7C] **18.** $\frac{2t - 7}{50}$ [3.5B] **19.** $6\frac{2}{5}$ [3.7B] **20.** $11\frac{1}{2}$ [3.8B] **21.** $-3z + 8$ [4.1B] **22.** $-a + 4$ [4.1B] **23.** $14\frac{4}{5}$ [3.9A] **24.** $b = 1\frac{1}{2}$ [4.4A] **25.** $a = -2$ [4.4A] **26.** [4.9A]

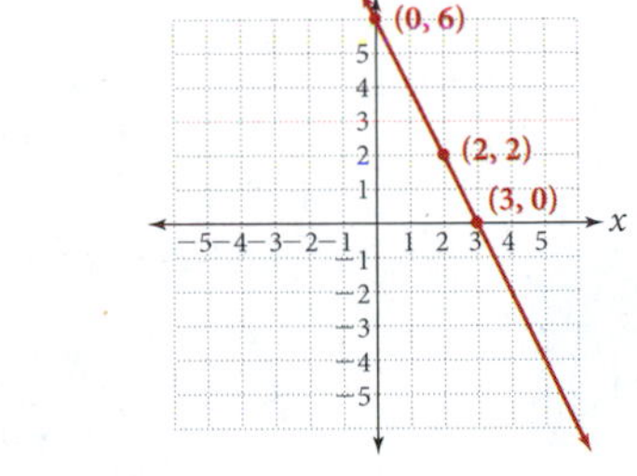

27. 72 in., 207 in² [1.2D, 1.8A]

28. $2\frac{11}{12}$ cm [3.3B] **29.** 47 [2.3A] **30.** 11 [4.6B] **31.** 9,261 cm³ [1.8B] **32.** $2 \cdot 3^2 \cdot 7$ [3.2B] **33.** 42 [4.5A] **34.** 14 [4.6B] **35.** (10, 4), (8, 2), (4, -2) [4.7A] **36.** -2 [4.9B] **37.** 220 in. [4.3B] **38.** $4\frac{1}{2}$ mi [4.4C] **39.** 10,368 in³ [1.8B]

40. a.

NEW YORK CITY MARATHON

Year	Number of Finishers	Rounded to the Nearest Hundred
1975	339	300
1980	12,512	12,500
1985	15,881	15,900
1990	23,774	23,800
1995	26,754	26,800

b.

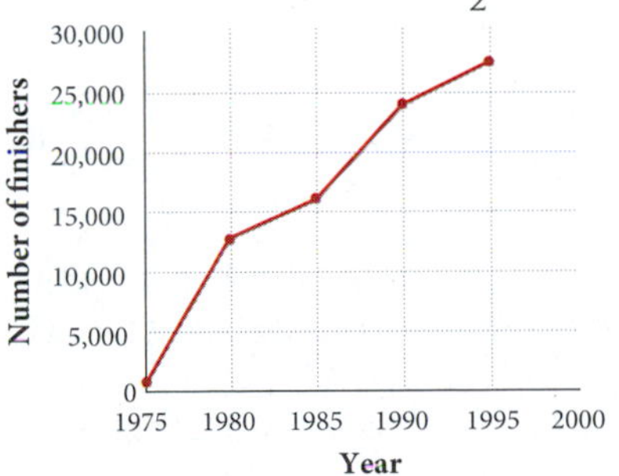

c. Answers will vary, but they should be around 30,000.

Chapter 4 Test

1. $6x - 5$ [4.1B] **2.** $3b - 4$ [4.1B] **3.** -3 [4.1C] **4.** 9 [4.1C] **5.** Yes [4.2A] **6.** 4 [4.2B] **7.** 27 [4.2B] **8.** -4 [4.3A] **9.** 1 [4.3A] **10.** $\frac{2}{3}$ [4.3A] **11.** 1 [4.3A] **12.** The number is -8. [4.5A] **13.** 22 mi/hr [4.5A] **14.** The length is 9 cm and the width is 5 cm. [4.5C] **15.** Susan is 11 years old and Karen is 6 years old. [4.5D] **16.** 8 [4.6B]

17. [4.8A]

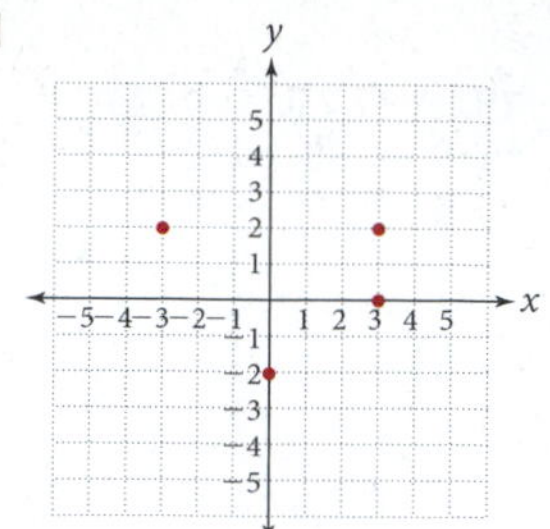

18.

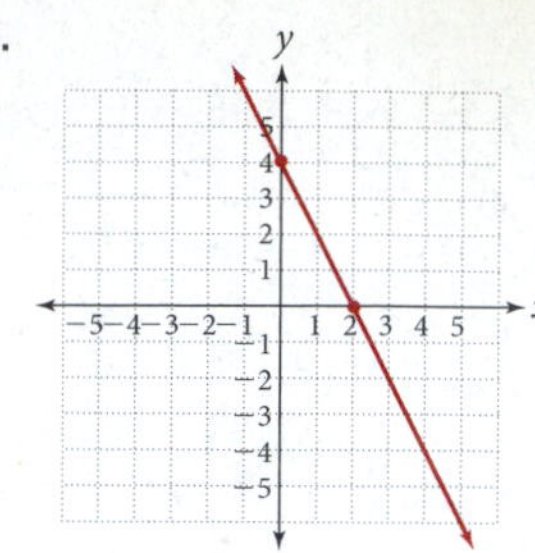

19.

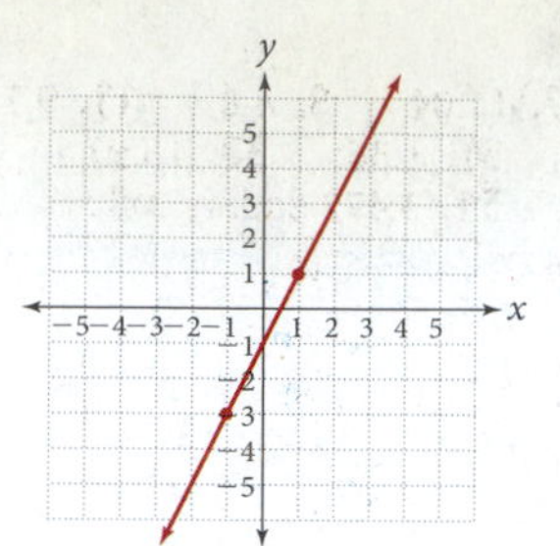

20.

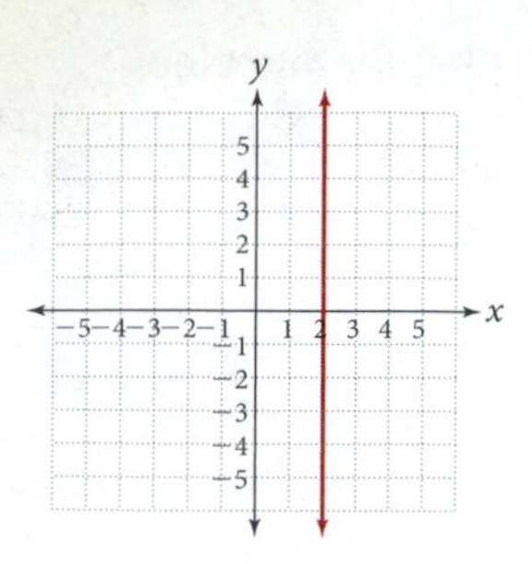

Chapter 5

Getting Ready for Chapter 5

1. 407,927 **2.** 25,576 **3.** 436 **4.** 663 **5.** 132,980 **6.** 728 **7.** 12,768 **8.** 96 **9.** 1,848 **10.** 298 R 14 **11.** 74
12. $\frac{1}{16}$ **13.** $\frac{3}{10}$ **14.** $1\frac{41}{64}$ **15.** 9,200 **16.** $\frac{19}{50}$ **17.** $2^4 \cdot 3$ **18.** $2^2 \cdot 3^2 \cdot 5$

Problem Set 5.1

1. Three tenths **3.** Fifteen thousandths **5.** Three and four tenths **7.** Fifty-two and seven tenths **9.** $405\frac{36}{100}$ **11.** $9\frac{9}{1{,}000}$
13. $1\frac{234}{1{,}000}$ **15.** $\frac{305}{100{,}000}$ **17.** Tens **19.** Tenths **21.** Hundred thousandths **23.** Ones **25.** Hundreds **27.** 0.55 **29.** 6.9
31. 11.11 **33.** 100.02 **35.** 3,000.003
37. **a.** $<$ **b.** $>$ **39.** 0.002 0.005 0.02 0.025 0.05 0.052 **41.** 7.451, 7.54 **43.** $\frac{1}{4}$ **45.** $\frac{1}{8}$ **47.** $\frac{5}{8}$ **49.** $\frac{7}{8}$ **51.** 9.99
53. 10.05 **55.** 0.05 **57.** 0.01

		Rounded to the Nearest			
	Number	**Whole Number**	**Tenth**	**Hundredth**	**Thousandth**
59.	47.5479	48	47.5	47.55	47.548
61.	0.8175	1	0.8	0.82	0.818
63.	0.1562	0	0.2	0.16	0.156
65.	2,789.3241	2,789	2,789.3	2,789.32	2,789.324
67.	99.9999	100	100.0	100.00	100.000

69. Hundredths **71.**

PRICE OF 1 GALLON OF REGULAR GASOLINE	
Date	**Price (Dollars)**
5/5/08	3.903
5/12/08	3.919
5/19/08	3.952
5/26/08	4.099

73. Three and eleven hundredths; two and five tenths

75. Fifteen hundredths **77.** $6\frac{31}{100}$ **79.** $6\frac{23}{50}$ **81.** $18\frac{123}{1{,}000}$ **83.** $\frac{3}{16} < \frac{3}{10} < \frac{3}{8} < \frac{3}{4}$ **85.** $<$ **87.** $>$

Problem Set 5.2

1. 6.19 **3.** 1.13 **5.** 6.29 **7.** 9.042 **9.** 8.021 **11.** 11.7843 **13.** 24.343 **15.** 24.111 **17.** 258.5414 **19.** 666.66
21. 11.11 **23.** 3.57 **25.** 4.22 **27.** 120.41 **29.** 44.933 **31.** 7.673 **33.** 530.865 **35.** 27.89 **37.** 35.64
39. 411.438 **41.** 6 **43.** 1 **45.** 3.1 **47.** 5.9 **49.** 3.272 **51.** 4.001 **53.** 1.47 seconds **55.** \$1,571.10 **57.** 4.5 in.
59. \$5.43 **61.** 6.42 sec **63.** 2 in. **65.** \$3.25; three \$1 bills and a quarter **67.** 3.25 **69.** $\frac{3}{100}$ **71.** $\frac{51}{10{,}000}$ **73.** $1\frac{1}{2}$
75. 1,400 **77.** $\frac{3}{20}$ **79.** $\frac{147}{1{,}000}$ **81.** 132,980 **83.** 2,115 **85.** 12 **87.** 16 **89.** 20 **91.** 68

Problem Set 5.3

1. 0.28 **3.** 0.028 **5.** 0.0027 **7.** 0.78 **9.** 0.792 **11.** 0.0156 **13.** 24.29821 **15.** 0.03 **17.** 187.85 **19.** 0.002
21. 27.96 **23.** 0.43 **25.** 49,940 **27.** 9,876,540 **29.** 1.89 **31.** 0.0025 **33.** 5.1106 **35.** 7.3485 **37.** 4.4
39. 2.074 **41.** 3.58 **43.** 187.4 **45.** 116.64 **47.** 20.75 **49.** 371.34 meters **51.** 0.126 **53.** Moves it two places to the right
55. \$1,381.38 **57.** **a.** 83.21 mm **b.** 551.27 mm² **c.** 1,102.53 mm³ **59.** 1.18 in² **61.** $C = 25.12$ in.; $A = 50.24$ in²
63. $C = 24{,}492$ mi **65.** 168 in. **67.** 100.48 ft³ **69.** 50.24 ft³ **71.** 1,879 **73.** 1,516 R 4 **75.** 298 **77.** 34.8 **79.** 49.896
81. 825 **83.** $\frac{3}{10} < \frac{2}{5} < \frac{1}{2} < \frac{4}{5}$ **85.** $\frac{3}{2} < 1\frac{2}{3} < 1\frac{5}{6} < \frac{25}{12}$ **87.** No

Problem Set 5.4

1. 19.7 **3.** 6.2 **5.** 5.2 **7.** 11.04 **9.** 4.8 **11.** 9.7 **13.** 2.63 **15.** 4.24 **17.** 2.55 **19.** 1.35 **21.** 6.5 **23.** 9.9 **25.** 0.05 **27.** 89 **29.** 2.2 **31.** 1.35 **33.** 16.97 **35.** 0.25 **37.** 2.71 **39.** 11.69 **41.** 3.98 **43.** 5.98 **45.** 0.77778 **47.** 307.20607 **49.** 0.70945 **51.** 3,472 square miles **53.** 7.5 mi **55.** $6.65/hr **57.** 22.4 mi **59.** 5 hr **61.** 7 min

63.

Rank	Name	Number of Events	Total Earnings	Average per Event
1.	Lorena Ochoa	25	$1,838,616	$73,545
2.	Annika Sorenstam	13	$1,295,585	$99,660
3.	Paula Creamer	24	$891,804	$37,159
4.	Seon Hwa Lee	28	$656,313	$23,440
5.	Jeong Jang	27	$642,320	$23,790

65. 2.73 **67.** 0.13 **69.** $\frac{3}{4}$ **71.** $\frac{2}{3y}$ **73.** $\frac{3xy}{4}$ **75.** $\frac{19}{50}$ **77.** $\frac{6}{10}$ **79.** $\frac{60}{100}$ **81.** $\frac{12x}{15x}$ **83.** $\frac{60}{15x}$ **85.** $\frac{18}{15x}$ **87.** 0.75 **89.** 0.875 **81.** 19 **93.** 3

Problem Set 5.5

1. 0.125 **3.** 0.625

5.

Fraction	$\frac{1}{4}$	$\frac{2}{4}$	$\frac{3}{4}$	$\frac{4}{4}$
Decimal	0.25	0.5	0.75	1

7.

Fraction	$\frac{1}{6}$	$\frac{2}{6}$	$\frac{3}{6}$	$\frac{4}{6}$	$\frac{5}{6}$	$\frac{6}{6}$
Decimal	$0.1\overline{6}$	$0.\overline{3}$	0.5	$0.\overline{6}$	$0.8\overline{3}$	1

9. 0.48 **11.** 0.4375 **13.** 0.92 **15.** 0.27 **17.** 0.09 **19.** 0.28

21.

Decimal	0.125	0.250	0.375	0.500	0.625	0.750	0.875
Fraction	$\frac{1}{8}$	$\frac{1}{4}$	$\frac{3}{8}$	$\frac{1}{2}$	$\frac{5}{8}$	$\frac{3}{4}$	$\frac{7}{8}$

23. $\frac{3}{20}$ **25.** $\frac{2}{25}$ **27.** $\frac{3}{8}$ **29.** $5\frac{3}{5}$ **31.** $5\frac{3}{50}$ **33.** $1\frac{11}{50}$

35. 2.4 **37.** 3.98 **39.** 3.02 **41.** 0.3 **43.** 0.072 **45.** 0.8 **47.** 1 **49.** 0.25 **51.** $4\frac{4}{10} = 4\frac{2}{5}$ **53.** $16.22 **55.** $52.66 **57.** 9 in.

59.

CHANGE IN STOCK PRICE

Date	Gain ($)	As a Decimal ($) (To the Nearest hundredth)
Monday, March 6, 2000	$\frac{3}{4}$	0.75
Tuesday, March 7, 2000	$\frac{9}{16}$	0.56
Wednesday, March 8, 2000	$\frac{3}{32}$	0.09
Thursday, March 9, 2000	$\frac{7}{32}$	0.22
Friday, March 10, 2000	$\frac{1}{16}$	0.06

61. 104.625 calories **63.** $10.38

65. Yes **67.** 33.49 mi³ **69.** 248.35 in³ **71.** 22.28 in² **73.** −9 **75.** −49 **77.** −6 **79.** −3 **81.** 2 **83.** 1 **85.** $\frac{1}{81}$ **87.** $\frac{25}{36}$ **89.** 0.25 **91.** 1.44 **93.** 25 **95.** 100 **97.** 852 **99.** 20,675

Problem Set 5.6

1. −1.5 **3.** 0.77 **5.** 0.15 **7.** −0.35 **9.** −0.8 **11.** 2.05 **13.** −0.064 **15.** 12.03 **17.** 0.44 **19.** −0.175 **21.** 2 **23.** 3,000 **25.** $1.90 per bushel **27.** The car was driven 110 miles. **29.** The car was driven 147 miles. **31.** Mary has 8 nickels and 18 dimes. **33.** You have 16 dimes and 32 quarters. **35.** The call was 16 minutes long. **37.** Katie has 7 nickels, 10 dimes, and 12 quarters. **39.** 125 **41.** 9 **43.** $\frac{1}{81}$ **45.** $\frac{25}{36}$ **47.** 0.25 **49.** 1.44 **51.** −24 **53.** 28 **55.** $-\frac{1}{2}$ **57.** −16 **59.** −25 **61.** 24

Problem Set 5.7

1. 8 **3.** 9 **5.** 6 **7.** 5 **9.** 15 **11.** 48 **13.** 45 **15.** 48 **17.** 15 **19.** 1 **21.** 78. **23.** 9 **25.** $\frac{4}{7}$ **27.** $\frac{3}{4}$ **29.** False **31.** True **33.** 10 in. **35.** 13 ft **37.** 6.40 in. **39.** 17.49 m **41.** 1.1180 **43.** 11.1803 **45.** 3.46 **47.** 11.18 **49.** 0.58 **51.** 0.58 **53.** 12.124 **55.** 9.327 **57.** 12.124 **59.** 12.124 **61.** 6.7 miles **63.** 30 ft **65.** 25 ft **67.** 5 miles

69.

Height h(feet)	Distance d(miles)
10	4
50	9
90	12
130	14
170	16
190	17

71. $25 \cdot 3$ **73.** $25 \cdot 2$ **75.** $4 \cdot 10$ **77.** 2^5 **79.** $2 \cdot 7^2$ **81.** $2^4 \cdot 3$ **83.** $\frac{2}{5}$ **85.** $9\frac{8}{25}$ **87.** 8 **89.** $2\frac{4}{5}$ **91.** $\frac{7}{16}$ **93.** 19

Problem Set 5.8

1. $2\sqrt{3}$ **3.** $2\sqrt{5}$ **5.** $6\sqrt{2}$ **7.** $7\sqrt{2}$ **9.** $2\sqrt{7}$ **11.** $10\sqrt{2}$ **13.** $2x\sqrt{3}$ **15.** $5x\sqrt{2}$ **17.** $5x\sqrt{3x}$ **19.** $5x\sqrt{2x}$ **21.** $4xy\sqrt{2y}$ **23.** $9x^2\sqrt{3}$ **25.** $6xy^2\sqrt{2}$ **27.** $2xy\sqrt{3xy}$ **29.** $3\sqrt{2}$ in. **31.** $4\sqrt{2}$ in. **33.** $x\sqrt{2}$ in. **35.** 1.25 sec **37.** See chapter introduction **39.** 8.4852814 **41.** 11.18034 **43.** 55.901699

45.

x	$\sqrt{x}$	$2\sqrt{x}$	$\sqrt{4x}$
1	1	2	2
2	1.414	2.828	2.828
3	1.732	3.464	3.464
4	2	4	4

47.

x	$\sqrt{x}$	$3\sqrt{x}$	$\sqrt{9x}$
1	1	3	3
2	1.414	4.243	4.243
3	1.732	5.196	5.196
4	2	6	6

49. $23x$ **51.** $27y$ **53.** $7ab$ **55.** $-7xy + 50x$ **57.** -3 **59.** -5 **61.** -2 **63.** 2 **65.** 0 **67.** 5 **69.** 10

Problem Set 5.9

1. $10\sqrt{3}$ **3.** $4\sqrt{5}$ **5.** $7\sqrt{x}$ **7.** $9\sqrt{7}$ **9.** $6\sqrt{y}$ **11.** $7\sqrt{2}$ **13.** $8\sqrt{3}$ **15.** $-2\sqrt{3}$ **17.** $8\sqrt{10}$ **19.** $x\sqrt{2}$ **21.** $17x\sqrt{5x}$ **23.** $7\sqrt{2}$ ft **25.** $\sqrt{2} + \sqrt{3} \approx 3.1463$; $\sqrt{5} \approx 2.2361$

27.

x	$\sqrt{x^2+9}$	$x+3$
1	3.162	4
2	3.606	5
3	4.243	6
4	5	7
5	5.831	8
6	6.708	9

29.

x	$\sqrt{x+3}$	$\sqrt{x}+\sqrt{3}$
1	2	2.732
2	2.236	3.146
3	2.450	3.464
4	2.646	3.732
5	2.828	3.968
6	3	4.182

31.

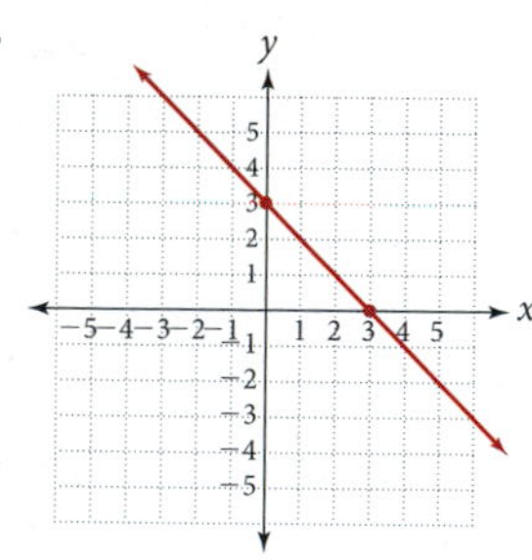

33.

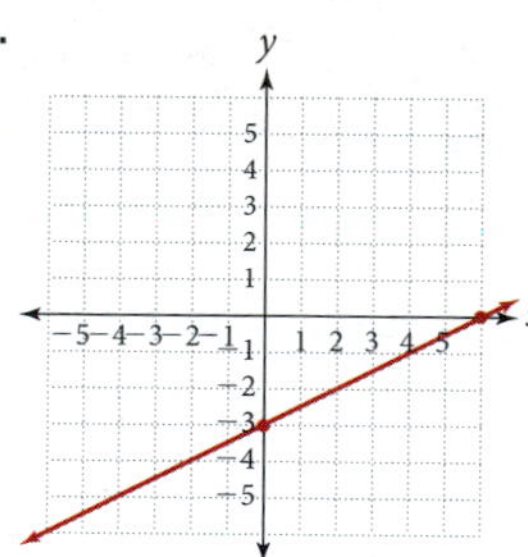

35.

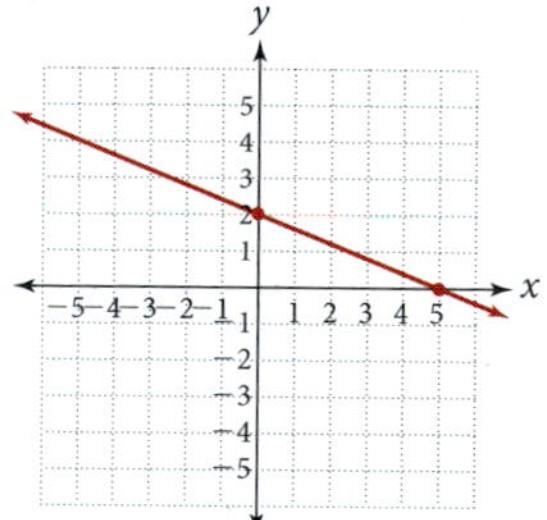

Chapter 5 Review

1. Thousandths **3.** 37.0042 **5.** 98.77 **7.** 7.779 **9.** -21.5736 **11.** 1.23 **13.** $\frac{141}{200}$ **15.** 2.42 **17.** 79 **19.** -5.9 **21.** 6.4 **23.** 15 **25.** 47 **27.** 17.4 in. **29.** 5.4 ft **31.** $P = 12.4$ ft; $A = 5$ ft² **33.** 100.48 in. **35.** $2\sqrt{3}$ **37.** $2x\sqrt{5}$ **39.** $3xy\sqrt{2y}$ **41.** $10\sqrt{3}$ **43.** $5\sqrt{3}$ **45.** $5\sqrt{6}$ **47.** $-9\sqrt{3}$

Chapter 5 Cumulative Review

1. $\frac{6x+5}{x}$ [3.8A] **2.** 0 [1.6B] **3.** 16,072 [1.5A] **4.** 30 [1.5B] **5.** $-2x - 8$ [2.3A] **6.** $-6x - 5$ [2.3A] **7.** 55.728 [4.1A] **8.** 76 [1.6B] **9.** $\frac{1}{2}$ [3.7B] **10.** 464,000 [1.3A] **11.** $15\frac{3}{4}$ [3.6B] **12.** -11, 11 [2.1B, 2.1C] **13.** No [4.2A]

14. [5.5B]

Decimal	0.125	0.250	0.375	0.500	0.625	0.750	0.875	1
Fraction	$\frac{1}{8}$	$\frac{1}{4}$	$\frac{3}{8}$	$\frac{1}{2}$	$\frac{5}{8}$	$\frac{3}{4}$	$\frac{7}{8}$	$\frac{8}{8}$

15. -9 [1.6A] **16.** Commutative and associative properties of multiplication [1.6C] **17.** $3(13 + 4) = 51$ [4.5A] **18.** $\frac{12a}{7}$ [3.2C] **19.** $x = 5$ [4.6B] **20.** 32 [2.5C] **21.** 7 [2.5C] **22.** $6xy\sqrt{2x}$ [5.8A] **23.** -0.34 [5.5C] **24.** $-2b + 1$ [4.1B] **25.** $14\frac{7}{8}$ [3.9A] **26.** -20 [5.9A] **27.** 28 [3.7A] **28.** 0.15 [5.6A] **29.** 1 [4.1A] **30.** $\frac{5}{17}$ [4.1A] **31.** $1\frac{1}{8}$ [3.5B] **32.** Lost $20 [2.2C]

33. 84 [1.7D] **34.** 10 in. [5.7C] **35.** 38.8 cm [5.2B] **36.** $8\frac{1}{4}$ cups [3.7C] **37.** 105 in³ [1.8B] **38.** $9.60 [1.6A]
39. 32 students [3.7C]

Chapter 5 Test

1. Five and fifty-three thousandths [5.1B] **2.** Thousandths [5.1A] **3.** 17.0406 [5.1A] **4.** 46.75 [5.1D] **5.** 8.18 [5.2A]
6. 6.056 [5.2A] **7.** 35.568 [5.3A] **8.** 8.72 [5.4A] **9.** 0.92 [5.1C] **10.** $\frac{14}{25}$ [5.1C] **11.** 14.664 [5.6A] **12.** 4.69 [5.6A]
13. 17.129 [5.6A] **14.** 0.26 [5.6A] **15.** 0.19 [5.6A] **16.** 36 [5.9A] **17.** $\frac{5}{9}$ [5.8A] **18.** $11.53 [5.2B] **19.** $24.47 [5.3B]
20. $6.55 [5.3B] **21.** 5 in. [5.7C] **22.** $6\sqrt{2}$ in. **23.** $4x\sqrt{3y}$ **24.** $8\sqrt{7}$ in. **25.** $-8\sqrt{3}$ in.

Chapter 6

Getting Ready for Chapter 6

1. $\frac{1}{3}$ **2.** 2 **3.** 0.25 **4.** 0.125 **5.** 65 **6.** 1.2 **7.** 297.5 **8.** 4 **9.** 8 **10.** 12 **11.** 62.5 **12.** 0.695 **13.** 3.98
14. $\frac{3}{2}$ **15.** 8.7 **16.** 6.5

Section 6.1

1. $\frac{4}{3}$ **3.** $\frac{16}{3}$ **5.** $\frac{2}{5}$ **7.** $\frac{1}{2}$ **9.** $\frac{3}{1}$ **11.** $\frac{7}{6}$ **13.** $\frac{7}{5}$ **15.** $\frac{5}{7}$ **17.** $\frac{8}{5}$ **19.** $\frac{1}{3}$ **21.** $\frac{1}{10}$ **23.** $\frac{3}{25}$ **25. a.** $\frac{1}{2}$ **b.** $\frac{1}{3}$ **c.** $\frac{2}{3}$
27. $\frac{11}{14}$ **29.** $\frac{20 \text{ mg}}{1 \text{ mL}}$ **31.** $\frac{75 \text{ mg}}{2 \text{ mL}}$ **33. a.** $\frac{13}{8}$ **b.** $\frac{1}{4}$ **c.** $\frac{3}{8}$ **d.** $\frac{13}{3}$ **35. a.** $\frac{6}{1}$ **b.** $\frac{5}{1}$ **c.** $\frac{4}{1}$ **d.** $\frac{3}{1}$ **e.** $\frac{4}{1}$ **37. a.** $\frac{44}{39}$ **b.** $\frac{1}{1}$ **c.** $\frac{6}{7}$
39. 18 **41.** 62.5 **43.** 0.615 **45.** 176 **47.** 184 **49.** 0.087 **51.** 0.048 **53.** $\frac{5}{8}$ **55.** $\frac{11}{2} = 5\frac{1}{2}$ **57.** $\frac{5}{14}$ **59.** 96

Section 6.2

1. 55 mi/hr **3.** 84 km/hr **5.** 0.2 gal/sec **7.** 14 L/min **9.** 19 mi/gal **11.** $4\frac{1}{3}$ mi/L **13.** The Midwest **15.** 480 mL/hr
17. 4.95¢ per oz **19.** 34.7¢ per diaper, 31.6¢ per diaper; Happy Baby **21.** $1.00 = 0.646 Euros **23.** $1.00 = 0.509 pounds
25. The 18 oz box is the best buy at $0.277 per ounce. **27.** 54.03 mi/hr **29.** 9.3 mi/gal **31.** $64 **33.** $16,000 **35.** $n = 6$
37. $n = 4$ **39.** $n = 4$ **41.** $n = 65$ **43.** $\frac{7}{8}$ **45.** $\frac{1}{40}$ **47.** $\frac{1}{60}$ **49.** $\frac{11}{6} = 1\frac{5}{6}$ **51.** The 100 ounce size is the better value.

Section 6.3

1. Means: 3, 5; extremes: 1, 15; products: 15 **3.** Means: 25, 2; extremes: 10, 5; products: 50 **5.** Means: $\frac{1}{2}$, 4; extremes: $\frac{1}{3}$, 6; products: 2
7. Means: 5, 1; extremes: 0.5, 10; products: 5 **9.** 10 **11.** $\frac{12}{5}$ **13.** $\frac{3}{2}$ **15.** $\frac{10}{9}$ **17.** 7 **19.** 14 **21.** 18 **23.** 6 **25.** 40
27. 50 **29.** 108 **31.** 3 **33.** 1 **35.** $\frac{1}{4}$ **37.** 108 **39.** 65 **41.** 41 **43.** 108 **45.** 20 **47.** 297.5 **49.** 450
51. 5 **53.** Tens **55.** 26.516 **57.** 0.39

Section 6.4

1. 329 mi **3.** 360 points **5.** 15 pt **7.** 427.5 mi **9.** 900 eggs **11.** 35 mg/pill **13.** 5.5 mL **15.** 2.5 tablets
17. 435 in. = 36.25 ft **19.** $313.50 **21.** 265 g **23.** 91.3 L **25.** 60,113 people **27.** 78 teachers **29.** 850 meters
31. 15 mL **33.** 900 mg/day **35.** 2 **37.** 147 **39.** 20 **41.** 147 **43.** 1.35 **45.** 3.816 **47.** 4 **49.** 160 **51.** 183.79

Section 6.5

1. 9 **3.** 14 **5.** 12 **7.** 25 **9.** 32 **11.**

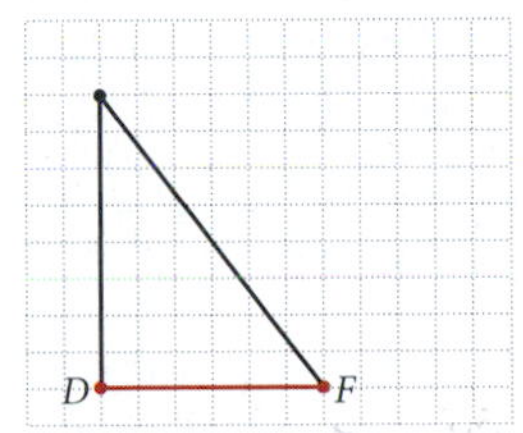

13.

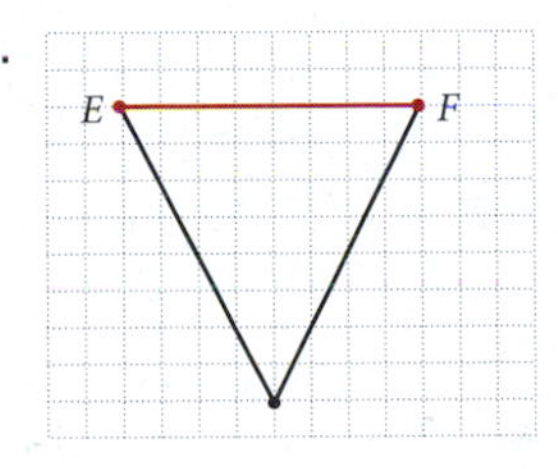

15.

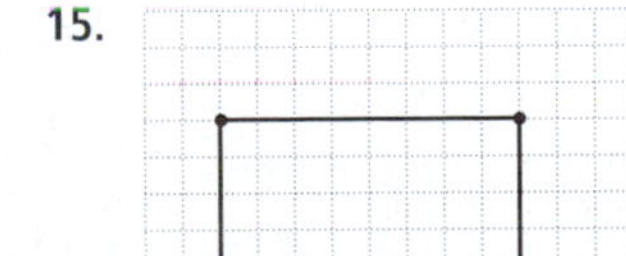

17.

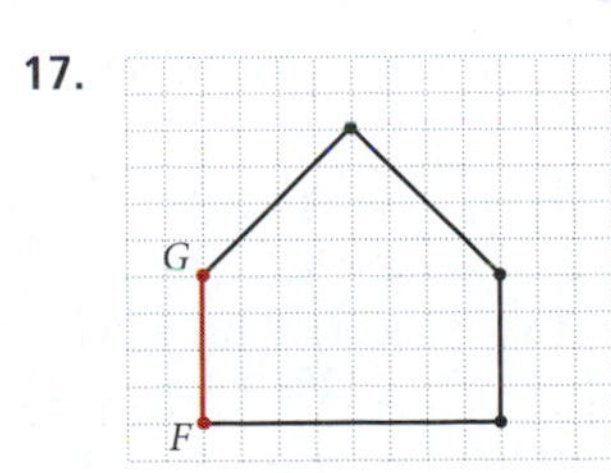

19. 45 in. **21.** 960 pixels **23.** 1,440 pixels **25.** 57 ft

27. 177 ft **29.** 13.99 **31.** 40.999 **33.** 0.10545 **35.** 18 **37.** $4\frac{1}{6}$ **39.** 4
41. a. $\frac{1}{2}$ **b.** 18 rectangles should be shaded. **c.** $\frac{1}{3}$

Chapter 6 Review

1. $\frac{3}{10}$ **3.** $\frac{3}{4}$ **5.** $\frac{7}{5}$ **7.** $\frac{1}{2}$ **9.** $\frac{1}{3}$ **11.** $\frac{9}{2}$ **13.** $\frac{1}{4}$ **15.** 19 mi/gal
17. 9.05¢/ounce; 8.41¢/ounce; 32-ounce carton is the better buy. **19.** 49 **21.** $\frac{1}{10}$ **23.** 1,500 mL **25.** 5 weeks **27.** 3 tablets
29. 9 **31.** 160

Chapter 6 Cumulative Review

1. $-2x + 12$ [4.1B] **2.** $\frac{6x-5}{6}$ [2.6B] **3.** 316 [1.4B] **4.** $13\frac{5}{12}$ [3.9A] **5.** $-2x - 6$ [4.3A] **6.** 42,771 [1.5A] **7.** $-\frac{1}{16}$ [3.4A]
8. 37.65 [1.3A] **9.** $\frac{39}{8}$ [3.6A] **10.** Thirty-eight thousand, six hundred nine [1.1C] **11.** Distributive property of multiplication [1.5B]
12. 126 [1.7B] **13.** $\frac{6}{7}$ [5.8A] **14.** $-10\sqrt{3}$ [5.9A] **15.** 3 [1.7B] **16.** -1 [2.5B] **17.** 6 [2.1C] **18.** $\frac{2}{5}$ [6.1A] **19.** $\frac{8}{9}$ [6.1A]
20. -3 [4.6B] **21.** $2(x + 9) - 4x$ [4.5A] **22.** 56 [6.2B] **23.** -0.15 [4.2A] **24.** 30 [4.2A] **25.** 5 [4.1A]
26. $P = 42$ in.; $A = 72$ in² [1.2D, 1.8B] **27.** 66 cm; $A = 150$ cm² [1.2D] **28.** 0.56 [5.5A] **29.** -2 [4.6B] **30.** $\frac{11y}{4}$ [3.2C]
31. 17° [2.2C] **32. a.** 18.84 ft **b.** 565.2 ft³ [4.5C] **33.** 20 women [6.1B] **34.** $80\frac{9}{16}$ in. [3.7C] **35.** 97 mi [1.7D]
36. 150 sections [6.2A] **37.** 8.5¢/ounce; 13.5¢/ounce; 72-ounce carton is the better buy [6.2C] **38.** 492 in. = 41 ft [6.4A]
39. 16 cm [6.5B] **40.** 35 ft [6.5D]

Chapter 6 Test

1. $\frac{4}{3}$ [6.1A] **2.** $\frac{9}{10}$ [6.1A] **3.** $\frac{3}{2}$ [6.1A] **4.** $\frac{3}{10}$ [6.1A] **5.** $\frac{3}{5}$ [6.1A] **6.** $\frac{24}{5}$ [6.1B] **7.** $\frac{6}{25}$ [6.1B] **8.** 23 mi/gal [6.2A]
9. 16-ounce can: 16.2¢/ounce; 12-ounce can: 15.8¢/ounce; 12-ounce can is the better buy [6.2C] **10.** 36 [6.3B] **11.** 8 [6.3B]
12. 24 hits [6.4A] **13.** 135 mi [6.4A] **14. a.** LGB **b.** twice as large [6.5B] **15.** $h = 16$ [6.5A] **16.** $h = 300$ [6.5B] **17.** 27 mg
18. 33.15 mg

Chapter 7

Getting Ready for Chapter 7

1. 141.44 **2.** 225 **3.** 1,477,432 **4.** $\frac{13}{40}$ **5.** 20 **6.** 489 **7.** 9.45 **8.** 0.352 **9.** 0.0362 **10.** 117.2 **11.** 4 **12.** $\frac{9}{25}$
13. $\frac{9}{200}$ **14.** $\frac{65}{2}$ **15.** 0.375 **16.** $0.41\overline{6}$ **17.** 2.5 **18.** 62.5 **19.** 15,300 **20.** 2,976.74

Section 7.1

1. $\frac{20}{100}$ **3.** $\frac{60}{100}$ **5.** $\frac{24}{100}$ **7.** $\frac{65}{100}$ **9.** 0.23 **11.** 0.92 **13.** 0.09 **15.** 0.034 **17.** 0.0634 **19.** 0.009 **21.** 23% **23.** 92%
25. 45% **27.** 3% **29.** 60% **31.** 80% **33.** 27% **35.** 123% **37.** $\frac{3}{5}$ **39.** $\frac{3}{4}$ **41.** $\frac{1}{25}$ **43.** $\frac{53}{200}$ **45.** $\frac{7{,}187}{10{,}000}$ **47.** $\frac{3}{400}$
49. $\frac{1}{16}$ **51.** $\frac{1}{3}$ **53.** 50% **55.** 75% **57.** $33\frac{1}{3}$% **59.** 80% **61.** 87.5% **63.** 14% **65.** 325% **67.** 150% **69.** 48.8%
71. a. 0.506 **b.** 0.527 **c.** 0.537 **73. a.** $\frac{27}{50}, \frac{7}{25}, \frac{3}{20}, \frac{3}{100}$ **b.** 0.54, 0.28, 0.15, 0.03 **c.** About two times as likely **75.** 78.4%
77. 11.8% **79.** 72.2% **81.** 18.5 **83.** 10.875 **85.** 0.5 **87.** 62.5 **89.** 0.5 **91.** 0.125 **93.** 0.625 **95.** 0.0625
97. 0.3125 **99.** 2 **101.** 2 **103.** 2 **105.** 2

Section 7.2

1. 8 **3.** 24 **5.** 20.52 **7.** 7.37 **9.** 50% **11.** 10% **13.** 25% **15.** 75% **17.** 64 **19.** 50 **21.** 925 **23.** 400 **25.** 5.568
27. 120 **29.** 13.72 **31.** 22.5 **33.** 50% **35.** 942.684 **37.** 97.8 **39.** What number is 25% of 350?
41. What percent of 24 is 16? **43.** 46 is 75% of what number? **45.** 4.8% calories from fat; healthy
47. 50% calories from fat; not healthy **49.** 0.80 **51.** 0.76 **53.** 48 **55.** 0.25 **57.** 0.5 **59.** $\frac{5}{8}, \frac{3}{4}, \frac{7}{8}$ **61.** $\frac{1}{4}$ **63.** 0.25
65. 0.75 **67.** 0.125 **69.** 0.375 **71.** $\frac{1}{8}, \frac{1}{4}, \frac{3}{8}, \frac{1}{2}, \frac{5}{8}, \frac{3}{4}, \frac{7}{8}$

Section 7.3

1. 70% **3.** 84% **5.** 45 mL **7.** 18.2 acres for farming; 9.8 acres are not available for farming **9.** $11.20
11. Bush 286; Kerry 251 **13.** 34,266,960 **15.** 3,000 students **17.** 400 students **19.** 1,664 female students **21.** 31.25%
23. 50% **25.** About 19.2 million **27.** 33 **29.** 8,685 **31.** 136 **33.** 0.05 **35.** 15,300 **37.** 0.15 **39.** $\frac{1}{5}$ **41.** $\frac{5}{12}$ **43.** $\frac{3}{4}$
45. $\frac{2}{3}$ **47.** 39.4% **49.** 94 hits **51.** At least 16 hits

Section 7.4

1. $52.50 **3.** $2.70; $47.70 **5.** $150; $156 **7.** 5% **9.** $420.90 **11.** $2,820 **13.** $200 **15.** 14% **17.** 26.9%
19. 62.8 cents or $0.628% **21.** $560 **23.** $11.93 **25.** 4.5% **27.** $3,995 **29.** 1,100 **31.** 75 **33.** 0.16 **35.** 4 **37.** 396
39. 415.8 **41.** $\frac{1}{2}$ **43.** $2\frac{2}{3}$ **45.** $1\frac{1}{2}$ **47.** $4\frac{1}{2}$ **49.** Sales tax = $3,180; luxury tax = $2,300 **51.** $2,300
53. You saved $1,600 off the sticker price and $150 in the luxury tax. If you lived in a state with a 6% sales tax rate, you saved an additional 0.06($1,600) = $96.

Section 7.5

1. \$24,610 **3.** \$7,020 **5.** \$13,200 **7.** 10% **9.** 20% **11.** 21% **13.** \$45; \$255 **15.** \$381.60 **17.** 13.9% **19.** 16% **21.** 21.8 in. to 23.2 in. **23.** \$46,595.88 **25. a.** 51.9% **b.** 7.8% **27.** 140 **29.** 4 **31.** 152.25 **33.** 3,434.7 **35.** 10,150 **37.** 10,456.78 **39.** 2,140 **41.** 3,210 **43.** 1 **45.** $\frac{3}{4}$ **47.** $1\frac{5}{12}$ **49.** 6

Section 7.6

1. \$2,160 **3.** \$665 **5.** \$8,560 **7.** \$2,160 **9.** \$5 **11.** \$813.33 **13.** \$406.34 **15.** \$5,618 **17.** \$8,407.56 **19.** \$974.59 **21. a.** \$13,468.55 **b.** \$13,488.50 **c.** \$12,820.37 **d.** \$12,833.59 **23.** 30% **25.** $16\frac{2}{3}\%$ **27.** 108 **29.** 162 **31.** 8 **33.** 3 **35.** $\frac{1}{2}$ **37.** $\frac{3}{8}$ **39.** $\frac{2}{3}$ **41.** $\frac{3}{2} = 1\frac{1}{2}$ **43.** $\frac{1}{12}$ **45.** $1\frac{1}{4}$ **47.** $\frac{16}{21}$ **49.** $\frac{64}{525}$ **51.** \$30.78

53. Percent increase in production cost: *Star Wars* 1 to 2 = 63.6%; *Star Wars* 2 to 3 = 80.6%; *Star Wars* 3 to 4 = 253.8%

Section 7.7

1. a. $\frac{41}{100}$ **b.** $\frac{38}{41}$ **c.** $\frac{31}{50}$ **d.** $\frac{31}{19}$ **3. a.** 78% **b.** 10% **c.** 18% **d.** 22%

5. a. 240 people **b.** 960 people **c.** 750 people **d.** 1,800 people

7. **9.**

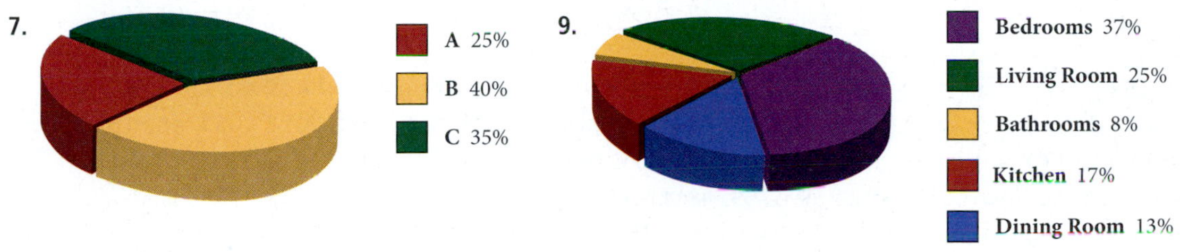

11. $\frac{8}{3}$ or $2\frac{2}{3}$ **13.** $\frac{1}{40}$ **15.** $\frac{73}{20}$ or $3\frac{13}{20}$ **17.** $\frac{62}{25}$ or $2\frac{12}{25}$ **19.** $\frac{4}{3}$ or $1\frac{1}{3}$

Chapter 7 Review

1. 0.35 **3.** 0.05 **5.** 95% **7.** 49.5% **9.** $\frac{3}{4}$ **11.** $1\frac{9}{20}$ **13.** 30% **15.** $66\frac{2}{3}\%$ **17.** 16.8 **19.** 50% **21.** 80 **23.** 75% **25.** \$477 **27.** \$4.93 **29.** 55.4% increase **31.** \$60; 20% off **33.** \$42

35.

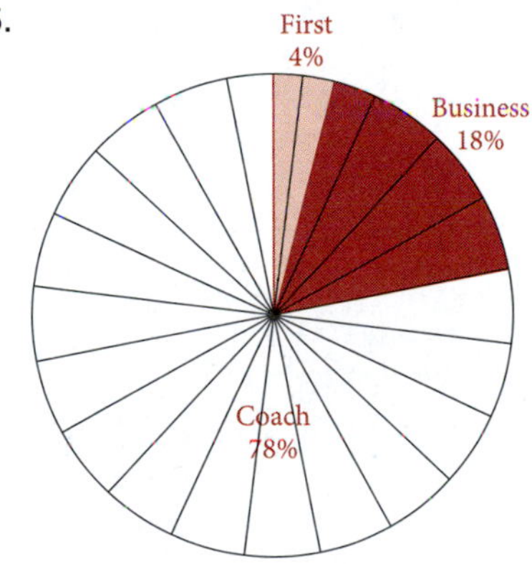

Chapter 7 Cumulative Review

1. 4,576 **2.** 1,605 **3.** 5,040 **4.** 90.5 **5.** 42 **6.** $\frac{2}{3}$ **7.** $-\frac{3}{28}$ **8.** $-\frac{8}{27}$ **9.** $17\frac{1}{2}$ **10.** 3.85 **11.** −9.648 **12.** $1\frac{3}{4}$ **13.** $1\frac{5}{7}$ **14.** $\frac{1}{20}$ **15.** 7.617 **16.** 0.94 **17.** 21 **18.** $9\frac{3}{4}$ **19.** $21x$ **20.** $23x + 7$ **21.** 25 **22.** −7 **23.** 32 cm, 35 cm^2 **24.** 91.06 mm **25.** $2 \cdot 2 \cdot 2 \cdot 3 \cdot 11$ **26.** 2 **27.** $\frac{5}{4}$ **28.** $\frac{3}{5}$ **29.** 350% **30.** 1.25 **31.** 0.6 **32.** 16 **33.** \$17 **34.** 7% **35.** \$24 **36.** Perimeter = 22.4 in., area = 31 in^2 **37.** 477 cm **38.** 20.8¢/soda; 25.0¢/soda; the store brand is less expensive **39.** 456.75 in.

Chapter 7 Test

1. 0.18 [7.1B] **2.** 0.04 [7.1B] **3.** 0.005 [7.1B] **4.** 45% [7.1C] **5.** 70% [7.1C] **6.** 135% [7.1C] **7.** $\frac{13}{20}$ [7.1A] **8.** $1\frac{23}{50}$ [7.1A] **9.** $\frac{7}{200}$ [7.1A] **10.** 35% [7.1E] **11.** 37.5% [7.1E] **12.** 175% [7.1E] **13.** 45 [7.2A] **14.** 45% [7.2A] **15.** 80 [7.2A] **16.** 92% [7.3A] **17.** \$960 [7.4B] **18.** \$40; 16% off [7.5C] **19.** \$220.50 [7.4A] **20.** \$100 [7.6A] **21.** \$2,520 [7.6B]

Chapter 8

Getting Ready for Chapter 8

1. $\frac{2}{5}$ **2.** $\frac{4}{1}$ **3.** 192 **4.** 12,500 **5.** 3,267,000 **6.** 3,600 **7.** 9.3375 **8.** 190.4 **9.** $2\frac{2}{3}$ **10.** 0.025 **11.** 450 **12.** 0.4 **13.** 50 **14.** 3.65 **15.** 750 **16.** 24 **17.** 122 **18.** 38.9 **19.** 0.75 **20.** Perimeter = 120 in., Area = 864 in^2

Section 8.1

1. 60 in. **3.** 120 in. **5.** 6 ft **7.** 162 in. **9.** $2\frac{1}{4}$ ft **11.** 13,200 ft **13.** $1\frac{1}{3}$ yd **15.** 1,800 cm **17.** 4,800 m **19.** 50 cm **21.** 0.248 km **23.** 670 mm **25.** 34.98 m **27.** 6.34 dm **29.** 5.3 miles **31.** 20 yd **33.** 80 in. **35.** 244 cm **37.** 2,960 chains **39.** 120,000 μm **41.** 7,920 ft **43.** 80.7 ft/sec **45.** 19.5 mi/hr **47.** 1,023 mi/hr **49.** 3,965,280 ft **51.** 179,352 in. **53.** 2.7 mi **55.** 18,094,560 ft **57.** 144 **59.** 8 **61.** 1,000 **63.** 3,267,000 **65.** 6 **67.** 0.4 **69.** 405 **71.** 450 **73.** 45 **75.** 2,200 **77.** 607.5 **79.** $\frac{1}{3}$ **81.** 6 **83.** $3\frac{1}{2}$ **85.** 6 **87.** 6 **89.** $\frac{7}{10}$ **91.** $10{,}000 \text{ steps} \cdot \frac{2.5 \text{ ft}}{1 \text{ step}} \cdot \frac{1 \text{ mi}}{5{,}280 \text{ ft}} = 4.7 \text{ mi}$

Section 8.2

1. 432 in^2 **3.** 2 ft^2 **5.** 1,306,800 ft^2 **7.** 1,280 acres **9.** 3 mi^2 **11.** 108 ft^2 **13.** 1,700 mm^2 **15.** 28,000 cm^2 **17.** 0.0012 m^2 **19.** 500 m^2 **21.** 700 a **23.** 3.42 ha **25.** 135 ft^2 **27.** 48 fl oz **29.** 8 qt **31.** 20 pt **33.** 480 fl oz **35.** 8 gal **37.** 6 qt **39.** 9 yd^3 **41.** 5,000 mL **43.** 0.127 L **45.** 4,000,000 mL **47.** 14,920 L **49.** 2,177,280 acres **51.** 30 a **53.** 5,500 bricks **55.** 16 cups **57.** 34,560 in^3 **59.** 48 glasses **61.** 20,288,000 acres **63.** 3,230.93 mi^2 **65.** 23.35 gal **67.** 21,492 ft^3 **69.** 192 **71.** 6,000 **73.** 300,000 **75.** 12.5 **77.** $\frac{5}{8}$ **79.** 9 **81.** $\frac{1}{3}$ **83.** $\frac{17}{144}$

Section 8.3

1. 128 oz **3.** 4,000 lb **5.** 12 lb **7.** 0.9 T **9.** 32,000 oz **11.** 56 oz **13.** 13,000 lb **15.** 2,000 g **17.** 40 mg **19.** 200,000 cg **21.** 508 cg **23.** 4.5 g **25.** 47.895 cg **27.** 1.578 g **29.** 0.42 kg **31.** 48 g **33.** 4 g **35.** 9.72 g **37.** 120 g **39.** 3 L **41.** 1.5 L **43.** 0.561 grams **45. a.** 3 capsules **b.** 1 capsule **c.** 2 capsules **47.** 20.32 **49.** 6.36 **51.** 50 **53.** 50 **55.** 122 **57.** 248 **59.** 40 **61.** $\frac{9}{50}$ **63.** $\frac{9}{100}$ **65.** $\frac{4}{5}$ **67.** $1\frac{3}{4}$ **69.** 0.75 **71.** 0.85 **73.** 0.6 **75.** 3.625 **77.** $\frac{1}{8}$ **79.** $6\frac{1}{4}$ **81.** 0.125 **83.** 6.25

Section 8.4

1. 15.24 cm **3.** 13.12 ft **5.** 6.56 yd **7.** 32,200 m **9.** 5.98 yd^2 **11.** 24.7 acres **13.** 8,195 mL **15.** 2.12 qt **17.** 75.8 L **19.** 339.6 g **21.** 33 lb **23.** 365°F **25.** 30°C **27.** 58°C **29.** 7.5 mL **31.** 3.94 in. **33.** 7.62 m **35.** 46.23 L **37.** 17.67 oz **39.** Answers will vary. **41.** 91.46 m **43.** 20.90 m^2 **45.** 88.55 km/hr **47.** 2.03 m **49.** 38.3°C **51.** 75 **53.** 82 **55.** 3.25 **57.** 22 **59.** 41 **61.** 48 **63.** 195 **65.** 3.25 **67.** $90 **69.** mean = 8, range = 6 **71.** mean = 6, range = 9 **73.** median = 21, range = 14 **75.** median = 40, range = 16 **77.** mode = 14, range = 7 **79. a.** 73 **b.** 71 **c.** 70 **d.** 23 **81.** 381.8 mg/day; 1.52 mL/day

Section 8.5

1. a. 270 min **b.** 4.5 hr **3. a.** 320 min **b.** 5.33 hr **5. a.** 390 sec **b.** 6.5 min **7. a.** 320 sec **b.** 5.33 min **9. a.** 40 oz **b.** 2.5 lb **11. a.** 76 oz **b.** 4.75 lb **13. a.** 54 in. **b.** 4.5 ft **15. a.** 69 in. **b.** 5.75 ft **17. a.** 9 qt **b.** 2.25 gal **19.** 11 hr **21.** 22 ft 4 in. **23.** 11 lb **25.** 5 hr 40 min **27.** 3 hr 47 min **29.** 52 min **31.** 7.5 seconds **33.** 8:18:18; 9:08:01 **35.** 00:06:15 **37.** $104 **39.** 10 hr **41.** $150 **43.** $6

45.

CAFFEINE CONTENT IN SOFT DRINKS

Drink	Caffeine (In Milligrams)
Jolt	100
Mountain Dew	55
Coca-Cola	45
Diet Pepsi	36
7 up	0

47. 1.79 sec

Chapter 8 Review

1. 144 in. **3.** 0.49 in. **5.** 435,600 ft^2 **7.** 576 in^2 **9.** 6 gal **11.** 128 oz **13.** 5,000 g **15.** 10.16 cm **17.** 7.42 qt **19.** 141.5 g **21.** 248°F **23.** 8.52 liters **25.** 4,383.23 mi^2 **27.** 402.5 km **29.** 600 bricks **31.** 80 glasses **33.** 117 mi/hr **35.** 322 km/hr **37.** 13 lb

Chapters 1–8 Cumulative Review

1. 5,555 **2.** 5,555 **3.** 73.8 **4.** 7.38 **5.** $\frac{64}{125}$ **6.** 8.936 **7.** 1.156 **8.** $-\frac{1}{6}$ **9.** $-\frac{1}{9}$ **10.** $-\frac{13}{27}$ **11.** $2\frac{1}{10}$ **12.** 36 **13.** 7 **14.** −9 **15.** 6 **16.** 6 **17.** $1\frac{1}{8}$ **18.** 2(9) + 2; 20 **19.** $\frac{1}{3}$ **20.** 13.6 mi/hr **21.** 1,600 acres **22.** 62.5% **23.** $\frac{11}{20}$ **24.** $\frac{7}{2} = 3.5$ **25.** 30,000 + 400 + 5 **26.** 210 **27.** $66\frac{2}{3}\%$ **28.** 1 **29.** 22 lbs **30.** 80.5 km/hr **31.** 122°F **32.** $75 **33.** $450 **34.** 84.9 ft **35. a.** 2,160 people **b.** 785 ft **c.** 39.25 ft/min **d.** 0.45 mi/hr

Chapter 8 Test

1. 21 ft [8.1A] **2.** 0.75 km [8.1B] **3.** 130,680 ft² [8.2A] **4.** 3 ft² [8.2A] **5.** 10,000 mL [8.2D] **6.** 8.05 km [8.2B] **7.** 10.6 qt [8.4A] **8.** 26.7° C [8.4B] **9.** 0.26 gal [8.4A] **10.** 20,844 in. [8.1A] **11.** 0.24 ft³ [8.2A] **12.** 70.75 liters [8.4A] **13.** 74.70 m [8.4A] **14.** 7.55 liters [8.4A] **15.** 90 tiles [8.5C] **16.** 64 glasses [8.5C] **17. a.** 330 min **b.** 5.5 hr [8.4A] **18.** 11 lb [8.5A]

Chapter 9

Getting Ready for Chapter 9

1. 9 **2.** −3 **3.** −3 **4.** −7 **5.** −3 **6.** −8 **7.** −2 **8.** 2,000 **9.** 64 **10.** 9 **11.** 86,400 **12.** 3.76 **13.** 6.5 **14.** 3.4 **15.** x **16.** $-11y$ **17.** $14x - 21$ **18.** x^6 **19.** $27x^3y^3$ **20.** 16

Section 9.1

1. x^8 **3.** a^8 **5.** y^{15} **7.** x^5 **9.** x^9 **11.** a^{12} **13.** y^6 **15.** $14x^9$ **17.** $24y^8$ **19.** $21x^2$ **21.** $192x^{15}$ **23.** $24a^3$ **25.** $125y^{12}$ **27.** $6x^5y^8$ **29.** $56a^4b^5$ **31.** $24a^6b^6$ **33.** $28a^{12}b^{12}$ **35.** $-30x^5y^8$ **37.** a^{21} **39.** $125y^{12}$ **41.** $81x^8$ **43.** $8a^3b^3$ **45.** $625x^8y^{12}$ **47.** $32x^{10}y^{25}$ **49.** a^{29} **51.** $72x^{12}y^{22}$ **53.** $175a^{12}b^9$ **55.** $a^{18}b^{34}$

57.

Number x	Square x^2
1	1
2	4
3	9
4	16
5	25
6	36
7	49

59.

Number x	Square x^2
−2.5	6.25
−1.5	2.25
−0.5	0.25
0	0
0.5	0.25
1.5	2.25
2.5	6.25

61. 52 ft **63.** 84 ft **65.** 9 in.

Section 9.2

1. $6x^2 + 6x + 8$ **3.** $5a^2 - 9a + 8$ **5.** $6x^3 + 4x^2 - 3x + 2$ **7.** $13t^2 - 2$ **9.** $33x^3 - 10x^2 - 4x - 4$ **11.** $2x^2 - 2x + 1$ **13.** $4y^3 - 8y^2 + 3y - 12$ **15.** $-4x - 4$ **17.** $x^2 - 30x + 64$ **19.** $7y^2 + 3y + 25$ **21.** 9 **23.** 0 **25.** 1 **27.** 2, 5, 8, 11 **29.** 4, 9, 16, 25 **31.** 10, 13, 18, 25

33.

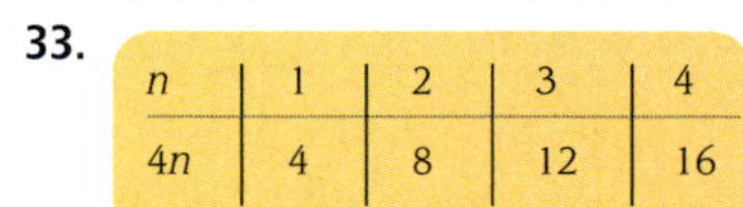

n	1	2	3	4
$4n$	4	8	12	16

35.

n	1	2	3	4
$n - 3$	−2	−1	0	1

37. 169 ft² **39.** 210 in² **41.** 480 yd²

Section 9.3

1. $x^6 + x^5$ **3.** $x^6 + x^5$ **5.** $6x^6 - 14x^4$ **7.** $12x^6y^3 + 8x^3y^6$ **9.** $6x^6 - 14x^4$ **11.** $12x^2y^2 - 27xy$ **13.** $12x^2y^2 - 27xy$ **15.** $6x^5y^7 + 4x^6y^6 + 10x^4y^5$ **17.** $x^2 + 5x + 6$ **19.** $x^2 + x - 6$ **21.** $6x^2 + 19x + 10$ **23.** $3x^2 + 6x - 24$ **25.** $16x^2 - 9$ **27.** $12x^2 - x - 1$ **29.** $6x^2 - 23x + 20$ **31.** $x^2 + 6x + 8$ **33.** $6x^2 + 13x + 6$ **35.** $21x^2 + 34x + 8$ **37.** $x^2 + 6x + 9$ **39.** $x^2 + 12x + 36$ **41.** $9x^2 + 12x + 4$ **43.** $4x^2 + 12x + 9$ **45.** $49x^2 + 84x + 36$ **47.** $16x^2 + 8x + 1$ **49.** $x^2 - 10x + 25$ **51.** $4x^2 - 16x + 16$

53.

x	$(x+3)^2$	$x^2 + 9$	$x^2 + 6x + 9$
1	16	10	16
2	25	13	25
3	36	18	36
4	49	25	49

55. $x^2 + 6x + 9$ **57.** $x^2 + 10x + 25$ **59.** $4x^2 + 12x + 9$ **61.** $V = 175$ in³; $SA = 190$ in² **63.** $V = 2{,}340$ ft³; $SA = 1{,}062$ ft²

Section 9.4

1. $\frac{1}{32}$ **3.** $\frac{1}{100}$ **5.** $\frac{1}{x^3}$ **7.** $\frac{1}{16}$ **9.** $-\frac{1}{125}$ **11.** $\frac{1}{4}$ **13.** 1,000 **15.** $\frac{1}{x^7}$ **17.** 9 **19.** $\frac{1}{8}$ **21.** 1,000 **23.** $\frac{1}{x^{12}}$ **25.** 2 **27.** 4 **29.** $\frac{1}{4}$ **31.** x **33.** $\frac{1}{x}$ **35.** $\frac{1}{10^7}$ **37.** $\frac{1}{10^7}$ **39.** 1,000 **41.** x^{24} **43.** 4 **45.** 10,000 **47.** $15x^5$ **49.** $\frac{6}{x^5}$ **51.** $-\frac{12}{x^7}$ **53.** $42x^2$ **55.** x^5 **57.** x^2 **59. a.** 405 sec **b.** 6.75 min **61. a.** 56 oz **b.** 3.5 lb **63. a.** 69 in. **b.** 5.75 ft

Section 9.5

1. 4.25×10^5 **3.** 6.78×10^6 **5.** 1.1×10^4 **7.** 8.9×10^7 **9.** 38,400 **11.** 57,100,000 **13.** 3,300 **15.** 89,130,000 **17.** 3.5×10^{-4} **19.** 7×10^{-4} **21.** 6.035×10^{-2} **23.** 1.276×10^{-1} **25.** 0.00083 **27.** 0.0625 **29.** 0.3125 **31.** 0.005 **33.** $\$4.2 \times 10^4$; 4,000,000 **35.** $\$5.75 \times 10^5$; 87,000,000 **37.**

Jupiter's Moon	Period (Seconds)	
Io	153,000	1.53×10^5
Europa	307,000	3.07×10^5
Ganymede	618,000	6.18×10^5
Callisto	1,440,000	1.44×10^6

39. 1.024×10^3 **41.** 1.074×10^9 **43.** $4x$ **45.** $-3x$ **47.** $9x$ **49.** $-6a$

Section 9.6

1. 6×10^{10} **3.** 1.5×10^{11} **5.** 6.84×10^{-2} **7.** 1.62×10^{10} **9.** 3.78×10^{-1} **11.** 2×10^3 **13.** 4×10^{-9} **15.** 5×10^{14} **17.** 6×10^1 **19.** 2×10^{-8} **21.** 4×10^6 **23.** 4×10^{-4} **25.** 2.1×10^{-2} **27.** 3×10^{-5} **29.** 6.72×10^{-1} **31.** $\$1.05 \times 10^{-2}$ **33.** $\$6.61 \times 10^{-3}$ **35.** 2,500,000 stones or 2.5×10^6 **37.** 7 **39.** -14 **41.** -14 **43.** 21

Chapter 9 Review

1. $15y^{14}$ **2.** $56y^{18}$ **3.** $54x^8y^9$ **4.** $84a^{11}b^{13}$ **5.** x^{20} **6.** y^{14} **7.** $625x^{12}y^8$ **8.** $72a^{14}b^9$ **9.** $9t^2 + 3t - 7$ **10.** $12y^2 - 14y + 3$ **11.** $3x^2 - 10x + 2$ **12.** $5y^2 - 3y + 1$ **13.** 16 **14.** -9 **15.** $6x^7 - 16x^3$ **16.** $15a^5b^5 - 6a^7b^9$ **17.** $3x^2 + 4x - 15$ **18.** $18x^2 + 24x + 8$ **19.** $x^2 - 6x + 9$ **20.** $16y^2 + 56y + 49$ **21.** $x^2 + 8x + 15$ **22.** $15x^2 + 41x + 14$ **23.** $\frac{1}{125}$ **24.** $\frac{1}{16}$ **25.** $\frac{1}{9}$ **26.** $\frac{1}{64}$ **27.** $\frac{1}{y^4}$ **28.** $\frac{1}{x}$ **29.** $\frac{1}{x^2}$ **30.** x^{16} **31.** $\frac{1}{8}$ **32.** 81 **33.** $-\frac{20}{x^8}$ **34.** x^2 **35.** 7.38×10^7 **36.** 2.935×10^{-3} **37.** 4,400,000,000 **38.** 0.000016 **39.** 1.728×10^3 **40.** 2.576×10^2 **41.** 2×10^{14} **42.** 5×10^{-1} **43.** 9×10^{-7} **44.** 2.8×10^{11} **45.** 4 music videos **46.** 940 songs, or 9.4×10^2

Chapters 1–9 Cumulative Review

1. 5,996 **2.** $\frac{8x + 5}{x}$ **3.** 4.559 **4.** $2\frac{3}{8}$ **5.** $18x^2 - 9x - 35$ **6.** $10a^5 - 35a^3$ **7.** 42 **8.** 1 **9.** 440,000 **10.** $\frac{12}{25}$ **11.** $6\frac{1}{3}$ **12.** 12 **13.** 80% **14.** 15.47 lb **15.** $1\frac{6}{25}$ **16.** 35% **17.** $4y\sqrt{2x}$ **18.** $-x - 5$ **19.** -7 **20.** 8 **21.** 4 **22.** 62 **23.** 56 **24.** $-\frac{4}{3}$ **25.** 5 **26.** $\frac{1}{8}$ **27.** $4x + 1$ **28.** Associative property of addition **29.** 82 in² **30.** Ben is 17; Ryan is 9 **31.** 16 mi/gal **32.** \$80; 20% **33.** 13 m **34.** \$14.40 **35.** \$21 **36.** 64 packets **37.** $-\$45$ **38.** \$784 **39.** 240 glasses **40.**

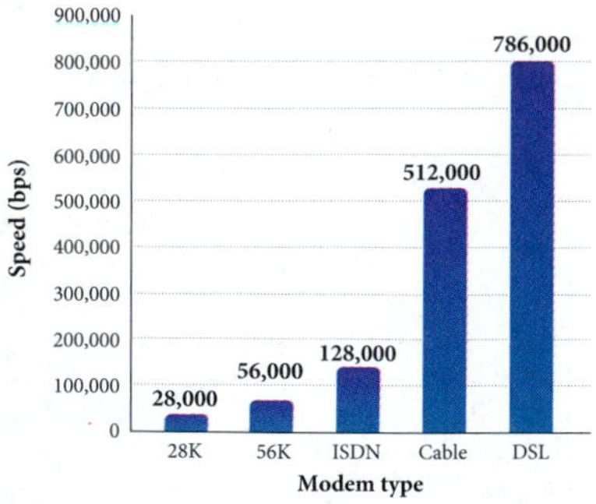

Speed (bps)
900,000
800,000
700,000
600,000
500,000
400,000
300,000
200,000
100,000
0
28,000
56,000
128,000
512,000
786,000
28K
56K
ISDN
Cable
DSL
Modem type

Chapter 9 Test

1. $12x^9$ [9.1A] **2.** $24x^7y^7$ [9.1A] **3.** a^{26} [9.1B] **4.** $216x^9y^6$ [9.1C] **5.** $256a^{16}b^{11}$ [9.1D] **6.** $13t^2 + t + 7$ [9.2A] **7.** $4a^2 + a - 9$ [9.2B] **8.** 36 [9.2C] **9.** 3 [9.2C] **10.** $12x^{10} - 18x^6$ [9.3A] **11.** $6x^2 + 8x - 30$ [9.3A] **12.** $16y^2 - 24y + 9$ [9.3A] **13.** $\frac{1}{16}$ [9.4A] **14.** $\frac{1}{x^4}$ [9.4A] **15.** 25 [9.4B] **16.** y^2 [9.4B] **17.** 3.85×10^{-5} [9.5A] **18.** 6,750,000 [9.5B] **19.** 1.701×10^3 [9.6A] **20.** 5.0×10^2 [9.6A] **21.** 4.0×10^{-4} [9.6A] **22.** 2.56×10^{-3} [9.6A] **23.** 12 music videos [9.6A] **24.** 4,000 songs, or 4.0×10^3 [9.6A]

Index